Fungi and Fungal Metabolites for the Improvement of Human and Animal Nutrition and Health

Fungi and Fungal Metabolites for the Improvement of Human and Animal Nutrition and Health

Editors

Laurent Dufossé
Mireille Fouillaud
Yanis Caro

MDPI • Basel • Beijing • Wuhan • Barcelona • Belgrade • Manchester • Tokyo • Cluj • Tianjin

Editors

Laurent Dufossé
Laboratoire CHEMBIOPRO
Université de La Réunion
Sainte-Clotilde
Reunion

Mireille Fouillaud
Laboratoire CHEMBIOPRO
Université de La Réunion
Sainte-Clotilde
Reunion

Yanis Caro
IUT-Laboratoire CHEMBIOPRO
Université de La Réunion
Saint-Pierre
Reunion

Editorial Office
MDPI
St. Alban-Anlage 66
4052 Basel, Switzerland

This is a reprint of articles from the Special Issue published online in the open access journal *Journal of Fungi* (ISSN 2309-608X) (available at: www.mdpi.com/journal/jof/special_issues/fungal_metabolites_health).

For citation purposes, cite each article independently as indicated on the article page online and as indicated below:

LastName, A.A.; LastName, B.B.; LastName, C.C. Article Title. *Journal Name* **Year**, *Volume Number*, Page Range.

ISBN 978-3-0365-1466-6 (Hbk)
ISBN 978-3-0365-1465-9 (PDF)

Contents

About the Editors

Laurent Dufossé

Laurent Dufossé, has held the position of Professor of Food Science and Biotechnology since 2006, at the Reunion Island University, which is located on a volcanic island in the Indian Ocean, near Madagascar and Mauritius. The island is one of France's overseas territories, with almost one million inhabitants, and the university has 15,000 students. Previously, Professor Dufossé was a researcher and senior lecturer at the Université de Bretagne Occidentale, Quimper, Brittany, France. He attended the University of Burgundy, D, where he received his PhD in Food Science in 1993 and has been involved in the field of biotechnology of food ingredients for more than 30 years. His main research over the last 20 years has focused on microbial production of pigments and studies are mainly devoted to aryl carotenoids, such as isorenieratene, C50 carotenoids, azaphilones, and anthraquinones.

Mireille Fouillaud

Mireille Fouillaud is a senior lecturer at the University of Reunion and a researcher at the Laboratory of Chemistry and Biotechnology of Natural Products (ChemBioPro). She graduated in 1990 as an industrial microbiology engineer. Following a Ph.D. in Cell Biology and Microbiology from the University of Aix-Marseille I, defended in 1994, she was recruited at the Faculty of Sciences and Technology of the University of Reunion. Between 2009 and 2020, she joined the Ecole Supérieure d'Ingénieurs Réunion Océan Indien (ESIROI), in the food engineering department. She taught biology and microbiology applied to agribusiness there. In 2020, she came back at the Faculty of Sciences of University of La Réunion to teach microbiology and project management. Her main research interests range from the microbial diversity of ecosystems and organisms to the production of metabolites of interest for industries, through biotechnology.

Yanis Caro

Dr Yanis Caro has held the position of Associate Professor of Chemistry and Biotechnology at University of La Reunion and senior researcher at Chemistry and Biotechnology of Natural Products Laboratory (CHEMBIOPRO) since 2010. He first joined the food engineering department of the engineering school ESIROI, and then the Department of Health, Safety and Environment (HSE) at the Institute of Technology (IUT) of La Reunion in 2017. Holder of a Master Degree in Agro-Resources Sciences (1997), Dr Caro obtained his Ph.D. (2001) in Lipid Chemistry and Biotechnology at the school INP of Toulouse (France) and his Habilitation at the University of La Reunion (2020). Dr Caro worked at CIRAD of Montpellier (1998 - 2001) and, from 2004 until 2010, he worked as graduate engineer and advisor technological development in charge of food safety innovation projects at the CRITT centre of La Reunion.

Preface to "Fungi and Fungal Metabolites for the Improvement of Human and Animal Nutrition and Health"

Fungi comprise 1, 2, 3, ..., or maybe around 5.1 million species? Even scientists do not currently agree on how many fungi species may be found on planet Earth, with only around 120,000 having been described so far. Fungi have been classified as a separate kingdom of organisms, as complex and diverse as plants and animals, of which only a few percent have been named and described. Fungal biomasses and fungal metabolites share a long common history with human and animal nutrition and health. Macrofungi and filamentous fungi constitute a large portfolio of proteins, lipids, vitamins, minerals, oligo elements, pigments, colorants, bioactive compounds, antibiotics, pharmaceuticals, etc. For example, industrially important enzymes and microbial biomass proteins have been produced from fungi for more than 50 years. Some start-ups convert byproducts and side streams rich in carbohydrates into a protein-rich fungal biomass. This biomass can then be processed into a vegan meat substitute for food applications. In recent years, there has also been a significant increase (in fact, a significant revival) in the number of publications in the international literature dealing with the production of lipids by microbial sources (the single-cell oils (SCO) that are produced by the so-called "oleaginous"microorganisms, including "oleaginous"fungi such as zygomycete species, e.g., *Cunninghamella echinulata* and *Mortierella isabellina*). Fungi are potential sources of polyunsaturated fatty acids (PUFA), as these microorganisms can accumulate large amounts of high-valued PUFAs, such as gamma-linolenic acid (GLA) and arachidonic acid (ARA).

The purpose of this Special Issue of Journal of *Fungi* (MDPI) was not to provide a comprehensive overview of the vast arena of how fungi and fungal metabolites are able to improve human and animal nutrition and health; rather, we, as Guest Editors, wished to encourage authors working in this field to publish their most recent work in this rapidly growing journal in order for the large readership to appreciate the full potential of wonderful and beneficial fungi. Thus, this Special Issue welcomed scientific contributions on applications of fungi and fungal metabolites, such as bioactive fatty acids, pigments, polysaccharides, alkaloids, terpenoids, etc., with great potential in human and animal nutrition and health.

Laurent Dufossé, Mireille Fouillaud, Yanis Caro
Editors

Journal of *Fungi*

MDPI

Editorial

Fungi and Fungal Metabolites for the Improvement of Human and Animal Nutrition and Health

Laurent Dufossé [1,2,*], Mireille Fouillaud [1,2] and Yanis Caro [1,3]

1 Laboratoire de Chimie et Biotechnologie des Produits Naturels—CHEMBIOPRO, Université de la Réunion, 15 Avenue René Cassin, CEDEX 9, CS 92003, F-97744 Saint-Denis, Ile de la Réunion, France; mireille.fouillaud@univ-reunion.fr (M.F.); yanis.caro@univ-reunion.fr (Y.C.)
2 Ecole Supérieure d'Ingénieurs Réunion Océan Indien—ESIROI, 2 Rue Joseph Wetzell, F-97490 Sainte-Clotilde, Ile de la Réunion, France
3 IUT de La Réunion, Département Hygiène, Sécurité, Environnement (HSE), 40 Avenue de Soweto, CEDEX 9, BP 373, F-97455 Saint-Pierre, Ile de la Réunion, France
* Correspondence: laurent.dufosse@univ-reunion.fr

Keywords: micro-fungi; macro-fungi; filamentous fungi; antioxidant; *Ganoderma*; kombucha; anti-cancer; carotenoid; medicinal mushroom; mycobiome; antimicrobial; antifungal; bioconversion; feed additive; cheese; dairy; *Sclerotinia*; secondary metabolite

check for updates

Citation: Dufossé, L.; Fouillaud, M.; Caro, Y. Fungi and Fungal Metabolites for the Improvement of Human and Animal Nutrition and Health. *J. Fungi* **2021**, *7*, 274. https://doi.org/10.3390/jof7040274

Received: 17 March 2021
Accepted: 1 April 2021
Published: 4 April 2021

Fungi: 1, 2, 3, ... 5.1 million species? Even scientists do not currently agree on how many fungi there might be on the planet Earth but only about 120,000 have been described so far [1]. They have been grouped into a separate kingdom of organisms, as complex and diverse as plants and animals, of which only a small percentage have been named and described. Fungal biomasses and fungal metabolites have a long common history with human and animal nutrition and health. Macro fungi and filamentous fungi bring a large portfolio of proteins, lipids, vitamins, minerals, oligo elements, pigments, colorants, bioactive compounds, antibiotics, pharmaceuticals, etc. For example, industrially important enzymes and microbial biomass proteins have been produced from fungi for more than 50 years. Some start-ups convert by-products and side-streams rich in carbohydrates into a protein-rich fungal biomass. The biomass is then processed into a vegan meat substitute for food applications. The last years there has also been a significant rise (in fact, a significant revival) in the number of publications in the international literature dealing with the production of lipids by microbial sources (the "single cell oils; SCOs" that are produced by the so-called "oleaginous" microorganisms, including "oleaginous" fungi, such as the zygomycete species, e.g., *Cunninghamella echinulata* and *Mortierella isabellina*). Fungi are potential sources of polyunsaturated fatty acids (PUFAs) as these microorganisms can accumulate large amounts of high-valued PUFAs, such as gamma-linolenic acid (GLA) and arachidonic acid (ARA).

The objective of the invitation to contribute to this MDPI Fungi Special Issue was not to give a complete coverage of how fungi and fungal metabolites are able to improve human and animal nutrition and health. We, as co-guest editors, simply wanted to encourage authors working in this field to publish their most recent work in a rapidly expanding journal, in order to make a large audience discover the full potential of wonderful and beneficial fungi. Thus, this Special Issue welcomes 22 scientific contributions (13 original research papers and 9 reviews) on applications of fungi and fungal metabolites, such as bioactive fatty acids, pigments, polysaccharides, alkaloids, terpenoids, etc., with great potential in human and animal nutrition and health.

Three original research papers investigate the advantages of using probiotics and living fungi in feed. Kanpiengjiai et al. [2] provide insights into a potential multifunctional probiotic yeast in aquaculture produced from *Sporidiobolus ruineniae* cultivated in tannin substrate. *Sporidiobolus ruineniae* A45.2, a carotenoid-producing yeast, was able to co-produce cell-associated tannase (CAT), gallic acid and viable cells with antioxidant activity when grown in a tannic acid substrate. Furthermore, viable cells were characterized by moderate hydrophobicity, high auto-aggregation and moderate co-aggregation with *Staphylococcus aureus*, *Salmonella ser. thyphimurium* and *Streptococcus agalactiae*. In another aspect of yeasts in animal feed, Keller et al. [3] investigate the gas-formation potential of yeasts present in liquid swine diets. Yeasts are said to cause sudden death in swine due to intestinal gas formation. As not all animals given high yeast content feed fall ill, growth and gas formation potential at body temperature were investigated as possible causally required properties. Most *Candida krusei* isolates formed high gas amounts within 24-h, whereas none of the *Candida lambica*, *Candida holmii* or most of the other isolates did. The gas pressure formed by yeast isolates varied more than tenfold. Only a minority of the yeasts were able to produce gas at temperatures common in the pig gut. Moving to dairy sheep milk performance, Mavrommatis et al. [4] evaluate the dietary administration of *Saccharomyces cerevisiae* live yeast on milk performance and composition, oxidative status of both blood plasma and milk, and gene expression related to the immune system of lactating ewes during the peripartum period. In conclusion, the dietary supplementation of ewes with *S. cerevisiae* improved the energy utilization and tended to enhance milk performance with simultaneous suppression of mRNA levels of pro-inflammatory genes during the peripartum period. Of course, fungi may also have an impact on human nutrition and health. Nguyen et al. [5] focus on *Monascus purpureus*, a filamentous fungus known for its fermentation of red yeast rice, that produces the metabolite monacolin K used in statin drugs to inhibit cholesterol biosynthesis. In their study, they show that active cultures of *M. purpureus* CBS 109.07, independent of secondary metabolites, use the mechanism of cholesterol assimilation to lower cholesterol in vitro. The findings demonstrate that active growing of *M. purpureus* CBS 109.07 can assimilate cholesterol, removing 36.38% of cholesterol after 48-h of incubation at 37 °C. The removal of cholesterol by resting or dead *M. purpureus* CBS 109.07 was not significant, with cholesterol reduction ranging from 2.75%–9.27% throughout a 72-h incubation. Cholesterol was also not shown to be catabolized as a carbon source.

The World Health Organization (WHO; Geneva, Switzerland) has established an urgent pathogen list of antibiotic-resistant bacteria to guide the research, discovery, and development of new antibiotics. This list includes carbapenem-resistant *Pseudomonas aeruginosa*, Enterobacteriaceae, and third generation cephalosporin-resistant bacteria as critical priorities as a result of the continuous and indiscriminate use of antibiotics, not only in the treatment of human diseases, but also in animals. This list includes antifungal compounds. Aguilar-Pérez et al. [6] explore and compare the antibacterial activity and chemical diversity of two endophytic fungi isolated from the plant *Hyptis dilatata* and cultured under different conditions by the addition of chemical elicitors, changes in the pH, and different incubation temperatures. A total of 34 extracts were obtained from both *Pestalotiopsis mangiferae* and *Pestalotiopsis microspora* and were tested against a panel of pathogenic bacteria. Three active extracts obtained from *P. mangiferae* were analyzed by Liquid Chromatography-Electrospray Ionization-Quadrupole-Time of Flight-Mass Spectrometry (LC-ESI-Q-TOF-MS) to screen the chemical diversity and the variations in composition. This allows the proposal of structures for some of the determined molecular formulas, including the previously reported mangiferaelactone, an antibacterial compound.

Arbuscular mycorrhizal fungi interact with plants, beneficial to humans or not. Habeeb et al. [7] highlight the integration of Arbuscular mycorrhizal fungi (AMF) and elevated CO_2 (eCO_2) into agricultural procedures as an ecofriendly approach to support the production and quality of plants. This study was conducted to investigate the effects of AMF and eCO_2, individually or in combination, on growth, photosynthesis, metabolism and functional food value of *Thymus vulgare*. Results revealed that both AMF and eCO_2 treatments improved photosynthesis and biomass production, however, much more positive impact was obtained by their synchronous application. Moreover, the levels of the majority of the detected sugars, organic acids, amino acids, unsaturated fatty acids, volatile compounds, phenolic acids and flavonoids were further improved as a result of the synergistic action of AMF and eCO_2, as compared to the individual treatments.

Elicitation effects of fungal products on plants and mushrooms are screened all over the world. Kim et al. [8] investigate the elicitation effects of alginate oligosaccharides extracted from brown algae (*Sargassum* species) on β-glucan production in cauliflower mushroom (*Sparassis latifolia*). Sodium alginate was refined from *Sargassum fulvellum*, *S. fusiforme*, and *S. horneri*, and characterized by proton nuclear magnetic resonance spectroscopy. The Solid Fraction (SF) of *S. fusiforme* and Liquid Fraction (LF) of *S. horneri* were chosen for elicitation on *S. latifolia*, yielding the highest β-glucan contents of 56.01 ± 3.45% and 59.74 ± 4.49% in the stalk, respectively. Total polyphenol content (TPC), antioxidant activities (2,2′-Azino-bis (3-ethylbenzthiazoline-6-sulfonic acid) (ABTS) radical scavenging and Superoxide dismutase (SOD)-like activity) of aqueous extracts of *S. latifolia* were also greatly stimulated by alginate elicitation.

Vegetal biomass, available in large amounts on the planet, could be a source of fungal substrate to produce useful compounds for humans, animals or plants. The project conducted by Ibarra-Cantún et al. [9] determines the quantity of secondary metabolites and the antioxidant activity of the extracts obtained by the solid-state fermentation of apple and agave mezcalero bagasse over 28 days, inoculated with the *Pleurotus ostreatus* strain. Solid-state fermentation (SSF) is used in enzyme and antibiotic production, bioethanol and biodiesel as an alternative energy source, biosurfactants with environmental goals, and the production of organic acids and bioactive compounds. The results showed a higher presence of phenolic compounds, flavonoids, total triterpenes and antioxidant activity in the apple bagasse from the SSF on day 21 in the extract of acetone and water (80:20 *v/v*), 100% methanol and aqueous; while the agave bagasse showed a significant presence of phenolic compounds and flavonoids only in the aqueous extract. In conclusion, the presence of secondary metabolites exhibiting antioxidant activities from the solid-state fermentation in the residues of the cider and mezcal industry is an alternative use for wasted raw material, and reduces the pollution generated from agro-industrial residues.

Fungal biotechnology brings much progress to humans, animals, and plants. Five research papers illustrate this aspect by the production of lipids and pigments or colorants. The biomass of *Mucor circinelloides*, a dimorphic oleaginous filamentous fungus, has a significant nutritional value and can be used for single cell oil production. Metal ions are micronutrients supporting fungal growth and metabolic activity of cellular processes. Dzurendova et al. [10] study the effect of 140 different substrates, with varying amounts of metal and phosphate ions concentration, on the growth, cell chemistry, lipid accumulation, and lipid profile of *M. circinelloides*. It was observed that Mg and Zn ions were essential for the growth and metabolic activity of *M. circinelloides*. An increase in Fe ion concentration inhibited fungal growth, while higher concentrations of Cu, Co, and Zn ions enhanced the growth and lipid accumulation. Lack of Ca and Cu ions, as well as higher amounts of Zn and Mn ions, enhanced lipid accumulation in *M. circinelloides*. Generally, the fatty acid profile of *M. circinelloides* lipids was quite constant, irrespective of media composition. In a second study dedicated to lipids, *Mortierella alpina* was used by Slaný et al. [11] in SSF for the bioconversion of animal fat by-products into high value fermented bioproducts enriched with arachidonic acid (ARA). Although in general the addition of an animal fat by-product caused a gradual cessation of ARA yield in the obtained fermented bioproduct, the content of ARA in fungal biomass was higher. Thus, *M. alpina* CCF2861 effectively transformed exogenous fatty acids from animal fat substrate to ARA. Maximum yield of 32.1 mg of ARA/g of bioproduct was reached when using cornmeal mixed with 5% (*w/w*) of an animal fat by-product as substrate.

Fungal pigments and colorants tend to replace synthetic, chemical ones as consumers request green and sustainable processing. Demand for microbial colorants is now becoming a competitive research topic for food, cosmetics, textile, and pharmaceutics industries. Textile dyeing is one of the most polluting aspects of the global fashion industry, devastating the environment and posing health hazards to humans. Many fungal genera/species are currently in use for large-scale, industrial-scale production of pigments and colorants. *Talaromyces albobiverticillius* (deep red pigment producer), *Emericella purpurea* (red pigment producer), *Paecilomyces marquandii* (yellow pigment producer) and *Trichoderma harzianum* (yellow-brown pigment producer) are discussed by Lebeau et al. [12]. Polyketide-based pigments from fungal submerged cultures are usually extracted by conventional liquid–liquid extraction methods requiring large volumes of various organic solvents and time. To address this question from a different angle, the authors proposed an investigation the use of three different aqueous two-phase extraction systems using either ammonium- or imidazolium-based ionic liquids. Their findings led them to conclude that (i) these alternative extraction systems using ionic liquids as the means for greener extractant worked well for this extraction of colored molecules from the fermentation broths of the filamentous fungi investigated; (ii) tetrabutylammonium bromide, [N4444]Br-, showed the best pigment extraction ability, with a higher putative affinity for azaphilone red pigments; (iii) the back extraction and recovery of fungal pigments from ionic liquid phases remained the limiting point of the method under their selected conditions for potential industrial applications. In another study, de Oliveira et al. [13] produce yellow-orange-red colorants with *Talaromyces amestolkiae* in a stirred-tank bioreactor. The effect of the pH-shift control strategy from 4.5 to 8.0 after 96-h of cultivation is evaluated at 500 rpm, resulting in an improvement of natural colorant production, with this increase being more significant for the orange and red examples, both close to four-fold. Next, the fermented broth containing the colorants is applied to the preparation of cassava starch-based films in order to incorporate functional activity in biodegradable films for food packaging. The presence of fermented broth did not affect the water activity and total solids of biodegradable films as compared with the standard. In the end, the films are used to pack butter samples (for 45 days) showing excellent results regarding antioxidant activity. It is demonstrated that the presence of natural colorants is obtained by a biotechnology process, which can provide protection against oxidative action, as well as be a functional food additive in food packing biomaterials. The last original paper and the third on fungal pigments and colorants deals with pigments derived from spalting fungi that have previously shown promise as textile dyes; however, their use has required numerous organic solvents with human health implications. Palomino Agurto et al. [14] explore the possibility of using linseed oil as a carrier for the pigment from *Scytalidium cuboideum* as a textile dye. Colored linseed oil effectively dyed a range of fabrics, with natural fibers showing better coloration. Scanning electron microscopy (SEM) revealed a pigment film over the fabric surface. While mechanical testing showed no strength loss in treated fabric, colorfastness tests showed significant changes in color in response to laundering and bleach exposure with variable effects across fabric varieties. SEM investigation confirmed differences in pigmented oil layer loss and showed variation in pigment crystal formation between fabric varieties. Heating of the pigmented oil layer was found to result in a bright, shiny fabric surface, which may have potential for naturally weatherproof garments.

A special issue is also an opportunity to go deeper into some scientific subjects via reviews. The first of six published in this issue describes how fungal metabolism meets modern demands for healthy aging, functional foods, and sustainability. Takahashi et al. [15] remind us that aging-associated, non-transmissible chronic diseases (NTCD) such as cancer, dyslipidemia, and neurodegenerative disorders have been challenged through several strategies including the consumption of healthy foods and the development of new drugs for existing diseases. Consumer health consciousness is guiding market trends toward the development of additives and nutraceutical products of natural origin. Fungi produce several metabolites with bioactivity against NTCD as well as pigments, dyes, antioxidants, polysaccharides, and enzymes that can be explored as substitutes for synthetic food additives. Research in this area has increased the yields of metabolites for industrial applications through improving fermentation conditions, application of metabolic engineering techniques, and fungal genetic manipulation. Several modern hyphenated techniques have impressively increased the rate of research in this area, enabling the analysis of a large number of species and fermentative conditions. Their review thus focuses on summarizing the nutritional, pharmacological, and economic importance of fungi and their metabolites resulting from applications in the aforementioned areas, examples of modern techniques for optimizing the production of fungi and their metabolites, and methodologies for the identification and analysis of these compounds.

The second review by Ye al. [16] focuses on selenium. Selenium (Se) is essential for human health, however, Se is deficient in soil in many places all around the world, resulting in human diseases, such as the notorious Keshan disease and Kashin–Beck disease. Therefore, Se biofortification is a popular approach to improve Se uptake and maintain human health. Beneficial microorganisms, including mycorrhizal and root endophytic fungi, dark septate fungi, and plant growth-promoting rhizobacteria (PGPRs), show multiple functions, especially increased plant nutrition uptake, growth and yield, and resistance to abiotic stresses. Such functions can be used for Se biofortification and increased growth and yield under drought and salt stress. The work summarizes the use of mycorrhizal fungi and PGPRs in Se biofortification, aiming to improving their practical use.

Another way to protect humans from chemical hazards is to find biological and sustainable alternatives to chemical fungicides. Zhang et al. [17] highlight this specific point for controlling postharvest decay of fruit. Fruit plays an important role in human diet. Whereas fungal pathogens cause huge losses of fruit during storage and transportation, abuse of chemical fungicides leads to serious environmental pollution and endangers human health. Antagonistic yeasts (also known as biocontrol yeasts) are promising substitutes for chemical fungicides in the control of postharvest decay owing to their widespread distribution, antagonistic ability, environmentally friendly nature, and safety for humans. Over the past few decades, the biocontrol mechanisms of antagonistic yeasts have been extensively studied, such as nutrition and space competition, mycoparasitism, and induction of host resistance. Moreover, combination of antagonistic yeasts with other agents or treatments was developed to improve biocontrol efficacy. Several antagonistic yeasts are used commercially. In the review of Zhang et al. [17], the application of antagonistic yeasts for postharvest decay control is summarized, including the antagonistic yeast species and sources, antagonistic mechanisms, commercial applications, and efficacy improvement. Issues requiring further study are also discussed.

In addition to the research papers dedicated to probiotics, two reviews illustrate the importance of this scientific field. Pais et al. [18] discuss how *Saccharomyces boulardii* became so successful. *Saccharomyces boulardii* is a probiotic yeast often used for the treatment of gastro-intestinal (GI) tract disorders such as diarrhea symptoms. It is genetically close to the model yeast *Saccharomyces cerevisiae* and its classification as a distinct species or a *S. cerevisiae* variant has long been discussed. Here, the authors review the main genetic divergencies between *S. boulardii* and *S. cerevisiae* as a strategy to uncover the ability to adapt to host physiological conditions by the probiotic. *S. boulardii*, which possesses discernible phenotypic traits and physiological properties that underlie its success as probiotic, such as optimal growth temperature, resistance to the gastric environment and viability at low pH. Its probiotic activity has been elucidated as a conjunction of multiple pathways, ranging from improvement of gut barrier function, pathogen competitive exclusion, production of antimicrobial peptides, immune modulation, and trophic effects. This work summarizes the participation of *S. boulardii* in these mechanisms and the multifactorial nature by which this yeast modulates the host microbiome and intestinal function. The second paper assigned to probiotics, written by Kunyeit et al. [19], summarize how probiotic yeasts can be applied against *Candida* species associated infections. Superficial and life-threatening invasive *Candida* infections are a major clinical challenge in hospitalized and immuno-compromised patients. Emerging drug-resistance among *Candida* species is exacerbated by the limited availability of antifungals and their associated side-effects. In the current review, authors discuss the application of probiotic yeasts as a potential alternative/combination therapy against *Candida* infections. Preclinical studies have identified several probiotic yeasts that effectively inhibit virulence of *Candida* species, including *Candida albicans*, *Candida tropicalis*, *Candida glabrata*, *Candida parapsilosis*, *Candida krusei* and *Candida auris*. However, *Saccharomyces cerevisiae* var. *boulardii* is the only probiotic yeast commercially available. In addition, clinical studies have further confirmed the in vitro and in vivo activity of the probiotic yeasts against *Candida* species. Probiotics use a variety of protective mechanisms, including posing a physical barrier, the ability to aggregate pathogens and to render them avirulent. Secreted metabolites such as short-chain fatty acids effectively inhibit the adhesion and morphological transition of Candida species. Overall, the probiotic yeasts could be a promising effective alternative or combination therapy for *Candida* infections.

Mushrooms have been valued as food and health supplements by humans for thousands of years. They are rich in dietary fiber, essential amino acids, minerals, and many bioactive compounds, especially those related to human immune system functions. Mushrooms contain diverse immunoregulatory compounds such as terpenes and terpenoids, lectins, fungal immunomodulatory proteins (FIPs) and polysaccharides. The distributions of these compounds differ among mushroom species and their potent immune modulation activities vary depending on their core structures and fraction composition chemical modifications. Zhao et al. [20] provide insight into the current status of clinical studies on immunomodulatory activities of mushrooms and mushroom products.

Mycotoxin is an important issue when dealing with human and animal health or nutrition. A specific outline by Chen et al. [21] discuss sphinganine-analog mycotoxins (SAMs). Sphinganine-analog mycotoxins (SAMs) include fumonisins from the *Fusarium* genus and *Alternaria alternata* f. sp. *lycopersici* (AAL) toxins. SAMs have shown diverse cytotoxicity and phytotoxicity, causing adverse impacts on plants, animals, and humans, and are a destructive force to crop production worldwide. The review summarizes the structural diversity of SAMs and encapsulates the relationships between their structures and biological activities. The toxicity of SAMs on plants and animals is mainly attributed to their inhibitory activity against the ceramide biosynthesis enzyme, influencing the sphingolipid metabolism and causing programmed cell death.

With the last articles, number 21 and 22 of this Journal of Fungi special issue, we return to fungal pigments. Fungi can produce a myriad of secondary metabolites, including pigments. Some of these pigments play a positive role in human welfare while others are detrimental. The review by Lin and Xu [22] summarizes the types and biosynthesis of fungal pigments, their relevance to human health, including their interactions with host immunity, and recent progress in studying their structure–activity relationships. Fungal pigments are grouped into carotenoids, melanin, polyketides, and azaphilones, etc. These pigments are phylogenetically broadly distributed. While the biosynthetic pathways for some fungal pigments are known, the majority remain to be elucidated.

Synthetic pigments/non-renewable coloring sources used normally in the textile industry release toxic substances into the environment, causing perilous ecological challenges. To be safer from such challenges, both academia and industry have explored the use of natural colorants such as microbial pigments. As explained by Venil et al. [23] such explorations have created a fervent interest among textile stakeholders in undertaking the dyeing of textile fabrics, especially with fungal pigments. The biodegradable and sustainable production of natural colorants from fungal sources is comparatively advantageous to synthetic dyes. The prospective scope of fungal pigments has emerged in the opening of many new avenues in textile colorants for wide ranging applications.

As a concluding remark, we, the Guest Editors, wish to thank all the authors and the reviewers for their significant contributions to this Special Issue and for making it a highly successful and timely collection of papers. Our acknowledgements also go to the whole MDPI team, i.e., assistant editors, editors, Editor-in-Chief, production office, website management, etc.

Funding: This research received no external funding.

Conflicts of Interest: The authors declare no conflict of interest.

References

1. Blackwell, M. The fungi: 1, 2, 3 ... 5.1 million species? *Am. J. Bot.* **2011**, *98*, 426–438. [CrossRef]
2. Kanpiengjai, A.; Khanongnuch, C.; Lumyong, S.; Kummasook, A.; Kittibunchakul, S. Characterization of *Sporidiobolus ruineniae* A45.2 Cultivated in Tannin Substrate for Use as a Potential Multifunctional Probiotic Yeast in Aquaculture. *J. Fungi* **2020**, *6*, 378. [CrossRef] [PubMed]
3. Keller, B.; Kuder, H.; Visscher, C.; Siesenop, U.; Kamphues, J. Yeasts in Liquid Swine Diets: Identification Methods, Growth Temperatures and Gas-Formation Potential. *J. Fungi* **2020**, *6*, 337. [CrossRef] [PubMed]
4. Mavrommatis, A.; Mitsiopoulou, C.; Christodoulou, C.; Karabinas, D.; Nenov, V.; Zervas, G.; Tsiplakou, E. Dietary Supplementation of a Live Yeast Product on Dairy Sheep Milk Performance, Oxidative and Immune Status in Peripartum Period. *J. Fungi* **2020**, *6*, 334. [CrossRef]
5. Nguyen, T.P.T.; Garrahan, M.A.; Nance, S.A.; Seeger, C.E.; Wong, C. Assimilation of Cholesterol by *Monascus purpureus*. *J. Fungi* **2020**, *6*, 352. [CrossRef] [PubMed]
6. Aguilar-Pérez, M.M.; Torres-Mendoza, D.; Vásquez, R.; Rios, N.; Cubilla-Rios, L. Exploring the Antibacterial Activity of *Pestalotiopsis* spp. under Different Culture Conditions and Their Chemical Diversity Using LC–ESI–Q–TOF–MS. *J. Fungi* **2020**, *6*, 140. [CrossRef] [PubMed]
7. Habeeb, T.H.; Abdel-Mawgoud, M.; Yehia, R.S.; Khalil, A.M.A.; Saleh, A.M.; AbdElgawad, H. Interactive Impact of Arbuscular Mycorrhizal Fungi and Elevated CO_2 on Growth and Functional Food Value of *Thymus vulgare*. *J. Fungi* **2020**, *6*, 168. [CrossRef]
8. Kim, Y.-W.; Wu, Y.; Choi, M.-H.; Shin, H.-J.; Li, J. Alginate-Derived Elicitors Enhance β-Glucan Content and Antioxidant Activities in Culinary and Medicinal Mushroom, *Sparassis latifolia*. *J. Fungi* **2020**, *6*, 92. [CrossRef]
9. Ibarra-Cantún, D.; Ramos-Cassellis, M.E.; Marín-Castro, M.A.; Castelán-Vega, R.d.C. Secondary Metabolites and Antioxidant Activity of the Solid-State Fermentation in Apple (*Pirus malus* L.) and Agave Mezcalero (*Agave angustifolia* H.) Bagasse. *J. Fungi* **2020**, *6*, 137. [CrossRef]
10. Dzurendova, S.; Zimmermann, B.; Tafintseva, V.; Kohler, A.; Horn, S.J.; Shapaval, V. Metal and Phosphate Ions Show Remarkable Influence on the Biomass Production and Lipid Accumulation in Oleaginous *Mucor circinelloides*. *J. Fungi* **2020**, *6*, 260. [CrossRef]
11. Slaný, O.; Klempová, T.; Shapaval, V.; Zimmermann, B.; Kohler, A.; Čertík, M. Biotransformation of Animal Fat-By Products into ARA-Enriched Fermented Bioproducts by Solid-State Fermentation of *Mortierella alpina*. *J. Fungi* **2020**, *6*, 236. [CrossRef]
12. Lebeau, J.; Petit, T.; Fouillaud, M.; Dufossé, L.; Caro, Y. Aqueous Two-Phase System Extraction of Polyketide-Based Fungal Pigments Using Ammonium- or Imidazolium-Based Ionic Liquids for Detection Purpose: A Case Study. *J. Fungi* **2020**, *6*, 375. [CrossRef] [PubMed]
13. De Oliveira, F.; de Azevedo Lima, C.; Lopes, A.M.; de Araújo Viana Marques, D.; Druzian, J.I.; Júnior, A.P.; Santos-Ebinuma, V.C. Microbial Colorants Production in Stirred-Tank Bioreactor and Their Incorporation in an Alternative Food Packaging Biomaterial. *J. Fungi* **2020**, *6*, 264. [CrossRef] [PubMed]
14. Agurto, M.E.P.; Gutierrez, S.M.V.; Van Court, R.C.; Chen, H.-L.; Robinson, S.C. Oil-Based Fungal Pigment from *Scytalidium cuboideum* as a Textile Dye. *J. Fungi* **2020**, *6*, 53. [CrossRef] [PubMed]
15. Takahashi, J.A.; Barbosa, B.V.R.; Martins, B.d.A.; Guirlanda, C.P.; Moura, M.A.F. Use of the Versatility of Fungal Metabolism to Meet Modern Demands for Healthy Aging, Functional Foods, and Sustainability. *J. Fungi* **2020**, *6*, 223. [CrossRef]
16. Ye, Y.; Qu, J.; Pu, Y.; Rao, S.; Xu, F.; Wu, C. Selenium Biofortification of Crop Food by Beneficial Microorganisms. *J. Fungi* **2020**, *6*, 59. [CrossRef]
17. Zhang, X.; Li, B.; Zhang, Z.; Chen, Y.; Tian, S. Antagonistic Yeasts: A Promising Alternative to Chemical Fungicides for Controlling Postharvest Decay of Fruit. *J. Fungi* **2020**, *6*, 158. [CrossRef] [PubMed]
18. Pais, P.; Almeida, V.; Yılmaz, M.; Teixeira, M.C. *Saccharomyces boulardii*: What Makes It Tick as Successful Probiotic? *J. Fungi* **2020**, *6*, 78. [CrossRef]
19. Kunyeit, L.; Anu-Appaiah, K.A.; Rao, R.P. Application of Probiotic Yeasts on *Candida* Species Associated Infection. *J. Fungi* **2020**, *6*, 189. [CrossRef]
20. Zhao, S.; Gao, Q.; Rong, C.; Wang, S.; Zhao, Z.; Liu, Y.; Xu, J. Immunomodulatory Effects of Edible and Medicinal Mushrooms and Their Bioactive Immunoregulatory Products. *J. Fungi* **2020**, *6*, 269. [CrossRef]
21. Chen, J.; Li, Z.; Cheng, Y.; Gao, C.; Guo, L.; Wang, T.; Xu, J. Sphinganine-Analog Mycotoxins (SAMs): Chemical Structures, Bioactivities, and Genetic Controls. *J. Fungi* **2020**, *6*, 312. [CrossRef] [PubMed]
22. Lin, L.; Xu, J. Fungal Pigments and Their Roles Associated with Human Health. *J. Fungi* **2020**, *6*, 280. [CrossRef] [PubMed]
23. Venil, C.K.; Velmurugan, P.; Dufossé, L.; Devi, P.R.; Ravi, A.V. Fungal Pigments: Potential Coloring Compounds for Wide Ranging Applications in Textile Dyeing. *J. Fungi* **2020**, *6*, 68. [CrossRef] [PubMed]

Journal of *Fungi*

MDPI

Article

Characterization of *Sporidiobolus ruineniae* A45.2 Cultivated in Tannin Substrate for Use as a Potential Multifunctional Probiotic Yeast in Aquaculture

Apinun Kanpiengjai [1,2,*], Chartchai Khanongnuch [3], Saisamorn Lumyong [2,4,5], Aksarakorn Kummasook [6] and Suwapat Kittibunchakul [7]

1 Division of Biochemistry and Biochemical Technology, Department of Chemistry, Faculty of Science, Chiang Mai University, Chiang Mai 50200, Thailand
2 Research Center of Microbial Diversity and Sustainable Utilization, Faculty of Science, Chiang Mai University, Chiang Mai 50200, Thailand
3 Division of Biotechnology, Faculty of Agro-Industry, Chiang Mai University, Mae-Hia, Chiang Mai 50100, Thailand; chartchai.k@cmu.ac.th
4 Division of Microbiology, Department of Biology, Faculty of Science, Chiang Mai University, Chiang Mai 50200, Thailand; scboi009@gmail.com
5 Academy of Science, The Royal Society of Thailand, Sanam Suea Pa, Dusit, Bangkok 10300, Thailand
6 Unit of Excellence in Infectious Disease, Department of Medical Technology, School of Allied Health Sciences, University of Phayao, Muang, Phayao 56000, Thailand; aksarakorn.ku@up.ac.th
7 Institute of Nutrition, Mahidol University, 999 Phutthamonthon 4 Rd., Nakhon Pathom 73170, Thailand; suwapatkt@gmail.com
* Correspondence: ak.apinun@gmail.com or apinun.k@cmu.ac.th

Received: 4 November 2020; Accepted: 14 December 2020; Published: 18 December 2020

check for updates

Abstract: At present, few yeast species have been evaluated for their beneficial capabilities as probiotics. *Sporidiobolus ruineniae* A45.2, a carotenoid-producing yeast, was able to co-produce cell-associated tannase (CAT), gallic acid and viable cells with antioxidant activity when grown in a tannic acid substrate. The aim of this research study was to identify the potential uses of *S. ruineniae* A45.2 obtained from a co-production system as a potential feed additive for aquaculture. *S. ruineniae* A45.2 and its CAT displayed high tolerance in pH 2.0, pepsin, bile salts and pancreatin. Furthermore, its viable cells were characterized by moderate hydrophobicity, high auto-aggregation and moderate co-aggregation with *Staphylococcus aureus*, *Salmonella* ser. Thyphimurium and *Streptococcus agalactiae*. These attributes promoted *S. ruineniae* A45.2 as a multifunctional probiotic yeast. In addition, the intact cells possessed antioxidant activities in a 100–150 µg gallic acid equivalent (GAE)/mL culture. Remarkably, the fermentation broth demonstrated higher antioxidant activity of 9.2 ± 1.8, 9.0 ± 0.9, and 9.8 ± 0.7 mg GAE/mL culture after FRAP, DPPH and ABTS assays, respectively. Furthermore, higher antimicrobial activity was observed against *Bacillus cereus*, *Staphylococcus aureus* and *Strep. agalactiae*. Therefore, cultivation of *S. ruineniae* A45.2 with a tannic acid substrate displayed significant potential as an effective multifunctional feed additive.

Keywords: feed additive; probiotic; yeast; *Sporidiobolus ruineniae*; tannase

---

1. Introduction

Aquaculture is the most rapidly growing sector of food production throughout the world. Its global demand continues to rise as it is applied to farming practices in ever-increasing proportions [1]. Industrial-scale aquaculture production is recognized as a challenge within the farming sector as it requires sustainable and efficient technologies [2] that address concerns of potential human exposure to microbial diseases and the possibility of severe economic losses across the industry [3]. The use of antibiotics is a simple approach in the control of diseases, however, excess antibiotics that are discharged into the environment are known to be responsible for the spread of antibiotic-resistant genes of pathogenic and commensal bacteria, all of which can lead to increases in drug resistance among animal and human populations. The addition of additives to antibiotics, vaccines, immunostimulants, prebiotics and probiotics, in particular, is an environmentally friendly alternative and recognized as a sustainable strategy [2]. The Food and Agriculture Organization (FAO)/World Health Organization (WHO) defined probiotics as live microorganisms that, when administered in adequate amounts, confer a health benefit upon the host [4]. In terms of their potential applications in aquaculture, yeasts are considered a second leading source of probiotics after bacteria. However, the use of probiotic yeasts is not as popular as bacteria. Indicative of a greater potential for profit than bacterial probiotics, yeasts are not affected by antibacterial compounds and are known to contain various immunostimulant compounds [3,5]. However, their applicable use among a wide variety of animals is limited due to the fact that the normal body temperature of animals is higher than the temperature for the optimal growth of yeast. To date, two yeast species, namely *Saccharomyces cerevisiae* and *Debaryomyces hansenii*, are widely recognized as potential probiotic yeasts [5]. Additionally, yeasts isolated from fish microbiota exhibit certain probiotic properties. These yeast species include *Candida deformans*, *Rhodotorula mucilaginosa*, *Yarrowia lipolytica*, *Metschnikowia viticola* and *Cryptococcus laurentii* [3].

Sporidiobolus ruineniae A45.2, isolated from fermented tea-leaves of northern Thailand, namely Miang, is a pigment-producing and tannin-tolerant type of yeast [6]. It is among the range of yeasts commonly found in the intestines of humans [7]. Based on evidence established by previous studies [6,8], *S. ruineniae* A45.2 is assumed to have a unique cell wall structure that serves its tannin-tolerance and may promote the organism as a potential probiotic yeast. On the other hand, its carotenoid pigment is considered highly valuable in terms of the enhancement of some aquaculture pigmentations [9]. Moreover, *S. ruineniae* A45.2 is capable of producing thermostable and pH-stable cell-associated tannase (CAT) and can degrade tannic acid to gallic acid [10]. Tannase is a feed additive enzyme that plays an important role in the reduction of tannins, an antinutritional factor in animal feed. The enzymatic degradation of tannins releases gallic acid that can be used as both an antimicrobial and an antioxidant agent. In aquaculture feed ingredients, tannins come from plant-derived, alternate fish feed ingredients that are used as protein sources, such as soybean meal, rapeseed meal, pea seed meal and mustard oil cake [11]. This drawback leads to significantly decreased levels of cumulative feed intake and digestibility [12]. However, this can be overcome by the addition of tannase.

In previous studies involving the co-production of gallic acid and CAT derived from tannins, both cells and the culture broth rich in gallic acid may be used as a potential source of tannase and gallic acid in the feed industry, respectively. The aims of this research study were to evaluate *S. ruineniae* A45.2 for its potential to be used as a probiotic in aquaculture. Our objectives were to also investigate the potential for fermented broth cultivated in a tannic acid substrate to be further applied as a multifunctional feed additive. Further carotenoids produced by this yeast were also characterized.

2. Materials and Methods

2.1. Chemicals

Bile salts, 40× pancreatin, pepsin, methyl gallate, gallic acid, rhodanine, 2,4,6-Tris (2-pyridyl)-*s*-triazine (TPTZ), 2,2-diphenyl-1-picrylhydrazyl (DPPH), 2,2′-azino-bis (3-ethylbenzothi-azoline-6-ysulfonic acid (ABTS) and potassium persulfate were all of analytic grade and of the highest quality available from Sigma and Sigma-Aldrich (St. Louis, MO. USA). All media used in this research study, including yeast–malt extract broth (YMB), nutrient broth (NB), trypticase soy broth (TSB) and agar, were purchased from HiMedia (Nashik, India).

2.2. Microorganisms and Culture Conditions

S. ruineniae A45.2 was maintained on yeast-malt extract agar (YMA) at 4 °C for further use. To prepare the seed inoculum, a single colony of yeast was inoculated in YMB and incubated at 30 °C on a 150 rpm rotary shaker for 15–18 h or until the optical density at 600 nm reached 2.0–3.0. *Escherichia coli* TISTR 527, *Salmonella* ser. Thyphimurium TISTR 1472, *Staphylococcus aureus* TISTR 746 and *Bacillus cereus* TISTR 747 were maintained on nutrient agar (NA) and were grown in a nutrient broth (NB) at 37 °C on a 100 rpm rotary shaker when necessary. Furthermore, *Listeria monocytogenes* DMST 17303 and *Streptococcus agalactiae* DMST 11366 were maintained on trypticase soy agar (TSA). These pathogenic bacteria were grown in TSB at 37 °C on a 100 rpm rotary shaker when necessary.

2.3. Co-Production of Gallic Acid, CAT and Viable Cells of S. ruineniae A45.2

A total of 10% (*v/v*) of the prepared seed inoculum was transferred into a 1 L stirred tank fermenter (B.E. Marubishi Co. Ltd., Tokyo, Japan) with a 60% working volume of the optimized medium [10] that contained 12.3 g/L tannic acid, 6.91 g/L glucose, 10 g/L yeast extract, 2 g/L $(NH_4)_2SO_4$, 0.5 g/L tween 80 and 1 g/L glutamate (pH 6.0). Culture conditions were administered at 30 °C with an aeration rate of 1 vvm and an agitation speed of 250 rpm. After 48 h of cultivation, the cells were harvested by centrifugation at 8000 rpm, 4 °C for 10 min and washed twice with phosphate buffer saline (PBS) supplemented with 0.1% (*v/v*) triton X-100. The cell pellets were then resuspended in 0.85% (*w/v*) NaCl to obtain a concentration of 10^8 cells/mL for further experimentation.

2.4. Tolerance of S. ruineniae A45.2 and Stability of Tannase at Low pH Values

A total of 0.5 mL of the prepared cell suspension (10^8 cells/mL) was transferred to a 125 mL Erlenmeyer flask containing 49.5 mL of 0.85% (*w/v*) NaCl adjusted to pH 2.0 and 3.0 by 0.1 N HCl. The cell suspension in PBS was then used as a control. All mixtures were incubated at 30 °C for 4 h. Samples were periodically taken for measurement of viable cells by plate count technique and residual tannase activity. Initial cell concentration and tannase activity without incubation were set to 100%.

2.5. Tolerance of S. ruineniae A45.2 and Stability of Tannase in Simulated Gastric Juice

A total of 0.5 mL of the prepared cell suspension (10^8 cells/mL) was transferred to a 125 mL Erlenmeyer flask containing 49.5 mL of simulated gastric juice (0.3% (*w/v*) pepsin, 0.85% (*w/v*) NaCl, pH 2.0). Cell suspension in PBS was used as a control. All mixtures were incubated at 30 °C for 4 h. Samples were periodically taken for measurement of viable cells by plate count technique and to determine residual tannase activity. Initial cell concentration and tannase activity without incubation was set to 100%.

2.6. Bile Salt Tolerance of S. ruineniae A45.2 and Stability of Tannase

A total of 0.5 mL of the prepared cell suspension (10^8 cells/mL) was transferred to a 125 mL Erlenmeyer flask containing 49.5 mL of solution that consisted of 0.85% (*w/v*) NaCl and 0.3% (*w/v*) bile salts. Cells suspended with PBS were used as a control. All mixtures were incubated at 30 °C for 6 h. Samples were periodically taken for measurement of viable cells by the plate count technique to determine residual tannase activity. Initial cell concentration and tannase activity without incubation were set to 100%.

2.7. Tolerance of S. ruineniae A45.2 and Stability of Tannase in Simulated Intestinal Fluid

A total of 0.5 mL of the prepared cell suspension (10^8 cells/mL) was transferred to a 125 mL Erlenmeyer flask containing 49.5 mL of simulated intestinal fluid (0.3% (*w/v*) bile salts, 0.3% (*w/v*) pancreatin and 0.85% (*w/v*) NaCl). Cells suspended with PBS were used as a control. All mixtures were incubated at 30 °C for 8 h. Samples were periodically taken for measurement of viable cells by plate count technique to determine residual tannase activity. Initial cell concentration and tannase activity without incubation were set to 100%.

2.8. Assay of Tannase

Tannase activity was assayed using the previously described method [13]. Briefly, 50 µL of enzyme solution was mixed with 50 µL of 12.5 mM methyl gallate in 100 mM citrate–phosphate buffer pH 6.5 and incubated at 30 °C. After the incubation procedure, the reaction was stopped by adding 60 µL of 0.667% (*w/v*) methanolic rhodanine and the mixture was left at room temperature (25 °C) for 5 min. Subsequently, 40 µL of 0.5 M KOH was added to the mixture, which was then left at room temperature for 5 min prior to adding 800 µL of distilled water. Absorbance of the mixture was measured at 520 nm. One unit of tannase was defined as the amount of enzyme that liberated 1 µmol of gallic acid per minute under assay conditions.

2.9. Cell Surface Hydrophobicity

Yeast adherence was determined by cell surface hydrophobicity. The cell suspension (3 mL) ($A_{initial}$) was transferred to a glass tube (12 × 100 mm) containing 1 mL of chloroform, agitated using a vortex mixer for 2 min and allowed to stand at room temperature for 30 min. The optical density of the aqueous phase (A_{final}) was measured at a wavelength of 600 nm. The hydrophobicity index (HPBI) was calculated using the following equation:

$$\text{HPBI } (\%) = (1 - \frac{A_{final}}{A_{initial}}) \times 100$$

2.10. Auto-Aggregation Assay

A total of 3 mL of the yeast suspension in PBS ($A_{initial}$) was transferred to a glass tube (12 × 100 mm), vortexed for 10 s and incubated at 30 °C for 2 h. Absorbance of the upper part of the mixture (approximately 1 mL) was measured at 600 nm (A_{final}). Auto-aggregation was calculated using the following equation:

$$\text{Auto} - \text{aggregation}(\%) = (1 - \frac{A_{final}}{A_{initial}}) \times 100$$

2.11. Co-Aggregation Assay

Equal volumes (1.5 mL) of the yeast suspension (A_{yeast}) and pathogenic bacterium ($A_{pathogen}$) were transferred into a glass tube (12 × 100 mm), vortexed for 30 s and incubated at 30 °C for 2 h. Absorbance of the upper part of the mixture (A_{mix}) was measured at 600 nm. Co-aggregation was calculated using the following equation:

$$\text{Co} - \text{aggregation}(\%) = (1 - \frac{A_{mix}}{(A_{yeast} + A_{pathogen})/2}) \times 100$$

2.12. Adherence of Bacteria onto Yeast Cells

Adhesion of bacteria onto yeast cells was accomplished by mixing 1 mL of the yeast suspension (10^8 cells/mL) in PBS and 1 mL of each pathogenic bacteria (10^8 cells/mL). The specimens were then incubated at 30 °C. After 2 h of incubation, 10 µL of the mixture was smeared onto a microscopic slide for Gram-staining [14]. The Gram-stained slide was then used to visualize the adherence of the bacteria onto the yeast cells under a phase-contrast light microscope. The pathogenic bacteria used in the adherence test were *B. cereus*, *E. coli*, *Sal.* Typhimurium, *Staph. aureus*, *L. monocytogenes* and *Strep. agalactiae*.

2.13. Determination of Antimicrobial Activity

The agar well diffusion method was used to determine antagonistic activity of culture broth obtained from the co-fermentation and fermentation in YMB. Briefly, an overnight culture (approximately 10^6–10^8 CFU/mL) of the pathogenic bacteria was swabbed onto an NA plate for *B. cereus*, *E. coli*, *Sal.* Thyphimurium and *Staph. aureus* and a TSA plate for *L. monocytogenes* and *Strep. agalactiae*. The wells were prepared by being punched with a 6 mm diameter sterile cork-borer and were filled with 50 μL of sterile culture broth or 50 μg/mL of chloramphenicol as the positive control. The plates were incubated at 30 °C for 18 h.

2.14. Analysis of Carotenoids

The previously obtained cell pellet of *S. ruineniae* A45.2 was lyophilized into a dry cell for carotenoid extraction. The freeze-dried cell (0.25 g) was then placed into a screw-cap glass tube (25 × 150 mm) containing 10 mL of acetone and 30 g of glass beads. Cell disruption was carried out through vigorous mixing for 10 min at room temperature. The mixture was filtered through filter paper to collect the cell extract and then centrifuged at 10,000 rpm for 10 min for the purposes of clarification. To quantify total carotenoids, absorbance of the clear extract was measured at 485 nm. The total carotenoid content of the yeast cells was calculated based on the extinction coefficient ($E^{1\%}_{1\ cm}$) of 2680 and expressed in terms of total carotenoids (μg)/g dry cell weight. To determine the carotenoid composition, individual carotenoids were separated by Mightysil RP-18 GP prepacked column (150 × 2.0 mm; Kanto Chemical Co., Inc., Tokyo, Japan) equilibrated with a solution of methanol/acetonitrile (90:10 *v/v*). The conditions were carried out at 30 °C with a flow rate of 1.0 mL/min. The separated carotenoids were detected using a UV detector at 485 nm. Meanwhile, the evaporated cell extract was resuspended in acetone. This was then spotted on a thin layer chromatography (TLC) Silica gel plate (Silica gel 60 F_{254}, Merck Millipore, Germany) and developed in a chamber containing hexane/acetone (70:30 *v/v*). The pigments separated by TLC were recovered and dissolved in acetone for measurement of the wavelength of maximum absorbance (λ_{max}) using a UV-visible spectrophotometer.

2.15. Assay of Antioxidants

A culture of *S. ruineniae* A45.2 obtained from the co-production system was harvested by centrifugation at 8000 rpm for 10 min. The cell pellet was washed twice with PBS solution, suspended in the same solution and the fermented broth was then collected. Both intact cells and cell-free extract were determined for antioxidant activity using three different methods, including ferric-reducing antioxidant power (FRAP), DPPH free-radical-scavenging activity and ABTS free-radical-scavenging activity.

For the FRAP assay, the FRAP reagent consisted of 300 mM acetate buffer pH 3.6, a solution of 10 mM TPTZ in 40 mM HCl and 20 mM $FeCl_3$ at a ratio of 10:1:1 (*v/v/v*). The sample solution (0.10 mL) was mixed thoroughly with 3.40 mL of the FRAP reagent for 30 min prior to measuring the absorbance at 593 nm. A standard curve was prepared using different concentrations of gallic acid. The results were expressed in terms of milligram gallic acid equivalent (GAE)/mL culture.

For the DPPH assay, a sample (0.25 mL) was mixed with freshly prepared 40 ppm methanolic DPPH (2.25 mL) and allowed to stand in the dark at room temperature. A decrease in absorbance at 517 nm was determined after 10 min of the incubation process. The concentration of the sample that produced between 20% and 80% inhibition of the blank absorbance was determined and adapted. Radical scavenging activity was expressed as the concentration of the extract required for reduction of the initial concentration of DPPH by 50% (EC_{50}) under specified experimental conditions. DPPH radical scavenging activity was expressed in terms of mg GAE/mL culture.

For the ABTS assay, 0.0384 g of ABTS was prepared in 10 mL of water. Subsequently, 5 mL of the solution was mixed with 88 μL of 140 mM potassium persulfate and adjusted to 25 mL with deionized water in a volumetric flask for further experimentation. A total of 1.75 mL of the ABTS solution was mixed thoroughly with 0.25 mL of the sample and allowed to stand in the dark at room temperature for 10 min. A decrease in absorbance at 734 nm was measured. Radical scavenging activity was expressed as the concentration of the extract required for reduction of the initial concentration of ABTS by 50% (EC_{50}) under specified experimental conditions. ABTS radical scavenging activity was expressed in terms of mg GAE/mL culture.

3. Results

3.1. Survival of Yeast and CAT Stability under Simulated Gastrointestinal Tract (GIT) Conditions

Gastric and intestinal conditions were simulated for evaluation of the probiotic properties of *S. ruineniae* A45.2. Under simulated gastric conditions, cell viability of *S. ruineniae* A45.2 and its residual CAT activity were evaluated at pH 2.0 and pH 3.0 (Figure 1a,b) and carried out with and without (Figure 1c,d) the supplementation of pepsin. The results revealed that the values of both pH and pepsin did not significantly affect cell viability of the yeast or its CAT. At a pH value of 2.0, both in the presence and in the absence of pepsin, approximately 90% of the initial viable cells and residual CAT activity were retained. Furthermore, at a pH value of 3.0, the supplementation of pepsin slightly decreased tannase activity but had no effect on cell viability.

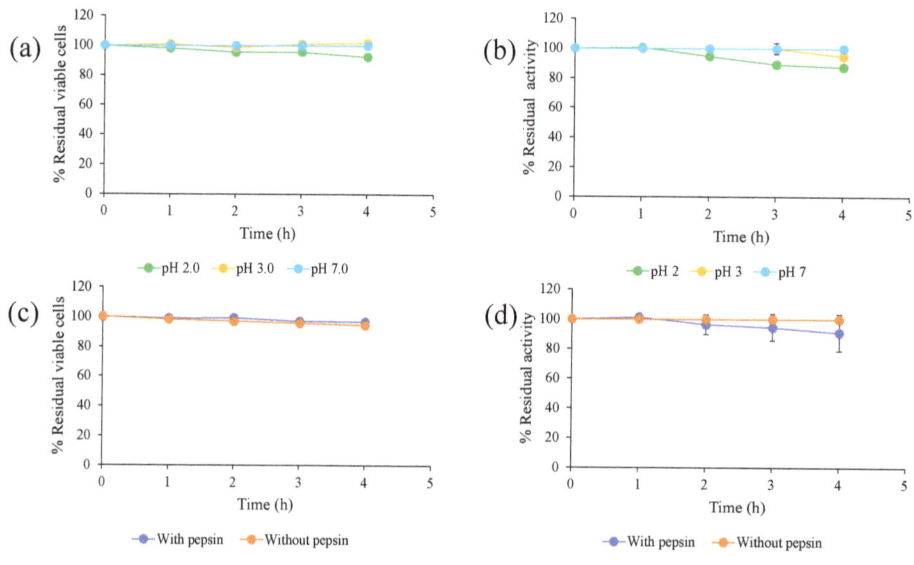

Figure 1. Effect of pH 2.0 and pH 3.0 on (**a**) survival of *S. ruineniae* A45.2 and (**b**) residual CAT activity and effect of pepsin on (**c**) survival of *S. ruineniae* A45.2 and (**d**) residual CAT activity.

The same results were also obtained when *S. ruineniae* A45.2 and CAT were incubated under simulated intestinal conditions, in which bile salts and pancreatin acted as key factors. It was revealed that *S. ruineniae* A45.2 and CAT were resistant to bile salts (Figure 2a,b) and pancreatin (Figure 2c,d), as they retained 100% residual viable cells and tannase activity after incubation.

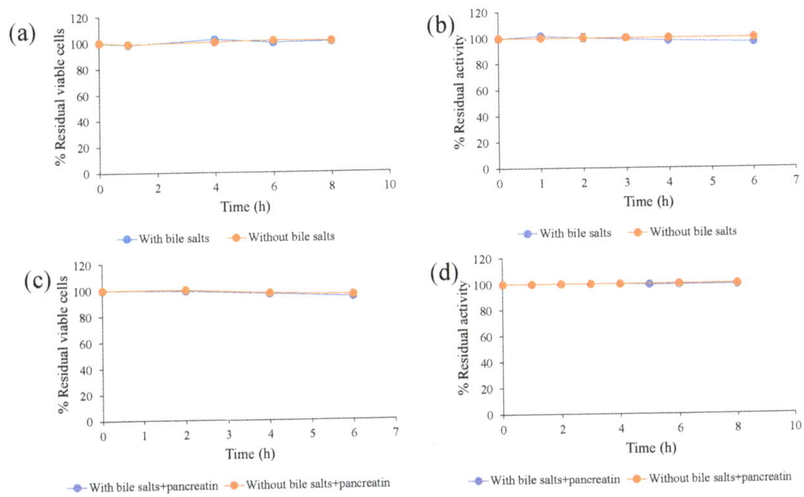

Figure 2. Effect of bile salts on (**a**) survival of *S. ruineniae* A45.2 and (**b**) residual CAT activity and effect of pancreatin in combination with bile salts on (**c**) survival of *S. ruineniae* A45.2 and (**d**) residual CAT activity.

3.2. Cell Surface Hydrophobicity, Auto-Aggregation and Co-Aggregation

Cell surface hydrophobicity, auto-aggregation and co-aggregation of *S. ruineniae* A45.2 are presented in Table 1. *S. ruineniae* A45.2 displayed 58.4 ± 2.7% cell surface hydrophobicity. Significantly high values of auto-aggregation were observed at up to 88.2 ± 1.2% along with the ability to be co-aggregated with pathogenic bacteria, including *B. cereus, E. coli, Staph. aureus, Sal.* Thyphimurium, *L. monocytogenes* and *Strep. agalactiae* at different percentages of co-aggregation. *S. ruineniae* A45.2 displayed a stronger co-aggregation ability with *Strep. agalactiae, Sal.* Thyphimurium and *Staph. aureus* than other tested pathogenic bacteria. This evidence was in agreement with their adherence ability (Figure 3), wherein pathogenic bacteria obviously adhered to the yeast cells.

Table 1. Cell surface hydrophobicity, auto-aggregation and co-aggregation against pathogenic bacteria.

Properties	%
Cell surface hydrophobicity	58.4 ± 2.7
Auto-aggregation	88.2 ± 1.2
Co-aggregation	
B. cereus	36.2 ± 2.8
E. coli	33.8 ± 0.7
Staph. aureus	44.0 ± 0.8
Sal. Thyphimurium	45.8 ± 0.1
L. monocytogenes	37.8 ± 2.1
Strep. agalactiae	51.5 ± 2.6

3.3. Antimicrobial Activity

The antagonistic effects of *S. ruineniae* A45.2 were observed after being exposed to various indicator microorganisms, including *B. cereus, E. coli, Sal.* Thyphimurium, *Staph. aureus, L. monocytogenes* and *Strep. agalactiae*. Culturing periods of 24 and 48 h of the co-production system were tested in comparison with specimens cultured in YMB. Only the supernatant obtained from co-production exhibited antimicrobial activity against some pathogenic bacteria (Figure 4), i.e., *B. cereus, Staph. aureus* and *Strep. agalactiae*.

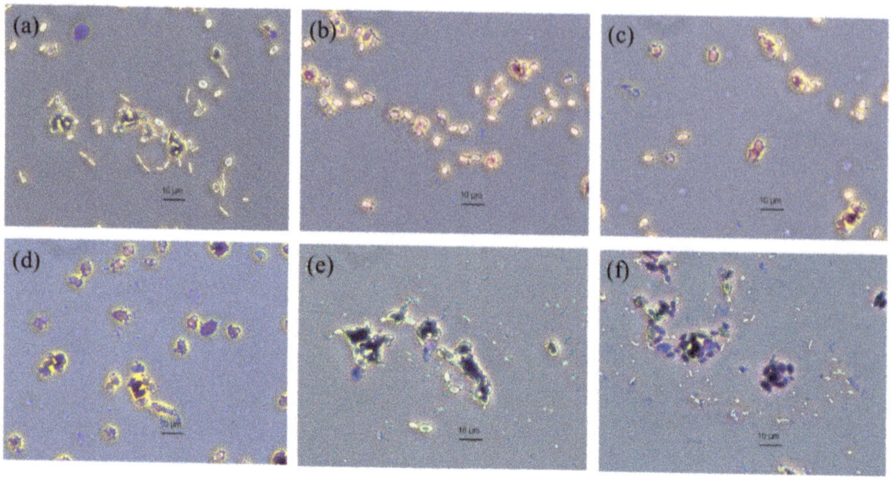

Figure 3. Adherence of (**a**) *B. cereus*, (**b**) *E. coli*, (**c**) *Staph. aureus*, (**d**) *Sal.* Thyphimurium, (**e**) *L. monocytogenes* and (**f**) *Strep. agalactiae* on yeast cell walls observed under a phase-contrast light microscope at 100× magnification.

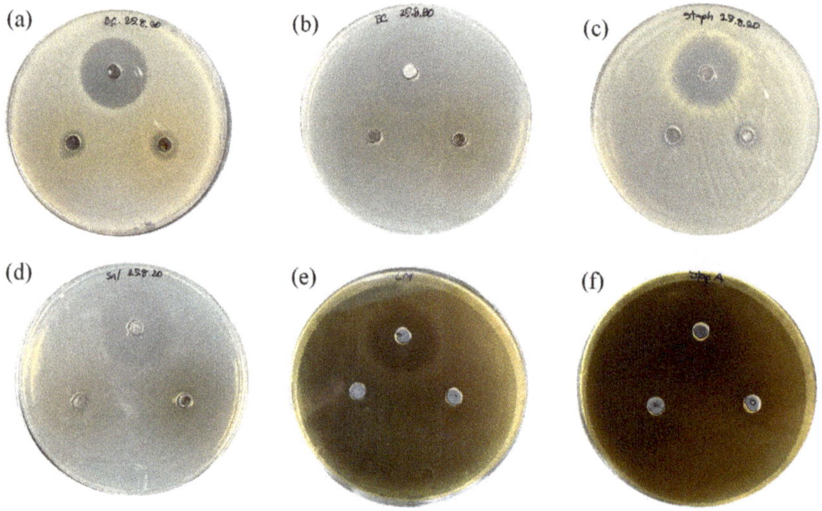

Figure 4. Antimicrobial activity of cell-free extract of *S. ruineniae* A45.2 cultivated in tannic acid substrate at 30 °C for 24 (left) and 48 h (right) of cultivation against (**a**) *B. cereus*, (**b**) *E. coli*, (**c**) *Staph. aureus*, (**d**) *Sal.* Thyphimurium, (**e**) *L. monocytogenes* and (**f**) *Strep. agalactiae* compared to control (top) (50 μg/mL chloramphenicol)

3.4. Carotenoids Produced by S. ruineniae A45.2

S. ruineniae A45.2 grown in tannic acid were harvested after 48 h of cultivation, lyophilized and used for carotenoid extraction. Identification and characterization of the carotenoid pigment was performed by HPLC. Three main peaks were separated from the carotenoid extracts (Figure 5a). These peaks corresponded to the three spots that were isolated and visualized by TLC (Figure 5b). The first spot from the bottom was rosy-red in color and migrated with slower mobility than the second spot, which appeared orange to red in color, while the third spot was yellow in color and migrated with the same degree of mobility as β-carotene. These pigments were identified based on their Visible absorbance maxima (Figure 5c). The least degree of polar pigment was identified as β-carotene due to similar absorbance maxima values recorded at 429 nm and 485 nm, with the maximal degree of absorbance recorded at 456 nm. The most notable polar pigment revealed a spectrum with three absorption maxima at wavelengths of 474, 527 with the absorption optimum at 499 nm, thereby identified as torularhodin. The second most polar pigment was torulene with three absorbance maxima values recorded at 462 nm and 516 nm, with an absorption maximum value recorded at 488 nm. Total carotenoids produced by *S. ruineniae* A45.2 after cultivation at 30 °C for 48 h were recorded at 88.0 ± 0.2 μg/g dry cell weight, equivalent to 500 ± 130 μg/L culture.

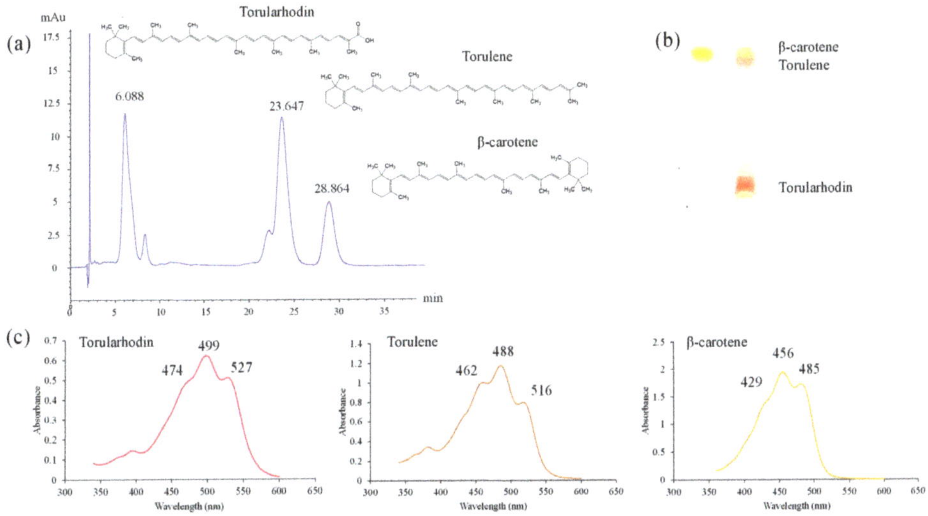

Figure 5. Characterization of pigments produced by *S. ruineniae* A45.2 (**a**) Separation of pigments by HPLC and (**b**) TLC and (**c**) Visible absorption spectra of the major pigments.

3.5. Antioxidant Activity

Antioxidant activities of intact cells of *S. ruineniae* A45.2 and cell-free extract obtained from the co-production system were measured using FRAP, DPPH and ABTS assays. Overall, there were no significant differences in the antioxidant activities among the different methods. Antioxidant activities ranging from 100–120 μg GAE/mL culture and 110–150 μg GAE/mL culture were detected from the intact cells obtained from 24-h and 48-h periods of cultivation, respectively (Table 2). The cell-free extract revealed significantly higher antioxidant activities than the intact cells. The antioxidant activities of 5.6–6.8 mg GAE/mL culture and 9.0–9.8 mg GAE/mL culture were obtained from the cell-free extract fraction after 24-h and 48-h periods of cultivation, respectively. Activities of both intact cells and cell-free extract increased relative to the incubation time of the co-production system.

Table 2. FRAP, DPPH and ABTS antioxidant activity of cell-free extract and intact cells obtained from co-production of gallic acid and viable cells of *S. ruineniae* A45.2.

Time (h)	Cell-Free Extract (mg GAE/mL)			Intact Cells (μg GAE/mL)		
	FRAP	**DPPH**	**ABTS**	**FRAP**	**DPPH**	**ABTS**
24	5.6 ± 0.8	6.8 ± 0.6	6.4 ± 0.4	122.1 ± 9.3	108.5 ± 2.4	104.7 ± 0.8
48	9.2 ± 1.8 *	9.0 ± 0.9 *	9.8 ± 0.7 *	143.6 ± 3.4 *	114.6 ± 1.4 *	111.1 ± 0.5 *

* significant difference within a column ($p < 0.05$).

4. Discussion

In this study, *S. ruineniae* A45.2 and its culture broth obtained from the co-production of cells, gallic acid and tannase were characterized for their potential use in animal feed, specifically in feed prepared for fish and other aquatic organisms. *S. ruineniae* A45.2 was isolated from Miang, which is rich in tannins and considered a microbial inhibitor [6,8]. The cell wall structure and composition of *S. ruineniae* A45.2 are believed to promote its growth along with high concentrations of tannic acid, yet this yeast was found to be a promising probiotic. When used as a functional probiotic yeast, growth temperature is a crucial limitation for the application of probiotics in animals, since the yeast must be able to survive and grow at the animal's normal body temperature in order to enhance the animal's growth performance and promote the health of the animal [15]. Typically, the growth temperatures of yeasts range from 0 to 47 °C with an optimal temperature between 25 and 30 °C [16], yet probiotic yeasts might actively function when they are used in aquaculture. Probiotic yeasts are less popular than bacteria but can offer some major physiological contributions over bacteria. These include their cell volume and the production of a wide spectrum of simple and more complex compounds that may be beneficial to the health of aquatic organisms. However, only a few varieties of probiotic yeasts have been isolated for aquaculture applications. It was reported that marine and other aquatic environments, along with the gut microbiota of aquatic organisms, are potential sources of probiotic yeasts. In addition to *S. cerevisiae*, *D. hansenii* is a ubiquitous yeast that is frequently associated with fish and marine environments [7]. As of yet, no reports of using *S. ruineniae* as a probiotic yeast have been identified.

To be a good probiotic yeast, it must be able to successfully survive under gastrointestinal tract (GIT) conditions and provide beneficial conditions for the enhancement of the health of the host. In this study, both *S. ruineniae* A45.2 and its CAT were exposed to GIT conditions in order to assess the degree of residual cell viability and CAT. The temperature used in this research study was 30 °C, as it was identified as an optimal temperature of *S. ruineniae* A45.2 (data not shown). Considering cell viability, *S. ruineniae* A45.2 resisted low pH values ranging from 2.0 to 3.0, which were within the range found in the stomachs of fish. The degree of acidity in the stomach of a fish can vary depending on the fullness of the stomach and the species of the fish [17,18]. Moreover, *S. ruineniae* A45.2 was not found to be affected by the digestive enzymes we tested, namely pepsin and pancreatin (a mixture of amylase, protease and lipase). These attributes are considered important selection criteria for a good probiotic yeast [19]. On the other hand, the CAT of *S. ruineniae* A45.2 exhibited a good degree of thermostability and pH stability. Surprisingly, positive stability values were observed under simulated GIT conditions by retaining more than 90% of initial activity after treatment. This indicates that the yeast species could be applicable in the aquafeed industry. Plant-based products in fish diets contain valuable proteins used to replace fishmeal. These plant feed ingredients contain considerable amounts of tannins that can have an adverse effect on animals by reducing the nutritional value of the feed [11]. This circumstance can also decrease the palatability of the feed due to an unpleasant taste caused by a high concentration of tannins [12]. The results of this study indicate that both cells of *S. ruineniae* A45.2 might be able to survive in transit through the stomach and small intestines and function effectively in the large intestines. However, its CAT might be stable in stomach environments and could be active in the intestines, as the environments are similar to the known optimal values for pH and temperature.

Cell surface hydrophobicity is defined as a nonspecific interaction in adhesion between probiotic microorganisms onto GIT epithelial cells, where they may provide prophylactic and therapeutic benefits [20]. Colonization in the intestinal epithelial cell wall and mucosal surfaces can prevent pathogenic bacteria adhesion and inflammatory reactions [21]. Yet, hydrophobicity is an important attribute for selecting potential probiotics. *S. ruineniae* A45.2 showed high cell surface hydrophobicity toward chloroform and was comparable with those reported in *Bacillus subtilis* [19,21], various strains of *Lactobacillus* sp. [22] and *Sac. unisporus* [20].

Auto-aggregation is defined as aggregation among yeast cells to form flocs and colonize the intestinal environment of the host when the cells approach harmful conditions [20,23]. Probiotic microorganisms should be associated with higher auto-aggregation than pathogenic microorganisms [22], specifically *Strep. agalactiae*, a representative fish pathogen. Under the same experimental conditions as this study, the percentage auto-aggregation of pathogenic bacteria ranged between 15–35% for *L. monocytogenes*, *Sal.* Thyphimurium and *Staph. aureus* [22]. Within 2 h of the auto-aggregation test, *S. ruineniae* reported 88.2 ± 1.2%, which was higher than previously reported probiotic yeasts, namely *P. kluyveri*, *Issatchenkia orientalis*, *P. kudriavzevii* [24], *Yarrowia lipolytica*, *Wickerhamomyces anomalus* and *Sac. cerevisiae* [23]. Auto-aggregation capacity is strain-specific, while a capacity greater than 50% displayed the potential to prevent the invasion of various other pathogenic microorganisms through film formation.

Co-aggregation is defined as the close interaction between probiotics and different pathogenic bacteria [23]. It was reported that adherence of enteric bacteria onto yeast cells is irreversible, thus transient passage of the bacteria occurs through GIT and subsequent flushing out in the feces [20]. The co-aggregation ability of *S. ruineniae* A45.2 agreed with its adherence ability. This could be explained by the specific fimbriae present on bacteria with mannan on yeast cells and the electrostatic and hydrophobic nonspecific bindings [20].

No antimicrobial activity of *S. ruineniae* A45.2 against the tested pathogenic bacteria was detected when it was cultivated in YMB. It is therefore implied that no antimicrobial metabolite was produced by the organisms typically identified in various yeasts [20,24–26]. Most yeasts scavenge pathogenic infection by indirect mechanisms such as auto-aggregation, co-aggregation and adherence ability [27]. On the contrary, the growing of *S. ruineniae* A45.2 in tannic acid containing medium led to the release of gallic acid, which enhanced the antimicrobial activity of the culture broth against *B. cereus*, *E. coli*, *Staph. aureus* and *Strep. agalactiae*. The results suggest that production of the yeast should be performed in the presence of tannic acid to promote gallic acid production and CAT, thereby gaining antimicrobial activity. Supplementation of gallic acid in animal feed, especially aquatic feed, was scarcely reported. Current research found that the supplementation of gallic acid in broiler diets at levels ranging from 75 to 100 mg/kg improved the performance of broiler chicks in terms of feed utilization, breast muscle yield and oxidative stability, while positively modulating jejunum intestinal morphology [28]. Hence, our results provide supplemental, supportive evidence for the use of gallic acid as an alternative to antibiotics in animal feed or for the determination of synergistic interactions of gallic acid that could enhance the effects of antibiotics.

S. ruineniae A45.2 is a basidiomycetous yeast that forms a natural pink-red pigment made up of carotenoids. The pigments extracted from the yeast were separated into three types of carotenoids based on the separation by HPLC and TLC. These pigments displayed distinctively different visible spectra. The most polar pigment showed a rosy-red color and had a similar visible spectrum to torularhodin, while the others displayed a similar spectrum to torulene and β-carotene as the second most polar and the least polar pigments, respectively [29–31]. However, structural elucidation of these compounds must be confirmed. Currently, carotenoid-producing yeasts are mainly represented by the genera *Rhodosporidium*, *Xanthophylomyces Rodotorula* and *Sporobomyces*. The latter genus has a close relationship to the genus *Sporidiobolus* and represents the main source of torulene and torularhodin [32]. The quantity of total carotenoids produced by *S. ruineniae* A45.2 was in the range of those produced by the yeast studied in previously published reports [31,33].

After cell wall components, some probiotic yeasts exhibit multifunctional potential in the production of bioactive compounds with certain antioxidant properties, such as carotenoids, organic acids and glutathione [24]. As *S. ruineniae* A45.2 is a carotenoid-producing yeast, it is likely that it possesses antioxidant capacity. The intact cells and cell-free extract obtained from YMB were evaluated for their antioxidant activity (data not shown). No antioxidant activity was detected in the cell-free extract, while the intact cells possessed approximately 10 times lower the degree of antioxidant activity than that obtained from cultivation in tannic acid. This may have resulted from the presence of β-glucan as a component of yeast cell wall composition [34,35]. Cultivation of *S. ruineniae* A45.2 in tannic acid could potentiate the antioxidant activity of not only intact cells but also cell-free extracts. It was determined that the fermentation of S. *ruineniae* A45.2 induced the production of CAT which strongly affected the degradation of tannic acid, resulting in gallic acid production. During the degradation of tannic acid, large amounts of gallic acid were released into the fermentation broth and attached to the yeast cell surface, reported in previous studies [10,36]. The antioxidant activity in terms of gallic acid equivalent is likely a consequence of gallic acid content, as reported in previously published studies [10]. This result agrees with previously reported evidence published on the fermentation of plant-based foods [37], including grape seed flour and extracts [38], as well as Miang [39], as sources of *S. ruineniae* A45.2.

Overall, *S. ruineniae* A45.2 may be capable of exhibiting the beneficial characteristics attributed to a probiotic yeast that can be used for aquaculture. Cultivation of the yeast in tannic acid substrate might provide a number of benefits. These benefits include the assertion that yeast cells can be a source of antioxidant agents, tannase and carotenoids for aquatic organisms. Furthermore, it is believed that the resulting culture broth can display strong antioxidant activity as well as the potential to display antimicrobial activity against some pathogenic bacteria, especially fish pathogens. Therefore, this research study described and verified an alternative integrative strategy for the production of feed additives. To our knowledge, this is the first report to suggest that *S. ruineniae* exhibits probiotic properties.

5. Conclusions

S. ruineniae A45.2 was tolerant to simulated GIT conditions, displaying tolerance to pH 2.0, pepsin, bile salts and pancreatin. A high percentage of auto-aggregation was observed, and this species co-aggregated various pathogenic bacteria, specifically *Strep. agalactiae*, and adhered to some specific strains of pathogenic bacteria. These are considered beneficial attributes that support the use of *S. ruineniae* as a probiotic yeast. The fermentation of tannic acid to gallic acid has resulted in the co-production of CAT, gallic acid and viable yeast cells. Moreover, CAT was found to be stable and may be able to function under simulated GIT conditions, while the cells possessed antioxidant activity. Thus, *S. ruineniae* A45.2 as a carotenoid- and CAT-producing yeast could be labeled as a multifunctional probiotic yeast suitable for the feed of animals, particularly aquatic animals. In addition, its cell-free extract derived from the co-production system could be a potential alternative source of natural antioxidants and antimicrobial agents.

Author Contributions: Conceptualization, A.K. (Apinun Kanpiengjai); funding acquisition, A.K. (Apinun Kanpiengjai); investigation, A.K. (Apinun Kanpiengjai), C.K., S.L., A.K. (Aksarakorn Kummasook) and S.K.; methodology, A.K. (Apinun Kanpiengjai), A.K. (Aksarakorn Kummasook) and S.K.; project administration, A.K. (Apinun Kanpiengjai); supervision, A.K. (Apinun Kanpiengjai), C.K. and S.L.; writing—original draft, A.K. (Apinun Kanpiengjai); writing—review and editing, A.K. (Apinun Kanpiengjai). All authors read and approved of the final version of this manuscript.

Funding: This research study was partially funded by the Thailand Research Fund through the Research Grant for New Scholars (MRG6280057), and Chiang Mai University.

Conflicts of Interest: The authors declare that they have no conflict of interest.

References

1. Gephart, J.A.; Golden, C.D.; Asche, F.; Belton, B.; Brugere, C.; Froehlich, H.E.; Fry, J.P.; Halpern, B.S.; Hicks, C.C.; Jones, R.C.; et al. Scenarios for global aquaculture and its role in human nutrition. *Rev. Fish. Sci. Aquac.* **2020**, *10*, 1–17. [CrossRef]
2. Adel, M.; Lazado, C.C.; Safari, R.; Yeganeh, S.; Zorriehzahra, M.J. Aqualase®, a yeast-based in-feed probiotic, modulates intestinal microbiota, immunity and growth of rainbow trout Oncorhynchus mykiss. *Aquac. Res.* **2017**, *48*, 1815–1826. [CrossRef]
3. Caruffo, M.; Navarrete, N.; Salgado, O.; Díaz, A.; López, P.; García, K.; Feijóo, C.G.; Navarrete, P. Potential probiotic yeasts isolated from the fish gut protect zebrafish (Danio rerio) from a Vibrio anguillarum challenge. *Front. Microbiol.* **2015**, *6*, 1093. [CrossRef] [PubMed]

4. Hill, C.; Guarner, F.; Reid, G.; Gibson, G.R.; Merenstein, D.J.; Pot, B.; Morelli, L.; Canani, R.B.; Flint, H.J.; Salminen, S.; et al. The International Scientific Association for Probiotics and Prebiotics consensus statement on the scope and appropriate use of the term probiotic. *Nat. Rev. Gastroenterol. Hepatol.* **2014**, *11*, 506–514. [CrossRef] [PubMed]
5. Banu, M.R.; Akter, S.; Islam, R.; Mondol, N.; Hossain, A. Probiotic yeast enhanced growth performance and disease resistance in freshwater catfish gulsa tengra, Mystus cavasius. *Aquac. Rep.* **2020**, *16*, 100237. [CrossRef]
6. Kanpiengjai, A.; Chui-Chai, N.; Chaikaew, S.; Khanongnuch, C. Distribution of tannin-'tolerant yeasts isolated from Miang, a traditional fermented tea leaf (Camellia sinensis var. assamica) in northern Thailand. *Int. J. Food Microbiol.* **2016**, *238*, 121–131. [CrossRef] [PubMed]
7. Navarrete, P.; Tovar-Ramírez, D. Use of yeasts as probiotics in fish aquaculture. In *Sustainable Aquaculture Techniques*; Hernandez-Vergara, M.P., Perez-Rostro, C.I., Eds.; IntechOpen: London, UK, 2014.
8. Okada, S.; Daengsubha, W.; Uchimura, T.; Ohara, N.; Kozaki, M. Flora of lactic acid bacteria in Miang produced in northern Thailand. *J. Gen. Appl. Microbiol.* **1986**, *32*, 57–65. [CrossRef]
9. De Carvalho, C.C.C.R.; Caramujo, M.J. Carotenoids in aquatic ecosystems and aquaculture: A colorful business with implications for human health. *Front. Mar. Sci.* **2017**, *4*, 93. [CrossRef]
10. Kanpiengjai, A.; Khanongnuch, C.; Lumyong, S.; Haltrich, D.; Nguyen, T.-H.; Kittibunchakul, S. Co-production of gallic acid and a novel cell-associated tannase by a pigment-producing yeast, Sporidiobolus ruineniae A45.2. *Microb. Cell Fact.* **2020**, *19*, 95. [CrossRef]
11. Francis, G.; Makkar, H.P.; Becker, K. Antinutritional factors present in plant-derived alternate fish feed ingredients and their effects in fish. *Aquaculture* **2001**, *199*, 197–227. [CrossRef]
12. Omnes, M.-H.; Le Goasduff, J.; Le Delliou, H.; Le Bayon, N.; Quazuguel, P.; Robin, J.H. Effects of dietary tannin on growth, feed utilization and digestibility, and carcass composition in juvenile European seabass (*Dicentrarchus labrax* L.). *Aquac. Rep.* **2017**, *6*, 21–27. [CrossRef]
13. Kanpiengjai, A.; Unban, K.; Nguyen, T.-H.; Haltrich, D.; Khanongnuch, C. Expression and biochemical characterization of a new alkaline tannase from Lactobacillus pentosus. *Protein Expr. Purif.* **2019**, *157*, 36–41. [CrossRef] [PubMed]
14. Kanpiengjai, A.; Mahawan, R.; Lumyong, S.; Khanongnuch, C. A soil bacterium Rhizobium borbori and its potential for citrinin-degrading application. *Ann. Microbiol.* **2016**, *66*, 807–816. [CrossRef]
15. Pais, P.; Almeida, V.; Yılmaz, M.; Teixeira, M.C. Saccharomyces boulardii: What makes it tick as successful probiotic? *J. Fungi* **2020**, *6*, 78. [CrossRef] [PubMed]
16. Bullerman, L.B. Spoilage fungi in food–An overview. In *Encyclopedia of Food Sciences and Nutrition*, 2nd ed.; Caballero, B., Ed.; Academic Press: Oxford, UK, 2003; pp. 5511–5522. [CrossRef]
17. Hlophe, S.N.; Moyo, N.A.G.; Ncube, I. Postprandial changes in pH and enzyme activity from the stomach and intestines of Tilapia rendalli (Boulenger, 1897), Oreochromis mossambicus (Peters, 1852) and Clarias gariepinus (Burchell, 1822). *J. Appl. Ichthyol.* **2014**, *30*, 35–41. [CrossRef]
18. Solovyev, M.; Kashinskaya, E.; Izvekov, E.I.; Glupov, V.V. pH values and activity of digestive enzymes in the gastrointestinal tract of fish in Lake Chany (West Siberia). *J. Ichthyol.* **2015**, *55*, 251–258. [CrossRef]
19. Khan, I.R.; Choudhury, T.G.; Kamilya, D.; Monsang, S.J.; Parhi, J. Characterization of Bacillus spp. isolated from intestine of Labeo rohita—Towards identifying novel probiotics for aquaculture. *Aquac. Res.* **2020**. [CrossRef]
20. Gut, A.M.; Vasiljevic, T.; Yeager, T.; Donkor, O. Characterization of yeasts isolated from traditional kefir grains for potential probiotic properties. *J. Funct. Foods* **2019**, *58*, 56–66. [CrossRef]
21. Kavitha, M.; Raja, M.; Perumal, P. Evaluation of probiotic potential of Bacillus spp. isolated from the digestive tract of freshwater fish Labeo calbasu (Hamilton, 1822). *Aquac. Rep.* **2018**, *11*, 59–69. [CrossRef]
22. Xu, H.; Jeong, H.S.; Lee, H.Y.; Ahn, J. Assessment of cell surface properties and adhesion potential of selected probiotic strains. *Lett. Appl. Microbiol.* **2009**, *49*, 434–442. [CrossRef]
23. Suvarna, S.; Dsouza, J.; Ragavan, M.L.; Das, N. Potential probiotic characterization and effect of encapsulation of probiotic yeast strains on survival in simulated gastrointestinal tract condition. *Food Sci. Biotechnol.* **2018**, *27*, 745–753. [CrossRef] [PubMed]
24. Ogunremi, O.R.; Sanni, A.; Agrawal, R. Probiotic potentials of yeasts isolated from some cereal-based Nigerian traditional fermented food products. *J. Appl. Microbiol.* **2015**, *119*, 797–808. [CrossRef] [PubMed]

25. Yildiran, H.; KiliÇ, G.B.; Çakmakçi, A.G.K. Characterization and comparison of yeasts from different sources for some probiotic properties and exopolysaccharide production. *Food Sci. Technol.* **2019**, *39*, 646–653. [CrossRef]
26. Di Cagno, R.; Filannino, P.; Cantatore, V.; Polo, A.; Celano, G.; Martinovic, A.; Cavoski, I.; Gobbetti, M. Design of potential probiotic yeast starters tailored for making a cornelian cherry (*Cornus mas* L.) functional beverage. *Int. J. Food Microbiol.* **2020**, *323*, 108591. [CrossRef]
27. Tiago, F.C.P.; Martins, F.S.; Souza, E.L.S.; Pimenta, P.F.P.; Araujo, H.R.C.; Castro, I.M.; Brandão, R.L.; Nicoli, J.R. Adhesion to the yeast cell surface as a mechanism for trapping pathogenic bacteria by Saccharomyces probiotics. *J. Med. Microbiol.* **2012**, *61*, 1194–1207. [CrossRef]
28. Samuel, K.; Wang, J.; Yue, H.Y.; Wu, S.; Zhang, H.; Duan, Z.; Qi, G. Effects of dietary gallic acid supplementation on performance, antioxidant status, and jejunum intestinal morphology in broiler chicks. *Poult. Sci.* **2017**, *96*, 2768–2775. [CrossRef]
29. Libkind, D.; Diéguez, M.D.C.; Moliné, M.; Pérez, P.; Zagarese, H.; Van Broock, M. Occurrence of photoprotective compounds in yeasts from freshwater ecosystems of northwestern Patagonia (Argentina). *Photochem. Photobiol.* **2006**, *82*, 972–980. [CrossRef]
30. Park, P.; Kim, E.Y.; Chu, K. Chemical disruption of yeast cells for the isolation of carotenoid pigments. *Sep. Purif. Technol.* **2007**, *53*, 148–152. [CrossRef]
31. Zoz, L.; Carvalho, J.C.; Soccol, V.T.; Casagrande, T.C.; Cardoso, L.A.D.C. Torularhodin and Torulene: Bioproduction, properties and prospective applications in food and cosmetics–A review. *Braz. Arch. Biol. Technol.* **2015**, *58*, 278–288. [CrossRef]
32. Zhao, Y.; Guo, L.; Xia, Y.; Zhuang, X.; Chu, W. Isolation, identification of carotenoid-producing rhodotorula sp. from marine environment and optimization for carotenoid production. *Mar. Drugs* **2019**, *17*, 161. [CrossRef]
33. Kot, A.M.; Błażejak, S.; Gientka, I.; Kieliszek, M.; Bryś, J. Torulene and torularhodin: "new" fungal carotenoids for industry? *Microb. Cell Fact.* **2018**, *17*, 49. [CrossRef] [PubMed]
34. Jaehrig, S.C.; Rohn, S.; Kroh, L.W.; Fleischer, L.-G.; Kurz, T. In vitro potential antioxidant activity of (1→3), (1→6)-β-d-glucan and protein fractions from saccharomyces cerevisiae cell walls. *J. Agric. Food Chem.* **2007**, *55*, 4710–4716. [CrossRef] [PubMed]
35. Kogan, G.; Pajtinka, M.; Babincova, M.; Miadoková, E.; Rauko, P.; Slamenova, D.; Korolenko, T.A. Yeast cell wall polysaccharides as antioxidants and antimutagens: Can they fight cancer? *Neoplasma* **2008**, *55*, 387–393. [PubMed]
36. Seth, M.; Chand, S. Biosynthesis of tannase and hydrolysis of tannins to gallic acid by Aspergillus awamori —Optimisation of process parameters. *Process. Biochem.* **2000**, *36*, 39–44. [CrossRef]
37. Hur, S.J.; Lee, S.Y.; Kim, Y.-C.; Choi, I.; Kim, G.-B. Effect of fermentation on the antioxidant activity in plant-based foods. *Food Chem.* **2014**, *160*, 346–356. [CrossRef]
38. Cho, Y.-J.; Kim, D.; Jeong, D.; Seo, K.; Jeong, H.S.; Lee, H.G.; Kim, H. Characterization of yeasts isolated from kefir as a probiotic and its synergic interaction with the wine byproduct grape seed flour/extract. *LWT* **2018**, *90*, 535–539. [CrossRef]
39. Unban, K.; Khatthongngam, N.; Shetty, K.; Khanongnuch, C. Nutritional biotransformation in traditional fermented tea (Miang) from north Thailand and its impact on antioxidant and antimicrobial activities. *J. Food Sci. Technol.* **2019**, *56*, 2687–2699. [CrossRef]

Journal of *Fungi*

MDPI

Article

Yeasts in Liquid Swine Diets: Identification Methods, Growth Temperatures and Gas-Formation Potential

Birgit Keller [1,*], Henrike Kuder [1], Christian Visscher [1], Ute Siesenop [2] and Josef Kamphues [1]

1 Institute for Animal Nutrition, University of Veterinary Medicine Hannover, Foundation, 30173 Hannover, Germany; henrike.kuder@uzh.ch (H.K.); christian.visscher@tiho-hannover.de (C.V.); josef.kamphues@tiho-hannover.de (J.K.)

2 Institute for Microbiology, University of Veterinary Medicine Hannover, Foundation, 30173 Hannover, Germany; ute.siesenop@tiho-hannover.de

* Correspondence: birgit.keller@tiho-hannover.de

Received: 30 September 2020; Accepted: 2 December 2020; Published: 4 December 2020

check for updates

Abstract: Liquid feed is susceptible to microbiological growth. Yeasts are said to cause sudden death in swine due to intestinal gas formation. As not all animals given high yeast content feed fall ill, growth and gas formation potential at body temperature were investigated as possible causally required properties. The best identification method for these environmental yeasts should be tested beforehand. Yeasts derived from liquid diets without (LD − S) and liquid diets with maize silage (LD + S) were examined biochemically (ID32C-test) and with MALDI-TOF with direct smear (DS) and an extraction method (EX). Growth temperature and gas-forming potential were measured. With MALDI-EX, most yeast isolates were identified: *Candida krusei* most often in LD − S, and *C. lambica* most often in LD + S, significantly more than in LD − S. Larger colonies, 58.75% of all yeast isolates, were formed at 25 °C rather than at 37 °C; 17.5% of all isolates did not grow at 37 °C at all. Most *C. krusei* isolates formed high gas amounts within 24 h, whereas none of the *C. lambica*, *C. holmii* and most other isolates did. The gas pressure formed by yeast isolates varied more than tenfold. Only a minority of the yeasts were able to produce gas at temperatures common in the pig gut.

Keywords: yeasts; liquid swine diets; MALDI-TOF; biochemical identification; growth temperature Ancom Gas Production System; *Candida krusei*; *Candida lambica*

---

1. Introduction

Yeasts, about 600 species of which are known [1], are ubiquitous in nature and can also be found on feedstuffs [2]. They pose a risk factor regarding hygiene in liquid diets associated with off-flavor and loss of nutrients [3–5]. Depending on the species or strain, as well as on the growth conditions like temperature, substrate and its aw-value (activity of water), yeasts are able to metabolize numerous sugars, starch, protein, amino acids or even fats, and therefore lead to a loss of nutrients and energy in the feed [3–5]. In pig fattening, these energy losses in the feed are particularly undesirable [4,5]. In addition, the flavor and smell of the feed can be negatively affected [6,7]. High cell counts of yeasts in liquid swine diets due to pronounced metabolic activity are often seen in the presence of easily fermentable, low molecular weight sugars [5]. Choosing maize silage for pig feed was used with the aim of feeding the pigs to increase the feeling of satiety without making them fat [8]. The relatively high initial yeast flora of the feed has to be taken into account [8]. Therefore, feed hygiene related to yeast content was of special concern.

In liquid feeds, mostly microflora develops, which is dominated by lactic acid-producing bacteria [7]. A pH-value lower than 5.0, which significantly reduces several bacteria, is often achieved in a shorter time with the use of starter cultures for fermented liquid feeds [3,9,10]. Yeasts are not only able to stay alive but also continue growing in fermented feeds [11], even if the pH-value is 4.5 [12].

Besides these complications concerning feed composition and quality, animal health may be affected due to the yeast content in the feed [11,13–15]. Hemorrhagic bowel syndrome (HBS), mainly caused by yeasts [13], is supposed to be causally responsible for gastric torsion and gastrointestinal tympani [16], being sometimes associated with liquid feeding [15]. HBS preferentially affects fattened pigs in the second half of the fattening period [16]. Those animals most affected are, as a rule, the better developed pigs in the group [14]. The fact that the affected animals are in excellent health makes this disease of particular economic importance [15].

In feed analyses, yeasts, irrespective of the species, are classified as spoilage indicators in animal feed [17]. A liquid diet with more than 10^6 cfu yeasts/g original substance (OS) is considered as significantly increased, while less than 10^5 cfu/g feed OS in liquid feed is considered as normal [18]. On the other hand, selected yeasts are authorized feed additives in human nutrition and animal feedstuffs as they synthesize vitamin B1, B2, B6, B12, folic acid, niacin, pantothenic acid and biotin, as well as containing some minerals (potassium, sodium, calcium, zinc and iron) [1]. For swine diets, viable *S. cerevisiae* is authorized as a feed additive as intestinal flora stabilizers, digestibility enhancers and microorganisms with a minimum concentration of 1×10^9 cfu/kg complete feed (88% DM) [19].

Pathogenicity factors of yeasts have been analyzed to identify high-risk yeasts and their effects on humans and animals. In their study about potential virulence of food-borne yeasts, Rajkowska et al. [20] stated that the ability to grow at 37 °C was crucial; hence, they referred to this as preliminary criterion for pathogenicity. Adaptation to pH-value was also suggested to be a key to pathogenicity, especially important for yeasts entering the digestive tract where the pH-value changes from pH 2 to pH 8 [21,22]. The ability to form biofilms also on abiotic surfaces [21] or even to colonize them is a prerequisite for colonizing the liquid feeding system, which allows the yeasts to stay alive even if the hygiene of the liquid feed was improved [23]. Stalljohann et al. [3] distinguished yeasts according to their ability to produce high or low amounts of CO_2 with regard to their pathogenicity for swine, but did not mention which species produced the high gas amounts. Such detailed information on these possible indicators of pathogenicity is provided in the present paper.

The hypothesis of this study was that different yeast species could be found in different feedstuffs. For this reason, a comparison of biochemical differentiation and identification with MALDI-TOF was performed to determine the method with the most reliable identification. Presumably, only distinct species would be able to grow and to produce high amounts of gas at 37 °C.

A comparison between gas amounts produced from yeasts measured with the Ancom Gas Production System under defined conditions in a standardized Sabouraud glucose bouillon, had, to the best of the authors' knowledge, never been carried out previously. This permits a comparison of yeast isolates not only to see whether but also how much gas can be produced by yeasts within a certain time period regardless of feedstuffs. Further studies must clarify whether and to what extent these properties have an influence on the development of diseases such as HBS. These new aspects could then allow to make better predictions concerning the ability of high yeast cell counts in liquid diets to cause clinical problems.

2. Materials and Methods

2.1. Sample Origin

For our project, samples from farms with liquid feeding common to all samples were collected. We obtained these samples either by contacting farms in the area, or from our own studies, which were also carried out on similar farms. In total, 42 liquid feed samples were analyzed. Of these samples, 33 were submitted for diagnostic purposes to the Institute for Animal Nutrition, University of Veterinary Medicine Hannover, Foundation, Germany. These included common liquid feeds to which no silage had been added, referred to as liquid diets without silage (LD − S). The remaining nine samples obtained from field trials carried out by the Institute for Animal Nutrition, additionally contained whole plant maize-silage (up to 66% DM; liquid diets with silage, LD + S). For collecting the liquid feed samples, a standard laboratory protocol was used for both the submissions and the samples from the studies. The protocol required that the samples were taken fresh, packed directly into a sterile, unbreakable vessel, filled to 2/3 at most, immediately cooled and not sent before the weekend. All samples were processed directly, or in case they arrived late in the afternoon, refrigerated and processed the following morning.

2.2. Detection Techniques

Yeasts were isolated and morphologically characterized on Sabouraud glucose agar (SAB-Agar, PO 5096A, Thermo Fisher Scientific GmbH, Bremen, Germany) and then incubated at 30 °C. Only yeasts that grew at the highest decimal dilution levels of the agar plates were considered.

2.2.1. Biochemical Differentiation

Biochemical differentiation of the yeast isolates was performed by ID 32 C strip (bioMérieux SA, Marcy-l'Ètoile, France). This was performed in accordance with the manufacturer's instructions. The strip consisted of 32 cavities, each containing a dehydrated carbohydrate substrate, testing the assimilation by the yeast. Pure culture yeast material of 44–48 h-grown subcultures was suspended in 3 mL aqua destillatum. Turbidity was set in accordance with a McFarland standard of 2.0 using a Densitometer DEN-1B (BioSan, Riga, Lettland). From this solution, 250 µL were added to the API C medium included in the test kit. After careful vortexing, 135 µL were transferred from this liquid medium to each well of the test strip. The strip was incubated at 30 °C for 44–48 h. Yeast growth resulted in turbidity of the liquid medium in the cupules, which was visually evaluated. The obtained results were noted on a result sheet. The values corresponding to the positive reactions were then added up within groups. Three results each were added up for a group. Group values were coded into a numerical profile. This was analyzed by means of identification software (APIWEB^{TM}, bioMérieux). The results of two of these carbohydrates, *N*-acetylglucosamine (NAG) and lactic acid (LAT), are examined in more detail below. Only results that received good, very good or excellent (classified as "Very good identification") ratings were evaluated. Rice agar (Thermo Fisher Scientific GmbH), incubated at 25 °C for 44–48 h, was selected for some isolates if the identification software required this deficiency medium, with a cover glass placed over the inoculum for an oxygen-reduced atmosphere.

2.2.2. MALDI-TOF

In the MALDI-TOF analysis, the sample (e.g., bacteria or yeasts) was ionized by a laser beam. These ions were then accelerated differently depending on their mass and charge. The time required to pass through the length of the flight tube was determined [24]. In this way, a characteristic spectrum can be generated for bacteria or fungi, which usually allows a species diagnosis [24,25]. As an advantage, less time is required for this method compared to biochemical methods [24].

MALDI-TOF analyses were performed on a Microflex LT/SH MALDI-MS Biotyper (Bruker Daltonik GmbH, Bremen, Germany) with the direct smear method (MALDI-DS) and with a formic acid–ethanol extraction (MALDI-EX). The latter is used for hardly soluble bacteria or yeasts.

For MALDI-DS, direct on-plate smearing was performed with yeasts incubated 44–48 h on an SAB-plate at 30 °C. Small amounts of colony material of every isolate were evenly applied with a toothpick to two circles of the target plate (8280800 MSP 96 Target polished steel BC, Bruker Daltonik). After air drying the sample material at room temperature for five to ten minutes, 1 µL of an α-cyano-4-hydroxycinnamic acid (HCCA, 19182, Sigma Aldrich Inc., St. Louis, MO, USA) matrix solution was applied to each circle of the target plate and dried again at room temperature.

For MALDI-EX, a 1 µL loop of 20–24 h-grown yeast material on an SAB-plate was vortexed in 300 µL deionized water at 30 °C; 900 µL of ethanol absolute, HPLC-grade, was added and it was vortexed again. The samples were pelleted by centrifuging for two minutes (13,000 U/min;

approximately 3000× *g*); the supernatant was discarded, centrifuged and then the supernatant was discarded again. The pellet in the Eppendorf tube was air dried for three to five minutes and resuspended in 40 µL 70% formic acid. After adding 40 µL acetonitrile (ACN, Acetonitrile HPLC Gradient Grade, 20060.320 VWR International Inc., Radnor, PA, USA), this was followed by the same centrifugation step as above. Eppendorf tubes were taken carefully out of the centrifuge and 1 µL of the supernatant was applied to a circle of the target and air dried for 5 min at room temperature. Immediately afterwards, 1 µL of the same matrix solution that had been previously used for the MALDI-DS method was applied to a circle of the target and air dried for 5 min at room temperature.

For the control of both methods, 1 µL BTS (Bacterial Test-Standard, 8290190 Bruker IVD Bacterial Test-Standard) was placed on every target plate and, additionally, a control strain, *E. coli* DH5 α, was tested on every target plate in two circles.

Each sample was analyzed by a Microflex LT/SH MALDI-TOF MS in the linear mode across a mass-to-charge ratio range between 2000 and 20,000. The obtained data were analyzed automatically by using the MBT Compass Library BDAL and MBT Flex Control software, BTyp2.0-Sec.Library, 1.0. Every strain was tested in two circles, with a decreased concentration in the second circle. The result from the two circles that achieved the higher score value was used for analysis.

The identification cut-off scores were interpreted as per Bruker's recommendation scores as follows: obtaining scoring thresholds between 2.30 and 3.0 suggested highly probable species identification; 2.00–2.29, probable species identification; 1.70–1.99, identification at the genus level was postulated; whereas cut-off scores <1.70 indicated no reliable identification.

2.2.3. Method Comparison

The results "excellent identification" and "very good identification" adopted in the software APIWEB^{TM} were equated to "highly probable species identification" from the MALDI-TOF analysis. "Good identification" was equated to "probable species identification"; "probable genus identification" was equated to "good identification at genus level"; and "doubtful profile" and "no identification" were equated to "unacceptable profile".

2.3. *Temperature Comparison*

After cultivating the yeasts from liquid feeds on SAB-agar and subcultivating a single colony, a subculture on two SAB-agar plates was produced for the temperature comparison. One agar plate each was cultivated in an incubator at either 25 °C or 37 °C for 48 h. The colony growth (diameter) was compared visually.

2.4. *PH-Value*

Using a calibrated glass electrode (HI 2211 pH/ORP Meter, Hanna Instruments Inc., Woonsocket, RI, USA), pH-values were measured. Results of 25/33 LD − S and 16/17 LD + S were obtained immediately after dividing the samples for further microbiological testing.

2.5. *Gas Pressure Measurement*

To measure yeast activity in gas production, 40 selected yeast isolates were examined for 24 h at 37 °C with the ANKOM RF Gas Production System (ANCOM Technology, Macedon, NY, USA), which remotes pressure under controlled pressure measurements and records these on a standard Excel spreadsheet. Gas production curves were generated. In 100 mL glass bottles containing 100 mL Sabouraud glucose broth (SAB-B; CM 0147 B, Thermo Fisher Scientific GmbH), a 10 µL yeast suspension, McFarland standard 0.3 (Densitometer DEN-1B, Biosan, Riga, Lettland), in a physiologic salt solution, was added. SAB-B conforms to the parameters from the harmonized EP/USP/JP Microbial Limit Testing for the microbial enumeration tests and tests for specified microorganisms. The bottles were placed on magnetic stirring panels (MIXdrive magnetic e motion with Mixcontrol 20, 2mag AG, Munich, Germany) for permanent mixing at 210 rpm. Gas pressure was measured over a 24-h period, taking into account that feed normally does not normally stay longer in animals' gastrointestinal tract. Line charts of the cumulated gas production were generated with the Ancom Gas Pressure Monitor. Each isolate was tested at least twice.

The results were divided into two groups. A very small pressure increase (<100 mbar) at the beginning was also observed if no further gas was produced thereafter. If less than 100 mbar of cumulated pressure within 24 h was observed, the result was determined as negative. If a yeast was able to produce more than 800 mbar within 24 h, the result was determined as gas production. This value was defined based on the results found, because no yeast produced gas amounts between 100 and 800 mbar.

2.6. Statistics

Data were statistically analyzed using the SAS® Enterprise Guide® (version 7.1, Fa. SAS Institute Inc. Cary, NC, USA). Pearson's chi-square homogeneity test and Fisher's exact test, used to analyze qualitative analytical characteristics, were applied to check if a yeast was found significantly more with one feed; if one of the identification tests found significantly more reliable results; whether yeasts grew better at a certain temperature; or if yeasts built up a distinct pressure at 37 °C within 24 h. Fisher's exact test was used especially for low absolute frequencies.

3. Results

In total, 95 morphologically different yeast colonies (color, size and surface structure) were isolated from a total of 42 feed samples. In each feed sample, one to four different yeast species were found.

3.1. Identification

In spite of a different morphology, yeast identification led to the same result for 15 yeast isolates. Yeasts diagnosed twice in the same sample were not considered in the evaluation of the number of yeast species or species of yeast-like organisms that were found in the respective feed sample.

The isolates originated from six genera (*Candida* (76.25%; n = 61), *Geotrichum* (7.5%, n = 6), *Trichosporon* (3.75%, n = 3), *Saprochaete* (2.5%, n = 2), *Rhodotorula* (1.25%, n = 1), *Pichia* (1.25%, n = 1) and non-identified yeasts (7.5%, n = 6)). A total of 19 different yeast species were identified.

In Table 1, the most often isolated species are listed. Less frequently isolated (one to three times) were *C. pelliculosa* (n = 3, 2× LD + S, 1× LD − S), *C. valida* (n = 3; 2× LD − S, 1× LD + S), *Sap. suaveolens* (n = 2, both from LD − S), *C. rugosa* (n = 2, both from LD − S), *C. kefyr* (n = 1, LD − S), *C. variabilis* (n = 1, LD − S), *C. spherica* (n = 1, LD − S), *T. asahii* (n = 1, LD + S), *T. coremiiforme* (n = 1, LD − S), *T. laibachii* (n = 1, LD − S), *P. manshurica* (n = 1, LD − S), *Rhodotorula mucilaginosa* (n = 1, LD − S) and *Candida* spp. (n = 1, LD − S). Six isolates (5× in LD − S; 1× in LD + S) were not reliably identified.

Table 1. Often isolated yeast species.

Liquid Diet	Often Isolated Yeast Species in %
LD − S (n = 63)	*C. krusei* (23.8), *C. holmii* (12.7), *C. humilis* (7.9), isolates not identified (7.9), *C. lambica* (6.3), *S. cerevisiae* (6.3), *G. silvicola* (6.3)
LD + S (n = 17)	*C. lambica* (29.4), *C. krusei* (23.5), *C. holmii* (17.6), *C. pelliculosa* (11.8), *S. cerevisiae* (5.9), *C. valida* (5.9), isolates not identified (5.9)
Total (n = 80)	*C. krusei* (23.75), *C. holmii* (13.75), *C. lambica* (11.25), isolates not identified (7.5), *C. humilis* (6.25), *S. cerevisiae* (6.25), *G. silvicola* (6.25)

C. = *Candida*; S. = *Saccharomyces*; G. = *Geotrichum*.

Only *C. lambica* was determined to have a significantly higher incidence in LD + S ($p < 0.0126$). The occurrence in LD − S or LD + S did not significantly change (Figure 1).

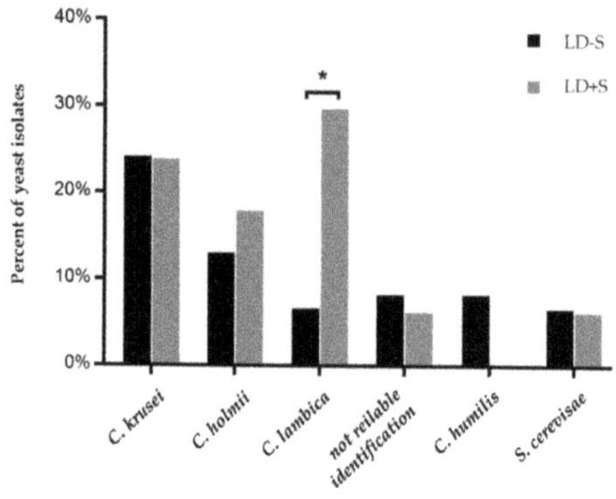

Figure 1. Incidence of yeasts according to their feed origin. An asterisk (*) indicates a significant difference (* $p < 0.05$). *C.* = *Candida*; *S.* = *Saccharomyces*.

If, despite different colony morphology, two identical yeast species were diagnosed in one feed sample, only one yeast species was further examined. In 33 LD − S, 7/70 yeast colonies with different morphologies were selected, but identified as the same yeast species within the same diet. In LD + S, 8/25 morphologically different yeasts were identified as the same yeast within the same diet. Therefore, 63 yeasts from LD − S and 17 yeast from LD + S were further examined. As a result of this, significantly more different yeast morphologies were observed from the LD + S samples ($p < 0.0211$).

On the other hand, no differences between the two feeding groups were found regarding the actual (not morphologically) different yeasts. In both diet groups, an average of 1.9 different yeasts per feed sample were diagnosed: 63 different yeasts of 33 LD − S diets, and 17 different yeasts of nine LD + S diets.

3.1.1. Method Comparison

The comparison of methods revealed the most reliable results with MALDI-EX (78.75% reliable species identification), closely followed by ID32C with 75.0% reliable results. These test results did not differ statistically significantly ($p < 0.6762$). Among the reliable results, ID32C provided differentiation of ten isolates (8%), seven isolates with species identification and three isolates were at least identified up to genus level, which MALDI-EX was not able to differentiate. Hence, taking the results of both methods together, 71 (88.75%) of all isolates were identified up to species level. A probable identification or rather identification only at genus level was possible in three (3.75%) isolates. No identification with any of the three diagnostic methods was made for 6/80 (7.5%) yeast isolates. Only few results (37.5%) were provided by MALDI-DS (Table 2). When evaluating the reliable results, both MALDI-EX and ID32C differed significantly (both $p < 0.001$) from the MALDI-DS. No identification with any of the three diagnostic methods was made for 6/80 yeast isolates (7.5%). For better comparison, the results of all three methods were presented together in one table. However, the yeast names (according to their teleomorphic or anamorphic growth) given in the results varied sometimes according to the evaluation software.

Some yeasts were diagnosed more accurately with one or the other method. Five isolates differentiated with MALDI-EX as *C. humilis* were diagnosed as *C. holmii* in ID 32 C. *C. humilis* was not included in the database used to evaluate the ID 32 C. Both of these species had very similar biochemical reactions. Therefore, identification with MALDI-EX was chosen to be more accurate.

The two yeasts identified as *C. pararugosa* in MALDI-EX were identified as *C. rugosa* in ID 32 C. Results in MALDI-EX were only 1.88 and 1.79, respectively. Therefore, they had to be named according to ID 32 C, where the results for both isolates revealed very good identification scores, namely, 99.8% and 99.5% for *C. rugosa*. On the other hand, *C. pararugosa* was not included in the database used to evaluate the ID 32 C.

Table 2. Comparison of performance of methods with 80 yeast isolates in %.

Evaluation Score of MALDI-TOF	MALDI-EX		MALDI-DS		ID-32C	
Highly probable species identification (>2.3)	38.75 (n = 31)	78.75 *,a (n = 63)	5.0 (n = 4)	37.5 *,b (n = 30)	47.5 (n = 38)	75.0 **,a (n = 60)
Probable species identification (2.0–2.29)	40.0 (n = 32)		32.5 (n = 26)		27.5 (n = 22)	
Probable genus identification (1.7–1.99)	7.5 (n = 6)	21.25 *** (n = 17)	33.75 (n = 27)	62.5 *** (n = 70)	3.75 (n = 3)	25.0 *** (n = 20)
No identification (<1.7)	13.75 (n = 11)		28.75 (n = 23)		21.25 (n = 17)	

[a,b] Different superscripts differ significantly in a row. * Reliable identification: Highly probable species identification and probable species identification; ** Reliable identification: Good and very good identification together; ***: No species identification.

Even with the formic acid extraction method, the slimy and red growing *Rhodotorula* (*R.*) *mucilaginosa* could not be detected with MALDI-TOF; there was no reliable identification, although it was registered in the database.

In the ID 32 C, two isolates were diagnosed as *Cryptococcus* (*C. curvatus and C. laurentii*), which showed no mucus capsule in the Indian ink preparation, but showed hyphal growth, arthro- and blastosporogenesis on rice agar under the cover glass (Figure 2), and were identified in MALDI-EX as *T. coremiiforme* and *T. laibachii* (Table 3). On the other hand, two *T. asahii* isolates could be recognized well or very well by both methods.

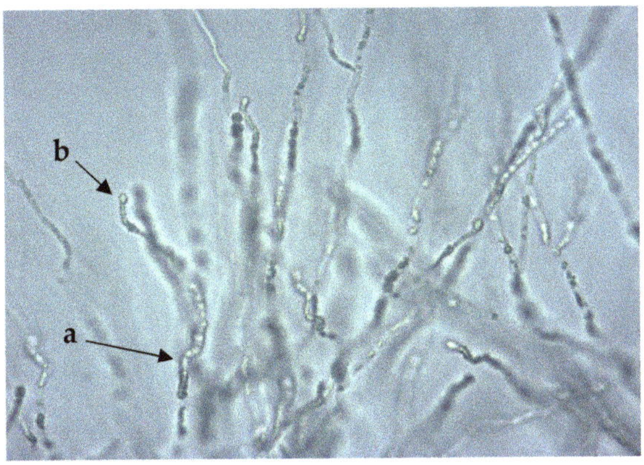

Figure 2. *Trichophyton coremiiforme* on rice agar: arthrospores (**a**) and blastospores (**b**). 400× magnification.

Table 3. Isolates differentially diagnosed with MALDI-EX and ID 32C.

Number of Isolates	MALDI-EX	ID 32C
5	***Candida humilis***	*Candida holmii*
2	*Candida pararugosa*	***Candida rugosa***
1	*No identification*	***Rhodotorula mucilaginosa***
1	***Trichosporon coremiiforme***	*Cryptococcus curvatus*
1	***Trichosporon laibachii***	*Cryptococcus laurentii*

Yeasts highlighted in bold represent the selected diagnoses.

Six isolates had a score between 1.79 and 1.98 in the MALDI-EX. Of these isolates, *S. cerevisiae* (score 1.98) and *C. holmii* (score 1.97) had the same result in ID32C (see Table 4), with very good identification. Furthermore, two isolates were identified as *C. pararugosa* in the MALDI-EX (with scores

1.88 and 1.63). *C. pararugosa* was not included in the identification software (APIWEB^{TM}, bioMérieux). The remaining two isolates consisted of *Saprochaete (Sap.) suaveolens* (score 1.88), which was diagnosed as a *Geotrichum* spp. in ID32C, and *Pichia occidentalis* (score 1.83), diagnosed with 99.7% as *C. krusei* in ID32C. Bearing in mind that *Sap. suaveolens* was formerly diagnosed as *Geotrichum fragrans*, the diagnosis made by MALDI-EX was most likely the one with the currently correct name. The name of the yeast in ID32C was probably out of date; the yeast was still correctly identified. Only one isolate was differently identified by the two methods as *P. occidentalis* (MALDI-TOF) and *C. krusei* (ID32C; Table 4).

Table 4. Isolates in MALDI-EX rated as probable genus identification (1.70 and 1.99).

MALDI-EX		ID32C		
Diagnosis	**Score**	**Diagnosis**	**Identification**	**%**
C. pararugosa	1.88	***C. rugosa***	very good	99.8
Sap. suaveolens	1.88	***Geotrichum spp.***	very good	99.7
P. occidentalis	1.83	***C. krusei***	very good	99.7
C. holmii	1.97	***C. holmii***	very good	99.2
C. pararugosa	1.79	***C. rugosa***	very good	99.5
S. cerevisiae	1.98	***S. cerevisiae***	very good	99.7

Yeasts in bold represent the selected diagnoses *C.* = *Candida*; *Sap.* = *Saprochaete*; *P.* = *Pichia*; *S.* = *Saccharomyces*.

3.1.2. Biochemical Reactions

In total, all the investigated yeasts were able to assimilate glucose and no yeast grew in the cupule where no substrate was present (cupule F). These reactions were considered as the positive growth control or negative control (no contamination). From the large number of biochemical reactions, two of them will be examined in more detail in the following section, since the ability of the yeasts to metabolize them could be an advantage, especially in maize silage.

Metabolization of N-Acetylglucosamine (NAG)

In the ID 32 C-Test, 27 yeasts from a total of 63 yeasts in the LD − S samples were able to metabolize *N*-acetylglucosamine and 36 yeasts were not. In LD + S, 11 yeasts were able to build *N*-acetylglucosamine and six yeasts were not. Despite the fact that this is insignificant ($p = 0.0788$), the ability to build NAG was more often seen in yeasts from LD + S.

Metabolization of Lactic Acid (LAT)

In the ID 32 C-Test, 36 yeasts of a total of 63 yeasts in the LD − S samples were able to metabolize lactic acid and 27 yeasts were not. In LD + S, 12 yeasts were able to metabolize lactic acid and five yeasts were not. These results were insignificant ($p = 0.4075$). Nevertheless, the ability to metabolize lactic acid could be found more often with yeasts that had to stay alive or even grow in maize silage than for yeasts in LD − S.

3.2. *Temperature Comparison*

Most yeast isolates ($n = 47$; 58.75%) formed larger colonies at 25 °C than at 37 °C, among them, 14 isolates (17.5%) did not grow at 37 °C at all. These included ten isolates from LD − S (3× *C. holmii*, 1× each for *C. humilis*, *C. lambica*, *T. laibachii*, *C. pelliculosa*, *Geotrichum* spp., *S. cerevisiae* and an isolate not identified) and four from LD + S (3× *C. holmii* and 1× isolate not identified). Among all yeasts, which grew better or only at 25 °C, many isolates of *C. holmii*, *C. humilis* and *C. lambica* were found. Only 23.75% of isolates grew better at 3 °C compared to 25 °C; this was often the case for *C. krusei* (12/19 isolates) and *S. cerevisiae* (4/5 isolates). Additionally, *C. kefyr* (1/1), *C. holmii* (1/11 isolates) and one isolate that could not be identified (1/6) showed better growth at 37 °C. All *S. cerevisiae* isolates were harvested from LD − S samples. Larger colonies at 37 °C were formed from nine *C. krusei* isolates gained from LD − S and three from LD + S, while three isolates from LD − S and one from LS + S formed larger colonies at 25 °C; one isolate showed equal colony growth at 25 °C or 37 °C. Hence, for *C. krusei*, no difference was observed, regardless of which feed it was isolated from. When comparing both feeds, it was noticeable that especially yeasts isolated from LD + S grew poorly at 37 °C (see Table 5). Better growth at 37 °C than at 25 °C for yeasts harvested from LD + S was only seen for three isolates, all of which were *C. krusei*. However, there was no statistically significant difference ($p < 0.3862$) between the two feed sources concerning growth performance (colony size) of the yeasts at either of the temperatures.

Table 5. Growth performance of the yeasts at different temperatures depending on feed.

Yeast Isolates	No Growth at 37 °C	Better Growth at 25 °C	25 °C = 37 °C	Better Growth at 37 °C
LD − S (*n* = 63)	10 (15.9%)	24 (38.1%)	13 (20.6%)	16 (25.4%)
LD + S (*n* = 17)	4 (23.5%)	9 (52.9%)	1 (5.9%)	3 (17.6%)
Total (*n* = 80)	14 (17.5%)	33 (41.25%)	14 (17.5%)	19 (23.75%)

3.3. pH-Value in Liquid Swine Diets

The pH-values of LD − S (n 25/33) ranged from 3.87–5.78, while the pH-values of LD + S (*n* = 8/9) achieved higher pH-values ranging from pH 4.79 to pH 5.61. Related to the feed origin, 44 yeasts isolated from LD − S were harvested from liquid feed, with an average pH-value of 4.59. Yeasts gained from LD + S were harvested from liquid feed, with an average pH-value of 5.51. *C. krusei* was isolated from liquid swine diets with the lowest (pH 3.87) and highest pH-values (pH 5.78) as well. *C. humilis and C. holmii* were found in diets with lower pH-values (pH 3.9 to pH 5.11), whereas *C. lambica* was isolated once from a diet with a pH-value of 4.45. However, other isolates were harvested from diets with higher pH-values (ranging from pH 4.97 to pH 5.61).

3.4. Gas Production

The results of the duplicate testing of each isolate showed small deviations, possibly caused by small differences in cell counts at the beginning as well as differences in replication time and counts of spores formed by each yeast cell during the 24-h incubation period.

Only two groups were formed: yeasts that produced virtually no gas within 24 h at 37 °C and yeasts that produced more than 800 mbar. A further subdivision of the yeasts into groups producing little or a lot of gas was omitted, because too little information was available from the literature as to which quantities could be classified as a lot or little.

No yeast produced gas amounts between 100 mbar and 800 mbar. More yeasts harvested from LD − S produced gas than yeasts that were found in LD + S, but the quantity was not significant ($p < 0.2216$).

Gas production with more than 800 mbar was observed for a total of 13 (40.6%) isolates (Table 6): 10/11 *C. krusei*-isolates, 2/3 *S. cerevisiae*-isolates, 1/1 *C. kefyr* and 1/1 *C. humilis*. *C. kefyr* formed the highest gas pressure, with 10,419 mbar, followed by the *C. krusei* isolates (7134.5, 7073, 6659, 6487, 6383, 6164.5, 4839, 4147.5, 3954.5 and 3940.5 mbar), both isolates of *S. cerevisiae* (1160.5 and 1466.5 mbar) and *C. humilis* (888 mbar). Eleven of these isolates grew better at 37 °C than at 25 °C within 48 h (see Section 3.2). Nonetheless, two isolates, which also grew better at 37 °C, were not able to produce more than 100 mbar gas in 24 h. These two yeasts were one *C. holmii* and one *C. krusei* isolate harvested from LD + S. The latter one produced these high gas quantities only after a 40 h incubation time. On the other hand, one isolate of *C. krusei*, harvested from LD − S, which grew better at 25 °C than at 37 °C, nevertheless produced 3954.5 mbar gas within 24 h. Only one yeast isolate from LD + S could produce significant quantities of gas under the abovenamed circumstances within 24 h (Table 4).

Table 6. Gas production (mbar) at 37 °C within 24 h.

Sample	Number	<100 mbar	>800 mbar
LD − S	$n = 32$	19 (59.4%)	13 (40.6%)
LD + S	$n = 8$	7 (87.5%)	1 (12.5%)
total	$n = 40$	26 (65.0%)	14 (35.0%)

Gas production less than 100 mbar was demonstrated in 5/5 *C. lambica*, 4/4 *C. holmii*, 3/3 *Trichosporon* spp., 3/3 *G. silvicola*, 2/2 *C. pelliculosa*, 2/2 *C. rugosa*, 1/1 *Sap. suaveolens*, 1/1 *Candida* spp., 1/1 *C. valida*, 1/1 *P. manshurica*, 1/1 *C. spherica*, 1/3 *S. cerevisiae* and 1/11 *C. krusei*. Yeasts that showed some signs of growth or grew particularly well at 37 °C within 48 h showed different reactions. Some isolates needed more than 24 h to produce high amounts of gas (Figure 3). Some isolates did not produce amounts greater than 100 mbar, even within a given 60-h period. In Figure 3 such yeasts are *C. valida* and *C. lambica*. Therefore, their curves in Figure 3 are so close to the x-axis they are hardly visable, just like the curve of the control (sterile SAB-bouillon without yeast isolate).

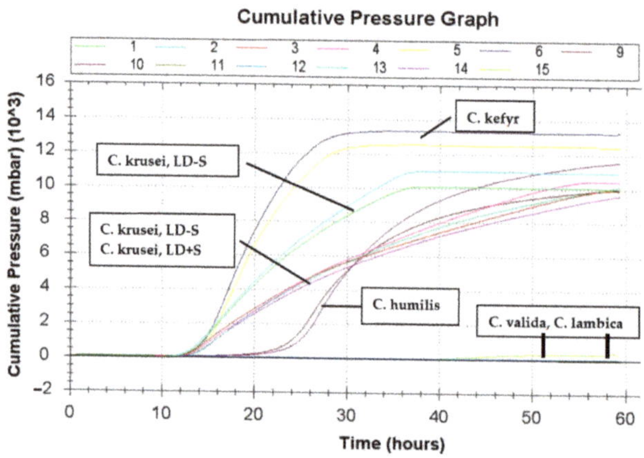

Figure 3. Cumulative pressure graph C. = *Candida; C. krusei*, LD − S = isolate of *C. krusei* from LD − S; *C. krusei*, LD-+ = isolate of *C. krusei* from LD + S.

The highest correlation between yeast growth at 25 °C and 37 °C within 48 h (see Section 3.2) and gas formation was found in yeasts that did not grow at 37 °C at all. None of these yeasts (8/8) were able to produce gas during the 24-h incubation period at 37 °C.

4. Discussion

Increased numbers of yeasts in liquid feed for pigs has been the subject of some publications in previous years [10–12,16,26–28]. Some publications compared yeasts in liquid feed for pigs with different feed composition taken from different stables or with and without the addition of starter cultures. However, none have yet compared the yeasts in liquid swine diets with and without maize silage, with the identification results derived from two methods.

4.1. Identification

As in most other studies, the genus *Candida* (C.) was found most often in our research study. *C. krusei* was found most often in LD − S and in all samples as a whole, whereas *C. lambica* was found most often in LD + S. In addition to these two species, another 17 yeast species from six genera were also diagnosed in this study.

4.1.1. *C. krusei*

Overall, the most often found yeast in this study, *C. krusei*, was isolated from feed samples with the lowest and highest pH-values. From the literature, it is known to grow at low pH-values [29], ferment up to a pH-value of 3.6 [30] and can also form biofilms [31]. These properties are likely to be beneficial for yeast persistence in the liquid feed and the feeding system. In part, *C. krusei* is capable of pseudohyphae formation and mostly of growing at 37 °C, both characteristics that could contribute to HBS.

C. krusei is responsible for about 2% of yeast infections caused by *Candida* species in humans [32]. *Pichia kudriavzevii, Issatchenkia orientalis* and *Candida glycerinogenes* are proven to be the same yeast with collinear genomes 99.6% identical in DNA sequence. Under these names, the yeast is used for industrial-scale production of glycerol and succinate, and is also used to make some fermented foods [32]. The latter use in fermented foods also explains the frequent occurrence in liquid feed for pigs, which also has a low pH value (see Section 3.3).

4.1.2. *C. lambica*

The significantly higher presence of *C. lambica* in LD + S is possibly due to maize silage in the feed but could also be due to the lower storage temperature of the maize silage outdoors during winter [33]. As LD + S samples were gained from the institute's own research projects, it is known that animals did not develop HBS or any other disease and that they ate a lot more with the ad libitum feeding of LD + S in comparison to the previous feed intake with commercial feed (Jörling, 2017) [8]. This was observed, although the yeast content of the feed was temporarily more than 1×10^8 cfu/g feed (Jörling, personal observations, results of which have not yet been published). Olstorpe et al. [26] discovered that *Pichia fermentans* (*C. lambica*) was dominant in all their experiments. They assumed that *C. lambica* was able to improve palatability as it has been described to improve the flavor composition during wine- and cheese-making [26]. Presumably, the yeast species within the diet might be more important than the orientation values, which are the same for all yeasts when the hygiene status of the liquid diets is under debate.

4.1.3. Yeasts from Liquid Diets for Pigs

In the present study, mostly *C. krusei*, *C. holmii*, *C. lambica*, *S. cerevisiae*, *C. humilis* and *Geotrichum* spp. were identified (see Section 3.1, Table 1). Together, they accounted for 66.5% of all yeast isolates. Other species could not be identified (7.5%, Table 1) or were only detected in lower proportions (26%; see Section 3.1).

Middelhoven et al. [33] observed, in whole-crop maize ensiled for two weeks, similar yeast species compared to those found in LD + S. They predominately found *C. holmii*, *C. lambica*, *C. milleri* (current name: *C. humilis*), *Hansenula anomala* (current name: *Wickerhamomyces anomalis*, anamorph: *C. pelliculosa*) and *Saccharomyces dairensis* [33], whereby only the latter yeast did not occur in our study. The comparability of the yeast species in both studies could indicate that it is not so much the storage over winter but rather the substrate that influences the yeast occurrence.

The biochemical profiles of both yeasts are very similar and therefore sometimes misidentified [34]. Both yeasts are able to assimilate mostly glucose, lactose, glycerol, inositol and N-acetylglucosamine, while xylose is only metabolized from *C. lambica*. On the other hand, the next frequently identified yeasts, *C. holmii*, *C. humilis* and *S. cerevisiae*, cannot perform this metabolic function, with the exception of glucose. Instead, they metabolize galactose and raffinose, and, in part, trehalose (*C. holmii and C. humilis*), sucrose (*C. holmii* and partly also *S. cerevisiae*) as well as maltose (*S. cerevisiae*). This could suggest that the assimilative capacities of the yeasts are not essential for their presence or absence in different liquid feeds for pigs.

Likewise, many different yeasts were identified in studies on yeast determination from liquid feed samples for pigs, and different species dominated in the different feed samples [6,26–28].

Similarities to the isolated yeasts in the present study were observed in the studies by Olstorpe et al. [26], who examined liquid feeds based on a cereal grain mix and wet wheat distiller's grain with and without starter cultures. Without starter cultures, they observed *P. fermentans* (*C. lambica*), *C. pararugosa*, *C. rugosa*, *P. galeiformis* (current name: *P. mandshurica*), *T. asahii*, *Issatchenkia orientalis* (*C. krusei*, *P. kurdriavzevii*), *C. ethanolica* and *C. vini*. These yeasts were found in the present study as well, except the last three mentioned ones.

Olstorpe et al. [26] isolated *C. kefyr* from wheat-based liquid feed as the dominant yeast, which was isolated only once in the present study. This previous publication also found *C. krusei*, *C. pelliculosa* and *Pichia membranaefaciens* (*C. valida*). Plumed-Ferrer and Wright [35] most frequently observed *K. exigua* (*C. humilis*), *Debaromyces hansenii* and *Pichia derserticola* in fresh batches of liquid feed, of which only *K. exigua* was often observed in the present study. On the other hand, in this previous study, other yeasts were also isolated, such as *Pichia kurdriavzevii* (*C. krusei*), *S. exiguous* (*C. holmii*), *Pichia membranaefaciens* (*C. valida*) and *Wickerhamomyces anomalus* (*C. pelliculosa*), which were identified in the present study, too.

Significantly more morphologically different yeasts were observed in LD + S (see Section 3.1), which had lower pH-values than conventional feed and were stored outdoors during winter, which, as a consequence, were exposed to changing temperatures. These different colonial morphologies could be a result of changing environmental conditions, as explained in previous publications [36,37].

The genera *Geotrichum* and *Trichosporon* are classified as yeast or yeast-like organisms, but *Geotrichum* was formerly classified as a mold [38–40]. The colony morphology is very similar to other yeasts and therefore was described in many previous studies concerning yeasts in liquid swine diets [25,27,28], so that comparability with other studies is possible. *Saprochaete suaveolens*, formerly classified as *Geotrichum fragrans*, is also classified as a yeast or yeast-like organism (mycobank.org [41]). Hereafter, for the sake of simplicity, all genera are referred to as yeasts, even if the term yeast or yeast-like organism would be more accurate.

4.1.4. Method Comparison

The present study compared different methods for identifying yeasts to find the best method for the chosen substrate and the yeasts contained in it. From previous studies [6,10,26,28,42], it was known that many tests for identifying yeasts from the environment produce fewer results than those from clinical material [43–45]. Additionally, different databases on which different test procedures are based also influence the obtained results [45,46].

For clinical samples consisting mainly of *Candida* spp., the method of MALDI-TOF outperformed the diagnosis capacities of the phenotypic tests by reducing the delay in results and improving the reliable identification rate at species level [43]. On the other hand, this method requires significantly higher acquisition costs for the equipment. Therefore, this method was compared with the ID 32 C test, which has virtually no purchase costs.

ID32C

In our study, 5.2% less reliable results were observed with the ID 32 C test in comparison to MALDI-EX. Nevertheless, in individual cases, correct identification could only be made with this simple biochemical method (see Table 4). *C. rugosa* was twice identified with more than 99.5% accuracy as "very good identification", while MALDI-EX identified these two yeast isolates as *C. pararugosa*. The latter was not included in the ID32C-database. Considering the fact that several authors [44,45,47] proposed a lower identification score for the yeast identification with MALDI (see below), perhaps the MALDI results are the correct ones. In the case of *Geotrichum* spp. and *Saprochaete suaveolens*, the situation is similar. *Saprochaete suaveolens* was not included in the database of the ID32C test. Comparable to the finding in the present study, namely that *Rhodotorula* was better identified with ID32C, Olstorpe et al. [6] reported that, with the applied PCR fingerprinting, two *Rhodotorula glutinis* isolates were incorrectly classified as *Cryptococcus satoi* or *Pichia membranaefaciens*, but correctly identified with ID32C.

The ID32C test can be easily performed in any laboratory and does not require an expensive device. In addition to species identification, the biochemical test has the advantage of showing which enzymes can be produced by the respective yeast isolate. This, in turn, could allow or exclude opportunities for identifying which feed components could be metabolized by the yeast.

Selected Biochemical Reactions of the ID32C-Test

Metabolization of *N*-Acetylglucosamine (NAG): More yeasts from LD + S, even if not significant, were able to metabolize the amino sugar NAG. This is the monomeric constituent of chitin, which is one of the most abundant renewable resources found in nature [48]. The uptake of NAG into the yeast cell, its metabolites in the cell and conversion to cell wall formation have already been described for various yeasts [48]. The cell wall reinforced by NAG (chitin) offers protection against low pH-values in the environment [22]. Although the pH-values in the LD − S were not significantly lower than those of LD + S, the prolonged period of survival in silage (see below) may have led to the ability of yeasts to metabolize NAG.

Metabolization of Lactic Acid (LAT): More yeasts from LD + S were able to metabolize lactic acid. As this finding is not statistically significant, the ability to metabolize lactic acid obviously is not a prerequisite for yeasts in LD + S. Lactic acid bacteria are the predominant group of bacteria found in maize silages, and are able to multiply in liquid feed, lactic acid being a main product of their metabolism [49]. Maize silage used for LD + S in the present study was kept outdoors during winter and early spring until it was fed to the animals in late spring and early summer. Being able to use a substrate present in the environment is presumed to be an advantage for yeasts [50,51], which have to survive in these conditions for a long time.

MALDI-DS: The less time-consuming and less expensive MALDI-DS reduced the identification rate significantly ($p < 0.001$) by more than half compared to MALDI-EX (37.5% vs. 78.75% reliable identification). Thus, the use of this method is clearly limited, at least if different yeasts are to be identified from environmental samples. In contrast to bacteria, yeasts possess a thick and chitinous cell wall, which might lead to the difficulties encountered with the MALDI-DS method [52].

MALDI-EX: In our study, 78.75% of the 80 different yeasts could be identified by MALDI-EX and 75.0% by the ID 32 C. While 11 isolates were not identified at all, six isolates achieved only probable results at the genus level (see Table 4). An incorrect diagnosis was observed only once, mistakenly identifying *P. occidentalis* instead of *C. krusei* (see Section 3.1.1).

A comparison of identification of 96 foodborne yeasts with MALDI-TOF and two conventional tests, of which one was ID 32 C, was made by Pavlovic et al. 2014 [53]. In their study, more yeast isolates could be identified with MALDI-TOF than with the ID32C test, too.

The identification rate of the different methods in the present study was comparable to those of others in which yeasts were isolated from the environment rather than from clinical material [6,54–56]. As already shown by these and other authors [1,57], none of the methods were capable of reliably detecting all yeast isolates from the liquid feed. Many authors attribute these differentiation failures to the background of the ID32C, MALDI-TOF and other commercially available tests, as these were developed and established for clinically relevant yeasts in human beings and not for yeasts in animal feed [1,45,57]. None of the available methods can be considered as the golden standard for the differentiation of yeasts from liquid feeds. With respect to the low examination costs, low workload, fast availability of results and available databases, which means the highest rate of correct identification, the different methods exhibit advantages and disadvantages.

Although MALDI-EX was the best method for gaining the most reliable identification results in this study, it has to be considered that results from this method are only as good as the underlying database [46]. Vlek et al. [46] identified 61.5% of their yeasts from human patients using the Bruker Daltonic database (BDAL), but improved their identification rate up to 86.8% by adding their database with the in-house database from the Centraalbureau voor Schimmelcultures (Central Bureau for Fungal Cultures) (BDAL + CBS in-house). This allows the assumption to be made that even more yeasts will be identified with this method in the future, if correspondingly relevant data continue to be added, especially for the non-clinical yeasts found in the surroundings. An improvement in the identification results of 845 environmental yeasts by one third was also described by Augustini et al. [45] after developing a supplementary database.

Besides the databases as reason for missing reliable yeast identification, Augustini et al. [45] stated that identification scores <2.00 are not able to unequivocally affirm that the identification at species level is unreliable. They cited studies that showed identification results under 2.00, but with correct identifications. This observation was underlined in the studies by Tan et al. [44]. Repeating MALDI-TOF attempts in 10.2% of the yeast isolates, which had indicated spectral scores as being unacceptable on the first attempt (scores < 2.00), resulted in acceptable scores (>2.00). Most of these achieved a correct identification on the first attempt [44]. The authors concluded that lowering the identification score from <2.00 to <1.70 could reduce the repetition rate [44]. With a cut-off of <1.70, Lee et al. [50] also improved the identification rate of their 284 pathogenic yeasts from clinical samples compared to the required cut-off value of >2.00 [49]. When comparing the results of two different MALDI-TOF systems (Biotyper from Bruker and ASTA MALDI-TOF MS), Lee at al. [50] found that only 39.5% of the isolates with confirmed identification with molecular sequencing met the cut-off score in both systems. The majority of the isolates (58.6%) ranged between 1.70 and 2.00 when using the Bruker Biotyper and scores > 140 using ASTA MALDI-TOF.

Lee et al. [52] performed a formic acid extraction with a shorter protocol. Most of the yeasts obtained from samples of clinically infected humans were identified correctly, but the method failed to identify the slimy *Cryptococcus* spp. Considering the fact that in our study no *Cryptococcus* spp. were found, possibly this shorter, easier and inexpensive method could have provided as good results as MALDI-EX. On the other hand, different *Cryptococcus* spp. were isolated from liquid swine diets in studies by Olstorpe et al. [6]. Therefore, MALDI-EX seemed to be the best method to reliably identify as many yeasts species as possible.

Extending databases, lowering the identification score for yeasts as well as shorter protocols could improve the ratio of reliable results of environmental yeasts with MALDI-EX in the future, so that the results of this method could be highlighted even more.

Various molecular biological methods described in the literature were not included in this study, although previous authors achieved good results [27]. Gori et al. [27] had difficulties in separating the two most commonly occurring yeasts in their study with 26S rRNA sequencing: *C. humilis* (formerly named *C. milleri*; 58.4%) and *C. holmii* (*Kasachstania exigua*; 17.5%), together accounting for 75.9% of all results (n = 766 yeasts). They distinguished the two yeasts biochemically according to their sucrose and raffinose metabolism [27]. In the present study, *C. humilis* and *C. holmii* accounted together for 20% of all results (n = 16). In retrospect, it can be assumed that the 26S rRNA method would not have been advantageous in these cases.

4.2. Temperature

In the present study, clearly more than half of the yeasts (52.9%) grew better at 25 °C than at 37 °C or did not grow at 37 °C at all (23.5%). Those yeasts that did not grow at 37 °C at all will presumably not grow in the intestines of pigs, where the internal body temperature normally still exceeds 37 °C.

Considering only yeasts isolated from LD + S, there are even more isolates that prefer cooler temperatures (see Table 5). An explanation for these yeasts preferring lower temperatures than yeasts from LD − S could be the chosen time of sampling of LD + S in late spring and early summer in the two projects, when the liquid diets were composed. After harvesting the maize plants and making silage in the fall in the respective projects, this was stored outdoors during winter, where yeasts had to cope with low temperatures. Thus, yeasts may have adapted to these temperatures or died. Storing feed materials or liquid diets at cool temperatures possibly reduces the yeast species, which prefer 37 °C, and as a result have little or no impact on gut health.

The present results could also indicate an adaptation of the yeasts to their feed origin and storage temperature. These results were obtained directly after the cultivation of the yeasts from the respective feed (see Section 2.3). Therefore, yeasts had little or no opportunity to adapt to the new temperatures. This is in the broadest sense comparable with the climatic conditions during the long period between the fall and spring. On the other hand, the possibility to adapt would exist at warmer outside temperatures and in case of the pre-fermentation of the liquid feed (24 h, 38 °C), as is sometimes practiced, especially with controlled fermentation [58]. Suutari et al. [59] reported morphological changes in some yeasts that had to adapt to cooler or very warm temperatures in a bouillon. The investigations in this previous study on growth performance at different temperatures was made on agar plates. The possible easier adaptation to new temperature conditions in a bouillon could also be an explanation for the observations that some yeasts only produce gas at 37 °C after a longer period of time (see Section 3.4, *C. humilis* in Figure 3).

Margesin et al. [60] isolated yeasts and bacteria from cold-adapted habitats and classified 60% of the yeasts but only 8% of bacteria to be true psychrophils, which showed no growth above 20 °C, indicating that the remaining microorganisms are able to adapt to warmer temperatures. Yeasts that do not grow or grow very poorly at 37 °C are thought to have little or no effect on gut health [20]. As a result, the lack of or partly low clinical symptoms on farms with a high yeast load in the feed are explicable. Correspondingly, yeasts that did not grow at all or worse at 37 °C than at 30 °C were also found on yeasts obtained from swab samples from milking machines [55]. A large majority of them could not be recovered from the milk collected with these milking machines.

In both groups (LD − S and LD + S), 1.9 different yeasts were identified. On the other hand, significantly more different colonial morphologies of the yeasts were found in LD + S, possibly indicating that temperature could have an influence on morphology, as was also observed by Nadeem et al. [37].

4.3. pH-Value

In the present study, the LD + S had on average slightly higher pH-values than LD − S and they contained significantly more *C. lambica*. Whether this connection is accidental or related to the higher pH-value can only be suspected due to the small number of farms of origin. Lack of growth at 37 °C [34], a good smell/taste [26] but no described ability of biofilm formation, as found by the Olostorpe et al. [26], could mean that this yeast is expected to be less harmful as a feed contaminant and for gut health than other yeasts. On the other hand, some yeasts are known to adapt to pH-values, to temperature and to different media [37], so that the safety of *C. lambica* in liquid swine diets still needs to be tested.

For fungi as well as bacteria, one of the most important environmental conditions is ambient pH. Changes in external pH result in phenotypic, metabolic and physical changes of the microorganisms [22]. The low pH-values in liquid feeds, especially fermented ones or those containing silage compared to normal feed for pigs, in general favor yeasts. This is due to the fact that at pH-values < 5.0, many bacteria are not able to stay alive or to grow as fast as they do at higher pH-values [61]. Molds depend on oxygen, but yeasts are able to grow at low pH-values with and without oxygen [3]. Some yeasts are known to be able to adapt to low pH-values in their surroundings by forming a thicker cell wall with chitin (*N*-acetyl glucuronidase) [22]: the high buffering capacity in the cytosol, high H^+-ATP-ase and/or high endogenous energy reserves of *C. krusei* [29]. Therefore, fermented liquid feeds, especially after controlled fermentation, always poses a certain risk of increased yeast content.

4.4. Gas Production

Quantitatively comparing gas production of different yeasts under standardized conditions with Ancom RF Gas Production System was, to the best of our knowledge, performed for the first time. Investigations in a bouillon, produced in accordance with European and US Pharmacopoeia guidelines, allows for a comparison of gas-producing yeasts irrespective of feed or water. The SAB-bouillon provides ideal conditions for yeasts and contains high amounts of glucose (20 g/L). However, the total gas quantities measured do not describe quantities that would also be produced in the feed or in the animal, since the competing flora is always different, and feed is not composed like a bouillon or an agar for yeasts.

Different generation times, sizes and numbers of buds make it difficult to precisely calculate the yeast quantity with density determination or even with quantitative cell counting. Hence, the amounts of gas production were not precisely determined but categorized to two major groups, as described above. Additionally, not the exact yeast numbers per milliliter were determined but only the density by means of the McFarland standard. Exemplarily, for some samples with the density of McFarland 0.3, the yeasts were counted, resulting in 1–4 $\times$ 10^5 cfu yeasts per mL. Thus, these yeast counts are just about acceptable regarding the requirements in liquid feed according to Kamphues et al. [17].

In the present study, slightly increasing gas pressures were also measured for yeasts that did not produce gas at the beginning of the experiment. This could be explained by the rising room temperature during processing to the 37 °C in the incubator.

The gas formation capacity of the yeasts differed very clearly between 888 mbar and 10,419 mbar. *C. kefyr* formed over ten times more gas than one of the *S. cerevisiae* isolates. Only one yeast isolate from LD + S was able to produce higher amounts of gas at 37 °C within 24 h. This was partly caused by its preference for cooler temperatures, as described above. The reason for the differing amounts of gas production of yeasts may to some extent be seen in the lack of oxygen produced in the Ancom Gas Production System, which is also found in the pig's colon. Some yeasts like *C. sphaerica, C. variabilis, C. kefyr, C. lambica, C. krusei, S. cerevisiae* and *C. pelliculosa* are known to metabolize glucose under anaerobic conditions; variable metabolization is expected from other yeasts like *C. valida, G. candidum* and *G. capitatum*, while *C. rugosa* and *Rhodotorula* spp. are mostly not capable of fermentation [61]. The latter cannot be expected to produce gas amounts under the conditions available in the present study as well as those found in the gastrointestinal tract of pigs. As such, they cannot be expected to cause a disease such as HBS. Comparing the growth of a yeast from liquid feed for pigs at 37 °C and 25 °C can give a good indication of whether a yeast is likely to cause HBS. However, it is not possible to make an accurate prediction because yeasts are partially capable of adapting to temperatures and some yeasts hardly ferment under anaerobic conditions despite growth at 37 °C. On the other hand, no high gas production within 24 h was observed in the present study when a yeast isolate did not grow at 37 °C. Presumably, those yeasts are not supposed to cause HBS. A test of growth at 37 °C would be easy to perform in every laboratory and could give a hint at whether a yeast would be able to grow in a pig's alimentary tract. Further studies will be needed to clarify which amount of gas production can generally be called high or low. Apart from this, it has to be considered that a yeast, even if it is not able to form a biofilm itself, may colonize the biofilm of the lines of the feeding system. Such yeasts could potentially be capable of adapting to warmer temperatures, especially in the summer months.

4.5. Summary

In several studies of liquid feed, samples for pigs' yeasts were identified, which were also found in the present study. The most commonly detected yeast in our study was *C. krusei*. This is the first study of liquid feed with and without maize silage. In liquid feed with maize silage (LD + S), significantly more *C. lambica* was found.

MALDI-EX provided the most reliable results (78.75%), but the ID 32 C-test, easy to perform in every laboratory, was sufficient for confirming 75.0% of the identified yeasts. Both tests together identified 88.75% of the yeasts because some yeasts were only reliably identified with one or the other test. The quicker MALDI-DS-method provided only 37.5% reliable results, this being significantly less than the other two methods. Thus, a formic acid/acetonitrile extraction (MALDI-EX) before analysis should be preferred.

Clearly more than half of all yeast isolates grew better at 25 °C than at 37 °C. Fourteen isolates showed no growth at all at 37 °C. Gas amounts produced by the different yeast isolates differed more than tenfold within a 24-h incubation period at 37 °C in SAB-bouillon measured with the Ancom

Gas Production System. Most of the tested *C. krusei* and *S. cerevisiae* but none of the tested *C. holmii*, *Trichosporon* spp., *G. silvicola* and *C. pelliculosa* were able to produce gas. While only one yeast from LD + S was able to produce gas within 24 h, more yeasts (40.6%) from LD − S were able to do so. None of the yeasts that did not grow on the SAB-agar at 37 °C were able to produce high amounts of gas within a 24-h incubation period at 37 °C in the bouillon, presuming that those yeasts could only slightly affect the animals' health.

Due to the fact that the majority of *C. krusei* isolates were able to grow at 37 °C, produce high amounts of gas, grow in low pH conditions and form biofilms, as is known from the literature, this yeast species seems to be predestinated to grow in liquid diets and to remain in a biofilm in the pipelines serving the liquid diet. Therefore, special interest should be given to this yeast species. The evaluation of yeast levels in liquid feed for pigs has so far only been determined on the basis of the number of yeasts per gram feed. Laboratory values alone could possibly incorrectly estimate the influence of yeasts on the health of the animals as either being too low or too high. Additional investigations are needed to further characterize the effect of each yeast species on pig health. Moreover, investigating the effect of having the storage temperature of the feed significantly below body temperature could be interesting.

Author Contributions: Conceptualization, B.K.; methodology, B.K.; software, B.K. and U.S.; validation, B.K., H.K., U.S. and C.V.; formal analysis, B.K. and C.V.; investigation, H.K. and B.K.; resources, J.K. and C.V.; writing—original draft preparation, B.K.; writing—review and editing, B.K., C.V. and U.S.; visualization, B.K. and C.V.; supervision, C.V.; project administration, B.K.; funding acquisition, J.K. All authors have read and agreed to the published version of the manuscript.

Funding: This publication was supported by the German Research Foundation (Deutsche Forschungsgemeinschaft) and the University of Veterinary Medicine Hannover, Foundation, Hannover, Germany within the funding program Open Access Publishing.

Acknowledgments: We would like to thank Frances Sherwood-Brock for proof-reading the manuscript to ensure correct English.

Conflicts of Interest: The funders had no role in the design of the study; in the collection, analyses, or interpretation of data; in the writing of the manuscript, or in the decision to publish the results.

References

1. Fiedler, B. *Hefen: Lebensmittelassoziierte Mikroorganismen Fermentationsleistung und Verderbspotential*, 2nd ed.; Behr's Verlag GmbH: Hamburg, Germany, 2017; pp. 7, 12, 56.
2. Büchl, N.R. Identifizierung von Hefen durch Fourier-transform Infrarotspektroskopie und künstlich neuronale Netzen. Ph.D. Thesis, Faculty Sciene Center Weihenstephan, Weihenstephan, Germany, 19 June 2009. Available online: http://mediatum.ub.tum.de/?id=684162 (accessed on 2 December 2020).
3. Stalljohann, G. Tiergesundheitsmanagement auf betrieblicher Ebene. In *Tiergesundheit Schwein*; Brede, W., Blaha, T., Hoy, S., Eds.; DLG-Verlags-GmbH: Frankfurt am Main, Germany, 2010; pp. 236–241.
4. Santos, M.C.; Golt, C.; Joerger, R.D.; Mechor, G.D.; Murao, G.B.; Kung, L., Jr. Identification of the major yeasts isolated from high moisture corn and corn silages in the United States using genetic and biochemical methods. *J. Dairy Sci.* **2017**, *100*, 1151–1160. [CrossRef] [PubMed]
5. Nagel, M. Hygiene in Flüssigfütterungsanlagen. In Themen zu Tierernährung, Fachtagung Deutsche Vilomix Tierernährung GmbH 2004/2005. 2005. Available online: http://docplayer.org/32990535-Hygiene-in-fluessigfuetterungsanlagen.html (accessed on 2 December 2020).
6. Olstorpe, M.; Lyberg, K.; Lindberg, J.E.; Schnürer, J.; Passoth, V. Population diversity of yeasts and lactic acid bacteria in pig feed fermented with whey, wet wheat distillers' grains or water at different temperatures. *Appl. Environ. Microbiol.* **2008**, *74*, 1696–1703. [CrossRef] [PubMed]
7. Brooks, P.H. Fermented liquid feed for pigs. *CAB Rev. Perspect. Agric. Veter- Sci. Nutr. Nat. Resour.* **2008**, 3. [CrossRef]
8. Jörling, U. Untersuchung zum Ansatzverhalten und zur Mikroflora im Kot von Mastschweinen unter den Bedingungen einer Konditionsbasierten Energie- und Nährstoffversorgung und Ad-libitum-fütterungsbedingungen. Master Thesis, University of Veterinary Medicine Hannover, Hanover, Germany, 2017.

9. Jensen, B.B. The impact of feed additives on the microbial ecology of the gut in young pigs. *J. Anim. Feed Sci.* **1998**, *7*, 45–64. [CrossRef]
10. Canibe, B.; Jensen, B.B. Fermented liquid feed—Microbial and nutritional aspects and impact on enteric diseases in pigs. *Anim. Feed Sci. Technol.* **2012**, *173*, 17–40. [CrossRef]
11. Missotten, J.A.; Michiels, J.; Ovyn, A.; De Smet, S.; Dierick, N.A. Fermented liquid feed for pigs. *Arch. Anim. Nutr.* **2010**, *64*, 437–466. [CrossRef]
12. Canibe, N.; Jensen, B.B. Fermented and nonfermented liquid feed to growing pigs: Effect on aspects of gastrointestinal ecology and growth performance. *J. Anim. Sci.* **2003**, *81*, 2019–2031. [CrossRef]
13. Johannsen, U.; Strijkstra, G.; Klarmann, D.; Janthur, I. Untersuchungen zum Enterohämorrhagischen Syndrom (EHS) der Schweine. *Prakt. Tierarzt.* **2000**, *12*, 440–451.
14. Reiner, G. Enterohaemorrhagisches Syndrom. In *Krankes Schwein—Kranker Bestand*; Eugen Ulmer Verlag: Stuttgart, Germany, 2015; pp. 132–134.
15. Straw, B.; Dewey, C.; Kober, J.S.; Henry, C. Factors associated with death due to hemorrhagic bowel syndrome in two large commercial swine farms. *J. Swine Health Prod.* **2002**, *10*, 75–79.
16. Missotten, J.; Michiels, A.; Degroote, J.; De Smet, S. Fermented liquid feed for pigs: An ancient technique for the future. *J. Anim. Sci. Biotechnol.* **2015**, *6*, 4. [CrossRef]
17. Kamphues, J.; Wolf, P.; Coenen, M.; Eder, K.; Iben, C.; Kienzle, E.; Liesegang, A.; Männer, K.; Zebeli, Q.; Zentek, J. Beurteilung der Mikrobiologischen Qualität von FM. In *Supplemente zur Tierernährung für Studium und Praxis*, 12th ed.; Schaper Verlag: Hannover, Germany, 2014; p. 206.
18. Nagel, M. Mikrobiologische Vorgänge in Flüssigfutter für Schweine. In *Handbuch der Tierischen Veredlung*; Kamlage-Verlag: Osnabrück, Germany, 1998; Volume 23, pp. 189–200.
19. *Grüne Broschüre 2020 Das geltende Futtermittelrecht. (Band FZ: Futtermittelzusatzstoffe)*, 2nd ed.; Allround Media Service e. K: Rheinbach, Germany, 2019; pp. 310–315, 322–323, 326–327, 338–339, 352–353.
20. Rajkowska, K.; Kunicka-Styczyńska, A. Typing and virulence factors of food-borne *Candida* spp. isolates. *Int. J. Food Microbiol.* **2018**, *279*, 57–63. [CrossRef]
21. Mayer, F.L.; Wilson, D.; Hube, B. *Candida albicans* pathogenicity mechanisms. *Virulence* **2013**, *4*, 119–128. [CrossRef] [PubMed]
22. Sherrington, S.L.; Sorsby, E.; Mahtey, N.; Kumwenda, P.M.; Lenardon, D.; Brown, I.E.; Ballou, R.D.; MacCallum, M.; Hall, R.A. Adaptation of *Candida albicans* to environmental pH induces cell wall remodelling and enhances innate immune recognition. *PLoS Pathog.* **2017**, *13*. [CrossRef] [PubMed]
23. Puligundla, P.; Mok, C. Potential applications of nonthermal plasmas againstbiofilm-associated micro-organisms in vitro. *J. Appl. Microbiol.* **2017**, *122*, 1134–1148. [CrossRef] [PubMed]
24. Singhal, N.; Kumar, M.; Kanaujia, P.K.; Jugsharan, S.; Virdi, J.S. MALDI-TOF mass spectrometry: An emerging technology for microbial identification and diagnosis. *Front. Microbiol.* **2015**, *6*, 791. [CrossRef] [PubMed]
25. Lee, H.; Park, J.H.; Oh, J.; Cho, S.; Koo, J.; Park, I.C.; Kim, J.; Park, S.; Choi, J.S.; Shin, S.Y.; et al. Evaluation of a new matrix-assisted laser desorption/ionization time-of-flight mass spectrometry system for the identification of yeast isolation. *J. Clin. Lab. Anal.* **2018**, *33*, e22685. [CrossRef]
26. Olstorpe, M.; Axelsson, L.; Schnürer, J.; Passoth, V. Effect of starter culture inoculation on feed hygiene and microbial population development in fermented pig feed composed of a cereal grain mix with wet wheat distillers' grain. *J. Appl. Microbiol.* **2010**, *108*, 129–138. [CrossRef]
27. Gori, K.; Kryger Bjørklund, M.; Canibe, N.; Petersen, A.O.; Jespersen, L. Occurrence and Identification of Yeast Species in Fermented Liquid Feed for Piglets. *Microbial. Ecol.* **2011**, *61*, 146–153. [CrossRef]
28. Urubschurow, V.; Janczyk, P.; Pieper, R.; Souffrant, W.B. Biological diversity of yeasts in the gastrointestinal tract of weaned piglets under different farm conditions. *FEMS Yeast Res.* **2009**, *8*, 1349–1356. [CrossRef]
29. Halm, M.; Hornbaek, T.; Arneborg, N.; Sefa-Dedeh, S.; Jespersen, L. Lactic acid tolerance determined by measurement of intracellular pH of single cells of *Candida krusei* and *Saccharomyces cerevisiae* isolated from fermented maize dough. *Int. J. Food Microbiol.* **2004**, *94*, 97–103. [CrossRef]
30. Spicher, G.; Schröder, R. Die Mikroflora des Sauerteiges. *Z. Lebensm. Unters. Forsch.* **1980**, *170*, 119–123. [CrossRef]
31. Silva, S.; Rodrigues, C.F.; Araùjo, D.; Rodrigues, M.E.; Henriques, M. *Candida* species biofilms' antifungal resistance. *J. Fungi* **2017**, *3*, 8. [CrossRef]

32. Douglass, A.P.; Offei, B.; Braun-Galleani, S.; Coughlan, A.Y.; Martos, A.A.R. Population genomics shows no distinction between pathogenic *Candida krusei* and environmental *Pichia kudriavzevii*: One species, four names. *PLoS Pathog.* **2018**, *14*, e1007138. [CrossRef] [PubMed]
33. Middelhoven, W.J.; de Jong, I.M.; de Winter, M. Yeasts and fungi occurring in ensiled whole-crop maize a7d other ensiled vegetable crops. *Antonie Van Leeuwenhoek* **1990**, *57*, 153–158. [CrossRef]
34. Vervaeke, S.; Vandamme, K.; Boone, E.; De Laere, E.; Swinne, D.; Surmont, I. A case of *Candida lambica* fungemia misidentified as *Candida krusei* in an intravenous drug abuser. *Med. Mycol.* **2008**, *46*, 853–856. [CrossRef] [PubMed]
35. Plumed-Ferrer, C.; von Wright, A. Antimicrobial activity of weak acids in liquid feed fermentations, and its effects on yeasts and lactic acid bacteria. *J. Sci. Food Agric.* **2011**, *91*, 1032–1040. [CrossRef]
36. Kuthan, M.; Devaux, F.; Janderová, B.; Jacq, C.I.; Palkov, Z. Domestication of wild *Saccharomyces cerevisiae* is accompanied by changes in gene expression and colony morphology. *Mol. Microbiol.* **2003**, *47*, 745–754. [CrossRef]
37. Nadeem, S.G.; Shafik, A.; Hakim, S.T.; Anjum, Y.; Kazm, S.U. Effect of growth media, pH and temperature on yeast to hyphal transition in Candida albicans. *Open J. Med. Microbiol.* **2013**, *3*, 185–192. [CrossRef]
38. Bauer, J.; Schwaiger, K. Spezielle Mykologie und Prototheken. In *Tiermedizinische Mikrobiologie, Infektions- und Seuchenlehre*; Selbitz, H.-J., Truyen, U., Valentin-Weigand, P., Eds.; Enke Verlag: Stuttgart, Germany, 2015; p. 365.
39. Pohlmann, L.M.; Chengappa, M.M. Yeasts—*Cryptococcus, Malassezia and Candida.* In *Veterinary Microbiology*, 3rd ed.; McVey, S.D., Kennedy, M., Chengappa, M.M., Eds.; Willey-Blackwell: Ames, IA, USA, 2013; pp. 319–320.
40. mycobank.org. Available online: www.mycobank.org/BioloMICS.aspx?TableKey=14682616000000067&Rec=39099 (accessed on 14 August 2020).
41. mycobank.org. Available online: www.mycobank.org/name/Geotrichumfragrans (accessed on 14 August 2020).
42. Canibe, N.; Pedersen, A.O.; Jensen, B.B.; Jespersen, L. Microbiological and biochemical characterization of fermented liquid feed samples from 40 Danish farms. *Livestock Sci.* **2010**, *134*, 158–161. [CrossRef]
43. Stefaniuk, E.; Baraniak, A.; Fortuna, M.; Hryniewicz, W. Usefulness of CHROMagar Candida medium, biochemical methods—API ID32C and VITEK 2 Compact and two MALDI-TOF MS Systems for *Candida* spp. Identification. *Polish J. Microbiol.* **2016**, *65*, 111–114. [CrossRef]
44. Tan, B.; Ellis, C.; Lee, R.; Stamper, P.D.; Zhang, S.X.; Carroll, K.C. Prospective Evaluation of a Matrix-Assisted Laser Desorption Ionization–Time of Flight Mass Spectrometry System in a Hospital Clinical Microbiology Laboratory for Identification of Bacteria and Yeasts: A Bench-by-Bench Study for Assessing the Impact on Time to Identification and Cost-Effectiveness. *J. Clin. Microbiol.* **2012**, *50*, 3301–3308. [CrossRef] [PubMed]
45. Augustini, B.; Silva, L.P.; Bloch, C. Evaluation of MALDI-TOF mass spectrometry for identification of environmental yeasts and development of supplementary database. *Appl. Microbiol. Biotechnol.* **2014**, *98*, 5645–5654. [CrossRef] [PubMed]
46. Vlek, A.; Koecka, A.; Khayhan, K.; Theelen, B.; Groenewald, M.; Boel, E.; Multicenter Study Group; Boekhout, T. Interlaboratory comparison of sample preparation methods, database expansions, and cutoff values for identification of yeasts by matrix-assisted laser desorption ionization-time of flight mass spectrometry using a yeast test panel. *J. Clin. Microbiol.* **2014**, *52*, 3023–3029. [CrossRef] [PubMed]
47. Koht, P.D.; Farrance, C.E. Evaluation of MALDI-TOF Mass Spectrometry for Identification of Yeasts Commonly Found During Environmental Monitoring. *Am. Pharm. Rev.* **2016**. Available online: https://www.americanpharmaceuticalreview.com/Featured-Articles/331621-Evaluation-of-MALDI-TOF-Mass-Spectrometry-for-Identification-of-Yeasts-Commonly-Found-During-Environmental-Monitoring/ (accessed on 3 December 2020).
48. Inokuma, K.; Matsuda, M.; Sasaki, D.; Hasunuma, T.; Kondo, A. Widespread effect of *N*-acetyl-d-glucosamine assimilation on the metabolisms of amino acids, purines, and pyrimidines in *Scheffersomyces stipitis*. *Microb. Cell Factoriest* **2018**, *17*, 153. [CrossRef] [PubMed]
49. Dellaglio, F.; Torriani, S. DNA-DNA homology, physiological characteristics and distribution of lactic acid bacteria isolated from maize silage. *J. Appl. Bacteriol.* **1986**, *60*, 83–92. [CrossRef]
50. Drihuis, F.; Oude Elferink, S.J.W.H. The impact of the quality of silage on animal health and food safety: A review. *Vet. Q.* **2000**, *22*, 212–217. [CrossRef] [PubMed]

51. Ponomarova, O.; Gabrielle, N.; Séverin, D.C.; Mülleder, M.; Zimgibl, K.; Bulyha, K.; Andrejev, S.; Kafkia, E.; Typas, A.; Sauer, U.; et al. Yeast creates a niche for symbiotic lactic acid, bacteria through nitrogen overflow. *Cell Syst.* **2017**, *5*, 345–357. [CrossRef]
52. Lee, H.S.; Shin, J.H.; Choi, M.J.; Won, E.J.; Kee, S.J.; Kim, S.H.; Shin, M.G.; Suh, S.P. Comparison of the Bruker Biotyper and VITEK MS Matrix-Assisted Laser Desorption/Ionization Time-of-Flight Mass Spectrometry Systems Using a Formic Acid Extraction Method to Identify Common and Uncommon Yeast Isolates. *Ann. Lab. Med.* **2017**, *37*, 223–230. [CrossRef]
53. Pavlovica, M.; Mewesa, A.; Maggipintoa, M.; Schmidta, W.; Messelhäußera, U.; Balsliemkeb, J.; Hörmansdorfer, S.; Buscha, U.; Huber, I. MALDI-TOF MS based identification of food-borne yeast isolates. *J. Microbiol. Methods* **2014**, *106*, 123–128. [CrossRef]
54. Latouche, G.N.; Daniel, H.-M.; Lee, O.K.C.; Mitchell, T.G.; Sorrell, T.C.; Meyer, W. Comparison of use of phenotypic and genotypic characteristics for identification of species of the anamorph genus Candida and related teleomorph yeast species. *J. Clin. Microbiol.* **1997**, *35*, 3171–3180. [CrossRef]
55. Keller, B.; Scheibl, P.; Bleckmann, E.; Hoedemaker, M. Differenzierung von Hefen in Mastitismilch. *Mycoses* **2000**, *43* (Suppl. 1), 17–19. [PubMed]
56. Hoedemaker, M.; Schmidt, A.; Keller, B.; Bleckmann, E.; Böhm, K.H. Isolation von Hefen aus Milch von Kühen mit Mastitis und aus Tupferproben von der Melkanlage. *Prakt. Tierarzt.* **2006**, *87*, 890–898.
57. Buchan, B.W.; Ledeboer, N.A. Advances in identification of clinical yeast isolates by use of Matrix-Assisted Laser Desorption Ionization-Time of Flight mass spectrometry. *J. Clin. Microbiol.* **2013**, *51*, 1359–1366. [CrossRef] [PubMed]
58. Bunte, S. The Fermentation of Liquid Diets on Feeding Concept in Fattening Pigs—Potentials, but also Risks from the View of Animal Nutrition and Veterinary Medicine. Master Thesis, University of Veterinary Medicine Hannover, Hanover, Germany, 2018.
59. Suutari, M.; Liukkonen, K.; Laakso, S. Temperature adaptation in yeasts: The role of fatty acids. *J. Gen. Microbiol.* **1990**, *136*, 1469–1474. [CrossRef] [PubMed]
60. Margesin, R.; Gander, S.; Zacke, G.; Gounot, A.M.; Schinner, F. Hydrocarbon degradation and enzyme activities of cold-adapted bacteria and yeasts. *Extremophiles* **2003**, *7*, 451–458. [CrossRef]
61. Rüschendorf, A. *Medizinische Mykologie. Bestimmung und Differenzierung von Sproßpilzen, Schimmelopilzen, Dermatophyten und dimorphen Pilzen*, 3rd ed.; Lehmanns Media: Berlin, Germany, 2014; pp. 30, 33, 37, 60, ISBN 978-3-86541-629-2.

Journal of *Fungi*

Article

Dietary Supplementation of a Live Yeast Product on Dairy Sheep Milk Performance, Oxidative and Immune Status in Peripartum Period

Alexandros Mavrommatis [1], Christina Mitsiopoulou [1], Christos Christodoulou [1], Dimitris Karabinas [1], Valentin Nenov [2], George Zervas [1] and Eleni Tsiplakou [1,*]

[1] Laboratory of Nutritional Physiology and Feeding, Department of Animal Science, School of Animal Biosciences, Agricultural University of Athens, Iera Odos 75, GR-11855 Athens, Greece; mavrommatis@aua.gr (A.M.); chr_mitsiopoulou28@hotmail.com (C.M.); c.christodoulou@aua.gr (C.C.); vdkarabinas@vkarabinas-sa.gr (D.K.); gzervas@aua.gr (G.Z.)

[2] Phileo Lesaffre Animal Care, Marcq en Baroeul, 59700 Nord, France; info@phileo.lesaffre.com

* Correspondence: eltsiplakou@aua.gr; Tel.: +30-2105294435

Received: 10 November 2020; Accepted: 1 December 2020; Published: 3 December 2020

Abstract: This study evaluated the dietary administration of *Saccharomyces cerevisiae* live yeast on milk performance and composition, oxidative status of both blood plasma and milk, and gene expression related to the immune system of lactating ewes during the peripartum period. Chios ewes were fed either a basal diet (BD) (Control, $n = 51$) or the BD supplemented with 2 g of a live yeast product/animal (ActiSaf, $n = 53$) from 6 weeks prepartum to 6 weeks postpartum. Fatty acid profile, oxidative, and immune status were assessed in eight ewes per treatment at 3 and 6 weeks postpartum. The β-hydroxybutyric acid concentration in blood of ActiSaf fed ewes was significantly lower in both pre- and postpartum periods. A numerical increase was found for the milk yield, fat 6% corrected milk (Fat corrected milk ($FCM_{6\%}$)), and energy corrected milk yield (ECM) in ActiSaf fed ewes, while daily milk fat production tended to increase. The proportions of $C_{15:0}$, $C_{16:1}$, $C_{18:2n6t}$, and $C_{18:3n3}$ fatty acids were increased in milk of ActiSaf fed ewes, while $C_{18:0}$ was decreased. Glutathione reductase in blood plasma was increased ($p = 0.004$) in ActiSaf fed ewes, while total antioxidant capacity measured by 2,2′-Azino-bis (3-ethylbenzthiazoline-6-sulfonic acid) (ABTS) method was decreased ($p < 0.001$). Higher ABTS values were found in the milk of the treated group. The relative transcript levels of *CCL5*, *CXCL16*, and *IL8* were suppressed, while that of *IL1B* tended to decrease ($p = 0.087$) in monocytes of ActiSaf fed ewes. In conclusion, the dietary supplementation of ewes with *S. cerevisiae*, improved the energy utilization and tended to enhance milk performance with simultaneous suppression on mRNA levels of pro-inflammatory genes during the peripartum period.

Keywords: *Saccharomyces cerevisiae*; livestock; ewes; energy; antioxidant; cytokines

1. Introduction

Both meat and dairy products consumption are expected to increase in 2050 by 73 and 58%, respectively, compared to their 2010 levels [1,2], due to the rapid population growth rate. Ruminants' milk (67%) and meat (33%) cover 51% of proteins derived from the livestock sector and have a dominant role in food security, which is linked to how efficient animals utilize feed. Ruminants' feed efficiency depends upon the microbes residing within the rumen that ferment and transform feeds into volatile fatty acids (VFAs), proteins, and vitamins which are exploited by the host [3]. This multikingdom ecosystem's efficiency is dependent on various factors, the most prominent being that of diet. The improvement of the rumen microbiome habitat through the advancement of feed efficiency technologies entails a fundamental stepping stone in the overall improvement of livestock systems sustainability and food security concerns.

In intensive farming systems, high genetic merit animals require higher amounts of concentrate to fulfil their energy and nutrient demands, resulting in metabolic imbalances in rumen function and their microbiome governance. Probiotic yeasts are currently popular and widely used in ruminant feeding systems, especially since some of them have been officially authorized as feed additives in Europe [4]. The main purpose for using such additives in ruminant diets is to prevent rumen flora disorders and disturbances [5]. Dietary supplementation with live yeast (LY), *Saccharomyces cerevisiae*, improves rumen function through several modes of action [6]. This improvement is related to the oxygen scavenging properties of yeast in rumen (anaerobiosis mechanism), which upgrades bacterial viability and therefore the animal production [6]. Amongst the favorable bacteria are cellulolytics, which through the increase in their activity enhance fiber digestion. Moreover, LY can also stabilize the ruminal pH [7], not only after feeding, but also during the peripartum period where animals often find themselves in a negative energy balance and are further sensitive to metabolic diseases. It has been proven that even a low-grade energy deficiency weakens the animals' antioxidant system, which fails to neutralize the formation of Reactive Oxygen Species (ROS) and triggers the pro-inflammatory response [8,9].

By improving ruminants' feed utilization and both energy and nutrient availability during the peripartum period, not only can milk performance and chemical composition be enhanced, but furthermore a downregulation in the immunostimulation response can be achieved through the limitation of lipomobilization metabolites [10,11]. Although LY supplementation in ruminants' diets is a well-established nutritional strategy, previous works have only focused on district parameters instead of a holistic approach. Specifically, except for milk performance [7,10–14], scarce information has been linked to the potential improvement of energy balance and oxidative status and therefore to the immune response under the influence of dietary yeasts inclusion in ruminants.

Taking into account the aforementioned information, the objective of this work was to evaluate the effect of LY *S. cerevisiae* (CNCM I-4407, 10^{10} CFU/g, ActiSaf; Phileo Lesaffre Animal Care, France) in dairy sheep during the transition and early lactation period (6 weeks prepartum and 6 weeks postpartum) on milk performance and composition, antioxidant status (determined by Glutathione transferase (GST), Glutathione reductase (GR), Superoxide dismutase (SOD), Glutathione peroxidase (GSH-Px), Catalase (CAT) and Lactoperoxidase (LPO) activities, antioxidant capacity with 2,2′-Azino-bis (3-ethylbenzthiazoline-6-sulfonic acid) (ABTS) and Ferric reducing ability of plasma (FRAP) methods and oxidative stress indicators such as Malondialdehyde (MDA) and protein carbonyls (PCs)) on both milk and blood plasma and key-gene expression (*CCL5, CXCL16, INFG, IL1B, IL2, IL6, IL8, IL10, TNF, NFKB*) in monocytes and neutrophils which are associated with cytokine production.

2. Materials and Methods

2.1. Location and Environmental Conditions

The experiment was conducted from November 2019 to March 2020 on a commercial dairy sheep farm in the region of Chiliomodi in Korithia, Greece. This region has a typical Mediterranean climate with hot dry summers and relatively mild wet winters. During the experimental period, the mean temperatures in November, December, January, February, and March were 12.2, 8, 9.3, 11.1 and 13.1 °C, respectively. The selected farm represents the typical intensive dairy sheep production system of Greece.

2.2. Animals and Diets

Animals' housing, management, handling, and care complied with the latest European Union Directive on the protection of animals used for scientific purposes [15], while taking into account an extended experimental design report, the Bioethical Committee of Faculty of Animal

Science (currently known as the Agricultural University of Athens Ethical Committee in Research; FEK 38/A/2-3-2018, eide AUA) approved the experimental protocol. One hundred and twenty (120), 1- to 3-year-old dairy ewes (*Ovis aries*), of pure Chios breed, were physically selected from a flock of six hundred. At approximately 6 weeks before parturition, the ewes were divided into two homogenous groups based on their body weight (BW), number of parturition, and the milk yield from the previous year only for the case of multiparous ewes (2.1 ± 0.68 kg). Both groups had the same number of primi- (n = 20) and multiparous (n = 40) animals. More specifically, the ewes mean BW in the Control group (n = 60) was 61.5 ± 10.70 (SD) kg while in the ActiSaf group (n = 60) was 61.5 ± 11.02 (SD) kg. The Control group was fed a basal diet comprising of concentrate mix, alfalfa hay, and oat hay, while the ActiSaf group consumed the same basal diet supplemented with 2 g of *S. cerevisiae* LY/day/ewe (CNCM I-4407, 10^{10} CFU/g, ActiSaf; Phileo Lesaffre Animal Care, France) (Table 1). The animals were housed in two pens based on the dietary treatment. Both diets were isonitrogenous and isocaloric and were designed to meet ewes' requirements in the transition period and early lactation according to the flock fat (6%) corrected milk yield [16,17]. The animals were fed on a group basis while forages were offered separately from the concentrate in three equal portions after milking. Diet selectivity did not occur, no refusals of forages and/or concentrate were observed, and all animals had free access to fresh water. The experimental procedure lasted 6 weeks started from each ewes' parturition. After this, each ewe was returned to the commercial farm flock and the experiment ended when the final ewe had completed its 6th week on lactation. Since milk performance was recorded at the same time points, lactation stage had no effect on milk performance. Control ewes (n = 60) gave birth to 141 lambs (prolificacy = 2.35; 69 females and 72 males) while those of the ActiSaf (n = 60) gave birth to 142 (prolificacy = 2.36; 65 females and 77 males). In addition, since the experimental trial took place on farm-scale conditions, few ewes were unable to be exploited for data curation due to abortions (4), mastitis (10) or dystocia (2), hence the final number of subjects was re-adjusted to 51 and 53 for the Control and ActiSaf groups, respectively.

Table 1. Concentrates composition (g/kg), diet intake (g), daily nutrients intake (g/ewe), and feeds chemical composition and fatty acid profile (%).

Ingredients (g/kg)	**Concentrates**			
	Control		**ActiSaf**	
Maize grain	575		575	
Wheat middlings	180		178	
Soybean meal	220		220	
Mineral and vitamin	25		25	
ActiSaf	-		2	
Daily feed intake (g/day/ewe)				
	Prepartum	Postpartum	Prepartum	Postpartum
Oat hay	700	700	700	700
Alfalfa hay	700	700	700	700
Con. Mix	1000	2000	-	1000
ActiSaf Mix	-	-	1000	1000
Nutrients intake (g/day/ewe)				
Dry matter	2093	2956	2091	2953
Crude protein	318	480	318	481
Ether extract	46	69	44	66
NDF	835	991	835	992
ADF	562	626	561	624

Table 1. *Cont.*

Diet chemical composition (%)				
	Con. Mix	ActiSaf Mix	Alfalfa hay	Oat hay
Dry matter	86.28	86.02	88.21	87.63
Organic matter	81.70	81.16	80.87	80.91
Crude protein	16.23	16.29	14.72	7.38
Ether extract	2.22	2.01	1.38	2.06
NDF	15.62	15.68	43.25	53.68
ADF	6.32	6.19	32.56	38.74
Ash	4.58	4.86	73.4	6.72
Fatty acids composition (%)				
$C_{14:0}$	0.1	0.1	0.6	5.6
$C_{16:0}$	14.3	14.3	23.9	3.7
$C_{18:0}$	3.9	3.9	3.1	19.5
cis-9 $C_{18:1}$	36.1	36.1	3.1	16.2
$C_{18:2\ n\text{-}6}$	42.2	42.2	21.3	54.0
$C_{18:3\ n\text{-}3}$	1.7	1.7	41.9	1.2

NDF = Neutral detergent fiber; ADF = Acid detergent fiber.

2.3. Feed Samples Analyses

Samples of the alfalfa hay, oat hay, and concentrate were analyzed for organic matter (OM; Official Method 7.009), dry matter (DM; Official Method 7.007), and crude protein (CP; Official Method 7.016) according to the Association of Official Analytical Chemists (1984) using a Kjeldahl Distillation System (FOSS Kjeltec 8400, Demark). Neutral detergent fiber (NDF) and acid detergent fiber (ADF) expressed exclusive of residual ash according to the method of Van Soest using an ANKOM 2000 Fiber Analyzer (USA) as described by Tsiplakou et al. [18] (Table 1).

2.4. Milk Samples Collection

The sheep were milked three times per day (at 0700, 1300 and 2000 h) with a milking machine equipped with a digital milk meter and an electronic identification system (Sylco, Greece); thus, milk yield was recorded daily, and software (Sylco, Greece) was set to provide weekly averages. Milk samples were collected from each ewe weekly (at 7, 14, 21, 28, 35 and 42 days from parturition) for a 6-week period with sampling bottles (Sylco, Greece) of 200 mL appropriately for the milking parlor, to receive a representative sample of the milked quantity. Each of the milk samples from the mix of three subsamples was derived from each milking time (at 0700, 1300 and 2000 h) by taking 5% of the milked quantity.

2.5. Milk Chemical Composition

The milk samples were analyzed for fat, protein, lactose, total solids, and total solids no-fat by IR spectrometry (MilkoScan 120; FOSS, Hillerød, Demark) after proper calibration according to the methods of Gerber [19] and Kjeldahl [20].

2.6. Blood Metabolic Biomarker (B-HBA) Determination

Four weeks before the expected parturition, blood B-HBA was individually determined (before the morning feeding, 0700 h) once every three days until the lambing to ensure that 15 days before parturition a measurement would be recorded (Table S1). Two weeks postpartum, the sample collection for B-HBA was repeated. Blood ketone concentrations were measured using an electrochemical capillary blood monitoring device (FreeStyle Precision Neo, Abbott Laboratories Hellas S.A) with the corresponding individual foil-wrapped test strips for B-HBA. This method of B-HBA determination possesses 98.4% accuracy for the prediction of both toxemias' pregnancy and ketosis in Chios ewes [21]. After the insertion of a test strip into the device, a drop of blood was applied to the assigned spot, and the B-HBA concentration was recorded. Data were interpreted using 208 determinations in the two aforementioned sampling time points.

2.7. Antioxidant Status, Immune Response, and Milk Fatty Acid Profile

Eight (n = 8) ewes of each group with comparable weights (Control: 60.2 ± 5.11 kg; ActiSaf: 60.3 ± 4.88 kg), ages (Control: 1.84 ± 0.16 kg; ActiSaf: 1.85 ± 0.18 kg), milk performance (Fat corrected milk 6% ($FCM_{6\%}$) data used up 14th day in milk (DIM), Control: 2.3 ± 0.15 kg; ActiSaf: 2.3 ± 0.21 kg), prolificacy (Control: 2; ActiSaf: 2), and same lactation stage (up to 3 days deviation between animals) were selected for determining the antioxidant status of both milk and blood, for immune system gene response, and for milk fatty acid profile. Milk samples were collected (as mentioned above) in the 3rd and 6th week postpartum and stored at −80 °C. Blood samples were also collected, before the morning feeding (0700 h), at the same time points in heparin contained tubes for cell extraction and plasma isolation.

2.8. Enzyme Assays, Oxidative Stress Biomarkers, and Total Antioxidant Capacity

The enzyme activities, oxidative stress biomarkers, and the total antioxidant capacity were measured spectrophotometrically (Helios alpha, UNICAM, Cambridge, UK) as previously described by Tsiplakou et al. [22]. Briefly, Glutathione transferase (GST) activity in blood plasma was measured according to the method described by Labrou et al. [23] by measuring the conjunction of reduced glutathione to 1-chloro-2,4-dinitrobenzene at 340 nm. Catalase (CAT) activity in blood plasma and milk were assessed using a continuous spectrophotometric rate for the determination of H_2O_2 at 520 nm, according to the Sigma-Aldrich Catalase Assay Kit (CAT100). Glutathione peroxidase (GSH-Px) activity in blood plasma was measured according to the method of Paglia and Valentine [24] at 340 nm. Glutathione reductase (GR) activities in both blood plasma and milk were measured according to the method of Mavis and Stellwagen [25] by measuring the reduction in oxidized glutathione at 340 nm. Superoxide dismutase (SOD) activities in both blood plasma and milk were assayed using the method of McCord and Fridovich [26] by measuring the inhibition of cytochrome c oxidation at 550 nm. Lactoperoxidase (LPO) activity in milk was performed according to the methods of Keesey [27] by measuring the oxidation of ABTS present in hydrogen peroxide at 340 nm. Malondialdehyde (MDA) was determined according to the method of Nielsen et al. [28] with some modifications. More specifically, 100 µL blood plasma was added to 700 µL ortho-phosphoric acid (Panreac ITW Companies) and 200 µL aquarius thiobarbituric acid (TBA, Sigma-Aldrich CO USA) and then the samples were heated at 100 °C for 60 min. In milk samples, 1 mL of raw milk was added to 7 mL ortho-phosphoric acid (Panreac ITW Companies) and 2 mL of aquarius TBA (thiobarbituric acid, Sigma-Aldrich CO USA) and then incubated at 100 °C for 60 min. After that, absorbance was recorded at 532 nm. The protein carbonyl (PC) content was determined according to the method of Patsoukis et al. [29] by measuring the conjunction of 2,4-dinitrophenylhydrazine (DNPH) on protein carbonyls at 375 nm. The 2,2′-Azino-bis (3-ethylbenzthiazoline-6-sulfonic acid) (ABTS) radical scavenging assay was based on the published methods [30,31]. Ferric reducing ability of plasma (FRAP) assay was used to measure total antioxidant potential according to the method described by Benzie and Strain [32].

2.9. Milk Fatty Acid Profile

Milk fatty acid profile was determined using Gas Chromatography (Agilent 6890 N GC, Agilent 7683 B autosampler injector), equipped with an HP-88 capillary column (60 m × 0.25 mm i. d. with 0.20 µm film thickness, Agilent Technologies, USA) and a flame ionization detector (FID) as previously described by Mavrommatis and Tsiplakou [33].

2.10. Monocytes and Neutrophils Immune Genes Expression

Blood monocytes and neutrophils were isolated and then total RNA was extracted as previously described by Tsiplakou et al. [34]. Pure RNA (500 ng) from 64 individual (monocytes (32) and neutrophils (32)) samples was reverse transcribed with the PrimeScript First Strand cDNA Synthesis Kit (Takara, Japan) according to the manufacturer's instructions using a mix of random hexamers and oligo-dT primers. A pair of primers specific for each target gene was designed using Geneious software (Biomatters, New Zealand) according to the respective *Ovis aries* gene coding sequences (CDSs in GenBank) (Table S7). The specificity of each pair of primers was tested against genomic DNA (positive control) to confirm that a single amplicon would emerge after quantitative real-time PCR. In addition, dissociation curves were generated, and the amplification products were subjected to agarose gel electrophoresis to confirm the production of a single amplicon per reaction. The relative expression levels of the target genes were calculated as $(1 + E)^{-\Delta Ct}$, where ΔCt is the difference between the geometric mean of the two housekeeping genes' Cts and the Ct of the target gene, and the primer efficiency is the mean of each amplicon's efficiency per primer, which was calculated by employing the linear regression method on the log (fluorescence) per cycle number (ΔRn) using the LinRegPCR software [35]. Glyceraldehyde 3-Phosphate Dehydrogenase (*GAPDH*) and Tyrosine 3-monoxygenase/tryptophan 5-monooxygenase activation protein, zeta polypeptide (*YWHAZ*) were used as housekeeping genes to normalize the cDNA template concentrations; the RT-PCR protocols are described in Tsiplakou et al. [18].

2.11. Statistical Analysis

Experimental data were analyzed using the SPSS.IBM statistical package (version 20.0) and results are presented as mean ± mean standard error (SEM). Dietary treatment effects were determined using a general linear model (GLM) for a repeated measures analysis of variance (ANOVA). With the dietary treatments (D = Control and ActiSaf) used as the fixed factor and the sampling time (S) as the repeated measure, while including their interactions (D*S) to evaluate differences over time, according to the model:

$$Yijkl = \mu + Di + Sj + Ak + (D \times S)ij + eijkl$$

where Υijk is the dependent variable, μ the overall mean, Di the effect of dietary treatment (i = 2; Control and ActiSaf), Sj the effect of sampling time (j = 6 for milk performance, 2 for B-HBA concentration, fatty acids profile, antioxidant and immune system), Ak the animal's random effect, (DxS)ij the interaction between dietary treatments and sampling time, and eijk the residual error. Posthoc analysis was performed when appropriate using a Tukey's multiple range test [36]. For all tests, the significance level was set at $p = 0.05$. In order to simplify the visualization of the results, GraphPad Prism 6.0 (2012) was used for interleaved bars while error bars represent the mean standard error (SEM).

Moreover, discriminant analysis was also applied to pooled data to establish those variables capable of distinguishing and classifying samples among the two dietary treatments. Wilk's lambda (λ) criterion was used for selecting discriminant variables [37]. Forty variables were entered to develop a model to discriminate the thirty-two samples of each case. Specifically, five variables were used for grouped fatty acids in milk, ten and 10 for immune system genes' relative expression in monocytes and neutrophils, respectively, and seven and eight in antioxidant indices in milk and blood, respectively.

3. Results

3.1. Animal Performance

Dietary supplementation with LY ActiSaf significantly reduced the B-HBA concentrations in ewes' blood by 27% (0.86 ± 0.07 vs. 0.63 ± 0.06 mmol/L, $p = 0.018$) in the prepartum period, and by 17% (0.67 ± 0.04 vs. 0.56 ± 0.03 mmol/L, $p = 0.028$) in the postpartum period. Overall, the B-HBA was reduced by 24% (0.77 vs. 0.59 mmol/L, SEM = 0.04, $p = 0.003$) in the whole experimental period (Figure 1; Tables S1 and S2). Mean BW did not differ between the dietary treatments in the whole experimental period (Tables S3 and S4). However, ewes' BW recovered between lambing and 6th week of lactation tended to increase by 75% (2.00 ± 0.31 vs. 3.44 ± 0.28 kg, $p = 0.092$) in the ActiSaf compared with the Control group (Figure 2; Table S5).

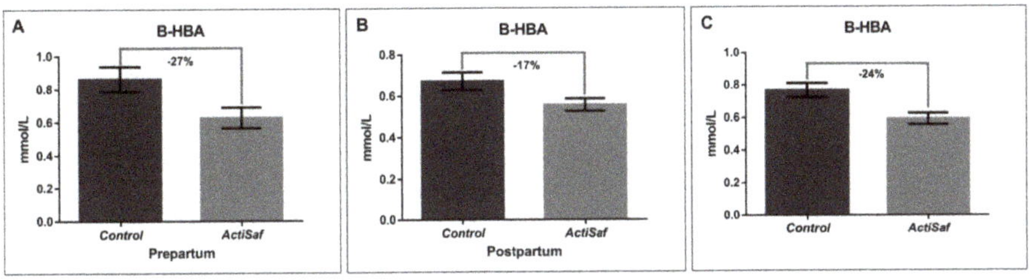

Figure 1. Graphical representation of (**A**) β-hydroxybutyric acid (B-HBA) in ewes' blood prepartum (mean ± SE), (**B**) β-hydroxybutyric acid in blood postpartum (mean ± SE), and (**C**) β-hydroxybutyric acid in blood of ewes in Control (black, n = 51 ewes) and ActiSaf (grey, n = 53 ewes) groups in experimental period of 12 weeks (mean ± mean standard error (SEM)).

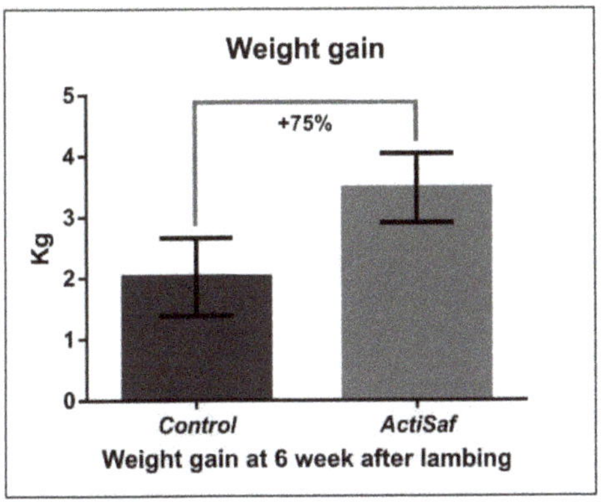

Figure 2. Graphical representation of body weight gain (recovery) between lambing and 6th week of lactation of ewes in Control (black, n = 51 ewes) and ActiSaf (grey, n = 53 ewes) groups (mean ± SE).

Milk, fat corrected milk 6% ($FCM_{6\%}$), and energy corrected milk yield (ECM) were numerically increased by 7.6 (2.50 vs. 2.69 kg/day, SEM = 0.159, $p = 0.395$), 12 (2.07 vs. 2.32 kg/day, SEM = 0.126, $p = 0.161$), and 10% (1.93 vs. 2.13 kg/day, SEM = 0.116, $p = 0.231$), respectively, in the ActiSaf compared to the Control group (Figure 3; Table S6; Figure S1). Concerning milk chemical composition, fat and protein contents were slightly decreased by 1.2 ($p = 0.740$) and 3% ($p = 0.381$), respectively, in the ActiSaf group, due to higher daily milk yield. However, both daily milk fat (114 vs. 131 g/day, SEM = 7.188, $p = 0.104$) and milk protein production (133 vs. 143 g/day, SEM = 8.288, $p = 0.434$) were increased by 15 and 7.5%, respectively, in the ActiSaf group (Figure 3; Supplementation Table S6). The milk yield in the treated ewes showed a moderate increase after the third week in lactation, and a peak in the fourth week, indicating a more intense milk persistence.

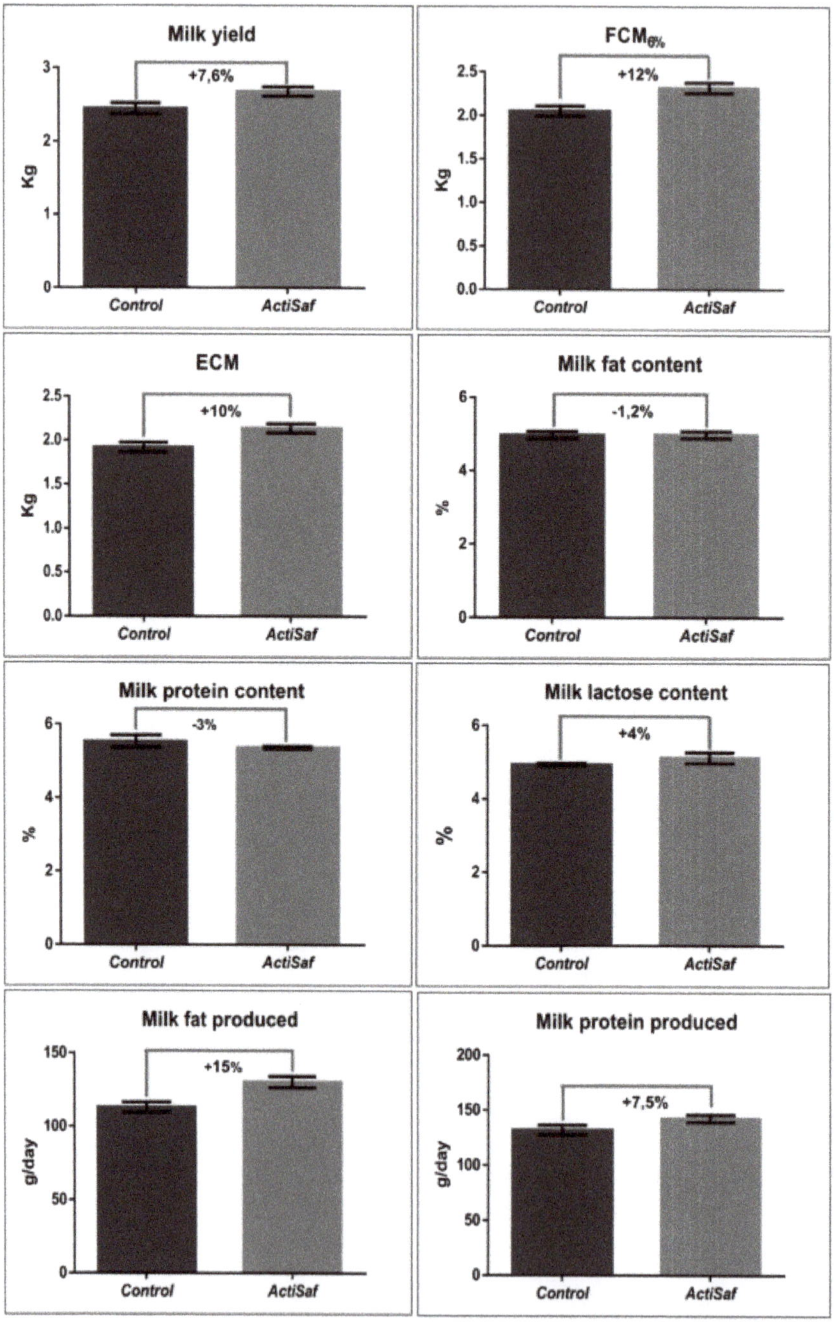

Figure 3. Graphical representation of milk yield and chemical composition of ewes in Control (black, $n = 51$ ewes) and ActiSaf (grey, $n = 53$ ewes) groups (mean ± SEM). FCM: Fat corrected milk in 6% according to the equation Y6% = (0.28 + 0.12F) M, where F = fat% and M = milk yield in kg. ECM: Energy corrected milk = milk yield × (0.071 × fat (%) + 0.043 × protein (%) + 0.2224) [38].

3.2. Milk Fatty Acid Profile

Milk fatty acid profile was not altered among dietary supplementation except for certain minor differences. Specifically, pentadecanoic acid ($C_{15:0}$), palmitoleic acid ($C_{16:1}$), trans linoleic acid ($C_{18:2n6t}$), and linolenic acid ($C_{18:3n3}$) were increased in ActiSaf milk by 15 (0.82 vs. 0.95%, SEM = 0.045, $p = 0.042$), 13 (0.29 vs. 0.33%, SEM = 0.014, $p = 0.033$), 9 (0.19 vs. 0.22%, SEM = 0.008, $p = 0.049$), and 20% (0.40 vs. 0.48%, SEM = 0.027, $p = 0.075$), respectively, while stearic acid ($C_{18:0}$) decreased by 5% (8.91 vs. 8.44%, SEM = 0.490, $p = 0.029$) (Table 2).

Table 2. The mean individual fatty acids (FAs) (% of total FA), FA groups and Saturated Fatty Acids (SFAs)/Unsaturated Fatty Acids (UFAs) of milk from ewes fed Control (n = 8 ewes) and ActiSaf (n = 8 ewes) diet throughout the experimental period (21 and 42 experimental days).

	Diets (D)			Sampling Time (T) in Weeks			Effect *		
	Control	ActiSaf	SEM †	3	6	SEM †	D	T	DxT
$C_{4:0}$	4.81	4.59	0.288	4.79	4.61	0.205	0.522	0.230	0.989
$C_{6:0}$	3.75	3.60	0.191	3.42	3.93	0.181	0.788	0.005	0.589
$C_{8:0}$	3.17	3.12	0.171	2.90	3.89	0.125	0.922	0.010	0.489
$C_{10:0}$	9.08	9.14	0.639	8.05	10.17	0.451	0.789	0.001	0.367
$C_{12:0}$	4.88	5.01	0.395	4.33	5.56	0.285	0.422	0.000	0.148
$C_{14:0}$	10.84	11.16	0.325	10.41	11.58	0.251	0.398	0.000	0.257
$C_{14:1}$	0.24	0.30	0.030	0.26	0.28	0.028	0.152	0.890	0.789
$C_{15:0}$	0.82	0.95	0.045	0.83	0.94	0.037	0.042	0.004	0.174
$C_{15:1}$	0.24	0.29	0.032	0.28	0.25	0.027	0.259	0.520	0.258
$C_{16:0}$	23.94	24.90	0.596	24.10	24.42	0.428	0.189	0.890	0.478
$C_{16:1}$	0.29	0.33	0.014	0.309	0.314	0.011	0.033	0.621	0.585
$C_{17:0}$	0.60	0.65	0.031	0.701	0.544	0.020	0.698	0.003	0.984
$C_{17:1}$	0.31	0.34	0.030	0.40	0.25	0.022	0.980	0.000	0.970
$C_{18:0}$	8.91	8.44	0.490	8.86	8.48	0.352	0.029	0.045	0.368
Σ_{trans} $C_{18:1}$	1.07	1.01	0.110	1.05	1.02	0.088	0.893	0.880	0.489
‡ $_{trans\text{-}11}$ $C_{18:1}$	1.26	1.16	0.158	1.22	1.20	0.125	0.358	0.499	0.984
cis_{-9} $C_{18:1}$	20.99	20.05	1.260	23.09	17.97	0.785	0.639	0.002	0.874
$C_{18:2n6t}$	0.19	0.22	0.008	0.19	0.22	0.008	0.049	0.019	0.321
$C_{18:2n6c}$	3.02	3.09	0.213	2.84	3.26	0.168	0.890	0.009	0.284
$C_{20:0}$	0.12	0.12	0.004	0.11	0.13	0.006	0.980	0.033	0.678
$C_{18:3n3}$	0.40	0.48	0.027	0.45	0.43	0.020	0.075	0.459	0.574
$C_{20:3n3}$ + $C_{22:1}$	0.31	0.28	0.015	0.34	0.25	0.012	0.784	0.000	0.348
trans-11, cis-9 $C_{18:2}$	0.76	0.77	0.110	0.74	0.80	0.081	0.899	0.269	0.635
§SCFA	20.81	20.45	0.896	19.16	22.10	0.678	0.678	0.002	0.354
¶MCFA	41.07	42.07	0.858	40.69	43.04	0.689	0.201	0.008	0.528
††LCFA	9.03	8.56	0.489	8.97	8.62	0.361	0.302	0.099	0.598
‡‡MUFA	24.40	23.48	1.140	26.61	21.28	0.651	0.522	0.008	0.789
§§ PUFA	4.68	4.84	0.308	4.57	4.95	0.212	0.622	0.028	0.654
¶¶ SFA	70.91	71.67	1.046	68.82	73.75	0.721	0.589	0.007	0.354
†††UFA	29.09	28.33	1.046	31.18	26.25	0.712	0.453	0.007	0.352
‡‡‡SFA/UFA	2.43	2.53	0.143	2.21	2.80	0.086	0.870	0.006	0.123
§§§AI	2.56	2.68	0.133	2.33	2.92	0.099	0.256	0.002	0.099

* Effect: The dietary treatment (D), time (T), and the interaction between dietary treatment × time (DxT) effects were analyzed by analysis of variance (ANOVA) using a general linear model (GLM) for repeated measures and posthoc analysis was performed when appropriate using Tukey's multiple range test. †SEM = Standard error of the mean. ‡ $_{trans\text{-}11}$ $C_{18:1}$ = these values are not included in the Σ $_{trans}$ $C_{18:1}$ content. §SCFAs: Short-Chain Saturated Fatty Acids = $C_{6:0}$ + $C_{8:0}$ + $C_{10:0}$ + $C_{11:0}$; ¶MCFAs: Medium-Chain Saturated Fatty Acids = $C_{12:0}$ + $C_{13:0}$ + $C_{14:0}$ + $C_{15:0}$ + $C_{16:0}$ + $C_{17:0}$. ††LCFAs: Long-Chain Saturated Fatty Acids = $C_{18:0}$ + $C_{20:0}$. ‡‡MUFAs: Mono-Unsaturated Fatty Acids = $C_{14:1}$ + $C_{15:1}$ + $C_{16:1}$ + $C_{17:1}$ + $C_{18:1\ cis\text{-}9}$ + $_{trans\text{-}11}$ $C_{18:1}$ + $_{trans}$ $C_{18:1}$; §§PUFAs: Poly-Unsaturated Fatty Acids = $_{cis\text{-}9,\ trans\text{-}11}$ $C_{18:2}$ (CLA) + $C_{18:2n\text{-}6c}$ + $C_{18:2n\text{-}6t}$ + $C_{18:3n\text{-}3}$ + $C_{18:3n\text{-}6}$ + $C_{20:3n\text{-}3}$; ¶¶SFAs: Saturated Fatty Acids = SCFA + MCFA + LCFA; †††UFAs: Unsaturated Fatty Acids = PUFA + MUFA; ‡‡‡S/U: Saturated/Unsaturated = (SCFA + MCFA + LCFA)/(PUFA + MUFA), and §§§AI: Atherogenicity index = ($C_{12:0}$ + 4 * $C_{14:0}$ + $C_{16:0}$)/(PUFA + MUFA).

3.3. Oxidative Status

Amongst the dietary treatments, we did not report any significant differences in both blood and milk antioxidant enzymes. However, Glutathione Reductase (GR) in blood plasma was significantly increased by 13% (0.067 vs. 0.076 units/mL, SEM = 0.002, $p = 0.004$) in ActiSaf fed ewes. A numerical increase in lactoperoxidase (LPO) and catalase (CAT) activities by 20 and 10%, respectively, in milk of ActiSaf fed ewes was observed. The total antioxidant capacity measured by ABTS assay was significantly higher by 16.7% (37.563 vs. 43.850% inhibition, SEM = 3.564, $p = 0.001$) in the milk of ActiSaf fed ewes (Table 3). A negative correlation between the total antioxidant capacity determined by FRAP assay and the proportions of MUFA and oleic acid ($C_{18:1\ cis\text{-}9}$) in milk was found. The same trend was reported between Glutathione Peroxidase (GPx) activity in blood plasma and the aforementioned fatty acids of milk. The correlation between blood malondialdehyde (MDA) content and the proportions of milk's PUFA was also negative. On the other hand, the correlations between blood MDA content and the proportions of MUFA and oleic acid, respectively, were positive (Figure 4).

Table 3. Enzymes activities (Units/mL), total antioxidant capacity, and oxidative status biomarkers in blood plasma and milk of ewes fed the two diets (Control, $n = 8$ and ActiSaf, $n = 8$) at two sampling times.

	Diets (D)			Sampling Time (T) in Weeks			Effect *		
	Control	ActiSaf	SEM †	3	6	SEM †	D	T	DxT
				Blood plasma					
GST	0.150	0.132	0.011	0.148	0.134	0.011	0.282	0.336	0.903
GR	0.067	0.076	0.002	0.067	0.076	0.002	0.004	0.003	0.458
GSH-Px	0.398	0.423	0.025	0.353	0.468	0.021	0.476	0.000	0.847
SOD	18.544	20.304	0.987	18.409	20.440	1.170	0.227	0.301	0.427
CAT	4.969	5.008	0.146	4.893	5.084	0.131	0.855	0.255	0.477
FRAP	0.641	0.764	0.045	0.675	0.730	0.036	0.099	0.230	0.409
ABTS	36.888	33.045	0.497	35.577	34.356	0.617	0.000	0.246	0.195
MDA	0.601	0.566	0.067	0.604	0.562	0.055	0.729	0.497	0.304
PC	4.065	4.015	0.085	3.886	4.193	0.110	0.685	0.114	0.981
				Milk					
LPO	0.504	0.602	0.059	0.628	0.478	0.047	0.265	0.006	0.467
SOD	3.677	3.569	0.173	3.577	3.669	0.157	0.664	0.660	0.859
CAT	29.284	32.359	3.564	33.558	28.085	2.836	0.552	0.054	0.750
FRAP	3.013	3.090	0.179	2.855	3.247	0.158	0.766	0.078	0.445
ABTS	37.563	43.85	1.014	40.254	41.158	1.057	0.001	0.572	0.726
MDA	0.363	0.369	0.018	0.363	0.369	0.015	0.843	0.693	0.687
PC	2.916	2.877	0.106	2.925	2.867	0.108	0.801	0.718	0.541

† SEM = Standard error of the mean. GST: Glutathione transferase. GR: Glutathione reductase. SOD: Superoxide dismutase. GSH-Px: Glutathione peroxidase. CAT: Catalase. ABTS: 2,2′-Azino-bis (3-ethylbenzthiazoline-6-sulfonic acid) (inhibition%). FRAP: Ferric Reducing Ability of Plasma (μM ascorbic acid). MDA: Malondialdehyde (μM MDA). PCs: Protein carbonyls (nmol/mL). LPO: Lactoperoxidase. Significance level below 0.05 indicates significant difference. * Effect: The dietary treatment (D), time (T), and the interaction between dietary treatment × time (DxT) effects were analyzed by ANOVA using a general linear model (GLM) for repeated measures and posthoc analysis was performed when appropriate using Tukey's multiple range test.

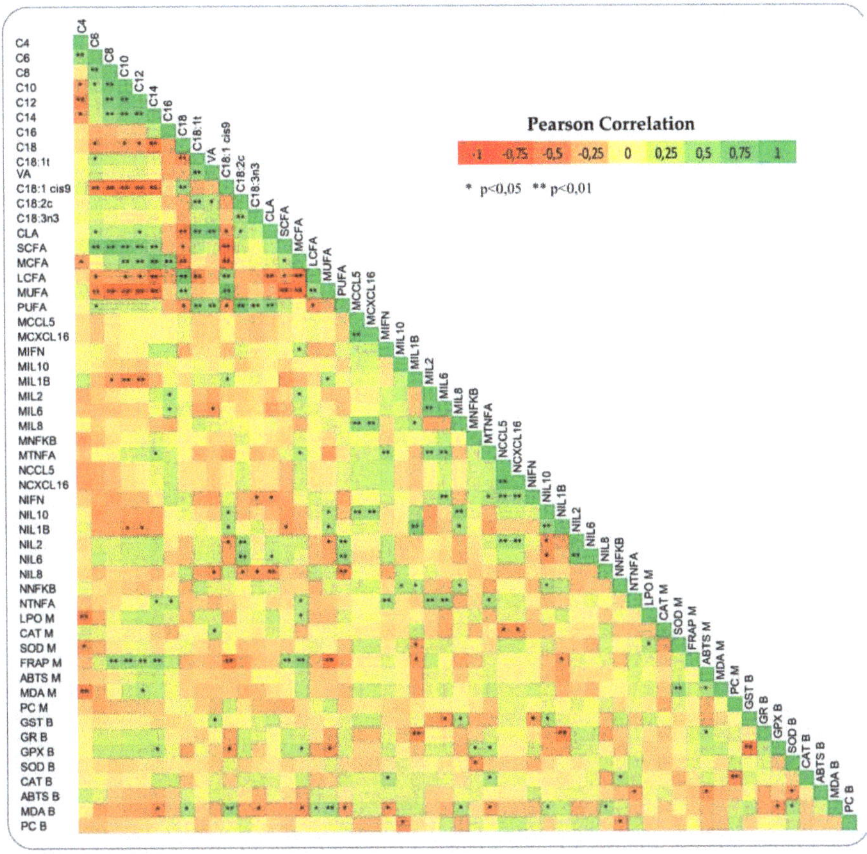

Figure 4. Heat-map represents a Pearson correlation of milk fatty acids, immune system gene expression in monocytes and neutrophils, antioxidant enzymes activities, total antioxidant capacity, and oxidative indices in both blood plasma and milk of ewes. In immune system genes, *M = monocytes and N = neutrophils*, while in antioxidants B = blood plasma and M = milk. CCL5: C-X-C motif chemokine 5, CXCL16: C-X-C motif chemokine ligand 16, INFG: Interferon γ, IL1B: Interleukin-1 beta, IL2: Interleukin-2, IL6: Interleukin-6, IL8: Interleukin-8, IL10: Interleukin-10, TNF: Tumor Necrosis Factor, NFKB: Nuclear Factor kappa B, GST: Glutathione transferase, GR: Glutathione reductase, SOD: Superoxide dismutase, GPx: Glutathione peroxidase, CAT: Catalase, ABTS: 2,2′-Azino-bis (3-ethylbenzthiazoline-6-sulfonic acid), FRAP: Ferric Reducing Ability of Plasma, MDA: Malondialdehyde, PCs: Protein carbonyls, LPO: Lactoperoxidase, VA: Vaccenic acid, CLA: Conjugated linoleic acid, SCFAs: Short-Chain saturated fatty acids, MCFAs: Medium-Chain saturated fatty acids, LCFAs: Long-Chain saturated fatty acids, MUFAs: Mono-unsaturated fatty acids, and PUFAs: Poly-unsaturated fatty acids.

3.4. Immune Status

The relative transcript levels of both *CCL5* and *CXCL16* in monocytes of ActiSaf fed ewes were significantly suppressed by 30% (0.053 vs. 0.037, SEM = 0.003, $p = 0.007$ and 0.042 vs. 0.029, SEM = 0.008, $p = 0.008$, respectively) (Table 4). Amongst cytokines, Interleukin 8 (*IL8*) relative transcript levels were significantly decreased by 80% (0.0020 vs. 0.0004, SEM = 0.0000, $p = 0.031$), while Interleukin 1β (*IL1B*) showed a tendency to decrease by 43% (0.007 vs. 0.004, SEM = 0.001, $p = 0.087$) in ActiSaf fed ewes (Table 4). A significant downregulation in the relative expression of Interleukin 10 (*IL10*) by 30% (0.014 vs. 0.010, SEM = 0.002, p = 0.047) in the neutrophils of the treated ewes was observed (Table 4). In addition, the relative transcript levels of *IL1B* in monocytes were positively correlated with the proportions of both MUFA and oleic acid in milk. The same trend was found between the relative expression of the Interleukin 2 gene and the proportion of palmitic acid in milk (Figure 4). Monocytes and neutrophil relative transcript levels of *IL1B* were negatively correlated with GR and Glutathione Transferase (GST) activities in blood plasma, respectively (Figure 4).

Table 4. Relative transcript levels of several genes in blood monocytes and neutrophils of ewes fed the two diets (Control, n = 8 and ActiSaf, n = 8) at two sampling times.

	Diets (D)			Sampling Time (T) in Weeks			Effect *		
	Control	ActiSaf	SEM †	3	6	SEM †	D	T	DxT
Monocytes									
CCL5	0.053	0.037	0.003	0.047	0.043	0.004	0.007	0.621	0.030
CXCL16	0.042	0.029	0.008	0.037	0.034	0.004	0.008	0.599	0.032
IFNG	0.009	0.008	0.001	0.006	0.010	0.001	0.301	0.034	0.856
IL1B	0.007	0.004	0.001	0.007	0.004	0.001	0.087	0.010	0.016
IL2	0.005	0.006	0.001	0.004	0.007	0.001	0.484	0.147	0.837
IL6	0.002	0.002	0.000	0.001	0.002	0.000	0.896	0.061	0.786
IL8	0.0020	0.0004	0.0000	0.0020	0.0004	0.0000	0.031	0.007	0.007
IL10	0.019	0.018	0.002	0.020	0.017	0.002	0.727	0.126	0.383
TNF	0.053	0.048	0.003	0.040	0.061	0.003	0.266	0.000	0.886
NFKB	0.251	0.245	0.022	0.254	0.242	0.024	0.842	0.727	0.708
Neutrophils									
CCL5	0.083	0.079	0.014	0.067	0.095	0.014	0.854	0.179	0.356
CXCL16	0.069	0.065	0.012	0.055	0.079	0.012	0.856	0.190	0.355
IFNG	0.026	0.015	0.007	0.018	0.023	0.005	0.289	0.424	0.719
IL1B	0.014	0.010	0.002	0.017	0.008	0.002	0.171	0.003	0.022
IL2	0.023	0.021	0.004	0.011	0.033	0.003	0.703	0.003	0.975
IL6	0.007	0.006	0.002	0.001	0.012	0.002	0.821	0.005	0.642
IL8	0.003	0.004	0.001	0.004	0.003	0.001	0.781	0.142	0.119
IL10	0.014	0.010	0.002	0.016	0.008	0.002	0.047	0.007	0.032
TNF	0.298	0.442	0.052	0.262	0.478	0.062	0.075	0.061	0.578
NFKB	0.182	0.185	0.028	0.239	0.0128	0.025	0.935	0.018	0.359

Significance level below 0.05 indicates significant difference. † SEM = Standard error of the mean. *CCL5*: C-X-C motif chemokine 5. *CXCL16*: C-X-C motif chemokine ligand 16. *INFG*: Interferon γ. *IL1B*: Interleukin-1 beta. *IL2*: Interleukin-2. *IL6*: Interleukin-6. *IL8*: Interleukin-8. *IL10*: Interleukin-10. *TNF*: Tumor Necrosis Factor. *NFKB*: Nuclear Factor kappa B. * Effect: The dietary treatment (D), time (T), and the interaction between dietary treatment × time (DxT) effects were analyzed by ANOVA using a general linear model (GLM) for repeated measures and posthoc analysis was performed when appropriate using Tukey's multiple range test.

3.5. Holistic Statistics

Discriminant analysis was applied to pooled data of two sampling times (3rd and 6th week postpartum) according to fatty acids in milk, immune gene expression in both monocytes and neutrophils, and antioxidant indices in blood plasma and milk (Figure 5) to investigate if the samples can be distinguished according to the type of the diet (Control and ActiSaf). The percentages of the samples that were classified into the correct group, according to the dietary treatment, were 100%. Wilks' lambda was observed at 0.001 for Function 1 (p = 0.159), while the relative transcript levels of *CCL5, CXCL16, and IL6* in monocytes', *IL10, IL6,* and *IL8* in neutrophils, and the GR activity in blood plasma were the variables that contributed the most.

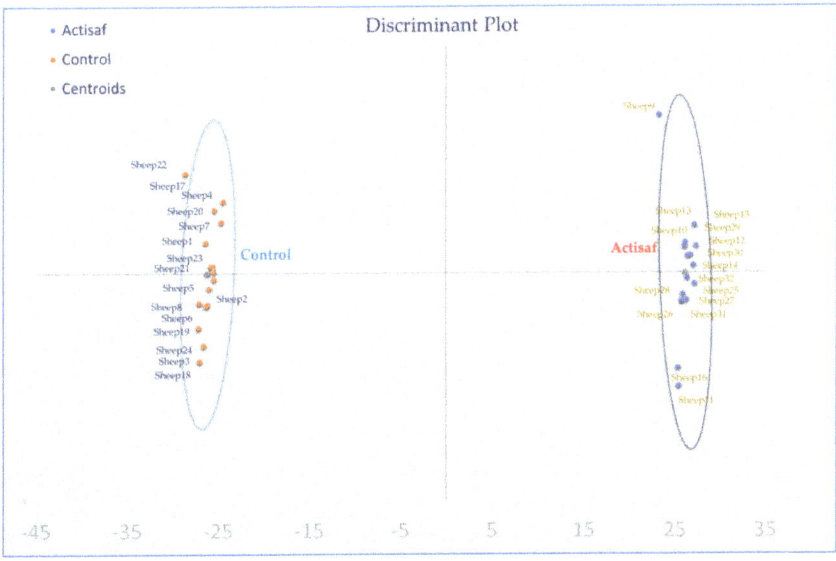

Figure 5. Discriminant plots separating the Control and ActiSaf fed ewes according to their fatty acid grouped values, immune gene expression in both monocytes and neutrophils, and antioxidant indices in blood and milk.

4. Discussion

Blood B-HBA concentrations reflect the magnitude of negative energy balance (NEB) and lipid mobilization and are a diagnostic marker for subclinical (SCK) and clinical ketosis (CK) in ruminants. The B-HBA content in the blood of sheep with SCK ranges from 0.5 to 1.6 mmol/L [39–42], while in those with CK from 1.6 to 7 mmol/L [39,40,43]. However, in the case of healthy pregnant sheep these values could be around 0.8–0.9 mmol/L [44,45]. Nonetheless, the values of the Control group, first and foremost during the prepartum period, may indicate a moderate NEB (0.86 mmol/L). On the other hand, results regarding the B-HBA concentration in the blood of ActiSaf fed ewes indicate an improvement in the energetic status of the animals. It should be underlined here that the prolificacy between the two groups was the same (around 2.35), which means that although the number of fetuses affects B-HBA content [46], it had minimum impact in our trial. The same levels of B-HBA content in blood of healthy ewes of the same breed in early lactation have been previously reported [18,38]. An increase in the host energy availability might be due to a better rumen function and microbiome homeostasis. During the peripartum period, the energy and nutrient demands increase exponentially while the dry matter intake decreases. Thus, the optimal rumen function and the balance between VFA for a maximum feed efficiency are momentous in the transition period. The mechanism underling LY contribution in rumen may be down to yeast's oxygen scavenging properties (anaerobiosis). Specifically, the improvement of anaerobiosis in rumen increases the bacteria viability and thus, microbial protein synthesis and fiber digestibility [6]. Energy balance might be improved as a result of the dry matter (DM) and NDF digestion enhancement as have been reported by Plata et al. [47]. Furthermore, Panda et al. [48] also found that crude protein (CP) digestibility was also increased by 4.8% with dietary supplementation of yeast culture on male calves. In agreement with our findings, ActiSaf dietary inclusion (5g/day) in early lactating dairy cows, significantly decreased serum B-HBA and non-esterified fatty acids (NEFA) concentrations [9], while in mid-lactating cows, 4 g/day of LY supplementation did not affect B-HBA concentration since animals were not prone to NEB [49]. In addition, ewes in the ActiSaf group recovered their body weight from lambing until the sixth week postpartum in a more efficient manner, considering the increased available energy, as concluded by B-HBA concentration.

In compliance with our findings (12% $FCM_{6\%}$, $p = 0.161$), Stella et al. [12], reported a significant increase in goats' milk yield by 14% when their diet was supplemented with *S. cerevisiae*. It is worth mentioning that, in the same study, treated goats showed an upward trend in milk yield after the fourth week postpartum which decreased slower compared to the control group, showing a persistence in milk similar to our study. The dietary inclusion of *S. cerevisiae* enhanced cows' milk yield in early [13] mid- [14] and late [7] lactation. However, Dehghan-Banadaky et al. [49], showed that the milk yield was not affected in *S. cerevisiae* supplemented cows after the 145 DIM, possibly due to the absence of NEB. The results from 22 studies with more than 9039 lactating dairy animals showed an increase in their milk production by 7.3% (ranging from 2 to 30%) when their diets were supplemented with Yea-Sacc®1026 yeast [50].

Interestingly, in a meta-analysis study, Dehghan-Banadaky [49] concluded that an enhancement in milk yield was accompanied by an increase in feed intake in supplemented animals with yeast products. Moreover, yeast administration in prepartum cows' diets improved DMI [51]. Additionally, Habeeb et al. [52] reported that an enhancement of animal performance by the inclusion of yeast in their diet was mainly attributed to an increase in feed intake rather than feed digestibility improvement.

The milk fatty acid's profile was not holistically modified; however, certain interesting results, related to the biohydrogenation process (BH), were unveiled. Julien et al. [53] first observed the impact of LY administration on ruminal biohydrogenation processes. Specifically, LY promotes growth and activity of rumen lactate-utilizing bacteria, such as *Megasphera elsdenii* or *Selenomonas ruminantium*, *Actinobacteria*, including *Propionibacterium acnes* as well as fibrolytic bacteria. Consequently, LY could be involved at different stages of BH; firstly, by altering biohydrogenating microorganisms, i.e., improving growth of either t11 or t10 isomer producing bacteria, and secondly by modulating the ruminal biotope, i.e., by stabilizing ruminal pH or favoring stronger reducing conditions. In addition, Julien et al. [53] reported that LY supplementation increased the accumulation of trans $C_{18:1}$ in vitro and decreased the proportion of $C_{18:0}$, suggesting an inhibition of the last step of BH of $_{c9c12}$-$C_{18:2}$ fatty acids. Thus, in our study it could be hypothesized that the improved rumen conditions by LY administration may favor the isomerisation of $_{c9c12}$-$C_{18:2}$ and consequently increased the production of intermediate fatty acids in the rumen, which induced an inhibition or a saturation of the enzyme activity of bacteria involved in the second reduction step [54].

During the peripartum period, animals' augmented requirement for energy and nutrient results in lipid mobilization and blood hyperketonemia which induce oxidative stress [55]. Optimizing nutrition requirements by improving rumen efficiency may suppress the concentration of such trigger metabolites and improve the oxidative status. Glutathione reductase (GR) has a central role in the antioxidant defense system since it catalyzes the conversion of oxidized glutathione disulfide to the reduced form of glutathione, which is a critical molecule in resting oxidative stress [56]. It is known that glutathione is extremely important since it acts as substrate or co-substrate in enzymatic reactions (e.g., the glutathione-S-transferase or glutathione-shuttle enzymes), reacts directly with free radicals and lipid peroxides, and protects cells [57]. The mechanism under which GR increased its activity remains unclear, thus our assumptions are oriented toward a prudent liver function where glutathione is de novo synthesized due to the lower B-HBA concentration. Another possible mechanism that can increase GR activity might be the Flavin Adenine Dinucleotides' (FADs) co-substrate. Specifically, yeasts are sources of B-complex vitamins that act as precursors of the essential co-enzymes NAD and FAD that are responsible for biological oxidation [58]. In addition, high genetic merit dairy animals often burden their metabolism since they require increased levels of energy in order to meet their demands, leading to ROS production and later to the annihilation of the milk oxidative stability [59]. Hence, total antioxidant capacity enhancement in the milk of ActiSaf fed ewes may be important for the dairy industry.

Concerning chemokines, CCL-5 is involved in the activation of T cells, macrophages, eosinophils and basophils, and its enhancement is related to an inflammation response [60]. On the other hand, CXCL-16, a transmembrane protein, is detached from the membrane by metalloproteinase

ADAM10 induced chemotaxis [61]. In this study, the downregulation in the relative transcript levels of both *CCL5* and *CXCL16* in monocytes of ActiSaf treated ewes, indicates a lower inflammatory response during the first 6 weeks of lactation. The *IL1B* which was suppressed in our study, regulates B-cell maturation and proliferation, activates the Natural Killer (NK) cells and is generally related to the acute manifestation of inflammation in immune cells [62]. Interleukin-8, on the other hand, has a chemotaxis-inducer effect mainly in neutrophils. Pro-inflammatory chemokines and cytokine downregulations is directly attributed to B-HBA mitigation. Specifically, blood ketones derived from ketogenesis through acetyl-CoA metabolization have been shown to act as stimulants in chemokines and cytokines in cows' mammary epithelia cells [63].

5. Conclusions

In conclusion, supplementing dairy sheep diets with 2 g of the ActiSaf live yeast/day/ewe during the transition and early lactation periods have a beneficial impact on animals' performance whilst simultaneously portraying an improvement on pro-inflammatory responses attributed to a lower lipomobilization. This overall stress suppression during this turning point for the ruminants' period may unveil the potential of live yeasts as health modulators towards the collective effort of reducing antibiotic dependance at the farm scale. However, further research is needed to deeply understand the mechanism under the enhancement of energy supply in small ruminants.

Supplementary Materials: The following are available online at http://www.mdpi.com/2309-608X/6/4/334/s1, Table S1. Analysis of variance in blood β-hydroxybutyric acid in prepartum and postpartum period; Table S2. Repeated measure analysis of variance in blood β-hydroxybutyric acid in overall experimental period using prepartum and postpartum sampling time as repeated factor; Table S3. Analysis of variance in body weight at the start of the experiment, at lambing, and at the end; Table S4. Repeated measure analysis of variance in body weight in overall experimental period using the weighing at the start of the experiment, at lambing and at the end, as repeated factor.; Table S5. Analysis of variance in body weight recovery from lambing to 6 weeks postpartum. Table S6. Milk yield and milk chemical composition of ewes fed the Control and ActiSaf diets in the six sampling times; Table S7. Sequences of primers for target genes used in real-time qPCR; Figure S1. Graphical representation of milk chemical composition of the Control and ActiSaf groups.

Author Contributions: Conceptualization, G.Z. and V.N.; methodology, A.M., C.M. and C.C.; software, A.M.; validation, A.M., C.M., C.C., V.N. and E.T.; formal analysis, A.M.; investigation, G.Z. and E.T.; resources, V.N.; data curation, A.M. and D.K.; writing—original draft preparation, A.M.; writing—review and editing, E.T., C.C. and G.Z.; visualization, A.M.; supervision, E.T.; project administration, G.Z.; funding acquisition, G.Z. All authors have read and agreed to the published version of the manuscript.

Funding: The study was partially funded by Phileo Lesaffre Animal Health, Lille, France.

Acknowledgments: Authors would like to thank Flessas Dairy Farm for their collaboration.

Conflicts of Interest: The authors declare no conflict of interest.

References

1. Alexandratos, N.; Bruinsma, J. *World Agriculture towards 2030/2050: The 2012 Revision*; FAO: Rome, Italy, 2012.
2. Gerber, P.J.; Steinfeld, H.; Henderson, B.; Mottet, A.; Opio, C.; Dijkman, J.; Falcucci, A.; Tempio, G. *Tackling Climate Change Through Livestock—A Global Assessment of Emissions and Mitigation Opportunities*; FAO: Rome, Italy, 2013; ISBN 978-92-5-107920-1.
3. McLoughlin, S.; Spillane, C.; Claffey, N.; Smith, P.E.; O'Rourke, T.; Diskin, M.G.; Waters, S.M. Rumen Microbiome Composition Is Altered in Sheep Divergent in Feed Efficiency. *Front. Microbiol.* **2020**, *11*, 1981. [CrossRef] [PubMed]
4. Council Directive 96/51/EC of 23 July 1996 Amending Directive 70/524/EEC Concerning Additives in Feeding Stuffs. Available online: https://core.ac.uk/download/pdf/76783458.pdf (accessed on 15 September 2020).
5. Fonty, G.; Chaucheyras-Durand, F. Effects and modes of action of live yeasts in the rumen. *Biologia* **2006**, *61*, 741–750. [CrossRef]
6. Newbold, C.J. *I.D.* Probiotics: Principles for use in ruminant nutrition. In *Role of Probiotics in Animal Nutrition and their link to the demands of European Consumers*; Van Vuuren, A.M., Rochet, B., Eds.; Lelystad Report on European Probiotic Association Seminar: Lelystad, The Netherlands, 2003; Volume 03/0002713, pp. 29–39.

7. Moallem, U.; Lehrer, H.; Livshitz, L.; Zachut, M.; Yakoby, S. The effects of live yeast supplementation to dairy cows during the hot season on production, feed efficiency, and digestibility. *J. Dairy Sci.* **2009**, *92*, 343–351. [CrossRef] [PubMed]
8. Lapointe, J.; Roy, C.; Lavoie, M.; Bergeron, N.; Beaudry, D.; Blanchet, I.; Petit, H.V.; Palin, M.F. Negative Energy Balance Is Associated with Inflammatory and Oxidative Stress Conditions in Early Lactating Dairy Cows. *Free Radic. Biol. Med.* **2015**, *87*, S42. [CrossRef]
9. Wathes, D.C.; Cheng, Z.; Chowdhury, W.; Fenwick, M.A.; Fitzpatrick, R.; Morris, D.G.; Patton, J.; Murphy, J.J. Negative energy balance alters global gene expression and immune responses in the uterus of postpartum dairy cows. *Physiol. Genom.* **2009**, *39*, 1–13. [CrossRef]
10. Kumprechtová, D.; Illek, J.; Julien, C.; Homolka, P.; Jančík, F.; Auclair, E. Effect of live yeast (Saccharomyces cerevisiae) supplementation on rumen fermentation and metabolic profile of dairy cows in early lactation. *J. Anim. Physiol. Anim. Nutr.* **2018**, *103*, 447–455. [CrossRef]
11. Wankhade, P.R.; Manimaran, A.; Kumaresan, A.; Jeyakumar, S.; Ramesha, K.P.; Sejian, V.; Rajendran, D.; Varghese, M.R. Metabolic and immunological changes in transition dairy cows: A review. *Vet. World* **2017**, *10*, 1367–1377. [CrossRef]
12. Stella, A.V.; Paratte, R.; Valnegri, L.; Cigalino, G.; Soncini, G.; Chevaux, E.; Dell'Orto, V.; Savoini, G. Effect of administration of live Saccharomyces cerevisiae on milk production, milk composition, blood metabolites, and faecal flora in early lactating dairy goats. *Small Rumin. Res.* **2007**, *67*, 7–13. [CrossRef]
13. Rihma, E.; Kärt, O.; Mihhejev, K.; Henno, M.; Jõudu, I.; Kaart, T. Effect of dietary live yeast on milk yield, composition and coagulation properties in early lactation of Estonian holstein cows. *Agraarteadus* **2007**, *XVIII*, 37–41.
14. Rossow, H.A.; Riordan, T.; Riordan, A. Effects of addition of a live yeast product on dairy cattle performance. *J. Appl. Anim. Res.* **2017**, *46*, 159–163. [CrossRef]
15. Directive 2010/63/EU of the European Parliament and of the Council of 22 September 2010 on the Protection of Animals Used for Scientific Purposes Text with EEA Relevance. Available online: http://data.europa.eu/eli/dir/2010/63/oj (accessed on 15 September 2020).
16. National Research Council. *Nutrient Requirements of Dairy Cattle*, 7th Revised ed.; The National Academies Press: Washington, DC, USA, 2001. [CrossRef]
17. Zervas, G. *Ration Formulation*; Stamoulis: Athens, Greece, 2007; ISBN 978-960-351-676-7.
18. Tsiplakou, E.; Mavrommatis, A.; Skliros, D.; Sotirakoglou, K.; Flemetakis, E.; Zervas, G. The effects of dietary supplementation with rumen-protected amino acids on the expression of several genes involved in the immune system of dairy sheep. *J. Anim. Physiol. Anim. Nutr.* **2018**, *102*, 1437–1449. [CrossRef] [PubMed]
19. The Royal Society of Chemistry; British Standards Institution. Available online: https://pubs.rsc.org/en/Content/ArticleLanding/1952/AN/an952770546a#!divAbstract (accessed on 15 September 2020).
20. IDF: Bulletin of the IDF No. 285/1993—Reference materials and Interlaboratory collaborative studies (third series), by various Groups of Experts (See also Bulletins Nos 207/1986, 235/1988). ISSN 0250-5118.
21. Panousis, N.; Brozos, C.; Karagiannis, I.; Giadinis, N.D.; Lafi, S.; Kritsepi-Konstantinou, M. Evaluation of Precision Xceed® meter for on-site monitoring of blood β-hydroxybutyric acid and glucose concentrations in dairy sheep. *Res. Vet. Sci.* **2012**, *93*, 435–439. [CrossRef] [PubMed]
22. Tsiplakou, E.; Mitsiopoulou, C.; Mavrommatis, A.; Karaiskou, C.; Chronopoulou, E.G.; Mavridis, G.; Sotirakoglou, K.; Labrou, N.E.; Zervas, G. Effect of under- and overfeeding on sheep and goat milk and plasma enzymes activities related to oxidation. *J. Anim. Physiol. Anim. Nutr.* **2017**, *102*, 288–298. [CrossRef] [PubMed]
23. Labrou, N.; Mello, L.; Clonis, Y. Functional and structural roles of the glutathione-binding residues in maize (Zea mays) glutathione S-transferase I. *Biochem. J.* **2001**, *358*, 101–110. [CrossRef]
24. Paglia, D.E.; Valentine, W.N. Studies on the quantitative and qualitative characterization of erythrocyte glutathione peroxidase. *J. Lab. Clin. Med.* **1967**, *70*, 158–169. [CrossRef]
25. Mavis, R.D.; Stellwagen, E. Purification and subunit structure of glutathione reductase from bakers' yeast. *J. Biol. Chem.* **1968**, *243*, 809–814.
26. McCord, J.M.; Fridovich, I. The utility of superoxide dismutase in studying free radical reactions I. radicals generated by the interaction of sulfite, dimethyl sulfoxide, and oxygen. *J. Biol. Chem.* **1969**, *244*, 6056–6063.
27. Keesey, J. *Biochemica Information*, 1st ed.; Boehringer Mannheim Biochemicals: Indianapolis, IN, USA, 1987; p. 49.

28. Nielsen, F.; Mikkelsen, B.B.; Nielsen, J.B.; Andersen, H.R.; Grand-Jean, P. Plasma malondialdehyde as bio-marker for oxidative stress: Reference interval and effects of life-style factors. *Clin. Chem.* **1997**, *43*, 1209–1214. [CrossRef]
29. Patsoukis, N.; Zervoudakis, G.; Panagopoulos, N.T.; Georgiou, C.D.; Angelatou, F.; Matsokis, N.A. Thiol redox state (TRS) and oxidative stress in the mouse hippocampus after pentylenetetrazolinduced epileptic seizure. *Neurosci. Lett.* **2004**, *357*, 83–86. [CrossRef]
30. Pellegrini, N.; Serafini, M.; Colombi, B.; Del Rio, D.; Salvatore, S.; Bianchi, M.; Brighenti, F. Total antioxidant capacity of plant foods, beverages and oils consumed in Italy assessed by three different in vitro assays. *J. Nutr.* **2003**, *133*, 2812–2819. [CrossRef]
31. Li, P.; Huo, L.; Su, W.; Lu, R.; Deng, C.; Li, L.; He, C. Free radical scavenging capacity, antioxidant activity and phenolic content of Pouzolzia zeylanica. *J. Serbian Chem. Soc.* **2011**, *76*, 709–717. [CrossRef]
32. Benzie, I.F.; Strain, J.J. The ferric reducing ability of plasma (FRAP) as a measure of "antioxidant power": The FRAP assay. *Anal. Biochem.* **1996**, *239*, 70–76. [CrossRef]
33. Mavrommatis, A.; Tsiplakou, E. The impact of the dietary supplementation level with *Schizochytrium* sp. on milk chemical composition and fatty acid profile, of both blood plasma and milk of goats. *Small Rumin. Res.* **2020**, *193*, 106252. [CrossRef]
34. Tsiplakou, E.; Mavrommatis, A.; Skliros, D.; Righi, F.; Flemetakis, E. The impact of rumen-protected amino acids on the expression of key- genes involved in the innate immunity of dairy sheep. *PLoS ONE* **2020**, *15*, e0233192. [CrossRef]
35. Ramakers, C.; Ruijter, J.M.; Deprez, R.H.; Moorman, A.F. Assumption-free analysis of quantitative real-time polymerase chain reaction (PCR) data. *Neurosci. Lett.* **2003**, *339*, 62–66. [CrossRef]
36. McHugh, M.L. Multiple comparison analysis testing in ANOVA. *Biochem. Med.* **2011**, *21*, 203–209. [CrossRef]
37. Alrawashdeh, M.; Radwan, T. "Wilk's lambda based on robust method". *AIP Conf. Proc.* **2017**, *1842*, 030032. [CrossRef]
38. Tsiplakou, E.; Mavrommatis, A.; Kalogeropoulos, T.; Chatzikonstantinou, M.; Koutsouli, P.; Sotirakoglou, K.; Zervas, G. The effect of dietary supplementation with rumen-protected methionine alone or in combination with rumen-protected choline and betaine on sheep milk and antioxidant capacity. *J. Anim. Physiol. Anim. Nutr.* **2016**, *101*, 1004–1013. [CrossRef]
39. Andrews, A.H.; Holland-Howes, V.E.; Wilkinson, J.I.D. Naturally occurring pregnancy toxaemia in the ewe and treatment with recombinant bovine somatotropin. *Small Rumin. Res.* **1997**, *23*, 191–197. [CrossRef]
40. Balikci, E.; Yildiz, A.; Gurdogan, F. Investigation on some biochemical and clinical parameters for pregnancy toxemia in Akkaraman ewes. *J. Anim. Vet. Adv.* **2009**, *8*, 1268–1273.
41. Anoushepour, A.; Mottaghian, P.; Sakha, M. The comparison of some biochemical parameters in hyperketonemic and normal ewes. *Eur. J. Exp. Biol.* **2014**, *4*, 83–87.
42. Feijó, J.O.; Schneider, A.; Schmitt, E.; Brauner, C.C.; Martins, C.F.; Barbosa-Ferreira, M.; Del Pino, F.A.B.; Faria Junior, S.P.; Rabassa, V.R.; Corrêa, M.N. Prepartum administration of recombinant bovine somatotropin (rBST) on adaptation to subclinical ketosis of the ewes and performance of the lambs. *Arq. Bras. Med. Vet. Zootec.* **2015**, *67*, 103–108. [CrossRef]
43. Lacetera, N.; Franci, O.; Scalia, D.; Bernabucci, U.; Ronchi, B.; Nardone, A. Effects of nonesterified fatty acids and BHB on functions of mononuclear cells obtained from ewes. *Am. J. Vet. Res.* **2002**, *63*, 414–418. [CrossRef]
44. Durak, M.H.; Altiner, A. Effect of energy deficiency during late pregnancy in Chios Ewes on free fatty acids, β-hydroxybutyrate and urea metabolites. *Turk. J. Vet. Anim. Sci.* **2006**, *30*, 497–502.
45. Al-Qudah, K.M. Oxidant and antioxidant profile of hyperketonemic ewes affected by pregnancy toxemia. *Vet. Clin. Pathol.* **2011**, *40*, 60–65. [CrossRef]
46. Moallem, U.; Rozov, A.; Gootwine, E.; Honig, H. Plasma concentrations of key metabolites and insulin in late-pregnant ewes carrying 1 to 5 fetuses. *J. Anim. Sci.* **2012**, *90*, 318–324. [CrossRef]
47. Plata, P.F.; Mendoza, M.G.D.; Gama, J.R.B.; Gonzalez, M.S. Effect of a yeast culture (Saccharomyces cerevisiae) on neutral detergent fiber digestion in steers fed oat straw based diets. *Anim. Feed Sci. Technol.* **1994**, *49*, 203–210. [CrossRef]
48. Panda, A.K.; Singh, R.; Pathak, N.N. Effect of dietary inclusion of Saccharomyces cerevisiae on growth performance of crossbred calves. *J. Appl. Anim. Res.* **1995**, *7*, 195–200. [CrossRef]

49. Dehghan-Banadaky, M.; Ebrahimi, M.; Motameny, R.; Heidari, S.R. Effects of live yeast supplementation on mid-lactation dairy cows performances, milk composition, rumen digestion and plasma metabolites during hot season. *J. Appl. Anim. Res.* **2013**, *41*, 137–142. [CrossRef]
50. Dawson, K.A.; Tricarico, J. The evolution of yeast cultures- 20 years of research. In: Navigating from Niche Markets to Mainstream. In Proceedings of the Alltech's European, Middle Eastern and African Lecture Tour, Stamford, UK, 20 November 2011; pp. 26–43.
51. Dann, H.M.; Drackley, J.K.; McCoy, G.C.; Hutjens, M.F.; Garrett, J.E. Effects of yeast culture (Saccharomyces cerevisiae) on prepartum intake and postpartum intake and milk production of Jersey cows. *J. Dairy Sci.* **2000**, *83*, 123–127. [CrossRef]
52. Habeeb, A.A.M. Importance of Yeast in Ruminants Feeding on Production and Reproduction. *Ecol. Evol. Biol.* **2017**, *2*, 49–58. [CrossRef]
53. Julien, C.; Marden, J.P.; Enjalbert, F.; Bayourthe, C.; Troegeler, A. Live yeast as a possible modulator of polyunsaturated fatty acid biohydrogenation in the rumen. *Rev. Med. Vet.* **2010**, *8–9*, 391–400.
54. Troegeler-Meynadier, A.; Bret-Bennis, L.; Enjalbert, F. Rates and efficiencies of reactions of ruminal biohydrogenation of linoleic acid according to pH and polyunsaturated fatty acids concentrations. *Rep. Nutr. Dev.* **2006**, *46*, 713–724. [CrossRef]
55. Li, Y.; Ding, H.Y.; Wang, X.C.; Feng, S.B.; Li, X.B.; Wang, Z.; Liu, G.W.; Li, X.W. An association between the level of oxidative stress and the concentrations of NEFA and BHBA in the plasma of ketotic dairy cows. *J. Anim. Physiol. Anim. Nutr.* **2016**, *100*, 844–851. [CrossRef]
56. Tsiplakou, E.; Chatzikonstantinou, M.; Mitsiopoulou, C.; Karaiskou, C.; Mavrommatis, A.; Sotirakoglou, K.; Zervas, G. Effect of soya bean and fish oil inclusion in diets on milk and plasma enzymes from sheep and goat related to oxidation. *J. Anim. Physiol. Anim. Nutr.* **2016**, *101*, 733–742. [CrossRef]
57. Bernabucci, U.; Ronchi, B.; Lacetera, N.; Nardone, A. Influence of Body Condition Score on Relationships Between Metabolic Status and Oxidative Stress in Periparturient Dairy Cows. *J. Dairy Sci.* **2005**, *88*, 2017–2026. [CrossRef]
58. Habeeb, A.A.M. Current View of the Significance of Yeast for Ruminants a Review 1- Role of Yeast and Modes of Action. *Am. J. Libr. Inf. Sci.* **2017**, *1*, 53–59. [CrossRef]
59. Wullepit, N.; Raes, K.; Beerda, B.; Veerkamp, R.F.; Fremaut, D.; De Smet, S. Influence of management and genetic merit for milk yield on the oxidative status of plasma in heifers. *Livest. Sci.* **2009**, *123*, 276–282. [CrossRef]
60. Aldinucci, D.; Colombatti, A. The Inflammatory Chemokine CCL5 and Cancer Progression. *Mediat. Inflamm.* **2014**, 1–12. [CrossRef]
61. Matsumura, S.; Demaria, S. Up-regulation of the Pro-inflammatoryChemokine CXCL16 is a Common Response of Tumor Cells to Ionizing Radiation. *Radiat. Res.* **2010**, *173*, 418–425. [CrossRef]
62. Galli, C.; Calder, P.C. Effects of fat and fatty acid intake on inflammatory and immune responses: A critical review. *Ann. Nutr. Metab.* **2009**, *55*, 123–139. [CrossRef]
63. Hillreiner, M.; Flinspach, C.; Pfaffl, M.W.; Kliem, H. Effect of the Ketone Body Beta-Hydroxybutyrate on the Innate Defense Capability of Primary Bovine Mammary Epithelial Cells. *PLoS ONE* **2016**, *11*, e0157774. [CrossRef]

Publisher's Note: MDPI stays neutral with regard to jurisdictional claims in published maps and institutional affiliations.

Journal of *Fungi*

Article

Assimilation of Cholesterol by *Monascus purpureus*

Theresa P. T. Nguyen *, Margaret A. Garrahan †, Sabrina A. Nance †, Catherine E. Seeger † and Christian Wong †

Department of Chemistry & Biochemistry, Loyola University Maryland, Baltimore, MD 21210, USA; magarrahan@loyola.edu (M.A.G.); sanance@loyola.edu (S.A.N.); ceseeger@loyola.edu (C.E.S.); ckwong@loyola.edu (C.W.)

* Correspondence: tptnguyen@loyola.edu; Tel.: +1-410-617-2862

† These authors contributed equally to this work.

Received: 31 October 2020; Accepted: 8 December 2020; Published: 9 December 2020

Abstract: *Monascus purpureus*, a filamentous fungus known for its fermentation of red yeast rice, produces the metabolite monacolin K used in statin drugs to inhibit cholesterol biosynthesis. In this study, we show that active cultures of *M. purpureus* CBS 109.07, independent of secondary metabolites, use the mechanism of cholesterol assimilation to lower cholesterol in vitro. We describe collection, extraction, and gas chromatography-flame ionized detection (GC-FID) methods to quantify the levels of cholesterol remaining after incubation of *M. purpureus* CBS 109.07 with exogenous cholesterol. Our findings demonstrate that active growing *M. purpureus* CBS 109.07 can assimilate cholesterol, removing 36.38% of cholesterol after 48 h of incubation at 37 °C. The removal of cholesterol by resting or dead *M. purpureus* CBS 109.07 was not significant, with cholesterol reduction ranging from 2.75–9.27% throughout a 72 h incubation. Cholesterol was also not shown to be catabolized as a carbon source. Resting cultures transferred from buffer to growth media were able to reactivate, and increases in cholesterol assimilation and growth were observed. In growing and resting phases at 24 and 72 h, the production of the mycotoxin citrinin was quantified via high-performance liquid chromatography-ultraviolet (HPLC-UV) and found to be below the limit of detection. The results indicate that *M. purpureus* CBS 109.07 can reduce cholesterol content in vitro and may have a potential application in probiotics.

Keywords: *M. purpureus*; red yeast rice; filamentous fungi; cholesterol reduction; probiotic potential

1. Introduction

Monascus purpureus is a filamentous fungus that produces a variety of secondary metabolites, including pigments, lipids, and monacolins. *M. purpureus* is most widely known for the fermentation of white rice to produce a deep red rice known as angkak or beni koji [1–3]. More commonly, the fermented product is called "red yeast rice", though *Monascus* species are more accurately molds. *M. purpureus* fermented rice is used in food preparation for flavoring, coloring, and preservation, and is also consumed in traditional Chinese medicine to improve ailments of circulation and heart health [2,4–7]. Modern research explored the health claims, and found a plausible cause: *M. purpureus* can synthesize monacolins, naturally occurring compounds capable of decreasing cholesterol levels by inhibiting HMG-CoA reductase, the rate-limiting step in cholesterol biosynthesis [7,8]. The most potent of the *Monascus* monacolins, monacolin K, was isolated and patented as lovastatin and is widely prescribed to treat hypercholesterolemia [8]. Statins are effective treatments for high cholesterol; however, side effects, low tolerance, and the cost of the drug have led patients to pursue alternative options to lower their cholesterol levels [9]. As lyophilized red yeast rice (RYR) supplements emerged as a naturopathic alternative in the U.S., the Food and Drug Administration (FDA) restricted the market, determining that without standardization and quality control for the amount of monacolin K, RYR was not a dietary supplement, but rather an unauthorized new drug [5,10]. As a result, current U.S. supplements marketed as RYR may contain, at most, only trace amounts of monacolin K [5,10–12]. Nevertheless, several clinical studies report that RYR supplements may be an effective treatment option for hypercholesterolemia as significant reductions in low-density lipoprotein (LDL) cholesterol and total cholesterol levels were observed in patients taking RYR [4,10–14].

Two-thirds of cholesterol required for cell membranes and biosynthesis of steroid hormones and bile acids is endogenously synthesized; however, an excess of cholesterol is a major risk factor for cardiovascular disease [15–17]. The World Health Organization (WHO) projects that by 2030, nearly 23.6 million people will die from cardiovascular disease, which includes heart disease and stroke as the leading causes of death [18]. Research into the efficacy of therapies such as RYR supplements in the treatment of elevated cholesterol levels is ongoing and critical for treating cardiovascular disease.

M. purpureus strains administered via RYR supplements in clinical trials often have unspecified viability [19,20]. This led us to ask if living *M. purpureus*, and not simply its secondary metabolites, have cholesterol-lowering properties. Several studies have demonstrated that active probiotic microorganisms introduced into intestinal microbiota can improve the overall lipid content in human blood serum [15,21–25]. Probiotics are defined as live microorganisms that, when administered in adequate amounts, can confer a health benefit on the host [26]. The fermented product of *M. purpureus* can be labeled as "contains live and active cultures," but clinical studies on the safety, viability, and dosage of live *M. purpureus* strains are still necessary before characterizing strains of *Monascus* as probiotic [26–28]. Additionally, factors like mycotoxins should considered; *Monascus* species, like *Aspergillus* and *Penicillium*, naturally synthesize the cytotoxic citrinin at low levels [29,30]. In the European Union (EU) and US, *Monascus* pigments are prohibited from use in food industries [3,31,32]. Still, many *Monascus* strains are generally regarded as safe (GRAS) in many Asian countries and the commercial interest in fermentation has led to easy access of RYR outside of Asia [6,11,33].

While the cholesterol-lowering benefits of probiotics have been highlighted in in vivo studies, the mechanisms of probiotics remain not fully understood [34–36]. Potential mechanisms have been proposed in vitro, such as the removal of cholesterol from media by the assimilation or uptake of cholesterol by probiotic strains of *Lactobacilli* and *Bifidobacteria* [35,37–41]. The ability of cholesterol removal appears to be growth- and strain-specific, with nearly all research focused on bacterial strains [35–41]. We anticipate that as beneficial fungi are characterized, greater attention will be drawn to the roles of fungi in health, nutrition, and the mycobiome. Currently, *Saccharomyces boulardii* is the only fungus with a strain commercially labeled in the U.S. as a probiotic, as it can survive the passage through the gastrointestinal tract, present good growth at 37 °C, complement treatment of gastrointestinal diseases, and also assimilate cholesterol [37,42–46].

In this study, we evaluated the ability of the fungus *M. purpureus* CBS 109.07 to assimilate cholesterol in vitro. We developed new sample collection methods and used gas chromatography to quantify the levels of cholesterol remaining after incubation with *M. purpureus* CBS 109.07. Our data indicate that this strain of *M. purpureus* is capable of cholesterol assimilation, a cholesterol-lowering mechanism separate from its ability to produce monacolins and which has not been previously reported. The present study does not establish the safety and efficacy of CBS 109.07 as a therapeutic agent, however, these results lay the groundwork for the possibility that active *M. purpureus* CBS 109.07 may have probiotic potential based on its ability to assimilate cholesterol.

2. Materials and Methods

2.1. Strains and Media Conditions

M. purpureus Went teleomorphic type strain CBS 109.07 (ATCC 16365) was obtained from the American Type Culture Collection (ATCC) strain bank and chosen for its ability to grow at 30 °C. *M. purpureus* CBS 109.07 has been used in food-grade studies for human and animal food applications [6,47,48]. *M. purpureus* was grown in a malt extract media (MEA) containing: 2% soluble Bactomalt extract (BD Bioscience), 2% glucose, 1% peptone, and pH adjusted to pH 7. Plate media contained 2% agar. Phosphate-buffered saline (PBS) solution contained 8.0 g NaCl, 0.2 g KCl, 1.44 g Na_2HPO_4, 0.24 g KH_2PO_4 for every 1.0 L solution and was pH adjusted to pH 7. Where indicated, PBS was supplemented with 6.72 g/L of yeast nitrogen base without amino acids (BD Difco) or 5 g/L ammonium sulfate, and pH was adjusted to 7 before sterilization. Bile salt supplemented media contained 0.3% (*w*/*v*) oxgall (BD Difco).

2.2. Submerged Culture Preparation

A sterilized 5 mm cork-borer was used to remove an agar plug of *M. purpureus* CBS 109.07 grown on MEA plate media. The agar plug was subcultured into 4 mL of liquid MEA media and incubated at 30 °C. After four day incubation at 30 °C at 150 rpm, a colorless, spherical fungal pellet was formed. The pellet was transferred into new media with 0.3% oxgall at 37 °C and 60 rpm for growth curve, cholesterol assimilation, and citrinin production experiments.

2.3. Cholesterol Assimilation

2.3.1. Cholesterol Reagents

A stock solution of cholesterol (Lipids Cholesterol Rich from adult bovine serum; Sigma-Aldrich, St. Louis, MO, USA) at 10 mg/mL was used to prepare cholesterol assimilation assays and to prepare a 6-point calibration curve as described in Section 2.3.6 [37]. A stock solution of 5-α-cholestane (Sigma-Aldrich, St. Louis, MO, USA) at 2.5 mg/mL was used as an internal standard in the lipid extractions.

2.3.2. Culture Preparation for Growing, Resting, Dead, and Control Conditions

Growing, resting, and *M. purpureus* control conditions contained *M. purpureus* CBS 109.07 pellets that were homogenized using a sterilized glass douncer in a sterile 50 mL conical tube and divided into replicates. Dead culture conditions contained *M. purpureus* CBS 109.07 pellets that were autoclaved at 121 °C for 20 min under 15 psi pressure and transferred to fresh media. Growing and dead cultures contained 10 mL MEA media in sterile 50 mL borosilicate tubes; resting cultures contained 10 mL of PBS in sterile 50 mL borosilicate tubes; resting cultures supplemented with nitrogen sources contained 10 mL of PBS with ammonium sulfate or 10 mL of PBS with yeast nitrogen base. Cholesterol assimilation and dry weight experiments were supplemented with 0.3% oxgall and incubated at 37 °C and 60 rpm. With the exception of the *M. purpureus* control, all conditions were incubated with 120 μg/mL cholesterol. Media control without *M. purpureus* contained 120 μg/mL cholesterol and 0.3% (*w*/*v*) oxgall in 10 mL MEA.

2.3.3. Cholesterol Assimilation and Dry Weight Growth Curve

Cholesterol assimilation and dry weight experiments for each growth condition were prepared in triplicate, where three independent sets of culture were homogenized and divided into six sterile 50 mL borosilicate tubes to account for six timepoints (0, 24, 36, 48, 60, and 72 h). At designated time point, a 1.0 mL aliquot of culture supernatant was collected, centrifuged at 2000× *g* for 15 min, and stored in a 50 mL borosilicate glass tube with a PTFE-lined cap at −20 °C. The remaining 9.0 mL of the culture was then harvested and filtered via vacuum flask and a pre-weighed Whatman filter #1. The contents were allowed to air dry for five days and dry weight was measured on an analytical balance.

2.3.4. *M. Purpureus* Dormancy Experiment

Cholesterol assimilation and dry weight experiments were prepared in triplicate, where three independent sets of cultures were homogenized and divided into five sterile 50 mL borosilicate tubes to account for five timepoints (0, 24, 48, 72, and 96 h). *M. purpureus* was incubated in PBS buffer, pH 7 with 0.3% oxgall and 120 μg/mL cholesterol. At designated time points, a 1.0 mL aliquot of culture supernatant was collected, centrifuged at 2000× *g* for 15 min, and stored in a 50 mL borosilicate glass tube with a PTFE-lined cap at −20 °C. *M. purpureus* samples were then washed twice under sterile conditions with 10 mL of PBS + 0.3% oxgall, and transferred to 10 mL MEA media with 0.3% oxgall and 120 μg/mL cholesterol. A 1.0 mL aliquot of culture supernatant was collected at the starting point (t = 0 h), and also at day 4 and day 7 of incubation in MEA media. At day 7, the culture was harvested and filtered via vacuum flask and a pre-weighed Whatman filter #1. The contents were allowed to air dry for five days and dry weight was measured on an analytical balance.

2.3.5. Resting *M. purpureus* Supplemented with Nitrogen Sources Experiment

Cholesterol assimilation and dry weight experiments were prepared in duplicate, where two independent sets of cultures were homogenized and divided into three sterile 50 mL borosilicate tubes to account for three timepoints (0, 24, and 72 h). *M. purpureus* was incubated in PBS buffer supplemented with either ammonium sulfate or yeast nitrogen base without amino acids, pH 7 with 0.3% oxgall and 120 μg/mL cholesterol. At designated time points, a 1.0 mL aliquot of culture supernatant was collected, centrifuged at 2000× *g* for 15 min, and stored in a 50 mL borosilicate glass tube with PTFE lined cap at −20 °C. *M. purpureus* samples were then washed twice under sterile conditions with 10 mL of PBS + 0.3% oxgall, and transferred to 10 mL MEA media with 0.3% oxgall. At day 4, the culture was harvested and filtered via vacuum flask and a pre-weighed Whatman filter #1. The contents were allowed to air dry for five days and dry weight was measured on an analytical balance.

2.3.6. Cholesterol Extraction

Cholesterol assimilation samples were thawed and a stock solution of cholesterol (10 mg/mL) was used to prepare cholesterol standards in a range from 10–150 μg/mL in a borosilicate glass tube. To each sample and standard, 20 μL of 2.5 mg/mL internal standard 5-α-cholestane was added. Direct saponification was carried out on all samples and standards based on the method described by Fletouris et al. [49]. Four millimeters of methanolic 0.5 M KOH solution was added to each tube which was then capped and vortexed for 15 s. The samples and standards were heated for a total of 15 min in an 80 °C water bath, and removed every 5 min to vortex for 10 s. After cooling to room temperature, 4 mL of hexane was added for lipid extraction and vortexed for 1 min. After incubating at room temp for 10 min to permit phase separation, the entire hexane layer of each sample was transferred to a clean test tube. The hexane layer was evaporated using speed vacuum at −109 °C. Dried samples and standards were resuspended in 0.6 mL of hexane and transferred to autosampler vials for gas chromatography (GC) analysis.

2.3.7. Gas Chromatography Methods

Cholesterol was determined using gas chromatography (Shimadzu GC-2014, Kyoto, Japan) with a flame ionized detector (FID) and an autosampler [49]. The separation was completed using an SPB-1 column (15 m × 0.32 mm i.d.; film thickness 1.0 mm) (Supelco Inc., Bellefonte, PA, USA) using helium as a carrier gas at a flow rate of 2 mL/min. The oven temperature was set at 285 °C, injection port temperature at 300 °C, and flame ionization detector temperature at 300 °C. The injection volume was 1 μL with a split ratio of 20:1. Matrix effects were addressed by the addition of an internal standard of 5-α-cholestane to all samples and standards. In addition, extracting the standards for each set of experiments using the same process as for the samples helped account for any errors in the preparation process. This allowed for the determination of the limits of detection (LOD) and quantitation (LOQ) for the experimental conditions of 8.31 μg/mL and 27.71 μg/mL, respectively.

2.3.8. Calculations for Cholesterol Assimilation

The integrated peak areas for cholesterol and the internal standard 5-α-cholestane were used to determine a 6-point calibration curve for cholesterol and used to extrapolate cholesterol recovered.

The experimental calibration curve to determine LOD and LOQ were created by combining the calibration curves from five experiments to generate a linear calibration curve with R^2 = 0.9832. Cholesterol assimilated and % cholesterol assimilated were calculated as follows, where Cholesterol_i represents cholesterol content recovered at t = 0 and Cholesterol_f represents cholesterol content recovered at given time point.

$$\text{Cholesterol assimilated} = \text{Cholesterol}_i - \text{Cholesterol}_f \tag{1}$$

$$\%\ \text{Cholesterol assimilated} = \frac{\text{Cholesterol assimilated}}{\text{Cholesterol}_i} \times 100\% \tag{2}$$

2.4. Citrinin Production

2.4.1. Citrinin Reagents

Citrinin, high-performance liquid chromatography (HPLC)-grade, was purchased from Sigma-Aldrich. A stock solution of citrinin at 100 ug/mL was prepared in HPLC-grade methanol (J.T. Baker) and used to construct a 7-point calibration curve as described in Section 2.4.3. Acetonitrile and water used in chromatography were HPLC grade (J.T. Baker), and trifluoroacetic acid (TFA) (Sigma-Aldrich, St. Louis, MO, USA) was analytical grade.

2.4.2. Culture Preparation and Extraction for Citrinin Production

A 5 mm agar plug of *M. purpureus* was pre-cultured at 150 rpm and 30 °C for 4 days, and transferred to 10 mL of MEA + 0.3% or PBS + 0.3% oxgall and grown at 60 rpm and 37 °C, as described in culture preparations for cholesterol assimilation assays Section 2.3.2. At 24 h, 72 h, and 14 days, cultures were extracted for citrinin as described in Liu and Xu, with some modifications [50]. Briefly, 10 mL cultures were dounced and extracted with 10 mL of ethanol (1:1). Samples were then vortexed for 5 min and sonicated for 20 min. Samples were spun down at 4200× *g* for 10 min. Supernatant was collected, dried down, and resuspended in 1 mL HPLC-grade methanol. Extraction method was validated with recovery controls, where *M. purpureus* grown in MEA + 0.3% oxgall at 24 h and 72 h were spiked with citrinin at 10 µg/mL and extracted as described previously.

2.4.3. High-Performance Liquid Chromatography Methods

Citrinin was determined using high-performance liquid chromatography, HPLC (Agilent 1100 liquid chromatograph) with a diode array detector (DAD) and an autosampler. The separation was completed using a Discovery C18 column (5 µm, 150 × 4.6 mm column) (Supelco Inc., Bellefonte, PA, USA) and an isocratic elution. The mobile phase consisted of acetonitrile:water containing 0.05% TFA and the volume ratio was 35:65 [50,51]. All samples, standards, and solvents were filtered through 0.22 µm membrane filters prior to HPLC analysis. The flow rate was 1 mL/min, and 20 µL sample was injected. The UV-DAD detection was monitored at 254 nm and 334 nm. The integrated peak areas at 334 nm for standard citrinin were used to determine a 7-point calibration curve and used to extrapolate citrinin recovered. The LOD and LOQ for the experimental conditions were 1.11 µg/mL and 3.70 µg/mL, respectively. Recovery of citrinin was determined by dividing citrinin concentration recovered by known citrinin concentration injected.

2.5. Statistical Analysis

For cholesterol assimilation and growth curves, growing, resting, and dead *M. purpureus* cultures and controls were conducted in triplicate for each time point. For citrinin assays, growing and resting *M. purpureus* cultures were conducted in duplicate for each time point. All GC-FID and high-performance liquid chromatography-ultraviolet (HPLC-UV) samples were measured in duplicate. Two-way ANOVA was carried out to examine the effect of *M. purpureus* × incubation time interaction on growing, resting, or dead conditions. Tukey's test was used to compare means. Significance was defined at $p < 0.05$ or $p < 0.01$. Standard deviation is calculated as either absolute error, or percent error through propagation of uncertainty from Equation (2) of Section 2.3.8. All statistical analyses were carried out using GraphPad Prism 8.0.

3. Results

3.1. Cholesterol Assimilation

The in vitro removal of cholesterol by the filamentous fungi *M. purpureus* CBS 109.07 (hereon referred to in the Results section as *M. purpureus*) was analyzed at 37 °C in media containing 0.3% (*w*/*v*) oxgall and 120 μg/mL cholesterol. Three growth phases were assessed: growing, where culture is active in MEA media; resting, where culture is dormant in PBS buffer; and dead, where culture has been heat-killed and incubated in MEA media. At indicated time points, an aliquot of spent media was collected from three independent replicates.

After 36 h, *M. purpureus* removed 18.69 μg/mL or 18.78% of the cholesterol in spent media, which is a significant decrease from the initial concentration (Table 1, Figure 1, $p < 0.01$). The rate of cholesterol removal was most dramatic from 36 to 60 h, and cholesterol removed increased from 18.78% to 50.27% ($p < 0.05$). At 72 h, 69.65% cholesterol was removed.

Table 1. Cholesterol assimilated in *M. purpureus* CBS 109.07 at different growth phases. All cultures were incubated at 37 °C with 120 μg/mL cholesterol and 0.3% (*w*/*v*) oxgall bile salts. Cholesterol assimilated was calculated from initial cholesterol, and determined from three independent trials conducted for each growth phase at each time point, and measured in duplicate via gas chromatography-flame ionized detection (GC-FID). Standard deviation calculated is absolute error (μg/mL) and percent error (%).

Time (h)	Cholesterol Assimilated					
	Growing *M. purpureus*		Resting *M. purpureus*		Dead *M. purpureus*	
	(μg/mL)	(%)	(μg/mL)	(%)	(μg/mL)	(%)
0	–	–	–	–	–	–
24	10.89 ± 5.62 $^{+}$	10.91 ± 7.47	6.78 ± 2.79	6.07 ± 6.97	5.68 ± 4.40	5.59 ± 4.95
36	18.69 ± 4.53 $^{a,+}$	18.78 ± 7.03 $^{a,+}$	7.20 ± 3.33 ^{b}	6.40 ± 6.02 ^{b}	6.04 ± 4.54 ^{b}	5.94 ± 5.02 ^{b}
48	36.18 ± 3.84 $^{a,+}$	36.38 ± 5.38 $^{a,+}$	10.53 ± 7.94 $^{b,+}$	9.29 ± 7.25 ^{b}	2.88 ± 4.45 ^{c}	2.75 ± 4.69 ^{b}
60	49.86 ± 5.30 $^{a,+}$	50.27 ± 13.5 $^{a,+}$	4.54 ± 2.48 ^{b}	4.03 ± 6.03 ^{b}	4.21 ± 4.63 ^{b}	4.08 ± 4.71 ^{b}
72	69.13 ± 3.95 $^{a,+}$	69.65 ± 12.5 $^{a,+}$	6.70 ± 6.33 ^{b}	5.91 ± 7.23 ^{b}	4.44 ± 4.67 ^{b}	4.33 ± 5.16 ^{b}

$^{a, b, c}$ Means within a row are significantly different ($p < 0.01$). $^{+}$ Means significantly different from the initial value at t = 0 ($p < 0.01$).

In contrast, *M. purpureus* cultures dormant in PBS for resting phase or heat-killed in dead phase removed a negligible amount of cholesterol after 72 h (Table 1) ($p > 0.05$). Non-growing *M. purpureus* conditions removed less than 10% of cholesterol, with a high percent error accounting for the propagation of the standard deviation. The media control containing MEA or PBS incubated without *M. purpureus* similarly showed no significant change in cholesterol content or cholesterol assimilation over 72 h (Table S1). We note that cholesterol content of resting and dead *M. purpureus* and media controls did not change significantly from the initial concentration of cholesterol within each trial ($p < 0.01$).

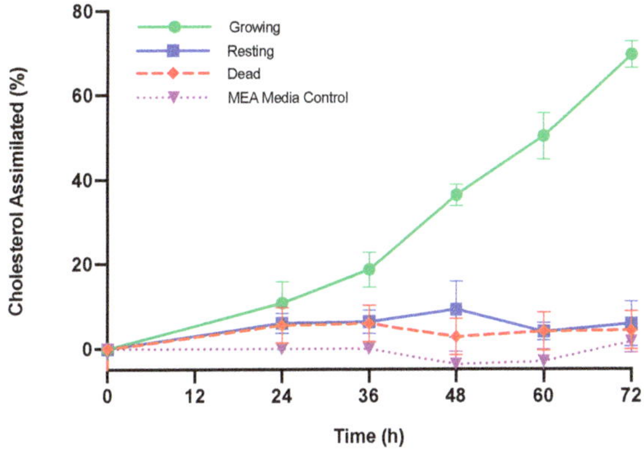

Figure 1. Cholesterol assimilated (%) by *M. purpureus* CBS 109.07 at different growth phases. Growing, resting, and dead *M. purpureus* cultures were incubated with 120 μg/mL cholesterol and 0.3% (*w*/*v*) oxgall bile salts at 37 °C. Malt extract media (MEA) control contained 120 μg/mL cholesterol and 0.3% (*w*/*v*) oxgall bile salts incubated without *M. purpureus*. Three independent trials were conducted for each condition at each time point, and cholesterol in samples were measured in duplicate via GC-FID. Standard deviation in cholesterol assimilated is percent error.

3.2. Growth of M. purpureus

The morphology of *M. purpureus* in submerged media is a compact, smooth, and spherical pellet consisting of intertwined hyphae [52–54]. Consistent with studies on other filamentous fungi, we found that dry weight measurement is the most reproducible method of quantifying *M. purpureus* growth [55–61].

After an aliquot of spent media was collected from three independent replicates, the remaining culture was harvested to measure dry weight (Table 2, Figure 2). Growth phases growing, resting, and dead were assessed and conditions included 0.3% oxgall and 120 μg/mL cholesterol. *M. purpureus* control was grown in MEA media without cholesterol. The dry weight of growing *M. purpureus* was significantly different from that of resting or dead *M. purpureus* ($p < 0.05$). The presence of cholesterol did not significantly enhance or inhibit the growth of *M. purpureus* [62] ($p > 0.05$). In both resting and dead conditions, *M. purpureus* had no significant growth ($p > 0.05$).

Table 2. Dry weight of *M. purpureus* CBS 109.07. Growing, resting, or dead *M. purpureus* cultures were incubated in 120 μg/mL cholesterol and 0.3% (*w*/*v*) oxgall bile salts at 37 °C. *M. purpureus* incubated without cholesterol served as a control. Three independent trials, corresponding to samples used for cholesterol assimilation, were conducted for each growth phase at each time point.

	Dry Weight (mg)			
Time (h)	Growing *M. purpureus*	Resting *M. purpureus*	Dead *M. purpureus*	*M. purpureus* without Cholesterol
0	17.54 ± 2.5	18.40 ± 1.5	18.60 ± 1.3	19.91 ± 2.9
24	30.77 ± 4.9 $^{a,+}$	24.17 ± 2.0 ^{a}	21.17 ± 1.5 ^{b}	37.05 ± 0.5 $^{c,+}$
36	40.77 ± 0.9 $^{a,+}$	21.07 ± 3.5^{b}	16.93 ± 5.6 ^{b}	40.45 ± 4.6 $^{a,+}$
48	45.40 ± 4.2 $^{a,+}$	22.93 ± 0.5 ^{b}	25.00 ± 3.6 $^{b,+}$	43.72 ± 2.5 $^{a,+}$
60	45.17 ± 2.7 $^{a,+}$	24.60 ± 3.3 ^{b}	24.17 ± 1.3 ^{b}	50.87 ± 5.6 $^{a,+}$
72	48.20 ± 3.9 $^{a,+}$	22.27 ± 1.7 ^{b}	21.50 ± 2.2 ^{b}	54.63 ± 1.8 $^{a,+}$

$^{a, b, c}$ Means within a row are significantly different ($p < 0.05$). $^{+}$ Means significantly different from the initial value at t = 0 ($p < 0.05$).

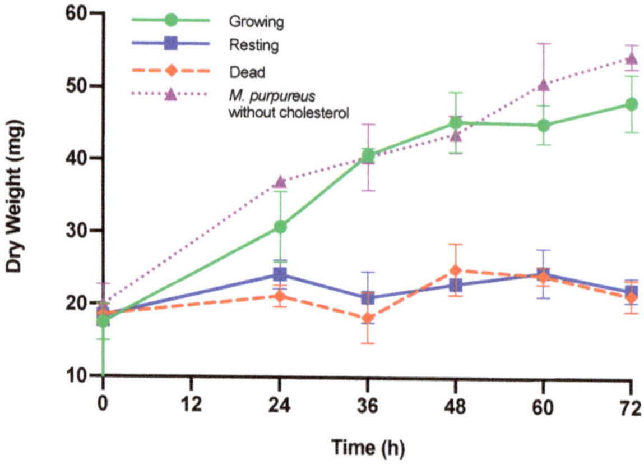

Figure 2. Dry weight of *M. purpureus* CBS 109.07. Growing, resting, or dead *M. purpureus* cultures were incubated in 120 µg/mL cholesterol and 0.3% (*w/v*) oxgall bile salts at 37 °C. *M. purpureus* incubated without cholesterol served as a control. Three independent trials were conducted for each growth phase at each time point. Standard deviation is represented by error bars.

3.3. Reactivating Dormant M. purpureus

M. purpureus incubated in PBS buffer, pH 7 with 0.3% oxgall and 120 µg/mL cholesterol does not assimilate cholesterol during a 96 h incubation (Table 3). To examine if resting conditions correspond to a dormant *M. purpureus*, cultures incubated in PBS were washed and transferred to MEA media with 0.3% oxgall and 120 µg/mL cholesterol. After four days of incubation in MEA, cholesterol assimilation was measured, but was not significantly different than MEA media control ($p > 0.05$). After seven days of incubation, cholesterol assimilation was initiated and cholesterol content is comparable to growing phase *M. purpureus* (Table S1). Dry weight of rescued *M. purpureus* was increased from resting *M. purpureus* (Table 2). The length of time incubated in PBS before the transfer to MEA did not have a significant effect on the ability to assimilate cholesterol ($p < 0.05$).

Table 3. Cholesterol content (µg/mL) and dry weight (mg) of *M. purpureus* CBS 109.07 incubated in phosphate-buffered saline (PBS) and rescued in MEA. All cultures were incubated at 37 °C with 120 µg/mL cholesterol and 0.3% (*w/v*) oxgall bile salts. Resting *M. purpureus* was washed in PBS + 0.3% oxgall and then transferred to MEA with 0.3% oxgall and 120 µg/mL cholesterol, and sample collected after 4 and 7 days. Data were determined from three independent trials conducted at each time point. Cholesterol content was measured in duplicate via GC-FID. Standard deviation in cholesterol content is absolute error.

PBS [1]			Day 4 Incubation in MEA [1]	Day 7 Incubation in MEA [1]	Day 7 Incubation in MEA [1]
Time (h)	Cholesterol Content (µg/mL)	Time [2] (h)	Cholesterol Content (µg/mL)	Cholesterol Content (µg/mL)	Dry Weight (mg)
24	114.46 ± 3.73 [a]	24	97.73 ± 2.66 [b]	38.83 ± 9.95 [c,+]	32.87 ± 2.9 [++]
48	118.30 ± 3.12 [a]	48	97.12 ± 4.40 [b]	40.88 ± 9.51 [c,+]	31.70 ± 2.7 [++]
72	117.36 ± 6.84 [a]	72	98.36 ± 2.10 [b]	37.24 ± 11.47 [c,+]	31.60 ± 4.9 [++]
96	114.81 ± 3.86 [a]	96	98.63 ± 8.19 [b]	52.23 ± 5.30 [c]	25.33 ± 2.0

[1] Media contained 0.3% oxgall and 120 µg/mL cholesterol. [2] Time corresponds to duration incubated in PBS prior to rescue in MEA media. [a, b, c] Cholesterol content means within a row are significantly different ($p < 0.05$). [+] Means significantly different from the MEA media control Table S1 ($p < 0.05$). [++] Means significantly different from the resting *M. purpureus* dry weight Table 2 ($p < 0.05$).

To determine if cholesterol is catabolized by *M. purpureus* as a carbon source, resting phase cultures in PBS were incubated with a nitrogen source, either ammonium sulfate or yeast nitrogen base without amino acids, at concentrations found in minimal media. We observed that resting *M. purpureus* in PBS supplemented with nitrogen sources had no significant cholesterol assimilation throughout a 24 and 72 h incubation (Table 4, $p > 0.05$).

Table 4. Cholesterol content (μg/mL) of *M. purpureus* CBS 109.07 incubated in PBS with nitrogen sources. All cultures were incubated at 37 °C with 120 μg/mL cholesterol and 0.3% (*w*/*v*) oxgall bile salts. Cholesterol content was determined from two independent trials conducted for each growth phase at each time point, and measured in duplicate via GC-FID. All means were not significantly different from the initial value at t = 0 ($p < 0.05$). Standard deviation in cholesterol content is absolute error.

	Cholesterol Content of *M. purpureus* Incubated in PBS with 0.3% Oxgall (μg/mL)	
Time (h)	**With Yeast Nitrogen Base without Amino Acids**	**With Ammonium Sulfate**
0	105.82 ± 2.20	105.46 ± 4.59
24	97.91 ± 2.03	89.64 ± 2.59
72	97.57 ± 1.60	93.45 ± 8.03

Fungi samples were then washed twice and transferred to MEA + 0.3% oxgall. The dry weight at day 4 in MEA of *M. purpureus* previously incubated in yeast nitrogen base was a significant increase from the initial weight before rescue (Table 5, $p < 0.05$), supporting the observation that dormant *M. purpureus* can be reactivated.

Table 5. Dry weight (mg) of *M. purpureus* CBS 109.07 after incubation in PBS with 120 μg/mL cholesterol and 0.3% (*w*/*v*) oxgall bile salts and rescued in MEA with 0.3% (*w*/*v*) oxgall bile salts. *M. purpureus* is washed in PBS + 0.3% oxgall and then transferred to MEA with 0.3% oxgall, and sample collected after 4 days. Two independent trials, corresponding to samples used in PBS cholesterol assimilation, were conducted for each growth phase at each time point. Standard deviation in cholesterol content is absolute error.

	Dry Weight of *M. purpureus* (mg)			
	PBS + 0.3% Oxgall		**Day 4 Incubation in MEA + 0.3% Oxgall**	
Time [1] (h)	**With Yeast Nitrogen Base**	**With Ammonium Sulfate**	**With Yeast Nitrogen Base**	**With Ammonium Sulfate**
0	19.4 ± 3.1 *	18.8 ± 3.1 *		
24			37.5 ± 2.8 +	34.7 ± 3.3 +
72	25.8 ± 5.3	23.3 ± 3.2	39.3 ± 0.4 +	30.8 ± 1.8 +

[1] Time corresponds to duration incubated in PBS prior to rescue in MEA media. * Initial weight at t = 0. + Means significantly different from the initial weight at t = 0 ($p < 0.05$).

3.4. Citrinin Production in M. purpureus

The production of the mycotoxin citrinin was measured in *M. purpureus* grown under conditions used in cholesterol assimilation assays, where a 5 mm agar plug of *M. purpureus* was precultured in MEA at 150 rpm and 30 °C and then transferred to fresh media with oxgall and incubated at 60 rpm and 37 °C (Table 6). Samples collected at 24 h and 72 h did not have citrinin production above the limit of detection. The culture broth in these samples were also colorless. After 14 days, *M. purpureus* grown in MEA + 0.3% oxgall began to produce red pigment and culture broth turned reddish. *M. purpureus* in resting phase did not become pigmented. Though 14 days is outside the incubation period for this study's cholesterol assimilation experiments, we extracted the red cultures, and measured 6.77 ± 1.02 μg citrinin per mL of culture broth. To validate extraction methods and show effectiveness, a set of *M. purpureus* cultures grown in MEA + 0.3% oxgall at 24 h and 72 h were spiked with citrinin at 10 μg/mL and extracted. Recovery of citrinin was 72.9% ± 3.8.

Table 6. Citrinin production under experimental conditions. Citrinin concentration (μg/mL) of *M. purpureus* CBS 109.07 under experimental conditions of growing and resting phases was measured at 24 and 72 h via HPLC-UV. Two independent trials were conducted for each time point. Standard deviation in cholesterol content is absolute error.

Citrinin Production of *M. purpureus* (μg Citrinin/mL of Culture Broth)			
Media Conditions	**After 24 h [1]**	**After 72 h [1]**	**After 14 Days**
MEA + 0.3% oxgall	ND	ND	6.77 ± 1.02
PBS + 0.3% oxgall	ND	ND	

[1] ND is not detected with a peak below the limit of detection (LOD) of 1.11 μg/mL.

4. Discussion

As a natural source for monacolins, fermentation products of *Monascus purpureus* are widely used as alternative treatments for hypercholesterolemia. To the best of our knowledge, our data is the first to show that a strain of *M. purpureus* is capable of a cholesterol-lowering mechanism separate from its ability to produce monacolins and other secondary metabolites. We observed that active growing *M. purpureus* CBS 109.07 can assimilate cholesterol in vitro, and after 48 h incubation at 37 °C and high bile salt conditions, 36.38% of cholesterol content was removed. The removal of cholesterol by resting or dead *M. purpureus* CBS 109.07 was not statistically significant, and cholesterol was not catabolized as a carbon source. When resting cultures were washed and transferred to MEA media, *M. purpureus* CBS 109.07 became active and cholesterol assimilation and growth were observed. Citrinin production of *M. purpureus* CBS 109.07 incubated in growing or resting phase conditions at 24 h and 72 h was lower than the limit of detection, and we note that CBS 109.07 produced citrinin under our experimental conditions at day 14 when red pigment production was observed.

The ability of microorganisms to remove cholesterol in vitro from growth media is an indicator of probiotic potential, and the range of reduction percentage is wide and dependent on strain. We note that as with any therapeutic dosage, the concentration of microorganisms present will play a major role in cholesterol assimilation percentage. Miremadi et al. tested strains of *Lactobacilli* and *Bifidobacteria* and found 14 strains capable of removing cholesterol with a range of 34–65% assimilation after 24 h [41]. Eukaryotes capable of lowering cholesterol include strains of *S. boulardii*, *S. cerevisiae*, and *I. orientalis*, which after 48 h of incubation was observed to assimilate 90.6%, 96.8%, and 88.1% of cholesterol, respectively [37]. Strains of *P. kudriazevii*, *Galactomyces* sp., and *Y. lipolytica* were observed to assimilate 45.7%, 36.3%, and 30.9% of cholesterol, respectively, after 48 h of incubation in Chen et al. [45]. In the same study, the commercially available yeast probiotic *S. boulardii* lowered cholesterol by 36.5% at 48 h, and 41.5% cholesterol at 72 h. In this study, active growing *M. purpureus* CBS 109.07 was comparable to *S. boulardii* and was able to lower cholesterol from the media by 36.38% at 48 h, and 69.65% cholesterol at 72 h (Table 1). We note that at higher aeration and agitation, *M. purpureus* CBS 109.07 was able to assimilate a higher percentage of cholesterol (Table S2). When *M. purpureus* CBS 109.07 is resting or dead, cholesterol removal is not significant (Figure 1) and ranged from 2.75–9.29% removal after 72 h of incubation (Table 1). Other studies observed similar trends where resting and heat-killed cultures did not significantly reduce cholesterol [41,63–65], suggesting a mechanism where actively growing strains are more efficient at removing cholesterol.

To eliminate the possibility that cholesterol assimilation by *M. purpureus* was an artifact of starvation and the uptake of available carbon sources, we allowed cultures to incubate undisturbed until the entire culture was collected at the designated time point. This procedure differed from other cholesterol assimilation studies, where one-tenth of the culture volume was removed at each time point and, thus, could significantly impact nutrient availability [38,39,41,63,65–67]. We note that in our methods, the presence of cholesterol at 120 μg/mL did not enhance or inhibit the growth of *M. purpureus* CBS 109.07 as measured by dry weight (Figure 2).

We investigated the ability of *M. purpureus* CBS 109.07 to transition out of microbial dormancy after one to four days of incubation in PBS. Resting phase cultures incubated in PBS with 120 μg/mL cholesterol did not show significant cholesterol assimilation (Table 3). When washed and transferred to MEA media with 120 μg/mL cholesterol, previously resting phase *M. purpureus* CBS 109.07 cultures were able to restore cholesterol assimilation and growth at day 7 of rescue (Table 3) at levels comparable to growing cultures (Table S1, Table 2). There was no significant difference in reactivation of cholesterol assimilation between cultures that were incubated for one day or four days in PBS. We also observed that *M. purpureus* CBS 109.07 incubated in PBS with cholesterol and supplemented with nitrogen sources showed insignificant cholesterol assimilation between 24 to 72 h (Table 4), and were able to grow after transfer into MEA media for four days (Table 5). These results on resting phase reveal that *M. purpureus* CBS 109.07 incubated in PBS is indeed dormant and that cholesterol is not significantly taken up as a carbon source during dormancy. Additionally, the absence of cholesterol assimilation in PBS with nitrogen sources supports the assertion that *M. purpureus* CBS 109.07 does not metabolize cholesterol (Table 4). We posit that such an absence of cholesterol removal may reflect a mechanism where growing *M. purpureus* CBS 109.07 is more efficient at assimilating cholesterol.

Microorganisms can utilize the cholesterol-lowering mechanisms of active assimilation and passive adhesion to decrease host absorption of intestinal cholesterol [21,36,38]. Our results suggest that *M. purpureus* CBS 109.07 is capable of an in vitro active assimilation mechanism by growing cells. The dense pellet morphology of *M. purpureus* CBS 109.07 has made it difficult to measure the cholesterol content of the cell membrane, as similarly noted in biosorbent studies on other filamentous fungi such as *Aspergillus niger* and *Penicillium* sp. L1 strains [55–57]. Follow-up experiments will be conducted to lyse the *Monascus* membrane and examine membrane cholesterol content. Other cholesterol-lowering mechanisms by probiotic microorganisms include modulation of lipid metabolism and deconjugation of bile salts [36]. *M. purpureus* is capable of directly modulating lipid metabolism, as it synthesizes monacolins that directly inhibit HMG-CoA reductase, the committed step of cholesterol biosynthesis in the liver [7,8]. In future studies, we will assay *M. purpureus* CBS 109.07 for bile salt hydrolase (BSH) activity, the enzyme responsible for the deconjugation of bile salts found in many probiotic strains [68,69].

Like many strains within *Monascus*, *Aspergillus*, and *Penicillium* genera, *M. purpureus* CBS 109.07 can biosynthesize citrinin, with levels highly dependent on the growth conditions and amount of microorganisms used. In this study, the citrinin production in *M. purpureus* CBS 109.07 under growing and resting phase conditions replicated from our cholesterol assimilation experiments was below our limit of detection of 1.11 μg/mL (Table 6). Our results at 24 h and 72 h are consistent with published studies on other *M. purpureus* strains which measured citrinin production in different growth conditions over time. These studies observed delays in citrinin production, with detection of citrinin beginning as early as day 5 or late as day 10 [70–72]. Notably, the commencement of citrinin production corresponded to the commencement of red pigment production, and increases in agitation and aeration increased citrinin production [72,73]. *M. purpureus* CBS 109.07 studies in particular did not measure citrinin at early time points. However using thin-layer chromatography (TLC), they reported citrinin level to be 5 μg/mL after 14 day incubation in glucose media and unspecified agitation [30], and 65 μg/mL after 7 day incubation in ethanol media and 220 rpm agitation [74,75]. We used HPLC-UV to quantify citrinin production after 14 day incubation in MEA + 0.3% oxgall and 60 rpm. At day 14 under our conditions, the culture broth began to turn reddish, and we measured a citrinin concentration of 6.77 μg/mL [30,74–76]. The differences in citrinin production between CBS 109.07 studies highlight how critical growth conditions are to the control of citrinin levels in *Monascus* strains [77,78]. In future studies, we can target the citrinin issue as many studies have successfully eliminated or reduced the levels of citrinin by disrupting the citrinin biosynthetic genes *pksCT* or *ctnA* in *M. purpureus* [70,79–81].

To be beneficial for human health, probiotic microorganisms must be capable of surviving transit through the human gastrointestinal tract. Absorption of dietary cholesterol into the bloodstream occurs predominantly in the duodenum of the small intestine, where the pH varies from pH 6 to 7 and bile salts excreted from the bile duct assist in solubilizing cholesterol [82,83]. *M. purpureus* CBS 109.07 was cultured in media at physiological temperature and pH, and with a high bile salt concentration and low aeration and agitation. However we recognize the limitations of an in vitro study in reproducing gastrointestinal conditions. Additionally, the clinical safety of *M. purpureus* CBS 109.07 needs to be established—a potentially complicated issue if the restrictive regulations on *Monascus* pigments and red yeast rice supplements by the FDA and European Food Safety Authority are any indication [4,5,31,84]. In the current study, we are only beginning to raise the possibility of an application for *M. purpureus* CBS 109.07 in probiotics; we recognize that additional safety and gastrointestinal survival experiments are required, and that such advancement in understanding *Monascus* biochemistry may improve the restrictions on their usage in the US and EU [85]. We also recognize that other candidate strains of *M. purpureus* or other *Monascus* species may be found [86], and that CBS 109.07 may not be unique or exemplary. However, we note that the human consumption of *M. purpureus* CBS 109.07 has precedents, as several food-grade studies have considered CBS 109.07 an edible filamentous fungus and used it as the representative *Monascus* strain in human and animal food applications of mycoprotein [6,47,48].

5. Conclusions

Our findings demonstrate that *M. purpureus* CBS 109.07, which can biosynthesize statin-like monacolins, can also reduce cholesterol content in vitro via a mechanism of cholesterol assimilation at 37 °C with a high concentration of bile salts. The most effective removal of cholesterol occurred in growing *M. purpureus* CBS 109.07 cultures, while non-growing *M. purpureus* CBS 109.07 minimally adhered to cholesterol and did not metabolize cholesterol. Dormant cultures, once transferred from buffer to nutrient rich media, were able to be resume cholesterol assimilation at levels observed in active cultures. Citrinin production under our experimental conditions was not detected. Our results show that it is valuable to continue examining the cholesterol-lowering potential of active *M. purpureus* CBS 109.07 cultures, as further research may provide a possible insight in the treatment of hypercholesterolemia and will draw attention to the significance of filamentous fungi in human health and nutrition.

Supplementary Materials: The following are available online at http://www.mdpi.com/2309-608X/6/4/352/s1, Table S1: Cholesterol content of *M. purpureus* CBS 109.07, Table S2. Cholesterol content, cholesterol assimilated, and dry weight of growing *M. purpureus* CBS 109.07.

Author Contributions: Conceptualization, T.P.T.N.; methodology, T.P.T.N.; validation, T.P.T.N.; M.A.G.; S.A.N.; C.E.S.; C.W.; formal analysis, T.P.T.N.; investigation, T.P.T.N.; M.A.G.; S.A.N.; C.E.S.; C.W.; resources, T.P.T.N.; data curation, T.P.T.N.; M.A.G.; S.A.N.; C.E.S.; C.W.; writing—original draft preparation, T.P.T.N.; writing—review and editing, T.P.T.N.; M.A.G.; S.A.N.; C.E.S.; C.W.; visualization, T.P.T.N.; supervision, T.P.T.N.; project administration, T.P.T.N.; funding acquisition, T.P.T.N. All authors have read and agreed to the published version of the manuscript.

Funding: This research received no external funding.

Acknowledgments: We thank the members of the Chemistry & Biochemistry Department and the Natural & Applied Sciences College at Loyola University Maryland for support to carry out this project. We are especially grateful to Elizabeth E. Dahl for assistance with GC-FID and quantitative and statistical analyses, Courtney J. Hastings for assistance with HPLC-UV, Heather R. Schmidt for assistance with materials, and Timothy J. McNeese for helpful discussions.

Conflicts of Interest: The authors declare no conflict of interest.

References

1. Song, J.; Luo, J.; Ma, Z.; Sun, Q.; Wu, C.; Li, X. Quality and authenticity control of functional red yeast rice—a review. *Molecules* **2019**, *24*, 1944. [CrossRef] [PubMed]
2. Ma, J.; Li, Y.; Ye, Q.; Li, J.; Hua, Y.; Ju, D.; Zhang, D.; Cooper, R.; Chang, M. Constituents of red yeast rice, a traditional Chinese food and medicine. *J. Agric. Food Chem.* **2000**, *48*, 5220–5225. [CrossRef] [PubMed]

3. Caro, Y.; Venkatachalam, M.; Lebeau, J.; Fouillaud, M.; Dufossé, L. *Fungal Metabolites. Reference Series in Phytochemistry*; Mérillon, J.-M., Ramawat, K.G., Eds.; Springer International Publishing: Cham, Switzerland, 2017; ISBN 9783319250014.
4. Patel, S. Functional food red yeast rice (RYR) for metabolic syndrome amelioration: A review on pros and cons. *World J. Microbiol. Biotechnol.* **2016**, *32*, 87. [CrossRef] [PubMed]
5. National Center for Complementary and Integrative Health. Red Yeast Rice. 2013. Available online: https://nccih.nih.gov/health/redyeastrice (accessed on 31 October 2020).
6. Souza Filho, P.F.; Nair, R.B.; Andersson, D.; Lennartsson, P.R.; Taherzadeh, M.J. Vegan-mycoprotein concentrate from pea-processing industry byproduct using edible filamentous fungi. *Fungal Biol. Biotechnol.* **2018**, *5*, 1–10. [CrossRef] [PubMed]
7. Xiong, Z.; Cao, X.; Wen, Q.; Chen, Z.; Cheng, Z.; Huang, X.; Zhang, Y.; Long, C.; Zhang, Y.; Huang, Z. An overview of the bioactivity of monacolin K/lovastatin. *Food Chem. Toxicol.* **2019**, *131*, 110585. [CrossRef]
8. Endo, A. The origin of the statins. *Atheroscler. Suppl.* **2004**, *5*, 125–130. [CrossRef]
9. Subhan, M.; Faryal, R.; Macreadie, I. Exploitation of *Aspergillus terreus* for the production of natural statins. *J. Fungi* **2016**, *2*, 13. [CrossRef] [PubMed]
10. Peng, D.; van Pelt, A. The Effects of Red Yeast Rice Supplementation on Cholesterol Levels in Adults. *Am. J. Nurs.* **2017**, *117*, 46–53. [CrossRef]
11. Nguyen, T.; Karl, M.; Santini, A. Red yeast rice. *Foods* **2017**, *6*, 19. [CrossRef] [PubMed]
12. Yang, C.W.; Mousa, S.A. The effect of red yeast rice (*Monascus purpureus*) in dyslipidemia and other disorders. *Complement. Ther. Med.* **2012**, *20*, 466–474. [CrossRef]
13. Gerards, M.C.; Terlou, R.J.; Yu, H.; Koks, C.H.W.; Gerdes, V.E.A. Traditional Chinese lipid-lowering agent red yeast rice results in significant LDL reduction but safety is uncertain—A systematic review and meta-analysis. *Atherosclerosis* **2015**, *240*, 415–423. [CrossRef] [PubMed]
14. Burke, F.M. Red Yeast Rice for the Treatment of Dyslipidemia. *Curr. Atheroscler. Rep.* **2015**, *17*, 495. [CrossRef] [PubMed]
15. Kriaa, A.; Bourgin, M.; Potiron, A.; Mkaouar, H.; Jablaoui, A.; Gérard, P.; Maguin, E.; Rhimi, M. Microbial impact on cholesterol and bile acid metabolism: Current status and future prospects. *J. Lipid Res.* **2019**, *60*, 323–332. [CrossRef] [PubMed]
16. Sanchis-Gomar, F.; Perez-Quilis, C.; Leischik, R.; Lucia, A. Epidemiology of coronary heart disease and acute coronary syndrome. *Ann. Transl. Med.* **2016**, *4*, 1–12. [CrossRef] [PubMed]
17. World Health Organization. Global Health Observatory (GHO) Data: Raised Cholesterol. Available online: http://www.who.int/gho/ncd/risk_factors/cholesterol_text/en/ (accessed on 31 October 2020).
18. World Health Organization. Cardiovascular Disease. 2017. Available online: https://www.who.int/cardiovascular_diseases/about_cvd/en/ (accessed on 31 October 2020).
19. Lin, C.C.; Li, T.C.; Lai, M.M. Efficacy and safety of *Monascus purpureus* Went rice in subjects with hyperlipidemia. *Eur. J. Endocrinol.* **2005**, *153*, 679–686. [CrossRef] [PubMed]
20. Klimek, M.; Wang, S.; Ogunkanmi, A. Safety and efficacy of red yeast rice (*Monascus purpureus*) as an alternative therapy for hyperlipidemia. *Pharm. Ther.* **2009**, *34*, 313–317.
21. Miremadi, F.; Sherkat, F.; Stojanovska, L. Hypocholesterolaemic effect and anti-hypertensive properties of probiotics and prebiotics: A review. *J. Funct. Foods* **2016**, *25*, 497–510. [CrossRef]
22. Kumar, M.; Nagpal, R.; Kumar, R.; Hemalatha, R.; Verma, V.; Kumar, A.; Chakraborty, C.; Singh, B.; Marotta, F.; Jain, S.; et al. Cholesterol-lowering probiotics as potential biotherapeutics for metabolic diseases. *Exp. Diabetes Res.* **2012**, *2012*, 902917. [CrossRef]
23. Hassan, A.; Din, A.U.; Zhu, Y.; Zhang, K.; Li, T.; Wang, Y.; Luo, Y.; Wang, G. Updates in understanding the hypocholesterolemia effect of probiotics on atherosclerosis. *Appl. Microbiol. Biotechnol.* **2019**, *103*, 5993–6006. [CrossRef]
24. Bordoni, A.; Amaretti, A.; Leonardi, A.; Boschetti, E.; Danesi, F.; Matteuzzi, D.; Roncaglia, L.; Raimondi, S.; Rossi, M. Cholesterol-lowering probiotics: In vitro selection and in vivo testing of bifidobacteria. *Appl. Microbiol. Biotechnol.* **2013**, *97*, 8273–8281. [CrossRef]
25. Fava, F.; Lovegrove, J.A.; Gitau, R.; Jackson, K.G.; Tuohy, K.M. The Gut Microbiota and Lipid Metabolism: Implications for Human Health and Coronary Heart Disease. *Curr. Med. Chem.* **2006**, *13*, 3005–3021. [CrossRef] [PubMed]

26. Hill, C.; Guarner, F.; Reid, G.; Gibson, G.R.; Merenstein, D.J.; Pot, B.; Morelli, L.; Canani, R.B.; Flint, H.J.; Salminen, S.; et al. Expert consensus document: The international scientific association for probiotics and prebiotics consensus statement on the scope and appropriate use of the term probiotic. *Nat. Rev. Gastroenterol. Hepatol.* **2014**, *11*, 506–514. [CrossRef] [PubMed]
27. World Health Organization; Food and Agriculture Organization Agriculture of the United Nations. Guidelines for the Evaluation of Probiotics in Food. 2002. Available online: https://www.who.int/foodsafety/fs_management/en/probiotic_guidelines.pdf (accessed on 31 October 2020).
28. De Melo Pereira, G.V.; de Oliveira Coelho, B.; Magalhães Júnior, A.I.; Thomaz-Soccol, V.; Soccol, C.R. How to select a probiotic? A review and update of methods and criteria. *Biotechnol. Adv.* **2018**, *36*, 2060–2076. [CrossRef] [PubMed]
29. Adrio, J.L.; Demain, A.L. Fungal biotechnology. *Int. Microbiol.* **2003**, *6*, 191–199. [CrossRef] [PubMed]
30. Blanc, P.J.; Loret, M.O.; Goma, G.; Rangueil, C.S. De Production of citrinin by various species of *Monascus*. *Biotechnol. Lett.* **1995**, *17*, 291–294. [CrossRef]
31. Younes, M.; Aggett, P.; Aguilar, F.; Crebelli, R.; Dusemund, B.; Filipič, M.; Frutos, M.J.; Galtier, P.; Gott, D.; Gundert-Remy, U.; et al. Scientific opinion on the safety of *monacolins* in red yeast rice. *EFSA J.* **2018**, *16*. [CrossRef]
32. Dufossé, L.; Fouillaud, M.; Caro, Y.; Mapari, S.A.S.; Sutthiwong, N. Filamentous fungi are large-scale producers of pigments and colorants for the food industry. *Curr. Opin. Biotechnol.* **2014**, *26*, 56–61. [CrossRef]
33. Srianta, I.; Ristiarini, S.; Nugerahani, I.; Sen, S.K.; Zhang, B.B.; Xu, G.R.; Blanc, P.J. Recent research and development of *Monascus fermentation* products. *Int. Food Res. J.* **2014**, *21*, 1–12.
34. Ishimwe, N.; Daliri, E.B.; Lee, B.H.; Fang, F.; Du, G. The perspective on cholesterol-lowering mechanisms of probiotics. *Mol. Nutr. Food Res.* **2015**, *59*, 94–105. [CrossRef]
35. Lye, H.S.; Rusul, G.; Liong, M.T. Removal of cholesterol by lactobacilli via incorporation and conversion to coprostanol. *J. Dairy Sci.* **2010**, *93*, 1383–1392. [CrossRef]
36. Reis, S.A.; Conceição, L.L.; Rosa, D.D.; Siqueira, N.P.; Peluzio, M.C.G. Mechanisms responsible for the hypocholesterolaemic effect of regular consumption of probiotics. *Nutr. Res. Rev.* **2017**, *30*, 36–49. [CrossRef] [PubMed]
37. Psomas, E.I.; Fletouris, D.J.; Litopoulou-Tzanetaki, E.; Tzanetakis, N. Assimilation of cholesterol by yeast strains isolated from infant feces and feta cheese. *J. Dairy Sci.* **2003**, *86*, 3416–3422. [CrossRef]
38. Kimoto-Nira, H.; Mizumachi, K.; Nomura, M.; Kobayashi, M.; Fujita, Y.; Okamoto, T.; Suzuki, I.; Tsuji, N.M.; Kurisaki, J.I.; Ohmomo, S. *Lactococcus* sp. as potential probiotic lactic acid bacteria. *Jpn. Agric. Res. Q.* **2007**, *41*, 181–189. [CrossRef]
39. Mishra, A.K.; Kumar, S.S.; Ghosh, A.R. Probiotic Enterococcus faecalis AG5 effectively assimilates cholesterol and produces fatty acids including propionate. *FEMS Microbiol. Lett.* **2019**, *366*, 1–9. [CrossRef] [PubMed]
40. Ahire, J.J.; Bhat, A.A.; Thakare, J.M.; Pawar, P.B.; Zope, D.G.; Jain, R.M.; Chaudhari, B.L. Cholesterol assimilation and biotransformation by *Lactobacillus helveticus*. *Biotechnol. Lett.* **2012**, *34*, 103–107. [CrossRef] [PubMed]
41. Miremadi, F.; Ayyash, M.; Sherkat, F.; Stojanovska, L. Cholesterol reduction mechanisms and fatty acid composition of cellular membranes of probiotic *Lactobacilli* and *Bifidobacteria*. *J. Funct. Foods* **2014**, *9*, 295–305. [CrossRef]
42. Kelesidis, T.; Pothoulakis, C. Efficacy and safety of the probiotic *Saccharomyces boulardii* for the prevention and therapy of gastrointestinal disorders. *Therap. Adv. Gastroenterol.* **2012**, *5*, 111–125. [CrossRef]
43. Gil-Rodríguez, A.M.; Carrascosa, A.V.; Requena, T. Yeasts in foods and beverages: In vitro characterisation of probiotic traits. *LWT Food Sci. Technol.* **2015**, *64*, 1156–1162. [CrossRef]
44. Fakruddin, M.; Hossain, M.N.; Ahmed, M.M. Antimicrobial and antioxidant activities of *Saccharomyces cerevisiae* IFST062013, a potential probiotic. *BMC Complement. Altern. Med.* **2017**, *17*, 1–11. [CrossRef]
45. Chen, L.S.; Ma, Y.; Maubois, J.L.; He, S.H.; Chen, L.J.; Li, H.M. Screening for the potential probiotic yeast strains from raw milk to assimilate cholesterol. *Dairy Sci. Technol.* **2010**, *90*, 537–548. [CrossRef]
46. Pais, P.; Almeida, V.; Yılmaz, M.; Teixeira, M.C. *Saccharomyces boulardii*: What makes it tick as successful probiotic? *J. Fungi* **2020**, *6*, 78. [CrossRef] [PubMed]
47. Ferreira, J.A.; Lennartsson, P.R.; Taherzadeh, M.J. Production of ethanol and biomass from thin stillage using food-grade *Zygomycetes* and *Ascomycetes* filamentous fungi. *Energies* **2014**, *7*, 3872–3885. [CrossRef]

48. Yudianto, D.; Nainggolan, E.A.; Millati, R.; Hidayat, C.; Lennartsson, P.; Taherzadeh, M.J.; Niklasson, C. Bioconversion of pretreated wheat straw to ethanol by monascus purpureus CBS 109.07 and fusarium venenatum ATCC 20334 using simultaneous saccharification and fermentation. *Biodiversitas* **2019**, *20*, 2229–2235. [CrossRef]
49. Fletouris, D.J.; Botsoglou, N.A.; Psomas, I.E.; Mantis, A.I. Rapid Determination of Cholesterol in Milk and Milk Products by Direct Saponification and Capillary Gas Chromatography. *J. Dairy Sci.* **1998**, *81*, 2833–2840. [CrossRef]
50. Liu, R.; Xu, B. Optimization of Extraction Conditions of Citrinin from Red Yeast Rice by Orthogonal Design and Quantification of Citrinin by High-Performance Liquid Chromatography. *Food Anal. Methods* **2013**, *6*, 677–682. [CrossRef]
51. Xu, B.; Wang, Q.; Lee, J.; Jia, X.; Sung, C. HPLC analysis of citrinin in red yeast rice. *Food Sci. Biotechnol.* **2003**, *12*, 376–380.
52. Pazouki, M.; Panda, T. Understanding the morphology of fungi. *Bioprocess Eng.* **2000**, *22*, 127–143. [CrossRef]
53. Veiter, L.; Rajamanickam, V.; Herwig, C. The filamentous fungal pellet—Relationship between morphology and productivity. *Appl. Microbiol. Biotechnol.* **2018**, *102*, 2997–3006. [CrossRef]
54. Mapari, S.A.S.; Meyer, A.S.; Thrane, U. Evaluation of Epicoccum nigrum for growth, morphology and production of natural colorants in liquid media and on a solid rice medium. *Biotechnol. Lett.* **2008**, *30*, 2183–2190. [CrossRef]
55. Castillo, N.A.; Valdez, A.L.; Fariña, J.I. Microbial production of scleroglucan and downstream processing. *Front. Microbiol.* **2015**, *6*, 1–19. [CrossRef]
56. Liu, Y.; Hu, T.; Zhao, J.; Lv, Y.; Ren, R. Simultaneous removal of carbon and nitrogen by mycelial pellets of a heterotrophic nitrifying fungus-Penicillium sp. L1. *J. Biosci. Bioeng.* **2017**, *123*, 223–229. [CrossRef] [PubMed]
57. Li, S.; Huang, J.; Mao, J.; Zhang, L.; He, C.; Chen, G.; Parkin, I.P.; Lai, Y. In vivo and in vitro efficient textile wastewater remediation by *Aspergillus niger* biosorbent. *Nanoscale Adv.* **2019**, *1*, 168–176. [CrossRef]
58. Dikshit, R.; Tallapragada, P. *Monascus purpureus*: A potential source for natural pigment production. *J. Microbiol. Biotechnol. Res.* **2011**, *1*, 164–174.
59. Witherden, E.A.; Shoaie, S.; Hall, R.A.; Moyes, D.L. The human mucosal mycobiome and fungal community interactions. *J. Fungi* **2017**, *3*, 56. [CrossRef]
60. Ajdari, Z.; Ebrahimpour, A.; Abdul Manan, M.; Hamid, M.; Mohamad, R.; Ariff, A.B. Nutritional requirements for the improvement of growth and sporulation of several strains of monascus purpureus on solid state cultivation. *J. Biomed. Biotechnol.* **2011**, *2011*, 487329. [CrossRef]
61. Nimnoi, P.; Lumyong, S. Improving Solid-State Fermentation of Monascus purpureus on Agricultural Products for Pigment Production. *Food Bioprocess Technol.* **2011**, *4*, 1384–1390. [CrossRef]
62. Madani, G.; Mirlohi, M.; Yahay, M.; Hassanzadeh, A. How much in vitro cholesterol reducing activity of Lactobacilli predicts their in vivo cholesterol function? *Int. J. Prev. Med.* **2013**, *4*, 404–413.
63. Kimoto, H.; Ohmomo, S.; Okamoto, T. Cholesterol removal from media by lactococci. *J. Dairy Sci.* **2002**, *85*, 3182–3188. [CrossRef]
64. Tomaro-Duchesneau, C.; Jones, M.L.; Shah, D.; Jain, P.; Saha, S.; Prakash, S. Cholesterol Assimilation by Lactobacillus Probiotic Bacteria: An In Vitro Investigation. *BioMed Res. Int.* **2014**, *2014*, 380316. [CrossRef]
65. Tok, E.; Aslim, B. Cholesterol removal by some lactic acid bacteria that can be used as probiotic. *Microbiol. Immunol.* **2010**, *54*, 257–264. [CrossRef]
66. Pereira, D.I.A.; Gibson, G.R. Cholesterol assimilation by lactic acid bacteria and bifidobacteria isolated from the human gut. *Appl. Environ. Microbiol.* **2002**, *68*, 4689–4693. [CrossRef] [PubMed]
67. Gilliland, S.E.; Nelson, C.R.; Maxwell, C. Assimilation of cholesterol by *Lactobacillus acidophilus*. *Appl. Environ. Microbiol.* **1985**, *49*, 377–381. [CrossRef] [PubMed]
68. Begley, M.; Hill, C.; Gahan, C.G.M. Bile salt hydrolase activity in probiotics. *Appl. Environ. Microbiol.* **2006**, *72*, 1729–1738. [CrossRef] [PubMed]
69. Jones, B.V.; Begley, M.; Hill, C.; Gahan, C.G.M.; Marchesi, J.R. Functional and comparative metagenomic analysis of bile salt hydrolase activity in the human gut microbiome. *Proc. Natl. Acad. Sci. USA* **2008**, *105*, 13580–13585. [CrossRef] [PubMed]
70. Shimizu, T.; Kinoshita, H.; Ishihara, S.; Sakai, K.; Nagai, S.; Nihira, T. Polyketide synthase gene responsible for citrinin biosynthesis in *Monascus purpureus*. *Appl. Environ. Microbiol.* **2005**, *71*, 3453–3457. [CrossRef] [PubMed]

71. Patrovsky, M.; Sinovska, K.; Branska, B.; Patakova, P. Effect of initial pH, different nitrogen sources, and cultivation time on the production of yellow or orange *Monascus purpureus* pigments and the mycotoxin citrinin. *Food Sci. Nutr.* **2019**, *7*, 3494–3500. [CrossRef]
72. Hajjaj, H.; Blanc, P.; Groussac, E.; Uribelarrea, J.L.; Goma, G.; Loubiere, P. Kinetic analysis of red pigment and citrinin production by *Monascus* ruber as a function of organic acid accumulation. *Enzyme Microb. Technol.* **2000**, *27*, 619–625. [CrossRef]
73. Hajjaj, H.; Klaébé, A.; Goma, G.; Blanc, P.J.; Barbier, E.; François, J. Medium-chain fatty acids affect citrinin production in the filamentous fungus *Monascus* ruber. *Appl. Environ. Microbiol.* **2000**, *66*, 1120–1125. [CrossRef]
74. Rasheva, T.V.; Nedeva, T.S.; Hallet, J.N.; Kujumdzieva, A.V. Characterization of a non-pigment producing *Monascus purpureus* mutant strain. *Antonie Leeuwenhoek* **2003**, *83*, 333–340. [CrossRef]
75. Pisareva, E.; Savov, V.; Kujumdzieva, A. Pigments and citrinin biosynthesis by fungi belonging to genus *Monascus*. *Z. fur Naturforsch. C* **2005**, *60*, 116–120. [CrossRef]
76. Fabre, C.E.; Santerre, A.L.; Loret, M.O.; Baberian, R.; Pareilleux, A.; Goma, G.; Blanc, P.J. Production and Food Applications of the Red Pigments of *Monascus* ruber. *J. Food Sci.* **1993**, *58*, 1099–1102. [CrossRef]
77. Kang, B.; Zhang, X.; Wu, Z.; Wang, Z.; Park, S. Production of citrinin-free *Monascus pigments* by submerged culture at low pH. *Enzyme Microb. Technol.* **2014**, *55*, 50–57. [CrossRef]
78. Feng, Y.; Shao, Y.; Zhou, Y.; Chen, F. Monacolin K production by citrinin-free *Monascus pilosus* MS-1 and fermentation process monitoring. *Eng. Life Sci.* **2014**, *14*, 538–545. [CrossRef]
79. Xu, M.J.; Yang, Z.L.; Liang, Z.Z.; Zhou, S.N. Construction of a *Monascus purpureus* mutant showing lower citrinin and higher pigment production by replacement of ctnA with pks1 without using vector and resistance gene. *J. Agric. Food Chem.* **2009**, *57*, 9764–9768. [CrossRef] [PubMed]
80. Jia, X.Q.; Xu, Z.N.; Zhou, L.P.; Sung, C.K. Elimination of the mycotoxin citrinin production in the industrial important strain *Monascus purpureus* SM001. *Metab. Eng.* **2010**, *12*, 1–7. [CrossRef] [PubMed]
81. Chen, W.; He, Y.; Zhou, Y.; Shao, Y.; Feng, Y.; Li, M.; Chen, F. Edible Filamentous Fungi from the Species Monascus: Early Traditional Fermentations, Modern Molecular Biology, and Future Genomics. *Compr. Rev. Food Sci. Food Saf.* **2015**, *14*, 555–567. [CrossRef]
82. Evans, D.F.; Pye, G.; Bramley, R.; Clark, A.G.; Dyson, T.J.; Hardcastle, J.D. Measurement of gastrointestinal pH profiles in normal ambulant human subjects. *Gut* **1988**, *29*, 1035–1041. [CrossRef]
83. Cohen, D.E. Balancing cholesterol synthesis and absorption in the gastrointestinal tract. *J. Clin. Lipidol.* **2008**, *2*, 1–5. [CrossRef]
84. Dufossé, L.; Galaup, P.; Yaron, A.; Arad, S.M.; Blanc, P.; Murthy, K.N.C.; Ravishankar, G.A. Microorganisms and microalgae as sources of pigments for food use: A scientific oddity or an industrial reality? *Trends Food Sci. Technol.* **2005**, *16*, 389–406. [CrossRef]
85. Dufossé, L. Red colourants from filamentous fungi: Are they ready for the food industry? *J. Food Compos. Anal.* **2018**, *69*, 156–161. [CrossRef]
86. Tsukahara, M.; Shinzato, N.; Tamaki, Y.; Namihira, T.; Matsui, T. Red yeast rice fermentation by selected *Monascus* sp. with deep-red color, lovastatin production but no citrinin, and effect of temperature-shift cultivation on lovastatin production. *Appl. Biochem. Biotechnol.* **2009**, *158*, 476–482. [CrossRef] [PubMed]

Journal of *Fungi*

MDPI

Article

Exploring the Antibacterial Activity of *Pestalotiopsis* spp. under Different Culture Conditions and Their Chemical Diversity Using LC–ESI–Q–TOF–MS

Madelaine M. Aguilar-Pérez [1,†], Daniel Torres-Mendoza [1,2,†], Roger Vásquez [1], Nivia Rios [3] and Luis Cubilla-Rios [1,*]

1 Laboratory of Tropical Bioorganic Chemistry, Faculty of Natural, Exact Sciences and Technology, University of Panama, Panama 0824, Panama; mad25aguilar@gmail.com (M.M.A.-P.); dtorresm.507@gmail.com (D.T.-M.); royi071123@gmail.com (R.V.)

2 Vicerrectoría de Investigación y Postgrado, University of Panama, Panama 0824, Panama

3 Department of Microbiology, Faculty of Natural, Exact Sciences and Technology, University of Panama, Panama 0824, Panama; toxogondii@gmail.com

* Correspondence: luis.cubilla@up.ac.pa; Tel.: +507-6676-5824

† These authors contributed equally to this work.

Received: 16 July 2020; Accepted: 17 August 2020; Published: 19 August 2020

Abstract: As a result of the capability of fungi to respond to culture conditions, we aimed to explore and compare the antibacterial activity and chemical diversity of two endophytic fungi isolated from *Hyptis dilatata* and cultured under different conditions by the addition of chemical elicitors, changes in the pH, and different incubation temperatures. Seventeen extracts were obtained from both *Pestalotiopsis mangiferae* (*man-1* to *man-17*) and *Pestalotiopsis microspora* (*mic-1* to *mic-17*) and were tested against a panel of pathogenic bacteria. Seven extracts from *P. mangiferae* and four extracts from *P. microspora* showed antibacterial activity; while some of these extracts displayed a high-level of selectivity and a broad-spectrum of activity, *Pseudomonas aeruginosa* was the most inhibited microorganism and was selected to determine the minimal inhibitory concentration (MIC). The MIC was determined for extracts *man-6* (0.11 μg/mL) and *mic-9* (0.56 μg/mL). Three active extracts obtained from *P. mangiferae* were analyzed by Liquid Chromatography-Electrospray Ionization-Quadrupole-Time of Flight-Mass Spectrometry (LC–ESI–Q–TOF–MS) to explore the chemical diversity and the variations in the composition. This allows us to propose structures for some of the determined molecular formulas, including the previously reported mangiferaelactone (**1**), an antibacterial compound.

Keywords: Endophytic fungi; *Hyptis dilatata*; *Pestalotiopsis mangiferae*; *Pestalotiopsis microspora*; chemical elicitors; antibacterial activity; LC–ESI–Q–TOF–MS

1. Introduction

The World Health Organization (WHO; Geneva, Switzerland) has established an urgent pathogen list of antibiotic-resistant bacteria to guide the research, discovery, and development of new antibiotics. This list includes carbapenem-resistant *Pseudomonas aeruginosa* and *Enterobacteriaceae* and third generation cephalosporin-resistant bacteria as critical priorities as a result of the continuous and indiscriminate use of antibiotics, not only in the treatment of human diseases, but also in animals [1,2]. This list includes antifungal compounds [3].

Pharmaceutical conglomerates have abandoned this field of research due to the high costs. Despite the efforts made in recent years, e.g., investment in research and development (R&D) as well as in scientific and technological research, the strategies for the search of new antibiotics and antifungals remain uncertain [4,5]. In this context, natural products produced by endophytic fungi provide an alternative to supply new molecules with antimicrobial activities [6–9].

Endophytic fungi spend a large part of their life cycle inside the tissue of the host organism without causing apparent damage [10]. In the last 15 years, interest in endophytic fungi has grown exponentially because of their ability to produce a wide range of secondary metabolites with diverse and important biological activities. Plant endophytes are considered one of the least studied groups of microorganisms and have proven to be a source of natural products and therefore provide a way to discover novel compounds with biological activities [11,12]. Two species of the genus *Pestalotiopsis*, isolated from *Hyptis dilatata* (Labiatae), a plant that is distributed in the north and east of the Republic of Panama and that is known for producing abietane and pimarane diterpenes [13] were selected for these studies.

The genus *Pestalotiopsis* is considered a vast source of natural products from which more than 300 compounds have been isolated and characterized, including terpenoids, polyketides, chromones, quinones, coumarins, lactones, and nitrogen-containing molecules with a wide range of biological activities such as antifungal, antibacterial, anticancer, antioxidant, antiparasitic, antihypertensive, anti-inflammatory, and neuroprotective activities [14–19]. Previously, our group reported the isolation of a set of eleven compounds (see Figure S1) from the crude extract of *P. mangiferae*, including a polyhydroxylated macrolide named mangiferaelactone [20]; the crude extract showed growth inhibition against *Listeria monocytogenes* (29 mm diameter inhibition zone), and showed a minimal inhibitory concentration (MIC) of 1.69 mg/mL and 0.55 mg/mL against *L. monocytogenes* and *Bacillus cereus*, respectively. This compound belongs to the nonalide class that is associated with important biological activities such anticancer, antifungal, antibacterial, and antiviral activities. Its synthesis has been developed by different research groups [21–23]. Compounds such as taxol, isopestacin, and some polyketides have been isolated from *P. microspora* [24–27].

The production of secondary metabolites by microorganisms could be impacted by environmental dynamics, such as growing conditions, which include biotic and abiotic factors [28]. Therefore, the selective variation of these parameters during the cultivation of fungi [29,30] and/or the induction of stress through competition with other microorganisms in a co-culture represent interesting ways to generate greater activity, chemical diversity, and novel active molecules [31–33]. Hence, the opportunity to modify culture conditions allows for the optimization of secondary metabolite production [34]. Therefore, given the growing interest in enhancing the production of secondary metabolites by endophytic fungi, the study of the methods and strategies to stimulate the gene clusters responsible for the biosynthesis of new molecules has been intensified and could include chemical or physical factors [35,36]. For example, the use of metallic ions, organic and inorganic compounds, pH, and incubation temperature to optimize the production of enzymes or secondary metabolites have been described [37–47].

For this study, we focused on modifying the conditions of the culture medium by varying abiotic parameters and through this, activate fungal silent gene clusters [48–50] in *P. mangiferae* Hd08 and *P. microspora* Hd18 in order to increase the chemical diversity and to detect new antibacterial activities.

2. Materials and Methods

2.1. Chemicals and Reagents

All of the following chemicals were acquired from Sigma–Aldrich® (Sigma–Aldrich, St. Louis, MO, USA): arginine, glutamic acid, $CuSO_4$, $CaCl_2$, $FeSO_4$, tri-sodium citrate dihydrate, dimethyl sulfoxide (DMSO), and formic acid (FA). Ethyl acetate, acetone, and methanol used for extraction were American Chemical Society grade (Tedia®, Tedia Company Inc., Fairfield, OH, USA). The methanol for the liquid chromatography-mass spectrometry (LC–MS) analysis was LC–MS grade (J. T. Baker®, Avantor Performance Materials, Inc., Center Valley, PA, USA).

2.2. Isolation and Identification of Fungal Isolates

A healthy specimen of *Hyptis dilatata* (Labiatae) was collected in La Mesa, Veraguas Province in the Republic of Panama, in November 2010. An exsiccate from the plant material was deposited in the Herbarium of the University of Panama (PMA 084861). Mature leaves were surface sterilized as we previously reported in [51], and small fragments were cultured on 2% malt-extract agar (MEA; Difco™, Becton, Dickinson and Co., Sparks, MD, USA) under sterile conditions. Strains Hd08 and Hd18 were further isolated from the collection plate and successively re-plated until pure strains were obtained. Pure fungal strains were stored at −80 °C in a cryoprotectant solution of 10% glycerol and were preserved in the collection of the International Cooperative Biodiversity Group (ICBG) at the University of Panama. The identification of endophytic fungi was carried out as described previously [20]. Briefly, the total genomic DNA of each strain was isolated from fresh mycelium following U'Ren et al. [52]. Polymerase chain reaction (PCR) was used to amplify the nuclear ribosomal internal transcribed spacers and 5.8s gene (ITS rDNA; ca. 600 bp), and the first ca. 600 bp was sequenced bidirectionally [52]. The entire sequences for each strain were compared to the nucleotide database of the National Center for Biotechnology Information (NCBI) using the Basic Local Alignment Search Tools (BLAST®) Website.

2.3. Media Preparation and Cultivation of Fungal Strains

Strains Hd08 and Hd18 were reactivated aseptically on Petri dishes containing potato dextrose agar (PDA; Difco™, Becton, Dickinson and Co., Sparks, MD, USA) and incubated at 26 °C for seven days. Then, the mycelium was removed using a sterile spatula and was placed in sterile water to obtain a homogeneous solution. This solution was poured on the surface of Petri dishes (145 × 90 mm) containing MEA for all of the experimental conditions. After 15 days of incubation, the material was extracted, and the amount of crude organic extract was measured. Sterile controls were established for all of the experiments.

Chemicals as elicitors. MEA was prepared as indicated on the label; the resulting pH was measured using a pH meter (Thomas Scientific, Swedesboro, NJ, USA). The medium was then sterilized at 121 °C. When it had cooled to 45 °C, each of the elicitors was added, mixed well, and poured on Petri dishes (145 × 90 mm), and the strains then were inoculated and incubated at 26 °C for 15 days.

pH as an elicitor. MEA was prepared as described above and buffered with a 50 mmol/L tri-sodium citrate dihydrate solution to set values of 4.0, 4.6, and 5.6. The chemical elicitors were $CaCl_2$ and $CuSO_4$. The strains then were inoculated, and the plates were incubated at 26 °C for 15 days.

Incubation temperature as an elicitor. MEA was prepared and buffered at pH = 4.0 by adding $CaCl_2$ or $CuSO_4$, as described above. The incubation temperatures were set at 24, 28, and 30 ± 2 °C in the incubation chamber (Sheldon Manufacturing, Inc., Cornelius, OR, USA) for 15 days.

2.4. Extraction and Sample Preparation

In all of the experiments, after the incubation time, the mycelium was cut into small pieces and placed into a 1 L beaker and 500 mL of ethyl acetate was added. After 30 min, the mixture was triturated and homogenized using a Polytron® (Brinkmann Instruments, Westbury, NJ, USA) and subsequently filtered through filter paper (Whatman No. 7) using a vacuum. The organic solvent was evaporated under a vacuum at 30 °C using a rotary evaporator. The resulting crude extract was re-dissolved in acetone and transferred to scintillation vials, which were previously labeled and weighed, and then evaporated on a Speed Vac® Plus (Thermo Savant™, Thermo Fisher Scientific, Waltham, MA, USA) for 24 h. Thereafter, the amount of extract was determined.

2.5. Antibacterial Activity

Tested microorganisms: Among the microorganisms used for the antimicrobial test, eight strains (*Bacillus cereus* CECT 5050, *Escherichia coli* CECT 433, *Kocuria rhizophila* CECT 241, *Legionella pneumophila*

CECT 7109, *Listeria monocytogenes* CECT 935, *Pasteurella multocida* CECT 962, *Salmonella enterica* CECT 7161, *Salmonela enterica* CECT 7160, and *Shigella flexneri* CECT 4804) were acquired from the Spanish Types Culture Collection of the University of Valencia, Spain, and eight strains (*Enterobacter cloacae* ATCC 13047, *Enterococcus faecalis* ATCC 19433, *Klebsiella pneumonia* ATCC 13883, *Klebsiella pneumonia ozaenae* ATCC 11296, *Proteus vulgaris* ATCC 9484, *Pseudomonas aeruginosa* ATCC 10145, *Staphylococcus aureus* ATCC 25923, and *Streptococcus oralis* ATCC 35037) were acquired from the American Type Culture Collection.

In vitro bacterial growth inhibition: The antibacterial activity of each extract was determined through the susceptibility test of the British Society for Antimicrobial Chemotherapy (BSAC) [53]. The turbidity standard (0.5 McFarland Turbidity Standard) was a $BaCl_2$ solution which absorbance (0.08–0.10 at 625 nm) was verified in a spectrophotometer (Spectronic 21, Bausch & Lomb, Rochester, NY, USA). The solution was stored in the dark at 24 ± 2 °C. The bacterial inoculum of the seventeen pathogenic strains were prepared in Trypticase-Soy Agar (TSA; Bacto^{TM}, Becton, Dickinson and Co., Sparks, MD, USA) for 24 h. Thereafter, five colonies were picked up and transferred into a tube containing a saline and isotonic solution, then visually compared to the turbidity standardas previously reported [54].

Evaluation of the minimal inhibitory concentration (MIC): A broth dilution susceptibility testing method was applied for the determination of the (MIC) [55], using a stock solution prepared by adding 15 mg of the organic extract in 3 mL of Trypticase-Soy Broth (TSB). Serial dilutions of the organic extract (3.33 μg/mL, 1.67 μg/mL, 0.56 μg/mL, 0.18 μg/mL, 0.061 μg/mL, 0.0021 μg/mL, and 0.007 μg/mL) and positive control (gentamycin sulfate: 104.5 μg/mL, 35.0 μg/mL, 11.6 μg/mL, and 3.87 μg/mL) were performed. Each solution was inoculated with 50 μL (0.5 McFarland) of a culture of *Pseudomonas aeruginosa* and incubated at 37 °C for 18 h. The negative control was DMSO. A sterile culture media control was also used. Each assay was performed in duplicate.

2.6. Analysis of Organic Extracts by LC–MS

LC–MS analysis was carried out in an Agilent 1290 Infinity LC System (Agilent Technologies, Santa Clara, CA, USA) using a Zorbax® Eclipse Plus (1.8 μm) C_{18} reverse phase LC column, 100 × 3 mm (Agilent Technologies, Santa Clara, CA, USA). The mass spectrometer was a micrOTOF-QIII (Bruker Daltonics, Billerica, MA, USA) supplied with an electrospray ionization (ESI) source. For the positive mode Electrospray Ionization-Quadrupole-Time of Flight-Mass Spectrometry (ESI+–Q–TOF–MS) analysis, extracts were re-dissolved in methanol and filtered through a 0.45 μm cellulose acetate membrane filter. Solutions of 0.5 μg/mL were prepared, and aliquots of 10 μL were injected. The chromatographic analysis was carried out using a 37 min step gradient (UHPLC) run using mixtures of methanol and acidified water (99.9% H_2O-0.1% FA) as mobile phase, starting from: (a) 5–95% MeOH-H_2O for 2 min; (b) a 25 min gradient from 5:95 methanol:H_2O to 100% methanol; (c) 100% methanol for 8 min. The column was returned to the initial condition for 2 min. Prior to collecting the data, two level of calibration were employed; before the analysis, an external calibration was performed using an Agilent ESI-L Low-Calibration Tuning Mix, and during the evaluation of each sample, we used hexakis (1H, 1H, 2H-difluoroethoxy)-phosphazene (*m/z* 622.0290 $[M + H]^+$; Synquest Laboratories, Alachua, FL, USA) as an internal reference, for the lock mass calibration.

3. Results

3.1. Culture Conditions

The results obtained from evaluating the changes in culture conditions are listed in Table S1. The addition of chemical elicitors impacted the amount of crude extract obtained. In *P. mangiferae*, the best result was achieved in presence of Fe^{2+} and Ca^{2+} ion (*man-3*, 256.0 mg and *man-4*, 232.0 mg); in *P. microspora*, under the presence of Cu^{2+} and Ca^{2+} ions (*mic-4*, 297.0 mg and *mic-5*, 329.0 mg).

Low pH values increased the amount of crude extract obtained. For both species, the best results were obtained at pH = 4.0 and Cu^{2+} as elicitor: in *P. mangiferae* (*man-9*, 264.3 mg); in *P. microspora* (*mic-9*, 260.0 mg).

The third factor to consider was the incubation temperature. Maintaining constant the pH at 4.0, we found that the highest amount of extract in *P. mangiferae* was at 30 °C for both elicitors (*man-14*, 528.0 mg; *man-17* 448.0 mg); the production of extract using Ca^{2+} was 34-fold higher and 1.7-fold higher using Cu^{2+} compared to the amounts obtained in phase II. For *P. microspora*, the highest amount of extract was obtained at 24 °C using Cu^{2+} as elicitor (*mic-15*, 1194.0 mg) that was 4.6-fold higher that the obtained amount in phase II.

3.2. Antibacterial Activity

A total of 34 extracts were assayed against a panel of pathogenic bacteria in a preliminary antibacterial test (disc diffusion method, mm), and only 11 extracts showed growth inhibition: seven from *P. mangiferae* and four from *P. microspora* (Table 1). These extracts were capable of inhibiting the growth of 13 of 17 pathogenic bacterial strains. Larger inhibition zones were observed against *P. aeruginosa* (12.5 mm) and *L. monocytogenes* (11.0 mm). Extract *man-6* displayed the highest broad-spectrum of antimicrobial activity (inhibited seven pathogenic strains), followed by extracts *man-9* and *mic-9* (both inhibited six pathogenic strains). Nevertheless, the culture condition for *man-6* ($CaCl_2$, pH = 4.0 and T = 26 °C) induced one of the lowest amounts of organic extract (15.4 mg). The culture condition for extracts *man-9* and *mic-9* ($CuSO_4$, pH = 4.0, T = 26 °C) were more favorable.

Extracts *man-15* and *mic-11* exhibited selectivity against *P. aeruginosa*, the most sensitive strain. To determine the MIC using *P. aeruginosa*, five of the eleven extracts were selected. In this experiment, only extracts *man-6* (0.11 µg/mL) and *mic-9* (0.56 µg/mL) demonstrated growth inhibition.

Hence, to correlate the chemical profile with the antibacterial activity against *P. aeruginosa*, three samples were selected from *P. mangiferae* to be analyzed by LC–MS: (1) one active, *man-7*; (2) one with broad-spectrum activity, *man-9* (12.5 mm inhibition zone); and (3) one with selective activity, *man-15* (9.5 mm inhibition zone).

To our knowledge, there are only two reports of secondary metabolites from *P. mangiferae* [20,56].

3.3. Evaluation of the Chemical Diversity

Table 2 lists the principal molecular ions present in the analyzed extracts and their determined molecular formulas. Extracts *man-7* and *man-15* showed a similar chemical composition; nevertheless, at least four compounds were only present in extract *man-15*. The molecular ions are linked, mainly to polyoxygenate compounds, but some nitrogenous are present too. In all three extracts (*man-7*, *man-9*, *man-15*), the presence of mangiferaelactone was determined (retention time t_R 21.55–21.59 min; *m/z* 401.2017 $[M + H]^+$); none of the other previously isolated compounds from *P. mangiferae* was detected as principal components of the analyzed samples (see Figure S1). The pseudo molecular ion *m/z* 338.341 $[M + H]^+$, which appeared at t_R 28.3 min, has been established as a possible molecular formula (calculated for $C_{22}H_{44}NO$). This compound appeared in the chromatograms of extracts *man-9* and *man-15*, but it was absent in the *man-7* extract (Figure 1, Table 2), suggesting that it could be responsible for the antibacterial activity against *P. aeruginosa*.

Table 1. Antibacterial activity in the disk diffusion test [1] of organic extracts produced by *Pestalotiopsis* spp.

Fungi	Extract	*E. coli*	*P. aeruginosa*	*S. tiphy*	*S. flexneri*	*P. vulgaris*	*S. enterica*	*L. pneumophilla*	*E. faecalis*	*E. cloacae*	*P. multocida*	*K. rhizophila*	*L. monocytogenes*	*B. cereus*	*S. oralis*
P. mangiferae	*man-1*												8.5		9.5
	man-2		8.5			8.5							8.0		
	man-3		8.5		7.5				9.0	9.0			7.5		
	man-6		10.0	8.5	9.0		9.0				8.0		11.0	8.0	
	man-7							8.0	8.5	8.5		9.0			7.5
	man-9		12.5			8.5		8.5	8.0			7.5	7.5		
	man-15		9.5												
P. microspora	*mic-9*		8.0	8.5		7.5						8.0	7.5	8.5	
	mic-10	8.0										8.5	8.5		
	mic-11		9.5												
	mic-12		9.5			7.5								7.5	
Positive control	Gentamycin sulfate (10 µg/mL)	18.5	17.5	14.5	13.5	13.0	13.0	12.0	10.0	9.5	9.5	30.0	18.0	17.0	9.0

[1] Results are given in mm of inhibition.

Table 2. Molecular ions of secondary metabolites present in extracts obtained from *P. mangiferae*.

Retention Time [1] (t_R)	*m/z*	$[M + H]^+$	$[M + Na]^+$	Dimers	Molecular Formula	Extracts
19.44–19.50	588.36567	589.369		1177.720 $[2M + H]^+$	$C_{34}H_{52}O_8$	*man-7*, *man-15*
20.14–20.37	278.18765	279.194			$C_{17}H_{26}O_3$	*man-7*
20.78–20.87	588.36567	589.369		1177.720 $[2M + H]^+$	$C_{34}H_{52}O_8$	*man-7*, *man-15*
21.55–21.59	282.18524	283.193		565.375 $[2M + H]^+$	$C_{16}H_{26}O_4$	*man-7*, *man-9*, *man-15*
	400.20917	401.217	423.199	801.424 $[2M + H]^+$ 823.407 $[2M + Na]^+$	$C_{20}H_{32}O_8$	*man-7*, *man-9*, *man-15*
22.39–22.43	394.28663	395.295			$C_{27}H_{38}O_2$	*man-7*, *man-15*
	412.29317	413.303			$C_{22}H_{40}N_2O_5$	*man-7*, *man-15*
	470.33906	471.347			$C_{30}H_{46}O_4$	*man-7*, *man-15*
	488.34560	489.354		995.716 $[2M + H_2O + H]^+$	$C_{30}H_{48}O_5$	*man-7*, *man-15*
23.40–23.45	314.24516	315.256	337.236		$C_{38}H_{34}O_4$	*man-15*
23.72–23.81	428.30737		451.298		$C_{31}H_{40}O$	*man-7*, *man-15*
	428.31324	429.320			$C_{24}H_{44}O_6$	*man-15*
24.54	278.22403	279.231	301.211	557.446 $[2M + H]^+$	$C_{18}H_{30}O_2$	*man-7*, *man-15*
24.60	452.31324		475.304	927.610 $[2M + Na]^+$	$C_{26}H_{44}O_6$	*man-15*
26.86	281.27132	282.279	304.261		$C_{18}H_{35}NO$	*man-7*, *man-15*
27.52	283.28697	284.294	306.276		$C_{18}H_{37}NO$	*man-7*, *man-15*
27.83–27.87	390.27646	391.283	413.261	803.536 $[2M + Na]^+$	$C_{24}H_{38}O_4$	*man-7*, *man-9*, *man-15*
28–33	337.33392	338.341	360.320	675.670 $[2M + H]^+$	$C_{22}H_{43}NO$	*man-9*, *man-15*
28.40	418.28663	419.308	441.288	859.578 $[2M + Na]^+$	$C_{29}H_{38}O_2$	*man-7*
31.25–31.36	662.43884	663.444	685.40		$C_{40}H_{36}NO_4$	*man-7*, *man-9*, *man-15*
	721.51233	722.519			$C_{41}H_{71}NO_9$	*man-15*

[1] Time is given in minutes.

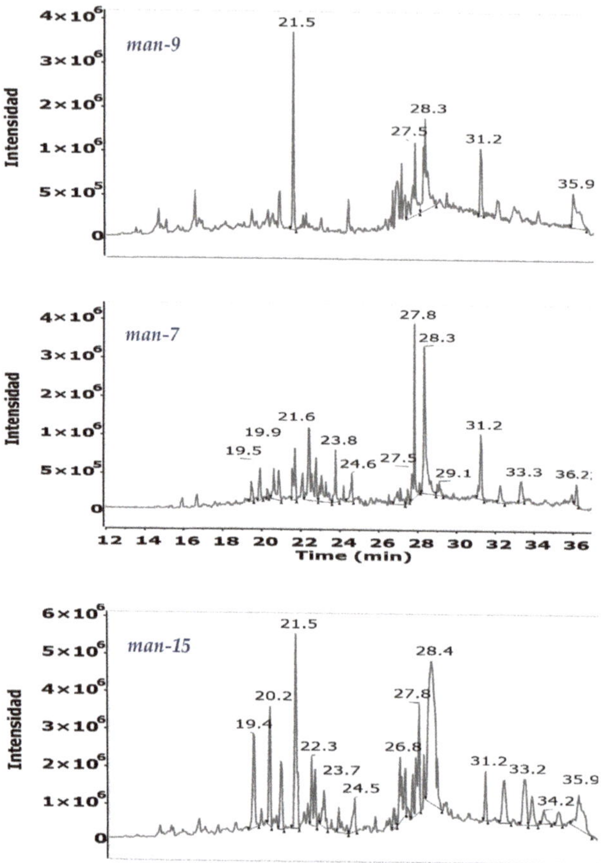

Figure 1. Total ion chromatograms (TICs) of extracts *man-9*, *man-7*, and *man-15*.

Chemical diversity of the three extracts from *P. mangiferae* (*man-7*, *man-9*, and *man-15*) was analyzed by LC–ESI–Q–TOF–MS. The total ion chromatograms for each sample are presented in Figure 1. The peak at t_R 21.5 min is common to all three analyzed samples. The MS spectrum of this peak showed ions $[M + H]^+$ and $[M + Na]^+$ at *m/z* 401.217 and 423.199, respectively. Additionally, two ion clusters were detected: M_2H^+ and M_2Na^+ at *m/z* 801.424 and 823.407, respectively (Figure 2A), which matched the MS data for mangiferaelactone, a previously characterized compound. This compound had a relatively lower concentration in sample *man-15*. Based on its selectivity against *P. aeruginosa* (9.5 mm inhibition zone), this compound could be proposed as the major component of the extract, for example, the peak at t_R 27.8 min with $[M + H]^+$, $[M + Na]^+$, and $[2M + Na]^+$ ions at *m/z* 391.283, 413.261, and 803.536, respectively (Figure 2D). The polar section of *man-15* was the most complex, indicating a higher level of chemical diversity then *man-7* and *man-9*.

Extract *man-7* showed a higher relative concentration among the components of the polar section (peaks at t_R 19.4, 20.2, 20.7, and 21.5 min). Its moderate polarity section included a peak at t_R 28.4 min with $[M + H]^+$,$[M + Na]^+$, and $[2M + Na]^+$ ions at *m/z* 419.308, 441.276, and 859.579, respectively (Figure 2B), compared with extracts *man-9* and *man-15* that had a peak at t_R 28.3 min with $[M + H]^+$, $[M + Na]^+$, and $[2M + H]^+$ ions at *m/z* 338.337, 360.319, and 675.670, respectively (Figure 2C).

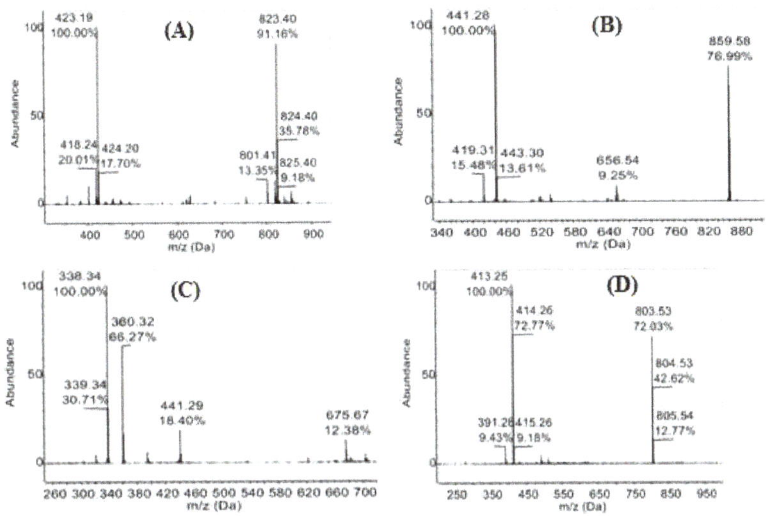

Figure 2. Mass spectra of selected peaks. (A) Mangiferaelactone at t_R 21.5 min in extracts *man-7*, *man-9*, and *man-15*. (B) Peak in extract *man-7* at t_R 28.4 min. (C) Peak in extracts *man-9* and *man-15* at t_R 28.3 min. (D) Peak in extracts *man-7* and *man-15* at t_R 27.8 min.

4. Discussion

As we mentioned above, there are only two reports related to the isolation and characterization of secondary metabolites from *P. mangiferae*; early culturing under two different conditions produced antibacterial compounds mangiferaelactone (**1**) and 4-(2,4,7-trioxa-bicyclo[4.1.0]heptan-3-yl) phenol (see Figure S1) [20,56]. Neither one was among the main components present in the extracts analyzed here by high resolution LC–MS, nor the other compounds with a low molecular weight.

Our results showed that in this initial study, *Pestalotiopsis* showed prolific antibacterial activity. The results of Table 2 indicate that each extract had a different biological activity profile. This means that the chemical composition changes in diversity and concentration.

The changes in the culture conditions played a role in differences in the chemical diversity of *P. mangiferae* and the genus *Pestalotiopsis*. This was confirmed by the wide range of the preliminary antibacterial activities determined for each of the extracts and through the LC–HRMS analysis of the three extracts. According to the revised reviews and recent publications (Table S2) on the secondary metabolites isolated from the genus *Pestalotiopsis* and the antibacterial activity previously determined, most of the molecular formulas for the metabolites reported here did not match with those reported earlier for the genus. Nonetheless, taking into account the previously isolated compounds from the genus *Pestalotiopsis* or from *P. mangiferae* and the molecular formula obtained through high-resolution mass spectrometry, we proposed some molecular structures; also, in most of the cases they are related with a previously isolated compound with antibacterial or antifungal activity.

Five major reviews on the chemistry and bioactivity of the genus *Pestalotiopsis* were published until 2017 [14–18]. An exhaustive exploration of the available compounds' structures allowed us to establish that the majority of the compounds present in the analyzed extracts of *P. mangiferae* have not been isolated from a species of the genus *Pestalotiopsis*. Nevertheless, it could be proposed that they belong to three of the main classes of compounds isolated from the genus, namely: (a) polyketides/polyols derivatives; (b) terpenoids/triterpenoids; and (c) nitrogen-containing compounds.

The major group of compounds present in the analyzed extracts could belong to polyketide/polyol derivatives. For example, we proposed the hydrolysis of mangiferaelactone (**1**) ($C_{20}H_{32}O_8$) that will result in the hypothetical trihydroxylactone (**1a**) (not-yet-detected by MS-analysis), its successive dehydration and methylation could lead to two lactones (**1b**) ($C_{16}H_{26}O_4$) and (**1d**) ($C_{17}H_{26}O_3$), respectively (Figure 3). Compound (**1b**) has the same molecular formula as koninginins B (**1e**; $C_{16}H_{26}O_4$) and E (**1f**; $C_{16}H_{26}O_4$) isolated from the genus *Trichoderma* [57–59], and they probably have the same precursor as compound (**1a**) ($C_{16}H_{28}O_5$).

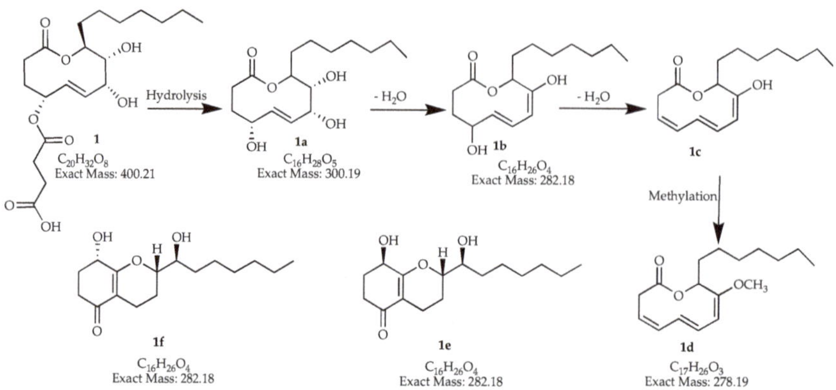

Figure 3. Proposed polyketide/polyol derivatives **1b,d** that could be present in extracts *man-7*, *man-9*, and *man-15*, having as a precursor compound **1** and the intermediates **1a**,**c**. Polyketides **1e**,**f** previously isolated from the genus *Trichoderma* with same molecular formula as **1b**.

Two molecular formulas could correspond to triterpenoids ($C_{30}H_{46}O_4$/ $C_{30}H_{48}O_5$). From the genus *Pestalotiopsis*, only oleanane- and ursane-type triterpenes have been isolated [14,17,60]. Ursane-type triterpenes have been reported when to the culture medium was added ursolic acid [61]. Nevertheless, oleanane-type were isolated (15α)-15- hydroxysoyasapogenol B (**2**), (7*β*, 15*α*)-7, 15-dihydroxysoyasapogenol B (**3**) and (7*β*)-7, 29-dihydroxysoyasapogenol B (**4**) from *Pestalotiopsis clavispora* [60], together with ursolic acid. These three compounds (**2–4**) could be synthesized by biological oxidation mechanism derived in one of the triterpenoids **2a**, **3a** or **4a**, respectively; they have the molecular formula $C_{30}H_{46}O_4$ or $C_{30}H_{48}O_5$ established by high resolution MS in extracts *man-7* and *man-15* (Figure 4).

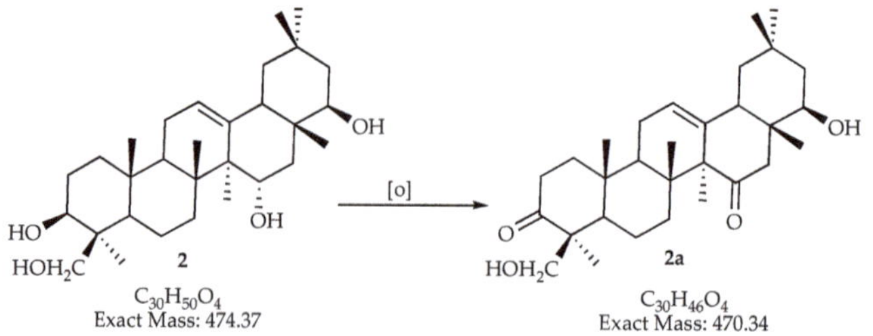

Figure 4. *Cont.*

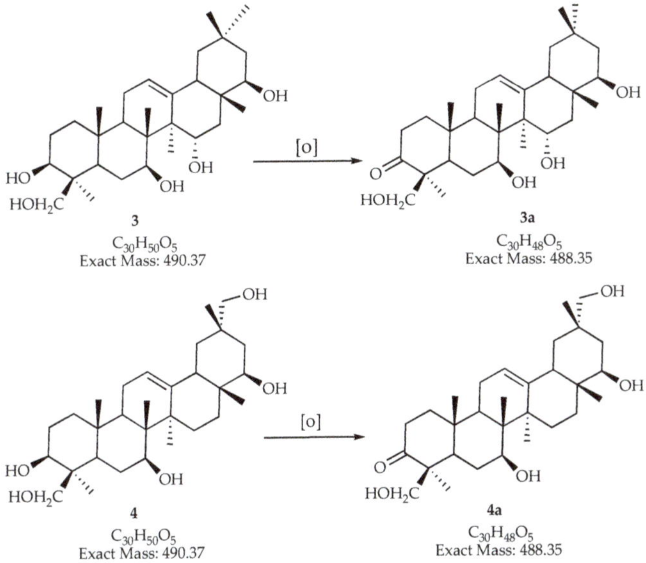

Figure 4. Proposed structure for compounds **2a**, **3a**, and **4a** based on the molecular formula determined by HRMS in extracts *man-7* and *man-15* and their proposed biosynthetic precursor.

In our results two molecular ions have the same molecular formula $C_{34}H_{52}O_8$, but they eluted at different time and are both present in extracts *man-7* and *man-15*. These isomers could be related to fusapirone (**5**; $C_{34}H_{54}O_9$) a compound with antifungal activity, previously, isolated from *Fusarium semitectum* [62], it possess multiple chiral centers and can derived into compound (**5a**; $C_{34}H_{52}O_8$) by dehydration (Figure 5).

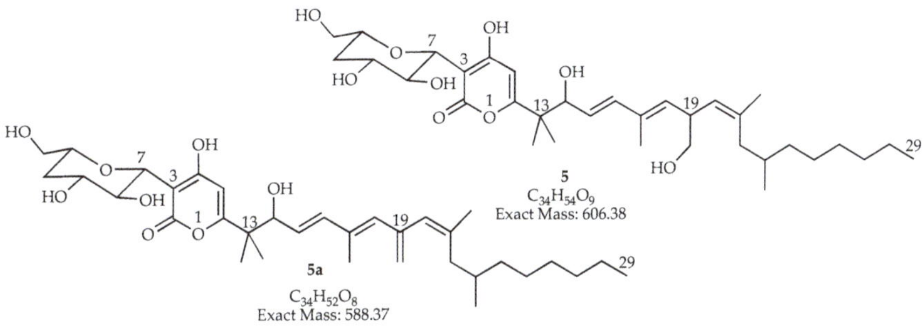

Figure 5. Dehydration of the polyketide derivative fusapirone **5** could produce compound **5a** with a molecular formula $C_{34}H_{57}O_8$, determined in extracts *man-7* and *man-15*.

The dehydrogenation of asperacine (**6**; $C_{40}H_{36}N_6O_4$) results in compound **6a** ($C_{40}H_{36}N_6O_4$) with an imine function (Figure 6), this molecular formula was determined for extracts *man-7*, *man-9*, and *man-15*.

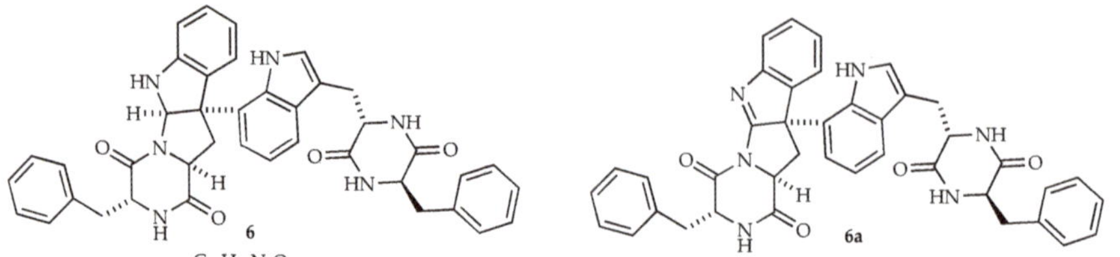

6
$C_{40}H_{36}N_6O_4$
Exact Mass: 664.28

6a
$C_{40}H_{34}N_6O_4$
Exact Mass: 662.26

Figure 6. Nitrogen derivative **6a**, that could be present in extracts *man-7*, *man-9*, and *man-15*.

5. Conclusions

For this study were selected two strains of *Pestalotiopsis* endophytic fungi (*P. microspora* Hd18 and *P. mangiferae*), that are capable of producing secondary metabolites with relevant biological activities. The strategy developed in this work included variations of the culture conditions, the determination of the antibacterial activity of the obtained extracts, together with the effective analysis of the chemical profile using LC-HRMS. This strategy could improve the discovery of new molecules with a pharmaceutical potential, in this case antibacterial. Hence, this work confirmed changes in the chemical diversity and biological activity of *P. microspora* Hd18 and, principally, *P. mangiferae* Hd08 under varying the culture conditions.

Taking into account the chemical diversity and the preliminary antibacterial activity displayed by *P. mangiferae,* further work will need to establish and confirm the chemical composition of each extract as well as the antibacterial activity of a single compound.

Supplementary Materials: The following are available online at http://www.mdpi.com/2309-608X/6/3/140/s1, Figure S1: Structure of compounds previously isolated from *P. mangiferae*. Table S1: Culture parameters and amount of organic extract produced by *Pestalotiopsis* spp., Table S2: Molecular ions and formulas of compounds isolated from the genus *Pestalotiopsis*

Author Contributions: M.M.A.-P., R.V., and N.R. designed and performed the experiments and analyzed the data. D.T.-M. analyzed the data and reviewed, edited, and wrote the paper. L.C.-R. designed the experiments, analyzed the data, and edited and wrote the paper. All of the authors have read and agreed to the published version of the manuscript.

Funding: This work was partially supported by the National Secretariat for Science, Technology and Innovation of Panama (SENACYT, grants COL10-060 and FID11-051), the Projects of Nagoya Protocols´ Application in Panama, and the National Research System of Panama (SNI).

Acknowledgments: We gratefully acknowledge the Government of Panama through the Ministerio de Ambiente (MiAMBIENTE) for granting the corresponding permits to collect the samples used in this study.

Conflicts of Interest: The authors declare no conflict of interest. The funders had no role in the design of the study; in the collection, analyses, or interpretation of data; in the writing of the manuscript; or in the decision to publish the results.

References

1. Escolà-Vergé, L.; Los-Arcos, I.; Almirante, B. New antibiotics for the treatment of infections by multidrug-resistant microorganisms. *Med. Clín. (Engl. Ed.)* **2020**, *154*, 351–357. [CrossRef]
2. Tacconelli, E.; Magrini, N. *Global Priority List of Antibiotic-Resistant Bacteria to Guide Research, Discovery, and Development of New Antibiotics*; World Health Organization: Geneva, Switzerland, 2017.
3. Mani Chandrika, K.V.S.; Sharma, S. Promising antifungal agents: A minireview. *Bioorg. Med. Chem.* **2020**, *28*, 115398. [CrossRef] [PubMed]
4. Outterson, K.; Rex, J.H. Evaluating for-profit public benefit corporations as an additional structure for antibiotic development and commercialization. *Transl. Res.* **2020**, *220*, 182–190. [CrossRef] [PubMed]
5. Singer, A.C.; Kirchhelle, C.; Roberts, A.P. (Inter)nationalising the antibiotic research and development pipeline. *Lancet Infect. Dis.* **2020**, *20*, e54–e62. [CrossRef]

6. Newman, D.J.; Cragg, G.M. Natural Products as Sources of New Drugs over the Nearly Four Decades from 01/1981 to 09/2019. *J. Nat. Prod.* **2020**, *83*, 770–803. [CrossRef]
7. Newman, D.J.; Cragg, G.M. Plant Endophytes and Epiphytes: Burgeoning Sources of Known and "Unknown" Cytotoxic and Antibiotic Agents? *Planta Med.* **2020**. [CrossRef]
8. Gupta, S.; Chaturvedi, P.; Kulkarni, M.G.; Van Staden, J. A critical review on exploiting the pharmaceutical potential of plant endophytic fungi. *Biotechnol. Adv.* **2020**, *39*, 107462. [CrossRef]
9. Hyde, K.D.; Xu, J.; Rapior, S.; Jeewon, R.; Lumyong, S.; Niego, A.G.T.; Abeywickrama, P.D.; Aluthmuhandiram, J.V.S.; Brahamanage, R.S.; Brooks, S.; et al. The amazing potential of fungi: 50 ways we can exploit fungi industrially. *Fungal Divers.* **2019**, *97*, 1–136. [CrossRef]
10. Sánchez-Fernández, R.E.; Sánchez-Ortiz, B.L.; Sandoval-Espinosa, Y.K.M.; Ulloa-Benítez, Á.; Armendáriz-Guillén, B.; García-Méndez, M.C.; Macías-Rubalcava, M.L. Hongos Endófitos: Fuente Potencial de Metabolitos Secundarios Bioactivos con Utilidad en Agricultura y Medicina. *Tip Rev. Espec. Cienc. Quím. Biol.* **2013**, *16*, 132–146. [CrossRef]
11. Pye, C.R.; Bertin, M.J.; Lokey, R.S.; Gerwick, W.H.; Linington, R.G. Retrospective analysis of natural products provides insights for future discovery trends. *Proc. Natl. Acad. Sci. USA* **2017**, *114*, 5601–5606. [CrossRef]
12. Torres-Mendoza, D.; Ortega, H.E.; Cubilla-Rios, L. Patents on Endophytic Fungi Related to Secondary Metabolites and Biotransformation Applications. *J. Fungi* **2020**, *6*, 58. [CrossRef]
13. Urones, J.G.; Marcos, I.S.; Diez, D.; Cubilla, L.R. Tricyclic diterpenes from Hyptys dilatata. *Phytochemistry* **1998**, *48*, 1035–1038. [CrossRef]
14. Yang, X.L.; Zhang, J.Z.; Luo, D.Q. The taxonomy, biology and chemistry of the fungal Pestalotiopsis genus. *Nat. Prod. Rep.* **2012**, *29*, 622–641. [CrossRef] [PubMed]
15. Wang, K.; Lei, J.; Wei, J.; Yao, N. Bioactive Natural Compounds from the Plant Endophytic Fungi Pestalotiopsis spp. *Mini Rev. Med. Chem.* **2012**, *12*, 1382–1393. [PubMed]
16. Kumar Deshmukh, S.; Prakash, V.; Ranjan, H. Recent advances in the discovery of bioactive metabolites from Pestalotiopsis. *Phytochem. Rev.* **2017**, *16*, 883–920. [CrossRef]
17. Xu, J.; Yang, X.; Lin, Q. Chemistry and biology of Pestalotiopsis-derived natural products. *Fungal Divers.* **2014**, *66*, 37–68. [CrossRef]
18. Xu, J.; Ebada, S.S.; Proksch, P. Pestalotiopsis a highly creative genus: Chemistry and bioactivity of secondary metabolites. *Fungal Divers.* **2010**, *44*, 15–31. [CrossRef]
19. Helaly, S.E.; Thongbai, B.; Stadler, M. Diversity of biologically active secondary metabolites from endophytic and saprotrophic fungi of the ascomycete order Xylariales. *Nat. Prod. Rep.* **2018**, *35*, 992–1014. [CrossRef]
20. Ortega, H.E.; Shen, Y.Y.; Tendyke, K.; Ríos, N.; Cubilla-Ríos, L. Polyhydroxylated macrolide isolated from the endophytic fungus Pestalotiopsis mangiferae. *Tetrahedron Lett.* **2014**, *55*, 2642–2645. [CrossRef]
21. Maram, L.; Das, B. The First Stereoselective Total Synthesis of Mangiferaelactone: An Antibacterial Fungal Nonanolide. *Synlett* **2014**, *25*, 2327–2330.
22. Kumar, R.; Rej, R.K.; Nanda, S. Asymmetric total synthesis of (-)-mangiferaelactone by using an appropriately substituted thiophene as a masked synthon for C-alkyl glycoside. *Tetrahedron Asymmetry* **2015**, *26*, 751–759. [CrossRef]
23. Reddy, B.V.S.; Reddy, P.S.; Babu, K.V.; Reddy, B.P.; Yadav, J.S. Stereoselective Total Synthesis of Mangiferaelactone using D-Mannose as a Chiral Pool. *Helv. Chim. Acta* **2015**, *98*, 1395–1402. [CrossRef]
24. Strobel, G.; Yang, X.; Sears, J.; Kramer, R.; Sidhu, R.S.; Hess, W.M. Taxol from Pestalotiopsis microspora, an endophytic fungus of Taxus wallachiana. *Microbiology* **1996**, *142*, 3–8. [CrossRef] [PubMed]
25. Strobel, G.; Ford, E.; Worapong, J.; Harper, J.K.; Arif, A.M.; Grant, D.M.; Fung, P.C.W.; Ming Wah Chau, R. Isopestacin, an isobenzofuranone from Pestalotiopsis microspora, possessing antifungal and antioxidant activities. *Phytochemistry* **2002**, *60*, 179–183. [CrossRef]
26. Liu, S.; Dai, H.; Makhloufi, G.; Heering, C.; Janiak, C.; Hartmann, R.; Mándi, A.; Kurtán, T.; Müller, W.E.G.; Kassack, M.U.; et al. Cytotoxic 14-Membered Macrolides from a Mangrove-Derived Endophytic Fungus, Pestalotiopsis microspora. *J. Nat. Prod.* **2016**, *79*, 2332–2340. [CrossRef]
27. Nalin Rathnayake, G.R.; Savitri Kumar, N.; Jayasinghe, L.; Araya, H.; Fujimoto, Y. Secondary Metabolites Produced by an Endophytic Fungus Pestalotiopsis microspora. *Nat. Prod. Bioprospect.* **2019**, *9*, 411–417. [CrossRef]

28. Bertrand, S.; Bohni, N.; Schnee, S.; Schumpp, O.; Gindro, K.; Wolfender, J.-L. Metabolite induction via microorganism co-culture: A potential way to enhance chemical diversity for drug discovery. *Biotechnol. Adv.* **2014**, *32*, 1180–1204. [CrossRef]
29. Bode, H.B.; Bethe, B.; Höfs, R.; Zeeck, A. Big Effects from Small Changes: Possible Ways to Explore Nature's Chemical Diversity. *Chembiochem* **2002**, *3*, 619–627. [CrossRef]
30. Takahashi, J.A.; Campos Teles, A.P.; De Almeida Pinto Bracarense, A.; Corrêia Gomes, D. Classical and epigenetic approaches to metabolite diversification in filamentous fungi. *Phytochem. Rev.* **2013**, *12*, 773–789. [CrossRef]
31. Frey-Klett, P.; Burlinson, P.; Deveau, A.; Barret, M.; Tarkka, M.; Sarniguet, A. Bacterial-Fungal Interactions: Hyphens between Agricultural, Clinical, Environmental, and Food Microbiologists. *Microbiol. Mol. Biol. Rev.* **2011**, *75*, 583–609. [CrossRef]
32. Pettit, R.K. Mixed fermentation for natural product drug discovery. *Appl. Microbiol. Biotechnol.* **2009**, *83*, 19–25. [CrossRef]
33. Shank, E.A.; Kolter, R. New developments in microbial interspecies signaling. *Curr. Opin. Microbiol.* **2009**, *12*, 205–214. [CrossRef] [PubMed]
34. Adelin, E.; Slimani, N.; Cortial, S.; Schmitz-Alfonso, I.; Ouazzani, J. Platotex: An innovative and fully automated device for cell growth scale-up of agar-supported solid-state fermentation. *J. Ind. Microbiol. Biotechnol.* **2011**, *38*, 299–305. [CrossRef] [PubMed]
35. Kjærbølling, I.; Mortensen, U.H.; Vesth, T.; Andersen, M.R. Strategies to establish the link between biosynthetic gene clusters and secondary metabolites. *Fungal Genet. Biol.* **2019**, *130*, 107–121. [CrossRef] [PubMed]
36. Ramesha, K.P.; Mohana, N.C.; Nuthan, B.R.; Rakshith, D.; Satish, S. Epigenetic modulations of mycoendophytes for novel bioactive molecules. *Biocatal. Agric. Biotechnol.* **2018**, *16*, 663–668. [CrossRef]
37. Chaichanan, J.; Wiyakrutta, S.; Pongtharangkul, T.; Isarangkul, D.; Meevootisom, V. Optimization of zofimarin production by an endophytic fungus, Xylaria sp. Acra L38. *Braz. J. Microbiol.* **2014**, *45*, 287–293. [CrossRef] [PubMed]
38. Pacheco Fill, T.; Pallini, H.F.; da Silva Amaral, L.; da Silva, J.V.; Lazarin Bidóia, D.; Peron, F.; Pelegrin Garcia, F.; Vataru Nakamura, C.; Rodrigues-Filho, E. Copper and Manganese Cations Alter Secondary Metabolism in the Fungus Penicillium brasilianum. *J. Braz. Chem. Soc.* **2016**, *27*, 1444–1451.
39. Fillat, Ú.; Martín-Sampedro, R.; Macaya-Sanz, D.; Martín, J.A.; Ibarra, D.; Martínez, M.J.; Eugenio, M.E. Screening of eucalyptus wood endophytes for laccase activity. *Process Biochem.* **2016**, *51*, 589–598. [CrossRef]
40. Goutam, J.; Sharma, V.K.; Verma, S.K.; Singh, D.K.; Kumar, J.; Mishra, A.; Kumar, A.; Kharwar, R.N. Optimization of Culture Conditions for Enhanced Production of Bioactive Metabolites Rich in Antimicrobial and Antioxidant Activities Isolated from Emericella quadrilineata an Endophyte of Pteris pellucida. *J. Pure Appl. Microbiol.* **2014**, *8*, 2059–2073.
41. Li, P.; Xu, L.; Mou, Y.; Shan, T.; Mao, Z.; Lu, S.; Peng, Y.; Zhou, L. Medium optimization for exopolysaccharide production in liquid culture of endophytic fungus Berkleasmium sp. Dzf12. *Int. J. Mol. Sci.* **2012**, *13*, 11411–11426. [CrossRef]
42. Mou, Y.; Luo, H.; Mao, Z.; Shan, T.; Sun, W.; Zhou, K.; Zhou, L. Enhancement of palmarumycins C12 and C13 production in liquid culture of endophytic fungus Berkleasmium sp. Dzf12 after treatments with metal ions. *Int. J. Mol. Sci.* **2013**, *14*, 979–998. [CrossRef] [PubMed]
43. Nomila Merlin, J.; Nimal Christhudas, I.V.S.; Praveen Kumar, P.; Agastian, P. Optimization of growth and bioactive metabolite production: Fusarium solani. *Asian J. Pharm. Clin. Res.* **2013**, *6*, 98–103.
44. Panuthai, T.; Sihanonth, P.; Piapukiew, J.; Sooksai, S.; Sangvanich, P.; Karnchanatat, A. An extracellular lipase from the endophytic fungi Fusarium oxysporum isolated from the Thai medicinal plant, Croton oblongifolius Roxb. *Afr. J. Microbiol. Res.* **2012**, *6*, 2622–2638.
45. Sorgatto, M.; Guimaraes, N.C.A.; Zanoelo, F.F.; Marques, M.R.; Peixoto-Nogueira, S.C.; Giannesi, G.G. Purification and characterization of an extracellular xylanase produced by the endophytic fungus, Aspergillus terreus, grown in submerged fermentation. *Afr. J. Biotechnol.* **2012**, *11*, 8076–8084. [CrossRef]
46. Zhao, X.-M.; Wang, Z.-Q.; Shu, S.-H.; Wang, W.-J.; Xu, H.-J.; Ahn, Y.-J.; Wang, M.; Hu, X. Ethanol and Methanol Can Improve Huperzine a Production from Endophytic Colletotrichum gloeosporioides ES026. *PLoS ONE* **2013**, *8*, e61777. [CrossRef] [PubMed]

47. Zhao, J.; Wang, X.; Sun, W.; Mou, Y.; Peng, Y.; Zhou, L. Medium optimization for palmarumycin C13 production in liquid culture of endophytic fungus Berkleasmium sp. Dzf12 using response surface methodology. *Electron. J. Biotechnol.* **2013**, *16*, 16. [CrossRef]
48. Kaeberlein, T.; Lewis, K.; Epstein, S.S. Isolating "Uncultivable" Microorganisms in Pure Culture in a Simulated Natural Environment. *Science* **2002**, *296*, 1127–1129. [CrossRef]
49. Wohlleben, W.; Mast, Y.; Stegmann, E.; Ziemert, N. Antibiotic drug discovery. *Microb. Biotechnol.* **2016**, *9*, 541–548. [CrossRef]
50. Netzker, T.; Fischer, J.; Weber, J.; Mattern, D.J.; König, C.C.; Valiante, V.; Schroeckh, V.; Brakhage, A.A. Microbial communication leading to the activation of silent fungal secondary metabolite gene clusters. *Front. Microbiol.* **2015**, *6*, 1–13. [CrossRef]
51. Molinar, E.; Rios, N.; Spadafora, C.; Elizabeth Arnold, A.; Coley, P.D.; Kursar, T.A.; Gerwick, W.H.; Cubilla-Rios, L. Coibanoles, a new class of meroterpenoids produced by Pycnoporus sanguineus. *Tetrahedron Lett.* **2012**, *53*, 919–922. [CrossRef]
52. U'Ren, J.M.; Lutzoni, F.; Miadlikowska, J.; Laetsch, A.D.; Elizabeth Arnold, A. Host and geographic structure of endophytic and endolichenic fungi at a continental scale. *Am. J. Bot.* **2012**, *99*, 898–914. [CrossRef] [PubMed]
53. Andrews, J.M. BSAC standardized disc susceptibility testing method (version 8). *J. Antimicrob. Chemother.* **2009**, *64*, 454–489. [CrossRef] [PubMed]
54. Vásquez, R.; Rios, N.; Solano, G.; Cubilla-Rios, L. Lentinoids A–D, New Natural Products Isolated from Lentinus strigellus. *Molecules* **2018**, *23*, 773. [CrossRef] [PubMed]
55. Jorgensen, J.H.; Turnidge, J.D. Susceptibility Test Methods Dilution and Disk Diffusion Methods. In *Manual of Clinical Microbiology*, 11th ed.; Jorgensen, J., Pfaller, M., Carrol, K., Funke, G., Landry, M., Richter, S., Warnock, D., Eds.; ASM Press: Washington, DC, USA, 2015; pp. 1253–1273.
56. Subban, K.; Subramani, R.; Johnpaul, M. A novel antibacterial and antifungal phenolic compound from the endophytic fungus Pestalotiopsis mangiferae. *Nat. Prod. Res.* **2013**, *27*, 1445–1449. [CrossRef]
57. Cutler, H.G.; Himmelsbach, D.S.; Jacyno, J.M.; Cole, P.D.; Yagen, B.; Arrendale, R.F.; Cox, R.H. Koninginin B: A Biologically Active Congener of Koninginin A from Trichoderma koningii. *J. Agric. Food Chem.* **1991**, *39*, 977–980. [CrossRef]
58. Ghisalberti, E.L.; Rowland, C.Y. Antifungal Metabolites from Trichoderma harzianum. *J. Nat. Prod.* **1993**, *56*, 1799–1804. [CrossRef]
59. Parker, S.R.; Cutler, H.G.; Schreiner, P.R. Koninginin E: Isolation of a Biologically Active Natural Product from *Trichoderma koningii*. *Biosci. Biotechnol. Biochem.* **1995**, *59*, 1747–1749. [CrossRef]
60. Luo, D.-Q.; Deng, H.-Y.; Yang, X.-L.; Shi, B.-Z.; Zhang, J.-Z. Oleanane-Type Triterpenoids from the Endophytic Fungus Pestalotiopsis clavispora Isolated from the Chinese Mangrove Plant Bruguiera sexangula. *Helv. Chim. Acta* **2011**, *94*, 1041–1047. [CrossRef]
61. Fu, S.B.; Yang, J.S.; Cui, J.L.; Meng, Q.F.; Feng, X.; Sun, D.A. Multihydroxylation of ursolic acid by Pestalotiopsis microspora isolated from the medicinal plant Huperzia serrata. *Fitoterapia* **2011**, *82*, 1057–1061. [CrossRef]
62. Evidente, A.; Amalfitano, C.; Pengue, R.; Altomare, C. High performance liquid chromatography for the analysis of fusapyrone and deoxyfusapyrone, two antifungal α-pyrones from Fusarium semitectum. *Nat. Toxins* **1999**, *7*, 133–137. [CrossRef]

Journal of *Fungi*

MDPI

Article

Interactive Impact of Arbuscular Mycorrhizal Fungi and Elevated CO_2 on Growth and Functional Food Value of *Thymus vulgare*

Talaat H. Habeeb [1], Mohamed Abdel-Mawgoud [2], Ramy S. Yehia [3,4], Ahmed Mohamed Ali Khalil [5], Ahmed M. Saleh [4,*] and Hamada AbdElgawad [6,*]

1 Biology Department, Faculty of Science at Yanbu, Taibah University, King Khalid Rd., Al Amoedi, Yanbu El-Bahr 46423, Saudi Arabia; thabeeb@yahoo.com

2 Department of Medicinal and Aromatic Plants, Desert Research Centre, Cairo 11753, Egypt; Mohamed_drc@yahoo.com

3 Department of Biological Sciences, College of Science, King Faisal University, Al-Ahsa 31982, Saudi Arabia; drramy4@hotmail.com

4 Department of Botany and Microbiology, Faculty of Science, Cairo University, Giza 12613, Egypt

5 Botany and Microbiology Department, Faculty of Science, Al-Azhar University, Cairo 13759, Egypt; khalilahmed_1980@hotmail.com

6 Botany and Microbiology Department, Faculty of Science, Beni-Suef University, Beni-Suef 62511, Egypt

* Correspondence: asaleh@sci.cu.edu.eg (A.M.S.); hamada.abdelgawad@uantwerpen.be (H.A.)

Received: 25 July 2020; Accepted: 4 September 2020; Published: 9 September 2020

Abstract: Arbuscular mycorrhizal fungi (AMF) and elevated CO_2 (eCO_2) have been effectively integrated to the agricultural procedures as an ecofriendly approach to support the production and quality of plants. However, less attention has been given to the synchronous application of AMF and eCO_2 and how that could affect the global plant metabolism. This study was conducted to investigate the effects of AMF and eCO_2, individually or in combination, on growth, photosynthesis, metabolism and the functional food value of *Thymus vulgare*. Results revealed that both AMF and eCO_2 treatments improved the photosynthesis and biomass production, however much more positive impact was obtained by their synchronous application. Moreover, the levels of the majority of the detected sugars, organic acids, amino acids, unsaturated fatty acids, volatile compounds, phenolic acids and flavonoids were further improved as a result of the synergistic action of AMF and eCO_2, as compared to the individual treatments. Overall, this study clearly shows that co-application of AMF and eCO_2 induces a synergistic biofertilization impact and enhances the functional food value of *T. vulgare* by affecting its global metabolism.

Keywords: mycorrhizae; elevated CO_2; *Thymus vulgare*; growth; photosynthesis; metabolites; biological activity

1. Introduction

Herbal plants have been widely used in traditional and folk medicine as an effectual solution to cure many diseases, being a big store for bioactive compounds, especially secondary metabolites [1]. They are known to have various biological activities such as antioxidant, antimicrobial, anti-inflammatory and anticancer properties [2]. Recently, a priority was given to herbal plants in terms of enhancing the production of the economically important phytochemicals through the application of cultivation procedures under stimulated growth conditions [3]. In this aspect, arbuscular mycorrhizal fungi (AMF) have been regarded as one of the most important beneficial microorganisms that are able to associate with almost two thirds of terrestrial plants improving their growth and stress tolerance [4].

In some cases, mycorrhizal symbiosis is essential as the host plant cannot grow normally and/or survive without it [5]. The beneficial effects of AMF symbiotic association with plants include enhanced levels of mineral nutrients and accumulation of primary and secondary metabolites [6]. From environmental point of view, AMF can keep the balance of soil aggregates, hence, able to fight erosion [7]. Accordingly, AMF represent a promising trend that has found its way in the sustainable agricultural productivity [8]. For instance, utilization of AMF for enhancing the production and quality of aromatic plants have been reported [8]. In this regard, several medicinal aromatic plants, such as pennyroyal and parsley, showed enhanced levels of bioactive metabolites when associated with AMF [6].

In another aspect, the exposure of plants to elevated CO_2 (eCO_2) has been regarded as an effective approach to improve the nutritional and medicinal values of herbal plants [9]. eCO_2 can increase plant growth and productivity either directly by enhancing photosynthesis [10] or indirectly by stimulating plant water use efficiency [11]. On the other hand, the effect of eCO_2 on belowground communities, including AMF, are still not fully understood [12]. What is well known is that the higher the photosynthetic activity, under eCO_2, the more the photosynthate transfer to plant roots and the higher release to the associated microbial communities [13]. Furthermore, as being dependent on their host plant for carbon, AMF may be sensitive to global climatic changes that influence their host [14]. Therefore, such triple effect resulting from interactions among plant, AMF and eCO_2 is expected to have beneficial roles in increasing the productivity and quality of crops and medicinal plants.

One of the well-known plants for both culinary and medicinal purposes is *Thymus vulgaris* L., a member of the family Lamiaceae, being widely used in folk medicine for treatment of several diseases like bedwetting, diarrhea, stomach ache, arthritis, sore throat, cough, bronchitis and chest congestion [15]. The biological activities of *T. vulgaris* are mainly ascribed to its content of secondary metabolites, particularly essential oils that have been extensively studied for antioxidant, antimicrobial and antitumor activities [16]. Thus, improving the accumulation of these phytochemicals in *T. vulgaris* could support its nutritional, medicinal and pharmacological properties. In this regard, previous studies have reported the positive impacts of both AMF and eCO_2 on the growth and quality of herbal plant [17,18], however, the complete picture on how AMF-eCO_2 combination affect primary and secondary metabolomes is not fully drawn [19,20]. In addition, metabolic profiling of the host plant is essential to understand the mechanisms behind the changes occurring in response to the individual and/or combined effect of AMF and eCO_2. So far, the detailed metabolic implications induced by the synchronous application of eCO_2 and AMF on plants are not investigated. Thus, the current study was conducted to explore, in details, the individual and combined impacts of eCO_2 and AMF on *T. vulgaris*, as a model herbal plant. We have assessed the changes in mycorrhizal colonization, plant biomass production, photosynthesis, respiration and levels of individual primary (sugars, amino acids, fatty acids and organic acids), secondary (phenolic acids and flavonoids) metabolites and volatile oils. Further, the associated changes in nutritional and medicinal values of *T. vulgaris* were investigated.

2. Material and Method

2.1. Experimental Setup, Plant Materials and Growth Conditions

Soil potting was mixed with sterilized sand (1:3) and inculcated with a pure commercial inoculum of *Rhizophagus irregularis* (MUCL 41,833 obtained from Glomeromycota in vitro collection (GINCO)) at a concentration of 50 spores per soil in a pot (25 × 15 cm). The control treatments were represented by non-inoculated soil. The seeds of *T. vulgaris* were disinfected then sown in both treated and non-treated soils. Plants were grown in a controlled greenhouse at 21/18 °C, 16/8 h day/night, and 60% humidity, they were regularly watered. The pots inn each of the control and AMF-inoculated groups were equally subdivided into two sub-groups, one subjected to 410 ppm CO_2 (ambient CO_2; aCO_2) and the other subjected to 620 ppm CO_2 (elevated CO_2; eCO_2,) through the time course of the experiment. The plants were harvested after 6 weeks, then the aerial parts were immediately frozen in liquid nitrogen and stored at −20 °C to be used in different plant analyses. For determination of dry matter and mineral elements, plant shoots were washed with distilled water and dried at 75 °C for 72 h.

2.2. Mycorrhizal Parameters

Mycorrhizal colonization was demonstrated following Phillips and Hayman [21]. About 0.5 g of fresh roots were clarified with potassium hydroxide (10% *w*/*v*) and potassium hydroxide (10%) + hydrogen peroxide (10% *v*/*v*) in a ratio of 1:1 (*v*/*v*), then stained with 0.05% trypan blue in lactoglycerol. A stereomicroscope (40×) was used to show the stained roots, while the colonization rate was calculated by using gridline intersect method [22].

2.3. Photosynthesis Parameters

The light-saturated photosynthetic rates (μmol CO_2 m^{-2} s^{-1}) of mature leaves were measured (LI-COR LI- 6400, LI-COR Inc., Lincoln, NE, USA), according to AbdElgawad et al. [23]. Dark respiration was determined as the absolute CO_2 exchange rate determined at photosynthetic photon flux density (μmolm^{-2} s^{-1}).

2.4. Metabolic Profiling

For extraction of sugars, plant tissues were homogenized in 50% (*v*/*v*) acetonitrile. The method described by Hamad et al. [24] was applied to identify the individual sugars in the plant extract by using high-performance liquid chromatography (HPLC), then comparing their retention time with those of a standard mixture. Quantification of the sugar samples was achieved based on peak area comparison with a calibration curve of the corresponding standards. Organic acids were extracted in phosphoric acid (0.1% *v*/*v*) supplemented with butylated hydroxyanisole (3 g/L) and then analyzed using HPLC with a SUPELCOGEL C-610H column coupled to a UV detection system set at 210 nm (LaChromL-7455 diode array, LaChrom, Tokyo, Japan). The concentration of each organic acid was calculated by using a calibration curve [24]. For extraction of amino acids, a known weight of plant tissues was vigorously homogenized in 80% aqueous ethanol. Amino acids were measured using a Waters Acquity UPLC-tqd system (Milford, Worcester County, MA, USA) equipped with a BEH amide 2.1 × 50 column. The lipophilic fraction of plant samples was obtained by extraction in chloroform/methanol (2:1, *v*/*v*). Thereafter, fatty acids were detected, according to Hassan et al. [25], by using GC/MS analysis (Hewlett Packard, Palo Alto, CA, USA) with an HP-5 MS column (30 m × 0.25 mm × 0.25 mm). Fatty acids were quantified using NIST 05 database and Golm Metabolome Database (http://gmd.mpimp-golm.mpg.de). Phenolic acids and flavonoids were extracted in acetone-water solution (4:1 *v*/*v*) for 24 h. The method outlined in Hamad et al. [24] was followed up for determination of Phenolic acids and flavonoids using an HPLC system (SCL-10A vp, Shimadzu Corporation, Kyoto, Japan). The concentration of each compound was calculated with a calibration curve of the corresponding standard. For extraction of volatile oils, two hundred gm of fresh plant material were subjected to steam distillation with about 500 mL of water, where the volatiles were collected [26]. The levels of volatiles were determined using gas chromatography–mass spectrometry (GC–MS) according to the method outlined by El Hattab et al. [27].

2.5. Determination of Biological Activities

Several methods were used to determine the total antioxidant capacities of the plant extract, including the ferric reducing antioxidant power (FRAP), oxygen radical absorbance capacity (ORAC), inhibition of LDL (low density lipoprotein) oxidation (TBARS and conjugated dienes) and inhibition of hemolysis assays [23,24]. For LDL oxidation, dialyzed LDL (100 μg protein/mL) was diluted in 10 mM PBS (phosphate buffered saline containing 0.01 Mphosphate-buffer and 0.15 M NaCl, pH 7.4) and incubated at 37 °C in presence or absence of 10 μM $CuSO_4$. Oxidation was performed with or without the sample solution of colostrum proteins. After incubation, lipid peroxidation of the LDL was measured. Thiobarbituric acid reactive substances (TBARS) was determined at 532 nm/600 nm, using 1,1,3,3-Tetramethoxypropane as standard for calibration curve, while conjugated diene formation was measured at 232 nm of LDL solution (100 µg protein/mL) in PBS incubated with $CuSO_4$ (10 µM) in the absence or presence of various concentrations of bovine colostrums protein [28].

2.6. Statistical Analysis

Experiments were carried out following a randomized complete block design. Data normality and the homogeneity of variances were checked using the Kolmogorove–Smirnov test and Levene's test, respectively. Each experiment was done in five replicates (n = 5). All the data was subjected to one-way analysis of variance (ANOVA). Student's *t*-test at probability levels of 0.05, 0.01 or 0.001 was used to test the difference between the treatment and control or between AMF alone and the combined AMF+eCO_2 treatment. All statistical tests were performed using the computer program PASW statistics 18.0 (SPSS Inc., Chicago, IL, USA).

3. Results and Discussion

3.1. AMF Colonization and Hyphal Growth

It is known that AMF are largely dependent on their host plant for carbon, so they are sensitive to climatic changes that affect their host plant [14]. In this sense, eCO_2 could have an indirect effect on mycorrhizal colonization by promoting carbon assimilation and allocation to roots [29]. Since AMF are attached to plant roots, they are lucky to receive higher amount of photosynthates under eCO_2 before other soil microbes [30]. Herein, the mycorrhizal growth was significantly enhanced in *T. vulgare* by AMF treatments (Table 1). Such mycorrhizal proliferation was much more stimulated under eCO_2 conditions. Several studies have demonstrated some positive effects for eCO_2 on AMF-plant association such as increased mycorrhizal root length [14] and increased extra-radical hyphae [31]. However, other studies did not show any beneficial effects for eCO_2 on AMF growth in host plants [32,33]. Therefore, the impact of eCO_2 levels on mycorrhizal growth seems to be dependent on plant species, AMF species and soil type [34]. In fact, as being a member of *Glomeraceae*, the ratio of *R. irregularis* has been reported to be more positively influenced by eCO_2 than others, e.g., *Gigasporaceae* [35].

Table 1. Mycorrhizal colonization and growth parameters in roots of *Thymus vulgare* grown under normal conditions (control) or the effect of eCO_2 (620 ppm), arbuscular mycorrhizal fungi (AMF) or their combination (AMF-eCO_2). Values are mean ± standard error of five independent replicates. Asterisks indicate significant changes (*** $p < 0.001$) between AMF alone and the combined AMF+eCO_2 treatment.

Metabolite	Control	eCO_2	AMF	AMF + eCO_2
Colonization (% root)	nd	nd	33.06 ± 2.39	54.04 ± 1.11 ***
Hyphal length (cm g^{-1} soil)	nd	nd	12.94 ± 5.82	19.10 ± 9.74 b ***
Number of arbuscules (no. cm^{-1} root)	nd	nd	4.78±0.27	5.03 ± 0.18

nd = not detected.

3.2. AMF and eCO_2 Acts Synergistically to Improve Photosynthetic Capacity and Biomass Production

It has been known that the photosynthetic rate, and consequently biomass production, could be improved under the effect of AMF inoculation, as a result of the expected increased nutrients uptake [36], and also under eCO_2 atmosphere due to the enhancement of the carboxylation reaction of rubisco [37]. Supporting this hypothesis, the current results revealed that eCO_2 and AMF independently, and to more extent in combination, promoted photosynthetic rate and biomass production in *T. vulgare* (Figure 1). Such increments were much more induced by the interaction between both treatments.

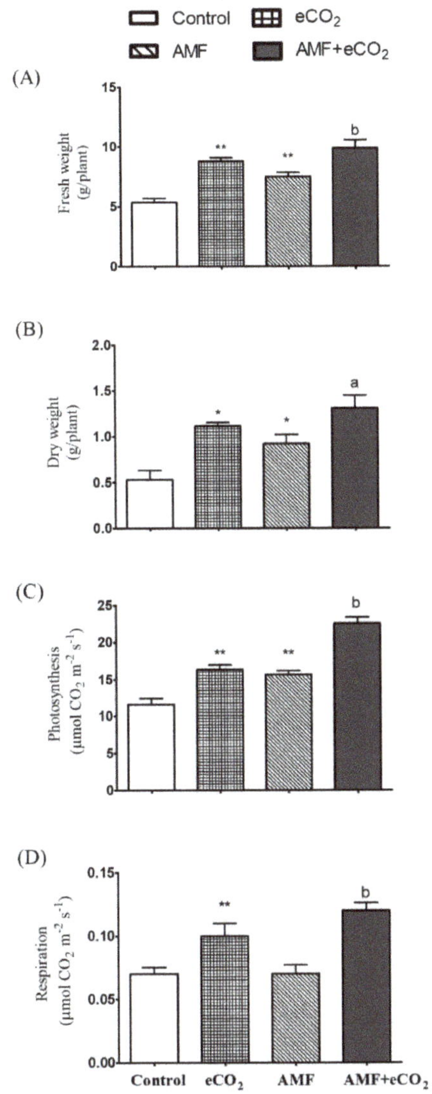

Figure 1. Fresh mass (**A**), dry mass (**B**), and rates of photosynthesis (**C**) and respiration (**D**) in *Thymus vulgare* grown under normal conditions (control) or the effect of eCO_2 (620 ppm), AMF or their combination (AMF-eCO_2). Values are mean ± standard error of five independent replicates. Asterisks indicate significant changes (* $p < 0.05$; ** $p < 0.01$) compared to control, as revealed by the Student's *t*-test. Lowercase letters indicate significant differences (^{a} $p < 0.05$; ^{b} $p < 0.01$) between AMF alone and the combined AMF+eCO_2 treatment.

Similar to our results, the positive effects of eCO_2 on biomass of *T. vulgaris* and some other medicinal plants, *Ocimum basilicum*, *Origanum vulgare*, *Mentha piperita* and *Mentha spicata*, have been previously investigated [38]. Moreover, the increased biomass production in plants inoculated with AMF was reported [39]. Regarding the interaction between eCO_2 and AMF, It is well known that eCO_2 stimulates the photosynthetic rate and plant growth [40], which in turn, affects the allocation of photosynthates to AMF, consequently makes more C available to AMF colonizing the roots [13], thus, increasing sink strength in mycorrhizal plants, and eventually this leads to increased C storage in soils [41]. Such effect is hypothesized to create a balance between carbon cost and nutrient benefits, besides reducing the negative effects of down regulation of photosynthesis caused by acclimation of plants to long-term exposure to eCO_2 [42]. In this regard, it has been found that both eCO_2 and AMF, when applied individually or in combination, improved biomass production of *Pisum sativum* and lettuce [32,43]. In addition, it has been indicated that mycorrhizal plants have higher photosynthetic rate [33] when grown under high CO_2 levels. However, such an effect might differ among variable cultivars [44]. In contrast, it was supposed that eCO_2 may impair the beneficial effects of AMF on plant biomass, especially when the fungal community is dominated by *Glomus* species [45]. This might be due to the difference among AMF taxa in their exchange of carbon and nutrients [46].

3.3. Application of AMF and eCO_2 Improves the Nutritional Value of *T. vulgare*

It was assumed that the nutritive value of plants is highly related to its content of primary metabolites, e.g., sugars, proteins and lipids [9]. In this regard, sugars and organic acids are related to taste and flavor [47], essential amino acids are involved in some biological processes, such as protein synthesis [48] and a lower saturated/unsaturated fatty acids (SFA/USFA) ratio is linked to cardio-protective effects [49]. It has been reported that the higher the CO2 levels, the higher the rate of photosynthetic activity, which is linked to the enhancement of the carboxylation reaction of rubisco, the enzyme responsible for CO_2 fixation [37]. As a consequence of photosynthesis improvement, sugars could be accumulated and also broken down via dark respiration, resulting in production of the precursors necessary for synthesis of different classes of primary and secondary metabolites [50]. Supporting such a concept, the individual AMF and eCO_2 treatments induced significant increases in the content of total soluble sugar of *T. vulgare,* about 1.6 folds, however, starch was significantly accumulated under eCO_2 only (Table 2, Figure 2). Further, the synchronous application of AMF and eCO_2 caused a significant accumulation in the levels of the majority of the measured sugars relative to AMF alone treatment. Similarly, CO_2 enrichment enhanced the accumulation of sucrose and starch in oil palm [51], and increased the accumulation of total soluble carbohydrates and starch in ginger varieties [52]. AMF treatments were found to induce the accumulation of total soluble sugars in lettuce [53]. Further, the interaction between AMF and CO_2 improved forage quality of alfalfa plants by increasing the levels of glucose, fructose and hemicellulose and decreasing that of lignin [54].

Table 2. Levels of primary metabolites (mg g^{-1} dry weight) in *Thymus vulgare* grown under normal conditions (control) or the effect of eCO_2 (620 ppm), AMF or their combination (AMF-eCO_2). Values are mean ± standard error of five independent replicates. Asterisks indicate significant changes (* $p < 0.05$; ** $p < 0.01$) compared to control, as revealed by the Student's t-test. Lowercase letters indicate significant differences (^{a} $p < 0.05$; ^{b} $p < 0.01$) between AMF alone and the combined AMF+eCO_2 treatment.

	Control	eCO_2	AMF	AMF + eCO_2
Sugars				
Glucose	1.34 ± 0.07	2.18 ± 0.12 **	1.32 ± 0.07	2.14 ± 0.01 ^{b}
Fructose	0.34 ± 0.05	0.45 ± 0.02	0.4 ± 0.02	0.76 ± 0.06 ^{b}
Sucrose	1.67 ± 0.15	2.07 ± 0.15	2.19 ± 0.04 *	2.85 ± 0.1 ^{b}
Soluble sugars	6.09 ± 0.31	10.29 ± 0.3 **	9.46 ± 0.35 **	11.44 ± 0.8
Starch	62.06 ± 1.89	79.26 ± 4.63 *	60.1 ± 1.7	85.22 ± 1.68 ^{b}
Total carbohydrates	105.38 ± 3.08	137.1 ± 6.75 *	105.62 ± 6.31	147.79 ± 2.01 ^{b}
Organic acids				
Oxalic	3.84 ± 0.33	2.97 ± 0.06	5.7 ± 0.34 *	7.75 ± 0.29 ^{b}
Malic	6.88 ± 0.3	6.25 ± 0.34	6.36 ± 0.29	7.38 ± 0.36
Isobutyric	3.46 ± 0.33	2.89 ± 0.46	6.34 ± 0.67 *	7.67 ± 0.53
Fumaric	0.93 ± 0.01	0.96 ± 0.08	0.67 ± 0.04 **	1.14 ± 0.07 ^{b}
Succinic	3.07 ± 0.33	4.21 ± 0.29	5.08 ± 0.39 *	6.31 ± 0.58
Citric	2.88 ± 0.33	5.32 ± 0.28 **	4.17 ± 0.54	5.9 ± 0.2 ^{a}

Table 2. *Cont.*

	Control	eCO_2	AMF	AMF + eCO_2
Essential amino acids (EAAs)				
Histidine	2.49 ± 0.19	2.75 ± 0.3	3.19 ± 0.23	2.95 ± 0.38
Isoleucine	0.15 ± 0.02	0.26 ± 0.03 *	0.28 ± 0.03 *	0.34 ± 0.01
Leucine	0.02 ± 0	0.03 ± 0	0.03 ± 0 *	0.04 ± 0
Lysine	4.2 ± 0.21	4.55 ± 0.29	4.97 ± 0.25	6.67 ± 0.5 [a]
Methionine	0.02 ± 0	0.02 ± 0 *	0.03 ± 0 *	0.03 ± 0
Phenylalanine	0.33 ± 0.04	0.68 ± 0.03 **	0.76 ± 0.1 *	1.31 ± 0.08 [a]
Valine	0.48 ± 0.05	0.58 ± 0.06	0.78 ± 0.08 *	1.72 ± 0.07 [b]
Threonine	0.11 ± 0.01	0.16 ± 0.02	0.19 ± 0.02 *	0.21 ± 0.01
Arginine	1.86 ± 0.13	2.35 ± 0.25	3.14 ± 0.34 *	3.76 ± 0.17
Total EAAs	9.66	11.38	13.37	17.03
Non-essential amino acids (NEAAs)				
Aspartate	0.03 ± 0	0.05 ± 0	0.06 ± 0.01 *	1.51 ± 0.1 [b]
Cystine	0.03 ± 0	0.2 ± 0.02 **	0.12 ± 0.01 **	0.62 ± 0.06 [b]
Glutamlic acid	77.02 ± 5.21	97.05 ± 3.71 *	82.21 ± 5.27	132.44 ± 6.78 [b]
Glutamine	96.66 ± 6.05	106.65 ± 0.98	159.99 ± 6.47 **	180.39 ± 9.12
Asparagine	1.23 ± 0.12	1.8 ± 0.19	1.84 ± 0.13 *	2.05 ± 0.09
Glycine	1.18 ± 0.13	2.03 ± 0.22 *	2.2 ± 0.24 *	2.67 ± 0.29
Ornithine	0.18 ± 0.03	0.21 ± 0.02	0.27 ± 0.04	0.32 ± 0.02
Proline	1.25 ± 0.06	2.24 ± 0.08 **	2.79 ± 0.15 **	3.52 ± 0.19 [a]
Serine	0.35 ± 0.04	0.59 ± 0.06 *	0.64 ± 0.07 *	0.78 ± 0.08
Tyrosine	0.99 ± 0.11	1.3 ± 0.02 *	1.64 ± 0.06 **	1.41 ± 0.09
Alanine	18.65 ± 0.97	22.22 ± 0.76 *	25.09 ± 0.67 **	28.06 ± 0.53 [a]
Total NEAAs	197.57	234.34	276.85	353.77
Saturated fatty acids (SFA)				
Lauric (C12:0)	1.42 ± 0.17	1.6 ± 0.15	1.31 ± 0.26	2.24 ± 0.19 [a]
Tetradecanoic (C14:0)	1.7 ± 0.15	2.6 ± 0.12 **	2.44 ± 0.14 *	3.05 ± 0.18
Hexadecanoic (C16:0)	10.18 ± 0.87	13.08 ± 0.73	13.38 ± 0.55 *	19.8 ± 1.14 [b]
Heptadecanoic (C17:0)	0.62 ± 0.06	0.81 ± 0.09	0.95 ± 0.09 *	1.62 ± 0.09 [b]
Octadecanoic (C18:0)	2.26 ± 0.21	2.96 ± 0.33	3.46 ± 0.34 *	5.9 ± 0.33 [b]
Eicosanoic (C20:0)	2 ± 0.19	2.16 ± 0.19	2.9 ± 0.14 *	4.06 ± 0.4
Total SFA	18.18 ± 1.15	23.2 ± 0.68 *	24.43 ± 0.94 *	36.68 ± 0.38 [b]
Unsaturated fatty acids				
Myristoleic (C14:1)	0.61 ± 0.06	0.8 ± 0.09	0.94 ± 0.09 *	1.6 ± 0.09 [b]
Palmitoleic (C16:1n-7)	1.61 ± 0.08	2.02 ± 0.1 *	1.48 ± 0.16	3.47 ± 0.19 [b]
Octadecenoic (C18:1)	7.45 ± 0.7	11.29 ± 0.35 **	10.51 ± 0.24 *	15.82 ± 0.23 [b]
Erucic acid (C22: 1)	12.19 ± 0.73	9.22 ± 0.28 *	12.93 ± 2.31	12.2 ± 0.54
Octadecadienoic (C18:2)	16.68 ± 1.57	24.65 ± 2.14 *	25.5 ± 1.48*	40.43 ± 0.78 [b]
Octadecatrienoic (C18:3)	5.36 ± 0.5	7.01 ± 0.28 *	6.1 ± 0.28	8.97 ± 0.28 [b]
USFA	43.9 ± 2.2	55 ± 2.55 *	57.46 ± 0.63 **	82.48 ± 0.68 [b]
SFA/USFA	0.35	0.34	0.35	0.35

Besides, the current results revealed that the combined AMF and eCO_2 treatment induced a significant increase in the majority of the detected organic acids, amino acids (including both essential and non-essential amino acids) and fatty acids in *T. vulgare*, relative to AMF alone (Table 2, Figure 2). Regarding the individual treatments, AMF was more efficient in inducing the accumulation of these primary metabolites than eCO_2. All AMF and/or eCO_2 treatments did not affect the SFA/USFA ratio. Similarly, it was reported that AMF-inoculated maize plants, under low temperature, had higher amino acid concentrations than non-mycorrizal ones, especially for Thr, Lys, Gly, Ala and His contents [55]. In contrast, proline content was reduced in mycorhizal *Capsicum annuum* grown under saline conditions [56]. Moreover, different effects of eCO_2 on amino acids were reported, which were reduced in barley [57], increased in spring wheat [58], or were not affected in maize [59]. It was also shown that organic acid levels were increased in mycorrhizal *Pinus sylvestris* grown under heavy metal concentrations [60], while they were not increased in *Portulacaria afra* under eCO_2 [61]. The concentration of most fatty acids of soybean was unchanged under higher levels of CO_2 [62]. On the other hand, a significant increase in the levels of individual fatty acids was reported in parsley and dill grown under eCO_2, which is more evident on UFA than SFA [9]. Therefore, the synchronous application of AMF and eCO_2 could be beneficial to avoid the negative impact of the individual treatment and/or to support their positive effects.

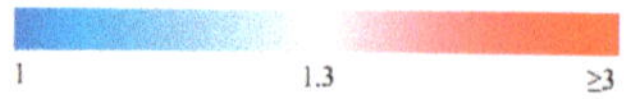

	eCO_2	AMF	AMF+eCO_2
Sugars			
Glucose	1.63**	0.99	1.60^{b}
Fructose	1.32	1.18	2.24^{b}
Sucrose	1.24	1.31*	1.71^{b}
Soluble sugars	1.69**	1.55**	1.88
Starch	1.28*	0.97	1.37^{b}
Total Carbohydrates	1.30*	1.00	1.40^{b}
Organic acids			
Oxalic	0.77	1.48*	2.02^{b}
Malic	0.91	0.92	1.07
Isobutyric	0.84	1.83*	2.22
Fumaric	1.03	0.72**	1.23^{b}
Succinic	1.37	1.65*	2.06
Citric	1.85**	1.45	2.05^{a}
Amino acids			
Lysine	1.08	1.18	1.59^{a}
Histidine	1.10	1.28	1.18
Arginie	1.26	1.69*	2.02
Isoleucine	1.73*	1.87*	2.27
Phenylalanine	2.06**	2.30*	3.97^{a}
Valine	1.21	1.63*	3.58^{b}
Leucine	1.50	1.50*	2.00
Methionine	1.00*	1.50*	1.50
Threonine	1.45	1.73*	1.91
Alanine	1.19*	1.35**	1.50^{a}
Asparagine	1.46	1.50*	1.67
Ornithine	1.17	1.50	1.78
Glycine	1.72*	1.86*	2.26
Serine	1.69*	1.83*	2.23
Proline	1.79**	2.23**	2.82^{a}
Aspartate	1.67	2.00*	5.33^{b}
Cystine	6.67**	4**	20.26b
Tyrosine	1.31*	1.66**	1.42
Glutamine	1.10	1.66**	1.87
Glutamlic acid	1.26*	1.07	1.72^{b}
Fatty acids			
C12:0 Lauric	1.13	0.92	1.58^{a}
C14:0 Myristic	1.53**	1.44*	1.79
C16:0 palmitic	1.28	1.31*	1.94^{b}
C17:0 heptadecanoic	1.32	1.53*	2.60^{b}
C18:0 stearic	1.30	1.53*	2.62^{b}

	eCO_2	AMF	AMF+eCO_2
C20:0 arachidic	1.08	1.45*	2.03
C14:1 myristoleic	1.31	1.54*	2.62^{b}
C16:1n-7 palmitoleic	1.25*	0.92	2.16^{b}
C18:1 oleic	1.52**	1.41*	2.12^{b}
C18:2 linoleic	1.48*	1.53*	2.42^{b}
C18:3 linolenic	1.31*	1.14	1.67^{b}
C22: 1 Erucic acid	0.76*	1.06	1.00
SFA	1.28*	1.34*	2.02^{b}
UFA	1.25*	1.31**	1.88^{b}
Gallic acid	1.41	1.15	1.47^{a}
Caffeic acid	1.54*	1.43*	1.59
p-Coumaric acid	1.06	0.87	1.41^{a}
Chicoric acid	1.13	0.96	1.46^{a}
Rosmarinic acid	1.10	0.92	1.45^{b}
Protocatechuic acid	1.09	1.07	1.56
Quercetin	1.00	1.14	1.14
Naringenin	1.2	1.00	1.20
Kaempferol	1.25*	1.00	1.63^{b}
Luteolin	1.43*	0.86	1.86^{b}
Apigenin	1.18	1.27	1.55^{a}
Naringenin	1.00	1.00**	1.50
Rutin	1.00	1.00	1.00
Chlorogenic acid	1.00	1.00	1.00
Essential oils			
alpha-pineno	1.30	1.23	1.83^{b}
Beta-pineno	1.28	1.18	1.84^{a}
p-cymene	1.33	1.31	1.87^{a}
limonene	0.8	1.00	1.20
1,8-cineol	1.04	1.17	1.03
linalol	1.04	0.93	1.16
Camphor	1.28	1.19	1.84^{a}
Borneol	1.03	0.92	1.15
Thymol	3.77**	1.65*	4.37^{b}
Carvacrol	1.28	1.18	1.84^{a}
alpha-Phellandrene	1.46	1.34	1.32
beta-Phellandrene	1.04	1.04	1.28
Humuline	1.21	7.57**	5.00^{b}
Myrcene	1.09	2.21**	1.43^{a}
alpha-thyjone	1.00	0.99	0.95
Tricyclene	2.2**	2.17*	4.03^{a}

Figure 2. Heatmap of fold change in the contents of primary and secondary metabolites of *Thymus vulgare* grown under the effect of eCO_2 (620 ppm), AMF or their combination (AMF-eCO_2). Asterisks indicate significant (* $p < 0.05$; ** $p < 0.01$) increased fold changes compared to control (untreated plants), as revealed by Student's t-test. Lowercase letters indicate significant differences (^{a} $p < 0.05$; ^{b} $p < 0.01$) between AMF alone and the combined AMF+eCO_2 treatment.

3.4. *AMF and eCO_2 Promote the Accumulation of Phenolic Compounds and Volatile Oils in T. vulgare*

Mycorrhizal symbiosis with medicinal plants has been recognized to induce the accumulation of secondary metabolites, especially phenolic compounds which play an important role in curing several ailments [63]. The present results showed that protocatechuic, p-coumaric and rosmarinic acids are the most abundant phenolic acids; while apigenin, kaempferol, quercetin and luteolin are the predominant flavonoid in *T. vulgare* (Table 3). Similarly, previous studies revealed the presence of some phenolic acids such as cinnamic, carnosic and rosmarinic acids, and also flavonoids such as luteolin and apigenin derivatives in *T. vulgare* [64]. There was a significant increment in the levels of the majority of the detected phenolic acids and flavonoids in *T. vulgare* under AMF and/or eCO_2 treatments, however the combined treatment was much more efficient than the individual ones (Figure 2). On the other hand, in consistence with the previous studies [65], the present results revealed the presence of 16 volatile oils in *T. vulgare*, whereas 1,8-cineol, carvacrol and p-cymene are the most dominant followed by less amounts of linalol, α- and β-pineno, α-Phellandrene, myrecene and thymol (Table 3). There is also a significant increase in the volatile oils of *T. vulgare*, under the individual and combined treatments.

Table 3. Levels of phenolic compounds and volatile oils (mg g^{-1} dry weight) and biological activities in *Thymus vulgare* grown under normal conditions (control) or the effect of eCO_2 (620 ppm), AMF or their combination (AMF-eCO_2). Values are mean ± standard error of five independent replicates. Asterisks indicate significant changes (* $p < 0.05$; ** $p < 0.01$) compared to control, as revealed by Student's t-test. Lowercase letters indicate significant differences (^{a} $p < 0.05$; ^{b} $p < 0.01$) between AMF alone and the combined AMF+eCO_2 treatment.

Metabolite	Control	eCO_2	AMF	AMF + eCO_2
Phenolic acids				
Caffeic acid	0.46 ± 0.04	0.71 ± 0.04 *	0.66 ± 0.03 *	0.73 ± 0.03
Chlorogenic acid	0.01 ± 0	0.01 ± 0	0.01 ± 0	0.01 ± 0 ^{a}
Protocatechuic acid	4.11 ± 0.32	4.47 ± 0.35	4.38 ± 0.45	6.42 ± 0.6
Gallic acid	0.34 ± 0.02	0.48 ± 0.06	0.39 ± 0.03	0.5 ± 0.02 ^{a}
p-Coumaric acid	2.66 ± 0.22	2.82 ± 0.35	2.32 ± 0.26	3.76 ± 0.36 ^{a}
Chicoric acid	1.28 ± 0.13	1.44 ± 0.15	1.23 ± 0.15	1.87 ± 0.16 ^{a}
Rosmarinic acid	1.72 ± 0.2	1.89 ± 0.1	1.59 ± 0.1	2.5 ± 0.11 ^{b}
Flavonoids				
Quercetin	0.07 ± 0	0.07 ± 0	0.08 ± 0.01	0.08 ± 0.01
Naringenin	0.05 ± 0	0.06 ± 0	0.05 ± 0.01	0.06 ± 0.01
Kaempferol	0.08 ± 0	0.1 ± 0.01 *	0.08 ± 0	0.13 ± 0.01 ^{b}
Luteolin	0.07 ± 0	0.1 ± 0.01 *	0.06 ± 0	0.13 ± 0.01 ^{b}
Apigenin	0.11 ± 0.02	0.13 ± 0	0.14 ± 0.01	0.17 ± 0.01 ^{a}
Rutin	0.01 ± 0	0.01 ± 0	0.01 ± 0	0.01 ± 0
Volatile oils				
alpha-pineno	1.62 ± 0.21	2.1 ± 0.08	2 ± 0.03	2.97 ± 0.1 ^{a,b}
Beta-pineno	2.79 ± 0.36	3.57 ± 0.27	3.29 ± 0.43	5.13 ± 0.23 ^{a}
p-cymene	4.45 ± 0.32	5.94 ± 0.48	5.85 ± 0.51	8.31 ± 0.37 ^{a}
limonene	0.1 ± 0.03	0.08 ± 0.01	0.1 ± 0	0.12 ± 0.02
1,8-cineol	8.23 ± 1.04	8.54 ± 1.19	9.61 ± 1.34	8.51 ± 0.73
linalol	3.42 ± 0.44	3.54 ± 0.49	3.19 ± 0.7	3.96 ± 0.23
Camphor	0.64 ± 0.08	0.82 ± 0.09	0.76 ± 0.1	1.18 ± 0.09 ^{a}
Borneol	0.39 ± 0.05	0.4 ± 0.06	0.36 ± 0.08	0.45 ± 0.09
Thymol	1.5 ± 0.19	5.66 ± 0.48 **	2.47 ± 0.28 *	6.55 ± 0.48 ^{b}
Carvacrol	4.97 ± 0.63	6.35 ± 0.68	5.85 ± 0.76	9.13 ± 0.15 ^{a}
alpha-Phellandrene	2.16 ± 0.21	3.15 ± 0.34	2.9 ± 0.38	2.86 ± 0.17
beta-Phellandrene	0.25 ± 0.03	0.26 ± 0.03	0.26 ± 0.04	0.32 ± 0.04
Humuline	0.14 ± 0.01	0.17 ± 0.01	1.06 ± 0.05 **	0.7 ± 0.04 ^{b}
Myrcene	1.84 ± 0.12	2 ± 0.4	4.07 ± 0.41 **	2.64 ± 0.12 ^{a}
alpha-thyjone	1.1 ± 0.05	1.1 ± 0.04	1.09 ± 0.08	1.04 ± 0.03
Tricyclene	0.3 ± 0.02	0.66 ± 0.04 **	0.65 ± 0.07 *	1.21 ± 0.14 ^{a}
Antioxidant capacity (FRAP)	17.08 ± 1.8	27.57 ± 1 **	22.85 ± 1.13	35.91 ± 0.91 ^{b}
Oxygen radical absorbance capacity (ORAC)	743.41 ± 33.19	1034.6 ± 108.25	983.99 ± 35.17 **	1680.57 ± 81.71 b
% inhibition of LDL oxidation				
TBARS)	14.08 ± 0.99	26.56 ± 0.76 **	24.55 ± 0.98 **	35.32 ± 2.25 ^{a}
conjugated dienes	17.7 ± 2.1	32.56 ± 0.51 **	28.51 ± 1.16 *	43.53 ± 1.29 ^{b}
% inhibition of hemolysis	13.4 ± 0.83	22.02 ± 1.97 *	16.45 ± 1.05	28.28 ± 1.08 ^{b}

Supporting our results, several studies have investigated the potential effects of eCO_2 and AMF, separately and in combination, on the levels of phenolic compounds and antioxidant activity in a variety of plant species. For instance, AMF treatments caused an increase in phenolic compounds content of lettuce [53,66], and in the antioxidant capacity of sweet basil [67]. Moreover, the flavonoids of some wild plants, such as *Libidibia ferrea*, were found to be accumulated by mycorrhizal association [68]. Similarly, eCO_2 induced the accumulation of some phenolic compounds in birch [69], and *Zingiber officinale* [52]. However, a low phenolic content was reported for some plants such as rice [70] under eCO_2. Regarding the interaction between eCO_2 and AMF, it was reported that the induction of secondary metabolites in lettuce and alfalfa by AMF was negatively affected under eCO_2, probably due to utilization of photoassimilates for increasing plant biomass and for AMF growth as well, at the expense of secondary metabolism [43,44]. Therefore, it could be suggested that climatic changes might have an impact on AMF, which in turn, affect the metabolic functions of their host plants.

Several scenarios have been proposed to explain the induction of secondary metabolites in response to AMF associations and eCO_2. It was found that AMF could affect secondary metabolites through improved photosynthesis and mineral content of the host plants, activation of pathways involved in synthesis of secondary metabolites, or higher expression of some genes related to secondary metabolism [68]. On the other hand, the eCO_2-induced changes in plant secondary metabolism have been attributed to either excess amount of non-structural carbohydrates, resulting in an increment in carbon-based secondary metabolites [9,37,51].

3.5. AMF and eCO_2-Induced Changes in Secondary Metabolites Support the Biological Activities of T. vulgare

Reactive oxygen species and free radicals have been recognized to induce harmful effects on living organisms. In this aspect, antioxidants, such as phenolic compounds and volatile oils, could act as free radical scavengers [71]. The present results showed an increase in the total antioxidant capacities of *T. vulgare*, tested by different methods (FRAP, ORAC, inhibition of LDL oxidation (TBARS and conjugated dienes) and inhibition of hemolysis), under the effects of AMF and/or eCO_2 (Table 3). It was previously reported that the antioxidant properties exhibited by *T. vulgare* extracts have been attributed to their content of volatile oils, especially carvacrol and thymol [72], flavonoids such as apigenin and luteolin derivatives and phenolic acids such as cinnamic and rosmarinic acids [64]. Moreover, some phenolic compounds were previously isolated from *T. vulgare* and proved to inhibit oxidative hemolysis [73]. The decreased levels of lipid peroxidation products such as TBARS and conjugated dienes might be ascribed to some protective effects of thymol [74].

4. Conclusions

Based on the above results, it is clear that the tested plant, *T. vulgare*, has benefited from the independent and combined effects of both AMF and eCO_2, however their synchronous application is much more beneficial. Such positive impacts are being reflected on improved biomass production and higher accumulation of primary (sugars, amino acids, fatty acids and organic acids), and secondary (phenolic acids, flavonoids and volatile oils) metabolites. *T. vulgare* plants grown under synchronous application of AMF and eCO_2 have taken much advantage over those grown under the individual effects of both factors in terms of improved growth and bioactive components. Thus, the current study clearly shows that co-application of AMF and eCO_2 is a promising approach to improve the growth and the nutritional and health promoting values of *T. vulgare*. Further, the robust monitoring of primary and secondary metabolites presented herein could support our understanding about the mechanisms behind the positive impacts of AMF and eCO_2 on plants.

Author Contributions: Conceived and designed the experiments: A.M.S., H.A., A.M.A.K. Conducted the main experiment: T.H.H., R.S.Y, A.M.A.K. Measured the fungal related parameters: R.S.Y., A.M.A.K. Performed the metabolite profiling: H.A., A.M.S., M.A.-M. Measured the biological activities: M.A.-M., Analyzed the data: A.M.S., M.A.-M., A.M.A.K., T.H.H. Wrote the original draft: M.A.-M., T.H.H., Reviewed, edited and prepared the MS for submission: A.M.S., H.A., A.M.A.K. All authors have read and agreed to the published version of the manuscript.

Funding: This research received no external funding.

Conflicts of Interest: The authors declare no conflict of interest.

References

1. Morales, F.; Padilla, S.; Falconí, F. Medicinal plants used in traditional herbal medicine in the province of Chimborazo, Ecuador. *Afr. J. Tradit. Complement. Altern. Med.* **2017**, *14*, 10–15. [CrossRef] [PubMed]
2. Kurian, A. Health benefits of herbs and spices. In *Handbook of Herbs and Spices*, 2nd ed.; Elsevier: Amsterdam, The Netherlands, 2012; Volume 2, pp. 72–88.
3. Abdelgawad, H.; Saleh, A.M.; Al, S.; Selim, S.; Hassan, M.O.; Wadaan, M.A.M.; Shuikan, A.M.; Mohamed, H.S.; Hozzein, W.N. Utilization of actinobacteria to enhance the production and quality of date palm (*Phoenix dactylifera* L.) fruits in a semi-arid environment. *Sci. Total Environ.* **2019**, *665*, 690–697. [CrossRef] [PubMed]
4. Wang, B.; Qiu, Y.-L. Phylogenetic distribution and evolution of mycorrhizas in land plants. *Mycorrhiza* **2006**, *16*, 299–363. [CrossRef] [PubMed]
5. Van der Heijden, M.G.A.; Boller, T.; Wiemken, A.; Sanders, I.R. Different arbuscular mycorrhizal fungal species are potential determinants of plant community structure. *Ecology* **1998**, *79*, 2082–2091. [CrossRef]
6. Gashgari, R.; Selim, S.; Abdel-Mawgoud, M.; Warrad, M.; Habeeb, T.H.; Saleh, A.M.; AbdElgawad, H. Arbuscular mycorrhizae induce a global metabolic change and improve the nutritional and health benefits of pennyroyal and parsley. *Acta Physiol. Plant.* **2020**, *42*, 1–11. [CrossRef]
7. Rillig, M.C. Arbuscular mycorrhizae, glomalin, and soil aggregation. *Can. J. Soil Sci.* **2004**, *84*, 355–363. [CrossRef]
8. Smith, S.E.; Read, D.J. *Mycorrhizal Symbiosis*, 3rd ed.; Academic Press: London, UK; Elsevier: Amsterdam, The Netherlands, 2008.
9. Saleh, A.M.; Selim, S.; Al Jaouni, S.; AbdElgawad, H. CO_2 enrichment can enhance the nutritional and health benefits of parsley (*Petroselinum crispum* L.) and dill (*Anethum graveolens* L.). *Food Chem.* **2018**, *269*, 519–526. [CrossRef]
10. Ainsworth, E.A.; Long, S.P. What have we learned from 15 years of free-air CO_2 enrichment (FACE)? A meta-analytic review of the responses of photosynthesis, canopy properties and plant production to rising CO_2. *New Phytol.* **2005**, *165*, 351–372. [CrossRef]
11. Morgan, J.A.; Mosier, A.R.; Milchunas, D.G.; LeCain, D.R.; Nelson, J.A.; Parton, W.J. CO_2 enhances productivity, alters species composition, and reduces digestibility of shortgrass steppe vegetation. *Ecol. Appl.* **2004**, *14*, 208–219. [CrossRef]
12. Van der Putten, W.H.; Bradford, M.A.; Pernilla Brinkman, E.; van de Voorde, T.F.J.; Veen, G.F. Where, when and how plant–soil feedback matters in a changing world. *Funct. Ecol.* **2016**, *30*, 1109–1121. [CrossRef]
13. Staddon, P.L.; Reinsch, S.; Olsson, P.A.; Ambus, P.; Lüscher, A.; Jakobsen, I. A decade of free-air CO_2 enrichment increased the carbon throughput in a grass-clover ecosystem but did not drastically change carbon allocation patterns. *Funct. Ecol.* **2014**, *28*, 538–545. [CrossRef]
14. Rillig, M.C.; Wright, S.F.; Shaw, M.R.; Field, C.B. Artificial climate warming positively affects arbuscular mycorrhizae but decreases soil aggregate water stability in an annual grassland. *Oikos* **2002**, *97*, 52–58. [CrossRef]
15. Fachini-Queiroz, F.C.; Kummer, R.; Estevao-Silva, C.F.; Carvalho, M.D.D.B.; Cunha, J.M.; Grespan, R.; Bersani-Amado, C.A.; Cuman, R.K.N. Effects of thymol and carvacrol, constituents of *Thymus vulgaris* L. essential oil, on the inflammatory response. *Evid. Based Complement. Altern. Med.* **2012**, *2012*. [CrossRef] [PubMed]
16. Nikolić, M.; Glamočlija, J.; Ferreira, I.C.F.R.; Calhelha, R.C.; Fernandes, Â.; Marković, T.; Marković, D.; Giweli, A.; Soković, M. Chemical composition, antimicrobial, antioxidant and antitumor activity of *Thymus serpyllum* L., *Thymus algeriensis* Boiss. and Reut and *Thymus vulgaris* L. essential oils. *Ind. Crops Prod.* **2014**, *52*, 183–190. [CrossRef]
17. Rajashekar, C.B. Elevated CO2 levels affect phytochemicals and nutritional quality of food crops. *Am. J. Plant Sci.* **2018**, *9*, 150. [CrossRef]
18. Selvaraj, T.; Sumithra, P. Effect of *Glomus aggregatum* and plant growth promoting rhizomicroorganisms on growth, nutrition and content of secondary metabolites in *Glycyrrhiza glabra* L. *Indian J. Appl. Pure Biol.* **2011**, *26*, 283–290.

19. Saleh, A.M.; Abdel-Mawgoud, M.; Hassan, A.R.; Habeeb, T.H.; Yehia, R.S.; AbdElgawad, H. Global metabolic changes induced by arbuscular mycorrhizal fungi in oregano plants grown under ambient and elevated levels of atmospheric CO_2. *Plant Physiol. Biochem.* **2020**, *151*, 255–263. [CrossRef]
20. French, K.E. Engineering mycorrhizal symbioses to alter plant metabolism and improve crop health. *Front. Microbiol.* **2017**, *8*, 1403. [CrossRef]
21. Phillips, J.M.; Hayman, D.S. Improved procedures for clearing roots and staining parasitic and vesicular-arbuscular mycorrhizal fungi for rapid assessment of infection. *Trans. Br. Mycol. Soc.* **1970**, *55*, 158–161. [CrossRef]
22. Giovannetti, M.; Mosse, B. An evaluation of techniques for measuring vesicular arbuscular mycorrhizal infection in roots. *New Phytol.* **1980**, *84*, 489–500. [CrossRef]
23. AbdElgawad, H.; De Vos, D.; Zinta, G.; Domagalska, M.A.; Beemster, G.T.S.; Asard, H. Grassland species differentially regulate proline concentrations under future climate conditions: An integrated biochemical and modelling approach. *New Phytol.* **2015**, *208*, 354–369. [CrossRef] [PubMed]
24. Hamad, I.; AbdElgawad, H.; Al Jaouni, S.; Zinta, G.; Asard, H.; Hassan, S.; Hegab, M.; Hagagy, N.; Selim, S. Metabolic analysis of various date palm fruit (*Phoenix dactylifera* L.) cultivars from Saudi Arabia to assess their nutritional quality. *Molecules* **2015**, *20*, 13620–13641. [CrossRef] [PubMed]
25. Hassan, M.O.; Saleh, A.M.; Abdelgawad, H. Sonchus oleraceus residue improves nutritive and health-promoting value of common bean (*Phaseolus vulgaris* L.): A metabolic study. *J. Agric. Food Chem.* **2018**, *66*, 2092–2100. [CrossRef] [PubMed]
26. Manzan, A.C.C.M.; Toniolo, F.S.; Bredow, E.; Povh, N.P. Extraction of essential oil and pigments from *Curcuma longa* [L.] by steam distillation and extraction with volatile solvents. *J. Agric. Food Chem.* **2003**, *51*, 6802–6807. [CrossRef] [PubMed]
27. El Hattab, M.; Culioli, G.; Piovetti, L.; Chitour, S.E.; Valls, R. Comparison of various extraction methods for identification and determination of volatile metabolites from the brown alga *Dictyopteris membranacea*. *J. Chromatogr. A* **2007**, *1143*, 1–7. [CrossRef]
28. Chen, C.-W.; Chang, C.-Y.; Chiang, S.-H. The inhibition effect of cell DNA oxidative damage and LDL oxidation by bovine colostrums. *Molecules* **2016**, *21*, 1378. [CrossRef]
29. Zhu, X.; Song, F.; Liu, S.; Liu, F. Role of arbuscular mycorrhiza in alleviating salinity stress in wheat (*Triticum aestivum* L.) grown under ambient and elevated CO_2. *J. Agron. Crop Sci.* **2016**, *202*, 486–496. [CrossRef]
30. Drigo, B.; Kowalchuk, G.A.; Knapp, B.A.; Pijl, A.S.; Boschker, H.T.S.; Van Veen, J.A. Impacts of 3 years of elevated atmospheric CO_2 on rhizosphere carbon flow and microbial community dynamics. *Glob. Chang. Biol.* **2013**, *19*, 621–636. [CrossRef]
31. Tingey, D.T.; Phillips, D.L.; Johnson, M.G. Elevated CO_2 and conifer roots: Effects on growth, life span and turnover. *New Phytol.* **2000**, *147*, 87–103. [CrossRef]
32. Gavito, M.E.; Curtis, P.S.; Mikkelsen, T.N.; Jakobsen, I. Atmospheric CO_2 and mycorrhiza effects on biomass allocation and nutrient uptake of nodulated pea (*Pisum sativum* L.) plants. *J. Exp. Bot.* **2000**, *51*, 1931–1938. [CrossRef]
33. Staddon, P.L.; Fitter, A.H.; Graves, J.D. Effect of elevated atmospheric CO_2 on mycorrhizal colonization, external mycorrhizal hyphal production and phosphorus inflow in *Plantago lanceolata* and *Trifolium repens* in association with the arbuscular mycorrhizal fungus *Glomus mosseae*. *Glob. Chang. Biol.* **1999**, *5*, 347–358. [CrossRef]
34. Compant, S.; van der Heijden, M.G.A.; Sessitsch, A. Climate change effects on beneficial plant–microorganism interactions. *FEMS Microbiol. Ecol.* **2010**, *73*, 197–214. [CrossRef] [PubMed]
35. Cotton, T.E.A.; Fitter, A.H.; Miller, R.M.; Dumbrell, A.J.; Helgason, T. Fungi in the future: Interannual variation and effects of atmospheric change on arbuscular mycorrhizal fungal communities. *New Phytol.* **2015**, *205*, 1598–1607. [CrossRef] [PubMed]
36. Fredeen, A.L.; Rao, I.M.; Terry, N. Influence of phosphorus nutrition on growth and carbon partitioning in *Glycine max*. *Plant Physiol.* **1989**, *89*, 225–230. [CrossRef]
37. Al Jaouni, S.; Saleh, A.M.; Wadaan, M.A.M.; Hozzein, W.N.; Selim, S.; AbdElgawad, H. Elevated CO_2 induces a global metabolic change in basil (*Ocimum basilicum* L.) and peppermint (*Mentha piperita* L.) and improves their biological activity. *J. Plant Physiol.* **2018**, *224–225*, 121–131. [CrossRef] [PubMed]

38. Tisserat, B. Influence of ultra-high carbon dioxide levels on growth and morphogenesis of Lamiaceae species in soil. *J. Herbs. Spices Med. Plants* **2002**, *9*, 81–89. [CrossRef]
39. Hartwig, U.A.; Wittmann, P.; Braun, R.; Hartwig-Räz, B.; Jansa, J.; Mozafar, A.; Lüscher, A.; Leuchtmann, A.; Frossard, E.; Nösberger, J. Arbuscular mycorrhiza infection enhances the growth response of Lolium perenne to elevated atmospheric p CO_2. *J. Exp. Bot.* **2002**, *53*, 1207–1213. [CrossRef]
40. Ainsworth, E.A.; Rogers, A. The response of photosynthesis and stomatal conductance to rising [CO_2]: Mechanisms and environmental interactions. *Plant. Cell Environ.* **2007**, *30*, 258–270. [CrossRef]
41. Gifford, R.M. The global carbon cycle: A viewpoint on the missing sink. *Funct. Plant Biol.* **1994**, *21*, 1–15. [CrossRef]
42. Makino, A.; Mae, T. Photosynthesis and plant growth at elevated levels of CO_2. *Plant Cell Physiol.* **1999**, *40*, 999–1006. [CrossRef]
43. Baslam, M.; Garmendia, I.; Goicoechea, N. Elevated CO_2 may impair the beneficial effect of arbuscular mycorrhizal fungi on the mineral and phytochemical quality of lettuce. *Ann. Appl. Biol.* **2012**, *161*, 180–191. [CrossRef]
44. Baslam, M.; Erice, G.; Goicoechea, N. Impact of arbuscular mycorrhizal fungi (AMF) and atmospheric CO_2 concentration on the biomass production and partitioning in the forage legume alfalfa. *Symbiosis* **2012**, *58*, 171–181. [CrossRef]
45. Johnson, N.C.; Wolf, J.; Reyes, M.A.; Panter, A.; Koch, G.W.; Redman, A. Species of plants and associated arbuscular mycorrhizal fungi mediate mycorrhizal responses to CO_2 enrichment. *Glob. Chang. Biol.* **2005**, *11*, 1156–1166. [CrossRef]
46. Pearson, J.N.; Jakobsen, I. Symbiotic exchange of carbon and phosphorus between cucumber and three arbuscular mycorrhizal fungi. *New Phytol.* **1993**, *124*, 481–488. [CrossRef]
47. Malundo, T.M.M.; Shewfelt, R.L.; Ware, G.O.; Baldwin, E.A. Sugars and acids influence flavor properties of mango (*Mangifera indica*). *J. Am. Soc. Hortic. Sci.* **2001**, *126*, 115–121. [CrossRef]
48. Al-Alawi, R.A.; Al-Mashiqri, J.H.; Al-Nadabi, J.S.M.; Al-Shihi, B.I. Date palm tree (*Phoenix dactylifera* L.): Natural products and therapeutic options. *Front. Plant Sci.* **2017**, *8*, 1–12. [CrossRef]
49. Livingstone, K.M.; Lovegrove, J.A.; Givens, D.I. The impact of substituting SFA in dairy products with MUFA or PUFA on CVD risk: Evidence from human intervention studies. *Nutr. Res. Rev.* **2012**, *25*, 193–206. [CrossRef]
50. Leakey, A.D.B.; Ainsworth, E.A.; Bernacchi, C.J.; Rogers, A.; Long, S.P.; Ort, D.R. Elevated CO_2 effects on plant carbon, nitrogen, and water relations: Six important lessons from FACE. *J. Exp. Bot.* **2009**, *60*, 2859–2876. [CrossRef]
51. Ibrahim, M.H.; Jaafar, H.Z.E. Impact of elevated carbon dioxide on primary, secondary metabolites and antioxidant responses of *Eleais guineensis* Jacq.(Oil Palm) seedlings. *Molecules* **2012**, *17*, 5195–5211. [CrossRef]
52. Ghasemzadeh, A.; Jaafar, H. Effect of CO_2 enrichment on synthesis of some primary and secondary metabolites in ginger (*Zingiber officinale* Roscoe). *Int. J. Mol. Sci.* **2011**, *12*, 1101–1114. [CrossRef]
53. Baslam, M.; Garmendia, I.; Goicoechea, N.; Unidad, V.; Icvv, Z. Arbuscular mycorrhizal fungi (AMF) improved growth and nutritional quality of greenhouse-grown lettuce. *J. Agric. Food Chem.* **2011**, *59*, 5504–5515. [CrossRef] [PubMed]
54. Baslam, M.; Antolín, M.C.; Gogorcena, Y.; Muñoz, F.; Goicoechea, N. Changes in alfalfa forage quality and stem carbohydrates induced by arbuscular mycorrhizal fungi and elevated atmospheric CO_2. *Ann. Appl. Biol.* **2014**, *164*, 190–199. [CrossRef]
55. Zhu, X.; Song, F.; Liu, F. Altered amino acid profile of arbuscular mycorrhizal maize plants under low temperature stress. *J. Plant Nutr. Soil Sci.* **2016**, *179*, 186–189. [CrossRef]
56. Beltrano, J.; Ruscitti, M.; Arango, M.C.; Ronco, M. Effects of arbuscular mycorrhiza inoculation on plant growth, biological and physiological parameters and mineral nutrition in pepper grown under different salinity and p levels. *J. Soil Sci. Plant Nutr.* **2013**, *13*, 123–141. [CrossRef]
57. Sicher, R.C. Effects of CO_2 enrichment on soluble amino acids and organic acids in barley primary leaves as a function of age, photoperiod and chlorosis. *Plant Sci.* **2008**, *174*, 576–582. [CrossRef]
58. Oehme, V.; Högy, P.; Zebitz, C.P.W.; Fangmeier, A. Effects of elevated atmospheric CO_2 concentrations on phloem sap composition of spring crops and aphid performance. *J. Plant Interact.* **2013**, *8*, 74–84. [CrossRef]

59. Leakey, A.D.B.; Uribelarrea, M.; Ainsworth, E.A.; Naidu, S.L.; Rogers, A.; Ort, D.R.; Long, S.P. Photosynthesis, productivity, and yield of maize are not affected by open-air elevation of CO_2 concentration in the absence of drought. *Plant Physiol.* **2006**, *140*, 779–790. [CrossRef]
60. Ahonen-Jonnarth, U.; Van Hees, P.A.W.; Lundström, U.S.; Finlay, R.D. Organic acids produced by mycorrhizal Pinus sylvestris exposed to elevated aluminium and heavy metal concentrations. *New Phytol.* **2000**, *146*, 557–567. [CrossRef]
61. Huerta, A.J.; Ting, I.P. Effects of various levels of CO_2 on the induction of Crassulacean acid metabolism in *Portulacaria afra* (L.) Jacq. *Plant Physiol.* **1988**, *88*, 183–188. [CrossRef]
62. Zheng, G.; Chen, J.; Li, W. Impacts of CO_2 elevation on the physiology and seed quality of soybean. *Plant Divers.* **2019**. [CrossRef]
63. Jurkiewicz, A.; Ryszka, P.; Anielska, T.; Waligórski, P. Optimization of culture conditions of *Arnica montana* L.: Effects of mycorrhizal fungi and competing plants. *Mycorrhiza* **2010**, *20*, 293–306. [CrossRef]
64. Roby, M.H.H.; Sarhan, M.A.; Selim, K.A.-H.; Khalel, K.I. Evaluation of antioxidant activity, total phenols and phenolic compounds in thyme (*Thymus vulgaris* L.), sage (*Salvia officinalis* L.), and marjoram (*Origanum majorana* L.) extracts. *Ind. Crops Prod.* **2013**, *43*, 827–831. [CrossRef]
65. Mancini, E.; Senatore, F.; del Monte, D.; de Martino, L.; Grulova, D.; Scognamiglio, M.; Snoussi, M.; de Feo, V. Studies on chemical composition, antimicrobial and antioxidant activities of five *Thymus vulgaris* L. essential oils. *Molecules* **2015**, *20*, 12016–12028. [CrossRef] [PubMed]
66. Baslam, M.; Esteban, R.; García-Plazaola, J.I.; Goicoechea, N. Effectiveness of arbuscular mycorrhizal fungi (AMF) for inducing the accumulation of major carotenoids, chlorophylls and tocopherol in green and red leaf lettuces. *Appl. Microbiol. Biotechnol.* **2013**, *97*, 3119–3128. [CrossRef] [PubMed]
67. Hristozkova, M.; Gigova, L.; Geneva, M.; Stancheva, I.; Vasileva, I.; Sichanova, M.; Mincheva, J. Mycorrhizal fungi and microalgae modulate antioxidant capacity of basil plants. *J. Plant Prot. Res.* **2017**. [CrossRef]
68. Dos Santos, E.L.; da Silva, F.A.; da Silva, F.S.B. Arbuscular mycorrhizal fungi increase the phenolic compounds concentration in the bark of the stem of libidibia ferrea in field conditions. *Open Microbiol. J.* **2017**, *11*, 283. [CrossRef]
69. Peltonen, P.A.; Vapaavuori, E.; Julkunen-Tiitto, R. Accumulation of phenolic compounds in birch leaves is changed by elevated carbon dioxide and ozone. *Glob. Change Biol.* **2005**, *11*, 1305–1324. [CrossRef]
70. Goufo, P.; Pereira, J.; Figueiredo, N.; Oliveira, M.B.P.P.; Carranca, C.; Rosa, E.A.S.; Trindade, H. Effect of elevated carbon dioxide (CO2) on phenolic acids, flavonoids, tocopherols, tocotrienols, γ-oryzanol and antioxidant capacities of rice (*Oryza sativa* L.). *J. Cereal Sci.* **2014**, *59*, 15–24. [CrossRef]
71. Santos-Sánchez, N.F.; Salas-Coronado, R.; Villanueva-Cañongo, C.; Hernández-Carlos, B. Antioxidant compounds and their antioxidant mechanism. In *Antioxidants*; IntechOpen: London, UK, 2019.
72. Chizzola, R.; Michitsch, H.; Franz, C. Antioxidative properties of Thymus vulgaris leaves: Comparison of different extracts and essential oil chemotypes. *J. Agric. Food Chem.* **2008**, *56*, 6897–6904. [CrossRef]
73. Haraguchi, H.; Saito, T.; Ishikawa, H.; Date, H.; Kataoka, S.; Tamura, Y.; Mizutani, K. Antiperoxidative components in Thymus vulgaris. *Planta Med.* **1996**, *62*, 217–221. [CrossRef]
74. Meeran, N.; Fizur, M.; Javed, H.; Al Taee, H.; Azimullah, S.; Ojha, S.K. Pharmacological properties and molecular mechanisms of thymol: Prospects for its therapeutic potential and pharmaceutical development. *Front. Pharmacol.* **2017**, *8*, 380. [CrossRef] [PubMed]

Journal of
Fungi

MDPI

Article

Alginate-Derived Elicitors Enhance β-Glucan Content and Antioxidant Activities in Culinary and Medicinal Mushroom, *Sparassis latifolia*

Yong-Woon Kim [1,†], Yuanzheng Wu [2,†], Moon-Hee Choi [1], Hyun-Jae Shin [1,*] and Jishun Li [2]

1 Department of Biochemical and Polymer Engineering, Chosun University, Gwangju 61452, Korea; ywkim1205@naver.com (Y.-W.K.); aamoony1222@naver.com (M.-H.C.)

2 Shandong Provincial Key Laboratory of Applied Microbiology, Ecology Institute, Qilu University of Technology (Shandong Academy of Sciences), Jinan 250103, China; wuyzh@sdas.org (Y.W.); yewu2@sdas.org (J.L.)

* Correspondence: shinhj@chosun.ac.kr; Tel.: +82-62-230-7518

† These authors contributed equally to this work.

Received: 9 June 2020; Accepted: 24 June 2020; Published: 25 June 2020

Abstract: This study aimed to investigate the elicitation effects of alginate oligosaccharides extracted from brown algae (*Sargassum* species) on β-glucan production in cauliflower mushroom (*Sparassis latifolia*). Sodium alginate was refined from *Sargassum fulvellum*, *S. fusiforme*, and *S. horneri*, and characterized by proton nuclear magnetic resonance spectroscopy (^{1}H NMR), resulting mannuronic acid to guluronic acid (M/G) rationes from 0.64 to 1.38. Three oligosaccharide fractions, ethanol fraction (EF), solid fraction (SF), and liquid fraction (LF), were prepared by acid hydrolysis and analyzed by Fourier transform infrared (FT-IR) spectra and high-performance anion-exchange chromatography with a pulsed amperometric detector (HPAEC-PAD). The samples of *S. fusiforme* resulted in the highest hydrolysate in SF and the lowest in LF, which was consistent with its highest M/G ratio. The SF of *S. fusiforme* and LF of *S. horneri* were chosen for elicitation on *S. latifolia*, yielding the highest β-glucan contents of 56.01 ± 3.45% and 59.74 ± 4.49% in the stalk, respectively. Total polyphenol content (TPC) and antioxidant activities (2,2′-Azino-bis(3-ethylbenzthiazoline-6-sulfonic acid) (ABTS) radical scavenging and Superoxide dismutase (SOD)-like activity) of aqueous extracts of *S. latifolia* were greatly stimulated by alginate elicitation. These results demonstrate that alginate oligosaccharides extracted from brown algae may be useful as an elicitor to enhance the nutritional value of mushrooms.

Keywords: Alginate; β-glucan; oligosaccharides; elicitation; *Sargassum* species; *Sparassis latifolia*; polyphenol; antioxidant

1. Introduction

Mushrooms have been recognized as medicine sources and functional foods since ancient times owing to their bioactive compounds and diverse health benefits [1]. Cauliflower mushrooms, species of *Sparassis* Fr., are culinary and medicinal mushrooms that primarily, but not exclusively, grow on the stumps of coniferous trees and are widely distributed throughout northern temperate forests [2]. The Asian *Sparassis* isolate was originally known as *S. crispa*, until morphological and molecular studies redefined it as *S. latifolia* [3,4]. Recently this mushroom has become cultivable in Japan and Korea using conifers [5]. The fruiting bodies of *S. crispa* and *S. latifolia* exhibit excellent effects for enhancing cytokine synthesis and preventing human diseases, such as gastric ulceration, oesophageal cancer, hypertension, and diabetes. Such effects are attributable to different compounds including polysaccharides, terpenoids, phenolic compounds, and glycoproteins [6–8].

As major constituents of fungal cell walls, β-glucans are present in all mushroom species and play important roles in their beneficial properties for human health [9]. Previous studies described a wide range of mushroom β-glucans with different structures, such as linear 1-3-β-glucans isolated from *Poria cocos*, linear 1-6-β-glucans from *Agaricus* spp., 6-branched 1,3-β-glucans from *Lentinus edodes* (designated lentinan) and *Grifola frondosa* (designated grifolan), and 3-branched 1,6-β-glucans from *Sarcodon aspratus* [10–12]. Among most studied fungal glucans, 6-branched 1,3-β-glucans were reported to be efficient as biological response modifiers (BRM) for the treatment of cancer and infectious diseases [13]. *Sparassis* species contain considerably higher contents of β-glucan than other mushrooms, up to 43.6% of dry weight in the fruiting bodies of *S. crispa* [14,15]. The primary structure of β-glucan isolated from *S. crispa* was 6-branched 1,3-β-glucan with one branch in every third main-chain unit, showing high water solubility and an estimated molecular weight of ca. 510 kDa [16,17]. Extraction and purification of β-glucans from mushroom mycelia and fruiting bodies have been established [18,19]. However, there are few studies about the enhancement of β-glucan contents in mushrooms.

To enhance the production of desirable compounds in plants or microorganisms, elicitation has been employed using biotic or abiotic elicitors to stimulate the accumulation of secondary metabolites [20,21]. For example, physical elicitation and enzyme treatments have been exploited for the growth enhancement and promotion of β-glucan contents in *S. latifolia* [22,23]. Alginate is a biotic elicitor that has been demonstrated to provide induced plant defense against pathogens, tolerance improvement to environmental stresses such as drought and salinity, and growth enhancement in *Eucomis autumnalis*, *Vitis vinifera*, *Phoenix dactylifera*, and *Olea europaea* [24–27]. Chemically, alginate is a linear copolymer consisting of homopolymeric blocks of residues of (1-4)-linked β-D-mannuronate (M) and its C-5 epimer α-L-guluronate (G), covalently linked together in different sequences [28]. As a biocompatible and immunogenic polysaccharide, alginate can be extracted from brown algae such as *Ecklonia* spp., *Laminaria* spp., or *Sargassum* spp. [29]

In this study, the physicochemical characteristics of alginate oligosaccharides extracted from *Sargassum* species from Korea and the enhancement of β-glucan content in *S. latifolia* by alginate elicitation were investigated. To the best of our knowledge, there has been no prior attempt to investigate alginate-derived elicitors for enhancement of β-glucan synthesis in mushrooms.

2. Materials and Methods

2.1. Collection of Marine Brown Algae

Three brown algae of *Sargassum* species were collected from the Korean coast. *S. fulvellum* (Turner) C. Agardh was collected in March 2017 at Jindo (34°22′18.0″ N, 126°08′09.4″ E), *S. fusiforme* (Harvey) Setchell was harvested in July 2017 at Wando (34°20′11.8″ N, 126°48′54.6″ E), and *S. horneri* (Turner) C. Agardh was collected in February 2018 at Jeju (33°31′08.3″ N, 126°31′16.7″ E). All *Sargassum* samples were cleaned by washed in distilled water to remove sand and excess salt. Samples were then oven dried at 60 °C. The samples were stored in plastic bags at room temperature (25 °C) until alginate extraction.

2.2. Extraction of Sodium Alginate

Sodium alginate was extracted as described by Davis et al. [30] In brief, 25 g of *Sargassum* samples were soaked in 800 mL of 2% formaldehyde for 24 h at room temperature, washed with distilled water, added to 800 mL of 0.2 M HCl, and left for 24 h. The samples were then washed with distilled water before extraction with 2% sodium carbonate (Na_2CO_3) for 3 h at 100 °C. The filtered extract was centrifuged (10,000 rpm, 30 min, 4 °C) and the supernatant was precipitated with 3 volumes of 95% ethanol. The precipitate was washed with acetone and dried at 60 °C.

2.3. Preparation of Oligosaccharide Fractions

Oligosaccharide separation from sodium alginate was performed as described by Asilonu et al. [31] This method gave three oligosaccharide fractions instead of two fractions from sodium alginate. In brief, 10 g of sodium alginate samples were acid hydrolyzed with 1 L of 0.3 M HCl in an autoclave at 121 °C for 1 h. The suspension was quickly cooled using ice water and neutralized with NaOH. Sodium chloride (NaCl) was added to a final concentration of 0.5% (*w/v*), and the solution mixture precipitated with 2 volumes of 95% ethanol and centrifuged (4000 rpm, 30 min, 4 °C). The supernatant was marked as ethanol fraction (EF). The precipitate was then washed with distilled water, centrifuged again, dissolved in distilled water, and adjusted to pH 2.85 with 1 M HCl. The solid fraction (SF) and liquid fraction (LF) were separated and lyophilized. Thus, three different oligosaccharide fractions (EF, SF, and LF) were obtained for further analysis.

2.4. Physicochemical Analysis of Oligosaccharide Fractions

2.4.1. Molecular Weight Analysis

The average molecular weight (M_W) of the EF, LF, and SF oligosaccharide fractions was determined using an SB-803 HQ column (6 μm, 8.0 mm × 300 mm, Shodex, Japan) and an SB-805 HQ column (13 μm, 8.0 mm × 300 mm, Shodex, Japan). A mobile phase of 0.1 mol/L NaCl solution was used after 0.22 μm filtration. The flow rate was 0.5 mL/min and analyses were performed at 50 °C. The sodium alginate sample (100 mg) was then resuspended in 10 mL distilled water. Molecular weights were determined by reference to a calibration curve using pullulan standards [32].

2.4.2. H NMR Spectroscopy Analysis

Proton nuclear magnetic resonance (^{1}H NMR) spectroscopy was acquired on alginate solution (1% *w/v*) in D_2O, with recordings at 80 °C using an AVANCE III HD 400 (Bruker Scientific Instruments, USA) spectrometer operating at a frequency of 400 MHz. The individual blocks of guluronic and mannuronic acids (F_G and F_M, respectively), the homopolymeric (F_{GG} and F_{MM}, respectively) blocks, and the heterogeneous (F_{GM} or F_{MG}) blocks of alginate were calculated using the areas of I, II, and III (A_I, A_{II}, and A_{III}, respectively) signals according to the Grasdalen [33] method and Equations (1)–(3):

$$F_G = \frac{A_I}{A_{II} + A_{III}} \quad (1)$$

$$F_M = 1 - F_G \quad (2)$$

$$M/G = \frac{1 - F_G}{F_G} = \frac{F_M}{F_G} \quad (3)$$

The double fractions (F_{GM} and F_{MM}) were deduced by referring to Equations (4)–(6):

$$F_{GG} = \frac{A_{II}}{A_{II} + A_{III}} \quad (4)$$

$$F_{GM} = F_{MG} = F_G - F_{GG} \quad (5)$$

$$F_{MM} = F_M - F_{MG} \quad (6)$$

2.4.3. FT-IR Spectroscopy Analysis

Fourier transform infrared (FT-IR) spectra were measured on a Nicolet 6700 FT-IR (Thermo Electron Co., Waltham, MA, USA). Spectra of the alginate samples in KBr pellets were recorded in the 4000–400 cm^{-1} range.

2.4.4. Monosaccharide Analysis

Monosaccharide analysis for the three oligosaccharide fractions was performed as follows [34]. The first acid hydrolysis was performed in 3 mL of 72% H_2SO_4 with 0.3 g of sample for 2 h at 30 °C. The second hydrolysis was carried out by adding 84 mL of distilled water to the first hydrolysate followed by acid hydrolysis at 121 °C for 1 h in an autoclave. Analysis was performed using high performance anion exchange chromatography with a pulsed amperometric detector (HPAEC-PAD, ICS-5000, Dionex Co., Sunnyvale, CA, USA) equipped with a current detector, with CarboPac PA-4 (250 mm × 4 mm, Dionex Co., USA) used as the column at room temperature. An 18 mM NaOH solution was chosen as the mobile phase with a flowrate of 1.0 mL/min.

2.5. Cultivation of Sparassis Latifolia

Sparassis latifolia JF02-06 was provided by Jeonnam Forest Research Institute (Naju, Korea). A sawdust-based medium was used for the cultivation of cauliflower mushrooms as adopted by Park et al. [23] Fermented sawdust of *Larix kaempferi*, corn, and wheat flour were mixed at a ratio of 8:1:1 (*w/w/w*), followed by the addition of 10% aqueous solution of starch syrup and an adjustment of the moisture content to 55–60%. The medium was packed in a 500 mL incubation bottle and sterilized at 121 °C for 90 min. The inoculum of *S. latifolia* was prepared in potato dextrose broth (PDB) for 3 weeks and 10 mL of liquid inoculum was inoculated into each bottle. The inoculated medium was then incubated at room temperature. Cultivation was completed after 50–70 days from inoculation. When mycelium was formed in the medium, the culture was transferred to the cultivation room at a temperature of 20 ± 2 °C and 95% humidity. Fruit-shaped fruiting bodies were formed after 1 week and then harvested after 40–45 days.

2.6. Elicitation by Alginate Oligosaccharide Fractions

Two of the oligosaccharide fractions, LF and SF (each 200 mg/L in distilled water), were applied to the surface of the sawdust medium for elicitation: 12 mL/bottle after sterilization of the medium (first application) and 20 mL/bottle after transfer to the cultivation room (second application). The same volume of distilled water was used as a control group. All treatments were arranged in a completely randomized block design with four replicates. Fruiting body and stalk of *S. latifolia* were harvested, frozen in liquid nitrogen, and stored in a freezer at −80 °C until extraction and analysis.

2.7. Assay of Glucan Content

Content of total and β-glucans was determined by a β-glucan assay kit (Cat. No. K-YBGL, Megazyme International, Wicklow, Ireland), following the H_2SO_4 acid hydrolysis procedure by McCleary and Draga [35]. Briefly, after being milled to pass through a 1.0 mm screen, 100 mg of the dried mushroom samples were added with 2 mL of ice-cold 12 M H_2SO_4, and then the mixture was stirred vigorously and incubated in an ice-cold water bath for 2 h. Then, 12 mL of distilled water was added to each sample and the suspension was kept in a boiling-water bath (~100 °C) for 2 h. After cooling to room temperature, 6 mL of 10 M KOH was added and the volume was adjusted to 100 mL with 200 mM sodium acetate buffer (pH 5.0). After centrifugation at 12,000 rpm for 10 min, an aliquot of the supernatant (0.1 mL) was mixed with 0.05 mL of exo-1,3-β-glucanase (20 U/mL) plus β-glucosidase (4 U/mL) and incubated at 40 °C for 60 min. Then, the mixture was incubated at 40 °C for 20 min with 3 mL of glucose-oxidase/peroxidase-reagent (GOPOD). Total glucan content was evaluated by a UV-Vis spectrophotometer (S-3100, SCINCO, Seoul, Korea) at λ = 510 nm with the reagent blank.

The α-glucan content was determined after incubation at 40 °C for 30 min of the suspension of 100 mg mushroom samples in 2 mL of 2 M KOH with an addition of 8 mL of 1.2 M sodium acetate buffer (pH 3.8) with 0.2 mL of amyloglucosidase (1630 U/mL) plus invertase (500 U/mL). Each sample was centrifuged at 12,000 rpm for 10 min, and 0.1 mL of the supernatant was analyzed for glucose by mixing with 0.1 mL of 200 mM of sodium acetate buffer (pH 5.0) and 3 mL of GOPOD. α-Glucan content was evaluated by a UV-Vis spectrophotometer. β-Glucan content was determined as the difference between total and α-glucan contents.

2.8. Determination of Total Polyphenol Content

Total polyphenol content (TPC) of aqueous extracts of *S. latifolia* after alginate elicitation was determined by the modified Folin-Ciocalteu method [36]. An ramount of 0.5 mL of the extract was mixed with 0.5 mL of 0.2 N Folin-Ciocalteu reagent and 0.5 mL of 2% (*w/v*) sodium carbonate. The mixture was vortexed and incubated at room temperature for 30 min. The absorbance of the mixture was measured at λ = 750 nm with a UV-Vis spectrophotometer. Total polyphenol content was expressed as of gallic acid equivalent (GAE) mg/100 g.

2.9. Measurement of Antioxidant Activities

The antioxidant activities of aqueous extracts of *S. latifolia* were evaluated by scavenging activity of 2,2′-azino-bis-3-ethylbenzothiazoline-6-sulphonic acid (ABTS) radical cation and superoxide dismutase (SOD)-like activity.

2.9.1. Assay of ABTS Radical Scavenging Activity

2,2′-Azino-bis(3-ethylbenzthiazoline-6-sulfonic acid) (ABTS) radical scavenging activity was measured using a modified method of Re et al. [37] The ABTS stock solution was prepared by mixing 7 mM ABTS with 2.45 mM potassium persulfate and then kept in the dark for 12 h. The stock solution was then diluted with phosphate buffered saline (PBS, pH 7.4) until absorbance reached 0.80 ± 0.02 at λ = 730 nm using a UV-Vis spectrophotometer. Of the extracts of *S. latifolia*, 0.2 mL was mixed with 1 mL of diluted ABTS solution and left for 15 min in the dark. ABTS radical scavenging activity (%) was calculated as in Equation (7) as follows:

$$\text{ABTS radical scavenging activity } (\%) = \frac{A_0 - A_1}{A_0} \times 100, \tag{7}$$

where A_0 is the absorbance of the blank sample using water, and A_1 is the absorbance of the aqueous extracts of *S. latifolia*.

2.9.2. Assay of SOD-Like Activity

Superoxide dismutase (SOD)-like activity was measured by reference using an SOD kit (Cat. No. 19160, Sigma-Aldrich, St. Louis, MO, USA) [38]. Of the aqueous extracts of *S. latifolia*, 20 μL was mixed with 200 μL of water-soluble tetrazolium salt (WST) working solution and 20 μL of enzyme working solution to each well in a 96-well microplate. Then the microplate was incubated at 37 °C for 20 min. The absorbance of each sample was measured at λ = 450 nm using a microplate reader. The SOD-like activity (%) was calculated as in Equation (8) as follows:

$$\text{SOD} - \text{like activity } (\%) = \frac{(\text{Blank 1} - \text{Blank 3}) - (\text{Sample} - \text{Blank 2})}{(\text{Blank 1} - \text{Blank 3})} \times 100 \tag{8}$$

where Blank 1 is the absorbance of the water with WST working solution and enzyme working solution, Blank 2 is the absorbance of sample with WST working solution and dilution buffer, Blank 3 is the absorbance of water with WST working solution and dilution buffer, and Sample is the absorbance of the aqueous extracts of *S. latifolia*.

2.10. Statistical Analysis

Data were analyzed using analysis of variance (ANOVA) followed by Duncan's multiple-range test ($p < 0.05$) using SPSS software (version 23, SPSS, Chicago, IL, USA). Error bars indicate the mean ± SD and different letters describe significant differences within the same application data group.

3. Results

3.1. Physicochemical Properties of Alginate

Sodium alginate was extracted and refined from three *Sargassum* species, namely *S. fulvellum*, *S. fusiforme*, and *S. horneri*. The composition of extracted alginate is summarized in Table 1. The highest alginate yield was obtained from *S. fusiforme* (37.84 ± 0.48%), followed by 34.11 ± 1.65% from *S. horneri* and 30.88 ± 1.51% from *S. fulvellum*. The size of the extracted alginates was relatively high in *S. fusiforme* and *S. fulvellum* with an average M_W of 504.65 and 461.07 kDa, respectively. However, the alginate from *S. horneri* had a much lower M_W (138.10 kDa) than that of the other two algae. This indicates that different ranges of alginates are obtained from different *Sargassum* species.

Table 1. Composition of sodium alginate extracted from *Sargassum* species.

Species	Yield (%)	Mw (kDa)	M/G	F_M	F_G	F_{MM}	F_{MG}	F_{GG}
Sargassum fusiforme	37.84 ± 0.48	504.65	1.38	0.58	0.42	0.55	0.03	0.39
S. fulvellum	30.88 ± 1.51	461.07	0.88	0.47	0.53	0.43	0.04	0.49
S. horneri	34.11 ± 1.65	138.10	0.64	0.39	0.61	0.30	0.09	0.52

M_W: average molecular weight; F_M: Mannuronic acid block fractions; F_G: Guluronic acid block fractions; F_{MM}: Homopolymeric mannuronic acid block fractions; F_{MG}: Heterogeneous block fractions (mannuronic acid and guluronic acid); F_{GG}: Homopolymeric guluronic acid block fractions.

The ratio of mannuronic acid to guluronic acid (M/G) of the structural blocks were determined by ^{1}H NMR analysis. As shown in Table 1, the M/G ratio of *Sargassum* species ranged from 0.64 to 1.38, with the highest ratio of 1.38 from *S. fusiforme* and similar M/G rationes between *S. fusiforme* and *S. horneri*. The guluronic acid anomeric proton (G-1) occurred at 5.56–5.58 ppm (peak I), guluronic acid H-5 (G-5) occurred at 4.96–4.98 ppm (peak III), and mannuronic acid anomeric proton (M-1) and the C-5 of alternating blocks (GM-5) overlapped at 5.21–5.23 ppm (peak II), as shown in Figure 1.

The FT-IR spectrum of sodium alginate of *Sargassum* species is presented in Figure 2. A broad band at 3466.25 cm^{-1} was assigned to the hydrogen-bonded (O-H) stretching vibrations, and a weak signal at 2928.43 cm^{-1} was attributed to C-H stretching vibrations. The peaks at 1619.79 cm^{-1} and 1427.52 cm^{-1} were attributed to asymmetric stretching of carboxylate O-C-O vibrations. The characteristic peaks of alginate include 1096.76 cm^{-1} assigned to β-mannuronic acid and 1029.66 cm^{-1} assigned to α-L-guluronic acid. Moreover, signals of glucuronic acid (1737.01, 1629.72, and 1144.66-937.31 cm^{-1}) could be detected in Figure 2 (LF and EF of a,b,c).

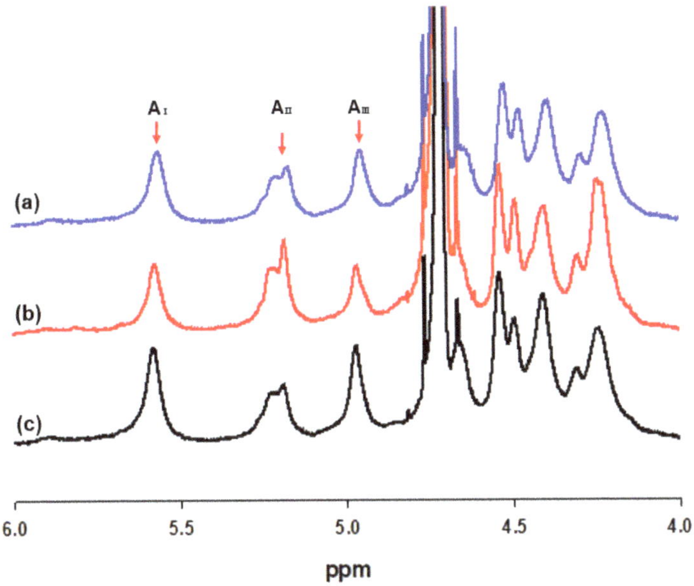

Figure 1. ^{1}H NMR spectra of sodium alginates extracted from (a) *Sargassum fulvellum*, (b) *S. fusiforme*, and (c) *S. horneri* (A_I: Area of guluronic acid anomeric proton peak I; A_{II}: Area of mannuronic acid anomeric proton and the C-5 of alternating blocks peak II; A_{III}: Area of guluronic acid peak III).

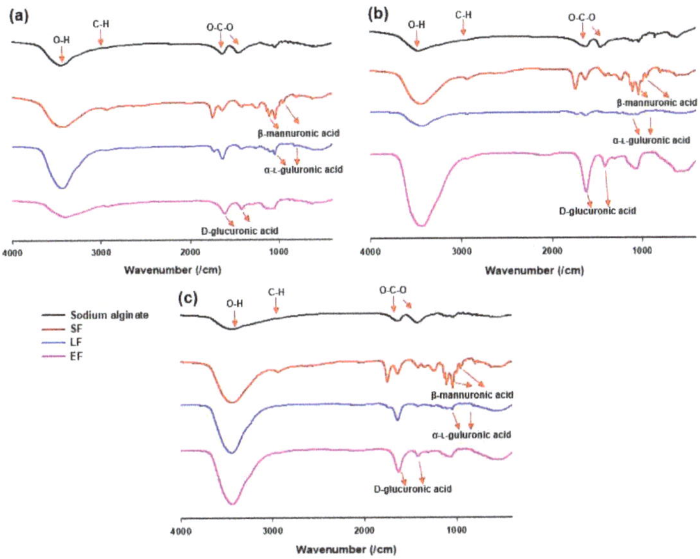

Figure 2. Fourier transform infrared (FT-IR) spectra of sodium alginate, solid fraction (SF), liquid fraction (LF), and ethanol fraction (EF) from (**a**) *Sargassum fulvellum*, (**b**) *S. fusiforme*, and (**c**) *S. horneri*.

3.2. Oligosaccharide Analysis

Sodium alginates extracted from *Sargassum* species were submitted to a two-step H_2SO_4 acid hydrolysis process to obtain partial hydrolysates for monosaccharide analysis. The block distribution of partial acid hydrolysates in EF, SF, and LF were 13.9–20.4%, 15.8–22.8%, and 61.1–65.0%, respectively.

The block distribution of this study was compared with that of previous reports by Haug et al. and Leal et al. [39,40], as summarized in Table 2. The comparison data showed a great difference in partial hydrolysis yield according to species used for the preparation of alginate oligosaccharides. The samples of *S. fusiforme* showed the highest hydrolysate yields in SF and the lowest in LF among *Sargassum* species, which was consistent with the highest M/G ratio of 1.38 and might be explained by the high content of polymannuronic acid and low content of polyguluronic acid. *S. fulvellum* and *S. horneri* both produced lower hydrolysate yields in SF and higher yields in LF, as indicated by their low M/G rationes. All three *Sargassum* species yielded similar percentages of hydrolysates in EF.

Table 2. Block distribution of sodium alginate from different species of brown algae.

Species	M/G	Partial Hydrolysis (Yields)			Reference
		EF (%)	SF (%)	LF (%)	
Sargassum fusiforme	1.38	63.8	20.4	15.8	This study
S. fulvellum	0.88	65.0	13.9	21.1	This study
S. horneri	0.64	61.1	16.1	22.8	This study
Ascophyllum nodosum	1.85	52	35	13	[39]
Chordaria flagelliformis	0.90	21	28	51	[39]
Desmarestia aculeata	0.85	27	23	50	[39]
Dictyosiphon foeniculaceus	0.85	25	25	50	[39]
Fucus serratus	1.30	35	34	31	[39]
Laminaria digitata	1.45	34	43	23	[39]
L. hyperborea, fronds	1.35	26	43	31	[39]
L. hyperborea, stripes	0.65	25	15	60	[39]
Pelvetia canaliculata	1.50	38	37	25	[38]
Pylaiella	0.75	40	18	42	[39]
Scytosiphon lomentaria	1.15	25	35	40	[39]
Spermatochnus paradoxus	1.30	35	32	33	[39]
Desmarestia ligulata	0.58	3.7	25.1	56.4	[40]
D. ligulata	0.77	3.5	37.0	47.1	[40]
Lessonia flavicans	1.03	8.5	41.3	22.2	[40]

EF: Ethanol fraction of acid-hydrolyzed sodium alginate; SF: Solid fraction at pH 2.85 (fraction enriched in polymannuronic acid); LF: Liquid fraction at pH 2.85 (fraction enriched in polyguluronic acid).

HPAEC-PAD analysis of the H_2SO_4 hydrolysate of sodium alginate showed that monosaccharides were dominated by the weight of fucose, rhamnose, arabinose, galactose, glucose, mannose, xylose, mannuronic acid, guluronic acid, and glucuronic acid (Table 3). The major alginate contents comprised mannuronic acid and guluronic acid, as expected. The content of uronic acids was almost the same as that of *S. turbinarioides* as reported by Fenoradosoa et al. [41] The main monosaccharides of alginate from EF, SF, and LF were mannuronic acid, guluronic acid, and glucuronic acid. The size of the extracted ethanol fraction ranged from 650 to 5500 Da.

Table 3. Monomeric carbohydrate contents from *Sargassum* species determined by high-performance anion-exchange chromatography with a pulsed amperometric detector (HPAEC-PAD) analysis after two-step sulfuric acid hydrolysis (mg·g^{-1}).

Species	Fraction	Fuc	Rham	Arab	Gal	Glu	Man	Xyl	Mannu	Gulur	Glucu
Sargassum fusiforme	SA	26.38 ± 0.88	ND	ND	6.21 ± 0.16	0.29 ± 0.02	7.17 ± 0.25	4.50 ± 0.18	179.64 ± 2.37	167.57 ± 1.88	10.67 ± 0.29
	SF	0.32 ± 0.01	ND	ND	0.10 ± 0.01	ND	ND	ND	54.29 ± 0.14	13.18 ± 0.51	ND
	LF	1.84 ± 0.03	ND	ND	0.69 ± 0.01	ND	0.082 ± 0.04	0.35	8.47 ± 0.09	107.57 ± 1.88	ND
	EF	7.68 ± 0.10	ND	ND	2.8 ± 0.04	ND	3.59 ± 0.04	0.87 ± 0.02	5.46 ± 0.02	11.38 ± 0.11	6.39 ± 0.06
S. fulvellum	SA	24.21 ± 0.60	0.12 ± 0.01	0.03 ± 0.00	15.07 ± 0.32	0.35 ± 0.00	8.59 ± 0.36	4.60 ± 0.05	136.79 ± 1.82	167.19 ± 1.10	20.27 ± 0.22
	SF	0.17 ± 0.00	ND	ND	ND	0.11 ± 0.00	0.48 ± 0.00	0.05 ± 0.00	42.07 ± 2.82	7.53 ± 6.79	ND
	LF	3.88 ± 0.05	0.03 ± 0.00	0.05 ± 0.00	2.40 ± 0.03	1.77 ± 0.02	2.09 ± 0.04	1.08 ± 0.03	26.18 ± 0.77	100.16 ± 0.54	4.11 ± 0.31
	EF	8.86 ± 0.06	0.05 ± 0.00	0.01 ± 0.00	4.96 ± 0.04	0.10 ± 0.03	4.42 ± 0.04	1.03 ± 0.04	9.11 ± 0.18	3.56 ± 0.20	4.24 ± 0.05
S. horneri	SA	20.14 ± 0.20	ND	ND	13.45 ± 0.09	ND	5.70 ± 0.13	4.40 ± 0.24	127.37 ± 0.39	189.99 ± 0.36	16.93 ± 0.32
	SF	0.24 ± 0.00	ND	ND	ND	0.19 ± 0.00	0.14 ± 0.00	0.09 ± 0.00	43.98 ± 5.70	7.89 ± 1.05	ND
	LF	3.26 ± 0.02	0.03 ± 0.00	0.01 ± 0.00	2.51 ± 0.02	0.16 ± 0.02	2.11 ± 0.05	0.34 ± 0.02	22.21 ± 1.08	135.61 ± 3.18	3.44 ± 0.31
	EF	9.26 ± 0.12	0.08 ± 0.00	ND	6.57 ± 0.04	0.16 ± 0.00	5.52 ± 0.08	1.03 ± 0.02	11.71 ± 0.39	4.77 ± 0.76	5.65 ± 0.32

Fuc: Fucose; Rham: Rhamnose; Arab: Arabinose; Gal: Galactose; Glu: Glucose; Man: Mannose; Xyl: Xylose; Mannu: Mannuronic acid; Gulur: Guluronic acid; Glucu: Glucuronic acid; SA: Sodium alginate; EF: Ethanol fraction; SF: Solid fraction; LF: Liquid fraction; ND: Not detected.

3.3. Elicitation by Alginate on β-Glucan Contents

From the results of oligosaccharide analysis of different fractions of *Sargassum* species, the solid fraction (SF) of *S. fusiforme* and liquid fraction (LF) of *S. horneri* were chosen as elicitors because of their high content of mannuronic acid and guluronic acid, respectively. The elicitation on *Sparassis latifolia* showed that SF treatment required a cultivation period of 47 days for mycelial growth, whereas LF treatment and the control group required a cultivation period of 43 days (Figure 3a). SF treatment provided higher mushroom production (197.30 ± 2.64 g) than the control group (192.04 ± 1.58 g), whereas LF treatment (191.05 ± 2.43 g) showed slightly lower results than the control group (Figure 3b). However, an additional spray of SF and LF elicitors resulted in lower yields of mushroom production, and this may be caused by excessive humidity as the control group also showed a decline with the second application.

These alginate-derived elicitors presented different effects on β-glucan content in the fruiting body and stalk of *S. latifolia*. SF elicitor was more effective for β-glucan enhancement in the fruiting body, whereas LF elicitor was more effective in the stalk (Figure 3c). For the second application, both SF (56.01 ± 3.45%) and LF (59.74 ± 4.49%) treatments showed obvious high increments of β-glucan contents in the stalk compared with the control (21.65%), which indicates that the additional spray of alginate oligosaccharide fractions apparently stimulated the β-glucan synthesis process in the stalk compared with that in the fruiting body (Figure 3d).

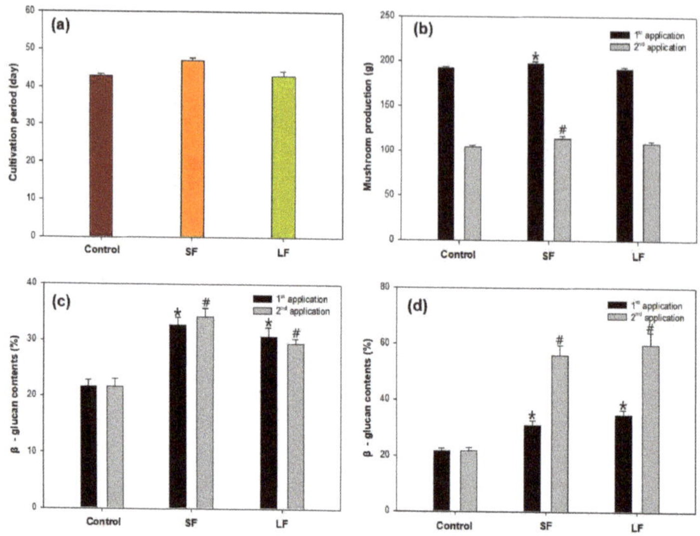

Figure 3. Elicitation of *Sparassis latifolia* by alginate oligosaccharide fractions on (**a**) cultivation period, (**b**) mushroom production, (**c**) β-glucan content of the fruiting body, and (**d**) β-glucan content of the stalk. SF: Solid fraction of *Sargassum fusiforme*; LF: Liquid fraction of *S. horneri*; Control: distilled water. (All data are presented as the mean ± SD. A *t*-test was used to compare the control and treated samples. * $p < 0.05$, vs. the control in first application; # $p < 0.05$, vs. the control in second application).

3.4. Alginate Effects on Total Polyphenol Content

After the second application of alginate oligosaccharide fractions, total polyphenol content (TPC) in the fruit body and in the stalk of *S. latifolia* was determined, and gallic acid was used as standard. The TPC of the samples ranged from 140.11 ± 6.72 to 420.87 ± 16.99 GAE mg/100g (Figure 4). There were significant differences in TPC ($p < 0.05$) depending on the mushroom part and treated elicitor. The TPC of stalk samples was higher than for the corresponding fruit body samples. LF treatment exhibited a higher TPC value than SF treatment, both in the fruit body and stalk samples.

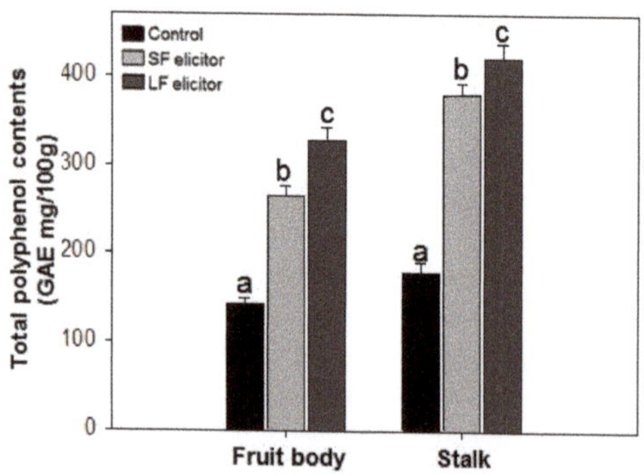

Figure 4. Total polyphenol contents of aqueous extracts of *Sparassis latifolia* after alginate elicitation. SF: Solid fraction of *Sargassum fusiforme*; LF: Liquid fraction of *S. horneri*; Control: distilled water. The data were analyzed using analysis of variance (ANOVA) followed by Duncan's multiple-range test ($p < 0.05$) using SPSS software (SPSS, Chicago, IL). Error bars indicate the mean ± SD and different letters describe greater differences within the same part of *S. latifolia* (fruit body and stalk).

3.5. Alginate Effects on Antioxidant Activities

The antioxidant activities of aqueous extracts of *S. latifolia* after alginate elicitation were measured by ABTS radical scavenging and SOD-like activity. ABTS radical scavenging ability was calculated by IC_{50} value. The results ranged from 926.67 ± 56.26 to 4002.19 ± 149.53 μg·mL^{-1} (Figure 5a). SOD-like activity was measured by a colorimetric method and calculated by IC_{50} value. IC_{50} values of SOD-like activity ranged from 6.74 ± 0.29 to 18.70 ± 0.57 mg·mL^{-1} (Figure 5b). The highest ABTS radical scavenging and SOD-like activity were recorded with LF elicitor treated stalk and the lowest were from the fruit body of the control group. Similar to the TPC results, both fruit body and stalk samples of LF treatment exhibited higher antioxidant activities than the corresponding samples treated by SF elicitor. The results of antioxidant activities were significantly ($p < 0.05$) stimulated with the alginate elicitation.

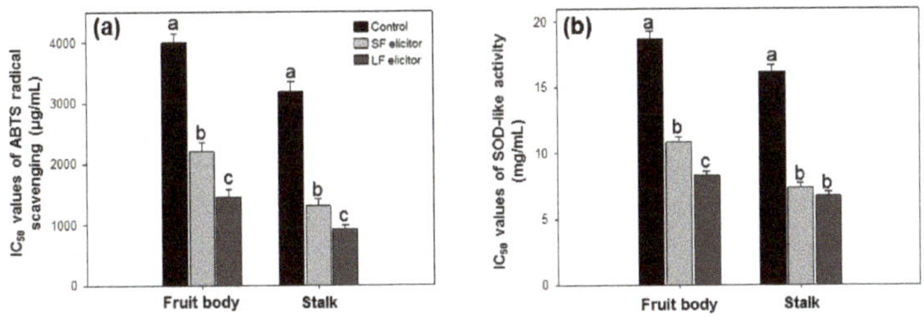

Figure 5. IC_{50} values of ABTS radical scavenging (**a**) and SOD-like activity (**b**) of aqueous extracts of *Sparassis latifolia* after alginate elicitation. SF: Solid fraction of *Sargassum fusiforme*; LF: Liquid fraction of *S. horneri*; Control: distilled water. The data were analyzed using analysis of variance (ANOVA) followed by Duncan's multiple-range test ($p < 0.05$) using SPSS software (SPSS, Chicago, IL, USA). Error bars indicate the mean ± SD and different letters describe great differences within the same part of *S. latifolia* (fruit body and stalk).

4. Discussion

Sparassis species are widely used medicinal mushrooms in traditional Chinese medicine. However, despite its commercial potential, cultivation of *S. latifolia* has been limited to a few high-tech commercial farms because of its slow mycelial propagation into solid medium [42]. In addition, its slow rate of growth is a major obstacle to the employment of *S. latifolia* as a producer of β-glucan. Biotic or abiotic elicitors have been employed to enhance the production of β-glucan in *S. latifolia*. Ryoo et al. exploited the effects of physical stimulation of UV irradiation and temperature shock on β-glucan contents in *S. latifolia* and found that β-glucan yields reached 41.36 ± 2.96% of flabella and 42.16 ± 2.90% of stipe after UV irradiation for 10 min [22]. Park et al. utilized chitinase, β-glucuronidase, and lysing enzyme complex as elicitors to enhance the β-glucan content of *S. latifolia* and produced an increase in β-glucan concentration of 31%, although the treatment caused a decrease in mushroom yield [23]. These results indicate that the elicitation technique needs to be evaluated for application in high-value mushroom cultivation.

This study aimed to increase the contents of β-glucan in *S. latifolia* through alginate elicitation, which was extracted from *Sargassum* species. Marine algae are an excellent biomass source used as a fertilizer for soil reformation because they are rich in carbohydrates and mineral content [43]. However, although *Sargassum* species contain a large amount of alginate and fucoidan, the high content of arsenic makes them unsuitable for consumption [44,45]. Polysaccharide-derived hydrocolloids in marine algae such as alginate, carrageenan, fucoidan, laminarin, ulvan, and glucuronan have been suggested as biotic elicitors to induce several different mechanisms including host defense mechanisms and growth enhancement in plant and fungi [46–49].

Alginate is the major structural polysaccharide of *Sargassum* species, consisting of MM, GG, and MG blocks arranged in various proportions in block units. The alginate M/G ratio and yield may vary in accordance with the algae harvest season and geographical location [41]. Leal et al. studied block fractions of alginate in the brown algae *Lessonia flavicans* and *Desmarestia ligulata* [40]. The ethanol fraction obtained by partial acid hydrolysis of alginate was mainly composed of a heteropolymer block (MG). In addition, the solid fraction was rich in polymannuronic acid (MM) and the liquid fraction was rich in polyguluronic acid (GG) [28]. These were consistent with the results found in this study (see Table 2).

Alginate elicitors have been employed to modify cell metabolism in order to enhance the productivity of useful metabolites in plants and microorganisms [50]. The addition of chitosan, chitosan oligosaccharide, and alginate oligosaccharide to a culture of *Panax ginseng* C. A. Mey hairy roots caused growth inhibition and rises in total ginseng saponin accumulation with elicitor concentration [51]. Mannuronic acid of alginate may have a potent stimulatory effect on cytokine production, and it appeared to affect β-glucan content because of increased specific immunity owing to immunological effects [52]. In contrast, guluronic acid suppresses tissue damage caused by cytokines produced in response to inflammatory stimuli [53]. Our group investigated the supplement of sodium alginate to *S. latifolia* for the induced expression of γ-aminobutyric acid (GABA) both in the mycelia and fruiting bodies, which inhibited the dendrite outgrowth of excitatory neurons, but not that of inhibitory neurons [54]. Genome sequencing and genomics studies are underway to determine the underlying effect of alginate elicitors on β-glucan content of *S. latifolia* [55].

The effects of alginate elicitors on antioxidant activity and polyphenol synthesis in plants and microorganisms were also verified. Chitosan, pectin, and alginate promoted accumulation of phenolic acids, particularly 3-*O*-glucosyl-resveratrol, in *Vitis vinifera*, which was positively correlated with increased accumulation of anthocyanin [25]. Ulvan, carrageenan, alginate, and laminarin were examined for any elicitation effect in twigs of olive trees to elicit phenolic metabolism and control against *Verticillium* wilt of olive caused by *V. dahlia*, and the results showed increased phenylalanine ammonia-lyase (PAL) activity and total polyphenol content combined with the decline of wilt symptoms [27]. Sodium alginate extracted from Moroccan brown algae *Fucus spiralis* and *Bifurcaria bifurcata* was evaluated for elicitation on phenolic metabolism including PAL activity and total polyphenol content in seedling roots of date palm, and the results showed that PAL activity and phenolic compound content were stimulated with 1 mg·mL^{-1} sodium alginate [26].

The present study suggested that treatment with *Sargassum* alginate exhibited elicitation effects on the growth and β-glucan contents in *S. latifolia*. Elicitors derived from algae alginate can be widely used in the cultivation process of other culinary and medicinal mushrooms.

5. Conclusions

Based on our results, the products of partial acid hydrolysis of alginate oligosaccharides could act as elicitors for stimulated growth of *S. latifolia* and production of useful metabolites such as β-glucan and polyphenols.

Author Contributions: Conceptualization and methodology, Y.-W.K. and H.-J.S.; formal analysis and investigation, Y.-W.K. and M.-H.C.; software and validation, Y.W. and J.L.; resources, M.-H.C.; data curation, Y.-W.K.; writing—original draft preparation, Y.-W.K.; writing—review and editing, Y.W.; supervision, H.-J.S.; project administration, H.-J.S.; funding acquisition, H.-J.S. and J.L. All authors have read and agreed to the published version of the manuscript.

Funding: This research was funded by a National Research Foundation of Korea (NRF) grant funded by the Korean government, Grant No. NRF-2017R1A2B4006204; Shandong Provincial Key Research and Development Program (International Science and Technology Cooperation), Grant No. 2019GHZ033; and International Science and Technology Cooperation Program of Shandong Academy of Sciences, Grant No. 2019GHPY05.

Conflicts of Interest: The authors declare no conflict of interest. The funders had no role in the design of the study; in the collection, analyses, or interpretation of data; in the writing of the manuscript; or in the decision to publish the results.

References

1. Roncero-Ramos, I.; Delgado-Andrade, C. The beneficial role of edible mushrooms in human health. *Curr. Opin. Food Sci.* **2017**, *14*, 122–128. [CrossRef]
2. Kimura, T. Natural products and biological activity of the pharmacologically active cauliflower mushroom *Sparassis crispa*. *BioMed Res. Int.* **2013**, *2013*, 982317. [CrossRef] [PubMed]
3. Ryoo, R.; Sou, H.D.; Ka, K.H.; Park, H. Phylogenetic relationships of Korean *Sparassis latifolia* based on morphological and its rDNA characteristics. *J. Microbiol.* **2013**, *51*, 43–48. [CrossRef] [PubMed]
4. Zhao, Q.; Feng, B.; Yang, Z.L.; Dai, Y.C.; Wang, Z.; Tolgor, B. New species and distinctive geographical divergences of the genus *Sparassis* (Basidiomycota): Evidence from morphological and molecular data. *Mycol. Prog.* **2013**, *12*, 445–454. [CrossRef]
5. Ryu, S.R.; Ka, K.H.; Park, H.; Bak, W.C.; Lee, B.H. Cultivation characteristics of *Sparassis crispa* strains using sawdust medium of *Larix kaempferi*. *Korean J. Mycol.* **2009**, *37*, 49–54. [CrossRef]
6. Chandrasekaran, G.; Oh, D.S.; Shin, H.J. Properties and potential applications of the culinary-medicinal cauliflower mushroom, *Sparassis crispa* Wulf.Fr. (Aphyllophoromycetideae): A review. *Int. J. Med. Mushrooms* **2011**, *13*, 177–183. [CrossRef]
7. Yoshikawa, K.; Kokudo, N.; Hashimoto, T.; Yamamoto, K.; Inose, T.; Kimura, T. Novel phthalide compounds from *Sparassis crispa* (Hanabiratake), Hanabiratakelide A-C, exhibiting anti-cancer related activity. *Biol. Pharm. Bull.* **2010**, *33*, 1355–1359. [CrossRef]
8. Bang, S.; Chae, H.S.; Lee, C.; Choi, H.G.; Ryu, J.; Li, W.; Lee, H.; Jeong, G.S.; Chin, Y.W.; Shim, S.H. New aromatic compounds from the fruiting body of *Sparassis crispa* (Wulf.) and their inhibitory activities on proprotein convertase subtilisin/kexin type 9 mRNA expression. *J. Agric. Food Chem.* **2017**, *65*, 6152–6157. [CrossRef]
9. Ruthes, A.C.; Smiderle, F.R.; Iacomini, M. D-Glucans from edible mushrooms: A review on the extraction, purification and chemical characterization approaches. *Carbohyd. Polym.* **2015**, *117*, 753–761. [CrossRef]
10. Synytsya, A.; Novák, M. Structural diversity of fungal glucans. *Carbohyd. Polym.* **2013**, *92*, 792–809. [CrossRef] [PubMed]
11. Smiderle, F.R.; Alquini, G.; Tadra-Sfeir, M.Z.; Iacomini, M.; Wichers, H.J.; van Griensven, L.J. *Agaricus bisporus* and *Agaricus brasiliensis* (1→6)-β-D-glucans show immunostimulatory activity on human THP-1 derived macrophages. *Carbohyd. Polym.* **2013**, *94*, 91–99. [CrossRef] [PubMed]
12. Morales, D.; Smiderle, F.R.; Villalva, M.; Abreu, H.; Rico, C.; Santoyo, S.; Iacomini, M.; Soler-Rivas, C. Testing the effect of combining innovative extraction technologies on the biological activities of obtained β-glucan-enriched fractions from *Lentinula edodes*. *J. Funct. Foods* **2019**, *60*, 103446. [CrossRef]
13. Chan, G.C.; Chan, W.K.; Sze, D.M. The effects of β-glucan on human immune and cancer cells. *J. Hematol. Oncol.* **2009**, *2*, 25. [CrossRef] [PubMed]
14. Ohno, N.; Miura, N.N.; Nakajima, M.; Yadomae, T. Antitumor 1,3-β-glucan from cultured fruit body of *Sparassis crispa*. *Biol. Pharm. Bull.* **2000**, *23*, 866–872. [CrossRef]
15. Harada, T.; Miura, N.N.; Adachi, Y.; Nakajima, M.; Yadomae, T.; Ohno, N. IFN-γ induction by SCG, 1, 3-β-D-glucan from *Sparassis crispa*, in DBA/2 mice in vitro. *J. Interf. Cytok. Res.* **2002**, *22*, 1227–1239. [CrossRef]
16. Harada, T.; Kawaminami, H.; Miura, N.N.; Adachi, Y.; Nakajima, M.; Yadomae, T.; Ohno, N. Comparison of the immunomodulating activities of 1,3-β-glucan fractions from the culinary-medicinal mushroom *Sparassis crispa* Wulf.: Fr. (Aphyllophoromycetideae). *Int. J. Med. Mushrooms* **2006**, *8*, 231–244. [CrossRef]
17. Tada, R.; Harada, T.; Nagi-Miura, N.; Adachi, Y.; Nakajima, M.; Yadomae, T.; Ohno, N. NMR characterization of the structure of a β-(1→3)-D-glucan isolate from cultured fruit bodies of *Sparassis crispa*. *Carbohyd. Res.* **2007**, *342*, 2611–2618. [CrossRef]
18. Nameda, S.; Harada, T.; Miura, N.N.; Adachi, Y.; Yadomae, T.; Nakajima, M.; Ohno, N. Enhanced cytokine synthesis of leukocytes by a β-glucan preparation, SCG, extracted from a medicinal mushroom, *Sparassis crispa*. *Immunopharm. Immunot.* **2003**, *25*, 321–335. [CrossRef]
19. Park, H.G.; Shim, Y.Y.; Choi, S.O.; Park, W.M. New method development for nanoparticle extraction of water-soluble β-(1→3)-D-glucan from edible mushrooms, *Sparassis crispa* and *Phellinus linteus*. *J. Agric. Food Chem.* **2009**, *57*, 2147–2154. [CrossRef]

20. Narayani, M.; Srivastava, S. Elicitation: A stimulation of stress in in vitro plant cell/tissue cultures for enhancement of secondary metabolite production. *Phytochem. Rev.* **2017**, *16*, 1227–1252. [CrossRef]
21. Pettit, R.K. Small-molecule elicitation of microbial secondary metabolites. *Microb. Biotechnol.* **2011**, *4*, 471–478. [CrossRef] [PubMed]
22. Ryoo, R.; Sou, H.D.; Ka, K.H.; Park, H. Elicitor-induced β-glucan contents in fruit body of cauliflower mushroom (*Sparassis latifolia*). *For. Sci. Technol.* **2018**, *14*, 119–125. [CrossRef]
23. Park, H.; Ka, K.H.; Ryu, S.R. Enhancement of β-glucan content in the cultivation of cauliflower mushroom (*Sparassis latifolia*) by elicitation. *Mycobiology* **2014**, *42*, 41–45. [CrossRef] [PubMed]
24. Salachna, P.; Grzeszczuk, M.; Meller, E.; Soból, M. Oligo-alginate with low molecular mass improves growth and physiological activity of *Eucomis autumnalis* under salinity stress. *Molecules* **2018**, *23*, 812. [CrossRef]
25. Cai, Z.; Kastell, A.; Mewis, I.; Knorr, D.; Smetanska, I. Polysaccharide elicitors enhance anthocyanin and phenolic acid accumulation in cell suspension cultures of *Vitis vinifera*. *Plant Cell Tissue Organ Cult.* **2012**, *108*, 401–409. [CrossRef]
26. Bouissil, S.; Alaoui-Talibi, E.; Pierre, G.; Michaud, P.; El Modafar, C.; Delattre, C. Use of alginate extracted from Moroccan brown algae to stimulate natural defense in date palm roots. *Molecules* **2020**, *25*, 720. [CrossRef]
27. Ben Salah, I.; Aghrouss, S.; Douira, A.; Aissam, S.; El Alaoui-Talibi, Z.; Filali-Maltouf, A.; El Modafar, C. Seaweed polysaccharides as bio-elicitors of natural defenses in olive trees against verticillium wilt of olive. *J. Plant Interact.* **2018**, *13*, 248–255. [CrossRef]
28. Pawar, S.N.; Edgar, K.J. Alginate derivatization: A review of chemistry, properties and applications. *Biomaterials* **2012**, *33*, 3279–3305. [CrossRef]
29. Silva, T.H.; Alves, A.; Ferreira, B.M.; Oliveira, J.M.; Reys, L.L.; Ferreira, R.J.; Sousa, R.A.; Silva, S.S.; Mano, J.F.; Reis, R.L. Materials of marine origin: A review on polymers and ceramics of biomedical interest. *Int. Mater. Rev.* **2012**, *57*, 276–306. [CrossRef]
30. Davis, T.A.; Ramirez, M.; Mucci, A.; Larsen, B. Extraction, isolation and cadmium binding of alginate from *Sargassum* spp. *J. Appl. Phycol.* **2004**, *16*, 275–284. [CrossRef]
31. Asilonu, E.; Bucke, C.; Keshavarz, T. Enhancement of chrysogenin production in cultures of *Penicillium chrysogenum* by uronic acid oligosaccharides. *Biotechnol. Lett.* **2000**, *22*, 931–936. [CrossRef]
32. Alban, S.; Schauerte, A.; Franz, G. Anticoagulant sulfated polysaccharides. I. Synthesis and structure-activity relationships of new pullulan sulfates. *Carbohyd. Polym.* **2001**, *47*, 267–276. [CrossRef]
33. Grasdalen, H. High-field, ^{1}H-n.m.r. spectroscopy of alginate: Sequential structure and linkage conformations. *Carbohyd. Res.* **1983**, *118*, 255–260. [CrossRef]
34. Manns, D.; Deutschle, A.L.; Saake, B.; Meyer, A.S. Methodology for quantitative determination of the carbohydrate composition of brown seaweeds (Laminariaceae). *RSC Adv.* **2014**, *4*, 25736–25746. [CrossRef]
35. McCleary, B.V.; Draga, A. Measurement of β-glucan in mushrooms and mycelial products. *J. AOAC Int.* **2016**, *99*, 364–373. [CrossRef]
36. Wootton-Beard, P.C.; Moran, A.; Ryan, L. Stability of the total antioxidant capacity and total polyphenol content of 23 commercially available vegetable juices before and after in vitro digestion measured by FRAP, DPPH, ABTS and Folin–Ciocalteu methods. *Food Res. Int.* **2011**, *44*, 217–224. [CrossRef]
37. Re, R.; Pellegrini, N.; Proteggente, A.; Pannala, A.; Yang, M.; Rice-Evans, C.A. Antioxidant activity applying an improved ABTS radical cation decolourising assay. *Free Radic. Biol. Med.* **1999**, *26*, 1231–1237. [CrossRef]
38. Odeyemi, S.; Dewar, J. Repression of acetaminophen-induced hepatotoxicity in HepG2 cells by polyphenolic compounds from *Lauridia tetragona* (L.f.) R.H. Archer. *Molecules* **2019**, *24*, 2118. [CrossRef]
39. Haug, A.; Larsen, B.; Smidsrød, O. Uronic acid sequence in alginate from different sources. *Carbohyd. Res.* **1974**, *32*, 217–225. [CrossRef]
40. Leal, D.; Matsuhiro, B.; Rossi, M.; Caruso, F. FT-IR spectra of alginic acid block fractions in three species of brown seaweeds. *Carbohyd. Res.* **2008**, *343*, 308–316. [CrossRef]
41. Fenoradosoa, T.A.; Ali, G.; Delattre, C.; Laroche, C.; Petit, E.; Wadouachi, A.; Michaud, P. Extraction and characterization of an alginate from the brown seaweed *Sargassum turbinarioides* Grunow. *J. Appl. Phycol.* **2010**, *22*, 131–137. [CrossRef]
42. Kim, S.R.; Kang, H.W.; Ro, H.S. Generation and evaluation of high β-glucan producing mutant strains of *Sparassis crispa*. *Mycobiology* **2013**, *41*, 159–163. [CrossRef] [PubMed]
43. Kumar, C.S.; Ganesan, P.; Suresh, P.V.; Bhaskar, N. Seaweeds as a source of nutritionally beneficial compounds-a review. *J. Food Sci. Technol.* **2008**, *45*, 1.

44. Sinha, S.; Astani, A.; Ghosh, T.; Schnitzler, P.; Ray, B. Polysaccharides from *Sargassum tenerrimum*: Structural features, chemical modification and anti-viral activity. *Phytochemistry* **2010**, *71*, 235–242. [CrossRef] [PubMed]
45. Yokoi, K.; Konomi, A. Toxicity of so-called edible hijiki seaweed (*Sargassum fusiforme*) containing inorganic arsenic. *Regul. Toxicol. Pharmacol.* **2012**, *63*, 291–297. [CrossRef] [PubMed]
46. Bi, F.; Iqbal, S.; Arman, M.; Ali, A.; Hassan, M.U. Carrageenan as an elicitor of induced secondary metabolites and its effects on various growth characters of chickpea and maize plants. *J. Saudi Chem. Soc.* **2011**, *15*, 269–273. [CrossRef]
47. Chandía, N.P.; Matsuhiro, B. Characterization of a fucoidan from *Lessonia vadosa* (Phaeophyta) and its anticoagulant and elicitor properties. *Int. J. Biol. Macromol.* **2008**, *42*, 235–240. [CrossRef]
48. Aziz, A.; Poinssot, B.; Daire, X.; Adrian, M.; Bézier, A.; Lambert, B.; Joubert, J.M.; Pugin, A. Laminarin elicits defense responses in grapevine and induces protection against *Botrytis cinerea* and *Plasmopara viticola*. *Mol. Plant Microbe Interact.* **2003**, *16*, 1118–1128. [CrossRef]
49. El Modafar, C.; Elgadda, M.; El Boutachfaiti, R.; Abouraicha, E.; Zehhar, N.; Petit, E.; El Alaoui- Talibi, Z.; Courtois, B.; Courtois, J. Induction of natural defense accompanied by salicylic acid-dependant systemic acquired resistance in tomato seedlings in response to biloelicitors isolated from green algae. *Sci. Hortic.* **2012**, *138*, 55–63. [CrossRef]
50. Chandía, N.P.; Matsuhiro, B.; Mejías, E.; Moenne, A. Alginic acids in *Lessonia vadosa*: Partial hydrolysis and elicitor properties of the polymannuronic acid fraction. *J. Appl. Phycol.* **2004**, *16*, 127–133. [CrossRef]
51. Jeong, G.T.; Park, D.H.; Ryu, H.W.; Hwang, B.; Woo, J.C.; Kim, D.; Kim, S.W. Production of antioxidant compounds by culture of *Panax ginseng* C.A. Meyer hairy roots: I. Enhanced production of secondary metabolite in hairy root cultures by elicitation. *Appl. Biochem. Biotechnol.* **2005**, *121*, 1147–1157. [CrossRef]
52. Yamamoto, Y.; Kurachi, M.; Yamaguchi, K.; Oda, T. Stimulation of multiple cytokine production in mice by alginate oligosaccharides following intraperitoneal administration. *Carbohyd. Res.* **2007**, *342*, 1133–1137. [CrossRef] [PubMed]
53. Iwamoto, M.; Kurachi, M.; Nakashima, T.; Kim, D.; Yamaguchi, K.; Oda, T.; Iwamoto, Y.; Muramatsu, T. Structure–activity relationship of alginate oligosaccharides in the induction of cytokine production from RAW264. 7 cells. *FEBS Lett.* **2005**, *579*, 4423–4429. [CrossRef] [PubMed]
54. Choi, M.H.; Ki, S.; Lee, S.E.; Lee, G.; Shin, H.J. Enhanced GABA content from sodium alginate-induced *Sparassis latifolia* influences dendrite development in primary cortical neurons. *J. Mushroom* **2019**, *17*, 275–283.
55. Kiyama, R.; Furutani, Y.; Kawaguchi, K.; Nakanishi, T. Genome sequence of the cauliflower mushroom *Sparassis crispa* (Hanabiratake) and its association with beneficial usage. *Sci. Rep.* **2018**, *8*, 16053. [CrossRef] [PubMed]

Journal of *Fungi*

MDPI

Article

Secondary Metabolites and Antioxidant Activity of the Solid-State Fermentation in Apple (*Pirus malus* L.) and Agave Mezcalero (*Agave angustifolia* H.) Bagasse

Diego Ibarra-Cantún [1], María Elena Ramos-Cassellis [2,*], Marco Antonio Marín-Castro [3] and Rosalía del Carmen Castelán-Vega [3]

1 Posgrado en Ciencias Ambientales, Benemérita Universidad Autónoma de Puebla, Col. Jardines de San Manuel, Edificio IC6, 72570 Puebla, Mexico; diego.ibarrac@alumno.buap.mx

2 Facultad de Ingeniería Química, Benemérita Universidad Autónoma de Puebla, Av. San Claudio y 18 Sur, 72570 Puebla, Mexico

3 Departamento de Investigación en Ciencias Agrícolas, Benemérita Universidad Autónoma de Puebla, 14 Sur 6301 Edificio IC1, 72570 Puebla, Mexico; marco.marin@correo.buap.mx (M.A.M.-C.); rosalia.castelan@correo.buap.mx (R.d.C.C.-V.)

* Correspondence: elena.ramos@correo.buap.mx; Tel.: +52-222-229-55-00 (ext. 7356)

Received: 30 June 2020; Accepted: 3 August 2020; Published: 18 August 2020

Abstract: Solid-state fermentation (SSF) is used in enzyme and antibiotic production, bioethanol and biodiesel as an alternative energy source, biosurfactants with environmental goals, and the production of organic acids and bioactive compounds. The present project determined the quantity of secondary metabolites and the antioxidant activity of the extracts obtained by the solid-state fermentation of apple and agave mezcalero bagasse over 28 days, inoculated with the *Pleurotus ostreatus* strain. The extraction was carried out with three solvents: acetone and water (80:20 *v/v*), 100% methanol and 100% water. The results showed a higher presence of phenolic compounds, flavonoids, total triterpenes and antioxidant activity in the apple bagasse from the SSF on day 21 in the extract of acetone and water (80:20 *v/v*), 100% methanol and aqueous; while the agave bagasse showed a significant presence of phenolic compounds and flavonoids only in the aqueous extract. In conclusion, the presence of secondary metabolites exhibiting antioxidant activities from the solid-state fermentation in the residues of the cider and mezcal industry is an alternative use for wasted raw material, plus, it reduces the pollution generated from the agroindustrial residues.

Keywords: agave mezcalero bagasse; apple bagasse; solid-state fermentation; secondary metabolites; *Pleurotus ostreatus*

1. Introduction

The residues generated by the agroindustry are, in most cases, not processed, causing environmental pollution and adverse effects on health [1]. In Mexico, this situation is mainly reflected in the production of the cider industry in Puebla and the production of mezcal in the state of Oaxaca. The former reported an annual production of 150 thousand bottles [2], where the processed fruit residues fluctuate between 15% and 30% of said production, depending on the state of maturation and the used technology [3]. The latter generated 122,696 tons of bagasse in 2012, which is usually dumped in rivers or streams, endangering the region's environment [4].

Most of these wastes degrade slowly due to their solid and organic nature, which corresponds to their lignocellulosic biomass, rich in cellulose, hemicellulose and lignin. In addition, they are not subjected to an adequate process of exploitation, giving as a result an utterly deficient disposal of the waste [5,6]. Therefore, the aim of the present investigation was to determine the content of the total phenolic compounds, total flavonoids and total triterpenes, and the antioxidant activity, by 2,2-diphenyl-1-picrylhydrazyl (DPPH•) and 2,2′-azino-bis-3-ethylbenzothiazoline-6-sulfonic acid radical cation (ABTS•+) test methods resulting from the solid-state fermentation of apple bagasse and agave mezcalero bagasse substrates inoculated with the *Pleurotus ostreatus* strain, in order to optimize the utilization of residues of the cider and mezcalero industry in the attaining of secondary metabolites.

Pleurotus ostreatus synthesizes bioactive compounds, such as: polysaccharides (alpha and beta glucan) and polyphenols [7,8]. The macromycetes, such as *Pleurotus* and *Agaricus*, contain tetracyclic triterpenoids, with the sterols being the most abundant ones. At the same time, in *Pleurotus mutilis* and *Clitopilus passeckerianus* (called *Pleurotus passeckerianus* in the past), pleuromutilin and diterpene were identified as having antibiotic action against microplasmatic infections in animals [9,10].

Singhania et al. [11] and Thomas et al. [12] describe solid-state fermentation (SSF) as a process that takes place in a solid matrix that is within an inert support or substrate, or near free water, with just enough to promote the growth and metabolic activity of microorganisms. The SSF reduces the production costs due to the lower energy consumption and production area (smaller reactors), proving that it is an economic, interesting and environmentally friendly process for the production of secondary metabolites from agroindustrial and forest residues [13]. The SSF has generated great interest in recent years due to its variety of uses, such as bioremediation, biofuels (biodiesel, bioethanol, biobutanol, and biohydrogen among others) and the production or extraction of bioactive compounds [14].

One of the main applications of the solid fermentation of agroindustrial wastes is the extraction of secondary metabolites, also called phytochemicals, which include carotenes, terpenes, sterols, tocopherols and polyphenols. These can be used in the treatment and prevention of various diseases [15], since they have antitumor, anti-inflammatory, antithrombotic, antimicrobial, immune system modulating, and hypoglycemic properties; also, they reduce the concentration of lipids in the blood and display antioxidant activities [16,17]. Waste from agribusiness, such as the peel of fruits and vegetables, has been employed recently as a potential source of production of bioactive compounds that are commonly known as phenolic compounds [18]. Therefore, the secondary metabolites obtained from the solid fermentation of residues can be used as natural bioactive triggers for the formulation of functional foods or serve as additives in food products to extend their shelf life [19].

2. Materials and Methods

2.1. Obtaining Plant Material

The apple bagasse was collected from Huejotzingo's cider industry, a town located in the state of Puebla, Mexico, and the agave mezcalero bagasse was collected from San Pablo Villa de Mitla's mezcal industry, a town located in the state of Oaxaca, Mexico.

These bagasses were transported in airtight containers at 20 °C in less than 4 h. They were subsequently bleached at 70 °C for 5 min and dried in a forced air dryer (HTP 72 Lumistell Celaya, Gto. México) for 8 h at 60 °C until the moisture content was less than 10%. Once dried, the bagasse was crushed in a mill to obtain a particle of 0.841 mm to be used during solid fermentation.

Physicochemical tests were performed on the crushed bagasse in order to discover its attributes and adjust its physicochemical properties to make them suitable for the growth of the *Pleurotus* fungus. The bagasse reconditioning was carried out according to Staments and Chilton [20] where it is mentioned that *Pleurotus ostreatus* culture must have the following conditions: a humidity percentage of 80–90%, temperature from 23 to 26 °C, and pH from 6.5 to 7.5. The temperature and relative humidity were controlled during the process. The physicochemical tests performed were: total soluble solids (°Brix) [21], pH [21], moisture percentage [21], water activity (*Aw*) [21], and titratable acidity percentage [21]; the meq value of malic acid were 0.067, the meq value of acetic acid were 0.060 and direct reducing sugars were also used [21].

2.2. Solid-State Fermentation

This process was accomplished with the contribution of the Mycology Laboratory of the Departamento de Investigación de Ciencias Agrícolas (Agricultural Sciences Research Department) of the Benemérita Universidad Autónoma de Puebla (Benemeritus Autonomous University of Puebla), which provided the *Pleurotus ostreatus* strains.

The solid-state fermentation was carried out in 9.7 × 2.6 cm glass tubes with a Bakelite cork, and each contained 10 g of each residue adjusted to a humidity percentage of 80% and a pH of 6.5, which was then sterilized in an autoclave ((CV300 AESA Tecámac, Edo. México, México)) at 120 °C and 15 psi for 45 min.

Then, a 1 cm diameter mycelium-agar circle of the *Pleurotus ostreatus* strain, corresponding to 3 mg of dry weight biomass, was incorporated into each tube and incubated in an oven (E51 Riossa Monterrey, N.L., México) at 25 °C. Samples were taken at 0, 7, 14, 21 and 28 days.

2.3. Solid Waste Extraction

After the fermentation days, 12 mL of each solvent was added to each glass tube for extraction. The used solvents were acetone and water (80:20 *v/v*), 100% methanol and distilled water. The samples were placed in a shaker (KJ-201BD Orbital Shaker Westtune Hangzhou, Zhejiang, China) at 80 rpm for 1 h, filtered and placed in a centrifuge (Z200A Hermle Wehingen, Germany) at 47 g, and the ensuing supernatant was stored in Eppendorf tubes at −20 °C for further analysis.

2.4. Quantification of Total Phenolic Compounds (TPC)

The analysis was performed using the method described by Singleton and Rossi [22], with some modifications. A total of 250 μL of 50% Folin–Ciocalteu reagent was added to 50 μL of the extract; then it was stirred and allowed to stand in the dark for 8 min. Subsequently, 1.25 mL of 5% $CaCO_3$ (*w/v*) was added and allowed to stand again for 30 min in the dark at room temperature. An absorbance at 725 nm was read afterwards on a UV-Vis spectrophotometer (7305 Jenway Staffordshire, UK). A calibration curve for gallic acid (Sigma Aldrich, CAS 149-91-7, Steinheim, Germany) was used in the range of 0 to 0.3 g/L. The results were expressed in mg gallic acid equivalent per g substrate on dry weight (mg GAE/g substrate dw). These analyses were performed in triplicate.

2.5. Total Flavonoid Quantification (TF)

The methodology proposed by Chang et al. [23] was used, with some modifications. A total of 500 μL of the extract, 1.5 mL of 80% ethanol (*v/v*), 100 μL of 10% hexahydrate aluminum chloride solution, 100 μL of 1 M potassium acetate, and 2.8 mL of distilled water were placed, stirred and incubated for 30 min at room temperature.

Subsequently, an absorbance at 415 nm was read on a UV-Vis spectrophotometer (7305 Jenway Staffordshire, UK). A quercetin calibration curve (Sigma Aldrich, CAS 117-39-5, Germany) was used at a concentration of 0 to 0.1 g/L. The results were expressed in mg quercetin equivalents per g substrate on dry weight (mg QE/g substrate dw). These analyzes were performed in triplicate.

2.6. Total Triterpenes Quantification (TT)

The vanillin-acetic acid method was used, with some modifications [24]. A total of 120 μL of each extract, 100 μL of vanillin (Sigma Aldrich, CAS 121-33-5, Germany) 5% (*w/v*), and 400 μL of perchloric acid in 0.1 N glacial acetic acid were placed together in a test tube at 60 °C for 15 min.

Then, the mixture was cooled down to room temperature and 2.5 mL of glacial acetic acid were added. The absorbance was measured at 550 nm in a UV-Vis spectrophotometer (7305 Jenway Staffordshire, UK). The results were expressed as mg ursolic acid per g substrate on dry weight (mg UA/g substrate dw). A calibration curve for uric acid (Sigma Aldrich, CAS 77-52-1, Germany) was used with a concentration from 0 to 120 g/L. These analyses were performed in triplicate.

2.7. Antioxidant Activity by 1,1-Diphenyl-2-Picril Hydracil (DPPH•)

The DPPH• (2,2-Diphenyl-1-picrilhydrazyl) radical in methanol solution was developed by Brand-Williams et al. [25], and was used with some alterations. A total of 990 μL of DPPH (Sigma Aldrich, CAS 1898-66-4, Germany) 0.188 mM was added to 10 μL of the extract. The mixture was then homogenized and kept in the dark for 30 min at room temperature.

The absorbance at 517 nm was measured in both the control (methanol (A0)) and the test samples (A1) on a UV-Vis spectrophotometer (7305 Jenway Staffordshire, UK). In order to calculate the percentage of inhibition, Equation (1) was used. These analyses were performed in triplicate.

Subsequently, the results were expressed as the inhibitory concentration (IC_{50}) and the percentage of inhibition of four different concentrations was calculated to obtain a linear regression curve. These analyses were performed in triplicate.

$$\text{Percent inhibition} = \left[\frac{A0 - A1}{A0}\right] * 100 \quad (1)$$

A0 = Control absorbance
A1 = Sample absorbance.

2.8. Antioxidant Activity by 2,2′-Azino-Bis-3-Ethylbenzothiazoline-6-Sulfonic Acid Radical Cation (ABTS•+)

The method proposed by Re et al. [26] was used. An ABTS•+ radical cation was formed from 0.0033 g of potassium persulfate and 0.0194 g of ABTS reagent (2,2′-azinobis-(3-ethylbenzothiazolin-6-sulfonic acid (Sigma Aldrich, CAS 30931-67-0, St. Louise, MO, USA) mixed in 5 mL of distilled water. The mixture was stirred and kept in the dark for 16 h at room temperature. After this, a mixture was made with absolute ethanol and the ABTS•+ radical cation until an absorbance of 0.70 ± 0.02 at 754 nm was obtained.

A total of 3920 μL of the ABTS radical solution were then added and the initial absorbance (Ai) was recorded; 80 μL of extract was afterwards added and mixed. After 7 min, the final absorbance (Af) was recorded with the use of a UV-Vis spectrophotometer (7305 Jenway Staffordshire, UK). The inhibition percentage was calculated according to Equation (2) to obtain the medium inhibitory concentration (IC_{50}) by a curve of lineal regression using the inhibitory percentage of the four different concentrations. These analyses were performed in triplicate.

$$\text{Percent inhibition } (\%) = \left[\frac{Ai - Af}{Ai}\right] * 100 \quad (2)$$

2.9. Statistic Analysis

The data analysis was performed through a completely randomized design. The experiment consisted of the inoculation of *Pleurotus ostreatus* in two bagasses (two treatments with five replicas) at 25 °C for 28 days. Samples of the substrates were taken at 0, 7, 14, 21 and 28 days of fermentation (5 periods) to obtain the extracts with three organic solvents: acetone and water (80:20 *v/v*), 100% methanol, and water at 100% (3 extracts); the secondary metabolite content and antioxidant activity were determined as well. Samples were analyzed in quadruplicate ($n = 4$).

The results were expressed as a mean and standard deviation, and were examined through the Statistical Analysis System (SAS) version 9.0 statistical package [27]. Additionally, an analysis of variance (ANOVA) means test (Tukey $\alpha = 0.05$) and Pearson correlation were performed.

3. Results

3.1. Physicochemical Characterization

Table 1 shows the physicochemical characterization of the residues before fermentation. The values in percent humidity and *Aw* are similar since both bagasses were subjected to the same drying process at 60 °C for 12 h and then they were crushed.

Table 1. Physicochemical characteristics of apple bagasse and agave mezcalero bagasse.

Agroindustrial Residue	°Brix	pH	Moisture Percentage (%)	*Aw*	Titratable Acid (%)	Direct Reducing Sugars [1] (%)	Direct Reducing Sugars [2] (%)
Apple bagasse	8.90 ± 0.28 [a]	4.04 ± 0.15 [b]	8.67 ± 0.20 [a]	0.98 ± 0.05 [a]	0.97 ± 0.23 [a]	86.20 ± 7.11 [a]	9.98 ± 0.01 [a]
Agave mezcalero bagasse	0.36 ± 0.20 [b]	5.00 ± 0.10 [a]	8.68 ± 0.20 [a]	0.97 ± 0.01 [a]	0.87 ± 0.21 [a]	6.79 ± 0.37 [b]	6.79 ± 0.37 [b]

Values are the mean ± standard deviation. Means with the same letters within each column do not differ statistically by Tukey's test ($p \leq 0.05$). [1] Direct reducing sugars before washing apple bagasse; [2] direct reducing sugars after washing apple bagasse.

3.2. Secondary Metabolites

3.2.1. Quantification of Total Phenolic Compounds (TPC)

In Figure 1, it is shown the total phenolic compounds resulting from the solid-state fermentation of agroindustrial wastes. The TPC had its highest concentration at the beginning of the experiment (day 0) in both residues, and after fermentation it decreased, increasing again after day 21.

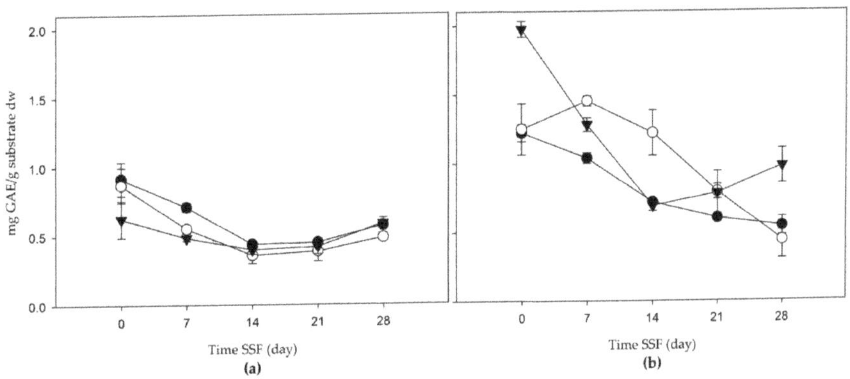

Figure 1. Total phenolic compounds (mg gallic acid equivalent per g substrate on dry weight (mg GAE/g substrate dw)) of the solid-state fermentation of agroindustrial residues: (**a**) Apple bagasse; (**b**) Agave mezcalero bagasse. Acetone and water (80:20 *v/v* (●)); methanol extract (○); aqueous extract (▼). Time SSF = time solid-state fermentation. Values are the mean ± standard deviation.

Apple bagasse extract (Figure 1a) increased its concentration 21 days after fermentation, with intervals of 0.385 and 0.587 mg gallic acid equivalent/g substrate dw. The highest concentration was present in the aqueous extract with a value of 0.587 ± 0.013 mg gallic acid equivalent/g substrate dw. The TPC content of agave mezcalero bagasse from the acetone and water (80:20 *v/v*) and methanol extracts tended to decrease during fermentation, while the aqueous extract increased its concentration after day 21, until it reached 0.976 ± 0.126 mg gallic acid equivalent/g substrate dw at 28 days.

3.2.2. Total Flavonoid Quantification (TF)

In Figure 2, the total flavonoids in apple bagasse and mezcalero agave are shown. First, the apple increased its TF content from day 21, giving final values (on day 28) of 0.015 ± 0.018, 0.013 ± 0.02, and 0.023 ± 0.012 mg QE/g substrate dw in extracts of acetone and water (80:20 *v/v*), methanol and the aqueous extract, respectively; the latter was the extract with the highest amount of flavonoids. Then, the agave mezcalero bagasse registered a similar tendency to the content of TPC. The total flavonoid concentrations in extracts of acetone and water (80:20 *v/v*) and methanol decreased during fermentation, while in the aqueous extract there was an increase in the TF content at the end of the fermentation time, with a final concentration of 0.012 mg QE/g substrate dw.

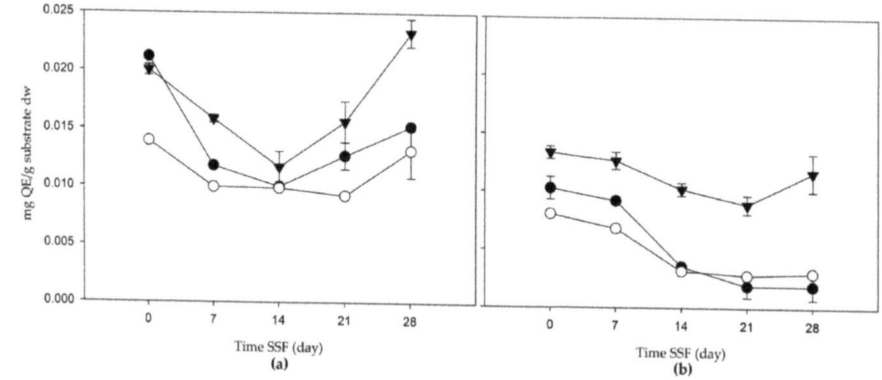

Figure 2. Total flavonoids (mg QE/g substrate dw) of the solid-state fermentation of agroindustrial residues: (**a**) apple bagasse; (**b**) agave mezcalero bagasse. Acetone and water (80:20 *v/v* (-●-)); methanol extract (-○-); aqueous extract (-▼-). Time SSF = time solid-state fermentation. Values are the mean ± standard deviation.

3.2.3. Total Triterpenes Quantification (TT)

In Figure 3, the total triterpenes content is shown. Apple bagasse presented TT contents which increased from day 21 in extracts of acetone and water (80:20 *v/v*) and methanol, with final values (on day 28) of 8.043 ± 0.696, and 9.411 ± 2.512 mg ursolic acid/g substrate dw, respectively. The aqueous extract showed steady growth from the beginning of the solid-state fermentation of triterpenes, and reached a concentration of 26.440 ± 0.949 mg ursolic acid/g substrate dw at 28 days. Despite the fact that the TT in agave mezcalero bagasse had the highest value at the beginning of the fermentation, its concentration decreased in all extracts during this process.

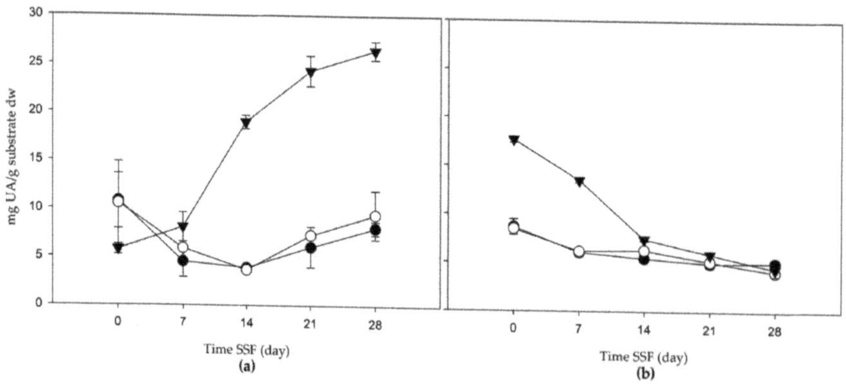

Figure 3. Total triterpenes (mg UA/g substrate dw) of the solid-state fermentation of agroindustrial residues: (**a**) apple bagasse; (**b**) agave mezcalero bagasse. Acetone and water (80:20 *v/v* (-●-)); methanol extract (-○-); aqueous extract (-▼-). Time SSF = time solid-state fermentation. Values are the mean ± standard deviation.

3.3. Antioxidant Activites

The antioxidant activity of the extracts obtained by the solid-state fermentation of *Pleurotus* from apple and agave mezcalero bagasses was expressed as the 50% inhibitory capacity of the radical in question (IC_{50}), which refers to the concentration of antioxidant required to reduce the initial amount of the DPPH and/or ABTS radical by 50%. This is the parameter used to measure the antioxidant properties of a substance [28], that is, the lower the IC_{50} value, the greater its antioxidant activity.

3.3.1. Antioxidant Activity by 1,1-Diphenyl-2-Picril Hydracil (DPPH•)

In apple bagasse, the IC_{50} of the DPPH assay ranged from 14.67 to 385.01 g/L residue. On the other hand, the agave mezcalero bagasse presented IC_{50} values from 61.80 to 377.75 g/L residue (Figure 4a,b). A tendency for IC_{50} to increase was observed from day 0 to day 21, however, a decrease was perceived from day 28 in different extracts.

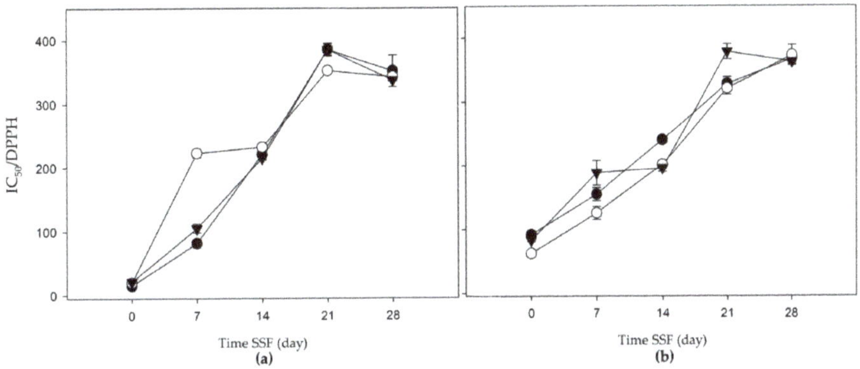

Figure 4. 50% inhibitory concentration in 2,2-diphenyl-1-picrylhydrazyl (DPPH (IC_{50}/DPPH)) of the solid-state fermentation of agroindustrial residues: (**a**) apple bagasse; (**b**) agave mezcalero bagasse. Acetone and water (80:20 *v/v* (–●–)); methanol extract (–○–); aqueous extract (–▼–). Time SSF = time solid-state fermentation. Values are the mean ± standard deviation.

3.3.2. Antioxidant Activity by 2,2′-Azino-Bis-3-Ethylbenzothiazoline-6-Sulfonic Acid Radical Cation (ABTS•+)

In the ABTS test, the apple bagasse presented a range of residue values from 60.20 to 423.09 g/L; meanwhile, the agave mezcalero bagasse ranged from 74.65 to 390.86 g/L (Figure 5a,b). The obtained results in the apple bagasse showed that the initial values (on day 0) were higher than the agave mezcalero bagasse with intervals of 61.80 to 91.42 g/L of residue.

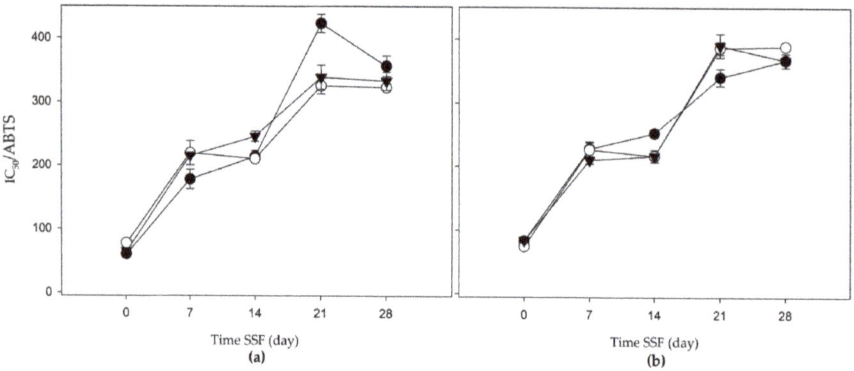

Figure 5. A 50% inhibitory concentration in 2, 2′-azino-bis (3-ethylbenzothiazoline)-6 ammonium sulfonate (ABTS (IC_{50}/ABTS)) of the solid-state fermentation of agroindustrial residues: (**a**) apple bagasse; (**b**) agave mezcalero bagasse. Acetone and water (80:20 *v/v* (—●—)); methanol extract (—○—); aqueous extract (—▼—). Time SSF = time solid-state fermentation. Values are the mean ± standard deviation.

The acetone and water (80:20 *v/v*) and aqueous extracts increased in IC_{50} during the days of fermentation. It should be noted that the methanol extract displayed a decrease in IC_{50} at day 28, which means that the secondary metabolites produced at the same time of solid fermentation denoted the antioxidant property of this residue.

4. Discussion

In both residues, the pH showed values lower than 5.0, lower than the value recommended for the adequate growth of *Pleurotus* (5.5–6.5) [29]. The pH was adjusted for proper fungus growth and fermentation development. The highest amounts of total soluble solids (°Brix), acidity percentage and reducing sugars were found in the apple bagasse because the residue had a significant amount of polysaccharides and organic acids [30]. This limited the growth of *Pleurotus*, thus, bagasse was subjected to a wash with tridestilated water. After washing the apple bagasse released sugars, it was observed that its content of reducing sugars decreased (Table 1), which was resulted in an adequate growth of the *Pleurotus* strain in the residue.

Apple bagasse presented a significant difference in the aqueous extract of the total triterpenes test, having the highest value; the other extracts showed no significant difference in the other tests performed. In the agave mezcalero bagasse, the significant difference between extracts was manifested in the tests of total flavonoids and total triterpenes; the aqueous extract had the highest value, while the others did not show significant differences in the rest of the performed tests.

The highest phenolic content was at the beginning, since the phenolic compounds are shown in the subtracts (bagasses). In addition, it contains lignin, which is a phenolic polymer made up of three principal monomerics: paracoumaryl alcohol, synapilic alcohol and conyferyl alcohol [31,32]. This acts as a carbon source for the *Pleurotus* fungi for its growth and development (trophophase) after the mycelial growth. However, the nutrient exhaustion produces the biosynthesis of secondary metabolites in the idiophase. In this period (T21) there is a metabolite increase according to Ferrer-Romero et al. [33] and Vamanu [34] which supports the presence of *Pleurotus* and its survival in the residue [35]. Therefore, to observe a higher antioxidant activity, idiophase time must be increased in order.

The TPC values generated from *Pleurotous* in other residues, such as the mixture of pineapple and rice straw residues [36], reported values of 0.179–0.650 mg GAE/g substrate dw. Other authors reported greater values than the ones obtained in this research: Hamdipour et al. [37] showed values of 5.53 to 11.6 mg GAE/g dry weight in the fermentation of apple residues with *Rhizopus oligosporus*, and Ajila et al. [38] obtained values of 4.6 to 16.12 mg GAE/g dry weight during the solid-state fermentation of apple residues with *Phanerocheate chrysosporium*. Meanwhile, Ferrer-Romero et al. [33] obtained 0.1 g of total phenols/g of glucose by submerged fermentation from the mycelial biomass of *Pleurotus ostreatus*.

The *Pleurotus ostreatus* strain of this study led to the degradation of lignin to obtain the necessary nutrients for its development, given that growth was observed in both lignocellulosic substrates: apple bagasse and mezcalero agave bagasse [39]. In addition, the genus *Pleurotus* has the capacity to produce laccase enzymes, which have a fundamental role both in the degradation of lignin and in the biosynthesis of phenolic compounds in white-rot fungi [40].

Boonsong et al. [41] conducted an investigation with the purpose of evidencing the high nutritional value of edible fungi. This investigation reported intervals from 2.29 to 2.61 mg QE/g in aqueous extracts of *Pleurotus* strains, which are higher values compared to the ones obtained in this project.

Furthermore, Braga et al. [42] reported significant concentrations of total flavonoids (1.76 mg QE/g) in grape marc residue, and 1.70 mg quercetin equivalent/g in mango residue; Dulf et al. [43] found slightly lower values of 0.29 and 0.36 mg QE/g, resulting from the solid-state fermentation of apricot residue with *Aspergillus niger* and *Rhizopus oligosporus* for 14 days; while Lessa et al. [44] found concentrations of TF with values of 31.5 mg QE/100 g in fermented cocoa flour by means of *Penicillium roqueforti*. Therefore, the low concentration of this type of metabolites maybe be determined, once again, by the used strain, that degrades lignin during solid-state fermentation.

Yang et al. [45] found a higher concentration of triterpenes, 47.10 mg/g, 30 days after the solid-state fermentation of citrus residues with the fungus *Antrodia cinnamomea*. In this investigation, the total triterpenes values have great importance, especially those found during solid fermentation in the apple bagasse because this type of secondary metabolite has antibacterial properties against plant pathogens; in addition, it has demonstrated effectiveness against oxidative damage through the elimination of free radicals and the modulation of enzymatic activity [46]. The authors mention solvents with low polarity for the extraction of triterpenes, such as ethyl acetate [47,48]. Nevertheless, aqueous extracts have been used to identify secondary metabolites, like triterpenes, in other macromycete fungi, *Phellinus* and *Ganoderma* [49]. As a consequence, it was decided to use this type of solvent, due to polyphenols, sterols, triterpenes and anthracenes being moderately abundant [50].

The production of secondary metabolites can be considered similar to the tendency of antioxidant activity, due to the fact that since day 21 there was an increase in antioxidant activity in the apple bagasse, which can be attributed to a higher concentration of polyphenols present in acetone and water (80:20 *v/v*) and aqueous extracts. As mentioned by Lu and Foo [51], the antioxidant activity is caused by polyphenols in apple residues. Meanwhile, Ajila et al. [38] report IC_{50} values of 12.24 ± 5.20 to 14.27 ± 2.50 μg in acetone and water (80:20 *v/v*) extracts and 40.16 ± 2.40 to 50.18 ± 2.00 μg in aqueous extracts in fermented apple residue.

While there is little information on antioxidant activity in agave mezcalero through solid-state fermentation, some authors have found this property in certain agave species. For example, Carmona et al. [52] mention the antioxidant activities of *Agave lechuguilla*, and Araldi et al. [53] mention it regarding *Agave sisalana*. Both species are used as a source of raw material in the production of biofuels from solid fermentation.

Although the values obtained can be considered low, the obtained methanol extract from the solid-state fermentation of the agave mezcalero can be used as an alternative for the production of compounds with antioxidant properties.

Overall, the IC_{50} values of apple bagasse are lower; they displayed greater inhibitory activity against the DPPH radicals and ABTS radicals in comparison to those of the agave mezcalero bagasse. On the other hand, Ignat et al. [54] express that the different results in the metabolite content may be due to the extraction conditions, type of solvent, and its susceptibility to degradation. This way, it is possible to indicate the variability of secondary metabolites and antioxidant activity in agroindustrial waste.

5. Conclusions

The growth of *Pleurotus ostreatus* was observed in both the apple bagasse and agave mezcalero bagasse residues due to their ability to secrete extracellular, non-specific ligninolytic enzymes during secondary metabolism. Spectrophotometric studies have demonstrated that, because of solid-state fermentation, there was presence of secondary metabolites—phenolic compounds, flavonoids and triterpenes.

The highest concentration of compounds of biological interest occurred in apple bagasse after 21 days, with the aqueous extract presenting the highest extraction of the different phytochemical compounds. This behavior was reflected in the antioxidant activity of the different extracts during the same fermentation time. Minor amounts of phenolic and flavonoid compounds were found in the aqueous extract of the agave bagasse from day 21 of fermentation.

The biosynthesis of secondary metabolites displaying antioxidant activities in the solid-state fermentation of *Pleurotus ostreatus* in the apple residues of the cider industry and in the agave residues of the mezcalera industry can contribute to the use of these wastes, as they can be used in functional food formulation or, due to their medical properties, in the pharmaceutical industry.

Author Contributions: Conceptualization, D.I.-C., M.A.M.-C., R.d.C.C.-V. and M.E.R.-C.; methodology, D.I.-C., M.A.M.-C., and M.E.R.-C.; validation, D.I.-C., M.A.M.-C., and M.E.R.-C.; formal analysis, D.I.-C., M.A.M.-C., R.d.C.C.-V. and M.E.R.-C.; investigation, D.I.-C.; resources, M.A.M.-C. and M.E.R.-C.; data curation, D.I.-C., M.A.M.-C. and M.E.R.-C.; writing—original draft preparation, D.I.-C.; writing—review and editing, D.I.-C., M.A.M.-C., R.d.C.C.-V. and M.E.R.-C.; visualization, D.I.-C., M.A.M.-C. and M.E.R.-C.; supervision, M.A.M.-C. and M.E.R.-C.; project administration, M.A.M.-C., R.d.C.C.-V. and M.E.R.-C.; funding acquisition, M.A.M.-C., R.d.C.C.-V. and M.E.R.-C. All authors have read and agreed to the published version of the manuscript.

Funding: This research was funded by Programa de Desarrollo Profesional Docente para el Tipo Superior (PRODEP) Folio PTC-BUAP.PTC-560.

Acknowledgments: The authors thank the Posgrado de Ciencias Ambientales—BUAP and CONACyT for the post-doctoral of Diego Ibarra-Cantún.

Conflicts of Interest: The authors declare no conflict of interest. The funders had no role in the design of the study; in the collection, analyses, or interpretation of data; in the writing of the manuscript, or in the decision to publish the results.

References

1. Vargas, Y.A.; Peréz, L.I. Aprovechamiento de Residuos Agroindustriales Para El Mejoramiento de La Calidad Del Ambiente. *Rev. Fac. Cienc. Básicas* **2018**, *14*, 59–72. [CrossRef]
2. SIAP. Servicio de Información Agroalimentaria y Pesquera SIAP. Available online: https://www.gob.mx/agricultura/articulos/la-sidra-mexicana-invitada-de-honor-en-las-fiestas-decembrinas (accessed on 8 January 2020).
3. Vicente, F.; Cueto, M.A.; de la Rosa, B.; Argamentería, A. Caracterización de Subproductos de La Manzana Para Su Uso En Nutrición Animal. *ITEA Inf. Tec. Econ. Agrar.* **2005**, *26*, 560–662.
4. Martínez Gutiérrez, G.A.; Iñiguez Covarrubias, G.; Ortiz-Hernández, Y.D.; López-Cruz, J.Y.; Bautista Cruz, M.A. Tiempos de Apilado Del Bagazo Del Maguey Mezcalero y Su Efecto En Las Propiedades Del Compost Para Sustrato de Tomate. *Rev. Int. Contam. Ambient.* **2013**, *29*, 209–216.
5. Mejía Giraldo, L.F.; Martínez Correa, H.A.; Betan court Gutiérrez, J.E.; Castrillón Castaño, C.E. Aprovechamiento de Residuo Agroindustrial Del Mango Común (*Mangifera indica* L.) Para Obtener Azúcares Fermentables. *Ing. Cienc.* **2007**, *3*, 41–62.
6. Sánchez Riaño, A.M.; Gutiérrez Morales, A.I.; Muñoz Hernández, J.A.; Rivera Barrero, C.A. Producción de Bioetanol a partir de Subproductos Agroindustriales Lignocelulósicos. *Tumbaga* **2010**, *1*, 61–91.
7. Unekwu, H.R.; Audu, J.A.; Makun, M.H.; Chidi, E.E. Phytochemical Screening and Antioxidant Activity of Methanolic Extract of Selected Wild Edible Nigerian Mushrooms. *Asian Pac. J. Trop. Dis.* **2014**, *4*, S153–S157. [CrossRef]
8. Palacios, I.; García-Lafuente, A.; Guillamón, E.; Villares, A. Novel Isolation of Water-Soluble Polysaccharides from the Fruiting Bodies of *Pleurotus ostreatus* Mushrooms. *Carbohydr. Res.* **2012**, *358*, 72–77. [CrossRef]

9. Suárez Arango, C.; Nieto, I.J. Cultivo biotecnológico de macrohongos comestibles: Unaalternativa en la obtención de nutracéuticos. *Rev. Iberoam. Micol.* **2013**, *30*, 1–8. [CrossRef] [PubMed]
10. Benkortbi, O.; Hanini, S.; Bentahar, F. Batch Kinetics and Modelling of Pleuromutilin Production by *Pleurotus mutilis*. *Biochem. Eng. J.* **2007**, *36*, 14–18. [CrossRef]
11. Singhania, R.R.; Sukumaran, R.K.; Patel, A.K.; Larroche, C.; Pandey, A. Advancement and Comparative Profiles in the Production Technologies Using Solid-State and Submerged Fermentation for Microbial Cellulases. *Enzym. Microb. Technol.* **2010**, *46*, 541–549. [CrossRef]
12. Thomas, L.; Larroche, C.; Pandey, A. Current Developments in Solid-State Fermentation. *Biochem. Eng. J.* **2013**, *81*, 146–161. [CrossRef]
13. Singhnee'Nigam, P.; Pandey, A. Biotechnology for Agro-Industrial Residues Utilisation: Utilisation of Agro-Residues. In *Biotechnology for Agro-Industrial Residues Utilisation*; Singhnee'Nigam, P., Pandey, A., Eds.; Springer: Dordrecht, The Netherlands, 2009. [CrossRef]
14. Lizardi-Jimenez, M.A.; Hernández-Martínez, R. Solid State Fermentation(SSF): Diversity of Applications to Valorize Waste and Biomass. *3Biotech* **2017**, *7*, 44. [CrossRef] [PubMed]
15. Yusuf, M. Agro-Industrial Waste Materials and Their Recycled Value-Added Applications: Review. In *Handbook of Ecomaterials*; Martínez, L.M.T., Kharissova, O.V., Kharisov, B.I., Eds.; Springer International Publishing: Cham, Switzerland, 2017; pp. 1–11. [CrossRef]
16. Wasser, S. Medicinal Mushrooms as a Source of Antitumor and Immunomodulating Polysaccharides. *Appl. Microbiol. Biotechnol.* **2002**, *60*, 258–274. [CrossRef] [PubMed]
17. Gregori, A.; Švagelj, M.; Pohleven, J. Cultivation Techniques and Medicinal Properties of *Pleurotus* spp. *Food Technol. Biotechnol.* **2007**, *45*, 238–249.
18. Gil-Chávez, G.J.; Villa, J.A.; Ayala-Zavala, J.F.; BasilioHeredia, J.; Sepulveda, D.; Yahia, E.M.; González-Aguilar, G.A. Technologies for Extraction and Production of Bioactive Compounds to Be Used as Nutraceuticals and Food Ingredients: An Overview. *Compr. Rev. Food Sci. Food Saf.* **2013**, *12*, 5–23. [CrossRef]
19. Kumar, K.; Yadav, A.N.; Kumar, V.; Vyas, P.; Dhaliwal, H.S. Food Waste: A Potential Bioresource for Extraction of Nutraceuticals and Bioactive Compounds. *Bioresour. Bioprocess.* **2017**, *4*, 18. [CrossRef]
20. Stamets, P.; Chilton, J.S. *The Mushroom Cultivator: A Practical Guide to Growing Mushrooms at Home*; Agarikon Press: Olympia, WA, USA, 1983.
21. Horwitz, W.; AOAC International (Eds.) *Official Methods of Analysis of AOAC International*, 18th ed.; AOAC International: Gaithersburg, MD, USA, 2006.
22. Singleton, V.L.; Rossi, J.A. Colorimetry of Total Phenolics with Phosphomolybdic-Phosphotungstic Acid Reagents. *Am. J. Enol. Vitic.* **1965**, *16*, 144–158.
23. Chang, C.-C.; Yang, M.-H.; Wen, H.-M.; Chern, J.-C. Estimation of Total Flavonoids Contentin Propolis by Two Complementary Colorimetric Methods. *J. Food Drug Anal.* **2002**, *10*, 178–182.
24. Fan, J.-P.; He, C.H. Simultaneous Quantification of Three Major Bioactive Triterpene Acids in the Leaves of Diospyros Kaki by High-Performance Liquid Chromatography Method. *J. Pharm. Biomed. Anal.* **2006**, *41*, 950–956. [CrossRef]
25. Brand-Williams, W.; Cuvelier, M.E.; Berset, C. Use of a Free Radical Method to Evaluate Antioxidant Activity. *LWT-Food Sci. Technol.* **1995**, *28*, 25–30. [CrossRef]
26. Re, R.; Pellegrini, N.; Proteggente, A.; Pannala, A.; Yang, M.; Rice-Evans, C. Antioxidant Activity Applying an Improved ABTS Radical Cation Decolorization Assay. *Free Radic. Biol. Med.* **1999**, *26*, 1231–1237. [CrossRef]
27. SAS Institute Inc. *SAS/STAT®User's Guide*; SAS Institute Inc.: Cary, NC, USA, 1999.
28. Rufino, M.S.M.; Fernandes, F.A.N.; Alves, R.E.; Debrito, E.S. Free Radical-Scavenging Behaviour of Some North-East Brazilian Fruitsina DPPH System. *Food Chem.* **2009**, *114*, 693–695. [CrossRef]
29. Philippoussis, A.N. Production of Mushrooms Using Agro-Industrial Residuesas Substrates. In *Biotechnology for Agro-Industrial Residues Utilisation*; Singhnee'Nigam, P., Pandey, A., Eds.; Springer: Dordrecht, The Netherlands, 2009; pp. 163–196. [CrossRef]
30. Damodaran, S.; Parkin, K.; Fennema, O.R.; BeMiller, J.N.; Huber, O.R. (Eds.) *Fennema's Food Chemistry*, 4th ed.; Food science and technology; CRC Press/Taylor&Francis: Boca Raton, FL, USA, 2008.
31. Vicente, A.R.; Manganaris, G.A.; Sozzi, G.O.; Crisosto, C.H. Nutritional Quality of Fruits and Vegetables. In *Postharvest Handling: A Systems Approach*; Florkowski, W.J., Shewfelt, R.L., Brueckner, B., Prussia, S.E., Eds.; Elsevier-Academic Press: San Diego, CA, USA, 2009; pp. 57–106. [CrossRef]

32. Zheng, H.; Hwang, I.-W.; Chung, S.-K. Enhancing Polyphenol Extraction from Unripe Apples by Carbohydrate-Hydrolyzing Enzymes. *J. Zhejiang Univ. Sci. B* **2009**, *10*, 912–919. [CrossRef] [PubMed]
33. Ferrer-Romero, J.C.; Mas-Diego, S.M.; Beltrán-Delgado, Y.; Morris-Quevedo, H.J.; Díaz-Fernández, U. Estudio Cinético de La Producción de Biomasa y Compuestos Fenólicos por *Pleurotus ostreatus* En Fase Sumergida. *Rev. Cub. Quim* **2019**, *31*, 1–17.
34. Vamanu, E. Antioxidant Properties of Mushroom Mycelia Obtained by Batch Cultivation and Tocopherol Content Affected by Extraction Procedures. *BioMed Res. Int.* **2014**, *2014*, 1–8. [CrossRef] [PubMed]
35. Liao, X.; Vining, L.C.; Doull, J.L. Physiological Control of Trophophase—Idiophase Separationin Streptomycete Cultures Producing Secondary Metabolites. *Can. J. Microbiol.* **1995**, *41*, 309–315. [CrossRef]
36. Rashad, M.M.; Abdou, H.M.; Mahmoud, A.E. Production of Some Bioactive Materials by *Pleurotus ostreatus* from Pineapple Residues and Rice Straw via Solid State Fermentation. *Res. J. Pharm. Biol. Chem. Sci.* **2016**, *7*, 2730–2736.
37. Hamdipour, S.; Rezazad, M.; Alizadeh, M. Production of Phenolic Antioxidants from Apple Residue Using *Rhizopus oligosporus*. *Int. J. Adv. Biol. Biomed. Res.* **2014**, *2*, 1937–1942.
38. Ajila, C.M.; Brar, S.K.; Verma, M.; Tyagi, R.D.; Valéro, J.R. Solid-State Fermentation of Apple Pomace Using *Phanerocheate chrysosporium*–Liberation and Extraction of Phenolic Antioxidants. *Food Chem.* **2011**, *126*, 1071–1080. [CrossRef]
39. Carlile, M.J.; Watkinson, S.C.; Gooday, G.W. *The Fungi*, 2nd ed.; Elsevier-Academic Press: San Diego, CA, USA, 2001.
40. Kirk, T.K.; Farrell, R.L. Enzymatic "Combustion": The Microbial Degradation of Lignin. *Annu. Rev. Microbiol.* **1987**, *41*, 465–501. [CrossRef]
41. Boonsong, S.; Klaypradit, W.; Wilaipun, P. Antioxidant Activities of Extracts from Five Edible Mushrooms Using Different Extractants. *Agric. Nat. Resour.* **2016**, *50*, 89–97. [CrossRef]
42. Braga, G.C.; Melo, P.S.; Bergamaschi, K.B.; Tiveron, A.P.; Massarioli, A.P.; de Alencar, S.M. Extraction Yield, Antioxidant Activity and phenolics from Grape, Mango and Peanut Agro-Industrial by-Products. *Ciênc. Rural* **2016**, *46*, 1498–1504. [CrossRef]
43. Dulf, F.V.; Vodnar, D.C.; Dulf, E.-H.; Pintea, A. Phenolic Compounds, Flavonoids, Lipids and Antioxidant Potential of Apricot (*Prunus armeniaca* L.) Pomace Fermented by Two Filamentous Fungal Strains in Solid State System. *Chem. Cent. J.* **2017**, *11*, 92. [CrossRef] [PubMed]
44. Lessa, O.A.; dos Santos Reis, N.; Leite, S.G.F.; Gutarra, M.L.E.; Souza, A.O.; Gualberto, S.A.; de Oliveira, J.R.; Aguiar-Oliveira, E.; Franco, M. Effect of the Solid State Fermentation of Cocoa Shellon the Secondary Metabolites, Antioxidant Activity, and Fatty Acids. *Food Sci. Biotechnol.* **2018**, *27*, 107–113. [CrossRef] [PubMed]
45. Yang, F.-C.; Ma, T.-W.; Lee, Y.-H. Reuse of Citrus Peel to Enhance the Formation of Bioactive Metabolite-Triterpenoidin Solid-State Fermentation of *A. cinnamomea*. *Biochem. Eng. J.* **2013**, *78*, 59–66. [CrossRef]
46. Lee, Y.-H.; Sun, Y.; Glickman, R. Ursolic Acid-Regulated Energy Metabolism—Relieveror Propeller of Ultraviolet-Induced Oxidative Stress and DNA Damage? *Proteomes* **2014**, *2*, 399–425. [CrossRef]
47. Chegwin-Angarita, C.; Nieto-Ramírez, I.J. Effect of Non-Conventional Carbon Sources on the Production of Triterpenoids in Submerged Cultures of *Pleurotus* Macrofungi. *J. Chil. Chem. Soc.* **2014**, *59*, 2287–2293. [CrossRef]
48. Lee, S.; Shiw, S.H.; Kim, J.S.; Shiw, K.H.; Kang, S.S. Aldose Reductase Inhibitors from the Fruiting Bodies of *Ganoderma applanatum*. *Biol. Pharm. Bull.* **2005**, *28*, 1103–1105. [CrossRef]
49. Trigos, Á.; Suárez, J. Biologically Active Metabolites of the genus *Ganoderma*: Three Decades of Myco-Chemistry Research. *Rev. Mex. Mic.* **2011**, *34*, 63–83.
50. Juliette-Ornely, O.B.; Eyi, N.H.C.; Rick-Léonid, N.M.M.; Sima Obiang, C.; Yembiyé, P.; Ondo, J.-P.; Ndong, A.G.R.; Obame-Engonga, L.-C. Phytochemical Screening, Antioxidant and Antiangiogenic Activities of *Daedaleopsis nitida*, *Pycnoporus sanguineus* and *Phellinus gilvus* Medicinal Mushrooms from Gabon. *Pharm. Chem. J.* **2019**, *6*, 71–80.
51. Lu, Y.; YeapFoo, L. Antioxidant and Radical Scavenging Activities of Polyphenols from Apple Pomace. *Food Chem.* **2000**, *68*, 81–85. [CrossRef]

52. Carmona, J.E.; Morales-Martínez, T.K.; Mussatto, S.I.; Castillo-Quiroz, D.; Ríos-González, L.J. Propiedades Químicas, Estructurales y Funcionales de la Lechuguilla (*Agave lechuguilla* Torr.). *Rev. Mex. Cienc. For.* **2017**, *8*, 100–122.
53. Araldi, R.P.; dos Santos, M.O.; Barbon, F.F.; Manjerona, B.A.; Meirelles, B.R.; de Oliva Neto, P.; da Silva, P.I.; dos Santos, L.; Camargo, I.C.C.; de Souza, E.B. Analysis of Antioxidant, Cytotoxic and Mutagenic Potential of *Agave sisalana* Perrine Extracts Using Vero Cells, Human Lymphocytes and Mice Polychromatic Erythrocytes. *Biomed. Pharmacother.* **2018**, *98*, 873–885. [CrossRef] [PubMed]
54. Ignat, I.; Volf, I.; Popa, V.I. A Critical Review of Methods for Characterisation of Polyphenolic Compounds in Fruits and Vegetables. *Food Chem.* **2011**, *126*, 1821–1835. [CrossRef] [PubMed]

Journal of *Fungi*

MDPI

Article

Metal and Phosphate Ions Show Remarkable Influence on the Biomass Production and Lipid Accumulation in Oleaginous *Mucor circinelloides*

Simona Dzurendova [1,*], Boris Zimmermann [1], Valeria Tafintseva [1], Achim Kohler [1], Svein Jarle Horn [2] and Volha Shapaval [1]

1 Faculty of Science and Technology, Norwegian University of Life Sciences, Drøbakveien 31, 1430 As, Norway; boris.zimmermann@nmbu.no (B.Z.); valeria.tafintseva@nmbu.no (V.T.); achim.kohler@nmbu.no (A.K.); volha.shapaval@nmbu.no (V.S.)

2 Faculty of Chemistry, Biotechnology and Food Science, Norwegian University of Life Sciences, Christian Magnus Falsens vei 1, 1433 As, Norway; svein.horn@nmbu.no

* Correspondence: simona.dzurendova@gmail.com or simona.dzurendova@nmbu.no

Received: 6 October 2020; Accepted: 27 October 2020; Published: 30 October 2020

Abstract: The biomass of *Mucor circinelloides*, a dimorphic oleaginous filamentous fungus, has a significant nutritional value and can be used for single cell oil production. Metal ions are micronutrients supporting fungal growth and metabolic activity of cellular processes. We investigated the effect of 140 different substrates, with varying amounts of metal and phosphate ions concentration, on the growth, cell chemistry, lipid accumulation, and lipid profile of *M. circinelloides*. A high-throughput set-up consisting of a Duetz microcultivation system coupled to Fourier transform infrared spectroscopy was utilized. Lipids were extracted by a modified Lewis method and analyzed using gas chromatography. It was observed that Mg and Zn ions were essential for the growth and metabolic activity of *M. circinelloides*. An increase in Fe ion concentration inhibited fungal growth, while higher concentrations of Cu, Co, and Zn ions enhanced the growth and lipid accumulation. Lack of Ca and Cu ions, as well as higher amounts of Zn and Mn ions, enhanced lipid accumulation in *M. circinelloides*. Generally, the fatty acid profile of *M. circinelloides* lipids was quite consistent, irrespective of media composition. Increasing the amount of Ca ions enhanced polyphosphates accumulation, while lack of it showed fall in polyphosphate.

Keywords: *Mucor circinelloides*; high-throughput screening; metal ions; phosphorus; lipids; biofuel; FTIR spectroscopy; bioremediation; co-production

1. Introduction

Mucor circinelloides is a dimorphic oleaginous filamentous fungus with a fully sequenced genome [1]. It has a versatile metabolism, allowing utilization of a variety of feedstocks, making this fungus widely applicable in a range of biotechnological processes [2]. *M. circinelloides* is well known as a robust cell factory, where extracellular products include enzymes (cellulases, lipases, proteases, phytases, and amylases) [2,3] and ethanol [4]. Further, *M. circinelloides* can synthesize and accumulate a number of valuable intracellular components, such as lipids, polyphosphates, carotenoids, and chitin/chitosan [5–9]. The biomass of *M. circinelloides* has a significant nutritional value and can be used as a feed ingredient [10]. Chitosan exhibits great chelating properties, mainly due to the low level of acetylation and the abundance of hydroxyl groups [9,11,12]. Due to the presence of chitin and chitosan in the cell wall of *M. circinelloides*, the biomass of this fungus can be used as bioabsorbent for heavy metals and applied as a bioremediation agent, for example in the wastewater treatment [13].

M. circinelloides has been extensively studied for the production of lipids for different applications [14–16]. The lipids are mainly in the form of triacylglycerides (TAGs) and contain palmitic, stearic, oleic, linoleic acid, and y-linoleic acids that make it particularly suitable for biodiesel production [17]. Therefore, the biomass of *M. circinelloides* could be considered as an important alternative feedstock for the biodiesel industry [18]. However, the cost of the *M. circinelloides* biomass production for biodiesel as a sole product is still too high compared to competitive bioprocesses. Thus, there is a need to further optimize lipid and biomass yield for *M. circinelloides*, and develop a coproduction concept, where other valuable components could be produced along with lipids in a single fermentation process [19,20].

Lipid accumulation, as well as biomass formation, can be affected by many different cultivation parameters, such as the nutrient composition of the growth medium, temperature, pH, aeration, or parameter shift during the fermentation (temperature/pH) [2,21,22]. Optimization of the nutrient composition of the growth medium is one of the most important aspects in improving fungal lipid production. In order to increase lipid and biomass yields, it is crucial to understand the role and effect of all media components. Many studies have assessed the effect of different carbon (C) and nitrogen (N) sources on the fungal lipid production in *Mucoromycota* fungi [23–29], where nitrogen limitation in carbon-rich media is the most frequently used strategy for inducing lipid accumulation and achieving high lipid yields. Macro- and micronutrients, such as phosphorus (P), potassium (K), sulfur (S), calcium (Ca), sodium (Na), iron (Fe), and magnesium (Mg), have previously been reported as essential for optimal fungal growth and metabolic activity [30].

Metal ions play an important role in fungal metabolism as they provide necessary redox and catalytic activities for the cellular processes [31]. The role of metal ions in yeast metabolism has been widely studied [32–35]. Bivalent metal ions are often reported as cofactors for different enzymes [36,37]. Metal ions are usually examined in the context of bioremediation capabilities of *M. cicrcinelloides* [9,38], while the role of metal ions in lipid accumulation of *Mucoromycota* fungi have been examined only to a limited extent. Different metal ions have shown strain-specific influence on lipid accumulation in *Mucoromycota* fungi, where either fatty acid profiles or total lipid content is affected. For example, manganese (Mn) has shown positive effects on the lipogenesis in *Mucor plumbeus* and *Mortierella* sp. [39,40]. Iron had an inhibiting effect on the arachidonic acid production in *Mortierella* sp. [39], while together with magnesium and zinc (Zn), it was enhancing lipid accumulation in *Cunninghamella* sp. Furthermore, zinc increased the gamma-linoleic acid production in *Cunninghamella* sp. [41], while iron, zinc, and copper (Cu) were reported as enhancers of arachidonic acid production in *Mortierella alpina* [42]. To the authors' knowledge, the effect of calcium and cobalt (Co) on the lipid accumulation in the oleaginous *Mucoromycota* fungi has not been investigated before. Moreover, there have been no studies reporting the effect of metal ions on the lipogenesis in *M. circinelloides*.

In our previous studies, we have assessed the chemical composition of nine different oleaginous filamentous fungi (including *M. circinelloides*) and revealed nutrient-induced coproduction of lipids, chitin/chitosan, and polyphosphates [43,44]. We reported that *M. circinelloides* has an ability to coproduce lipids, polyphosphate, and chitin/chitosan at different phosphorus concentrations and showed a versatile metabolism with a high adaptability level to different stress conditions. Thus, this fungus can be utilized in phosphorus recovery processes, while the co-production concept greatly contributes to the economic feasibility of such processes.

The aim of this study was to assess the effect of 140 different substrates, with varying amounts of metal and phosphate ions, on the growth, cell chemistry, and lipid production in *M. circinelloides*. Different concentrations of phosphorus source were used in order to study the effect of metal ions on the co-production of lipids, polyphosphates and cell wall polysaccharides, such as chitin/chitosan, triggered by phosphorus availability. Analogous to our previous studies, the study was performed in a high-throughput set-up using a Duetz microtiter plate system (Duetz-MTPs) combined with Fourier transform infrared (FTIR) spectroscopy [43–46]. FTIR spectroscopy was applied to obtain a biochemical fingerprint of the microbial cells [43,47–52], while gas chromatography was used to analyze the lipid yield and fatty acid profiles of the extracted lipids.

2. Materials and Methods

2.1. Growth Media and Cultivation Conditions

Fungal strain *M. circinelloides* VI04473, provided by the Veterinary Institute, Oslo, Norway was selected based on the previous study of 100 oleaginous fungal strains, as it showed the highest lipid and biomass yield of all tested *Mucor* strains [50]. Moreover, this strain has also shown coproduction potential for lipids, chitin/chitosan, and polyphosphates [43]. *M. circinelloides* was cultivated on malt extract agar (MEA) for 7 days at 25 °C in order to obtain fresh spores for the inoculation into nitrogen-limited broth media with various metal and phosphorus ion concentrations. Spores were collected from agar plates using 10 ml of saline solution and a bacteriological loop. The composition of the reference medium, used and modified in our previous studies [27,43–45], was the following: 80 g/L of glucose, 1.5 g/L of $(NH_4)_2SO_4$, 7 g/L of KH_2PO_4, 2 g/L of Na_2HPO_4, 1.5 g/L of MgSO4·7H2O, 0.1 g/L of CaCl2·2H2O, 0.008 g/L of FeCl3·6H2O, 0.001 g/L of ZnSO4·7H2O, 0.0001 g/L of CoSO4·7H2O, 0.0001 g/L of CuSO4·5H2O, 0.0001 g/L of MnSO4·5H2O, where the listed concentrations of the metal ions Ca, Cu, Co, Fe, Mg, Mn, and Zn were assigned as reference concentration and marked as "R" (Table 1). The higher—1000; 100; 10 × R and lower—0.1; 0.01 and 0 × R concentrations of the metal ions were assessed in the study as described in Table 1. The reference medium was modified by using four relative levels of metal and phosphate ions (Table 1). KH_2PO_4 and Na_2HPO_4 were used as phosphates substrate, and their total concentration is hereafter referred as "phosphates concentration" (Pi). Phosphate concentrations KH_2PO_4 7 g/L, Na_2HPO_4 2 g/L have been assigned as Pi1. In addition to Pi1 concentration, the higher—4 and 2 × Pi1 and lower—0.5 and 0.25 × Pi1 concentrations of phosphates were assessed in the study as described in Table 1. Broth media, with the lower than Pi1 amount of Pi—Pi0.5 and Pi0.25, contained KCl and NaCl in a corresponding concentration in order to have equal amounts of K^+ and Na^+ ions as in the Pi1 condition. Different media were prepared by modifying one metal ion concentration at a time for every level of Pi ions. The only exception was for the condition 0Mg10Ca, which was tested in order to examine a possibility of substitution of Mg by Ca. Thus, in total, 140 different media were prepared. The concrete concentrations of all media components can be found in the Supplementary Materials (Table S1).

Table 1. Overview over the relative levels of concentration of metal ions and inorganic phosphate in the media. The exact concentrations can be found in the Supplementary Materials (Table S1).

Ca	Mg	Cu	Co	Fe	Mn	Zn	Pi
0	0Mg 10Ca	0	0	0	0	0	0.25
0.01	0.01	R	R	R	R	R	0.5
0.1	0.1	10	10	10	10	10	Pi1
R	R	100	100	100	100	100	2
10		1000	1000	1000	1000	1000	4

Considering the high-throughput set-up of the study and consequently high number of samples, the reproducibility of the *M. circinelloides* growth was controlled by four biological replicates for the reference medium R with Pi1 concentration, two biological replicates for the reference metal ion concentrations, and the following phosphate ion concentrations: Pi4, Pi2, Pi0.5, and Pi0.25, and two biological replicates for the medium with 0Ca under all tested phosphate concentrations. The reproducibility of lipid accumulation was controlled under the reference metal concentrations by four biological replicates for the reference medium R-Pi1, two biological replicates for Pi4, Pi2, Pi0.5, and Pi0.25 media, and two biological replicates for the media with 0Ca-Pi2 and Pi1 levels. The biological variability is represented by error bars in Figure 1 and standard deviation in Table 3.

Cultivation was performed in Duetz-MTPS (Enzyscreen, Heemstede, The Netherlands) [43,45,47,50,53,54], consisting of 24-square well polypropylene deep well microtiter plates (MTPs), low evaporation sandwich covers with a clamp system. A total of 7 ml of sterile media broth was transferred into the autoclaved microtiter plates and each well was inoculated with 50 µL of the spore suspension. MTPs were placed on the shaking platform of the incubator MAXQ 4000 (Thermo Scientific, Oslo, Norway). Cultivations were performed for 7 days at 25 °C and 400 rpm agitation speed (1.9 cm circular orbit).

2.2. Lipid Extraction and GC-FID Analysis of Lipid Concentration and Fatty Acid Profile

Direct transesterification was performed according to Lewis et al. [55], with some modifications [44]: 2 mL screw-cap polypropylene (PP) tubes were filled with 30 ± 3 mg of freeze-dried biomass, 250 ± 30 mg of acid-washed glass beads, and 500 µL of methanol. Further, the fungal biomass was disrupted in a tissue homogenizer (Bertin Technologies Percellys Evolution, Montigny-le-Bretonneux, France). The disrupted fungal biomass was transferred into glass reaction tubes by washing the PP tube with 2400 µL of a methanol–chloroform–hydrochloric acid solvent mixture (7.6:1:1 *v/v*). Then, 1 mg of C13:0 TAG internal standard in 100 µL of hexane was added to the glass reaction tubes (100 µL from a 10.2 mg/mL^{-1} glyceryl tritridecanoate ($C_{42}H_{80}O_6$, C13:0 TAG (13:0/13:0/13:0), Sigma-Aldrich, St. Louis, Missouri, USA). Reaction tubes were incubated at 90 °C for 1 h, followed by cooling to room temperature and the addition of 1 mL distilled water. The fatty acid methyl esters (FAMEs) were extracted by the addition of 2 mL of a hexane–chloroform mixture (4:1 *v/v*) and applying 10 s of vortex mixing. The reaction tubes were centrifuged at 3000 g for 5 min at 4 °C and the upper hexane phase was collected in glass tubes. The extraction step was repeated three times for each sample. Subsequently, the solvent was evaporated under nitrogen at 30 °C and FAMEs were dissolved in 1.5 mL of hexane containing 0.01% of butylated hydroxytoluene (BHT, Sigma-Aldrich, St. Louis, Missouri, USA) and a small amount of anhydrous sodium sulfate (to remove traces of water in the sample). Samples were mixed by vortexing and, finally, dissolved FAMEs were transferred to the GC vials.

Fatty acid profile analysis was performed using a gas chromatography system with flame ionization detector (GC-FID) 7820A GC System, Agilent Technologies, controlled by Agilent OpenLAB software (Agilent Technologies, Santa Clara, CA, USA). Agilent J and W GC column 121–2323, DB-23, 20 m length; 0.180 mm diameter; 0.20 µm film was used for the separation of FAMEs. Then, 1 µL of the sample was injected in the 30:1 split mode with the split flow 30 mL/min. The inlet heater temperature was set on 250 °C and helium was used as the carrier gas. The flow of helium through the column was 1 mL/min. The total runtime for one sample was 36 min with the following oven temperature increase: initial temperature 70 °C for 2 min, after 8 min to 150 °C with no hold time, 230 °C in 16 min with 5 min hold time, and 245 °C in 1 min with 4 min hold time. For identification and quantification of fatty acids, the C4–C24 FAME mixture (Supelco, St. Louis, MO, USA) was used as an external standard, in addition to C13:0 TAG internal standard. The weight of individual FAs was calculated based on the peak areas, relative response factors (RRF), and C13 internal standard. The total lipids in the fungal biomass were the sum of FA (the weight of C13 IS was subtracted) divided by the weight of dry biomass.

2.3. Fourier Transform Infrared Spectroscopy of Fungal Biomass

Fourier transform infrared (FTIR) spectroscopy analysis of fungal biomass was performed according to Kosa et al. [45], with some modifications [43]. The biomass was separated from the growth media by centrifugation and washed with distilled water. Approximately 5 mg of washed biomass was transferred into a 2 ml polypropylene tube containing 250 ± 30 mg of acid washed glass beads and 0.5 ml of distilled water for further homogenization. The remaining washed biomass was freeze-dried for 24 h for determining biomass yield. The homogenization of fungal biomass was performed by using Percellys Evolution tissue homogenizer (Bertin Technologies, Aix-en-Provence, France) with the following set-up: 5500 rpm, 6 × 20 s cycle. Then, 10 µL of homogenized fungal biomass was pipetted onto an IR transparent 384-well silica microplate. Samples were dried at room temperature for 2 h. In total, 140 biomass samples were analyzed in three technical replicates by FTIR spectroscopy.

FTIR spectra were recorded in a transmission mode using the high throughput screening extension (HTS-XT) unit coupled to the Vertex 70 FTIR spectrometer (both Bruker Optik, Leipzig, Germany). Spectra were recorded as the ratio of the sample spectrum to the spectrum of the empty IR transparent microplate in the region between 4000 cm^{-1} and 500 cm^{-1}, with a spectral resolution of 6 cm^{-1}, a digital spacing of 1.928 cm^{-1}, and an aperture of 5 mm. For each spectrum, 64 scans were averaged. In total, 420 biomass spectra were obtained.

The OPUS software (Bruker Optik GmbH, Leipzig, Germany) was used for data acquisition and instrument control.

2.4. Data Analysis

The following software packages were used for the data analysis: Unscrambler X version 10.5.1 (CAMO Analytics, Oslo, Norway) and Matlab R2019a (The Mathworks Inc., Natick, MA, USA).

2.4.1. Analysis of FTIR Spectral Data

To evaluate the correlation of lipid content results obtained from FTIR and GC data and, further, investigate media associated changes of the lipid content in biomass, FTIR spectra were preprocessed by 2nd derivative using Savitzky–Golay algorithm with 2nd order polynomial and windows size 13, followed by spectral region of interest (SROI) selection 3050–2800 and 1800–1700 cm^{-1} and normalization by extended multiplicative signal correction (EMSC) [56] with linear and quadratic terms. Preprocessed FTIR data were then analyzed by principal component analysis (PCA) and score plots were used to compare the lipid related information in FTIR and GC data.

Different preprocessing was applied in order to evaluate the cell chemistry changes that occurred under different media conditions. To do so, SROI 3300–2800 and 1800–800 cm^{-1} was selected and normalized by EMSC using linear and quadratic terms and up-weighting the region 2800–1800 cm^{-1}. The up-weighting of the inactive region 2800–1800 cm^{-1} helped in reducing baselines by EMSC in the SROI. Afterwards, the dataset was split according to the concentrations of inorganic phosphorus (Pi) and separately analyzed by PCA. Correlation loading plots were obtained for Pi1, Pi2, and Pi4 in order to analyze the most pronounced correlation patterns in the data. To obtain such a plot, we used scores of each separate PCA model corresponding to one Pi concentration and projected on them variables of interest such as certain relevant peaks of FTIR data and other reference variables, such as pH, biomass yield, and lipid content from GC data, in addition to experimental design factors. The maxima of the corresponding chemical bonds selected for correlation loading plots based on the FTIR spectra of reference materials were pure sodium polyphosphate, chitin, and glyceryl trioleate (Table 2, Figure S3 in the Supplementary Materials). These compounds were the main cell components of interest. For plotting the peaks on the correlation loading plots, the preselected peaks of the preprocessed spectra were used (Table 2).

2.4.2. Analysis of GC Data

The detailed fatty acid profiles from the GC analysis were analyzed and compared to FTIR data. Fatty acid profiles were represented by both single fatty acids and sum of fatty acids—saturated fatty acids (SAT), monounsaturated fatty acids (MUFA), and polyunsaturated fatty acids (PUFA). Each data column was standardized (x/std (x)) and then analyzed by PCA. The scatter plot was used to compare the information related to total lipid content and profile obtained by GC and FTIR.

To find differences in fatty acid composition at different Pi concentrations, after standardization, the GC dataset was split into three datasets: Pi1, Pi2, Pi4, and analyzed separately by PCA. To learn about the correlation patterns in the data, the correlation loading plots for GC data were obtained using scores of GC based PCA models. The fatty acid profile of *M. circinelloides* is dominated by myristic (C14:0), palmitic (C16:0), palmitoleic (C16:1), stearic (C18:0), oleic (C18:1n9), linoleic (C18:2n6), and γ-linolenic (C18:3n6) acid, therefore these were presented in the correlation loading plots.

Table 2. The maxima of the corresponding chemical bonds selected for correlation loading plots based on FTIR data.

Cell Component	Peak Maxima	Molecular Vibration
Chitin/chitosan	3261	N-H stretching
	3105	N-H stretching
	2879	-C-H stretching
	1656	-C=O stretching (Amide I)
	1620	-C=O stretching (Amide I)
	1554	C-N-H deformation (Amide II)
	1375	$-CH_3$ deformation
	1305	C-N-H deformation (Amide III)
	1027	C-O-C str., C-O-H def. C-O-C def.
	950	$-CH_3$ def.
Lipids	3004	=C-H stretching
	2921	-C-H stretching
	2852	-C-H stretching
	1743	-C=O stretching
	1463	$-CH_2$ bending
	723	$>CH_2$ rocking
Polyphosphates	1263	P=O stretching
	885	P-O-P stretching

3. Results

3.1. Growth of Oleaginous M. Circinelloides in Metal Ion-Regulated Media with Different Pi Levels

In order to assess the variation between bioreplicates, cultivation of *M. circinelloides* in the reference medium R with Pi1 concentration was performed 11 times (part of the data were published previously [43]). Cultivations were performed at different timepoints and by using different MTPs. The results of statistical analysis of 11 bioreplicates show that the mean for the biomass production in the reference medium (R-Pi1) was 10.08 g/L (range: 9.35–11.52 g/L, median: 9.90 g/L), with 0.62 g/L standard deviation, and 0.19 g/L standard error. Therefore, it can be considered that all deviations higher than two standard deviations (13% of the average biomass concentration) are statistically significant and can be assigned as the effect of various metal and phosphates concentrations.

Growth of *M. circinelloides* in different media was strongly affected by the availability of phosphates. Low availability of phosphates (Pi0.25 and Pi0.5) led to a substantial decrease in pH (Table S2, Supplementary Materials) causing a significantly reduced growth of *M. circinelloides* (Figure 1, Table S3). Biomass yields in all Pi0.25 and Pi0.5 media were in the range from 0.51 g/L to 2.79 g/L and from 0.73 g/L to 4.70 g/L, respectively (Figure 1). The biomass production under moderate and high levels of phosphates (Pi1–Pi4) was substantially higher and in the range from 7.84 g/L to 12.90 g/L for Pi1, from 7.54 g/L to 12.47 g/L for Pi2, and from 7.89 g/L to 13.23 g/L for Pi4 (Figure 1). The biomass yield of *M. circinelloides* was, in several cases, higher than in the reference medium with Pi1 concentration (Figure 1, Table S3).

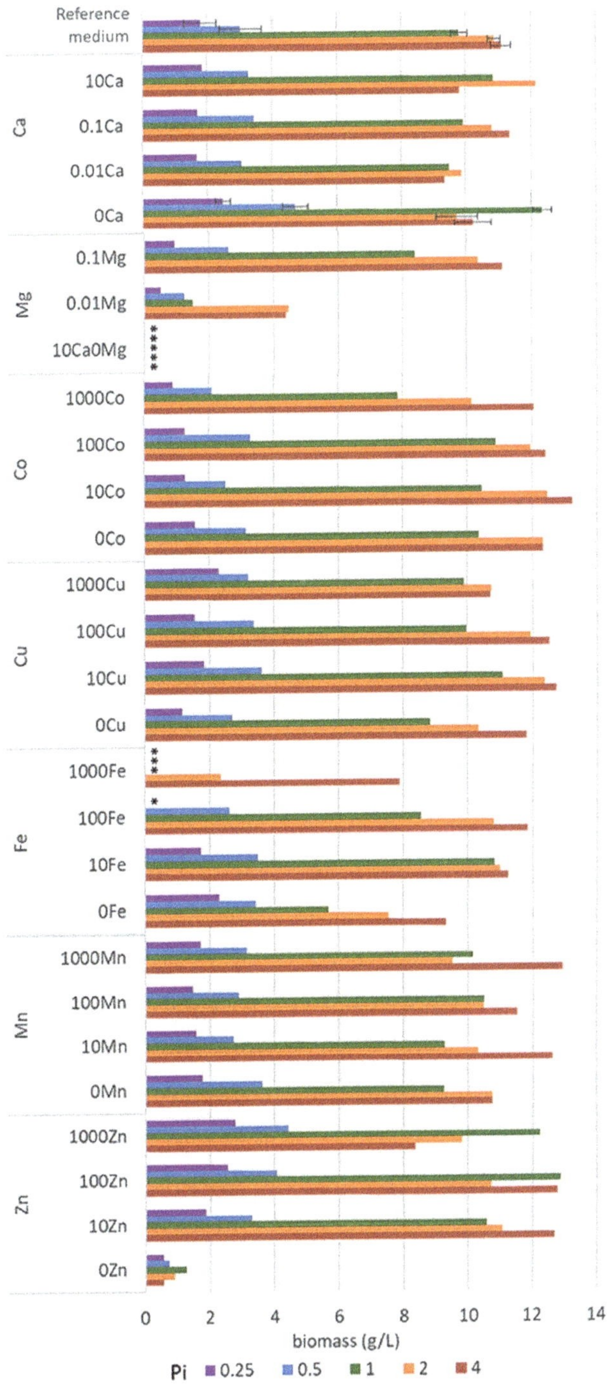

Figure 1. Final biomass concentrations after 7 days of incubation of *M. circinelloides* in the media with different concentrations of metal and phosphorus ions. * Empty slots indicate no growth.

Metal ions affected the growth of *M. circinelloides* differently and the strongest effect was observed at moderate and high levels of phosphates (Pi1, Pi2, and Pi4). Metal ions' starvation for most of the tested metal ions led to a reduced fungal growth. Thus, Zn and Mg starvation resulted in a low or no growth of *M. ciricinelloides* in all tested Pi conditions (Figure 1, Table S3). Further, removal of Cu ions led to a slight biomass decrease. Removal and low concentrations of Ca ions resulted in a slight biomass decrease at Pi2 and Pi4 and increase at Pi0.25, Pi0.5, and Pi1 levels (Figure 1, Table S3). Similar increase in the biomass yield was observed for the media without Fe ions at Pi0.25, and it was significantly decreased at higher Pi levels −44% for Pi1, −30% Pi2, and −16% for Pi4 compared to the reference amount of Fe (Figure 1, Table S3). Deficiency of Co ions did not result in any significant change in biomass formation of *M. circinelloides* in comparison with high concentrations of Co ions (100Co).

Generally, increased availability of metal ions showed either no effect or both growth-stimulating and -inhibiting effects, depending on the metal type and phosphates concentration. Thus, high metal ions concentration in the media with low Pi levels did not lead to any significant changes in the biomass formation of *M. ciricinelloides* (Figure 1), with the exception of Fe and Zn ions. High amount of Fe ions in a combination with the low concentration of phosphates showed a negative effect on the biomass formation, and no or very limited growth was observed for the following Fe conditions: 1000Fe-Pi 0.25, 1000Fe-Pi1, and 100Fe-Pi0.25. Low growth under these conditions can be connected to the acidic pH ranging from 2.79 for Pi4 to 1.62 for Pi0.25 (Table S2 in the supplementary materials). The opposite effect was observed for the media with elevated concentration of Zn ions, which enhanced growth of *M. circinelloides*, resulting in a higher biomass yield under Pi limitation, with the highest biomass yield of 4.44 g/L observed for 1000Zn condition (1 g/L $ZnSO_4{\cdot}7H_2O$) (Figure 1, Table S3).

A considerable effect of the increased metal ions availability was recorded for the media with moderate and high Pi levels (Figure 1). For example, high concentrations of Mn provided higher biomass at Pi4, with the highest yield of 12.93 g/L for 1000Mn condition, while the effect of other Mn conditions on the growth of M. *circinelloides* was generally negligible. Increasing amount of Fe ions up to the 100Fe condition positively affected the biomass production of *M. circinelloides* in the media with Pi4 and Pi2, and the highest yield of 11.84 g/L was observed for the condition 100Fe with Pi4 (Figure 1). However, very high iron concentration (1000Fe) showed an inhibiting effect at Pi1 and significantly decreased biomass at Pi2 and Pi4 (Figure 1). While increased concentration of Zn ions positively affected the growth of *M. circinelloides* in the media with low phosphates concentration (Pi0.5 and Pi0.25), it slightly decreased the growth in media with high phosphates concentration (Pi2 and Pi4). Generally, it can be concluded that increased concentrations of Zn ions (10Zn and 100Zn) in the media have beneficial effects, since under all Pi levels the biomass yield was increased compared to the standard conditions (R) (Figure 1). A similar effect can be seen for Cu and Co ions, where media with 10Cu condition provided the highest biomass yield for moderate and high Pi levels (Figure 1). Moreover, the highest biomass yield of 13.23 g/L of all tested conditions was observed for 10Co (0.001 g/L $CoSO_4{\cdot}7H_2O$) with Pi4 level of phosphorus substrate (Figure 1, Table S3).

3.2. *Effect of Metal Ions on Lipid Accumulation and Fatty Acid Profile of M. Circinelloides TAGs*

Lipid content in oleaginous *M. circinelloides* biomass grown in the different media is reported in Table 3. Due to the low growth and not sufficient amount of biomass for lipid extraction, samples Pi0.25 and Pi0.5 for all metal ions conditions, and samples 1000Fe, 0Mg, 0Zn, 10Ca0Mg were excluded from the lipid extraction and further data analysis.

Table 3. Lipid accumulation (% of lipids per dry cell weight) for *M. circinelloides* grown in nitrogen-limited metal ion-regulated media with different amounts of inorganic phosphorus substrates (Pi1, Pi2, and Pi4).

Metal Ion Condition	Pi1	Pi2	Pi4
Reference medium	41.13 ± 1.19	33.44 ± 1.28	33.15 ± 0.01
0Ca	61.16 ± 0.16	40.15 ± 2.31	31.51
0.01Ca	34.00	39.61	34.93
0.1Ca	60.55	37.22	43.70
10Ca	44.37	33.50	27.95
0.01Mg	11.43	20.57	22.80
0.1Mg	30.38	32.90	39.40
0Co	38.78	37.24	31.46
10Co	30.25	34.40	29.87
100Co	30.31	38.52	29.60
1000Co	31.49	35.08	29.38
0Cu	61.27	53.80	52.24
10Cu	47.11	37.75	35.63
100Cu	46.70	41.27	38.27
1000Cu	43.11	42.46	38.36
0Fe	37.27	37.00	30.62
10Fe	36.73	34.36	29.19
100Fe	30.77	33.00	27.11
0Mn	46.78	37.94	38.74
10Mn	34.52	31.67	33.21
100Mn	35.16	39.13	34.22
1000Mn	33.23	30.58	33.61
10Zn	49.78	38.31	37.65
100Zn	43.84	41.94	34.72
1000Zn	42.36	41.85	38.04

Lipid accumulation in *M. circinelloides* grown under the reference metal ion conditions reached approximately 41% for Pi1 and 33% for Pi2 and Pi4 (Table 3). Lack of several metals resulted in an increase of lipid content for several Pi conditions. For example, removal of Ca, Co, and Fe ions in the media with Pi2 and Pi1 and Cu and Mn ions in the media with Pi1, Pi2, and Pi4 resulted in an increase in lipid accumulation in *M. circinelloides*, compared to the reference conditions. The most significant increase in lipid accumulation was recorded for 0Ca-Pi1 and 0Cu-Pi1 conditions. Interestingly, the removal of some metal ions, such as Ca, Co, and Fe, enhanced lipid accumulation only at moderate phosphate concentrations in the media (Pi1 and Pi2), and decreased lipid accumulation at high phosphate concentrations (Pi4). Removal of Mn, and especially Cu, ions resulted in increased lipid accumulation at moderate and high phosphate concentrations (Pi1, Pi2, and Pi4) (Table 3).

Variation in the availability of metal ions showed diverse and metal-specific effects on the lipid accumulation in *M. circinelloides* (Table 3). Lack of Mn ions has resulted in relatively high lipid accumulation in *M. circinelloides* for all tested Pi levels. The inhibiting effect of higher Mn ion concentrations was more visible in the media with reference amounts of phosphates (Pi1). Similar results were recorded for the media with increased levels of Fe ions, where lipid yield was lower than under the reference Fe condition (R). Two tested concentrations of Mg ions provided low lipid yield in *M. circinelloides* with the lowest values of 11.43% at 0.01Mg-Pi1 condition, while lipid yield of 39.4% was observed in the biomass grown in the 0.1Mg-Pi4 condition. When increasing the amount of Co ions in the media with Pi4 and Pi2, a decrease in lipid accumulation was recorded, while the opposite effect was seen for the medium with Pi4. An increase in the concentration of Zn ions showed a triggering effect on lipid accumulation in *M. circinelloides* grown at different levels of phosphorus substrate, with the highest lipid yield of 49.78% at 10Zn-Pi1 (0.01 g/L $ZnSO_4 \cdot 7H_2O$), which was 9% higher than for the reference condition. A similar lipogenesis triggering effect was observed for increasing concentration of Cu ions at all tested Pi levels, while the highest lipid yield was recorded when Cu ions were removed. The most diverse effect on lipid accumulation in *M. circinelloides* was observed for different concentrations of Ca ions. Increase in Ca ion availability from 0Ca to 0.01Ca resulted in the decrease of lipid yield for condition 0.01Ca-Pi1, while a further increase in Ca ions to 0.1Ca resulted in the increase of lipid yield for biomass grown in the media with Pi1, Pi2, and Pi4. A high concentration of Ca ions (10Ca) resulted in the decrease in lipid accumulation in the medium with Pi4 and slight increase in the media with Pi2 and Pi1 (Table 3).

The fatty acid profile of *M. circinelloides* grown under the reference condition was dominated by oleic acid (C18:1n9; 38%), followed by palmitic (C16:0; 22%), linoleic (C18:2n6; 14%), and γ-linolenic (C18:3n6; 12%) acids. Further, stearic (C18:0; 5%), palmitoleic (C16:1; 1.75%), and myristic (C14:0; 1.5%) acids were recorded in smaller amounts (Figure 2, Table S4). An example chromatogram can be found in the Supplementary materials (Figure S1). The fatty acid profile of *M. circinelloides,* grown under reference metal ion conditions, slightly changed depending on the phosphorus availability in the media. Thus, we observed an increase in the unsaturation and amount of palmitoleic acid with the increasing amount of phosphorus (Figure 2, Table S4). An opposite effect of phosphorus availability (and the associated changes in pH of media) was recorded for the unsaturation of stearic acid into oleic and γ-linolenic acid, where decreasing unsaturation was evident with increasing Pi concentrations and higher pH. This pattern can be visible through all the samples, with some exceptions for 10Fe-Pi1/Pi2/Pi4, 100Fe-Pi1/Pi2, 1000Zn, 1000Co, 10Cu-Pi2, and 0.01Mg-Pi1conditions (Figure 2, Table S4). Minimal content of myristic acid (C14:0) was observed in 10Fe and 100Fe conditions, except for the 100Fe-Pi4 sample (Figure 2). Further, media with high amounts of Zn (1000Zn) and Co (1000Co) ions led to the synthesis of TAGs with the increased relative amount of stearic acid (Figure 2, Table S4).

To reveal underlying correlations among certain fatty acids, design variables, and reference variables, as well as sum of saturated (SAT), monounsaturated (MUFA), and polyunsaturated (PUFA) fatty acids, PCA analysis of fatty acid (FA) profiles was done for each Pi substrate level separately. The separation of data into different Pi concentrations was done in order to focus on the effect of metal ions only on FA profile, excluding the effect of phosphorus substrate availability. The results, in the form of correlation loading plots, are presented in Figure 3. Generally, the fatty acid profile of *M. circinelloides* was quite consistent, irrespective of media composition.

In the biomass obtained from the media with Pi1 and Pi2 amounts of phosphorus, high concentrations of Co (1000 Co) and Zn (1000 Zn) ions were positively correlated with the saturated fatty acids (SAT) (Figure 3A). This was also evident from the detailed FA profiles, where the relative amount of palmitic and stearic acid was increased under these conditions (Figure 2). Further, some tendency of positive correlation between increasing concentration of Fe ions and content of polyunsaturated fatty acids (PUFA) was observed in the media with Pi1 level (Figure 3A). In the media with high amounts of phosphorus substrate (Pi4 and Pi2), 1000 Mn and 0.01 Mg conditions were positively correlated with the polyunsaturated fatty acids (PUFAs) (Figure 3B,C).

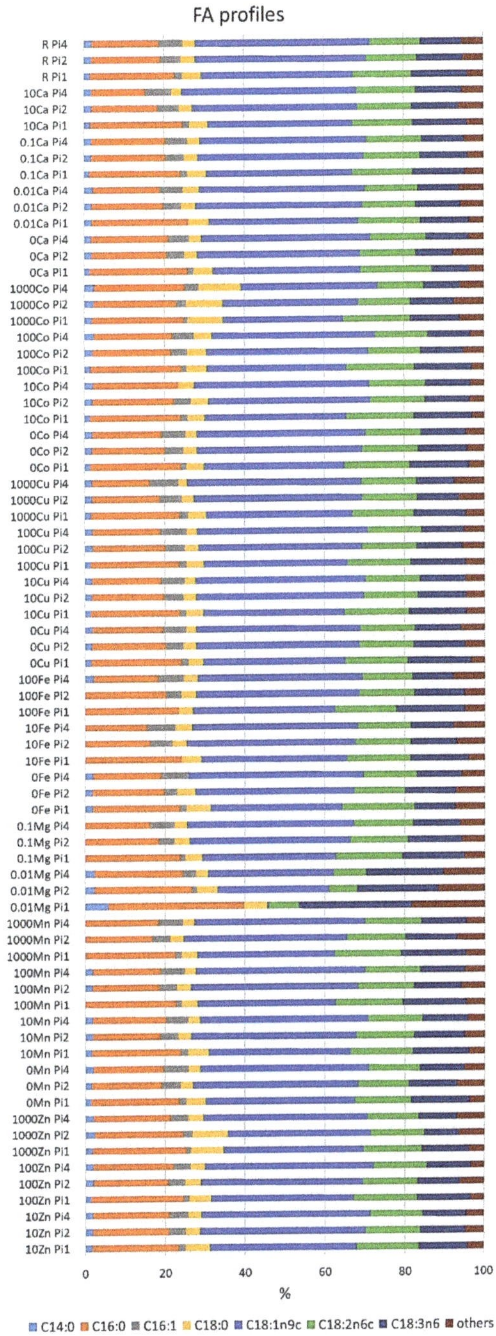

Figure 2. Fatty acid profile of lipids accumulated in *M. circinelloides* grown in media with Pi1, Pi2, and Pi4 levels of phosphorus. Only fatty acids present in amounts of more than 1% are displayed. The rest is summed up as 'others'. An example chromatogram can be found in the Supplementary Materials (Figure S1).

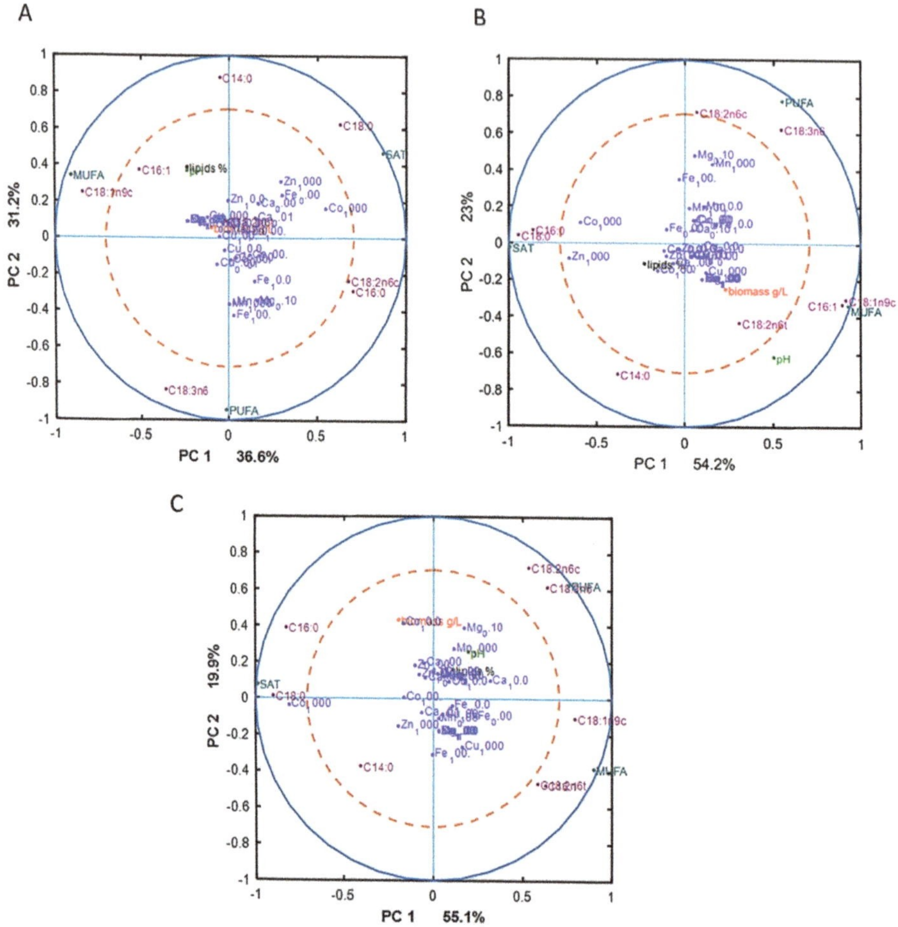

Figure 3. Correlation loading plots based on the PCA analysis of fatty acid (FA) profiles of lipids accumulated in *M. circinelloides* grown in metal ion-regulated media under Pi1 (**A**), Pi2 (**B**), and Pi4 (**C**) levels of phosphorus substrate.

3.3. Chemical Composition of M. Circinelloides Biomass

In order to study differences in the compositional profile of the *M. circinelloides* biomass, high-throughput Fourier transform infrared (FTIR-HTS) spectroscopy was used. Spectral regions and peaks related to three types of metabolites—lipids, chitin/chitosan, and polyphosphates were used in the analysis. The FTIR-HTS spectra (Figures S2 and S3 in the Supplementary Materials) showed that fungal biomass was dominated by signals of these intracellular metabolites. The spectra of reference materials can be found in the supplementary materials (Figure S2). The maxima of the peaks selected for the correlation loading plots are listed in Table 2. Due to the insufficient growth (Figure 1), the following samples have been disregarded from the FTIR-HTS spectral data analysis: (i) all samples grown under Pi0.25 and Pi0.5 levels; (ii) samples grown in the media with 1000Fe, 0Mg, 0.01Mg, 0Zn, and 10Ca0Mg.

First, we examined the lipid region of FTIR-HTS spectra (3050–2800 and 1800–1700 cm^{-1}) and analyzed the correspondence of it with GC data by PCA analysis (Figure 4). On the PCA score plots we can observe similar pattern for FTIR and GC data, indicating similarity in the obtained information about lipid yield and profile from the data of these analytical techniques.

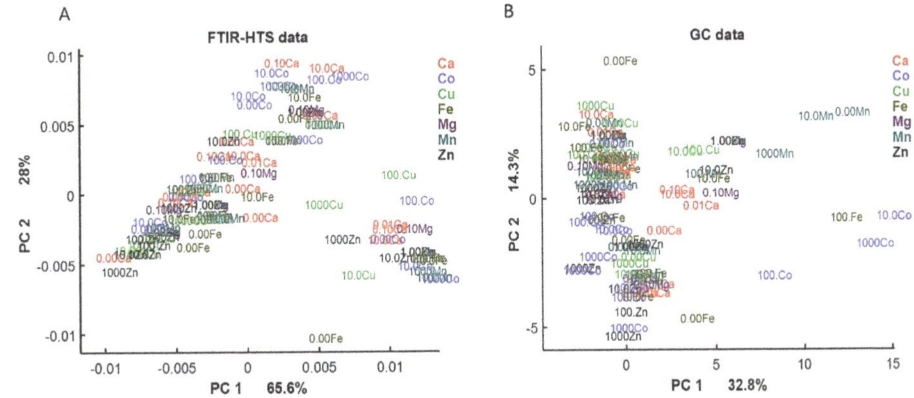

Figure 4. PCA score plots of FTIR-HTS (**A**) and GC (**B**) data. PCA analysis was performed on the preprocessed FTIR-HTS data (2nd derivative, polynomial order 2, window size 13; SROI: 3050–2800 and 1800–1700 cm^{-1}, EMSC) and normalized GC data.

The correlation loading plots from the PCA analysis visualize the relation between the presence of lipids, chitin/chitosan, and polyphosphate and different media (Figure 5). PCA analysis of EMSC preprocessed spectra was performed separately for different Pi concentrations in order to emphasize the effect of different metal ions and disconnect it from the effect of inorganic phosphorus substrate. The loading vectors and FTIR spectral scores are displayed in Figures S4–S6 in the Supplementary materials. The first principle component (PC1), which explained the highest variance in the FTIR data, was represented by lipids to proteins and to chitin and chitosan ratio. The second principle component (PC2) was represented by the polyphosphate peaks, which were strongly visible in the cases of polyphosphate accumulation triggering conditions when Pi4 and Pi2 phosphate concentrations were used in the media (Figures S5A and S6A), while no strong characteristic signals representing any of the studied metabolite were visible in the PC2 for Pi1 condition (Figure S4A).

Lipids and chitin/chitosan are both carbon-rich metabolites, therefore their synthesis processes are competing for the C source. In all correlation loading plots, we can observe that lipids and chitin/chitosan were anticorrelated, indicating that these metabolites cannot be produced simultaneously at high yields, while they still can be coproduced with one of them dominating (Figure 5). Further, we see that peaks 2879 cm^{-1} (-C-H stretching) and 950 cm^{-1} (C-O str, C-C str., C-O-H def. C-O-C def), responsible for chitin/chitosan, have been shown to be correlated with the lipid-related peaks (Figure 4). The reason is that the chemical bonds, represented by these peaks, are also present in lipids and the contribution of the lipid associated peaks was stronger than the contribution of chitin/chitosan peaks. The chitin/chitosan formation could also be negatively affected by the N-limitation. The biomass concentration was correlated with the lipid peaks, revealing that a good lipid accumulation can be achieved only with the optimal growth conditions providing good growth and biomass formation and that high biomass concentration was the result of the increased lipid accumulation (Figure 5).

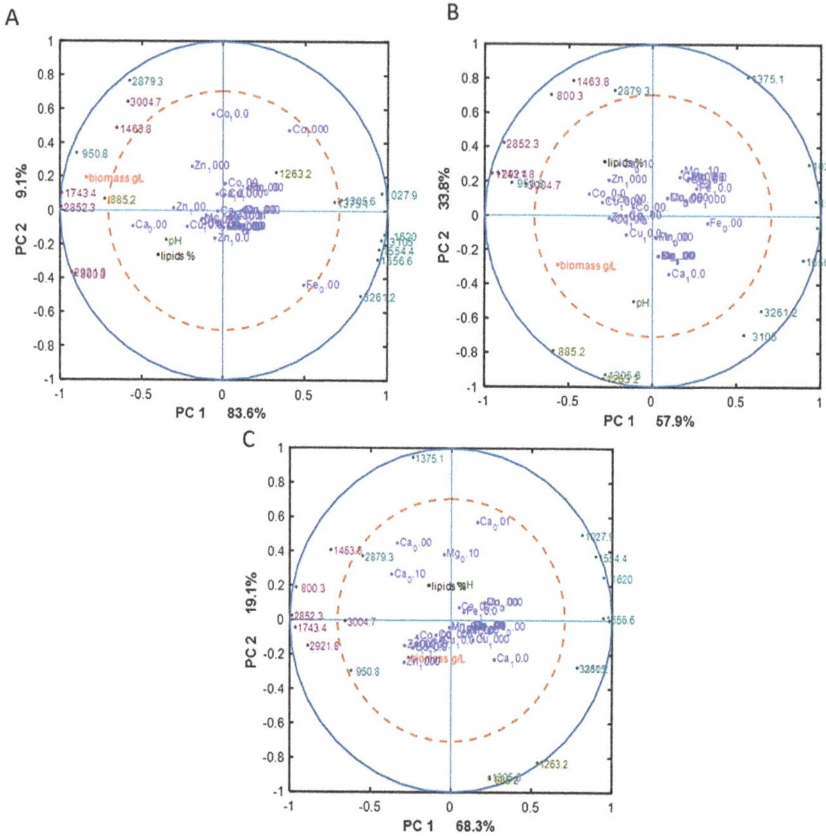

Figure 5. Correlation loading plots based on the PCA analysis of FTIR-HTS spectra of *M. circinelloides* biomass grown in metal ion-regulated media with Pi1 (**A**), Pi2 (**B**), and Pi4 (**C**) levels. Main peaks corresponding to the lipids (purple), chitin/chitosan (green), and polyphosphates (yellow) are presented.

The effect of metal ions on the *M. circinelloides* biomass composition in the media with reference level of phosphorus (Pi1) is displayed in a PCA score plot with loading vectors (Figure S4 in the Supplementary materials). Due to the fact that PC2 was not representing any clear relation between FTIR-HTS peaks and studied metabolites, we analyzed only PC1 representing the ratio of lipids to protein and to chitin and chitosan peaks. PCA score plots show that lipid content correlated with the absence of Ca ions (0Ca) (Figure S4B), which is in agreement with the lipid yield data presented in Table 3. Further, the PCA score plot shows that increasing Co amount displayed an inhibiting effect on the lipid accumulation (Figure S4B). The same results were observed for the lipid yield, where the highest lipid production was recorded for 0Co-Pi1 condition (Table 3). FTIR-HTS spectra of biomass grown in the media with reference level of inorganic phosphorus substrate (Pi1) did not exhibit significant absorbance for polyphosphate peaks. The correlation loading plot (Figure 5A) shows some effect of two metal ions—Fe and Co. Namely, a decrease in the concentration of Fe ions was correlated with the chitin/chitosan peaks, indicating that higher amounts of this metabolites is expected in the fungal biomass when Fe ions are in a low availability, while the absence of Fe ions (0Fe) was anticorrelated with the lipid peaks and biomass concentration (Figure 5A). This corresponds well to the biomass production results, where it is obvious that lack of Fe ions caused a significant decrease of biomass formation (Figure 1). A high concentration of Co ions (1000 Co) showed correlation with chitin/chitosan peaks of FTIR-HTS spectra, indicating that the relative content of this metabolite in the fungal biomass increased with the increased concentration of Co ions. Absence of Ca ions (0Ca) was correlated with the high lipid and biomass concentration. This is in agreement with the reference biomass and lipid concentration results, where 0Ca-Pi1condition provided the highest biomass and lipid production from all the tested Ca ions conditions (Figures 1 and 2).

The effect of metal ions on the synthesis of studied *M. circinelloides* metabolites in the media with Pi2 level of phosphorus is displayed on the Figure 5B, where we can see that increased amounts of Zn and Cu ions were corelated with the lipid yield. Similar correlation results were observed for Co ions, except for the condition of 1000Co-Pi2, which was slightly anticorrelated with lipid peaks. Correlation between polyphosphate spectral peaks and the highest tested Ca ions amount (10Ca) was observed (Figure 4B). Further, we can see correlation between amount of Zn ions and polyphosphate peaks. Finally, metal ions conditions 0.1Ca, 1000Zn, 100Zn, 10Zn, 10Co, 100Co, and 10Cu correlated with the lipid yield.

When examining the effect of metal ions under the high amounts of inorganic phosphorus substrate level (Pi4), on the PCA score plot it is seen that increasing amount of Ca ions corelated with the decrease of the relative lipid content and increase of the polyphosphate content (Figure S6 in the Supplementary materials). Further, in Figure 5C we observed that: (i) decreasing Cu ion availability and high concentration of Zn ions correlated with the increase of relative lipid content; (ii) low amount of Ca ions (0,1Ca) correlated with the lipid peaks and lipid yield and anticorrelated with polyphosphates peaks, while high amount of Ca ions (10Ca- 1 g/L $CaCl_2.2H_2O$) correlated with polyphosphate and chitin/chitosan peaks; (iii) low concentration of Mg ions (0,1Mg) correlated with lipid peaks and anticorrelates with polyphosphate peaks; (iv) there was no correlation observed for Co and Mn ions (Figure 5C).

4. Discussion

For half of a century, *M. circinelloides* has been studied as a microbial cell factory for production of a series of metabolites and valorization of different substrates. Today, this dimorphic oleaginous fungus is positioned as one of the most robust fungal cell factories for the biotech, biorefinery, and bioremediation industries [2].

Despite the deep understanding of *M. circinelloides* physiology and metabolic processes, the role and the effect of metal ions on the lipid accumulation and the cellular composition of this fungus have not been systematically investigated. The effect of metal ions on the growth and metabolic activity of *M. circinelloides* has, to the authors knowledge, only been assessed in the connection to bioremediation abilities of this fungus [11,12,57]. Therefore, in this study we performed an extensive screening of the growth, lipid accumulation, and compositional profile of *M. circinelloides* on 140 different media with variations in the concentrations of metals ions and phosphorus. Lipid accumulation and fatty acid profiles were determined by the GC-FID. The composition of the fungal biomass was investigated by the quantification of lipids, polyphosphates, and chitin/chitosan, as these components previously have been suggested for a coproduction concept involving *M. circinelloides* [43]. For the evaluation of biomass composition, the modern high-throughput analytical technique FTIR spectroscopy was applied. The main advantage of FTIR spectroscopy is that all biochemical components of the sample can be profiled in a single measurement run, without tedious extraction procedures [58–62]. FTIR spectroscopy provides detailed relative quantitative information about different chemical components of the samples and it has been previously utilized for the characterization of lipids [50,51,63], polyphosphates [64], and chitin/chitosan [65,66]. In this study, we have demonstrated that the FTIR analysis as a sole method coupled to multivariate data analysis can be applied for a fast and simple analysis of microbial biomass.

Efficiency of microbial biomass production, the yield of the targeted metabolite(s), and a coproduction potential are important assessment parameters in bioprocess development [19,20]. Biomass production is affected by factors such as pH, temperature, aeration, media composition, and cultivation mode [2]. For example, culture volume and mode of cultivation were reported by

Carvalho et al. as factors strongly affecting the final biomass yield of *M. circinelloides* [67]. The reported biomass production of *M. circinelloides*, depending on the culture conditions, varied greatly, from 5 g/L to 20 g/L [10,17,68]. In our previous studies, the biomass concentration for different *M. circinelloides* strains was between 10 g/L and 15 g/L for the cultivations performed in microtiter plates, and 15.8 g/L in bioreactors [54]. The biomass production of the *M. circinelloides* VI04473 strain in this study varied from 0.5 to 13.2 g/L (Figure 1). The standard growth medium containing reference amount of inorganic phosphorus substrate (Pi1) and the reference amounts of metal ions resulted in 9.8 g/L of biomass, significantly lower than in our previous screening study [50]. The reason for the lower biomass production in the reference medium in the present study was probably due to utilization of ammonium sulphate as a nitrogen source, instead of yeast extract as in the previous study [50]. Ammonium sulphate is a pure inorganic source of nitrogen, lacking any additional macro- and micronutrients, vitamins, and growth factors that are present in yeast extract. Buffering capacity of ammonium sulfate is lower than for yeast extract and, as it has been previously reported, the uptake of ammonium ions causes the release of H^+ by the fungal cells into the media, which accelerates pH lowering [69,70]. Further, possible formation of sulfuric acid during the uptake of ammonium could occur [71]. In addition, formation of organic acids by fungal cells either during exponential or during the stationary growth phase [72] significantly contributes to the acidity of the growth media. Thus, in the media with the low Pi levels we detected considerably low pH and suppressed growth and lipid accumulation in *M. circinelloides* (Figure 1). Acidic pH is a stress factor for many cellular organelles, especially for endoplasmic reticulum (ER), which is connected to protein folding and lipogenesis in fungal cells. It has been previously reported that acidic pH causes ER stress and induces unfolded protein response (UPR). This results in the accumulation of misfolded proteins in the ER and activation of the ER-stress sensor (Ire1p) and ER stress-responsive transcription factor (Hac1p), leading to the inhibition of growth and metabolic activity [73]. In our previous studies, we have showed that acidic pH affects cell wall and increasing chitin/chitosan production in *M. circinelloides* [43]. Cultivation in Duetz-MTPS does not allow continuous adjustment of pH and only start- and end-point measurements are possible, therefore the effect of low phosphorus concentrations was directly linked to drop in pH. Due to the fact, that acidic pH is quite an aggressive stress factor inhibiting fungal growth, the effect of metal ions on the growth and lipid accumulation under low phosphorus substrate availability was difficult to assess. Only two observations could be considered as significant—increase of biomass under higher Zn ion availability and Ca deficiency. Moreover, under Pi conditions lower than the reference (Pi0.5 and Pi0.25), K and Na ions were compensated with KCl and NaCl salts in order to provide the same Na and K amounts as in the reference Pi1 condition [43,44]. It has been reported that chlorides could have negative effect on the mycelium formation of some fungi [74,75]. Moreover, much higher concentrations of Cl (10–15% NaCl) than used in our study (KCl and NaCl in total below 5%) have shown some negative impact on fungal growth [76]. No negative impact of increased Cl^- on the biomass and lipid production has been observed when yeast extract was used as N-source [43,44]. Thus, we can hypothesize that in addition to pH-stress, increased Cl^- ions possibly negatively impacted the growth under low pH conditions. Therefore, these samples were excluded from further data analysis.

Variation in metal ion availability showed diverse and often metal- and pH-specific effects on biomass production and biomass composition of *M. circinelloides*. Growth of *M. circinelloides* was severely inhibited in media lacking Zn and Mg, indicating that these metal ions are essential for the growth and metabolic activity of the fungus. Inhibition of fungal growth in the media lacking Zn ions can be related to the fact that Zn plays an important role in the regulation of all genes in the eukaryotic cells [31]. Deficiency of Zn is detrimental for the fungal spore germination and further cell proliferation. Our study shows that elevated concentration of Zn ions has a beneficial effect on the biomass formation under phosphorus limitation. Low concentrations of Mg (0.01Mg condition) led to a decrease in biomass production and lipid yield, especially for Pi1 condition, where a lipid content of only 11.43% was reached (Figure 1, Table 3, Table S4). This can be explained by the fact that magnesium deficiency in eukaryotic cells can result in the decrease of glucose-6-phosphate, total content of phospholipids, and a remarkable decrease in oxygen and substrate delivery to the cells with further concomitant changes in membrane phospholipids, leading to the reduced cell growth, delay in the cell cycle, and metabolic activity [77]. It has been shown that long-term Mg deficiency for yeast may result in distortion of cell division, production of aberrant cell forms, and a decrease in viability that can lead to a delay or change of cell cycle [78]. Therefore, the difference in the FA profile of *M. circinielloides* grown under the Mg deficiency (0.01Mg condition) could be explained by disruption of the cell cycle [79].

In addition to Zn and Mg, Ca and Fe are known to be essential for fungal growth [31,80]. In our study, an absence of Fe ions in the medium suppressed the growth of *M. circinelloides* under conditions of moderate and high phosphorus concentrations. While these metals did not affect lipid accumulation. This is an interesting observation, due to the fact that Fe is an important cofactor of many enzymes, it is essential during DNA synthesis and cleavage, and, thus, Fe deficiency should strongly affect growth and metabolic activity of fungal cells.

An absence of Ca ions affected growth of *M. circinelloides* depending on the phosphorus concentration and associated pH of the growth media. A considerable increase in the biomass production of *M. circinelloides* was observed in the media lacking Ca ions and containing moderate (Pi1) and low concentrations of phosphates (Pi0.5 and Pi0.25). Elevated biomass production under the condition Ca0-Pi1 could be partially explained by the fact that the absence of Ca ions in the medium enhanced lipid yield up to 61% (*w*/*w*). Increase in lipid accumulation with the decrease of concentration of Ca ions was observed also for media containing Pi2 and Pi4 levels of phosphates. Calcium starvation enhancing lipid accumulation in oleaginous microorganisms has been reported for algae [81], where the lipid production was increased by 30% in Ca deprived media. To the authors knowledge, a similar effect of Ca ions deficiency on lipid accumulation has never been reported for oleaginous fungi. Currently, there is no clear understanding of the mechanisms behind Ca deficiency-induced lipid accumulation in oleaginous microorganisms, and the direct link between calcium and lipid accumulation and TAGs synthesis has not been clearly demonstrated yet. Similar observations have been reported for adipocyte cells, where low cellular availability of Ca ions mediated antilipolytic pathways through a calcium-sensing receptor (CaSR), resulting in enhancing of lipid content in adipose tissue [82]. Due to the fact that lipolytic pathways are functionally conserved from mammalian cells to fungi [83], we suggest that Ca deficiency is mediating similar antilipolytic pathways in oleaginous microorganisms. Further, Wang. W.A. et al. [83] showed that Ca ions are important for the basal sensitivity of the sterol sensing mechanism of the sterol response element binding proteins (SREBPs) pathway. Wang W.A. et al. discovered that reduction of Ca concentration in endoplasmic reticulum changes the distribution of intracellular sterol/cholesterol, resulting in the enhancement of SREBPs activation and triggering synthesis of neutral lipids. Sterol response element binding proteins (SREBP) are transcription factors that are synthesized on endoplasmic reticulum (ER) and considered as ER-associated integral membrane proteins [83]. SREBP were reported for eukaryotic cells, including mammalian and fungal cells [84]. The studies show that SREBP are involved in lipid homeostasis, while SREBP isoforms control the expression of genes responsible for the biosynthesis of sterol/cholesterol, fatty acids, triacylglycerols, and phospholipid in the cell [85]. Further, detailed studies would be needed to confirm if these two events are valid also for oleaginous fungi grown under calcium deficiency.

Increase in *M. circinelloides* biomass yield was observed also at high concentrations of Ca ions in the media with high phosphate concentrations (Pi2). Infrared spectra of *M. circinelloides* biomass grown in this medium showed strong absorbance values for polyphosphate peaks (Figure S5 in the Supplementary Materials). Thus, we can assume that increase in biomass production is associated with the intracellular accumulation of available inorganic phosphorus substrate in the form of polyphosphate. It has been previously reported that, in media with excess phosphorus source, *M. circinelloides* is able to perform so called luxury uptake of phosphorus and accumulate it in the form of polyphosphates either in the cell wall or in the form of intracellular polyphosphate granules [6]. Polyphosphate (polyP) is a polyanionic compound, and it has been reported by Kikuchi Y. et al. that in the fully dissociated form, polyP has one negative charge per Pi residue and two extra charges of terminal residues [85]. Therefore, accumulation of polyP in the cell results in the accumulation of a large amount of negative charge, which is probably compensated by an existence of a regulatory mechanism for maintaining charge neutrality of the cell. The studies involving temporal and quantitative analyses of cationic components of the fungal cells revealed that Na, K, Ca, and Mg ions were taken up by polyP, providing strong evidence that these ions play a major role in the neutralization of the negative charge of polyP in the fungal cell [85,86]. Thus, it is likely that with the higher availability of calcium ions in the medium, the neutralization of the polyP negative charge is more efficient and a higher amount of phosphorus can be stored intracellularly in the form of polyP. Due to fact that polyphosphate accumulation takes place in the exponential growth phase, while lipid accumulation in the stationary growth phase [87,88], it could be possible to perform a coproduction of these two components by manipulation of the availability of calcium and phosphorus substrate in the medium. Therefore, *M. circinelloides* can be utilized in the phosphorus recycling processes.

In addition to Ca-deprived media, lack of Cu and higher amounts of Zn and Mn considerably enhanced lipid accumulation in *M. circinelloides*. While elevated lipid production observed due to Ca deficiency could be explained by the above-mentioned hypothesis, there is no clear explanation of the high lipid accumulation under the copper deficiency condition that was significantly higher at all Pi levels. It has to be noted that the highest lipid yield was obtained under deficiency of Ca and Cu ions. In the literature, there has only been only one study, conducted on the liver cells, reporting Cu deficiency enhancing lipid storage [89], while metabolic pathways linking copper to lipid homeostasis have not been reported for fungal and any other microbial cells.

The FA profile of the accumulated in *M. circinelloides* TAGs was not significantly affected by the availability of metal ions and phosphorus. Only some tendency in increase of saturation with high Co and Zn amount was observed, but further enzymatic study would be needed to assess the activity of desaturases at these conditions.

By applying FTIR spectroscopy, we revealed that Ca, Co, and Zn ions at different concentrations correlated with lipid peaks; Ca and Zn correlated with polyphosphate, while Fe and Co with chitin/chitosan peaks of *M. circinelloides* biomass spectra. Thus, these ions could be considered as important components in optimizing and developing coproduction of lipids, polyphosphate, and chitin/chitosan by *M. circineloides*. However, further studies are needed to fully understand the role of these metal ions in the metabolic pathways of *M. circineloides* metabolites.

5. Conclusions

The aim of the study was to evaluate the effect of different metal ions and their concentration on biomass production, composition, and the lipid production in the oleaginous fungus *M. circinelloides*. Moreover, the growth experiments were conducted at different concentrations of phosphates. It can be concluded that, among tested metals, Mg and Zn are essential metals required for the optimal growth of *M. circinelloides*. Calcium availability is important for optimizing polyphosphate accumulation, while calcium and copper deficiency is important for lipid accumulation in *M. circinelloides*. Tested metal ions did not affect fatty acid profile of the accumulated TAGs. However, Ca, Co, Mg, and Zn ions have affected the cellular biochemical profile of *M. circinelloides*. Thus, metal ions are an important tool for optimizing lipid accumulation and coproduction of lipids, polyphosphate, and chitin/chitosan in *M. circinelloides*.

Supplementary Materials: The following are available online at http://www.mdpi.com/2309-608X/6/4/260/s1. Table S1: Concentrations of salts used for regulating metal ions and inorganic phosphorus levels in the growth media. Table S2: pH of culture supernatant. Table S3: Biomass concentration (g/L). Table S4: Fatty acid profiles (%). Figure S1: Example chromatogram, *Mucor circinelloides* grown in Pi1-R condition. Figure S2: EMSC corrected FTIR-HTS spectra of *Mucor circinelloides* biomass. Figure S3: FTIR spectra of reference materials. Adapted from Dzurendova et al. Figure S4: PCA analysis of FTIR-HTS spectra of *Mucor circinelloides* biomass grown under Pi1 level. The loadings of spectral PCA (A), the score plot (B). Figure S5: PCA analysis of FTIR-HTS spectra of *Mucor circinelloides* biomass grown under Pi2 level. The loadings of spectral PCA (A), the score plot (B). Figure S6: PCA analysis of FTIR-HTS spectra of *Mucor circinelloides* biomass grown under Pi4 level. The loadings of spectral PCA (A), the score plot (B).

Author Contributions: Conceived the research idea, V.S., B.Z., and A.K.; Designed the experiments, B.Z., S.D., V.S., and A.K.; methodology, V.S., B.Z., and A.K.; performed the experimental work, S.D.; analyzed the data, V.T. and S.D.; discussed the results, S.D., V.S., B.Z., A.K., and V.T.; wrote the manuscript, S.D.; discussed and revised the manuscript, S.D., V.S., B.Z., V.T., S.J.H., and A.K. All authors have read and agreed to the published version of the manuscript.

Funding: This research was funded by Research Council of Norway—FMETEKN Grant, project number 257622; BIONÆR Grant, project numbers 268305, 305215; DAAD Grant, project number 309220; HAVBRUK2 Grant, project number 302543/E40; MATFONDAVTALE Grant, project number 301834/E50.

Conflicts of Interest: The authors declare no conflict of interest. The funders had no role in the design of the study; in the collection, analyses, or interpretation of data; in the writing of the manuscript; or in the decision to publish the results.

References

1. Corrochano, L.M.; Kuo, A.; Marcet-Houben, M.; Polaino, S.; Salamov, A.; Villalobos-Escobedo, J.M.; Grimwood, J.; Álvarez, M.I.; Avalos, J.; Bauer, D. Expansion of signal transduction pathways in fungi by extensive genome duplication. *Curr. Biol.* **2016**, *26*, 1577–1584. [CrossRef] [PubMed]
2. Rodrigues Reis, C.E.; Bento, H.B.; Carvalho, A.K.; Rajendran, A.; Hu, B.; De Castro, H.F. Critical applications of *Mucor circinelloides* within a biorefinery context. *Crit. Rev. Biotechnol.* **2019**, *39*, 555–570. [CrossRef] [PubMed]
3. Carvalho, A.K.F.; Bento, H.B.; Reis, C.E.; De Castro, H.F. Sustainable enzymatic approaches in a fungal lipid biorefinery based in sugarcane bagasse hydrolysate as carbon source. *Bioresour. Technol.* **2019**, *276*, 269–275. [CrossRef]
4. Lübbehüsen, T.L.; Nielsen, J.; Mcintyre, M. Aerobic and anaerobic ethanol production by *Mucor circinelloides* during submerged growth. *Appl. Microbiol. Biotechnol.* **2004**, *63*, 543–548. [CrossRef]
5. Enrique, A.; Papp, T.; Breum, J.; Arnau, J.; Arturo, P. Strain and culture conditions improvement for β-carotene production with *Mucor*. In *Microbial Processes and Products*; Springer: Berlin/Heidelberg, Germany, 2005; pp. 239–256.
6. Ye, Y.; Gan, J.; Hu, B. Screening of phosphorus-accumulating fungi and their potential for phosphorus removal from waste streams. *Appl. Biochem. Biotechnol.* **2015**, *177*, 1127–1136. [CrossRef]
7. Carvalho, A.K.; Rivaldi, J.D.; Barbosa, J.C.; de Castro, H.F. Biosynthesis, characterization and enzymatic transesterification of single cell oil of *Mucor circinelloides*–A sustainable pathway for biofuel production. *Bioresour. Technol.* **2015**, *181*, 47–53. [CrossRef] [PubMed]
8. Fai, A.E.C.; Stamford, T.; Stamford-Arnaud, T.M.; Santa-Cruz, P.D.; Silva, M.C.; Campos-Takaki, G.M.; Stamford, T.L. Physico-chemical characteristics and functional properties of chitin and chitosan produced by *Mucor circinelloides* using yam bean as substrate. *Molecules* **2011**, *16*, 7143–7154. [CrossRef] [PubMed]
9. Hu, K.-J.; Hu, J.-L.; Ho, K.-P.; Yeung, K.-W. Screening of fungi for chitosan producers, and copper adsorption capacity of fungal chitosan and chitosanaceous materials. *Carbohydr. Polym.* **2004**, *58*, 45–52. [CrossRef]
10. Mitra, D.; Rasmussen, M.L.; Chand, P.; Chintareddy, V.R.; Yao, L.; Grewell, D.; Verkade, J.G.; Wang, T.; van Leeuwen, J.H. Value-added oil and animal feed production from corn-ethanol stillage using the oleaginous fungus *Mucor circinelloides*. *Bioresour. Technol.* **2012**, *107*, 368–375. [CrossRef]
11. Zhang, X.; Yang, H.; Cui, Z. *Mucor circinelloides*: Efficiency of bioremediation response to heavy metal pollution. *Toxicol. Res.* **2017**, *6*, 442–447. [CrossRef]
12. Cui, Z.; Zhang, X.; Yang, H.; Sun, L. Bioremediation of heavy metal pollution utilizing composite microbial agent of *Mucor circinelloides*, *Actinomucor* sp. and *Mortierella* sp. *J. Environ. Chem. Eng.* **2017**, *5*, 3616–3621. [CrossRef]
13. He, Q.; Rajendran, A.; Gan, J.; Lin, H.; Felt, C.A.; Hu, B.J. Phosphorus recovery from dairy manure wastewater by fungal biomass treatment. *Water Environ. J.* **2019**, *33*, 508–517. [CrossRef]
14. Tauk-Tornisielo, S.M.; Arasato, L.S.; Almeida, A.F.d.; Govone, J.S.; Malagutti, E.N. Lipid formation and γ-linolenic acid production by *Mucor circinelloides* and *Rhizopus* sp., grown on vegetable oil. *Braz. J. Microbiol.* **2009**, *40*, 342–345. [CrossRef] [PubMed]

15. Tang, X.; Chen, H.; Gu, Z.; Zhang, H.; Chen, Y.Q.; Song, Y.; Chen, W.J. Comparative proteome analysis between high lipid-producing strain *Mucor circinelloides* WJ11 and low lipid-producing strain CBS 277.49. *J. Agric. Food Chem.* **2017**, *65*, 5074–5082. [CrossRef] [PubMed]
16. Zhang, Y.; Adams, I.P.; Ratledge, C. Malic enzyme: The controlling activity for lipid production? Overexpression of malic enzyme in *Mucor circinelloides* leads to a 2.5-fold increase in lipid accumulation. *Microbiology* **2007**, *153*, 2013–2025. [CrossRef] [PubMed]
17. Vicente, G.; Bautista, L.F.; Rodríguez, R.; Gutiérrez, F.J.; Sádaba, I.; Ruiz-Vázquez, R.M.; Torres-Martínez, S.; Garre, V. Biodiesel production from biomass of an oleaginous fungus. *Biochem. Eng. J.* **2009**, *48*, 22–27. [CrossRef]
18. Ratledge, C.; Cohen, Z. Microbial and algal oils: Do they have a future for biodiesel or as commodity oils? *Lipid Technol.* **2008**, *20*, 155–160. [CrossRef]
19. Santek, M.I.; Beluhan, S.; Santek, B. Production of microbial lipids from lignocellulosic biomass. *Adv. Biofuels Bioenergy* **2018**, 137–164.
20. Meyer, V.; Basenko, E.Y.; Benz, J.P.; Braus, G.H.; Caddick, M.X.; Csukai, M.; de Vries, R.P.; Endy, D.; Frisvad, J.C.; Gunde-Cimerman, N.; et al. Growing a circular economy with fungal biotechnology: A white paper. *Fungal Biol. Biotechnol.* **2020**, *7*, 1–23. [CrossRef]
21. Mironov, A.A.; Nemashkalov, V.A.; Stepanova, N.N.; Kamzolova, S.V.; Rymowicz, W.; Morgunov, I.G. The Effect of pH and Temperature on Arachidonic Acid Production by Glycerol-Grown *Mortierella alpina* NRRL-A-10995. *Fermentation* **2018**, *4*, 17. [CrossRef]
22. Kamisaka, Y.; Kikutsugi, H.; Yokohichi, T.; Nakahara, T.; Suzuki, O. Studies on Production of Lipids in Fungi. XX. *J. Jpn. Oil Chem. Soc.* **1988**, *37*, 344–348. [CrossRef]
23. Papanikolaou, S.; Galiotou-Panayotou, M.; Fakas, S.; Komaitis, M.; Aggelis, G. Lipid production by oleaginous Mucorales cultivated on renewable carbon sources. *Eur. J. Lipid Sci. Technol.* **2007**, *109*, 1060–1070. [CrossRef]
24. Bellou, S.; Makri, A.; Sarris, D.; Michos, K.; Rentoumi, P.; Celik, A.; Papanikolaou, S.; Aggelis, G. The olive mill wastewater as substrate for single cell oil production by Zygomycetes. *J. Biotechnol.* **2014**, *170*, 50–59. [CrossRef]
25. Bellou, S.; Moustogianni, A.; Makri, A.; Aggelis, G. Lipids containing polyunsaturated fatty acids synthesized by Zygomycetes grown on glycerol. *Appl. Biochem. Biotechnol.* **2012**, *166*, 146–158. [CrossRef]
26. Vamvakaki, A.N.; Kandarakis, I.; Kaminarides, S.; Komaitis, M.; Papanikolaou, S. Cheese whey as a renewable substrate for microbial lipid and biomass production by Zygomycetes. *Eng. Life Sci.* **2010**, *10*, 348–360. [CrossRef]
27. Kavadia, A.; Komaitis, M.; Chevalot, I.; Blanchard, F.; Marc, I.; Aggelis, G. Lipid and γ-linolenic acid accumulation in strains of Zygomycetes growing on glucose. *J. Am. Oil Chem. Soc.* **2001**, *78*, 341–346. [CrossRef]
28. Fakas, S.; Papanikolaou, S.; Galiotou-Panayotou, M.; Komaitis, M.; Aggelis, G. Organic nitrogen of tomato waste hydrolysate enhances glucose uptake and lipid accumulation in *Cunninghamella Echinulata*. *J. Appl. Microbiol.* **2008**, *105*, 1062–1070. [CrossRef]
29. Dyal, S.D.; Bouzidi, L.; Narine, S.S. Maximizing the production of γ-linolenic acid in *Mortierella ramanniana* var. *ramanniana* as a function of pH, temperature and carbon source, nitrogen source, metal ions and oil supplementation. *Food Res. Int.* **2005**, *38*, 815–829. [CrossRef]
30. Moo-Young, M.; Chisti, Y. Biochemical engineering in biotechnology (Technical Report). *Pure Appl. Chem.* **1994**, *66*, 117–136. [CrossRef]
31. Gerwien, F.; Skrahina, V.; Kasper, L.; Hube, B.; Brunke, S. Metals in fungal virulence. *Fems Microbiol. Rev.* **2018**, *42*, fux050. [CrossRef]
32. Chandrasena, G.; Walker, G.M.; Staines, H.J. Use of response surfaces to investigate metal ion interactions in yeast fermentations. *J. Am. Soc. Brew. Chem.* **1997**, *55*, 24–29. [CrossRef]
33. Walker, G.M. Metals in yeast fermentation processes. *Adv. Appl. Microbiol.* **2004**, *54*, 197–230. [PubMed]
34. Birch, R.M.; Ciani, M.; Walker, G.M. Magnesium, calcium and fermentative metabolism in wine yeasts. *J. Wine Res.* **2003**, *14*, 3–15. [CrossRef]
35. Gadd, G.M.; Griffiths, A.J. Microorganisms and heavy metal toxicity. *Microb. Ecol.* **1977**, *4*, 303–317. [CrossRef] [PubMed]

36. Guchhait, R.B.; Polakis, S.E.; Dimroth, P.; Stoll, E.; Moss, J.; Lane, M.D. Acetyl coenzyme A carboxylase system of *Escherichia coli* purification and properties of the biotin carboxylase, carboxyltransferase, and carboxyl carrier protein components. *J. Biol. Chem.* **1974**, *249*, 6633–6645.
37. Gooday, G. Cell walls. In *The Growing Fungus*; Springer: Berlin/Heidelberg, Germany, 1995; pp. 43–62.
38. Albert, Q.; Baraud, F.; Leleyter, L.; Lemoine, M.; Heutte, N.; Rioult, J.-P.; Sage, L.; Garon, D. Use of soil fungi in the biosorption of three trace metals (Cd, Cu, Pb): Promising candidates for treatment technology? *Environ. Technol.* **2019**, *41*, 3166–3177. [CrossRef]
39. Šajbidor, J.; Koželouhova, D.; Ĉertik, M. Influence of some metal ions on the lipid content and arachidonic acid production by *Mortierella* sp. *Folia Microbiol.* **1992**, *37*, 404–406. [CrossRef]
40. Yoo, J.-Y.; Lee, H.-C.; Shin, D.-H.; Min, B.-Y. Production of Fungal Lipids-V. Effects of Vitamins, Metabolic Intermediates and Mineral Salts on the Growth and Lipid Accumulation of *Mucor plumbeus*. *Korean J. Food Sci. Technol.* **1982**, *14*, 151–155.
41. Muhid, F.; Nawi, W.; Kader, A.J.A.; Yusoff, W.M.W.; Hamid, A.A. Effects of metal ion concentrations on lipid and gamma linolenic acid production by *Cunninghamella* sp. 2A1. *Online J. Biol. Sci.* **2008**, *8*, 62–67. [CrossRef]
42. Kyle, D.J. Arachidonic Acid and Methods for the Production and Use Thereof. U.S. Patent 5,658,767, 19 August 1997. assigned to Martek Corporation.
43. Dzurendova, S.; Zimmermann, B.; Kohler, A.; Tafintseva, V.; Slany, O.; Certik, M.; Shapaval, V. Microcultivation and FTIR spectroscopy-based screening revealed a nutrient-induced co-production of high-value metabolites in oleaginous Mucoromycota fungi. *PLoS ONE* **2020**, *15*, e0234870. [CrossRef]
44. Dzurendova, S.; Zimmermann, B.; Tafintseva, V.; Kohler, A.; Ekeberg, D.; Shapaval, V. The influence of phosphorus source and the nature of nitrogen substrate on the biomass production and lipid accumulation in oleaginous Mucoromycota fungi. *Appl. Microbiol. Biotechnol.* **2020**, *104*, 8065–8076. [CrossRef]
45. Kosa, G.; Kohler, A.; Tafintseva, V.; Zimmermann, B.; Forfang, K.; Afseth, N.K.; Tzimorotas, D.; Vuoristo, K.S.; Horn, S.J.; Mounier, J.; et al. Microtiter plate cultivation of oleaginous fungi and monitoring of lipogenesis by high-throughput FTIR spectroscopy. *Microb. Cell Factories* **2017**, *16*, 101. [CrossRef]
46. Kosa, G.; Shapaval, V.; Kohler, A.; Zimmermann, B. FTIR spectroscopy as a unified method for simultaneous analysis of intra-and extracellular metabolites in high-throughput screening of microbial bioprocesses. *Microb. Cell Factories* **2017**, *16*, 195. [CrossRef] [PubMed]
47. Shapaval, V.; Møretrø, T.; Suso, H.P.; Åsli, A.W.; Schmitt, J.; Lillehaug, D.; Martens, H.; Böcker, U.; Kohler, A. A high-throughput microcultivation protocol for FTIR spectroscopic characterization and identification of fungi. *J. Biophotonics* **2010**, *3*, 512–521. [CrossRef] [PubMed]
48. Shapaval, V.; Schmitt, J.; Møretrø, T.; Suso, H.; Skaar, I.; Åsli, A.; Lillehaug, D.; Kohler, A. Characterization of food spoilage fungi by FTIR spectroscopy. *J. Appl. Microbiol.* **2013**, *114*, 788–796. [CrossRef] [PubMed]
49. Shapaval, V.; Afseth, N.K.; Vogt, G.; Kohler, A. Fourier transform infrared spectroscopy for the prediction of fatty acid profiles in *Mucor* fungi grown in media with different carbon sources. *Microb. Cell Factories* **2014**, *13*, 86. [CrossRef]
50. Kosa, G.; Zimmermann, B.; Kohler, A.; Ekeberg, D.; Afseth, N.K.; Mounier, J.; Shapaval, V. High-throughput screening of Mucoromycota fungi for production of low-and high-value lipids. *Biotechnol. Biofuels* **2018**, *11*, 66. [CrossRef]
51. Forfang, K.; Zimmermann, B.; Kosa, G.; Kohler, A.; Shapaval, V. FTIR spectroscopy for evaluation and monitoring of lipid extraction efficiency for oleaginous fungi. *PLoS ONE* **2017**, *12*, e0170611. [CrossRef]
52. Zimmermann, B.; Kohler, A. Optimizing Savitzky-Golay parameters for improving spectral resolution and quantification in infrared spectroscopy. *Appl. Spectrosc.* **2013**, *67*, 892–902. [CrossRef]
53. Duetz, W.A. Microtiter plates as mini-bioreactors: Miniaturization of fermentation methods. *Trends Microbiol.* **2007**, *15*, 469–475. [CrossRef]
54. Kosa, G.; Vuoristo, K.S.; Horn, S.J.; Zimmermann, B.; Afseth, N.K.; Kohler, A.; Shapaval, V. Assessment of the scalability of a microtiter plate system for screening of oleaginous microorganisms. *Appl. Microbiol. Biotechnol.* **2018**, *102*, 4915–4925. [CrossRef] [PubMed]
55. Lewis, T.; Nichols, P.D.; McMeekin, T.A. Evaluation of extraction methods for recovery of fatty acids from lipid-producing microheterotrophs. *J. Microbiol. Methods* **2000**, *43*, 107–116. [CrossRef]

56. Kohler, A.; Kirschner, C.; Oust, A.; Martens, H. Extended multiplicative signal correction as a tool for separation and characterization of physical and chemical information in Fourier transform infrared microscopy images of cryo-sections of beef loin. *Appl. Spectrosc.* **2005**, *59*, 707–716. [CrossRef] [PubMed]
57. Zhu, S.-C.; Tang, J.-X.; Zeng, X.-X.; Wei, B.-J.; Huang, B. Isolation of *Mucor circinelloides* Z4 and *Mucor racemosus* Z8 from heavy metal-contaminated soil and their potential in promoting phytoextraction with Guizhou oilseed rap. *J. Cent. South Univ.* **2015**, *22*, 88–94. [CrossRef]
58. Bağcıoğlu, M.; Kohler, A.; Seifert, S.; Kneipp, J.; Zimmermann, B. Monitoring of plant–environment interactions by high-throughput FTIR spectroscopy of pollen. *Methods Ecol. Evol.* **2017**, *8*, 870–880. [CrossRef]
59. Kenđel, A.; Zimmermann, B. Chemical analysis of pollen by FT-Raman and FTIR spectroscopies. *Front. Plant Sci.* **2020**, *11*, 352. [CrossRef] [PubMed]
60. Kohler, A.; Böcker, U.; Shapaval, V.; Forsmark, A.; Andersson, M.; Warringer, J.; Martens, H.; Omholt, S.W.; Blomberg, A. High-throughput biochemical fingerprinting of *Saccharomyces cerevisiae* by Fourier transform infrared spectroscopy. *PLoS ONE* **2015**, *10*, e0118052. [CrossRef]
61. Shapaval, V.; Brandenburg, J.; Blomqvist, J.; Tafintseva, V.; Passoth, V.; Sandgren, M.; Kohler, A. Biochemical profiling, prediction of total lipid content and fatty acid profile in oleaginous yeasts by FTIR spectroscopy. *Biotechnol. Biofuels* **2019**, *12*, 140. [CrossRef]
62. Byrtusová, D.; Shapaval, V.; Holub, J.; Šimanský, S.; Rapta, M.; Szotkowski, M.; Kohler, A.; Márová, I. Revealing the Potential of Lipid and β-Glucans Coproduction in Basidiomycetes Yeast. *Microorganisms* **2020**, *8*, 1034. [CrossRef]
63. Dean, A.P.; Sigee, D.C.; Estrada, B.; Pittman, J.K. Using FTIR spectroscopy for rapid determination of lipid accumulation in response to nitrogen limitation in freshwater microalgae. *Bioresour. Technol.* **2010**, *101*, 4499–4507. [CrossRef]
64. Khoshmanesh, A.; Cook, P.L.; Wood, B.R. Quantitative determination of polyphosphate in sediments using Attenuated Total Reflectance-Fourier Transform Infrared (ATR-FTIR) spectroscopy and partial least squares regression. *Analyst* **2012**, *137*, 3704–3709. [CrossRef]
65. Kasaai, M. A review of several reported procedures to determine the degree of N-acetylation for chitin and chitosan using infrared spectroscopy. *Carbohydr. Polym.* **2008**, *71*, 497–508. [CrossRef]
66. Cárdenas, G.; Cabrera, G.; Taboada, E.; Miranda, S.P. Chitin characterization by SEM, FTIR, XRD, and 13C cross polarization/mass angle spinning NMR. *J. Appl. Polym. Sci.* **2004**, *93*, 1876–1885. [CrossRef]
67. Carvalho, A.K.F.; da Conceição, L.R.V.; Silva, J.P.V.; Perez, V.H.; de Castro, H.F. Biodiesel production from *Mucor circinelloides* using ethanol and heteropolyacid in one and two-step transesterification. *Fuel* **2017**, *202*, 503–511. [CrossRef]
68. Andrade, V.S.; Sarubbo, L.A.; Fukushima, K.; Miyaji, M.; Nishimura, K.; Campos-Takaki, G.M.d. Production of extracellular proteases by *Mucor circinelloides* using D-glucose as carbon source/substrate. *Braz. J. Microbiol.* **2002**, *33*, 106–110. [CrossRef]
69. Peña, A.; Pardo, J.P.; Ramírez, J. Early metabolic effects and mechanism of ammonium transport in yeast. *Arch. Biochem. Biophys.* **1987**, *253*, 431–438. [CrossRef]
70. Torija, M.J.; Beltran, G.; Novo, M.; Poblet, M.; Rozès, N.; Mas, A.; Guillamón, J.M. Effect of organic acids and nitrogen source on alcoholic fermentation: Study of their buffering capacity. *J. Agric. Food Chem.* **2003**, *51*, 916–922. [CrossRef] [PubMed]
71. Beaulieu, M.; Beaulieu, Y.; Melinard, J.; Pandian, S.; Goulet, J. Influence of Ammonium Salts and Cane Molasses on Growth of *Alcaligenes eutrophus* and Production of Polyhydroxybutyrate. *Appl. Environ. Microbiol.* **1995**, *61*, 165–169. [CrossRef]
72. Papagianni, M.; Wayman, F.; Mattey, M. Fate and role of ammonium ions during fermentation of citric acid by *Aspergillus niger*. *Appl. Environ. Microbiol.* **2005**, *71*, 7178–7186. [CrossRef]
73. Kawazoe, N.; Kimata, Y.; Izawa, S. Acetic acid causes endoplasmic reticulum stress and induces the unfolded protein response in *Saccharomyces cerevisiae*. *Front. Microbiol.* **2017**, *8*, 1192. [CrossRef]
74. Diyaolu, S.; Adebajo, L. Effects of sodium chloride and relative humidity on growth and sporulation of moulds isolated from cured fish. *Food/Nahr.* **1994**, *38*, 311–317.
75. Babich, H.; Stotzky, G. Toxicity of zinc to fungi, bacteria, and coliphages: Influence of chloride ions. *Appl. Environ. Microbiol.* **1978**, *36*, 906–914. [CrossRef]
76. Al Tamie, M.S. Sodium chloride stress induced morphological changes in some halotolerant fungi. *Egypt. J. Hosp. Med.* **2016**, *62*, 109–126.

77. Gimenez, M.S.; Oliveros, L.B.; Gomez, N.N. Nutritional deficiencies and phospholipid metabolism. *Int. J. Mol. Sci.* **2011**, *12*, 2408–2433. [CrossRef]
78. Walker, G.M.; Duffus, J.H. Magnesium ions and the control of the cell cycle in yeast. *J. Cell Sci.* **1980**, *42*, 329–356.
79. Certik, M.; Shimizu, S. Kinetic analysis of oil biosynthesis by an arachidonic acid-producing fungus, *Mortierella alpina* 1S-4. *Appl. Microbiol. Biotechnol.* **2000**, *54*, 224–230. [CrossRef] [PubMed]
80. Aiba, S.; Humphrey, A.E.; Millis, N.F. *Biochemical Engineering*; Academic Press: New York, NY, USA, 1973; ISBN 0120450526.
81. Gorain, P.C.; Bagchi, S.K.; Mallick, N. Effects of calcium, magnesium and sodium chloride in enhancing lipid accumulation in two green microalgae. *Environ. Technol.* **2013**, *34*, 1887–1894. [CrossRef] [PubMed]
82. Wang, W.-A.; Liu, W.-X.; Durnaoglu, S.; Lee, S.-K.; Lian, J.; Lehner, R.; Ahnn, J.; Agellon, L.B.; Michalak, M. Loss of calreticulin uncovers a critical role for calcium in regulating cellular lipid homeostasis. *Sci. Rep.* **2017**, *7*, 1–15. [CrossRef]
83. Kurat, C.F.; Natter, K.; Petschnigg, J.; Wolinski, H.; Scheuringer, K.; Scholz, H.; Zimmermann, R.; Leber, R.; Zechner, R.; Kohlwein, S.D. Obese yeast: Triglyceride lipolysis is functionally conserved from mammals to yeast. *J. Biol. Chem.* **2006**, *281*, 491–500. [CrossRef]
84. Bien, C.M.; Espenshade, P.J. Sterol regulatory element binding proteins in fungi: Hypoxic transcription factors linked to pathogenesis. *Eukaryot. Cell* **2010**, *9*, 352–359. [CrossRef] [PubMed]
85. Kikuchi, Y.; Hijikata, N.; Yokoyama, K.; Ohtomo, R.; Handa, Y.; Kawaguchi, M.; Saito, K.; Ezawa, T. Polyphosphate accumulation is driven by transcriptome alterations that lead to near-synchronous and near-equivalent uptake of inorganic cations in an arbuscular mycorrhizal fungus. *New Phytol.* **2014**, *204*, 638–649. [CrossRef] [PubMed]
86. Allen, N.S.; Schumm, J.H. Endoplasmic reticulum, calciosomes and their possible roles in signal transduction. *Protoplasma* **1990**, *154*, 172–178. [CrossRef]
87. Beever, R.E.; Burns, D. Phosphorus uptake, storage and utilization by fungi. In *Advances in Botanical Research*; Elsevier: Amsterdam, The Netherlands, 1981; Volume 8, pp. 127–219.
88. Ratledge, C. Microorganisms for lipids. *Acta Biotechnol.* **1991**, *11*, 429–438. [CrossRef]
89. Burkhead, J.L.; Lutsenko, S. The role of copper as a modifier of lipid metabolism. In *Lipid Metabolism*; IntechOpen: London, UK, 2013.

Journal of *Fungi*

MDPI

Article

Biotransformation of Animal Fat-By Products into ARA-Enriched Fermented Bioproducts by Solid-State Fermentation of *Mortierella alpina*

Ondrej Slaný [1], Tatiana Klempová [1,*], Volha Shapaval [2], Boris Zimmermann [2], Achim Kohler [2] and Milan Čertík [1]

1 Institute of Biotechnology, Faculty of Chemical and Food Technology, Slovak University of Technology, Radlinského 9, 812 37 Bratislava, Slovakia; ondrej.slany@stuba.sk (O.S.); milan.certik@stuba.sk (M.Č.)
2 Faculty of Science and Technology, Norwegian University of Life Sciences, Postbox 5003, 1432 Ås, Norway; volha.shapaval@nmbu.no (V.S.); boris.zimmermann@nmbu.no (B.Z.); achim.kohler@nmbu.no (A.K.)
* Correspondence: tatiana.klempova@stuba.sk; Tel.: +421-949-582-157

Received: 25 September 2020; Accepted: 17 October 2020; Published: 21 October 2020

Abstract: Solid-state fermentation (SSF) is a powerful fermentation technology for valorizing rest materials and by-products of different origin. Oleaginous *Zygomycetes* fungi are often used in SSF as an effective cell factory able to valorize a wide range of hydrophilic and hydrophobic substrates and produce lipid-enriched bioproducts. In this study, for the first time, the strain *Mortierella alpina* was used in SSF for the bioconversion of animal fat by-products into high value fermented bioproducts enriched with arachidonic acid (ARA). Two cereals-based matrixes mixed with four different concentrations of animal fat by-product were evaluated for finding optimal conditions of a fat-based SSF. All obtained fermented bioproducts were found to be enriched with ARA. The highest substrate utilization (25.8%) was reached for cornmeal and it was almost double than for the respective wheat bran samples. Similarly, total fatty acid content in a fermented bioproduct prepared on cornmeal is almost four times higher in contrast to wheat bran-based bioproduct. Although in general the addition of an animal fat by-product caused a gradual cessation of ARA yield in the obtained fermented bioproduct, the content of ARA in fungal biomass was higher. Thus, *M. alpina* CCF2861 effectively transformed exogenous fatty acids from animal fat substrate to ARA. Maximum yield of 32.1 mg of ARA/g of bioproduct was reached when using cornmeal mixed with 5% (*w/w*) of an animal fat by-product as substrate. Furthermore, implementation of attenuated total reflectance Fourier transform infrared (ATR-FTIR) spectroscopy in characterization of obtained SSF bioproducts was successfully tested as an alternative tool for complex analysis, compared to traditional time-consuming methods.

Keywords: *Mortierella alpina*; solid-state fermentation; animal fat by-product; arachidonic acid; ATR-FTIR spectroscopy

1. Introduction

In Europe, approximately 16 million tonnes of animal fat (AF) by-products are processed annually by meat processors and renderers [1]. The total amount of AF by-products is expected to increase continuously, since consumers and health authorities in European countries focus on highly unsaturated fat diets [2]. In addition, due to the population growth, meat production in Europe is expected to grow. Even though some of these AF by-products comply with EU standards (EC No. 1069/2009) and can be served for human consumption, they are used almost exclusively as ingredients of animal feed. The biggest problem lies in the composition of AF by-products since they are rich in saturated and monounsaturated fatty acids. Long-term consumption of saturated fat elevates the risk of certain diseases such as diabetes or cardio-vascular problems. Thus, human food, as well as animal and aquaculture feeds, need to be supplied with the essential polyunsaturated ω-3 and ω-6 fatty acids (PUFAs), such as linoleic acid (LA; C18:2, ω-6), α-linolenic acid (ALA; C18:3, ω-3), arachidonic acid (ARA; C20:4, ω-6), eicosapentaenoic acid (EPA, C20:5, ω-3) and docosahexaenoic acid (DHA, C22:6, ω-3) [3]. Therefore, the development of sustainable processes for valorizing AF by-products into high-value bioproducts is urgently needed.

Fungal fermentation is an emerging technology for upgrading different types of by-products [4,5]. A certain group of filamentous fungi can accumulate up to 85% (*w/w*) of triacylglycerols (TAGs) containing high amount of essential PUFAs as a storage material of the cells [6]. Fungal biomass produced with these oleaginous fungi contains single-cell oils with a high share of essential ω-3 and ω-6 PUFAs, and is considered a high-value bioproduct with applications in food, feed and nutraceutical industry. Fungal biomass can be obtained by two fermentation processes: submerged or solid-state fermentation (SSF).

SSF is a fermentation technology using solid or semi-solid substrates in the absence of any water or using a low level of free-flowing water [7]. SSF has many advantages over more common submerged fermentation: (1) low energy requirements are accompanied by the high product yield; (2) downstream costs are significantly lower due to the high product applicability; (3) SSF conditions are more favorable for the microbial growth as they resemble the natural environment of microorganisms, resulting in a better performance of the fermentation process; (4) due to the high product/volume productivity of SSF, smaller fermentation volumes are possible [8]; (5) consumption of a lower amount of water and generation of low to almost zero amount of waste [9]. Solid substrates used in SSF frequently originate from agro-industrial waste materials and they serve as support materials for the optimal fungal growth, proliferation and single-cell oil production. Commonly used SSF substrates are cereal-based materials, such as barley flakes, millet grain, wheat bran, cornmeal or oat flakes [10,11]; fruit or vegetable-based materials, such as pulp, pomaces and peels (i.e., residues from plums, pineapple, carrot, papaya etc.) [12,13]; or they often come as residue from other types of industry, for example spent malt grain, groundnut fodder, forestry rests or wooden sawdust [7,14]. Several studies investigated the suitability of different oleaginous fungi, belonging to orders *Mucorales* and *Mortierellales*, for converting various waste materials and by-products based substrates by SSF into fermented BPs enriched with a wide range of essential PUFAs, such as γ-linolenic acid (GLA; C18:3, n-6), ARA or EPA [10,12,15–18]. Especially cereal-based solid matrixes are very suitable for these oleaginous fungi due to their chemical composition, such as the presence of easily accessible carbon, organic nitrogen and other macro- and micro-nutrients. Cereal-based matrixes provide a useful source for good fungal proliferation, hyphae penetration and stable lipid accumulation in fungal cells [10,15,16].

Oleaginous fungus *Mortierella alpina* is a well-known and thoroughly described species with a high capacity of lipid accumulation and good ability to produce industrially relevant essential PUFAs, especially ARA in both submerged fermentations and SSF [19–21]. It has also been reported that *M. alpina* is able to incorporate and transform exogenous fatty acids [22]. Thus, in order to improve the conversion of solid wastes into ARA-enriched bioproduct, supplementation with exogenous oils containing precursors of ARA is advised. Thus, the addition of vegetable oils, such as sunflower, rapeseed, corn, soybean or linseed oil containing individual fatty acid precursors of ARA led to a rapid increase of ARA yield in the final fermented bioproduct obtained by fungal SSF [10]. Such knowledge led to the hypothesis that the oleaginous fungus *M. alpina* should also be able to utilize and convert solid lipidic waste, such as AF by-products.

When developing a new SSF process and optimizing SSF processes, it is crucial to monitor key process parameters, such as humidity, airflow, oxygen transfer and controlling the quality of final fermented bioproduct. Quality parameters of the final fermented BP are fatty acid profile, total lipid content, amount of fungal biomass and total biochemical profile. For the determination of fatty acid profile and content, gas chromatography coupled with flame ionization detector (GC-FID) is the most commonly used technique [23], while the total amount of fungal biomass obtained by SSF is frequently analyzed by the glucosamine method [24]. Fourier transform infrared (FTIR) spectroscopy is a new emerging technique that has already been extensively used for characterizing biomass from different types of submerged cultivations [25–28]. FTIR spectroscopy is a biophysical technique allowing to measure the biomass in its native form. It is non-destructive and rapid and does not require extraction of single components. Therefore, the characterization of a bioproduct obtained in the process of SSF using FTIR spectroscopy could bring the new possibilities in complex monitoring of SSF process.

To the authors knowledge, this is the first study investigating the possibility to develop an efficient bioconversion of AF by-products into functional lipid-rich fermented bioproduct by SSF using *M. alpina*. Furthermore, the presented study introduces, describes and evaluates FTIR spectroscopy as an alternative method for accurate SSF monitoring.

2. Materials and Methods

2.1. Production Microorganism and Preparation of Spore Suspension

The oleaginous fungal strain *Mortierella alpina* CCF2861 used in this study was obtained from culture collection of fungi (CCF, Charles University, Prague, Czech Republic). The strain was kept on potato-dextrose agar media (PDA, Carl Roth, Germany) at 4 °C and regularly re-inoculated every 3 months.

The spore suspension for the inoculation of the fermentation substrate was prepared from a 14-day-old culture grown on PDA medium. Spores have been washed using an aqueous solution of 0.05% Tween®® 40 and suspension was diluted to achieve a final concentration of 10^6 spores/mL.

2.2. Conditions of Solid-State Fermentation

SSF cultivation was performed in the microporous high-density polyethylene bags (20 × 30 cm) containing 20 g of dry cereal-based matrix, to which various amounts of AF by-products (Norilia, Oslo, Norway) mixed with Tween®® 40 (Sigma-Aldrich, Darmstadt, Germany) were added. SSF experiments were performed using two waste cereal-based matrixes: cornmeal (CM; Amylum Slovakia, Boleráz, Slovakia) and wheat brans (WB; Mill Pohronský Ruskov, Pohronský Ruskov, Slovakia). The composition of AF used is summarized in Table 1.

Table 1. Fatty acid (FA) composition and percentage of individual lipid components of animal fat (AF) by-products used for the solid-state fermentation by *Mortierella alpina*.

FA	[%]
C14:0	2.23
C16:0	25.66
C16:1 n-7	2.18
C18:0	21.33
C18:1 n-9	38.34
C18:1 n-7	1.94
C18:2 n-6	5.47
C18:3 n-3	0.81
C20:0	0.23
Other fatty acids	1.81
Lipid Structure	
Polar lipids	0.14
Monoacylglycerols	0.28
Diacylglycerols	2.08
Sterol structures	9.38
Free fatty acids	7.20
Triacylglycerols	66.54
Esterified sterols	11.23
Other lipid structures	3.15

The cereal materials were soaked for 2 h in 20 mL (ratio solid to liquid 1:1) of either distilled water, Tween 40®® solution or AF emulsion, as per Table 2. After 2 h, all SSF substrates in bags were autoclaved (105 °C, 45 min). After cooling down, the substrates were inoculated with 4 mL of spore suspension of *M. alpina*. The fermentation ran for 10 days at 20 °C. In order to maintain the homogenous growth of fungi, substrates were mixed thoroughly after inoculation. During the fermentation process, the whole content of the fermentation bags was gently mixed once a day. All experiments were performed in three independent biological replicates.

Table 2. Amount of added Tween®® 40 and animal fat to cereal matrixes used for the solid-state fermentation by *Mortierella alpina*.

Cereal Matrix	Tween®® 40 [%(*w*/*w*)]	Animal Fat [%(*w*/*w*)]
cornmeal	0	0 [1]
	0.5	0 [2]
	1	0 [2]
	2	0 [2]
	3	0 [2]
	0.5	5 [3]
	1	10 [3]
	2	20 [3]
	3	30 [3]
wheat bran	0	0 [1]
	0.5	0 [2]
	1	0 [2]
	2	0 [2]
	3	0 [2]
	0.5	5 [3]
	1	10 [3]
	2	20 [3]
	3	30 [3]

[1] one-component samples (distilled water); [2] two-component samples (distilled water + Tween®® 40); [3] three-component samples (distilled water + Tween®® 40 + AF).

2.3. Preparation of AF Emulsion

Due to the fact that cereal-based matrixes have a hydrophilic nature while AF is hydrophobic, there is a need to perform the pretreatment of the fat materials before mixing it with cereal matrixes for obtaining homogenous substrates for SSF. The pretreatment of AF by-product was performed by preparing homogenous fat-water-Tween®® 40 emulsions, where the emulsifier Tween®® 40 formed a film around the dispersed lipid droplets and thereby reduced interfacial tension [29].

AF emulsions were prepared from distilled water, Tween®® 40 and AF using a combination of heating (80 °C, 10 min, stirring) and sonication (15 min) using VWR USC300T sonicator (VWR International, Radnor, PA, USA).

2.4. Humidity and Substrate Utilization Analysis

The obtained fermented bioproducts were collected and the humidity was measured. The humidity of substrates was measured by Moisture Analyzer Radwag 50/1. R (Radwag, Radom, Poland). Subsequently, the substrates were dried at 65 °C until a constant weight was achieved. The substrate utilization was calculated from the weight difference of the dry non-fermented substrate (control) and dry fermented bioproducts. Each sample was homogenized using a blender and stored in sterile Falcon tubes at laboratory temperature before the analysis was performed.

2.5. Estimation of Fungal Biomass in Fermented Bioproduct

To estimate the amount of fungal biomass in the fermented bioproducts, the method based on the estimation of glucosamine (GlcN) content was used [30,31]. In the first step, alkali insoluble material (AIM) was prepared according to Zamani et al. [24]: 0.5 M NaOH solution (3 mL) was added to 100 mg of each sample of fermented BP and the mixtures were heated at 90 °C for 16 h. Subsequently, samples were centrifuged (5000 rpm, 10 min) and washed 5 times with distilled water. Supernatants were removed and obtained AIMs were dried for 36 h at 75 °C. Further, AIM samples were hydrolyzed by adding 5 mL of 6M hydrochloric acid and incubating at 100 °C for 12 h. The analysis of glucosamine (GlcN) from the hydrolyzed AIM samples was performed spectrophotometrically according to Aidoo et al. [30] and GlcN content was calculated according to the standard calibration curve. Each analysis was performed in three independent technical replicates.

2.6. Microspopic Observation of Fungal Mycelia during SSF

The snapshots of fermented bioproducts were obtained by Dino-Lite Pro USB microscope. Fungal mycelia of *M. alpina* was cut out from the SSF matrix, suspended in 500 μL of sterile distilled water and observed by a confocal microscope Leica DM750 (Leica microsystems Ltd., Wetzlar, Germany), equipped with Leica HI PLAN 100x/1.25 oil objective lens (Leica microsystems Ltd., Wetzlar, Germany). Snapshots of individual hyphae were carried out with a Leica DFC290HD digital camera (Leica microsystems Ltd., Wetzlar, Germany).

2.7. Analysis of Fatty Acid Profile Andf Content in Fermented Bioproduct

Fatty acids (FA) from the SSF matrixes were converted into their methyl esters (FAMEs) by a modified method of Čertík and Shimizu [32]: 20 mg of dry homogenized bioproducts were mixed with 1 mL of dichloromethane containing 0.1 mg of heptadecanoic acid as an internal standard and 2 mL of anhydrous methanolic HCl solution. Samples were incubated at 50 °C for 3 h. After cooling down, 1 mL of distilled water was added and FAMEs were extracted with 1 mL of hexane. FAMEs were subsequently analyzed by gas chromatography according to the method described by Gajdoš et al. [33]. The identification of the FAMEs peaks was performed by comparison with authentic standards of C4-C24 FAME mixtures (Sigma Aldrich, USA). Quantitative evaluation of individual and total fatty acids was performed using an internal standard of heptadecanoic acid (C17:0, Sigma-Aldrich, Darmstadt, Germany) and calculated by ChemStation B 01 03 (Agilent Technologies, Santa Clara, CA, USA). Each analysis was performed in three independent technical replicates.

2.8. Lipid Isolation and Analysis of Lipid Classes by TLC

The lipids for the analysis of lipid classes were extracted with chloroform/methanol (2:1, *v/v*) according to original Folch et al. method [34] modified by Klempová et al. [15]. The lipid extracts were loaded on TLC silica plates 60 (Merck, Germany) using CAMAG TLC Sampler 4 (CAMAG, Muttenz, Switzerland) and TLC analysis was performed according to the method described by Gajdoš et al. [33]. Separated lipid fractions were analyzed densitometrically using CAMAG TLC Scanner 4 (CAMAG, Muttenz, Switzerland) and quantified using WinCATS software (CAMAG, Muttenz, Switzerland). Each analysis was performed in three independent technical replicates.

2.9. ATR-FTIR Spectroscopy Analysis

Fourier transform infrared (FTIR) spectroscopy was done employing an attenuated total reflectance (ATR) accessory for profiling the total biochemical composition of the obtained fermented bioproducts. ATR-FTIR measurements were performed using a Vertex 70 FTIR spectrometer (Bruker Optic, Billerica, MA, USA) with a single-reflection attenuated total reflectance (SR-ATR) accessory. The ATR-FTIR spectra were recorded with 32 scans using a horizontal SR-ATR diamond prism with 45° angle of incidence on a Specac (Slough, UK) High Temperature Golden Gate ATR Mk II. Of each homogenized dried fermented sample and of each non-fermented substrate sample, 10 mg of sample mass was prepared by homogenization with 2.8 mm stainless steel grinding balls (OPS Diagnostics, USA) using tissue homogenizer (5800 rpm, cycle 2 × 15 s with a 30 s pause) Percellys Evolution (Bertin Instruments, France). The samples were subsequently transferred to the surface of the ATR crystal. All ATR-FTIR measurements were performed in five technical replicates, resulting in 360 spectra in total. Spectra were recorded in a region between 7000 and 600 cm^{-1} with a resolution of 4 cm^{-1}. Each spectrum was recorded as the ratio of the sample spectrum to the spectrum of the empty ATR plate. The recording of spectra was performed using the OPUS 7.5 software (Bruker Optic, Billerica, MA, USA).

2.10. Data Analysis

The fatty acid GC-FID and GlcN data were analysed by ANOVA using Microsoft Excel (Microsoft Office 365 software pack) equipped with a data analysis tool. Post-hoc testing was performed for the ANOVA results using Tukey's HSD test in programming language R and in Python v. 3.7 using StatsModels libraries.

FTIR-ATR spectra were obtained for the region of 4000–600 cm^{-1}, selected as the spectral region containing bands distinctive for lipids, proteins and polysaccharides. The analysis of FTIR-ATR spectra was performed by first applying pre-processing and then principle component analysis (PCA). The pre-processing was performed by transforming to second-derivative spectra using the Savitzky–Golay algorithm with a polynomial of degree 2 and a window size of 11. The second-derivative spectra were pre-processed by extended multiplicative signal correction (EMSC) [35,36]. PCA was performed for three spectral regions, lipid (3050–2800 cm^{-1} combined with 1800–1700 cm^{-1}), protein (1700–1500 cm^{-1}) and polysaccharide (1200–700 cm^{-1}), using 11 principle components. The following software packages were used for data analysis: Unscrambler X version 10.5.1 (CAMO Analytics, Norway), Orange data mining toolbox version 3.24 (University of Ljubljana, Slovenia) [37].

3. Results

The present study is focused on utilization of AF by-products by the process of SSF employing fungus *M. alpina*. We evaluated two single-component control substrates, eight two-component control substrates containing Tween®® 40 and eight three-component animal fat-based substrate mixtures (Table 1). Since emulsifier Tween®® 40 may have impacted the fungal metabolism, the effect of Tween®® 40 itself was also investigated.

3.1. Fungal Growth and Aubstrate Utilization during SSF

It was observed that the presence of Tween 40 at different concentrations in cereal-based substrates had no significant impact on the substrate utilization, substrate humidity, pH, biomass growth and FA content (p-value > 0.4; F parameter > F_{crit}) (Table 2) and productivity was comparable with the control substrates based on cereal matrixes only.

It was also found that the utilization of control substrates and AF-based substrates in SSF varied depending on the type of cereal matrix used and the amount of AF added (Table 3). Generally, higher substrate utilization and fungal biomass growth was observed for the single-component and two-component substrates. The highest substrate utilization (25.8%) was reached for the substrates containing cornmeal (CM) and 1% of Tween (Table 3) and the highest fungal growth (289.3 mg/g of bioproducts) was obtained for the substrate containing CM with 3% of Tween. The utilization of CM single- and two-component control substrates was almost twice higher than for the respective wheat bran (WB) substrates (Table 3). Similarly, the fungal growth was 1.5-times higher using CM comparing to WB.

Table 3. Substrate utilization, substrate humidity, pH and fungal biomass (FBM) content in fermented bioproducts obtained from SSF of animal fat (AF) by-product by *Mortierella alpina*.

	Substrate Utilization [%]	Substrate Humidity [%]	pH	FBM [mg/g BP]
cornmeal (CM)	22.3 ± 0.9	60.0 ± 0.5	5.4 ± 0.1	241.3 ± 10.5
CM + 0.5% Tween 40	23.1 ± 0.5	58.1 ± 0.6	5.5 ± 0.0	244.0 ± 26.8
CM + 1% Tween 40	25.8 ± 0.5	60.1 ± 0.7	5.3 ± 0.1	272.8 ± 13.5
CM + 2% Tween 40	24.3 ± 0.6	59.0 ± 1.5	5.3 ± 0.3	286.2 ± 6.1
CM + 3% Tween 40	23.7 ± 1.0	58.6 ± 0.8	5.2 ± 0.1	289.3 ± 13.8
CM + 0.5% Tween 40 + 5% AF	21.1 ± 1.4	53.2 ± 0.6	5.2 ± 0.0	137.1 ± 15.6
CM + 1% Tween 40 + 10% AF	17.6 ± 0.4	48.8 ± 1.3	4.6 ± 0.1	120.7 ± 10.3
CM + 2% Tween 40 + 20% AF	13.1 ± 0.4	40.0 ± 1.1	4.5 ± 0.2	82.7 ± 14.8
CM + 3% Tween 40 + 30% AF	10.7 ± 0.4	36.5 ± 1.5	4.3 ± 0.1	73.7 ± 6.0
wheat bran (WB)	13.2 ± 0.5	58.7 ± 1.9	5.5 ± 0.1	188.8 ± 1.4
WB + 0.5% Tween 40	12.3 ± 0.6	58.9 ± 0.1	5.1 ± 0.5	174.2 ± 10.4
WB + 1% Tween 40	12.5 ± 0.4	56.8 ± 1.9	5.3 ± 0.2	193.6 ± 5.2
WB + 2% Tween 40	12.6 ± 0.7	58.4 ± 0.2	5.2 ± 0.2	168.1 ± 5.9
WB + 3% Tween 40	13.6 ± 0.6	54.2 ± 4.5	5.4 ± 0.3	166.0 ± 1.1
WB + 0.5% Tween 40 + 5% AF	10.4 ± 0.4	54.4 ± 1.0	5.1 ± 0.2	205.0 ± 1.1
WB + 1% Tween 40 + 10% AF	9.6 ± 0.2	48.7 ± 1.6	4.8 ± 0.1	189.9 ± 4.8
WB + 2% Tween 40 + 20% AF	9.3 ± 0.7	44.2 ± 2.2	4.5 ± 0.1	122.7 ± 5.1
WB + 3% Tween 40 + 30% AF	8.8 ± 0.5	39.1 ± 0.7	4.3 ± 0.0	92.2 ± 5.2

3.2. Microscopic Observation of SSF Process

Fungal growth during SSF was also observed microscopically (Figure 1a,b). Similar morphology characteristics of *M. alpina* grown on control single- and two-component substrates and AF supplemented substrates were found. *M. alpina* was able to completely cover the substrate surface and its hyphae also sufficiently penetrated inside the substrate. Moreover, the presence of lipid structures formed in individual fungal hyphae was detected (Figure 1c).

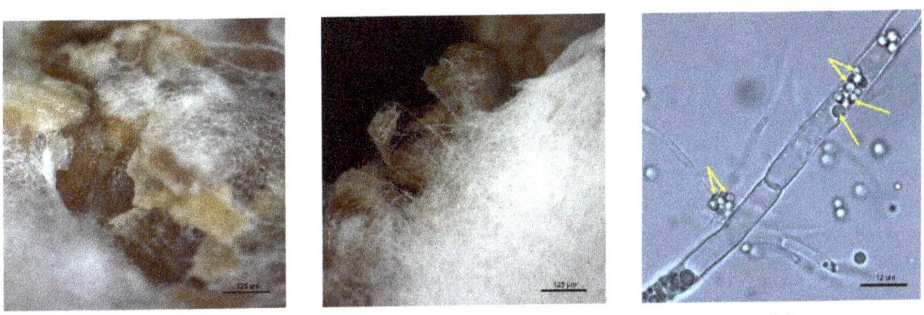

(a) (b) (c)

Figure 1. Microscopic images of *M. alpina* hyphae covering wheat bran (**a**), wheat bran with animal fat during solid-state fermentation (**b**) and lipid droplets accumulated inside *M. alpina* hyphae (**c**).

3.3. The Impact of AF Supplementation on the Humidity and pH of the Fermented Bioproducts

It was found that increasing the amount of AF in substrate mixtures reduced the humidity of the fermented bioproducts (Table 3). Fermented bioproducts obtained from the SSF of control substrates had pH 5.3 (CM) and pH 5.5 (WB), while fermented bioproducts obtained from SSF of AF-based substrates showed lower pH values. The increase in AF amounts in fermentation substrates strongly affected the pH values of fermented bioproducts and led to the gradual reduction of pH in the SSF system (Table 3). The lowest pH (4.3) was observed for both cereal-based fermented bioproducts with a 30% (*w*/*w*) AF supplementation.

3.4. Lipid Profile and Accumulation

We observed that for the production of fatty acid-rich fungal biomss the CM is a more suitable substrate compared to WB (Table 3). Although the total fatty acid content in both substrates was similar, the total fatty acids content in bioproducts derived from SSF of single and two-component control substrates based on CM was almost four times higher in contrast to WB-based bioproducts. The total fatty acid content in fermented WB-based bioproducts (single- and two- component) was always lower after fermentation (Table 4). In all three-component bioproducts, the accumulation of fatty acids was strongly affected by the presence of AF, since the added AF was homogenized with other substrate components (Table 4). In addition, it was observed that the total fatty acid content of fermented bioproducts still varied depending on the cereal matrix used. When CM was used as a matrix, the total fatty acid content in fermented bioproducts was elevated compared to the non-fermented substrate. When WB was used as a matrix, the total fatty acid content dropped in fermented bioproducts comparing to the non-fermented substrate. Nevertheless, these results suggest that fungal strain *M. alpina* CCF2861 is definitively able to utilize exogenous fatty acids of AF by-product added to the substrate not only for its own fatty acids biosynthesis, but for hyphae proliferation and growth as well.

Table 4. Total fatty acid (TFA) content and fatty acid composition of non-fermented (nf) and fermented (f) cereal substrates (cornmeal—CM, wheat bran—WB) with the addition of different amounts of Tween 40 and animal fat (AF). The results are average of three independent biological replicates with $\alpha < 5\%$.

	TFA	Fatty Acids (%)											
	(%/BP)	C14:0	C16:0	C16:1, n-7	C18:0	C18:1, n-9	C18:1, n-7	C18:2, n-6	C18:3, n-6	C18:3, n-3	C20:3, n-6	C20:4, n-6	Others
cornmeal													
CM nf	3.2	0.6	12.7	nd	2.5	27.3	0.7	54.7	nd	1.5	nd	nd	nd
CM f	13.8	0.8	14.5	0.1	5.6	13.7	0.7	23.8	2.3	0.5	2.5	26.4	9.0
+0.5%Tween 40 nf	3.7	0.7	14.3	nd	2.4	26.2	0.7	54.1	nd	1.5	nd	nd	nd
+0.5%Tween 40 f	13.7	0.6	15.6	0.1	5.8	13.2	1.0	23.5	2.4	0.3	2.5	27.5	7.6
+1%Tween 40 nf	3.4	0.8	16.6	nd	2.6	25.4	0.7	52.4	nd	1.6	nd	nd	nd
+1%Tween 40 f	13.4	0.8	16.0	0.1	5.7	12.8	0.6	22.6	2.3	0.4	2.6	27.1	8.9
+2%Tween 40 nf	4.3	1.0	20.3	nd	2.6	23.7	0.6	50.2	nd	1.5	nd	nd	nd
+2%Tween 40 f	13.8	0.8	17.3	0.1	5.8	13.0	0.6	22.7	2.3	0.5	2.5	26.1	8.4
+3%Tween 40 nf	4.7	0.6	23.9	nd	2.8	22.3	0.6	48.3	nd	1.3	nd	nd	nd
+3%Tween 40 f	13.7	0.8	18.2	0.1	5.7	12.7	0.6	23.0	2.3	0.5	2.5	25.4	8.2
+0.5%T + 5%AF nf	7.1	0.7	18.0	0.8	7.8	30.9	1.1	39.4	nd	1.3	nd	nd	nd
+0.5%T + 5%AF f	14.2	1.0	17.1	0.6	8.8	18.0	1.2	18.6	2.1	0.6	1.8	22.6	7.5
+1%T + 10%AF nf	11.3	1.2	22.2	1.7	12.0	33.4	1.5	26.9	nd	1.2	nd	nd	nd
+1%T + 10%AF f	16.1	1.1	18.8	1.0	11.3	21.1	1.5	16.4	2.0	0.6	1.4	18.3	6.4
+2%T + 20%AF nf	18.1	1.7	24.7	2.0	15.0	36.4	1.7	17.5	nd	1.1	nd	nd	nd
+2%T + 20%AF f	19.0	1.4	21.0	1.3	13.5	24.8	1.7	12.9	1.6	0.7	1.1	14.6	5.3
+3%T + 30%AF nf	24.3	1.9	25.8	2.1	16.6	37.4	1.8	13.3	nd	1.0	nd	nd	nd
+3%T + 30%AF f	24.7	1.6	22.4	1.6	14.7	30.4	1.7	12.7	1.2	0.8	0.7	8.6	3.6
wheat bran													
WB nf	3.5	1.7	17.1	nd	1.0	17.3	1.5	56.1	nd	4.5	nd	nd	0.7
WB f	3.1	0.2	12.6	0.1	1.6	15.3	1.1	38.6	2.2	2.3	0.6	20.4	5.2
+0.5%Tween 40 nf	3.6	nd	20.7	nd	1.1	16.6	1.5	54.7	nd	4.4	nd	nd	0.9
+0.5%Tween 40 f	3.4	0.1	13.9	nd	1.5	15.9	1.1	38.2	1.8	2.6	0.3	20.5	4.1
+1%Tween 40 nf	3.8	nd	21.8	0.2	1.2	16.7	1.4	53.4	nd	4.2	nd	nd	1.2
+1%Tween 40 f	3.3	0.3	15.7	nd	1.7	14.4	1.1	36.8	2.3	2.2	0.6	20.6	4.3
+2%Tween 40 nf	3.8	nd	26.7	nd	1.5	15.8	1.4	50.1	nd	3.9	nd	nd	0.7
+2%Tween 40 f	3.6	0.3	15.5	nd	2.1	13.8	1.1	34.9	2.4	2.1	0.2	22.3	5.5
+3%Tween 40 nf	4.6	0.2	33.6	0.2	1.8	13.7	1.2	44.9	nd	3.5	nd	nd	1.0
+3%Tween 40 f	4.0	0.3	18.8	0.2	2.3	13.0	1.0	31.5	2.1	2.0	0.7	23.1	5.1
+0.5%T + 5%AF nf	6.6	1.0	22.1	1.0	9.5	26.5	1.6	34.6	nd	2.9	nd	nd	0.8
+0.5%T + 5%AF f	5.6	0.5	17.2	0.7	7.1	18.2	1.9	21.8	3.5	1.2	0.8	19.6	7.4
+1%T + 10%AF nf	9.9	1.2	23.7	1.2	10.6	27.2	1.7	30.7	nd	2.6	nd	nd	1.1
+1%T + 10%AF f	7.6	0.8	17.7	1.0	8.6	22.9	1.91	17.7	3.6	1.2	0.8	18.1	5.9
+2%T + 20%AF nf	14.4	1.8	25.8	1.7	12.8	31.1	1.8	22.2	nd	2.0	Nd	nd	1.1
+2%T + 20%AF f	11.4	1.1	19.6	1.3	11.6	24.2	2.0	13.8	4.0	1.0	0.9	15.6	5.9
+3%T + 30%AF nf	23.4	1.3	26.1	1.6	16.0	33.3	1.8	17.1	nd	1.6	nd	nd	0.9
+3%T + 30%AF f	17.0	1.0	19.8	1.3	13.3	25.9	2.0	13.6	3.7	1.0	0.8	12.6	5.1

3.5. Arachidonic Acid Yield

Comparing the total fatty acid profile of the fermented bioproducts we made the following observations (Table 4): (i) the percentage of ARA of total fatty acid content in bioproducts derived from the CM single- and two-component substrates was higher than for WB-based bioproducts; (ii) the total fatty acid profile of bioproducts derived from AF containing substrates was strongly affected by the fatty acid composition of the added AF—higher content of palmitic and stearic acid provided lower content of linoleic acid; (iii) the addition of the AF by-product to the SSF substrates resulted in the decrease of ARA content in the fermented BPs, where the degree of decrease was higher in CM-based bioproducts than in WB-based bioproducts. The levels of ARA in total fatty acids of bioproducts derived from CM-based substrates dropped from 26.5% (average value of single- and two-component substrates) to 8.1%, depending on the AF addition. On the other hand, the percentage of ARA of total fatty acids derived from WB-based bioproducts varied from 21.4% (average of single- and two-component substrates) to 12.6%, depending on the AF addition (Table 3). Comparing of bioproducts derived from single- and two-components substrates, it is clear that CM-based substrates were much more suitable for ARA biosynthesis than WB-based substrates (Figure 2). The average yield of ARA achieved using CM-based substrates was 36.2 mg ARA/g of BP (136.9 mg of ARA/g of fungal biomass), that is almost five times higher than the content of ARA in WB-based bioproducts, where the average yield from single- and two-component bioproducts was 7.5 mg ARA/g of bioproducts (42.2 mg ARA/g of fungal biomass). The addition of AF affected the ARA yield significantly (Figure 2). It caused a gradual cessation of ARA yield in CM-based bioproduct from 32.1 mg of ARA/g of bioproduct (addition of 5% (*w*/*w*) of AF) down to 21.2 mg ARA/g BP (addition of 30% (*w*/*w*) of AF). However, the content of ARA in bioproducts from the three-component CM-based substrates was at least double comparing to single- and two-components substrates. On the other hand, the addition of AF increased the yield of ARA in WB-based bioproducts up to 21.4 mg ARA/g of bioproduct (234.4 mg ARA/g of fungal biomass) in comparison to the fermented bioproducts from single-component substrates (7.1 mg ARA/g of bioproduct; 36.2 mg ARA/g of fungal biomass (Figure 2)).

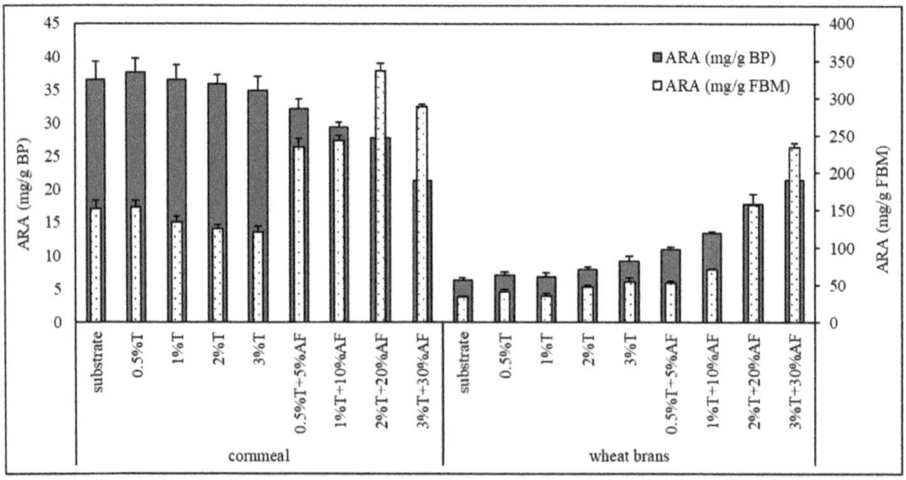

Figure 2. Content of arachidonic fatty acid (ARA) in fermented bioproducts (BP) and in fungal biomass (FBM). T—Tween 40, AF—animal fat by-product.

3.6. Biochemical Profile of the Fermented Bioproducts Obtained by FTIR-ATR Spectroscopy

FTIR-ATR spectroscopy allows rapid and non-destructive analysis of the total biochemical composition of different biological materials; in our case—bioproducts obtained by SSF of *M. alpina*.

Principal component analysis (PCA) of obtained FTIR-ATR spectra was used to study variations in the biochemical profile of the fermented bioproducts. The PCA scatter plot of the lipid-related spectral regions (3050–2800 cm^{-1} and 1800–1700 cm^{-1}) shows significant chemical differences in the lipid profile for the individual fermented bioproducts obtained in SSF of substrates with various addition of AF (Figure 3). It was observed that the FTIR lipid profile of bioproducts obtained from the WB-based substrates was quite similar to the lipid profile of bioproducts derived from the substrates with 5% and 10% (*w*/*w*) of AF, while the lipid profile of bioproducts from the substrates with 20% and 30% (*w*/*w*) of AF showed a very different lipid profile (Figure 3). On the other hand, in case of CM-based bioproduct, PCA score plot (Figure 3) shows large differences between all tested conditions. The protein and polysaccharide profile of the fermented bioproduct from the SSF of AF supplemented substrates was similar, while it was different from the control substrates (Figure 3). The complete biochemical profile of bioproducts from the SSF of the control CM substrates was more similar to the CM substrates supplemented with 5% and 10% (*w*/*w*) of AF than in the case of WB substrates (Figure 3).

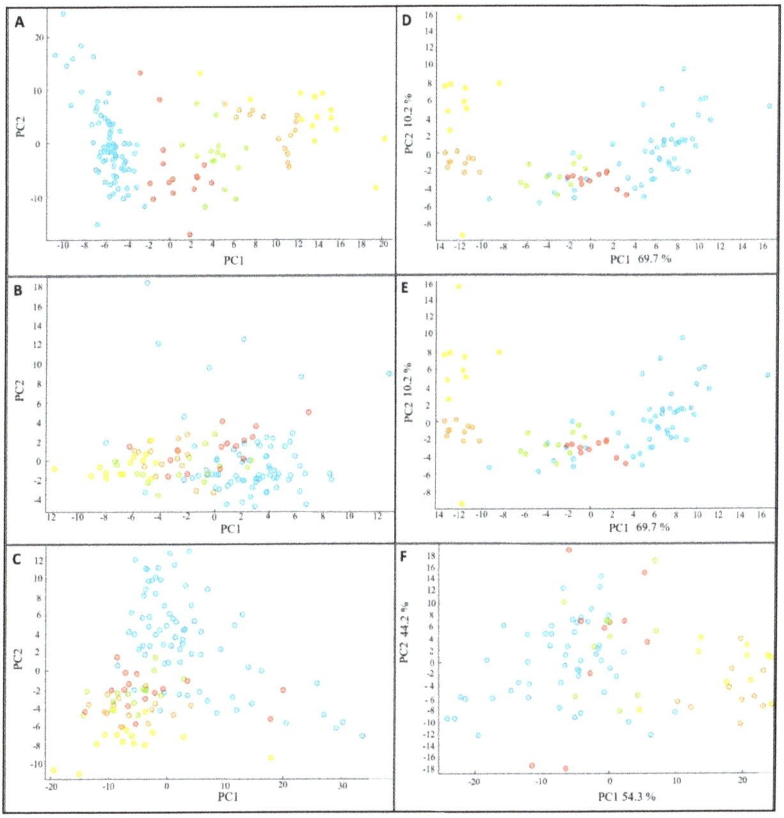

Figure 3. PCA scatter plots of FTIR-ATR spectra of fermented bioproducts obtained from the SSF of cornmeal-based and wheat bran-based substrates supplemented with animal fat at concentrations (blue) 0%, (red) 5%, (green) 10%, (orange) 20% and (yellow) 30%. (**A**)—lipid region of cornmeal-based fermented substrates, (**B**)—protein region of cornmeal-based fermented substrates, (**C**)—polysaccharide region of cornmeal-based fermented substrates, (**D**)—lipid region of wheat bran-based fermented substrates, (**E**)—protein region of wheat bran-based fermented substrates, (**F**)—polysaccharide region of wheat bran-based fermented substrates.

4. Discussion

Oleaginous fungus *Mortierella alpina* is a well-known and thoroughly described species with a high capacity for lipid accumulation and a good ability to produce industrially relevant essential PUFAs [19–21]. In order to verify growth ability and utilization of AF by-products by *M. alpina* in SSF, two types of cereal matrixes (wheat bran and cornmeal) were used in this study. Cereal-based matrixes are easily accessible and well-known as a good matrix for SSF substrates. Due to their chemical composition, such as presence of easily accessible carbon, organic nitrogen and other macro- and micro-nutrients, cereal-based matrixes provide a useful source for good fungal proliferation, hyphae penetration and stable lipid accumulation in fungal cells [10,15]. In order to improve the conversion of cereal-based carbon into PUFAs enriched bioproducts, supplementation with exogenous oils containing precursors of PUFAs is advised. Thus, the addition of vegetable oils, such as sunflower, rapeseed, corn, soybean or linseed oil containing individual fatty acid precursors of PUFAs, led to a rapid increase of PUFAs yield in the final fermented bioproducts obtained by fungal SSF [10]. It was also reported that *M. alpina* is able to incorporate and transform exogenous fatty acids [22]. In this study, an addition of animal fat by-product as a source of PUFA precursors (such as palmitic acid, stearic acid or oleic acid) to the cereal-based matrix was evaluated for the production of ARA-enriched fermented bioproducts in an SSF by *M. alpina* CCF2861.

Due to the fact that cereal-based matrixes have a hydrophilic nature while animal fat is hydrophobic, there is a need to perform pretreatment of the fat materials before mixing it with cereal matrixes for obtaining homogenous substrates for SSF. The pretreatment of the AF by-product was performed by preparing homogenous fat-water-Tween®® 40 emulsions, where emulsifier Tween®® 40 forms a film around the dispersed lipid droplets and thereby reduces interfacial tension [29].

While the total carbon content in both substrates is very similar (37.4% for CM and 37.9% for WB), the difference is in the profile of the dominant carbon source. The main carbon source in cornmeal is starch accounting for up to 85% (*w*/*w*) of the total carbon, whereas wheat bran contains only up to 24.5% of starch and up to 50% of cellulose and hemicellulose [38,39]. *M. alpina* is known to effectively use starch as a carbon source, however the cellulose and hemicellulose polymers are not suitable source of carbon for this strain [40]. This might be a reason for better fungal growth of *M. alpina* on CM in contrast to WB. Analysis of protein content using the automatic Kjeldahl method for nitrogen content showed that WB contained higher amount of proteins (17.4%) than CM (6.5%), indicating lower C/N ratio in WB than in CM. However, the amount of proteins was sufficient for fungal growth and proliferation in SSF process for both substrates.

Nevertheless, as mentioned above, fungal growth in the presence of AF was undoubtedly affected mostly by the carbon structure (CM—starch, WB—cellulose and hemicellulose polymers), significant changes in pH values were also noticed. Subsequent TLC analyses of lipid classes of extracted total lipid clearly indicated that AF contains relatively higher levels of free fatty acids (Table 1), which was probably the reason for lower pH of the fermented bioproducts after cultivation with AF addition.

Fatty acid analysis of all fermented bioproducts obtained in this study confirmed the previously described good ability of *M. alpina* to produce ARA (Figure 2, Table 3) [4,20]. High ARA yield in the fermented bioproducts obtained after SSF of AF supplemented substrates proves a high potential of applying *M. alpina* for the biotransformation of AF by-products into high-value ARA-enriched fermented bioproducts.

FTIR-ATR spectroscopy performed on all obtained fermented samples has proved possible high potential of this method for analysis of different types of materials. This method requires low sample amounts and does not involve extraction of any chemical components [26]. The infrared (IR) spectrum contains information about chemical bonds characteristic to all major biomolecules of the measured sample and different spectral regions representing information about lipids, proteins, phosphate containing molecules and polysaccharides. When analyzing cellular lipid profile based on FTIR spectra, the most important lipid associated peaks, used for the analysis, are: (i) peaks related to $-CH_3$ and $-CH_2$ stretching at 2947 cm^{-1}, 2925 cm^{-1}, 2855 cm^{-1}, 1465 cm^{-1} and 1377 cm^{-1}, indicating mainly the chain length of the carbon skeleton in lipid molecules; (ii) the peak related to the ester bond stretching at 1745 cm^{-1}, indicating the total lipid content in the cell; (iii) the peak related to the carboxyl bond vibrations in free fatty acids at 1710 cm^{-1}, and (iv) the peak related to =C-H stretching at 3010 cm^{-1}, indicating the unsaturation level of cellular lipids. Proteins are observed in the spectral region 1700–1500 cm^{-1} with the main peaks for amide I (1650 cm^{-1}) and amide II (1540 cm^{-1}) bonds and polysaccharides are observed in the region 1200–900 cm^{-1}. PCA of FTIR-ATR spectra proved distinct variations between individual samples in accordance with different composition of analyzed fermented samples. PCA score plots show large difference between all tested conditions, which could be caused by different composition of cereal substrate itself, as mentioned above. Higher similarity of complete biochemical profile in CM control samples with the CM substrates supplemented with 5% and 10% (*w*/*w*) of AF than in the case of WB substrates can be explained by the higher homogeneity of fungal growth and various cereal-based substrate utilization for the WB-based and CM-based substrates.

This is the first study on effective utilization of AF by-product into functional high-valued fermented bioproducts by SSF and obtained results represent broad fundamentals for subsequent research. Optimization of such fermentation process and extension for application of different fungal strains can lead to increased yields of desired metabolites or formatting a various high-value fungal specific product. Moreover, successful application of ATR-FTIR spectroscopy for rapid analysis of SSF bioproducts has been described, which can possibly extend industrial potential of the whole solid-state fermentation technology.

Author Contributions: Conceptualization, T.K.; methodology, O.S., T.K.; software, B.Z.; validation, T.K.; formal analysis, O.S., T.K., V.S.; investigation, O.S., T.K.; resources, T.K.; data curation, O.S., B.Z.; writing—original draft preparation, O.S.; writing—review and editing, T.K., V.S., M.Č.; visualization, O.S., T.K., A.K.; supervision, V.S., A.K., M.Č.; project administration, A.K., M.Č.; funding acquisition, V.S., A.K., M.Č. All authors have read and agreed to the published version of the manuscript.

Funding: This research was funded by Ministry of Education, Science, Research and Sports, Slovakia, grant VEGA 1/0323/19 and Council of Norway-FMETEKN grant, project number 257622; BIONÆR grant, project number 305215; HAVBRUK2 grant, project number 302543/E40; MATFONDAVTALE grant, project number 301834/E50 and DAAD grant, project number 309220. The funder had no role in the study design, data collection, and interpretation, or in the decision to submit the work for publication.

Acknowledgments: Authors would like to thank to Marije Oostindjer, Norilia AS (Oslo, Norway) for providing the animal fat by-products for experiments. Also, we would like to thank to Marek Jašurek and Pavel Lapeš for help with SSF experiments and to Margaréta Guttenová and Simona Dzurenová for assistance with ATR-FTIR analysis of SSF bioproducts.

Conflicts of Interest: The authors declare no conflict of interest.

References

1. Woodgate, L.S.; van der Veen, J.T. Fats and Oils—Animal Based. In *Food Processing: Principles and Applications*, 2nd ed.; Clark, S., Jung, S., Lamsal, B., Eds.; John Wiley and Sons: Hoboken, NJ, USA, 2004; pp. 481–499.
2. Nordic Council of Ministers. Nordic Nutrition Recommendations 2012. Available online: http://dx.doi.org/10.6027/Nord2014-002 (accessed on 20 August 2020).
3. Doppenberg, J.; van der Aar, P.J.; van Vuure, C. Animal fat: Nutritious ingredient for animal diets. *All Feed* **2015**, *23*, 9–11.
4. Antimanon, S.; Chamkhuy, W.; Sutthiwattanakul, S.; Laoteng, K. Efficient production of arachidonic acid of *Mortierella* sp. by solid-state fermentation using combinatorial medium with spent mushroom substrate. *Chem. Pap.* **2018**, *72*, 2899–2908. [CrossRef]
5. Szotkowski, M.; Byrtusova, D.; Haronikova, A.; Vysoka, M.; Rapta, M.; Shapaval, V.; Marova, I. Study of metabolic adaptation of red yeasts to waste animal fat substrate. *Microorganisms* **2019**, *7*, 578. [CrossRef] [PubMed]
6. Meng, X.; Yang, J.; Xu, X.; Zhang, L.; Nie, Q.; Xian, M. Biodiesel production from oleaginous microorganisms. *Renew. Energy* **2009**, *34*, 1–5. [CrossRef]
7. Soccol, C.R.; da Costa, E.S.F.; Letti, L.A.J.; Karp, S.G.; Woiciechowski, A.L.; Vandenberghe, L.P.S. Recent developments and innovations in solid state fermentation. *Biotech. Res. Innov.* **2017**, *1*, 52–71. [CrossRef]

8. Hölker, U.; Höfer, M.; Lenz, J. Biotechnological advantages of laboratory-scale solid-state fermentation with fungi. *Appl. Microbiol. Biotechnol.* **2004**, *64*, 175–186. [CrossRef] [PubMed]
9. Wang, L.; Liu, Z.; Duan, Y.; Chen, H. Relations between substrate morphological change and oxygen transfer in solid-state fermentation (SSF) using *Penicillium decumbens* JUA10. *J. Chem. Technol. Biotechnol.* **2014**, *89*, 1582–1589. [CrossRef]
10. Čertík, M.; Klempová, T.; Guothová, L.; Mihálik, D.; Kraic, J. Biotechnology for the functional improvement of cereal-based materials enriched with PUFA and pigments. *Eur. J. Lipid Sci. Technol.* **2013**, *115*, 1247–1256. [CrossRef]
11. Postemsky, P.D.; Curvetto, N.R. Solid-state fermentation of cereal grains and sunflower seed hulls by *Grifola gargal* and *Grifola sordulenta*. *Int. Biodeter. Biodegr.* **2015**, *100*, 52–61. [CrossRef]
12. Dulf, F.V.; Vodnar, D.C.; Toşa, M.I.; Dulf, E.-H. Simultaneous enrichment of grape pomace with γ-linolenic acid and carotenoids by solid-state fermentation with *Zygomycetes* fungi and antioxidant potential of the bioprocessed substrates. *Food Chem.* **2020**, *310*, 125927. [CrossRef]
13. Kaur, P.; Ghoshal, G.; Jain, A. Bio-utilization of fruits and vegetables waste to produce β-carotene in solid-state fermentation: Characterization and antioxidant activity. *Process Biochem.* **2019**, *76*, 155–164. [CrossRef]
14. Chandra, M.S.; Viswanath, B.; Reddy, R.B. Cellulolytic enzymes on lignocellulosic substrates in solid state fermentation by *Aspergillus niger*. *Indian J. Microbiol.* **2007**, *47*, 323–328. [CrossRef] [PubMed]
15. Klempová, T.; Slaný, O.; Šišmiš, M.; Marcinčák, S.; Čertík, M. Dual production of polyunsaturated fatty acids and beta-carotene with *Mucor wosnessenskii* by the process of solid-state fermentation using agro-industrial waste. *J. Biotechnol.* **2020**, *311*, 1–11. [CrossRef] [PubMed]
16. Slaný, O.; Klempová, T.; Marcinčák, S.; Čertík, M. Production of high-value bioproducts enriched with γ-linolenic acid and β-carotene by filamentous fungi *Umbelopsis isabellina* using solid-state fermentations. *Ann. Microbiol.* **2020**, *70*, 5. [CrossRef]
17. Asadi, S.Z.; Khosravi-Darani, K.; Nikoopour, H.; Bakhoda, H. Evaluation of the effect of process variables on the fatty acid profile of single cell oil produced by *Mortierella* using solid-state fermentation. *Crit. Rev. Biotechnol.* **2015**, *35*, 94–102. [CrossRef] [PubMed]
18. Qiao, W.; Tao, J.; Luo, Y.; Tang, T.; Miao, J.; Yang, Q. Microbial oil production from solid-state fermentation by a newly isolated oleaginous fungus, *Mucor circinelloides* Q531 from mulberry branches. *R. Soc. Open Sci.* **2018**, *5*, 180551. [CrossRef] [PubMed]
19. Cao, G.; Guan, Z.; Liu, F.G.; Liao, X.; Cai, Y. Arachidonic acid production by *Mortierella alpina* using raw crop materials. *Acta Sci. Pol. Technol. Aliment.* **2015**, *14*, 133–143. [CrossRef] [PubMed]
20. Kikukawa, H.; Sakuradani, E.; Ando, A.; Shimizu, S.; Ogawa, J. Arachidonic acid production by the oleaginous fungus *Mortierella alpina* 1S-4: A review. *J. Adv. Res.* **2018**, *11*, 15–22. [CrossRef]
21. Mironov, A.A.; Nemashkalov, V.A.; Stepanova, N.N.; Kamzolova, S.V.; Rymowicz, W.; Morgunov, I.G. The Effect of pH and Temperature on Arachidonic Acid Production by Glycerol-Grown *Mortierella alpina* NRRL-A-10995. *Fermentation* **2018**, *4*, 17. [CrossRef]
22. Čertík, M.; Sakuradani, E.; Shimizu, S. Desaturase-defective fungal mutants: Useful tools for the regulation and overproduction of polyunsaturated fatty acids. *Trends Biotechnol.* **1998**, *16*, 500–505. [CrossRef]
23. Zhang, H.; Wang, Z.; Liu, O. Development and validation of a GC–FID method for quantitative analysis of oleic acid and related fatty acids. *J. Pharm. Anal.* **2015**, *5*, 223–230. [CrossRef] [PubMed]
24. Zamani, A.; Jeihanipour, A.; Edebo, L.; Niklasson, C.; Taherzadeh, M.J. Determination of glucosamine and N-acetyl glucosamine in fungal cell walls. *J. Agric. Food Chem.* **2008**, *56*, 8314–8318. [CrossRef] [PubMed]
25. Forfang, K.; Zimmermann, B.; Kosa, G.; Kohler, A.; Shapaval, V. FTIR Spectroscopy for Evaluation and Monitoring of Lipid Extraction Efficiency for Oleaginous Fungi. *PLoS ONE* **2017**, *12*, e0170611. [CrossRef]
26. Kosa, G.; Zimmermann, B.; Kohler, A.; Ekeberg, D.; Afseth, N.K.; Mounier, J.; Shapaval, V. High-throughput screening of Mucoromycota fungi for production of low- and high value lipids. *Biotechnol. Biofuels* **2018**, *11*. [CrossRef]
27. Shapaval, V.; Afseth, N.K.; Vogt, G.; Kohler, A. Fourier transform infrared spectroscopy for the prediction of fatty acid profiles in Mucor fungi grown in media with different carbon sources. *Microb. Cell Fact.* **2014**, *13*, 86. [CrossRef] [PubMed]
28. Shapaval, V.; Brandenburg, J.; Blomqvist, J.; Tafintseva, V.; Passoth, V.; Sandgren, M.; Kohler, A. Biochemical profiling, prediction of total lipid content and fatty acid profile in oleaginous yeasts by FTIR spectroscopy. *Biotechnol. Biofuels* **2019**, *12*. [CrossRef]

29. Tamilvanan, S. Oil-in-water lipid emulsions: Implications for parenteral and ocular delivering systems. *Prog. Lipid Res.* **2004**, *43*, 489–533. [CrossRef]
30. Aidoo, K.; Hendry, R.; Wood, B.J.B. Estimation of fungal growth in solid state fermentation system. *Appl. Microbiol. Biot.* **1981**, *12*, 6–9. [CrossRef]
31. Katano, H.; Takakuwa, M.; Hayakawa, H.; Kimoto, H. Determination of Chitin Based on the Colorimetric Assay of Glucosamine in Acidic Hydrolysate. *Anal. Sci.* **2016**, *32*, 701–703. [CrossRef]
32. Čertík, M.; Shimizu, S. Kinetic analysis of oil biosynthesis by an arachidonic acid-producing fungus, Mortierella alpina 1S-4. *Appl. Microbiol. Biotechnol.* **2000**, *54*, 224–230. [CrossRef]
33. Gajdoš, P.; Nicaud, J.-M.; Rossignol, T.; Čertík, M. Single cell oil production on molasses by *Yarrowia lipolytica* strains overexpressing DGA2 in multicopy. *Appl. Microbiol. Biotechnol.* **2015**, *99*, 8065–8074. [CrossRef] [PubMed]
34. Folch, J.; Lees, M.; Sloane, G.H.S. A simple method for the isolation and purification of total lipides from animal tissues. *J. Biol. Chem.* **1957**, *226*, 497–509. [PubMed]
35. Kohler, A.; Kirschner, C.; Oust, A.; Martens, H. Extended multiplicative signal correction as a tool for separation and characterization of physical and chemical information in Fourier transform infrared microscopy images of cryo-sections of beef loin. *Appl. Spectrosc.* **2005**, *59*, 707–716. [CrossRef]
36. Zimmermann, B.; Kohler, A. Optimizing Savitzky-Golay parameters for improving spectral resolution and quantification in infrared spectroscopy. *Appl. Spectrosc.* **2013**, *67*, 892–902. [CrossRef]
37. Demšar, U.; Harris, P.; Brunsdon, C.; Fotheringham, A.; Mcloone, S. Principal Component Analysis on Spatial Data: An Overview. *Ann. Am. Assoc. Geogr.* **2013**, *103*, 106–128. [CrossRef]
38. Massarolo, K.C.; Ferreira, C.F.J.; de Borba, V.S.; Kupski, L.; Furlong, E.B. Particle size and physical-chemical characteristics of hydrothermally treated cornmeal on resistant starch content. *Food Chem.* **2019**, *283*, 39–45. [CrossRef] [PubMed]
39. Merali, Z.; Collins, S.R.A.; Elliston, A.; Wilson, D.R.; Käsper, A.; Waldron, K.W. Characterization of cell wall components of wheat bran following hydrothermal pretreatment and fractionation. *Biotechnol. Biofuels* **2015**, *8*. [CrossRef] [PubMed]
40. Dyal, S.D.; Narine, S.S. Implications for the use of *Mortierella* fungi in the industrial production of essential fatty acids. *Food Res. Int.* **2005**, *38*, 445–467. [CrossRef]

Publisher's Note: MDPI stays neutral with regard to jurisdictional claims in published maps and institutional affiliations.

Journal of *Fungi*

MDPI

Article

Aqueous Two-Phase System Extraction of Polyketide-Based Fungal Pigments Using Ammonium- or Imidazolium-Based Ionic Liquids for Detection Purpose: A Case Study

Juliana Lebeau [1], Thomas Petit [1,2], Mireille Fouillaud [1], Laurent Dufossé [1] and Yanis Caro [1,2,*]

1 Laboratoire de Chimie et de Biotechnologie des Produits Naturels (CHEMBIOPRO), Université de La Réunion, F-97490 Sainte-Clotilde, France; juliana.lebeau@gmail.com (J.L.); thomas.petit@univ-reunion.fr (T.P.); mireille.fouillaud@univ-reunion.fr (M.F.); laurent.dufosse@univ-reunion.fr (L.D.)

2 Département Hygiène Sécurité Environnement (HSE), Université de La Réunion—IUT La Réunion, F-97410 Saint-Pierre, France

* Correspondence: yanis.caro@univ-reunion.fr

Received: 14 October 2020; Accepted: 14 December 2020; Published: 18 December 2020

check for updates

Abstract: Demand for microbial colorants is now becoming a competitive research topic for food, cosmetics and pharmaceutics industries. In most applications, the pigments of interest such as polyketide-based red pigments from fungal submerged cultures are extracted by conventional liquid–liquid extraction methods requiring large volumes of various organic solvents and time. To address this question from a different angle, we proposed, here, to investigate the use of three different aqueous two-phase extraction systems using either ammonium- or imidazolium-based ionic liquids. We applied these to four fermentation broths of *Talaromyces albobiverticillius* (deep red pigment producer), *Emericella purpurea* (red pigment producer), *Paecilomyces marquandii* (yellow pigment producer) and *Trichoderma harzianum* (yellow-brown pigment producer) to investigate their selective extraction abilities towards the detection of polyketide-based pigments. Our findings led us to conclude that (i) these alternative extraction systems using ionic liquids as greener extractant means worked well for this extraction of colored molecules from the fermentation broths of the filamentous fungi investigated here; (ii) tetrabutylammonium bromide, [N4444]Br-, showed the best pigment extraction ability, with a higher putative affinity for azaphilone red pigments; (iii) the back extraction and recovery of the fungal pigments from ionic liquid phases remained the limiting point of the method under our selected conditions for potential industrial applications. Nevertheless, these alternative extraction procedures appeared to be promising ways for the detection of polyketide-based colorants in the submerged cultures of filamentous fungi.

Keywords: fungal pigment; natural colorant; extraction ability; marine fungi; *Talaromyces albobiverticillius*; aqueous two-phases system extraction; ionic liquids

---

1. Introduction

Improving sustainability and reducing the amounts of waste produced and discarded by chemical industries is a global challenge. The production of dyes is one of the industries which is tackling these issues, aiming to minimize these impacts by using viable and beneficial alternatives to both human health and the environment. The demand for natural colorants has been growing over the past two decades and is now becoming a competitive research topic for food, cosmetics, pharmaceutics and textiles industries [1]. The dyeing industry is currently suffering from the rising cost of raw materials due to the increasing demand for eco-friendly pigments to replace synthetic dyes (such as azo dyes). This is particularly true for the red-dye industries, which have no, or very few, natural red pigment alternatives with an adequate robustness meeting both the organoleptic expectations of consumers and the requirements of industrial processes. Unlike synthetic dyes, natural colorants are environmentally friendly as they do not require heavy and time- and energy-consuming benzene-based chemical synthesis, nor are they generating pollution and toxicity to human health wastewaters. Moreover, natural dyes often come as liquid solutions, while synthetic dyes tend to be dry powders, consequently requiring large amounts of water for dying purposes [2].

Recent literature has underlined the need to investigate new bio-pigments derived from bio-sources and notably pigments from microorganisms [3]. Indeed, the positive biological properties of some natural dyes, such as carotenoids and polyketide-based pigments of microbial origin, have made it possible to widen the potential area of specific applications of these natural colorants (such as for functional finishing of textiles, in functional foods, pharmaceuticals, etc.) [3]. Among unconventional sources, ascomycetous fungi of either marine or terrestrial origin are known to produce an extraordinary range of biomolecules such as bioactive polyketide pigments, which often tend to be more stable and water-soluble than plant pigments [4]. Thus, fungal colored polyketides are the most promising material in this respect. The production of natural colorants with a polyketide structure from filamentous fungi has already been reported, mostly using *Monascus*, *Talaromyces*, *Emericella*, *Paecilomyces* and *Trichoderma* species [3,4]. These aforementioned fungal species produce polyketide pigments with various hues, from yellow to reddish-purple, in the mycelium or as diffusing colorants in the fermentation broth (submerged culture). These fungal metabolites are structurally related (i.e., are biosynthesized by polyketide synthases, PKS) and are either azaphilone, naphthoquinone or hydroxyanthraquinone polyketide-based pigments [4].

However, to fully meet the demand of natural red colorants for the industries, there is an urgent need to develop and design green and sustainable detection, extraction and isolation techniques with higher recovery and greater selectivity in a faster and more benign manner than conventional methods. Indeed, the natural pigments available from plant, insect or microbial origins are currently essentially extracted using time-consuming, conventional liquid–liquid extraction methods that also require large volumes of various organic solvents. These conventional techniques usually involve prolonged contact with excess of organic solvents or a mixture thereof under high temperatures and/or pressures [5]. Moreover, these techniques sometimes require toxic solvents such as chlorinated or halogenated solvents, such as dichloromethane or toluene [6], and may impact on the final application (food, cosmetics, textiles, etc.) of the extracted colorants in regards to health and safety concerns. Furthermore, they often result in a relatively low extraction yield, poor selectivity and the need for further energy-intensive purification and wastewater recycling steps. Thus, any large-scale application raises environmental concerns, as well as uncertainties about their economic viability. Both scale-up and cost-effectiveness of the production of natural pigments at the industrial level, with a particular emphasis on red fungal pigments for food, cosmetic, pharmaceutical and fabric applications, still remain limited and require substantial optimization [4]. Ionic liquids, which are considered in this study, are new, unconventional solvents that have been receiving growing attention since the early 2000s because of their potential as green solvents in addition to their properties that are tunable to the target products and recyclability [7–13].

The aim of this study is the investigation of the potential of ionic liquids (ILs) as a more sustainable extraction technique for detection purpose of polyketide pigments from fermentation broths of filamentous fungi. Thus, we investigated the use of three different ammonium- or imidazolium-based ionic liquids, namely tetrabutylammonium bromide ($[N4444]Br^-$), 1-butyl-3-methylimidazolium bromide ($[C4Mim]Br^-$) and 1-butyl-3-methyl-imidazolium chloride ($[C4Mim]Cl^-$), as new types of solvents in aqueous two-phase system (ATPS) extraction. An IL-ATPS is formed by mixing a soluble IL (chaotropic agent) and a salt with a kosmotropic profile (Na_2CO_3 and $Na_2CH_3CO_2$). These systems allow the separation of two water-soluble compounds into two immiscible aqueous-rich phases. These specific ILs and salts were selected based on their water-miscibility profiles as well as their safety. It is worth mentioning that IL-based systems exhibit varying degrees of eco-toxicities driven both by the aromatic/alkyl chain of the IL itself as well as by the type of anionic salt it has been combined with [7,14,15]. As an example, tetrafluoroborate and hexafluorophosphate anions have been the commonly used salts with ILs due to their ability to enhance phase stability among other physicochemical advantages. However, fluorinated-based salts have the deleterious tendency to be hydrolyzed into toxic hydrofluoric acid [15]. Similarly, even the most hydrophobic ILs can be partially dispersed in water and become a potential polluting agent if not closely monitored. Only a few studies have addressed the environmental impact or potential risks associated with ILs' design, use and recycling. Thus, a rigorous selection of ILs and salt families and structures must be carried out beforehand based on the type of target biomolecules to be extracted, in order to make the most sustainable choice [14,16–18]. Imidazolium-based ILs such as $[C4Mim]Br^-$ and $[C4Mim]Cl^-$ are the most frequently used IL cores and are, therefore, well documented [19,20], hence we chose to study these in association with harmless salts. They were proven to have higher hydrophilicity than tetra-alkyl ammonium-based ILs ($[N4444]Br^-$) because of their aromatic characteristics allowing a stronger interaction with water due to the delocalization of the positive charge over the aromatic ring [21]. However, the combination with anions and kosmotropic salt also has a significant impact on the final ability to form ATPS [22–24], as well as on the polarity of the overall system, and thus potentially influences the final partition of fungal pigments. Moreover, the extraction capabilities of the IL-based system will be challenged by the fact that the target fungal pigments are contained in fermentation broths. These are complex biological matrices containing a myriad of other components such as proteins, lipids and other excretable secondary metabolites, which can most likely impact the chemical structure and availability of the target compounds. Therefore, both hydrophilic and hydrophobic IL cations combined with ATPS-forming anions and organic and inorganic salts were investigated here.

Within this scenario, we investigated this alternative IL-ATPS extraction technique using the three aforementioned ammonium- or imidazolium-based ionic liquids to selectively extract the pigmented polyketides from submerged cultures of four filamentous fungi producing a mixture of pigments, namely *Talaromyces albobiverticillius* (producing deep red pigments), *Emericella purpurea* (producing red pigments), *Paecilomyces marquandii* (producing yellow pigments) and *Trichoderma harzianum* (producing yellow to brown pigments). These four strains were selected because they are well known to produce biomolecules exhibiting a rich chemical and structural diversity with different colors, including red ones [25,26]. *Talaromyces* represents a good example of the diversity of fungal pigments. In our previous study, we highlighted the potential of the marine-derived fungus *T. albobiverticillius* 30548 to produce, in a submerged culture, azaphilone polyketide-based red colorants, such as *N*-threonine-monascorubramine, *N*-glutaryl-rubropunctamine and PP-O [26]. In other previous studies, some azaphilone (hexaketide) compounds with biological activities have also been isolated from the fermentation broth of *T. harzianum*, such as the yellow azaphilone pigment T22 [27] and the compounds harziphilone and fleephilone [28], with some other extrolites with different chemical structures (cyclonerodiol, trichosetin, harzianic acid, harzianopyridone and MR566A) [29]. Some species of *Trichoderma* are natural producers of anthraquinone (octaketide) pigments, whose biological activities are also well known. For example, several anthraquinone pigments have been reported earlier from a fermentation broth of *T. harzianum*—crysophanol (red), pachybasin (orange), emodin (orange), ω-hydroxypachybasin, 1,5-dihydroxy-3-hydroxymethyl-9,10-anthraquinone, 1,7-dihydroxy-3-hydroxymethyl-9,10-anthraquinone, 1,8-dihydroxy-3-methylanthraquinone and 1-hydroxy-3-methylanthraquinone [30,31]. Catenarin (red) is an anthraquinone derivative that has been isolated from a variety of fungi, including species of *Emericella* [4]. *Emericella* sp. is also known for its ability to produce other bioactive secondary metabolites with different chemical structures, including polyketide pigments. In particular, various extrolites such as epurpurin A–C, variecolins, versicolorins, albistrins, emerin, emindol PA, norsolorinic acid and shamixanthones have been isolated from *Emericella purpurea* [32]. Next, sorbicillinoids (e.g., sorbicillin, sorbicillinol, oxosorbicillinol and bisvertinol) are an important family of hexaketide compounds produced by terrestrial and marine fungi, including species of *Trichoderma* (e.g., *T. harzianum*) and *Paecilomyces* (e.g., *P. marquandii*) [33].

In regards to the diversity of the expected compounds, the selectivity of the IL-ATPS extraction for specific types of fungal pigments (e.g., red ones) was investigated. Thus, the main goal of our study here is the selective extraction of the fungal polyketide-based pigments with a red hue from the fermentation broths. Additionally, we looked into the reverse extraction of the colored molecules from the IL-ATPS by working on the ionizable profile of the polyketide molecules. Ultimately, this gave us tools to estimate the overall rate of pigment recovery from the fermentation broth and identify points of future improvements. It is worth mentioning that most of the studies conducted so far using IL-based extraction methods on biomaterials (fungal-, plant- or insect-originated) explored only either the extraction power of the IL on pure solution of compounds of interest or the stability and/or potential use of the said compounds in the IL phase [34]. Few and infrequent studies have been conducted on testing the recovery of the molecule of interest from the IL-based extractant as well as its reuse [34]. Nevertheless, it may be noted that the accurate composition of all the secondary metabolites produced by these aforementioned fungal strains has not been investigated here, because that is not the issue.

2. Materials and Methods

2.1. Selection of the Ionic Liquids and Formation of the Kosmotropic Salt Buffer

The compositions and structures of the three ionic liquids, ([N4444]Br^-), ([C4Mim]Br^- and ([C4Mim]Cl^-), used in this study are presented in Table 1. The reagents were purchased from Sigma Aldrich (Saint Louis, MO, USA). Alizarin (anthraquinone) purchased from Sigma Aldrich was used as control for IL-ATPS extraction purposes.

Table 1. Description of the selected ionic liquids (ILs) associated with the kosmotropic salt buffer, enabling a diphasic system.

ILs	Salt	Structure
[C4Mim]Br^- 1-Butyl-3-methylimidazolium bromide	Na_2CO_3 Sodium carbonate (Inorganic)	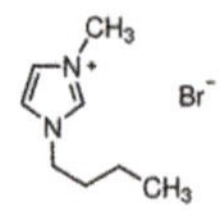
[C4Mim]Cl^- 1-Butyl-3-methylimidazolium chloride		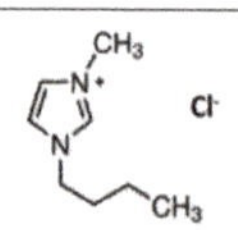
[N4444]Br^- Tetrabutylammonium bromide	$NaCH_3CO_2$ Sodium acetate (Organic)	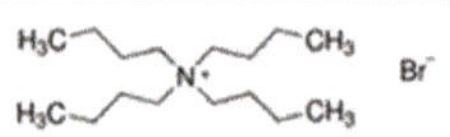

2.2. Fungal Strains and Culture Conditions

The marine-derived fungal strain *T. albobiverticillius* (collection number 30548, GenBank accession number MK937814) was isolated from the back reef-flat of Réunion Island [35]. The three other strains used in this study were isolated from terrestrial environments, including *T. harzianum* (commercial biological control strain T22), *E. purpurea* (collection reference LCP 3323) and *P. marquandii* (collection reference LCP 2271) and were bought from the fungal culture collection of the Muséum national d'Histoire naturelle (MNHN, Paris, France). For submerged cultures, potato dextrose broth

(PDB; Sigma Aldrich, Saint Louis, MO, USA) was used routinely as culture medium. The pH of the culture medium was adjusted to 6.0 ± 0.2 using 0.1 M HCl prior to sterilization. Pre-culture and cultivation were carried out in 250 mL Erlenmeyer flasks containing 100 mL of sterilized culture medium. The flasks were incubated at 26 °C for 7 days and agitated at 150 rpm using a rotary agitator (Infors Multitron, 50 mm excentration, Infors HT, Switzerland) as described by Lebeau and coworkers (2017) [25]. After 7 days of fermentation, the contents of each flask were collected and centrifuged at 14,000× *g* for 10 min; the resulting supernatant was filtered through a Whatman filter paper (GF/C) at a reduced pressure using a Büchner funnel to obtain the culture filtrate. The filtered colored supernatants were then freeze-dried at −84 °C in an ultra-low-temperature freezer (Sanyo, Guangzhou, China) for at least 2 h. The samples were then quickly transferred to a LABCONCO FreeZone 2.5 lyophilizer (LABCONCO, Kansas City, MO, USA) and lyophilized for 24 h. During freezing, the condenser temperature and vacuum pressure were maintained at −47 °C and 200 mbar, respectively (Figure 1).

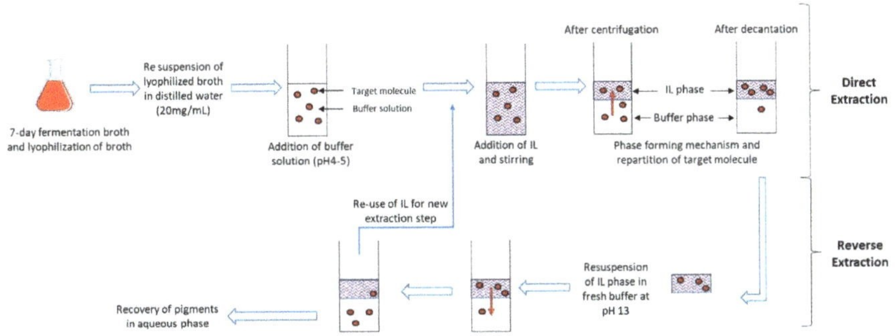

Figure 1. Brief description of the overall ideal IL aqueous two-phase system (IL-ATPS) workflow for direct detection and extraction of fungal pigments into ionic-rich phase, their back-extraction into salt-rich solution and re-injection of IL into the process.

2.3. Binodal Curves Determination

Ternary phase diagrams for each of the IL-ATPS-forming systems were determined using the Cloud Point titration method at room temperature and pressure [36]. Aqueous solutions of salts at 40% *w/w* and aqueous solutions of ILs at 30% *w/w* were also prepared at room temperature and pressure and were used for the determination of the binodal curves for each IL-ATPS. It was carried out by the drop-wise addition of the aqueous solution of the concentrated salt to the IL aqueous solution (or pure polymer) until the detection of a cloudy and biphasic solution. This was followed by the drop-wise addition of pure water (diluent) until the formation of a clear and limpid solution corresponding to the monophasic regime. The composition of the two-phase system was determined using the ratio of the weight of one component added to the total weight of all added components. The above procedure was repeated until the least amount of turbidity was observed to obtain sufficient data to generate binodal curves [37,38]. At the level of the biphasic region, the upper phase was composed by the IL-rich aqueous phase while the lower phase was the salt-rich phase for all ILs and salts investigated in this study.

2.4. Fungal Pigments Partitioning Using Ionic Liquids in an Aqueous Two-Phase System Extraction

Ternary mixtures within the biphasic region were prepared for each IL-ATPS and consisted of either (1) 15 wt.% of [C4Mim]Br^-, 15 wt.% Na_2CO_3 and 70 wt.% of the aqueous solution of pigmented fungal broth; (2) 15 wt.% of [C4Mim]Cl^-, 20 wt.% Na_2CO_3 and 65 wt.% pigmented fungal broth; and (3) 15 wt.% of [N4444]Br^-, 25 wt.% $NaCH_3CO_2$ and 60 wt.% pigmented fungal broth. The final volumes of the reaction mixtures were about 5 mL (final total weight of 5 g), with 0.75 g of the corresponding ionic liquid for 1 g of lyophilized fungal broth resuspended in corresponding weights of water and salt (i.e., 2.5 g of water + 0.75 g of Na_2CO_3, 2.25 g of water + 1 g of Na_2CO_3 or 2 g of water + 1.25 g of $NaCH_3CO_2$, respectively). The ternary mixtures' compositions were chosen based on the phase diagrams of each IL investigated (Figure 2). The extraction step was performed as follows: all components (lyophilized pigment broth previously obtained by fermentations, salts, water and IL) were mixed together, vigorously stirred and allowed to equilibrate for 30 min before centrifugation at 2455× *g* (4000 rpm) for 5 min. The mixture was then allowed to decant overnight (16 h). The next day, each fraction of the mixtures was carefully separated, weighted out and its volume was measured. These values were used to calculate the coefficient of partition of pigments ($K_{pigment}$), the phase ratio (R) and the selectivity of type of pigments ratio ($S_{pigment}$), as well as the efficiency of extraction (Effp%), as described in the following sections.

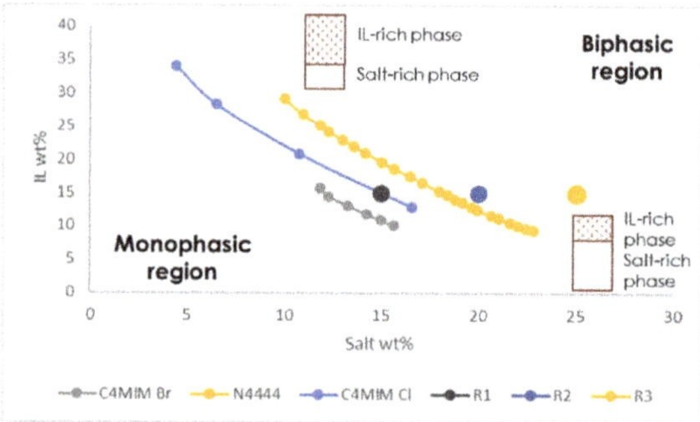

Figure 2. Binodal curve for the determination of weight ratio of IL, salt buffer (Na_2CO_3 or $Na_2CH_3CO_2$) and water to produce biphasic systems. R1, R2 and R3 display the composition of the ATPS made with $[C4Mim]Br^-$, $[C4Mim]Cl^-$ and $[N4444]Br^-$, respectively.

The recovery of the pigments extracted from the IL phase was carried out by resuspending the previously collected IL phase in concentrated aqueous salt solution and adjusting the pH to 13. The mixture was stirred and left to stand for 30 min before being centrifuged at 2455× *g* (4000 rpm) for 5 min and allowed to decant overnight (16 h). Similar measurements of weight, volumes and absorbance were carried out and the efficiency of pigment recovery was calculated as described below.

2.5. Determination of the Coefficients of Partition of Pigments ($K_{pigment}$)

After the extraction steps, samples of both IL-rich and salt-rich phases were diluted in distilled water and the pigment content of each phase was estimated by measuring the absorbance using a UV spectrophotometer (UV-1800, Shimadzu Corporation, Tokyo, Japan) at 490, 470 and 400 nm corresponding to red, orange and yellow hues, respectively. Relative repartitions ($K_{pigment}$) of each type of pigment (red, orange and yellow) were calculated according to the following Equations (1)–(3):

$$K_{red} = (A490_{IL} * FD_{IL})/(A490_{salt} * FD_{salt}) \quad (1)$$

$$K_{orange} = (A470_{IL} * FD_{IL})/(A470_{salt} * FD_{salt}) \quad (2)$$

$$K_{yellow} = (A400_{IL} * FD_{IL})/(A400_{salt} * FD_{salt}) \quad (3)$$

where $K_{pigment}$: coefficient of partition of the target shade of pigment (red, orange or yellow pigments); A_{IL}: absorbance (mAU) of the IL phase sample at the wavelength considered (490, 470 or 400 nm); FD_{IL}: dilution factor of the IL phase sample; A_{salt}: absorbance (mAU) of the salt phase at the wavelength considered (490, 470 or 400 nm); FD_{salt}: dilution factor of the salt phase.

2.6. Determination of the Selectivity of Type of Pigments Ratio ($S_{pigment}$)

In order to evaluate the higher selectivity of ILs tested against the isolation of either red, yellow or orange pigments, the selectivity of type of pigments ratio ($S_{pigment}$) was determined as follows:

$$S_{red/yellow} = K_{red}/K_{yellow} \quad (4)$$

$$S_{red/orange} = K_{red}/K_{orange} \quad (5)$$

where $S_{red/yellow}$: selectivity ratio of red to yellow pigments extracted; K_{red}: coefficient of partition of the red pigments extracted; K_{yellow}: coefficient of partition of the yellow pigments extracted; $Sr_{ed/orange}$: selectivity ratio of red to orange pigments extracted; K_{orange}: coefficient of partition of the orange pigments extracted.

2.7. Determination of the Extraction Efficiencies (Eff%)

The efficiency of the direct extraction was determined according to the following formula:

$$EffD\% = 100 \times [K_{pigment}/(1 + (1/R))] \quad (6)$$

where EffD%: efficiency of the direct extraction; $K_{pigment}$: coefficient of partition of the target pigments extracted; R: phase ratio between the IL- and salt-rich phase volumes.

The phase ratio (R) was calculated using the following equation:

$$R = V_{IL}/V_{salt}\ \text{phase} \quad (7)$$

where R: phase ratio; V_{IL}: volume (mL) of the IL phase sample; V_{salt}: volume (mL) of the salt phase.

3. Results

3.1. Determination of Biphasic Systems Using Binodial Curves

In the search for alternative extraction systems that are less hazardous for the environment and for human health, the properties of two different families of ionic liquids, namely the imidazolium- and the tetra-alkyl ammonium-based core families, were investigated specifically for extracting mixtures of fungal pigments. Subsequently, three IL-ATPSs, including $[C4Mim]Br^- + Na_2CO_3$, $[C4Mim]Cl^- + Na_2CO_3$ and $[N4444]Br^- + NaCH_3CO_2$, were investigated to evaluate their ability to selectively extract and detect mixtures of fungal pigments with different colors from fermentation broths [25]. Fungal fermentation broths are complex biological matrices containing various complex compounds, from proteins to lipids and other excretable secondary metabolites, including fungal pigments. Such diversity represents a burden for the development of fast, efficient and cost-effective detection and extraction processes, for which IL-ATPSs are expected to offer realistic options [19,21]. A brief description of the overall extraction process is shown in Figure 1. In this study, IL-ATPSs were applied to a lyophilized pigmented fermentation broth where the fungal pigments we are searching for (e.g., polyketide-derived pigments such as azaphilone red compounds, etc.) should have hydrophilic profiles.

Imidazolium cores have a stronger interaction with water than tetra-butyl ammonium and, consequently, tend to attract more hydrophilic systems. The combination with different anions and salts has an impact on the final overall hydrophilic/hydrophobic profile of the ionic liquid-rich phase, and this can be assessed by the determination of the binodal curves of each system [37–39]. Thus, binodal curves were produced for each of the IL-ATPSs in order to determine which ionic liquid/salt ratio was needed to generate a biphasic system. The resulting curves were produced using the Cloud Point titration method [28,36] and are displayed in Figure 2. We designed three IL-ATPSs that have in common the same ionic liquid content (15%), with the underlying strategy to avoid a large amount of ILs to be used due to both their cost and environmental impact. Although ionic liquids are considered as more ecological extraction solvents because of their higher extraction capacities for smaller volumes and recyclability, their final discard remains a grey area with regard to respect for the environment [14]. The corresponding amount of salt to be added for each specific ionic liquid system was established based on the binodal curves. Because clear and stable two-phase systems are desired, the less ATPS-forming a system is, the higher the amount of added salt (% *w/w*) was, in the limit of 25% *w/w* of the system. Both salts, sodium carbonate and sodium acetate, were adopted on the basis of their pH-buffering ability, as well as their biodegradability and lack of toxicity [40,41].

As shown in Figure 2, the phase-forming ability of the different ionic liquid systems can be ranked in the following order: $[C4Mim]Br^-$ > $[C4Mim]Cl^-$ > $[N4444]Br^-$. From the binodal curves, surprisingly, the most ATPS-forming systems were the imidazolium-based ATPS. Due to its longer alkyl chains, $[N4444]Br^-$ was expected to be the most prone to generate a two-phase systems. Among the systems investigated here, imidazolium-based ionic liquids were the least ATPS-forming options against tetra-alkyl ammonium-based ionic liquids, when taken under the same conditions (i.e., combination with the same anion and chaotropic/kosmotropic salts). Between the two classes of imidazolium-based ionic liquids chosen here, the only difference is the coupled anion (the salt being identical). $[C4Mim]Br^-$ showed higher ATPS-forming capacity than $[C4Mim]Cl^-$, and this is in line with previously reported results, where it was observed that bromide-based IL had greater aptitude to form ATPS than its chlorine anion-linked counterpart [37,42], when used under the same conditions (salts and ionic liquid core). This can be explained because bromide anions are more chaotropic than chlorine anions. The $[N4444]Br^-$ system involves a different salt, namely an organic salt of sodium acetate. A previous study comparing the effect of the salt on the ability to generate a two-phase system reported sodium carbonate, an inorganic salt, as significantly inclined to form biphasic systems [37]. It is then reasonable to assume that the lower ability of $[N4444]Br^-$ to form an ATPS, despite its longer alkyl chains, is mostly due to its combination with the organic salt of sodium acetate [43].

Those systems and their variation in profiles and behaviors have provided an unusual method for the selective extraction of pigments produced by submerged cultures of filamentous fungi from their fermentation broth. Indeed, we seek to selectively extract hydrophilic polyketide-derived pigments from ionic liquid-phase mixtures; therefore, we are investigating relatively hydrated ionic liquid phases that are still capable to form stable ATPSs.

3.2. Effect of the Ionic Liquids Aqueous Two-Phase System Extraction in the Partitioning of Fungal Pigments

In a previous published study, we identified the following four fungal strains: *T. albobiverticillius* of marine origin (deep red) and three strains of terrestrial origin, i.e., *E. purpurea* (red), *P. marquandii* (yellow) and *T. harzianum* (yellow-brown), based on their capacities to produce high amounts of polyketide-derived pigments in fermentation broth. One of the main limiting steps in the bioproduction of pigments is their inefficient and cost-effective extraction. We selected the aforementioned fungal strains to investigate the potential of IL-ATPS extraction for the selective detection and extraction of fungal pigments with different colors. Currently, there is a specific need for natural alternatives for red pigments for industrial use. We therefore focused our work on the potentialities of extracting fungal red pigments by IL-ATPS selected here and, more particularly, on the azaphilone-like red pigments from the fermentation broth of the marine-derived filamentous fungi *T. albobiverticillius* 30548, which was collected on the west coast of Réunion Island. Both tetra-butyl ammonium and imidazolium ionic liquids were chosen because they are considered more compatible (and less disruptive) with biomolecule structures and have proven large spectra of applications. Moreover, previous studies reported successful partitioning and recovery of hydrophobic pigments such as carotenoids using IL-ATPSs with similar polarity profiles as the target compounds [21,22,40,44,45]. Thus, the same strategy was applied here by investigating IL-ATPSs with greater hydrophilic profiles for the purpose of better azaphilone-like-pigment isolation. Based on these initial properties, it was assumed that the more hydrophilic the IL top phase, the higher the pigment partition coefficient and, consequently, the better the extraction would be. The results of the first direct extraction are shown in Figure 3. Globally speaking and across all three ATPSs tested here, the results of extraction showed that pigments tend to preferentially migrate towards the top ionic liquid-rich phase. Unanimously, [C4Mim]Br$^-$ with 15 wt.% Na_2CO_3 exhibited the smallest pigment partition in the four pigment-producing strains. Indeed, there was a significant loss of pigments remaining in the aqueous salt-rich phase, as shown in Figure 3. On the other hand, both systems, i.e., [C4Mim]Cl$^-$ + 20 wt.% Na_2CO_3 and [N4444]Br$^-$ + 25 wt.% $NaCH_3COO$, showed significantly higher pigment migration in the IL top phase, confirming their greater potential as an extraction system for hydrophilic fungal polyketide-derived pigments of various colors (red, orange and yellow).

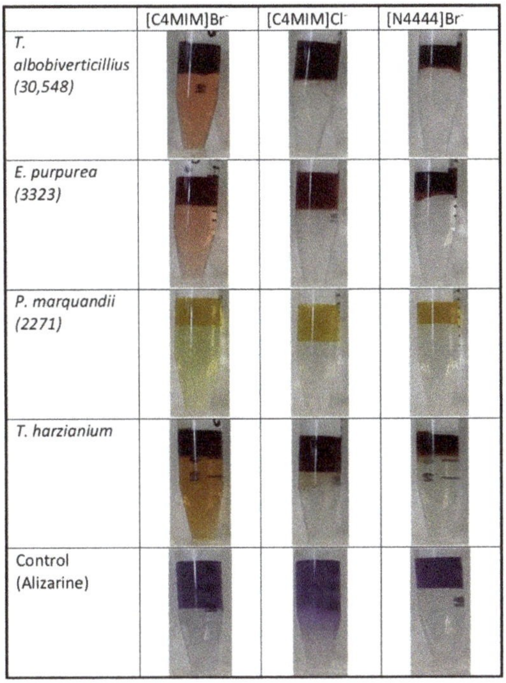

Figure 3. Visual aspect of the first step of extraction from the different IL-ATPSs tested and their extraction abilities on fungal fermentation broth and pure pigment control (alizarin). Lyophilized colored fermentation broths were mixed with the corresponding ionic liquid and salt. The mixtures were allowed to equilibrate for 30 min before centrifugation for 5 min at 4000 rpm and overnight decantation. Both ionic liquid-rich and salt-rich phases were collected and absorbances were measured. Top phase: ionic liquid-rich phase where most of the pigments migrated; Bottom phase: salt-rich phase.

The coefficients of partition of distinct pigment shades (K_{red}, K_{orange} and K_{yellow}) express the preference of each type of pigment to migrate in ionic liquid-rich phase. They were then calculated for each system and are displayed in Figure 4. In line with previous observations, [C4Mim]Br$^-$ + 15 wt.% Na_2CO_3 resulted in the lowest K values for all pigment shades for the four strains (Figure 4) and it was concluded as a non-suitable system for our purpose. Although the partition coefficient values obtained from the two different remaining ATPSs on the same fermentation broths suggested that the design of the IL-based ATPSs has an impact on pigment migration, it is important to consider that the composition in pigments and their respective polarity profiles will have an impact on the efficiency of the overall extraction of the color shade. We therefore investigated two red-pigment-producing fungal strains, namely *T. albobiverticillius* (30548) and *E. purpurea* (3323), which were submitted to the same extraction process. Interestingly, [N4444]Br$^-$ + 25 wt% $NaCH_3COO$ statistically showed significantly better pigment partition coefficients for red to orange pigments from *T. albobiverticillius* than those from *E. purpurea* and *T. harzianum*. On the other hand, the extraction efficiency was equally as good for both strains using [N4444]Br$^-$- and [C4Mim]Cl$^-$-based ATPSs (Figure 5). These observations led to the conclusion that quarterly ammonium-based IL-ATPS (i.e., N4444]Br$^-$) has better extraction capacities towards azaphilone-like red pigments. Moreover, it was noticed that the extraction yields tended to be less repeatable across the replicate experiments performed on pigments from *E. purpurea* for both [N4444]Br$^-$ and [C4Mim]Cl$^-$ based ATPSs. This led to the conclusion that the composition of the pigment mixture highly influenced the efficiency of one system. Previous studies have concluded on the production of azaphilone-like red compounds mainly by *T. albobiverticillius,* suggesting mostly hydrophilic molecules. [26]. The binodal curve initially established here highlighted the [N4444]Br$^-$ ATPS as the most hydrophilic system amongst our ATPSs, which could explain why and how this specific system appeared to be a more appropriate extraction agent for an apparent homogenous mixture of reddish hydrophilic pigments.

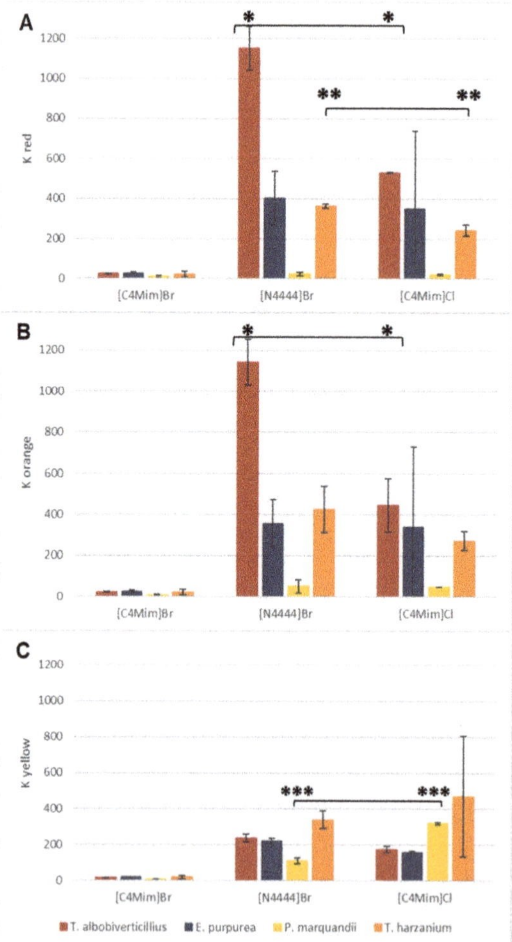

Figure 4. Partition coefficients of red (490 nm) (**A**), orange (470 nm) (**B**) and yellow (400 nm) (**C**) pigments for each strain in the four IL-ATPSs tested (*, ** and *** stand for statistical significance based on p-values < 0.05 performed for the same strain treated with different IL-ATPSs, n = 3).

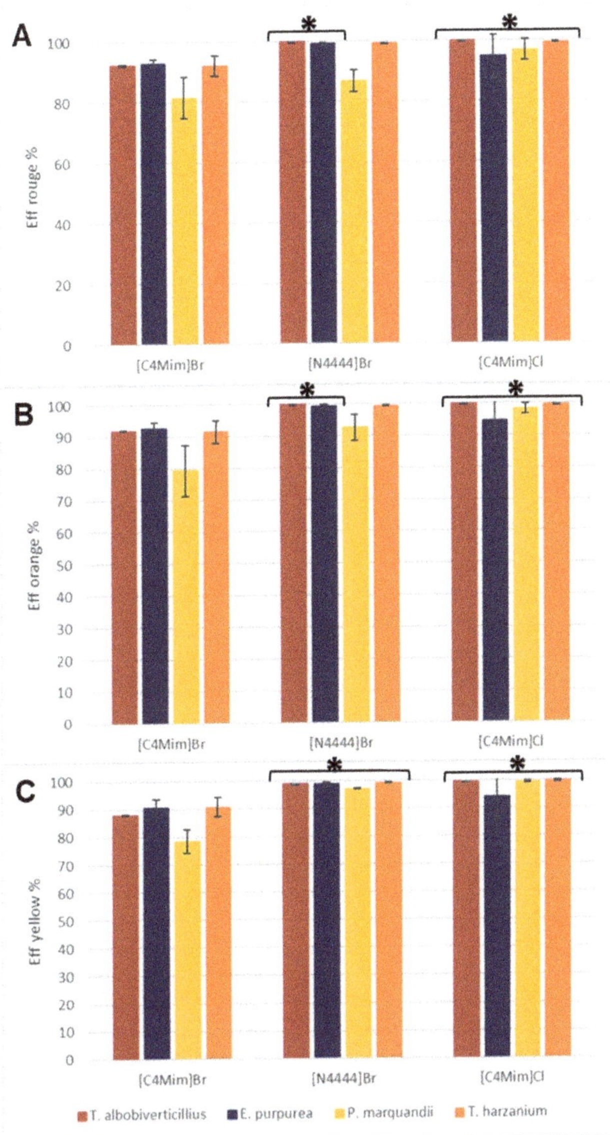

Figure 5. Direct extraction efficiency of red (490 nm) (**A**), orange (470 nm) (**B**) and yellow (400 nm) (**C**) pigments for each strain in the four IL-ATPSs tested (* indicate p-values < 0.05, n = 3).

In parallel, the repartition coefficient for yellow pigments remained drastically smaller than repartition coefficient for red, $K_{red,}$ across all ATPSs as well as amongst all fungal strain productions. Even the bright-yellow-pigment-producing fungal strain, namely *P. marquandii*, did not generate partition coefficient values as high as red-pigment-producing fungal strains (i.e., *T. albobiverticillius* and *E. purpurea*) for red to orange pigments. Such limited results for yellow compounds could be explained by the fact that reddish and orangish pigment-producing fungal strains may simply not produce many actual yellow molecules. In the case of the strong yellow-producing strain of *P. marquandii*, although the pigments clearly migrated to the ionic liquid-rich phase, the raw data of absorbance measured at 400 nm (yellow typical wavelength) were globally four times weaker in reddish strains, suggesting a smaller inherent yellow-pigment-production compared to red. Nevertheless, as shown in Figure 4C, no statistically significant differences were observed in yellow compound extractions from one IL-ATPS to another when used on broth containing a mixture of pigments produced by *T. albobiverticillius*, *E. purpurea* or *T. harzianum*. On the contrary, on the yellow-producing strain, namely *P. marquandii*, a statistically significantly better extraction capacity (i.e., higher K_{yellow}) was observed with $[C4Mim]Cl^-$.

Thus, from this first round of extraction, $[N4444]Br^-$ appeared as the more appropriate system amongst the ones investigated herein for consistent and efficient extractions of azaphilone-like red compounds from fungal resources. The yellow polyketide-based secondary metabolites produced by the investigated fungal strains were not the most preferred pigments extracted by either $[N4444]Br^-$ or $[C4Mim]Cl^-$; however, $[C4Mim]Cl^-$ showed relatively good robustness and efficiency with respect to the extraction of these yellow compounds (such as yellow azaphilone pigment T22 from *T. harzianum* and yellow sorbicillinoids (hexaketide compounds) from *P. marquandii*). It is worth mentioning that it seems more suitable to use this system on a strain generating a strongly yellow broth, as more inconsistent extraction results were observed on the yellow-brown mixture produced by *T. harzianum*. Thus, further optimizations, in particular on salt combinations and/or ratio thereof, could lead to better detection and extraction capacities.

3.3. *Selectivity of the IL-ATPS for Red Pigments in Various Fungal Broths*

As mentioned earlier, the selective isolation of red pigments from other fungal pigments and cellular components is the main goal of this study, due to the higher potential market of red colorants for industrial application [3,4,25]. To further determine the ability of each IL-ATPS to detect and isolate red colorants, two selectivity parameters were calculated as the ratio of the partition coefficients of red pigments to other fungal pigments and are displayed in Figure 6.

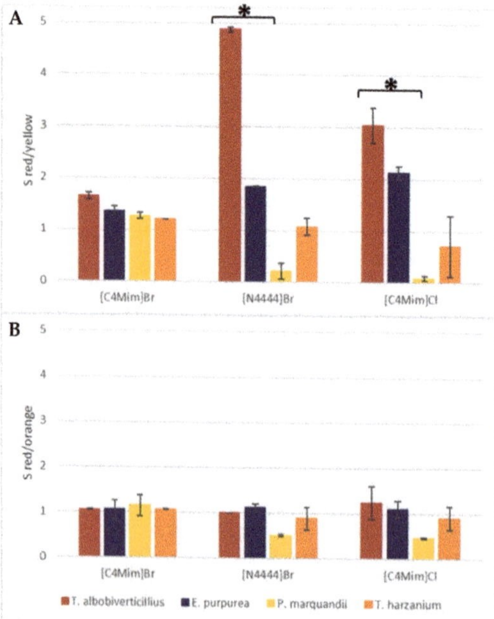

Figure 6. Extraction selectivity of red pigments over yellow pigments, $S_{red/yellow}$ (**A**) and over orange $S_{red/orange}$ (**B**) pigments for each strain in the four IL-ATPSs tested (* indicates p-values < 0.05, n = 3).

Overall, in red-producing strains, the selective extraction of red over yellow co-produced colorants was strongly favored, as suggested by the high $S_{red/yellow}$ values (i.e., ratio $S_{red/yellow}$ >>1) reported across the IL-ATPSs considered. In particular, we reported a statistically significantly better selectivity of red pigments' extraction from *T. albobiverticillius* when using [N4444]Br^- compared to [C4Mim]Cl^-, while no significant difference was observed from *E. purpurea*. This suggested that the migration and selectivity of mixtures of pigments are greatly impacted by the features of the ILs used. As far as we can see and without knowing the detailed composition of pigments in the different mixtures, we confirmed that the [N4444]Br^--based ATPS was the most suitable extraction system for the selective isolation of red colorants across different hue-producing strains. On the other hand, little or no selection occurred between red and orange pigments as shown in Figure 6B and in all IL-ATPSs and studied strains. Indeed, we reported a selective extraction ratio for red/orange pigments close to 1 for all selected strains (Figure 6B). We also noticed small distinction between red and orange or yellow colorants in all the IL-ATPSs used on *T. harzianum* (Figure 6A,B), suggesting a strong likeness of concomitant extraction of red and orange pigmented molecules from the fermentation broth showing a mixture of pigments of different shades. These observations led to the conclusion that our [N4444]Br^- system showed its highest potential under the conditions investigated here when applied on single-tone broth.

3.4. Reverse Extraction of the Fungal Pigments from the IL-ATPSs

The main objective of this study was to explore the potential of IL-ATPSs for the selective detection and extraction of red fungal pigments from a mixture of colored molecules. We succeeded in showing promising detection and extraction efficiencies of the [N4444]Br^- system for azaphilone-based compounds. This satisfied our initial scientific question. To further explore the potential applications of this technique, we investigated the back-extraction efficiency of the pigments from the IL system using a previously reported method. This was carried out with the purpose of rapidly estimating how straightforward and repeatable back extractions from IL-based systems across different types of biological matrixes are. Although several publications have reported the benefits of using IL-based extraction systems for the isolation of bioactive compounds [46], very few have addressed the corresponding challenges to develop efficient extraction of the said bio-compounds from IL-based solvents (referred to as back-extraction) [23,47]. Tan and coworkers [43] succeeded in back-extracting the hydrophilic anthraquinones of aloe-emodin from their IL-ATPS with [C4Mim]BF4 by adding an alkaline salt solution onto the IL-rich phase containing the pigment. The same approach was applied here using an alkaline (pH 13) solution of the respective salts initially used for each system, which were added to the separated IL-rich phase, shaken and left for decantation overnight. The visualization of the back-extraction steps before and after the alkalinization of the salt solution is shown in Figure S1. The visual aspects of the extracts (Figure S1) indicate that the reverse extractions remained extremely limited. The increase in pH up to 13 using drop-wise addition of NaOH did not allow a better recovery of any of the pigments tested under our experimental conditions. After the addition of the salt solution and before the alkalinization of the system, the phase separation between the IL- and salt-rich phases became blurry, especially for the [C4Mim]Cl^- system (Figure S1). Interestingly, in the [C4Mim]Cl^- and [C4Mim]Br^- ATPSs, before the pH increase, a slight repartition of the pigments in the salt phase for the red-producing strains (*T. albobiverticillius* and *E. purpurea*) and the brown-producing strain (*T. harzianum*) was observed. The increase in pH appeared to have counteracted the recovery of the pigments in the salt-rich phase, whilst further improving the partition in the IL-rich phase, as can be seen in particular with the clarification of the salt-rich phase in the case of the [C4Mim]Cl^- across the four fermentation broths tested (Figure S1).

The overall efficiencies and abilities of recovery of red pigments from the IL-ATPS after application of an alkaline salt solution are displayed in Figure S2. We concluded that the recovery of the pigments from any of the fermentation broth tested was very poor under these conditions. The highest overall recovery was obtained from the fermentation broth from *E. purpurea*, with a final recovery of about 20% (Figure S2B). We were also forced to conclude that we could not extract back, from the IL phase, the dark red pigments excreted by either *T. albobiverticillius* or *E. purpurea* because the pigments did not migrate to the alkaline salt-rich phase. Curiously, it seemed that increasing the pH reduced the migration of the pigments, which was in contradiction with the results obtained by Ventura et al. [40] on hydrophilic anthraquinone of aloe-emodin. However, this can be put into perspective by the fact that, firstly, we do not know the exact hydrophilic/hydrophobic equilibrium of all the fungal pigments produced in the fermentation broths, and secondly, we directly applied these IL-ATPSs onto complex fermentation matrices. Then, plenty of other metabolites were likely to interfere in the selective extraction of pigmented compounds. Globally speaking and despite the low overall recovery values, we could notice a statistically significant difference in the [N4444]Br^--based IL-ATPS on the extraction, back-extraction and overall recovery of red pigments. In conclusion, we were able to recover more red pigments from the fermentation broth of the second most red-pigment-producing strain, namely *E. purpurea,* by using the [N4444]Br^- ATPS.

To date, ionic liquids are considered to be a sustainable means of extraction, mainly because of their very high extraction power, perfectly illustrated here, revealing a promising potential as cost-effective, profitable and less hazardous solvents. However, it is important to recall that ionic liquids require heavy organic chemical synthesis, which can be counterintuitive against their reputation of being new-age green solvents. However, few arguments can be given here in favor of ionic liquids' green profile against conventional extraction solvents. First, the already well-known higher extraction power and recyclability of ionic liquids are their dire assets, as previously said. Second, a recent paper estimated and compared the environmental sustainability between a similar IL-based extraction system applied to carotenoid recovery and a conventional solvent extraction method (using acetone and ethanol) by calculating environmental factor (E-factor) and carbon footprint parameters [45]. The E-factor and carbon footprint calculations included all greenhouse gas (GHG) emissions from the production of all chemicals and the water and electricity consumed during the process. It was concluded that IL-based extraction offered a significantly reduced environmental impact compared to conventional solvents. Perspective must be taken though, as efficient recycling of both IL materials and water is paramount for ionic liquids to be properly referred to as green solvents.

The poor overall recovery obtained here highlights essential points of improvements to be made while perfectly illustrating how much case-by-case optimization is needed to use ionic liquids for detection and extraction purposes. It is worth mentioning that there are various options to explore in order to optimize back-extractions and enable multiple re-use of ionic liquids, such as using a different salt or polymer solutions as antisolvents, changing temperatures and applying ultrasounds or microwaves and a combination thereof [48]. Depending on the type and sensitivity of the target compounds trapped in IL phase, distillation or even adsorption can also be used. However, these options need to be carefully considered based on the additional energy, water, time and specific equipment expenditures they would require. However, this point was not in the scope of this paper.

4. Conclusions

Our findings suggested that the combination of an ATPS extraction system using ionic liquids as greener extraction solvents is a possible tool for the extraction and detection of colored molecules from the fermentation broth of filamentous fungi [49]. Among the IL-based systems investigated here, the [N4444]Br^- based system appeared to have the best ability to extract fungal pigments, with putative higher affinity for red-orange azaphilone-like pigments. Thus, the IL-ATPS consisting of 15% [N4444]Br^- and 25% $Na_2CH_3CO_2$ was found to be a promising solvent that is worth further investigation to expand its potential applications. Indeed, at this point, the most appropriate use of this system would be for the detection of red azaphilone-like pigments. More studies and optimizations, especially at the back-extraction level, are needed to use this [N4444]Br^- system as a powerful extraction technique. Thus, further studies would have to be carried out at larger scales in order to evaluate the obstacles and challenges associated with efficient target molecule recoveries and ionic liquids' recyclability.

Supplementary Materials: The following are available online at http://www.mdpi.com/2309-608X/6/4/375/s1, Figure S1: Visual aspect of the reverse extraction steps for each different IL-ATPS tested before (pH = 5) and after adjusting the pH (pH = 13), Figure S2: Back extraction efficiency (**A**) and overall efficiency (**B**) of red pigments extraction for each strain in the 4 IL-ATPS tested (n = 3).

Author Contributions: Conceptualization, Y.C.; formal analysis, J.L.; funding acquisition, M.F.; investigation, J.L. and Y.C.; methodology, J.L.; project administration, Y.C.; supervision, Y.C.; writing—original draft preparation, J.L.; writing—review and editing, J.L., T.P., M.F., L.D. and Y.C. All authors made critical revisions. All authors have read and agreed to the published version of the manuscript.

Funding: This research was funded by Conseil Régional de La Réunion (Réunion Island, France), grant No. DIRED 20140704 (COLORMAR Program).

Conflicts of Interest: The authors declare no conflict of interest.

References

1. Křížová, H. Natural dyes: Their past, present, future and sustainability. In *Recent Developments in Fibrous Material Science*; Křemenáková, D., Militký, J., Mishra, R., Eds.; Kanina-o.p.s: Kanina, Czech Republic, 2015; pp. 59–71.
2. Chaitanya Lakshmi, G. Food coloring: The natural way. *Res. J. Chem. Sci.* **2015**, *4*, 87–96.
3. Dufossé, L.; Fouillaud, M.; Caro, Y.; Mapari, S.A.S.; Sutthiwong, N. Filamentous fungi are large-scale producers of pigments and colorants for the food industry. *Curr. Opin. Biotechnol.* **2014**, *26*, 56–61. [CrossRef] [PubMed]
4. Caro, Y.; Venkatachalam, M.; Lebeau, J.; Fouillaud, M.; Dufossé, L. Pigments and colorants from filamentous fungi. In *Fungal Metabolites. Reference Series in Phytochemistry*; Merillon, J.M., Ramawat, K.G., Eds.; Springer: Cham, Switzerland, 2017; pp. 499–568.
5. Chemat, F.; Vian, M.A.; Cravotto, G. Green extraction of natural products: Concept and principles. *Int. J. Mol. Sci.* **2012**, *13*, 8615–8627. [CrossRef] [PubMed]
6. Joshi, D.R.; Adhikari, N. An overview on common organic solvents and their toxicity. *J. Pharm. Res. Int.* **2019**, *28*, 1–18. [CrossRef]
7. Han, D.; Row, K.H. Recent applications of ionic liquids in separation technology. *Molecules* **2010**, *15*, 2405–2426. [CrossRef]
8. Ghandi, K. A review of ionic liquids, their limits and applications. *Green Sustain. Chem.* **2014**, *4*, 44–53. [CrossRef]
9. Mussagy, C.U.; Santos-Ebinuma, V.C.; Gonzalez-Miquel, M.; Coutinho, J.A.P.; Pereira, J.F.B. Protic ionic liquids as cell-disrupting agents for the recovery of intracellular carotenoids from yeast *Rhodotorula glutinis* CCT-2186. *ACS Sustain. Chem. Eng.* **2019**, *7*, 16765–16776. [CrossRef]
10. Pereira, M.M.; Coutinho, J.A.P.; Freire, M.G. Ionic liquids as efficient tools for the purification of biomolecules and bioproducts from natural sources (Chapter 8). In *Ionic Liquids in the Biorefinery Concept: Challenges and Perspectives*; Bogel-Lukasik, R., Ed.; Royal Society of Chemistry: London, UK, 2016; pp. 227–257.
11. Egorova, K.S.; Gordeev, E.G.; Ananikov, V.P. Biological activity of ionic liquids and their application in pharmaceutics and medicine. *Chem. Rev.* **2017**, *117*, 7132–7189. [CrossRef]
12. Plechkova, N.V.; Seddon, K.R. Applications of ionic liquids in the chemical industry. *Chem. Soc. Rev.* **2008**, *37*, 123–150. [CrossRef]
13. Capela, E.V.; Coutinho, J.A.P.; Freire, M.G. Application of Ionic Liquids in Separation and Fractionation Processes. In *Green Chemistry and Chemical Engineering*; Han, B., Wu, T., Eds.; Encyclopedia of Sustainability Science and Technology Series; Springer: New York, NY, USA, 2019; pp. 637–665.
14. Ventura, S.P.M.; Gonçalves, A.M.M.; Sintra, T.; Pereira, J.L.; Gonçalves, F.; Coutinho, J.A.P. Designing ionic liquids: The chemical structure role in the toxicity. *Ecotoxicology* **2013**, *22*, 1–12. [CrossRef]
15. Freire, M.G.; Neves, C.M.S.S.; Marrucho, I.M.; Coutinho, J.A.P.; Fernandes, A.M. Hydrolysis of tetrafluoroborate and hexafluorophosphate counter ions in imidazolium-based ionic liquids. *J. Phys. Chem. A* **2010**, *114*, 3744–3749. [CrossRef] [PubMed]
16. Jastorff, B.; Störmann, R.; Ranke, J.; Mölter, K.; Stock, F.; Oberheitmann, B.; Hoffmann, W.; Hoffmann, J.; Nüchter, M.; Ondruschka, B.; et al. How hazardous are ionic liquids? Structure-activity relationships and biological testing as important elements for sustainability evaluation. *Green Chem.* **2003**, *5*, 136–142. [CrossRef]

17. Cserjési, P.; Bélafi-Bakó, K.; Nemestóthy, N.; Gubicza, L. Recent trends on application of ionic liquids in organic synthesis. *Hung. J. Ind. Chem. Veszprém* **2008**, *36*, 27–34.
18. Luís, A.; Dinis, T.B.V.; Passos, H.; Taha, M.; Freire, M.G. Good's buffers as novel phase-forming components of ionic-liquid-based aqueous biphasic systems. *Biochem. Eng. J.* **2015**, *101*, 142–149. [CrossRef]
19. Mcqueen, L.; Lai, D. Ionic liquid aqueous two-phase systems from a pharmaceutical perspective. *Front. Chem.* **2019**, *1*, 135. [CrossRef]
20. Gutowski, K.E.; Broker, G.A.; Willauer, H.D.; Huddleston, J.G.; Swatloski, R.P.; Holbrey, J.D.; Rogers, R.D. Controlling the aqueous miscibility of ionic liquids: Aqueous biphasic systems of water-miscible ionic liquids and water-structuring salts for recycle, metathesis, and separations. *J. Am. Chem. Soc.* **2003**, *125*, 6632–6633. [CrossRef]
21. Suarez Ruiz, C.A.; Emmery, D.P.; Wijffels, R.H.; Eppink, M.H.; van den Berg, C. Selective and mild fractionation of microalgal proteins and pigments using aqueous two-phase systems. *J. Chem. Technol. Biotechnol.* **2018**, *93*, 2774–2783. [CrossRef]
22. Freire, M.G.; Neves, C.M.S.S.; Marrucho, I.M.; Canongia Lopes, J.N.; Rebelo, L.P.N.; Coutinho, J.a.P. High-performance extraction of alkaloids using aqueous two-phase systems with ionic liquids. *Green Chem.* **2010**, *12*, 1715. [CrossRef]
23. Cláudio, A.F.M.; Marques, C.F.C.; Boal-Palheiros, I.; Freire, M.G.; Coutinho, J.A.P. Development of back-extraction and recyclability routes for ionic-liquid-based aqueous two-phase systems. *Green Chem.* **2014**, *16*, 259–268. [CrossRef]
24. Flieger, J.; Czajkowska-Zalazko, A. Ionic liquids in separation techniques. *J. Chromatogr. A* **2011**, *1184*, 6–18.
25. Lebeau, J.; Venkatachalam, M.; Fouillaud, M.; Petit, T.; Vinale, F.; Dufossé, L.; Caro, Y. Production and new extraction method of polyketide red pigments produced by ascomycetous fungi from terrestrial and marine habitats. *J. Fungi* **2017**, *3*, 34. [CrossRef] [PubMed]
26. Venkatachalam, M.; Zelena, M.; Cacciola, F.; Ceslova, L.; Girard-Valenciennes, E.; Clerc, P.; Dugo, P.; Mondello, L.; Fouillaud, M.; Rotondo, A.; et al. Partial characterization of the pigments produced by the marine-derived fungus *Talaromyces albobiverticillius* 30548. Towards a new fungal red colorant for the food industry. *J. Food Compos. Anal.* **2018**, *67*, 38–47. [CrossRef]
27. Vinale, F.; Marra, R.; Scala, F.; Ghisalberti, E.L.; Lorito, M.; Sivasithamparam, K. Major secondary metabolites produced by two commercial *Trichoderma* strains active against different phytopathogens. *Lett. Appl. Microbiol.* **2006**, *43*, 143–148. [CrossRef] [PubMed]
28. Qian-Cutrone, J.; Huang, S.; Chang, L.P.; Pirnik, D.M.; Klohr, S.E.; Dalterio, R.A.; Hugill, R.; Lowe, S.; Alam, M.; Kadow, K.F. Harziphilone and fleephilone, Two new HIV REV/RRE binding inhibitors produced by *Trichoderma harzianum*. *J. Antibiot.* **1996**, *49*, 990–997. [CrossRef]
29. Contreras-Cornejo, H.A.; Macías-Rodríguez, L.; Del-Val, E.; Larsen, J. Ecological functions of *Trichoderma* spp. and their secondary metabolites in the rhizosphere: Interactions with plants. *FEMS Microbiol. Ecol.* **2016**, *92*, 36. [CrossRef]
30. Lin, Y.R.; Lo, C.T.; Liu, S.Y.; Peng, K.C. Involvement of pachybasin and emodin in self-regulation of *Trichoderma harzianum* mycoparasitic coiling. *J. Agric. Food Chem.* **2012**, *60*, 2123–2128. [CrossRef]
31. Li, M.-F.; Li, G.-H.; Zhang, K.-Q. Non-volatile metabolites from *Trichoderma* spp. *Metabolites* **2019**, *9*, 58. [CrossRef]
32. Chen, A.J.; Frisvad, J.C.; Sun, B.D.; Varga, J.; Kocsubé, S.; Dijksterhuis, J.; Kim, D.H.; Hong, S.-B.; Houbraken, J.; Samson, R.A. *Aspergillus* section *Nidulantes* (formerly *Emericella*): Polyphasic taxonomy, chemistry and biology. *Stud. Mycol.* **2016**, *84*, 1–118. [CrossRef]
33. Meng, J.; Wang, X.; Xu, D.; Fu, X.; Zhang, X.; Lai, D.; Zhou, L.; Zhang, G. Sorbicillinoids from fungi and their bioactivities. *Molecules* **2016**, *21*, 715. [CrossRef]
34. Decay, R.K. Ionic Liquid Pre-Treatment of Microalgae and Extraction of Biomolecules. Ph.D. Thesis, Wageningen University, Wageningen, The Netherlands, 9 December 2016.
35. Venkatachalam, M.; Magalon, H.; Dufossé, L.; Fouillaud, M. Production of pigments from the tropical marine-derived fungi *Talaromyces albobiverticillius:* New resources for natural red-colored metabolites. *J. Food Compos. Anal.* **2018**, *70*, 35–48. [CrossRef]
36. Hatti-Kaul, R.; Kaul, A. The phase diagram. *Aqueous Two Phase Syst. Methods Protoc.* **2003**, *11*, 11–21.

37. Freire, M.G.; Cláudio, A.F.M.; Araújo, J.M.M.; Coutinho, J.A.P.; Marrucho, I.M.; Canongia Lopesac, J.N.; Rebelo, L.P.N. Aqueous biphasic systems: A boost brought about by using ionic liquids. *Chem. Soc. Rev.* **2012**, *41*, 4966–4995. [CrossRef] [PubMed]
38. Domínguez-Pérez, M.; Tomé, L.I.N.; Freire, M.G.; Marrucho, I.M.; Cabeza, O.; Coutinho, J.A.P. Extraction of biomolecules using aqueous biphasic systems formed by ionic liquids and aminoacids. *Sep. Purif. Technol.* **2010**, *72*, 85–91. [CrossRef]
39. Sintra, T.E.; Cruz, R.; Ventura, S.P.M.; Coutinho, J.A.P. Phase diagrams of ionic liquids-based aqueous biphasic systems as a platform for extraction processes. *J. Chem. Thermodyn.* **2014**, *77*, 206–213. [CrossRef]
40. Ventura, S.P.M.; Santos-Ebinuma, V.C.; Pereira, J.F.B.; Teixeira, M.F.S.; Pessoa, A.; Coutinho, J.A.P. Isolation of natural red colorants from fermented broth using ionic liquid-based aqueous two-phase systems. *J. Ind. Microbiol. Biotechnol.* **2013**, *40*, 507–516. [CrossRef]
41. Ferreira, A.M.; Coutinho, J.A.P.; Fernandes, A.M.; Freire, M.G. Complete removal of textile dyes from aqueous media using ionic-liquid-based aqueous two-phase systems. *Sep. Purif. Technol.* **2014**, *128*, 58–66. [CrossRef]
42. Flieger, J.; Grushka, E.-B.; Czajkowska-Zelazko, A. Ionic liquids in separation processes. *Chem. Thermodyn. Ind.* **2014**, *1*, 76–87.
43. Tan, Z.; Li, F.; Xu, X. Isolation and purification of aloe anthraquinones based on an ionic liquid/salt aqueous two-phase system. *Sep. Purif. Technol.* **2012**, *98*, 150–157. [CrossRef]
44. Montalvo-Hernández, B.; Rito-Palomares, M.; Benavides, J. Recovery of crocins from saffron stigmas (*Crocu sativus*) in aqueous two-phase systems. *J. Chromatogr. A* **2012**, *1236*, 7–15. [CrossRef]
45. De Souza Mesquita, L.M.; Martins, M.; Maricato, E.; Nunes, C.; Quinteiro, P.S.G.N.; Dias, A.C.R.V.; Coutinho, J.A.P.; Pisani, L.P.; de Rosso, V.V.; Ventura, S.P.M. Ionic liquid-mediated recovery of carotenoids from the bactris gasipaes fruit waste and their application in food-packaging chitosan films. *ACS Sustain. Chem. Eng.* **2020**, *8*, 4085–4095. [CrossRef]
46. Raja, S.; Murty, V.R.; Thivaharan, V.; Rajasekar, V.; Ramesh, V. Aqueous two phase systems for the recovery of biomolecules—A Review. *Sci. Technol.* **2012**, *1*, 7–16. [CrossRef]
47. Ventura, S.P.M.; Silva, F.A.E.; Quental, M.V.; Mondal, D.; Freire, M.G.; Coutinho, J.A.P. Ionic-liquid-mediated extraction and separation processes for bioactive compounds: Past, present, and future Trends. *Chem. Rev.* **2017**, *117*, 6984–7052. [CrossRef] [PubMed]
48. Zhou, J.; Sui, H.; Jia, Z.; Yang, Z.; He, L.; Li, X. Recovery and purification of ionic liquids from solutions: A review. *RSC Adv.* **2018**, *8*, 32832–32864. [CrossRef]
49. De Oliveira, F.; Rie Hirai, P.; Teixeira, M.F.S.; Pereira, J.F.B.; Santos-Ebinuma, V.C. *Talaromyces amestolkiae* cell disruption and colorant extraction using imidazolium-based ionic liquids. *Sep. Purif. Technol.* **2021**, *257*, 117759. [CrossRef]

Journal of *Fungi*

MDPI

Article

Microbial Colorants Production in Stirred-Tank Bioreactor and Their Incorporation in an Alternative Food Packaging Biomaterial

Fernanda de Oliveira [1], Caio de Azevedo Lima [1], André Moreni Lopes [2], Daniela de Araújo Viana Marques [3], Janice Izabel Druzian [4], Adalberto Pessoa Júnior [5] and Valéria Carvalho Santos-Ebinuma [1,*]

1 Department of Engineering Bioprocess and Biotechnology, School of Pharmaceutical Sciences, Universidade Estadual Paulista—UNESP, Araraquara 14800-903, Brazil; fernanda.oliveira1@unesp.br (F.d.O.); caio.a.lima@unesp.br (C.d.A.L.)

2 Faculty of Pharmaceutical Sciences, University of Campinas—FCF/UNICAMP, Campinas 13083-859, Brazil; amorenilopes@gmail.com

3 Laboratory of Biotechnology Applied to Infectious and Parasitic Diseases, Biological Science Institute, University of Pernambuco-ICB/UPE, Recife 50100-130, Brazil; daniela_viana@yahoo.com.br

4 Department of Bromatological Analysis, Faculty of Pharmacy, Postgraduate Program in Science of Food, Federal University of Bahia, Salvador 40170-115, Brazil; janicedruzian@hotmail.com

5 Department of Biochemical and Pharmaceutical Technology, University of São Paulo, São Paulo 05508-000, Brazil; pessoajr@usp.br

* Correspondence: valeria.ebinuma@unesp.br; Tel.: +55-16-3301-4647

Received: 28 August 2020; Accepted: 15 October 2020; Published: 2 November 2020

Abstract: Natural colorants from microbial fermentation have gained significant attention in the market to replace the synthetic ones. *Talaromyces* spp. produce yellow-orange-red colorants, appearing as a potential microorganism to be used for this purpose. In this work, the production of natural colorants by *T. amestolkiae* in a stirred-tank bioreactor is studied, followed by its application as additives in bio-based films. The effect of the pH-shift control strategy from 4.5 to 8.0 after 96 h of cultivation is evaluated at 500 rpm, resulting in an improvement of natural colorant production, with this increase being more significant for the orange and red ones, both close to 4-fold. Next, the fermented broth containing the colorants is applied to the preparation of cassava starch-based films in order to incorporate functional activity in biodegradable films for food packaging. The presence of fermented broth did not affect the water activity and total solids of biodegradable films as compared with the standard one. In the end, the films are used to pack butter samples (for 45 days) showing excellent results regarding antioxidant activity. It is demonstrated that the presence of natural colorants is obtained by a biotechnology process, which can provide protection against oxidative action, as well as be a functional food additive in food packing biomaterials.

Keywords: natural colorants; filamentous fungi; stirred-tank bioreactor; biodegradable films; antioxidant; food package

1. Introduction

Over the years, humans have used colorants to enhance or restore the appearance of products, especially in the food industry [1,2]. Nevertheless, most of the synthetic colorants used in food products are derived from petroleum and may have harmful effects on human health and the environment [3,4]. Therefore, there is a growing preference for natural additives in foods. Regarding the natural color market, a survey by the Grand View Research, Inc. estimated the world color market for 2025 to be

USD 37.49 billion. Specifically, for the food market, the natural colorant consumption will register a 5.9% compound annual growth rate (CAGR) in terms of revenue, in relation to the period from 2018 to 2025 [5].

There are several biological sources that produce natural colorants, such as insects, vegetables, and microorganisms [2,6]. Colorants from microbial origin, especially fungal ones, are of great interest mainly because of the recent improvements in biotechnology and bioprocessing technologies as well as a full control of the production conditions [2,7,8]. Fungus of the genus *Monascus* stands out for presenting good levels of colorant production. However, *Monascus* species co-produce citrinin, a mycotoxin [9]. Thus, other strains such as *Aspergillus*, *Trichoderma*, *Fusarium*, and *Talaromyces* (formerly *Penicillium* spp.) have been the subject of studies in order to produce colorants [8,10–12]. In this context, *Talaromyces amestolkiae* (former *Penicillium purpurogenum*) has aroused interest in the production potential of stable and non-toxic yellow, orange, and red colorants [10,13–15].

One of the main challenges of using filamentous fungi, such as *Talaromyces*, for metabolite production is the control of microorganism growth and form besides the mass transfer inside the bioreactor [16]. Hyphal concentration greatly influences the fermentation broth viscosity [15]. Meanwhile, the oxygen transfer in submerged cultures is an important factor for efficient fungal growth and an important parameter for a high yield of fungi colorants [17]. At a high cell concentration, the culture medium viscosity increases and it limits the oxygen supply for the cells [4,18]. Increasing the stirring speed can maintain dissolved oxygen concentration at high levels, however, potential damage to the fungal hyphae can limit the impeller speed and, consequently, the oxygen and nutrient transfer capability of a bioreactor [19,20]. In order to control cell growth and colorant biosynthesis, the well-directed process parameter shift represents a valuable control strategy [9,21]. In the production of *Monascus* colorants, the pH-shift strategy has been proven to be efficient in colorants biosynthesis [17,22]. Hence, great efforts to improve the bioreactor cultivation conditions are necessary for enhancing the large-scale production of *Talaromyces* colorants.

Biosynthetically, many fungi colorants are polyketides. Polyketide colorants range in structure from tetraketides to octaketides, which have four or eight C_2 units that contribute to the polyketide chain. The structure of the polyketides does not exhibit localized negatively-charged ions. These molecules often have a polyunsaturated function, i.e., a ring system, one or more carbonyl groups, carbolic acid, and functional ester or amide groups that absorb in the UV-visible spectrum [23]. One class of polyketide colorants are azaphilones, which are hexaketide colorants with pyrone-quinone structures and a chiral quaternary center [24]. Investigations focus on six major azaphilone colorants: Rubropunctamine and Monascorubramine (red); Rubropunctatin and Monascorubrin (orange); Monascin and Ankaflavin (yellow) [25]. One of the most interesting characteristics of azaphilone colorants is their potential use as a functional food colorant, because they may exert antioxidant and antimicrobial activities, relatively low cytotoxicity, as well as immunosuppressive, antiviral, anticancer, and cholesterol-reducing properties [26]. Moreover, our research group previously reported that *T. amestolkiae*-fermented broth presented low cytotoxicity against fibroblast cells and effective antimicrobial activity against *Staphylococcus aureus*, a representative food contaminant [13]. In this sense, bio-based materials with antioxidant properties are gaining attention in the food sector [27], mainly because non-ecofriendly packages are considered an environmental problem [28].

The application of natural compounds in active packaging has been increasingly applied in order to produce materials that interact with packaged foods, positively modifying their sensory and nutritional properties, preventing the deterioration of food, mainly caused by lipid oxidation and microbial growth [29]. Particularly, foods with high-fat content and especially those with a high degree of unsaturation are susceptible to deterioration by oxidation. In this way, alternative packaging technologies based on the inclusion of antioxidants compounds can create a barrier, improving the stability of packaged products sensitive to oxidation [30]. Therefore, the use of the *T. amestolkiae* natural colorants as an added component in bio-based films represents an alternative for increasing the properties of food packaging.

This work addresses the production of natural colorants by *T. amestolkiae* in a stirred-tank bioreactor. The effects of stirring speed and pH-shift as a control strategy on colorant accumulation are evaluated. After the selection of the best operational condition, the fermented broth containing the colorant is incorporated into cassava starch biodegradable films. The bio-based films prepared are characterized with respect to thickness, water activity, and total solid content. In the end, the cassava starch-based films are used to pack butter, supporting the applicability of natural colorants as additives in bio-based films. The effectiveness of the *T. amestolkiae* colorants protection against oxidative action is addressed on the basis of the peroxide content under accelerated oxidative conditions.

2. Materials and Methods

2.1. Materials

Sucrose and yeast extract were purchased from Synth (São Paulo, Brazil) and Acumedia (Lansing, MI, USA), respectively. Cassava starch (composed by 23.5% amylose and 64.2% amylopectin) was donated by Cargill Agrícola S.A. (Porto Ferreira, SP, Brazil). Commercial butter was obtained from Imperial (Bahia, Brazil). Low-density polyethylene (LDPE) film (0.020 mm thickness and 15.86×10^{-8} g H_2O·mm/m^2·h·kPa water vapor permeability) was purchased from local markets (Salvador, BA, Brazil). All of the other reagents were of analytical grade.

2.2. Microorganism Maintenance and Colorant Production

Talaromyces amestolkiae DPUA 1275 was generously provided by the Culture Collection of the Federal University of Amazonas (DPUA, Manaus, AM, Brazil). The cultures preserved in distilled water were reactivated in Czapeck Yeast Extract Agar (CYA) and maintained at 30 °C for 7 days. The CYA medium had the following composition (g/L in deionized water): K_2HPO_4 (1.0), yeast extract (5.0), sucrose (30.0), agar (15.0), and 10 mL/L of concentrated Czapeck [(g/100 mL of deionized water): $NaNO_3$ (30.0), KCl (5.0), $MgSO_4·7H_2O$ (5.0), and $FeSO_4·7H_2O$ (0.1)] [10].

The production process was composed of three phases: pre-inoculum, inoculum, and submerged culture. For the pre-inoculum preparation, a loop of fungus from stock culture was inoculated on a CYA plate and maintained in the same reactivation conditions. For inoculum, five mycelial agar discs (8 mm diameter) of *T. amestolkiae* were punched out from the pre-inoculum with a self-designed cutter and transferred to 50 mL of submerged culture medium in 250-mL Erlenmeyer flasks incubated at 30 °C and 150 rpm for 72 h. The inoculum medium was CYA liquid (without the addition of agar). Then, the entire volume obtained (0.2 L) was aseptically transferred to a single flask and then transferred to the bioreactor. The composition of the submerged culture medium was similar to that used for inoculum preparation, except for the concentration of sucrose and yeast extract, which were 48.50 and 11.80 g/L, respectively [10]. Both the inoculum medium and the culture broth had their pH adjusted to 4.5 with HCl (5 M) and were autoclaved at 121 °C for 15 min. The submerged culture was performed in a stirred-tank bioreactor Bioflo® 115 (New Brunswick, Edison, NJ, USA) with 3 L of working volume equipped with two Rushton impellers submersed into the bulk liquid. After sterilization, the bioreactor containing 1.8 L of culture medium was supplied with 0.2 L of the inoculum. The following operational conditions were kept constant by the bioreactor controllers: temperature at 30 °C and aeration rate at 2.0 vvm. The experiments were carried out for 240 h, and every 24 h, an aliquot was withdrawn. At the end of the bioprocess, the fermented broth was filtered first using filter paper Whatman #1 (Whatman, Marlborough, England) and later using a 0.45 μm filter acquired from Millipore. The filtrated samples were used to determine the production of yellow, orange, and red colorants as well as sucrose consumption. In fact, during the cultivation of *Talaromyces* spp., the three natural colorants (yellow, orange, and red) were produced at the same time. In order to guarantee that the colorants were produced, we analyzed the specific wavelength for each one of them, according to Section 2.3.

In the first set of experiments, the effects of the stirring speed from 100 to 600 rpm were evaluated. Next, pH-shift experiments were carried out at 500 rpm, adjusting the pH to 8.0 with NaOH (1 M) after the sucrose depletion (96 h of cultivation).

2.3. Analytical Methods

The sucrose concentration was determined according to the methodology described by Dubois et al. [31] and the pH was measured using a pH-meter (model MPC 227, Mettler Toledo, Columbus, OH, USA). The production of extracellular colorants was estimated by spectrophotometric analysis (model DU 640, Beckman, Irving, TX, USA) by reading the absorbance of supernatant at 410, 470, and 490 nm, which corresponds to the maximum absorbance for yellow, orange, and red colorants, respectively. The results were expressed in terms of absorbance units (AU) [10].

2.4. Kinetic Parameters

To calculate the substrate conversion factor (sucrose) in product ($Y_{P/S}$) and productivity (P), we employed Equations (1) and (2), respectively:

$$Y_{P/S} = \frac{Abs_f - Abs_0}{S_0 - S_f} \quad (1)$$

$$P = \frac{Abs_f}{t} \quad (2)$$

where: S_0 and S_f are the initial and final sucrose concentrations; Abs_f and Abs_0 are the final and initial colorant absorbance, respectively, and t is the time (h).

2.5. Preparation of Cassava Starch-Based Films and Incorporation of Natural Colorants

The cassava starch-based films were prepared according to the methodology described by Silva et al. [32]. In this way, the bio-based films were prepared by the casting technique, mixing cassava starch (4.0 wt%), plasticizers (0.7 wt% sucrose and 1.4 wt% inverted sugar), and the fermented broth containing the colorants (5.0 wt%). The fermented broth was used because of the synergy between the colorants molecules and the different metabolites present on fermented broth that increase the antioxidant power. Dispersions were heated (70 ± 2 °C), degassed for 30 min in ultrasonic bath to remove the bubbles, placed in polystyrene Petri dishes (150 × 15 mm), and dehydrated in an oven with airflow and circulation at 35 ± 2 °C for 24 h. A cassava starch-based film without addition of the fermented broth was prepared and used as control. Resulting bio-based films were stored in desiccators (at 23 ± 2 °C with 60 ± 2% relativity humidity) with a supersaturated solution of magnesium nitrate for 48 h before the characterization studies.

2.6. Cassava Starch-Based Films Characterization

The thickness was evaluated using a flat parallel surface micrometer (Mitutoyo model 103–137, precision 0.002 mm). Six measurements were taken at random positions around the film sample. The results were expressed as the mean ± respective standard deviations and used for the calculation of the contact area (mm).

The water activity (aw) was measured with a Decagon, Aqualab Lite, as calibration standards used pure water (aw of 1.000 ± 0.001%) and LiCl (aw of 0.500 ± 0.015%). Preconditioned samples (4 cm^2) were cut from the center of the films and evaluated in triplicate at room temperature (25 ± 2 °C).

The total solid content was determined by measuring the weight loss of films upon drying (105 °C) until constant weight. This determination was performed in triplicate. The results were expressed as the mean ± respective standard deviations.

2.7. Application of Cassava Starch-Based Films for Packaging Butter

Butter was packed in the cassava starch-based films with colorant additives. Square-shaped films (5 × 2 cm) of 0.164 and 0.212 mm in thickness were molded (Sealer Sulpack SM 400 TE, Caxias do Sul, Rio Grande do Sul, Brazil). Butter samples were homogenized and frozen in small pieces (with 3 × 2 cm and 10.00 ± 0.54 g), after they were involved with the film, the bubbles of oxygen were removed, and the film was sealed.

The samples of packaged butter were stored for different periods of time (0, 7, 15, 30, and 45 days) and under accelerated oxidative conditions (64% of relative humidity at 25 ± 2 °C). These analyses were carried out in a dark room to avoid the effects of light interference (i.e., color degradation) in the samples. For this, three types of packages were prepared: cassava starch-based films with colorant additives (CFC), cassava starch-based films without colorant additives (CF), and conventional plastic (CP). In addition, unpackaged butter (U) was used as control. To evaluate the antioxidant action of the films formulation, the oxidative stability of the butter stored under accelerated oxidative conditions was monitored for 45 days and the peroxide content (PC) was determined by titration method according to the methodology described by the Association of Official Analytical Chemists–AOAC (2000) [33].

2.8. Statistical Analysis

The data were analyzed by ANOVA using a StatSoft v.7 program (StatSoft, Inc., Tulsa, OK, USA). The Tukey test was used to evaluate the mean difference in results at the 95% level of significance.

3. Results and Discussion

3.1. Production of Natural Colorants in Bioreactor: Effect of Stirring Speed

The high viscosity of the medium can result in oxygen diffusion limitation, due to the combination of the high biomass concentration and the fungal morphology. Hence, efforts have been made in order to overcome mass transfer problems [16,34]. Therefore, the effects of the stirring speed and pH were investigated by our group. In this sense, the production of natural colorants by *T. amestolkiae* in a stirred-tank bioreactor was evaluated, we maintained the aeration rate constant at 2.0 vvm and varied the stirring speed from 100 to 600 rpm in order to achieve proper oxygen diffusion in our cultivations. The results are depicted in Figure 1.

As can be seen in Figure 1, the stirring speed influenced the colorants' yield. At 100 rpm, the colorants' production stopped after 120 h, probably because under this condition the microorganism distribution inside the bioreactor was not homogeneous. It is usually observed that submerged cultures of filamentous fungi present a pseudo plastic behavior (i.e., a non-Newtonian characteristic) [35], and low speed, associated with high medium viscosity, does not allow for system homogeneity. The main role of stirring in a bioreactor is to improve heat and mass transfer to achieve homogeneity in the system. A proper mixing of components in the culture medium is necessary to ensure an adequate flow in submerged aerobic culture [36]. The stirring of the culture medium may also cause different effects in filamentous microorganisms. Among them may be included: cell wall disruption, changes in filamentous morphology, variation in growth efficiency and growth rate, and the variation in the formation rate of the desired bioproduct [18,36]. The stirring process may damage both mycelial pellets and hyphae structures. The fragmentation of hyphae can result in the creation of small parts of these structures, promoting a new center for biomass growth, or may cause damage to the hypha allowing release of the cytoplasm. On the other hand, total pellet breakdown may occur, and, probably in this case, total autolysis or aggregates can occur [37].

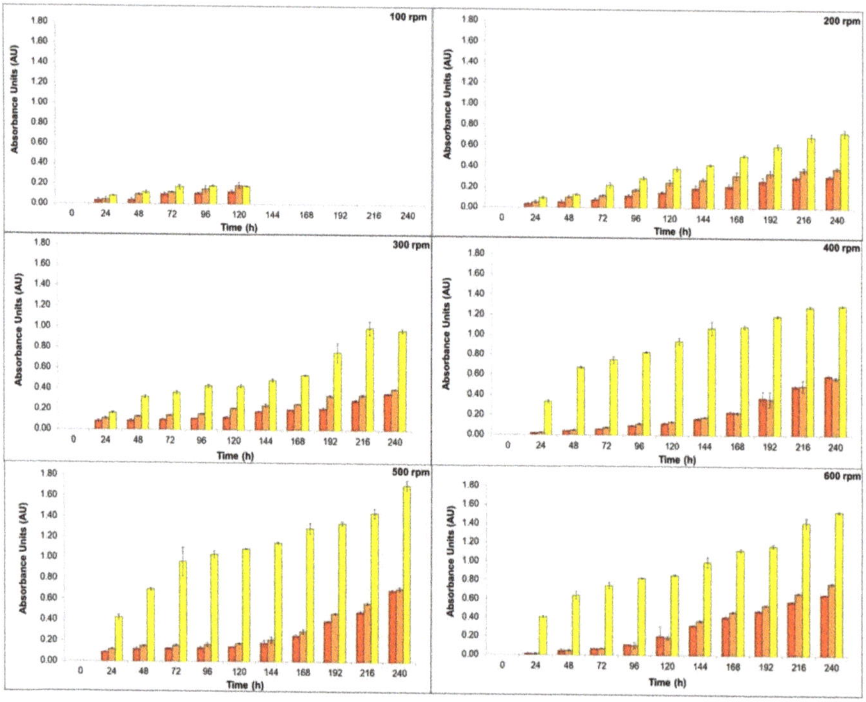

Figure 1. Production of yellow (yellow bars), orange (orange bars), and red colorants (red bars) by submerged culture of *T. amestolkiae* in stirred-tank bioreactor at 2.0 vvm and 30 °C by varying the stirring speed (100 to 600 rpm) as a function of time (0 to 240 h). The error bars represent 95% confidence levels for the mean of three independent assays.

Increasing the stirring speed could maintain dissolved oxygen concentration at high levels being a key parameter for high yield of *T. amestolkiae* colorants in the stirred-tank bioreactor cultivation. As the stirring speed increased, the homogeneity of the culture medium could be achieved, and it was possible to determine the natural colorant production. Since the oxygen transfer is directly proportional to the shear force and it is strictly related to the morphology of the filamentous fungus, stirring speeds above 600 rpm were not evaluated in this study. Considering the production of orange and red colorants, there was no statistical difference between the results achieved at 500 rpm and 600 rpm (statistical test: one-way ANOVA, $p = 0.0994$). However, as 500 rpm promote a lower shear stress, this stirring speed was chosen for the next experiments. In this condition, the absorbance of the yellow, orange, and red colorants was 1.71 AU_{400nm}, 0.71 AU_{470nm}, and 0.69 AU_{490nm}, respectively. Comparing the colorant production achieved with results previously reported by our research group [10], using the same culture media but in orbital shaker incubator, there was a decrease of 5-fold for the red colorant production in the bioreactor. This result confirms the statement that the scale-up from orbital shaker incubator to large fermenters is a difficult task. In fact, several conditions of cultivation can change in the bioreactor, mainly the oxygen transfer rate in the medium.

Figure 2 shows the results of sucrose consumption and pH over time of submerged culture of *T. amestolkiae* in bioreactor varying the stirring speed. All values are detailed in Table S1 in the supplementary material.

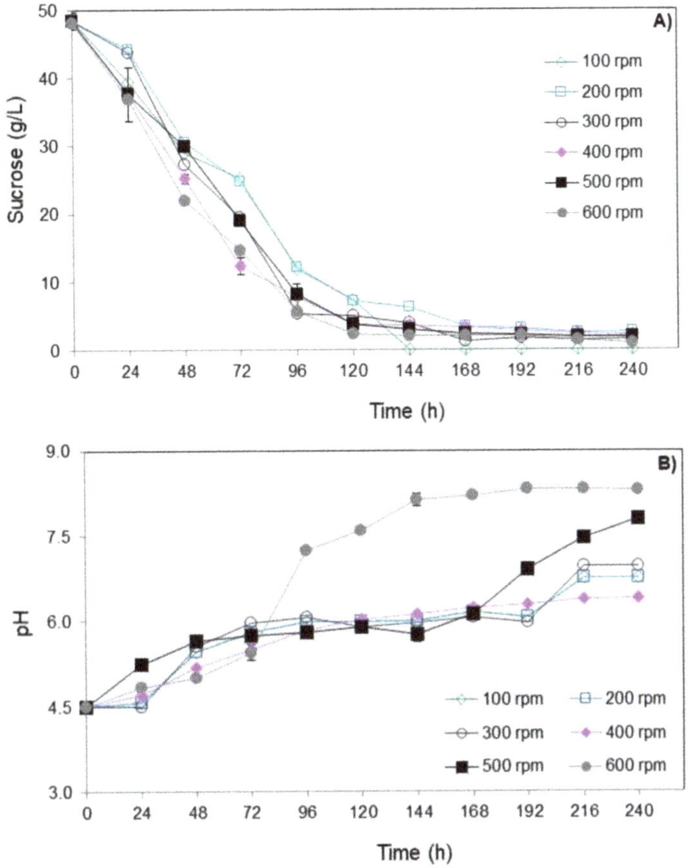

Figure 2. Consumption of sucrose (**A**) and pH (**B**) over time (0 to 240 h) of submerged culture of *T. amestolkiae* in stirred-tank bioreactor varying the stirring speed (100 to 600 rpm) at 2.0 vvm and 30 °C. The error bars represent 95% confidence levels for the measurements.

Regardless of the stirring speed evaluated, the sucrose consumption showed a similar profile (Figure 2A). In general, the concentration of sucrose decreased during the cultivation and a pronounced consumption of sucrose at 96 h occurred. At the end of cultivations, the final sucrose concentration was close to 2 g/L in all conditions studied.

With respect to pH variation, regardless of the stirring speed evaluated, there was an increase in its value throughout the cultivation, with the pH value of 4.5 reaching values close to 7.5 after 240 h of cultivation (Figure 2B). This profile was similar to what occurred in the submerged culture of *T. amestolkiae* (former *Penicillium purpurogenum*) in the orbital shaker incubator studied by Santos-Ebinuma et al. [10]. Therefore, pH 4.5 is suitable to provide good cell growth.

3.2. Production of Natural Colorants by Submerged Culture in Bioreactor: pH-Shift Strategy

The stirring speed of 500 rpm was selected to evaluate the pH-shift control strategy on *T. amestolkiae* colorant accumulation. At 500 rpm, the initial pH of 4.5 changed to approximately 8.0 at the end of cultivation (Figure 2B). According to Orozco and Kilikian (2008) [38], the production of red colorants by the submerged culture of *Monascus purpureus* in a bioreactor can be favored by a pH change. These authors selected a pH of 5.5 for the growth step and pH 8.0 for the production step. Thus, a similar strategy was used in this work in order to increase the colorant production in the bioreactor cultivation. In this way, at 500 rpm of stirring speed, a cultivation was carried out and, after 96 h, the pH was changed to 8.0 by the addition of 5 M NaOH. The time of 96 h was selected according to the experiments carried out that varied the stirring speed, because it was observed that, at this point, there was a marked decrease in the concentration of the primary carbon source (Figure 2B). Figure 3 depicts the production of yellow, orange, and red colorants, as well as the pH for the assays performed at 2.0 vvm, 30 °C, and 500 rpm.

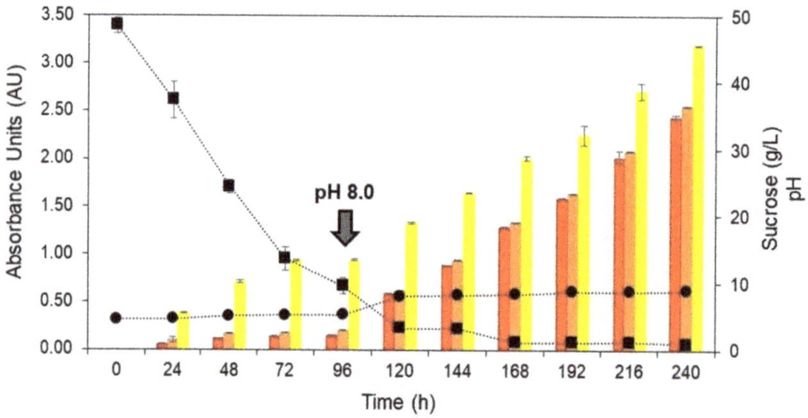

Figure 3. Production of yellow (yellow bars), orange (orange bars), and red colorants (red bars), pH (circle) and sucrose (square) as a function of time (0 to 240 h) during the submerged culture of *T. amestolkiae* in a stirred-tank bioreactor at 2.0 vvm, 30 °C, and 500 rpm. The black arrow indicates the pH-shift from 4.5 to 8.0 at 96 h for the production step. The error bars represent 95% confidence levels for the measurements.

After the pH-shift, it is possible to observe a significant increase in the production of all the colorants, with this increase being more significant for the orange and red ones, both close to 4-fold. Furthermore, the influence of the pH-shift on colorant production was evident. The maximum yield of yellow, orange, and red colorants was 3.20 AU_{400nm}, 2.56 AU_{470nm}, and 2.45 AU_{490nm}, respectively. These results corroborate those found by Orozco and Kilikian (2008) [38], who mention that different pH levels during the growth and production steps may lead to an improvement in the colorant production. Although the colorant production with the pH-shift strategy was greater than for the first set of experiments, the product yield was still lower than for the best condition in the orbital shaker incubator, which shows that further studies are necessary to improve the production of *T. amestolkiae* natural colorants in a stirred-tank bioreactor.

Compared with no pH change condition, all kinetic parameters evaluated improved with the pH-shift control strategy (Table 1). The rate of the substrate conversion to product increased from 0.037 to 0.067 AU·L/g for the yellow colorants and from 0.015 to 0.054 and 0.051 AU·L/g for the orange and red colorants, respectively. Similarly, the productivity of all colorants increased in this second experimental condition.

Generally, in bioreactor cultivation, reproducible kinetic growth of filamentous organisms is difficult to obtain. This phenomenon occurs frequently, due to mycelial aggregation in fermentation broths with long hairy mycelial morphologies that can adhere to surfaces and form a growing biofilm [39]. Many fungal strains grow preferentially on surfaces and will develop thick layers on walls and surfaces in a submerged fermentation process [40]. In the same way, reproducible samples to measure apparent viscosity are difficult to obtain, hence, it was also deemed appropriate to omit the time course for biomass and viscosity.

Table 1. Kinetic parameters calculated for substrate conversion factor (sucrose) in product ($Y_{P/S}$) and productivity during the submerged cultivation of *T. amestolkiae* in a stirred-tank bioreactor for the production of yellow (Y, AU_{400nm}), orange (O, AU_{470nm}), and red colorants (R, AU_{490nm}) without a pH change and after a pH-shift from 4.5 to 8.0 at 96 h of cultivation.

Experimental Condition	$Y_{P/S}$ (AU·L/g)			Productivity (AU/h)			Abs_O/Abs_Y	Abs_O/Abs_R
	Y	O	R	Y	O	R		
Without pH change	0.037	0.015	0.015	0.007	0.007	0.007	0.415	1.025
pH-shift strategy	0.067	0.054	0.051	0.013	0.011	0.010	0.800	1.049

Abs_O/Abs_Y: relationship between the production of orange and yellow colorants; Abs_O/Abs_R: relationship between the production of orange and red colorants.

Even though it is desirable to produce fungal colorants comprised of only one single color component [22], it is known that the fermented broth is composed of yellow, orange, and red colorants [13–15]. However, the careful selection of pH and nitrogen sources is a valid approach to produce extracellular extracts comprised of one predominant color component [14]. It has been widely accepted by the scientific community that orange colorants are the first biosynthetic product and the other colorants derive from the orange ones [41]. In this way, the extracellular colorant composition and concentration is dependent of the orange colorants produced that are transformed into red and yellow colorants under some specific fermentation conditions. For the *Monascus* genus, a low pH (pH 2.5 and 4.0) provides intracellular extracts composed mainly of orange colorants, independently of the nitrogen source employed. At acidic conditions, the secretion of orange colorants into the broth and the reaction with any amino unit for red colorants formation is limited [22]. Meanwhile, a change of pH to levels closer to neutral modify the extent of transformation of orange colorants into red ones [22]. Additionally, for *T. amestolkiae*, the synergistic effect of a low pH and nitrogen source is mainly important in the production and excretion of red colorants. A previous study of our research group presented the production of a glutamic acid-red colorant complex by *T. amestolkiae* in a chemically defined medium with monosodium glutamate (MSG) as the nitrogen source [14]. In the presence of MSG, deep yellow colorants were derived from neutral and basic pH, while deep red colors were derived from acidic pH. The glutamic acid–colorant complex seems to be more water-soluble than those produced in nitrogen complex media (meat extract and meat peptone) by the same strain [13].

In this work, yeast extract was used as complex nitrogen source and the synergistic effect of this nitrogen source with pH favored the increase of the relation between orange colorants and yellow/red ones. This is a reflection of the overall increased rate of substrate to product and productivity. It seems that a close to neutral pH favored the reduction of orange colorants to yellow ones (relation $Abs_O/Abs_Y < 1.0$ duplicated with pH-shifting), while the specific red conversion was limited (relation $Abs_O/Abs_R > 1.0$). Consequently, the extracellular extracts contained mainly yellow colorants. Therefore, extracellular red colorants derivatives were not mainly produced at a high pH. It can be considered that natural colorants can be obtained by the submerged culture of *T. amestolkiae* in a bioreactor, however, the culture conditions can be improved and the production increased by varying mainly the pH during cultivation. In order to study a possible application of the colorants produced, the incorporation of the fermented broth into an alternative food-packaging material based on its oxidative protection effect was evaluated.

3.3. Preparation and Characterization of Cassava Starch-Based Films

Biodegradable films were produced by a formulation of cassava starch, plasticizers, and the fermented broth containing the *T. amestolkiae* colorants produced in stirred-tank bioreactor. The incorporation of natural additives from plants for active packaging has been increasingly applied, such as: coffee-cocoa [42,43], carotenoid and yerba mate extract [44], green tea and palm oil extract [45], among others. However, natural additives produced by biotechnology is an innovative approach. The bio-based films produced were characterized in terms of thickness, water activity, and total solids content. The results are depicted in Table 2.

Table 2. Characterization of biodegradable films in the presence or absence of natural colorants in terms of thickness, water activity (aw), and total solids.

Characterization/Parameters Analyzed	Biodegradable Films Containing Natural Colorants *	Biodegradable Films Without Natural Colorants *
Thickness (mm)	0.147 ± 0.009 [a]	0.163 ± 0.013 [b]
aw (%)	0.612 ± 0.041 [a]	0.604 ± 0.050 [a]
Total solids	89.80 ± 0.540 [a]	88.96 ± 0.500 [a]

* Statistical test: one-way ANOVA (with Welch correction for unequal variances). The results were expressed as the mean ± respective standard deviations. Equal letters in the same line represent equal values to a significance level of 5%.

The physical properties of the films were determined to evaluate the influence of the colorant incorporation, since they could modify the structure of the polymer matrix, e.g., by weakening the inter-chain bonds. The incorporation of natural colorants in the film formulation caused a significant reduction of film thickness, around 10%, in relation to films with no additives. Other works also reported a reduction in thickness of films formulated with cassava starch and natural additives [43,46]. The control of the thickness in the biodegradable films is an important step, since variations in this parameter can affect some films properties, such as mechanical and barrier properties, which can compromise the package performance [45]. Thickness was also the only parameter addressed that was affected significantly.

The variations in the total solid content in the films after colorant incorporation were not significant, with values of 88.96 and 89.80%, respectively. In general, the incorporation of natural additives in cassava starch-based films does not affect significantly the solid content [43].

Additionally, no significant difference (<0.05) for water activity (aw) between the films produced was observed. These results demonstrate that the presence of natural colorants did not change the water activity of the films. Water activity brings information about the amount of free (also referred as unbound or active) water present in a sample [47] being a parameter used in food preservation. Low water activity reduces the availability of water to microorganisms, avoiding undesirable chemical changes for the storage of products [32]. As water migrates from areas of high aw to areas of low aw, it is important to have low values of aw (<0.600 according to Mathlouthi, 2001 [48]) for food product design. As the aw was around 0.6 for both films, they can be considered as food product packages. In this way, the cassava starch-based films were used to pack butter samples.

3.4. Oxidative Stability of the Packaged Butter during Storage

The oxidative stability of the butter packaged in different formulations, namely: cassava starch-based films with colorant additives (CFC), cassava starch-based films without colorant additives (CF), and conventional plastic (CP) was evaluated. Unpacked butter (U) was used as the control. To this purpose, the packed or unpackaged butter was stored for 45 days under accelerated oxidative conditions (64% of relative humidity and 25 °C) and monitored through peroxide content (PC) at 7, 15, 30, and 45 days of storage. Figure S1 from Supplementary Material shows an image of the butter packaged in each formulation evaluated. Figure 4 shows the results of PC for each condition over the 45 days of storage. All values are detailed in Table S2 in Supplementary Material.

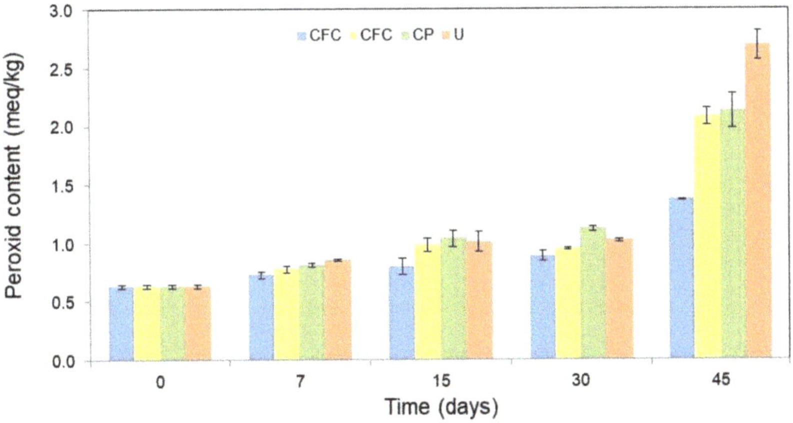

Figure 4. Peroxide content (PC, meq/kg) in the butter as a function of the packaging film used, such as: cassava starch-based films with colorant additives (CFC), cassava starch-based films without colorant additives (CF), conventional plastic (CP), and unpackaged butter (U), during different periods of time (0, 7, 15, 30, and 45 days) in storage. The error bars represent 95% confidence levels for the measurements.

From Figure 4 can be seen that, as expected, the PC of the U was higher than of the packaged one. The CF and CP showed higher peroxide indices than CFC. This demonstrates that natural colorants act as both a photooxidative protector and antioxidant agent, which allow a higher oxidative protection of the packaged product when compared to the other conditions evaluated.

After 45 days of storage, the PC increased 329.0% for the exposed butter, 239.7% for the CP, and 231.1% for the CF. While for the CFC, the PC increased only 118.5%. Therefore, even without antioxidant additives (CF), cassava starch-based films have a greater oxidative protection effect when compared to conventional package films (CP). However, the formulation with colorants incorporated in cassava starch-based films (CFC) enhanced the oxidative protection of butter, with a reduction in peroxide value. This shows a synergistic action of the bio-based film and the colorants, improving the shelf life of fatty products. In this way, *T. amestolkiae* colorants can be considered as potential natural antioxidants for the stabilization of lipid-containing foods for active packing formulation. However, additional studies on the concentration of colorants in the formulation should be performed in order to avoid a possible pro-oxidant effect. Since a high content of antioxidant additives can act as a pro-oxidant agent [44,45].

4. Conclusions

This work evaluated the influence of the stirring speed and pH on the production of natural colorants by *T. amestolkiae* in a stirred-tank bioreactor. The greatest colorant production occurred at 500 rpm, under the pH-shift strategy from 4.5 to 8.0 during the production phase. Although the regulatory mechanisms of the biosynthesis of *T. amestolkiae* colorants are not yet clear, the relationship between the culture conditions and the colorant formation it is of great importance. Despite the difficulties in terms of viscosity of the medium, during the submerged cultivation of filamentous fungi in a bioreactor, there is the possibility of using strategies to control growth and metabolite production, especially in terms of stirring and pH. Moreover, this work demonstrated the potential of natural colorants to be included in biodegradable film formulation. The incorporation of colorants in cassava starch-based films provides oxidative protection in packaged butter, by the decrease of peroxide index. Hence, *T. amestolkiae* colorants have great potential for use as functional food colorants. As future perspectives, our group intends to evaluate the surface morphologies of the films with or without colorants by scanning electron microscope in order to gather supplementary data about the characteristics and stability of our formulations extending, for example, its shelf-life application.

Supplementary Materials: The following are available online at http://www.mdpi.com/2309-608X/6/4/0264/s1, Figure S1: Butter packaged in cassava starch-based films with colorant additives (CFC), cassava starch-based films without colorant additives (CF), conventional plastic (CP), and unpackaged butter (U), Table S1: Consumption of sucrose and pH over time (0 to 240 h) of submerged culture of *T. amestolkiae* in stirred-tank bioreactor varying the stirring speed (100 to 600 rpm) at 2.0 vvm and 30 °C, Table S2: Peroxide content in the butter as a function of the packaging film used, such as: cassava starch-based films with colorant additives (CFC), cassava starch-based films without colorant additives (CF), conventional plastic (CP), and unpackaged butter (U), during different periods of time (0, 7, 15, 30, and 45 days) in storage.

Author Contributions: Conceptualization, F.d.O., C.d.A.L., V.C.S.-E.; methodology, F.d.O., J.I.D., A.P.J., V.C.S.-E.; software, F.d.O., V.C.S.-E.; validation, D.d.A.V.M., A.P.J., V.C.S.-E.; formal analysis, V.C.S.-E.; investigation, F.d.O., V.C.S.-E.; resources, A.P.J., V.C.S.-E.; data curation, A.P.J., V.C.S.-E.; writing—original draft preparation, F.d.O., C.d.A.L., A.M.L., V.C.S.-E.; writing—review and editing, F.d.O., A.M.L., D.d.A.V.M., V.C.S.-E.; visualization, A.P.J., V.C.S.-E.; supervision, J.I.D., A.P.J., V.C.S.-E.; project administration, V.C.S.-E.; funding acquisition, A.P.J., V.C.S.-E. All authors have read and agreed to the published version of the manuscript.

Funding: This work was developed with the financial support from FAPESP (São Paulo Research Foundation, Brazil) through the projects 2018/06908-8, 2014/01580-3, 2013/05039-2, 2008/58280-0. A.M. Lopes acknowledges the financial support from FAPESP through the projects 2017/10789-1 and 2018/10799-0. This study was financed in part by the Coordenação de Aperfeiçoamento de Pessoal de Nível Superior - Brasil (CAPES) - Finance Code 001. The authors also acknowledge the support from the CNPq (National Council for Scientific and Technological Development, Brazil).

Conflicts of Interest: The authors declare no conflict of interest.

References

1. Méndez, A.; Pérez, C.; Montañéz, J.C.; Martínez, G.; Aguilar, C.N. Red pigment production by *Penicillium purpurogenum* GH2 is influenced by pH and temperature. *J. Zhejiang Univ. Sci. B* **2011**, *12*, 961–968. [CrossRef]
2. Torres, F.A.E.; Zaccarim, B.R.; de Novaes, L.C.L.; Jozala, A.F.; Santos, C.A.d.; Teixeira, M.F.S.; Santos-Ebinuma, V.C. Natural colorants from filamentous fungi. *Appl. Microbiol. Biotechnol.* **2016**, *100*, 2511–2521. [CrossRef] [PubMed]
3. Carocho, M.; Barreiro, M.F.; Morales, P.; Ferreira, I.C.F.R. Adding molecules to food, pros and cons: A review on synthetic and natural food additives. *Compr. Rev. Food Sci. Food Saf.* **2014**, *13*, 377–399. [CrossRef]
4. Vendruscolo, F.; Bühler, R.M.M.; de Carvalho, J.C.; de Oliveira, D.; Moritz, D.E.; Schmidell, W.; Ninow, J.L. Monascus: A Reality on the Production and Application of Microbial Pigments. *Appl. Biochem. Biotechnol.* **2016**, *178*, 211–223. [CrossRef] [PubMed]
5. Global Info Research Global Natural Colorant (Natural Pigment) Consumption Market Report (2020). Available online: https://www.360marketupdates.com/enquiry/request-sample/13720643 (accessed on 29 September 2020).
6. Santos-Ebinuma, V.C.; Teixeira, M.F.S.; Pessoa, A. Submerged culture conditions for the production of alternative natural colorants by a new isolated Penicillium purpurogenum DPUA 1275. *J. Microbiol. Biotechnol.* **2013**, *23*, 802–810. [CrossRef] [PubMed]
7. Huang, Z.R.; Zhou, W.B.; Yang, X.L.; Tong, A.J.; Hong, J.L.; Guo, W.L.; Li, T.T.; Jia, R.B.; Pan, Y.Y.; Lin, J.; et al. The regulation mechanisms of soluble starch and glycerol for production of azaphilone pigments in Monascus purpureus FAFU618 as revealed by comparative proteomic and transcriptional analyses. *Food Res. Int.* **2018**, *106*, 626–635. [CrossRef]
8. Morales-Oyervides, L.; Ruiz-Sánchez, J.P.; Oliveira, J.C.; Sousa-Gallagher, M.J.; Méndez-Zavala, A.; Giuffrida, D.; Dufossé, L.; Montañez, J. Biotechnological approaches for the production of natural colorants by Talaromyces/Penicillium: A review. *Biotechnol. Adv.* **2020**, *43*, 107601. [CrossRef]
9. Yang, J.; Chen, Q.; Wang, W.; Hu, J.; Hu, C. Effect of oxygen supply on *Monascus* pigments and citrinin production in submerged fermentation. *J. Biosci. Bioeng.* **2015**, *119*, 564–569. [CrossRef]
10. Santos-Ebinuma, V.C.; Roberto, I.C.; Simas Teixeira, M.F.; Pessoa, A. Improving of red colorants production by a new Penicillium purpurogenum strain in submerged culture and the effect of different parameters in their stability. *Biotechnol. Prog.* **2013**, *29*, 778–785. [CrossRef]

11. Dufossé, L.; Fouillaud, M.; Caro, Y.; Mapari, S.A.S.; Sutthiwong, N. Filamentous fungi are large-scale producers of pigments and colorants for the food industry. *Curr. Opin. Biotechnol.* **2014**, *26*, 56–61. [CrossRef]
12. Tolborg, G.; Ødum, A.S.R.; Isbrandt, T.; Larsen, T.O.; Workman, M. Unique processes yielding pure azaphilones in Talaromyces atroroseus. *Appl. Microbiol. Biotechnol.* **2020**, *104*, 603–613. [CrossRef]
13. Zaccarim, B.R.; de Oliveira, F.; Passarini, M.R.Z.; Duarte, A.W.F.; Sette, L.D.; Jozala, A.F.; Teixeira, M.F.S.; de Santos-Ebinuma, V.C. Sequencing and phylogenetic analyses of *Talaromyces amestolkiae* from amazon: A producer of natural colorants. *Biotechnol. Prog.* **2018**, *35*. [CrossRef] [PubMed]
14. de Oliveira, F.; Pedrolli, D.B.; Teixeira, M.F.S.; de Santos-Ebinuma, V.C. Water-soluble fluorescent red colorant production by *Talaromyces amestolkiae*. *Appl. Microbiol. Biotechnol.* **2019**, *103*, 6529–6541. [CrossRef] [PubMed]
15. de Oliveira, F.; Ferreira, L.C.; Neto, Á.B.; Simas Teixeira, M.F.; de Santos Ebinuma, V.C. Biosynthesis of natural colorant by *Talaromyces amestolkiae*: Mycelium accumulation and colorant formation in incubator shaker and in bioreactor. *Biochem. Eng. J.* **2020**, *161*, 107694. [CrossRef]
16. Corrêia Gomes, D.; Takahashi, J.A. Sequential fungal fermentation-biotransformation process to produce a red pigment from sclerotiorin. *Food Chem.* **2016**, *210*, 355–361. [CrossRef] [PubMed]
17. Lv, J.; Zhang, B.B.; Liu, X.D.; Zhang, C.; Chen, L.; Xu, G.R.; Cheung, P.C.K. Enhanced production of natural yellow pigments from *Monascus purpureus* by liquid culture: The relationship between fermentation conditions and mycelial morphology. *J. Biosci. Bioeng.* **2017**, *124*, 452–458. [CrossRef]
18. Gibbs, P.A.; Seviour, R.J.; Schmid, F. Growth of filamentous fungi in submerged culture: Problems and possible solutions. *Crit. Rev. Biotechnol.* **2000**, *20*, 17–48. [CrossRef]
19. Cairns, T.C.; Zheng, X.; Zheng, P.; Sun, J.; Meyer, V. Moulding the mould: Understanding and reprogramming filamentous fungal growth and morphogenesis for next generation cell factories. *Biotechnol. Biofuels* **2019**, *12*, 1–18. [CrossRef]
20. Veiter, L.; Rajamanickam, V.; Herwig, C. The filamentous fungal pellet—relationship between morphology and productivity. *Appl. Microbiol. Biotechnol.* **2018**, *102*, 2997–3006. [CrossRef]
21. Kang, B.; Zhang, X.; Wu, Z.; Wang, Z.; Park, S. Production of citrinin-free *Monascus* pigments by submerged culture at low pH. *Enzyme Microb. Technol.* **2014**, *55*, 50–57. [CrossRef]
22. Shi, K.; Song, D.; Chen, G.; Pistolozzi, M.; Wu, Z.; Quan, L. Controlling composition and color characteristics of *Monascus* pigments by pH and nitrogen sources in submerged fermentation. *J. Biosci. Bioeng.* **2015**, *120*, 145–154. [CrossRef] [PubMed]
23. Mapari, S.A.S.; Meyer, A.S.; Thrane, U. Photostability of natural orange-red and yellow fungal pigments in liquid food model systems. *J. Agric. Food Chem.* **2009**, *57*, 6253–6261. [CrossRef] [PubMed]
24. Zhu, J.; Grigoriadis, N.P.; Lee, J.P.; Porco, J.A. Synthesis of the azaphilones using copper-mediated enantioselective oxidative dearomatization. *J. Am. Chem. Soc.* **2005**, *127*, 9342–9343. [CrossRef] [PubMed]
25. Mapari, S.A.S.; Thrane, U.; Meyer, A.S. Fungal polyketide azaphilone pigments as future natural food colorants? *Trends Biotechnol.* **2010**, *28*, 300–307. [CrossRef] [PubMed]
26. Keekan, K.K.; Hallur, S.; Modi, P.K.; Shastry, R.P. Antioxidant Activity and Role of Culture Condition in the Optimization of Red Pigment Production by Talaromyces purpureogenus KKP Through Response Surface Methodology. *Curr. Microbiol.* **2020**. [CrossRef] [PubMed]
27. Kurek, M.; Hlupić, L.; Elez Garofulić, I.; Descours, E.; Ščetar, M.; Galić, K. Comparison of protective supports and antioxidative capacity of two bio-based films with revalorised fruit pomaces extracted from blueberry and red grape skin. *Food Packag. Shelf Life* **2019**, *20*. [CrossRef]
28. Cazón, P.; Velazquez, G.; Ramírez, J.A.; Vázquez, M. Polysaccharide-based films and coatings for food packaging: A review. *Food Hydrocoll.* **2017**, *68*, 136–148. [CrossRef]
29. Pillai, S.K.; Maubane, L.; Sinha Ray, S.; Khumalo, V.; Bill, M.; Sivakumar, D. Development of antifungal films based on low-density polyethylene and thyme oil for avocado packaging. *J. Appl. Polym. Sci.* **2016**, *133*, 1–9. [CrossRef]
30. Mellinas, C.; Valdés, A.; Ramos, M.; Burgos, N.; Garrigós, M.d.C.; Jiménez, A. Active edible films: Current state and future trends. *J. Appl. Polym. Sci.* **2016**, *133*, 42631. [CrossRef]
31. DuBois, M.; Gilles, K.A.; Hamilton, J.K.; Rebers, P.A.; Smith, F. Colorimetric Method for Determination of Sugars and Related Substances. *Anal. Chem.* **1956**, *28*, 350–356. [CrossRef]
32. da Silva, J.B.A.; Pereira, F.V.; Druzian, J.I. Cassava Starch-Based Films Plasticized with Sucrose and Inverted Sugar and Reinforced with Cellulose Nanocrystals. *J. Food Sci.* **2012**, *77*, 14–19. [CrossRef]

33. Helrich, K. *Official Methods of Analysis of the AOAC*, 15th ed.; Helrich, K., Ed.; Wilson Boulevard: Arlington, VA, USA, 1990; Volume 1, ISBN 0935584420.
34. Papagianni, M. Fungal morphology and metabolite production in submerged mycelial processes. *Biotechnol. Adv.* **2004**, *22*, 189–259. [CrossRef]
35. Goudar, C.T.; Strevett, K.A.; Shah, S.N. Influence of microbial concentration on the rheology of non-Newtonian fermentation broths. *Appl. Microbiol. Biotechnol.* **1999**, *51*, 310–315. [CrossRef]
36. Smith, J.J.; Lilly, M.D.; Fox, R.I. The effect of agitation on the morphology and penicillin production of Penicillium chrysogenum. *Biotechnol. Bioeng.* **1990**, *35*, 1011–1023. [CrossRef] [PubMed]
37. Žnidaršič, P.; Pavko, A. The Morphology of Filamentous Fungi in Submerged Cultivations as a Bioprocess Parameter. *Food Technol. Biotechnol.* **2001**, *39*, 237–252.
38. Orozco, S.F.B.; Kilikian, B.V. Effect of pH on citrinin and red pigments production by Monascus purpureus CCT3802. *World J. Microbiol. Biotechnol.* **2008**, *24*, 263–268. [CrossRef]
39. Vecht-Lifshitz, S.E.; Magdassi, S.; Braun, S. Pellet formation and cellular aggregation in Streptomyces tendae. *Biotechnol. Bioeng.* **1990**, *35*, 890–896. [CrossRef]
40. Kim, H.J.; Kim, J.H.; Oh, H.J.; Shin, C.S. Morphology control of *Monascus* cells and scale-up of pigment fermentation. *Process. Biochem.* **2002**, *38*, 649–655. [CrossRef]
41. Chen, G.; Huang, T.; Bei, Q.; Tian, X.; Wu, Z. Correlation of pigment production with mycelium morphology in extractive fermentation of Monascus anka GIM 3.592. *Process. Biochem.* **2017**, *58*, 42–50. [CrossRef]
42. Calatayud, M.; López-De-Dicastillo, C.; López-Carballo, G.; Vélez, D.; Muñoz, P.H.; Gavara, R. Active films based on cocoa extract with antioxidant, antimicrobial and biological applications. *Food Chem.* **2013**, *139*, 51–58. [CrossRef] [PubMed]
43. Veiga-Santos, P.; Silva, L.T.; de Souza, C.O.; da Silva, J.R.; Albuquerque, E.C.C.; Druzian, J.I. Coffee-cocoa additives for bio-based antioxidant packaging. *Food Packag. Shelf Life* **2018**, *18*, 37–41. [CrossRef]
44. Moura, L.E.; de Souza, C.O.; de Oliveira, E.A.S.; Lemos, P.V.F.; Druzian, J.I. Bioactive efficacy of low-density polyethylene films with natural additives. *J. Appl. Polym. Sci.* **2018**, *135*. [CrossRef]
45. Perazzo, K.K.N.C.L.; De Conceição, A.C.V.; Dos Santos, J.C.P.; De Assis, D.J.; Souza, C.O.; Druzian, J.I. Properties and antioxidant action of actives cassava starch films incorporated with green tea and palm oil extracts. *PLoS ONE* **2014**, *9*, 1–13. [CrossRef] [PubMed]
46. De Souza, C.O.; Silva, L.T.; Druzian, J.I. Estudo comparativo da caracterização de filmes biodegradáveis de amido de mandioca contendo polpas de manga e de acerola. *Quim. Nova* **2012**, *35*, 262–267. [CrossRef]
47. Lewicki, P.P. Water as the determinant of food engineering properties. A review. *J. Food Eng.* **2004**, *61*, 483–495. [CrossRef]
48. Mathlouthi, M. Water content, water activity, water structure and the stability of foodstuffs. *Food Control.* **2001**, *12*, 409–417. [CrossRef]

Journal of
Fungi

MDPI

Article

Oil-Based Fungal Pigment from *Scytalidium cuboideum* as a Textile Dye

Mardonio E. Palomino Agurto [1], Sarath M. Vega Gutierrez [1], R. C. Van Court [1], Hsiou-Lien Chen [2] and Seri C. Robinson [1,*]

1 Department of Wood Science and Engineering, Oregon State University, Corvallis, OR 97331, USA; mardonio.palomino@oregonstate.edu (M.E.P.A.); sarathth@yahoo.co.uk (S.M.V.G.); ray.vancourt@oregonstate.edu (R.C.V.C.)

2 College of Business, Oregon State University, Corvallis, OR 97331, USA; hsiou-lien.chen@oregonstate.edu

* Correspondence: seri.robinson@oregonstate.edu; Tel.: +1-541-737-4233

Received: 31 March 2020; Accepted: 20 April 2020; Published: 22 April 2020

Abstract: Identification of effective natural dyes with the potential for low environmental impact has been a recent focus of the textile industry. Pigments derived from spalting fungi have previously shown promise as textile dyes; however, their use has required numerous organic solvents with human health implications. This research explored the possibility of using linseed oil as a carrier for the pigment from *Scytalidium cuboideum* as a textile dye. Colored linseed oil effectively dyed a range of fabrics, with natural fibers showing better coloration. Scanning electron microscopy (SEM) revealed a pigment film over the fabric surface. While mechanical testing showed no strength loss in treated fabric, colorfastness tests showed significant changes in color in response to laundering and bleach exposure with variable effects across fabric varieties. SEM investigation confirmed differences in pigmented oil layer loss and showed variation in pigment crystal formation between fabric varieties. Heating of the pigmented oil layer was found to result in a bright, shiny fabric surface, which may have potential for naturally weatherproof garments.

Keywords: fungal pigment; natural dye; spalting; *Scytalidium cuboideum*; dramada; sustainable clothing

1. Introduction

Natural dyes and pigments have a long history of use for coloring textiles, from ancient Egypt to the oldest South American cultures [1–5]. Artificial dyes currently dominate the market due to their ease of mass production, low price, and color variety [5]. However, most modern textile colorants are produced using hazardous chemicals and many contribute to water pollution through the production of effluents [6,7]. In recent decades, the low environmental impact and sustainability of natural dyes have become more desired by consumers, driving a shift in the market back to natural sources of coloration [8,9].

There are presently a number of natural colorant alternatives available, such as those derived from barks, fungal pigments, insects, and minerals [3]. Most have never been commercialized due to concerns about their sustainability, cost effectiveness, and profitability [10]. In addition, natural dyes often have problems with colorfastness, and, while methods have been developed to improve dye uptake [11–13], they are energy intensive and increase the production price [14–16].

One source of natural coloration that has been suggested as a replacement for synthetic dyes in fabric dyeing are pigments extracted from wood-rotting fungi. Research by Weber et al. [17] has shown that fungal pigments, when carried in dichloromethane (DCM), with and without mordants, showed promise for dyeing fabrics, using a dripping process. The pigments from four fungi were investigated: *Chlorociboria aeruginosa* (Oeder), Seaver, and *C. aeruginascens* (Nyl.) Kanouse, which produces a blue-green pigment known as xylindein, *Scytalidium cuboideum* (Sacc. & Ellis) Sigler & Kang, which produces a red pigment known as draconin red (Figure 1) that exists as both orange and red crystals, and *Scytalidium ganodermophthorum* Kang, Sigler, Y.W. Lee & S.H. Yun, which produces an unnamed yellow pigment. Further studies showed that polyester absorbed more pigment than other synthetic fabrics and that these colors were stable over time [17,18]. Hinsch [19] later confirmed that *S. cuboideum* and *C. aeruginosa* could be used to dye textiles with no mordant and attributed this to the carbon–carbon interaction of the main quinone structures.

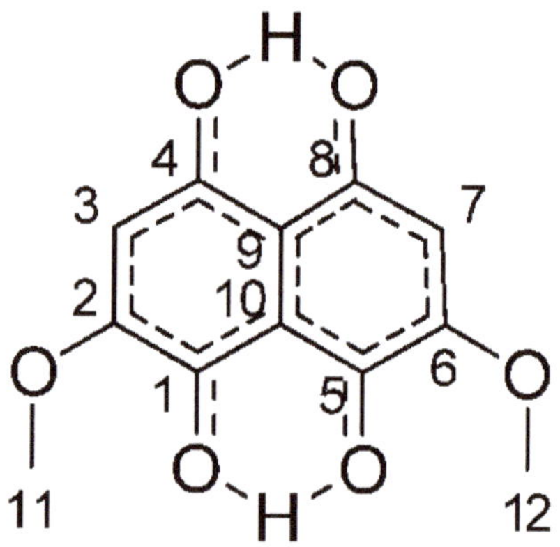

Figure 1. The structure of dramada, the orange and red crystal pigment from *Scytalidium cuboideum*.

Despite the success of early testing, the fungal pigments mentioned above never gained commercial traction. This is likely due to the need for the pigments to be carried in DCM, which is a potential human carcinogen and a known greenhouse gas [20–22]. Other solvents were tested for their ability to extract the pigments, and while tetrahydrofuran (THF), acetonitrile (ACN), and acetone were all moderately capable of extraction, all three interacted with the pigments and caused color change [23]. Due to these issues, an entirely new type of solvent was explored: natural oils. These oils proved to be highly successful at carrying the fungal pigments, though not at extracting them, with raw linseed oil allowing for the highest pigment stability of those tested [24]. However, of the tested pigments, only the red pigment from *S. cuboideum* was found to effectively color fabric, although it showed variation in coloration based on the material to which it was applied [25].

The purpose of study was to compare the dyeing capability, heat stability (in different washing temperatures), and pH stability (washing in different detergents), of three soft-rot spalting fungi, carried in raw linseed oil, on various fabrics. While it is already known that the pigments from these fungi are reasonable textile dyes when carried in DCM [17,18], this carrier is not practical for commercial scale use due to toxicity issues. An alternative carrier proposed is raw linseed oil, which is known to carry the pigments at high concentrations but has never been tested in textile applications. Using the oil carrier would result in a dyeing process with fewer environmental issues than the DCM process. Analysis via scanning electron microscopy (SEM) will further the understanding of the interaction between the fungal pigments and the fabrics. The research will assess if the oil changes the structure of the pigments, how the pigments bind to textiles, and if the reaction of the pigments changes due to temperature and pH changes based upon the oil carrier.

This work prepares the fungal pigments to fully enter into the commercial dye industry by assessing the viability of a more eco-friendly carrier and removing the largest hurdle currently facing the soft-rot fungal pigments, their DCM requirement for application.

2. Materials and Methods

All testing methodology followed either accepted the American Society of Testing Material (ASTM) standards (such as for fabric testing) or were based on testing used in previous work with these spalting pigments on textiles (see citations below).

2.1. Fungal Growth and Pigment Extraction

Cultures of *S. cuboideum* (UAMH 4802, isolated from oak lumber, location unknown) were used to inoculate malt agar plates amended with white rotted wood chips, following the procedure outlined in Robinson et al. [23]. Plates were allowed to grow for 2–3 weeks before they were dried for 24 h in a fume hood, extracted into DCM, and standardized to CIE $L^*a^*b^*$ values of $L = 82.32$, $a = 26.84$, and $b = 13.19$, as described in Robinson et al. [23]. After the standardization, raw linseed oil (Sunnyside) was mixed with the pigment carried in DCM and placed on a stir plate for 48 h in order to evaporate the solvent and leave the pigment suspended in the oil, following the methodology by Palomino Agurto et al. [25]. This pigmented oil was used for further testing.

2.2. Fabrics Tested

Unfinished 100% fabrics of polyester, nylon, cotton, and wool were used (see Table S1). Per each type of fabric, two different densities were used: low-density and high-density. To establish a difference between low-density and high-density, a minimum difference of 40 yarns per inch in the count number was established. Low-density nylon was discarded because its finishing process did not allow dying by the pigmented oil.

2.3. Mechanical Testing

Mechanical testing was done to evaluate the impact of the oil-pigments on tensile and tear strength of the fabrics, with samples for mechanical testing prepared according to the American Society of Testing Material (ASTM 2013). Submersion dying was carried out using borosilicate glass beakers (250 mL) (brand VWR, Randor, PA, USA), filled with 100 mL of pigmented oil from *S. cuboideum*. Samples from tested fabrics were separately submerged in the pigment solution in the uncovered beakers for 1 h. Once the time of exposure was achieved, samples were removed from the borosilicate beakers and placed over a metal mesh to dry for 48 h.

2.3.1. Tear Strength Test

The tear strength test was performed according to the ASTM D1424 using a falling-pendulum digital Elmendorf-type (produced by SDL ATLAS, Rock Hills, SC, USA) to measure the force required to propagate a single-rip tear. This test was performed on the samples of the four fabrics in filling directions. Twelve repetitions were performed on both dyed and undyed control samples (treated with only raw linseed oil), making a total of 36 samples per fiber fabric and 144 samples in total.

Estimated least squares means tests with Tukey adjustment was performed using SAS 9.8 (SAS Institute, Cary, NC, USA) to compare the effect of treatment and fabric type on stress (MPa) or force (N). The least squares means for main effects were reported if the main effect was statistically significant and not part of any statistically significant interaction or when the main effect was statistically significant and the effect of that factor was always in the same direction for the interacting variable(s). Selected samples were later evaluated using SEM analysis (below) to assess the behavior of pigment crystals inside the fabric structures.

2.3.2. Tensile Strength Test

The breaking (tensile strength) test was performed according to the ASTM D5034-09 (2013), using the Universal Testing Machine Instron Model 5582 (produced by Instron Company, Norwood, MA, USA). Samples were loaded into the tool and then exposed to tensile forces until breaking occurred.

The test was repeated for 30 samples for each fabric/density type, with 10 controls, 10 samples dyed with only raw linseed oil, and 10 samples dyed with pigmented oil. After the tensile strength test, affected samples were analyzed using SEM to evaluate any physical change in the attachment of the pigment to the fibers.

Estimated least squares means tests with Tukey adjustment were performed using SAS 9.8 (SAS Institute, Cary) to compare the effect of treatment and fabric type on f (N), and least squares means for main effects were reported if the main effect was statistically significant and not part of any statistically significant interaction or when the main effect was statistically significant and the effect of that factor was always in the same direction for the interacting variable(s). Selected samples were later evaluated using SEM analysis (below) to assess the behavior of pigment crystals inside the fabric structures.

2.4. *Analysis of Colorfastness*

2.4.1. Color Variation Across Fabrics

Difference in pigment uptake and coloration were compared between fabrics. Fabric samples were cut into 5.08 × 5.08 cm, and dripping was used to apply fungal pigment. This consisted of applying pigment suspended in raw linseed oil onto the surface of each fabric sample using a disposable pipette. Fifteen drops of the oil-solubilized pigments were applied onto each fabric sample using a 1 mL pipette (brand Gilson, Lewis Center, OH, USA), with drops weighing an average 0.02443 g.

Control samples were compared to samples colored using the dripping methodology, and color analysis was performed on the samples using a Konica Minolta Chroma Meter CR-5 colorimeter, using the CIE $L^*a^*b^*$ color space, with the ΔE calculation 2000. A one-way ANOVA was performed to test for the effect of fabric type on the response of color variation (ΔE) using SAS 9.8 (SAS Institute, Cary).

2.4.2. Laundry Test

Colorfastness to laundering with and without bleach was performed on the fabric samples dyed with the dripping method. The washing test was performed according to the AATCC method 61-2013 (Colorfastness to Laundering: Accelerated Using a Launder-O-Meter) in a Launderometer model LEF (produced by SDL ATLAS). Twelve 5.08 × 5.08 cm samples of each tested fabric type were prepared using dripping methodology (above), with six replicated exposed to bleach in testing and six without exposure. Samples were placed in a canister containing 50 stainless steel balls and AATCC standard detergent solution (with or without bleach). The sealed canisters were placed in the launder-O-meter with hot water (49 °C) for 45 min. When the process finished, samples inside the canister were rinsed and air-dried for 48 h. The color was measured in the Konica Minolta Chroma Meter CR-5, utilizing the CIE $L^*a^*b^*$ color space. A general linear model was used to analyze the interactions between the independent variables of laundry, fabric type, heat, and time of heating with a four-way ANOVA and Tukey–Kramer test ($p < 0.05$) to determine significant differences between groups. One sample for each fabric/treatment was randomly selected and was taken to be analyzed with SEM (below) to visualize any change from the control samples. The response of prepared pigmented fabrics to heating was then compared, with samples previously subjected to heat testing.

2.4.3. Heat Testing

Testing samples were prepared following methods described for the laundry test (using the same number of samples), then heated to one of three common dryer temperatures (low = 50 °C, medium = 65 °C, and high = 80 °C) in a forced air oven to simulate a home dryer. Initial color readings were taken after samples dried using a Konica Minolta Chroma Meter CR-5 colorimeter using the CIE $L^*a^*b^*$ color space, using the ΔE calculation 2000. Color change was then again measured after 30 and 60 min exposure to heat, and samples were removed with tweezers and transported in baggies to prevent dirtying. Once the color was measured, two samples were randomly selected from each fabric/treatment and analyzed via SEM (below) to visualize any change that occurred due to the heat. A repeated measurement ANOVA with a mixed model was performed to determine the difference in ΔE and any interactions between the independent variables fabric, time, heat, and treatment, and a Tukey Kramer test was performed to classify possible interactions.

2.4.4. Qualitative Analysis: SEM

The purpose of SEM was to visualize the physical interaction between the pigment when carried in raw linseed oil and the fabrics and how this varied between treatment groups and the control. All tested samples (noted above) were dried for 48 h at room temperature, then were mounted on aluminum studs 5 mm in diameter using carbon adhesive tape. Samples were coated following Vega Gutierrez (2016), with gold-palladium in a Cressington Sputter Coater 108 Auto (Cressington Scientific Instruments, Inc, Cranberry Twp, PA, USA) for 35 s to allow the samples to generate a coating that decreased the electron charging and increased the contrast on the sample images. Samples were analyzed with a FEI QUANTA 600F environmental SEM (FEI Co., Hillsboro, OR, USA). The electron spot used was 2 A and 2 kV (high voltage).

3. Results

3.1. *Influence of Fabric Type on Coloration of Samples*

The color intensity of dyed fabrics was found to vary significantly (df = 619, F = 150.15, $p < 0.001$) depending on the type of fabric, with Tukey groupings showing that low-density cotton had the most pigmentation ($\Delta E = 35.58 \pm 1.73$), followed by high-density cotton (33.21 ± 3.54), and low-density polyester (27.61 ± 4.22). Nylon, high-density polyester, high-density wool, and low-density wool were not found to differ significantly and showed lower coloration overall.

3.2. *Mechanical Testing*

3.2.1. Tensile Strength

No significant difference was seen in tensile strength between untreated fabric samples, fabric samples treated with linseed oil alone, and samples treated with pigmented oil (df = 2, F = 1.75, $p = 0.1806$). The fabric type was found to significantly affect the sample tensile strength (df = 6, F = 114.55, $p < 0.0001$), though no interaction was found between the interaction of fabric type and treatment, which would have indicated differences in strength changes for one or more of the fabrics due to the pigment (df = 12, F = 0.25, $p = 0.9944$).

SEM analysis showed differences in fiber morphology after tensile testing between tested fabric varieties. High-density cotton showed flattened or curvy twisted fibers close to the break area, with hook-like ends at the ends of fibers (Figure S1). Low density cotton also showed curvy, twisted fibers close to break point, with split fiber ends (Figure S2). Nylon showed flattening at the area of breakage and wide separation of fibers, with oil accumulation visible on their surface (Figure S3). High-density polyester showed uneven broken fiber ends and minor flattening of broken ends (Figure S4). Low-density polyester in contrast showed an irregular twisted surface and wavy fibers in the failure area, with oil accumulation evident (Figure S5). High-density wool showed flattened and elongated fibers with a sharp angle on failure sites, showcasing the separation of the cuticle from the fiber core (Figure S6). Low-density wool showed similar characteristics to high-density wool, with clear damage from the test (Figure S7).

3.2.2. Tear Strength

The tearing strength test showed a significant interaction between fabric type and treatment (df = 12, F = 9.76, $p < 0.0001$), with low density cotton showing a significant difference between untreated samples and samples treated with pigmented or non-pigmented linseed oil. Control samples tore at a significantly lower strength of 912.58 N compared to pigmented oil samples. Tukey groupings showed that the low-density wool with oil had the highest tear strength performance (5068.36 N), while high-density cotton treated with oil had the lowest performance (649.67 N) in this test.

High-density cotton showed clear flattened, broken ends with accumulated oil at the ends (Figure S8), while low-density cotton showed twisted sharp ends and wrinkled fibril bundles (Figure S9). High-density wool showed an even shape in the failure area with no loss of cuticle integrity, though a flattened core was present (Figure S10). Low-density wool showed cuticle loss and irregular surfaces, with breaking areas showing visible structural damage (Figure S11). Nylon showed un-flattened fibers with occasional bracket-like structures with accumulation of oil close to the break area on fiber ends, though most sections did not show this oil clumping (Figure S12). High-density polyester showed angled edges and flattened fibers, with oil accumulation and waviness seen in broken fiber ends (Figure S13). These was also seen in low-density polyester, which also showed twisting in the fibers (Figure S14).

3.3. *Laundry and Heat Testing*

Color change in response to laundry and heat processes was found to vary in response to testing conditions. The four-way ANOVA interaction between fabric type, laundry type (bleach/no bleach), temperature, and time was not found to be significant, nor were any three-way interactions. Significant two-way interactions included those between heating time and temperature (df = 4, f = 9.07, p = 0.0001), laundry type and temperature (df = 4, f = 7.66, p = 0.0005), heating time and temperature (df = 2, f = 9.07, p = 0.001), fabric type and temperature (df = 9, f = 1.88, p = 0.0349), and fabric type and laundry treatment (df = 8, f = 24.67, p <0.0001). Overall, these significant interactions suggested that fabrics responded differently to pigmentation, with bleach leading to discoloration and increased exposure to heat (either over time or at higher temperatures) leading to more intense color. Drying temperature seemed especially important, as it was present in all significant interactions, although its effects were mediated by other variables.

In general, the addition of bleach resulted in less color difference between pigmented samples and controls, although for low-density cotton, the addition of bleach resulted in a significantly higher ΔE (more color change). Polyester and nylon showed significant degradation of the pigments after laundry treatment, both with and without bleaching, as can be seen in Figure 1. However, in this case, the ΔE value does not tell the whole story. ΔE measures the total change in color, and, while in most cases in this study this change reflects the intensity of the red coloration, when bleach was applied on a number of fabrics, the hue of the pigmentation changed. In wool of both densities, the color changed from red to blue, and in cotton, the color changed from red to dark purple (Figure 2). Overall, bleach caused a more visible color change on natural fabrics over synthetic fibers. Synthetic fabrics followed a different pattern, presenting less color change after washing and drying.

In addition to the changes in color, as measured by ΔE, and the variation of this effect upon different fabrics, bleach seemed to have a profound effect on the applied pigment in oil, as was qualitatively seen in SEM. While fabrics did not appear degraded in SEM, pigments on samples showed evidence of breakdown. Figure 1 shows pigments looking different in the bleached sample compared to the unbleached sample on the same fabrics at the same drying temperature and time. The analysis of SEM images identified the loss of the pigment layer produced by the interaction of the fabric with the oil. As can be seen in Figure 3, the degradation of oil layers occurred regardless of whether bleach was used during laundry tests. When bleach was used, it severely altered the dye color. All cases showed a strong contrast in comparison with the initial coloration and/or a dramatic loss of the dyed area.

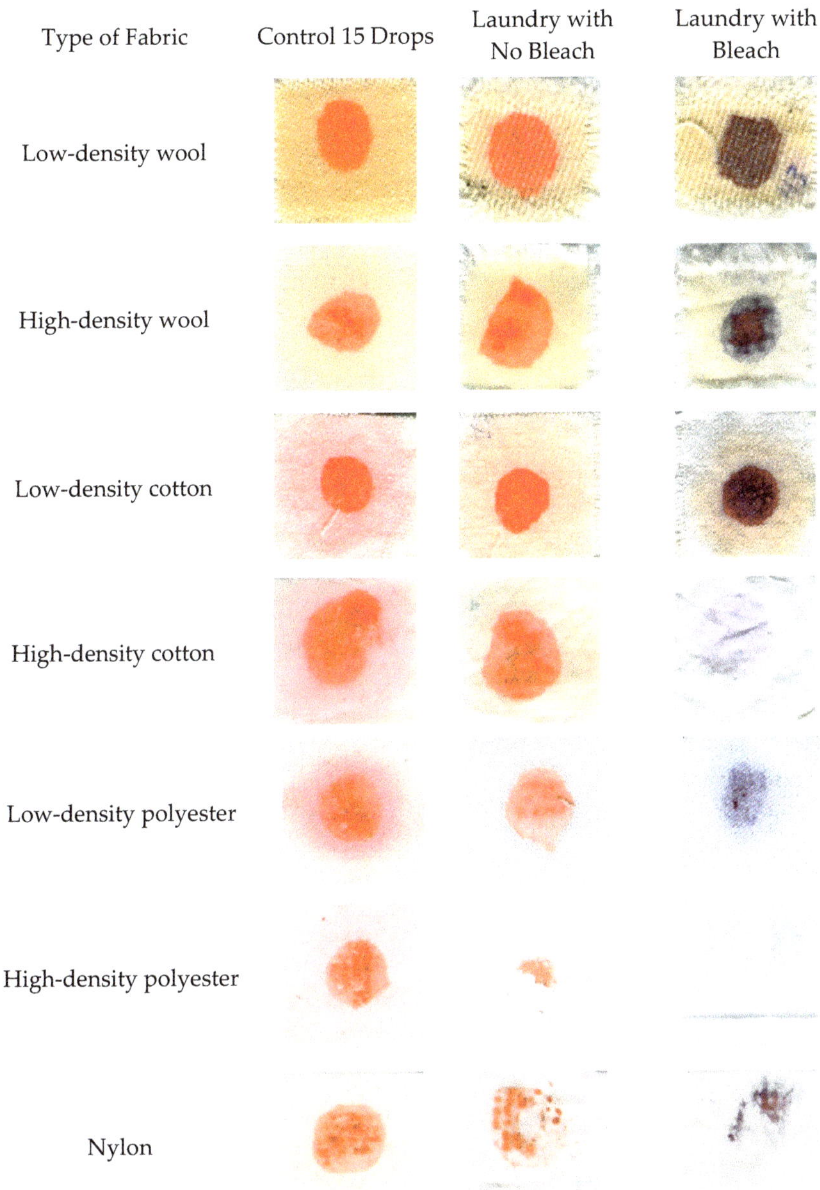

Figure 2. Differences in color and saturation between various fibers treated with 15 drops of dramada-pigmented oil across laundry without bleach, laundry with bleach, and the control (unwashed samples).

(a) (b) (c)

Figure 3. SEM images of a bleach/no bleach sample at the same temperature, showing degradation. (**a**) Low density polyester (control 15 drops); (**b**) low density polyester laundry without bleach; (**c**) low density polyester laundry with bleach.

No dramada crystals were seen in cotton fabrics, with high density cotton showing a smooth covering of oil with no presence of crystals, which became porous after laundry treatment without bleach, and, in the presence of bleach, no oil was found on the samples (Figure S15). Low density cotton showed a less smooth surface of applied pigment with fissures on the outer layer, likely due to retention in the most superficial layer (Figure S16b). After washing with and without bleach, low density cotton showed retention of a thick layer of pigmented oil with an irregular porous shape (Figure S16). Nylon also showed no crystals in the pigmented oil layer, with an irregular surface that was actually smoother after laundry without bleach. In contrast, laundry with bleach resulted in stripping of much of the oil layer remaining only in spaces between yarns (Figure S17).

High density wool also did not form visible crystals within the pigment layer, which formed a thick, even layer of oil that remained after laundry without bleach. However, laundry with bleach showed visible interstices and a porous layer (Figure S18). The low-density wool allowed for a higher degree of infiltration of the oil into the lower layer of the fabric than the high-density wool. Low density wool also showed a thick pigment layer covering the fibers, though interactions with inner layers of the wood fibers made it difficult to distinguish if a granular surface seen where pigment was directly applied was due to pigment/wool interaction or not. After laundry without bleach, the oil absorbance of the cotton became higher than wool or polyester, as seen through the interstices in the oil layer (indicating a dryer oil coat) (Figure S19c), with samples exposed to bleach also showing a reduction in oil and flaking along a porous surface.

In contrast to other fabrics, polyester showed high concentrations of dramada crystals in below oil layers. High-density polyester showed a somewhat more even oil distribution, with a smoother surface seen after laundry with bleach and a lack of visible coating after laundry with bleach (Figure 4). Low density polyester also showed a reduced oil layer after laundry, along with several facture points. Oil also seemed to be degraded by detergent and bleach, forming cube-like structures between fibers and an overall porous surface (Figure 5). Low density polyester also showed a reduced oil layer after laundry, along with several facture points (Figure S20).

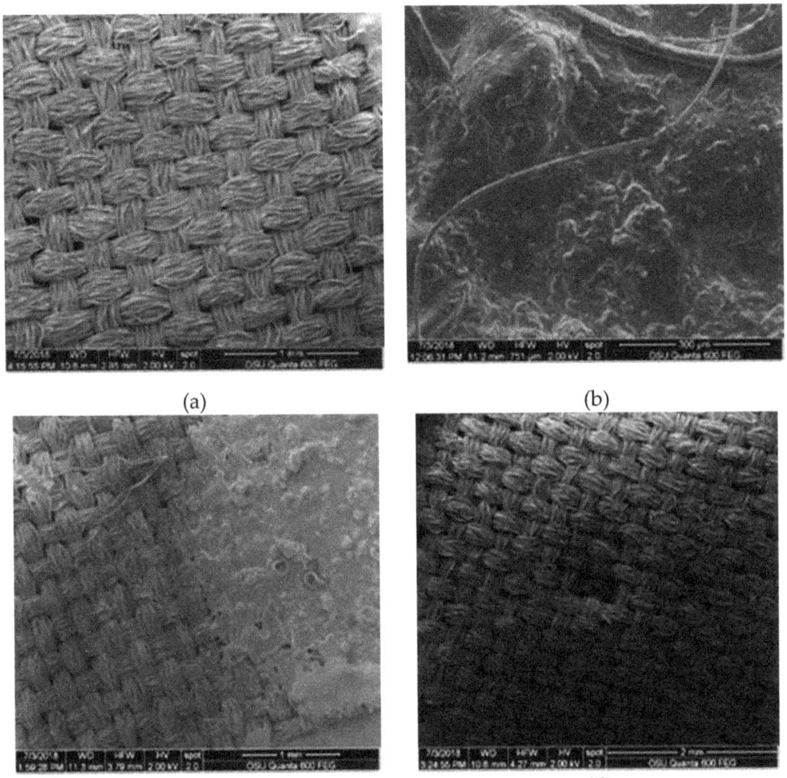

(c) (d)

Figure 4. SEM images of high-density polyester samples: (**a**) control; (**b**) sample treated with 15 drops of pigmented oil, showing presence of dramada crystals; (**c**) pigmented sample after laundry treatment with no bleach, showing a somewhat smoothed surface; (**d**) pigmented sample after laundry treatment with bleach showing no visible oil layer.

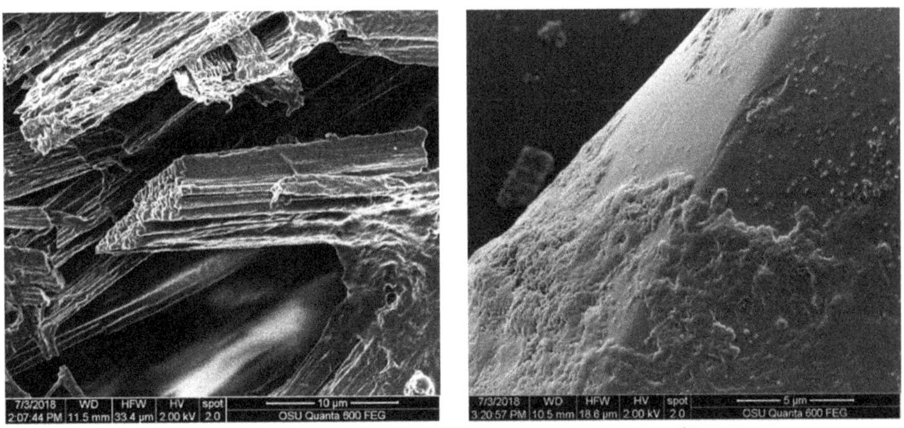

(a) (b)

Figure 5. Distinctive features seen in polyester. (**a**) High-density polyester sample with a dramada crystal; (**b**) SEM image showing low-density polyester sample after the laundry test with bleach.

3.4. Heat Testing

Significant variations of ΔE were seen between pigment treated fabrics exposed to drying conditions at either 50, 65, or 80 °C conditions for either 30 or 60 min. The four-way interaction between all these variables was not statistically significant (df = 12, F = 1.47, p = 0.1366), though two three-way interactions were found to be statistically significant. Fabric type and heating were found to influence color change after heating (df = 2, F = 7.41, p = 0.0008), with all fabrics having higher ΔE value after the heat treatment. In addition, the interaction between time, temperature, and treatment was significant (df = 12, F = 3.21, p = 0.0003), with an increase in ΔE after the heat process; a difference that can be seen in Figure 6. Tukey groups showed that high-density cotton and low-density polyester showed the highest (ΔE), while nylon showed the lowest (ΔE).

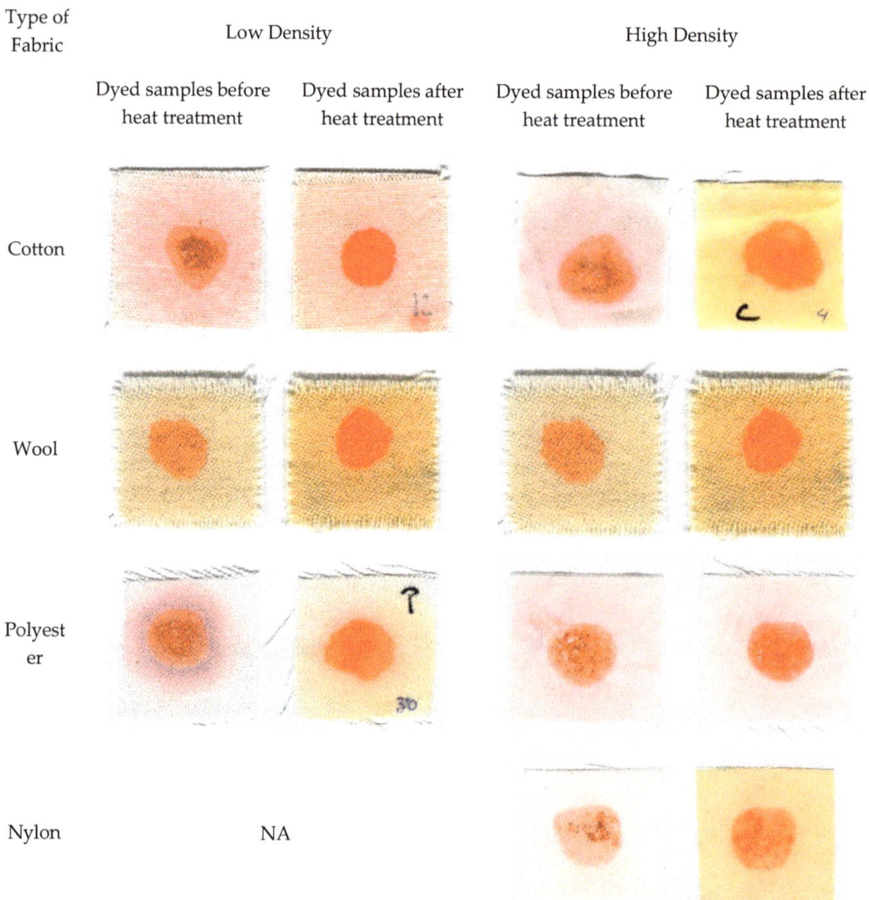

Figure 6. Fabrics dyed with pigmented oil before and after the heat treatment (30 min, 50 °C).

4. Discussion

The red pigment produced by *Scytalidium cuboideum*, carried in oil, was found to successfully dye a variety of fabric types without impact on tested mechanical properties, though wide variation in response to colorfastness treatments was seen. The color of the oil pigment varied based on fabric. These differences could be seen in the presence and morphology of the pigmented oil layer under SEM.

The presence of the pigment in oil layer did not affect material strength or tensile strength. As natural dyes have historically been associated with weakening of fabric and decreases in tear strength due their need to use mordants for effective dying [26–28], this is a significant benefit for use of spalting fungal pigments.

Pigmented low density cotton showed a significant difference in tear strength in comparison to untreated samples, but no significance was seen between pigmented oil and non-pigmented oil. This suggests that any differences might be due to the absorption of oil, which could have influenced slippage, leading to tearing. It is also possible that cellulose functional groups may have interacted with pigmented oil and formed a stronger bond; however, this would require further investigation. The fact that significance was only seen in this fabric type further suggests that this effect was not due to the pigment presence and instead suggests that factors, such as fabric thickness and diffusion, may influence dye absorption. Additionally, SEM showed that the pigmented oil did not stay in the internal structures of the high-density cotton but it did stay in the internal structures of low-density cotton, which had more space to absorb it. Both tensile strength and tear strength testing results mirrored those found by Hinsch [19], who reported no significant differences in strength values after application of *S. cuboideum* pigment in the DCM carrier.

While pigment presence was not seen to influence mechanical testing, clear differences in coloration of applied pigment were seen between fabric varieties with both cotton varieties and low-density polyester, showing significantly higher coloration than other fabric types ($p < 0.001$). It is likely that structural differences between fabric types are responsible for these differences in coloration. The more complex shapes created by natural fibers presented a different surface, leading to different interactions with the pigmented oil. A deep view with SEM showed that polyester fibers seemed to have the dye attached mostly in the outer areas of the sample. Polyester fabrics are normally hydrophobic materials; therefore, their affinity for oil-based compounds was supposed to be higher than that of cotton and wool (which are normally lipophobic materials). Previous research [1,19] did not show any relationship between this feature of the fabrics and their ability to be dyed; however, both studies found that polyester had a general affinity for natural dyes. This relationship still needs to be explored with other types of polyester and fabric blends.

While differences were seen between fabric varieties, density differences did not result in consistent patterns of pigment behavior. While it was speculated that the crystal structure contained in pigment from *S. cuboideum* could be entrapped differentially in fabrics of varying densities, SEM investigations showed that fabric density was also not related to crystal presence, with pigmented oil staying in the fibers without forming crystals. In most cases, a thin layer of oil formed on the surface and a marked accumulation was found in the intersections of each yarn. More crystals may have formed with time, as crystals from *S. cuboideum* have been shown to appear in oil after 60 days, potentially due to oxidation processes [29].

The natural fibers retained coloration better than the synthetics, with wool fabrics, in particular, showing stable color after laundering. This has been seen with other fungal dyes (although not from spalting fungi), specifically with *Talaromyces australis* (C.M. Visagie, N. Yilmaz & J.C. Frisvad) and *Penicillium murcianum* (C. Ramírez & A.T. Martínez) [30]. In most cases, laundry with detergent alone was sufficient to remove the oil layer from the applied pigment. It is likely that the detergent used in the laundry test affected the triester groups of the linseed oil through a complex formation with available -OH^- groups. However, complete removal was not seen on wool, and cotton samples though the coloration persisted, suggesting that the oil may influence pigment fixation, creating a mordant-like effect that has been described in other studies [31].

The pigmented oil layer degraded on exposure to bleach, likely through a saponification reaction. The strong alkalinity of bleach is also likely responsible for the color change seen in laundry samples. The pigment from *S. cuboideum* is known to turn blue under basic conditions, instead of its normal red coloration [32,33]. The mechanism behind this color change is under investigation but has not yet been described. It is possible that the variation in response to bleach seen may have been due to physical protection of pigment-functional groups pigment by the oil layer or by pigment binding to fabric surfaces.

While bonding mechanisms of fungal pigment attachment to different fabrics were not tested, these likely impacted the differential responses seen in colorfastness. It is possible that hydroxy groups in cotton, and potentially amino acids in wool, may form hydrogen bonds with the methoxy and hydroxy groups present in the fungal pigment dramada. For the polyester fabric, it is likely that the aromatic π-electron system of the phthalic unit may be interacting with the π-electron system of dramada, providing a strong bond between the fabric and the pigmented oil. This interaction has already been reported with an anthraquinone present in natural dyes [1]. It is also interesting that. in previous research studies, polyester fabrics were successfully dyed with *S. cuboideum* when using DCM as a carrier [18] and later with raw linseed oil [25]. However, there is a need to further investigate which interaction causes the bonding between the fabric and the dye.

Heating resulted in changes of fabric color, as it was present in all statistically significant interactions identified, with all fabrics brightening except nylon, which may have been due to low initial pigment uptake. Under SEM, a superficial layer could be seen on the dyed areas after heating. This layer was also seen in samples subjected to laundry testing and it appeared to be made of oil: it presented as a thin, compacted, smooth surface, and, in some cases, it showed interstices or pores. This layer was best seen in high-density wool and low-density cotton, which showed the best performance, as measured by ΔE after the heat test. The layer may have been formed due to evaporation, oxidation, or polymerization of the oil under the effect of the temperatures. The polymerization process of linseed oil is well known [34] and this process may have allowed for molecular stacking of the pigment into its crystal formation. Using high temperatures to accelerate the evaporation of the raw linseed oil has been previously reported in the preparation of oilcloths as a way to reduce drying time [35], although no oil layer, such as the one found in this research, has ever been reported. It is also notable that previous research has shown that temperatures higher than 70 °C have been shown to degrade the red color of dramada [29], although this effect was not seen during this research. It is possible that the oil provided some protection for the pigment.

While this research showed that a range of fabrics could be effectively dyed using fungal pigment from *Scytalidium cuboideum*, this method of dying presented colorfastness challenges, especially for synthetic fibers. In addition, the texture of completed cloth may be an obstacle, as the oil does not polymerize to a dry layer quickly, making samples oily. However, treatment of cotton with linseed oil to create a waterproof layer, known as an oilcloth, was commonly done in the past [36] and may be a sustainable waterproofing option. Testing other cellulosic fibers, such as bamboo, rayon, or hemp, should also be investigated, as should the water repellence of the pigmented oil layer. In addition, new technology would likely need to be developed to allow for industrial processing of quantities of this pigmented oil. The economics of this product would require further research. In addition, natural oils have been researched for protection of exterior wood in service [37,38], which may be an additional use.

The challenges seen using pigment from *S. cuboideum* in linseed oil may limit the ability of this technology to be adopted in the market. However, the use of spalting fungal pigments like that from *S. cuboideum* still have great potential for use in the textile industry, especially as they can be used without the need for mordants [18]. Other natural pigments, such as those from plant sources, have relied on mordants for their color stability [39–42], and the lack of mordants needed for fungal pigments represents a significant advantage. While it was hoped that linseed oil would provide a less toxic alternative to DCM, future work into identifying new solvents appropriate for use with dramada would be helpful, especially for use with synthetic fabrics. Recent work using pigment from *S. cuboideum* in inkjet printers may be a particularly effective method of textile dying with fungal pigments [43].

5. Conclusions

Red pigment from *Scytalidium cuboideum* carried in linseed oil was found to effectively dye a variety of fabric varieties, with natural fibers showing better coloration and colorfastness overall. Mechanical properties were found to be essentially unaffected by the application of pigment. Colorfastness was shown to vary among fabrics, with application of bleach associated with a color shift from red to blue of the applied pigment. Heating was found to result in the production of a bright red fabric, which may have potential as a sustainable waterproofing treatment.

Supplementary Materials: The following are available online at http://www.mdpi.com/2309-608X/6/2/53/s1, Figure S1: SEM images of high-density cotton exposed to tensile strength test; Figure S2: SEM images of low-density cotton exposed to tensile strength test; Figure S3: SEM images of nylon exposed to tensile strength test; Figure S4: SEM images of high-density polyester exposed to tensile strength test; Figure S5: SEM images of low-density polyester exposed to tensile strength test; Figure S6: SEM images of high-density wool exposed to tensile strength test; Figure S7: SEM images of low-density wool exposed to tensile strength test; Figure S8: SEM images of high-density cotton for tear strength test; Figure S9: SEM images of low-density cotton for tear strength test; Figure S10: SEM images of high-density wool for tear strength test; Figure S11: SEM images of low-density wool for tear strength test; Figure S12: SEM images of nylon for tear strength test; Figure S13: SEM images of high-density polyester for tear strength test; Figure S14: SEM images of low-density polyester for tear strength test; Figure S15: SEM images of high-density cotton samples from laundry test; Figure S16: SEM images of low-density cotton samples from laundry test; Figure S17: SEM images of nylon samples from laundry test; Figure S18: SEM images of high-density wool samples from laundry test; Figure S19: SEM images of low-density wool samples from laundry test; Figure S20: SEM images of low-density polyester samples.

Author Contributions: Conceptualization, S.C.R. and H.-L.C.; methodology, S.C.R. and H.-L.C.; validation, S.C.R. and H.-L.C.; formal analysis, M.E.P.A., S.M.V.G., and H.-L.C.; investigation, M.E.P.A.; resources, S.C.R. and H.-L.C.; data curation, M.E.P.A. and S.M.V.G.; writing—original draft preparation, M.E.P.A. and S.V.M.G.; writing—review and editing, R.C.V.C., S.M.V.G., H.-L.C., and S.C.R.; supervision, S.C.R. and H.-L.C.; project administration, S.C.R. and H.-L.C.; funding acquisition, S.C.R. and H.-L.C. All authors have read and agreed to the published version of the manuscript.

Funding: This research was funded by the Walmart Foundation Manufacturing Fund.

Acknowledgments: Authors gratefully acknowledge Oregon State University Electron Microscopy Facilities, Peter Eschbach, and Teresa Sawyer and would like to thank Ariel Muldoon for help with statistical analysis.

Conflicts of Interest: The authors declare no conflict of interest. The funders had no role in the design of the study; in the collection, analyses, or interpretation of data; in the writing of the manuscript, or in the decision to publish the results.

References

1. Räisänen, R.; Nousiainen, P.; Hynninen, P.H. Emodin and dermocybin natural anthraquinones as high-temperature disperse dyes for polyester and polyamide. *Text. Res. J.* **2001**, *71*, 922–927. [CrossRef]
2. Wouters, J.; Rosario-Chirinos, N. Dye analysis of pre-Columbian Peruvian textiles with high-performance liquid chromatography and diode-array detection. *J. Am. Inst. Conserv.* **1992**, *31*, 237–255. [CrossRef]
3. Kumar, J.K.; Sinha, A.K. Resurgence of natural colourants: A holistic view. *Nat. Prod. Res.* **2004**, *18*, 59–84. [CrossRef]
4. De Melo, J.; Rondão, R.; Burrows, H.D.; Melo, M.J.; Navaratnam, S.; Edge, R.; Voss, G. Spectral and photophysical studies of substituted indigo derivatives in their keto forms. *ChemPhysChem* **2006**, *7*, 2303–2311. [CrossRef]
5. Bechtold, T.; Mussak, R. *Handbook of Natural Colorants*; Wiley: Chichester, UK, 2009.
6. Brigden, K.; Santillo, D.; Johnston, P. *Nonylphenol Ethoxylates (NPEs) in Textile Products, and Their Release through Laundering*; Greenpeace Research Laboratories Technical Report: Exeter, UK, January 2012.
7. Luongo, G.; Thorsén, G.; Östman, C. Quinolines in clothing textiles—a source of human exposure and wastewater pollution? *Anal. Bioanal. Chem.* **2014**, *406*, 2747–2756. [CrossRef]
8. Taylor, G. Ancient textile dyes. *Chem. Br.* **1990**, *26*, 1155–1158.
9. Taylor, G. Natural dyes in textile applications. *Rev. Prog. Color. Relat. Top.* **1986**, *16*, 53–61. [CrossRef]
10. Glover, B. Are natural colorants good for your health? Are synthetic ones better? *Text. Chem. Color.* **1995**, *27*.
11. Woolf, J.A. Methods for Improving the Adhesion and/or Colorfastness of ink jet inks with Respect to Substrates Applied Thereto, and Compositions Useful Therefor. U.S. Patent 5,897,694, 27 April 1999.

12. Bhatti, I.A.; Adeel, S.; Jamal, M.A.; Safdar, M.; Abbas, M. Influence of gamma radiation on the colour strength and fastness properties of fabric using turmeric (Curcuma longa L.) as natural dye. *Radiat. Phys. Chem.* **2010**, *79*, 622–625. [CrossRef]
13. Batool, F.; Adeel, S.; Azeem, M.; Khan, A.A.; Bhatti, I.A.; Ghaffar, A.; Iqbal, N. Gamma radiations induced improvement in dyeing properties and colorfastness of cotton fabrics dyed with chicken gizzard leaves extracts. *Radiat. Phys. Chem.* **2013**, *89*, 33–37. [CrossRef]
14. Carr, C. *Chemistry of the Textiles Industry*; Springer Science & Business Media: Berlin/Heidelberg, Germany, 1995.
15. Yang, Y.; Huda, S. Comparison of disperse dye exhaustion, color yield, and colorfastness between polylactide and poly (ethylene terephthalate). *J. Appl. Polym. Sci.* **2003**, *90*, 3285–3290. [CrossRef]
16. Kadolph. *Textiles*, 11 ed.; Prentice Hall: Upper Saddle River, NJ, USA, 2010; p. 581.
17. Weber, G.; Chen, H.-L.; Hinsch, E.; Freitas, S.; Robinson, S. Pigments extracted from the wood-staining fungi Chlorociboria aeruginosa, Scytalidium cuboideum, and S. ganodermophthorum show potential for use as textile dyes. *Color. Technol.* **2014**, *130*, 445–452. [CrossRef]
18. Hinsch, E.M.; Weber, G.; Chen, H.-L.; Robinson, S.C. Colorfastness of Extracted Wood-staining Fungal Pigments on Fabrics: A new potential for textile dyes. *J. Text. Appar. Technol. Manag.* **2015**, *9*, 1–11.
19. Hinsch, E.M. A Comparative Analysis of Extracted Fungal Pigments and Commercially Available Dyes for Colorizing Textiles. Master's Thesis, Oregon State University, Corvallis, OR, USA, 2015.
20. Sherratt, P.J.; Williams, S.; Foster, J.; Kernohan, N.; Green, T.; Hayes, J.D. Direct comparison of the nature of mouse and human GST T1-1 and the implications on dichloromethane carcinogenicity. *Toxicol. Appl. Pharmacol.* **2002**, *179*, 89–97. [CrossRef] [PubMed]
21. Slater, K. *Environmental Impact of Textiles: Production, Processes and Protection*; Woodhead Publishing: Sawston, UK; Cambridge, UK, 2003; Volume 27.
22. Benbrahim-Tallaa, L.; Lauby-Secretan, B.; Loomis, D.; Guyton, K.Z.; Grosse, Y.; El Ghissassi, F.; Bouvard, V.; Guha, N.; Mattock, H.; Straif, K. Carcinogenicity of perfluorooctanoic acid, tetrafluoroethylene, dichloromethane, 1, 2-dichloropropane, and 1, 3-propane sultone. *Lancet Oncol.* **2014**, *15*, 924–925. [CrossRef]
23. Robinson, S.C.; Hinsch, E.; Weber, G.; Freitas, S. Method of extraction and resolubilisation of pigments from Chlorociboria aeruginosa and Scytalidium cuboideum, two prolific spalting fungi. *Color. Technol.* **2014**, *130*, 221–225. [CrossRef]
24. Robinson, S.C.; Vega Gutierrez, S.; Garcia, R.A.C.; Iroume, N.; Vorland, N.R.; McClelland, A.; Huber, M.; Stanton, S. Potential for carrying dyes derived from spalting fungi in natural oils. *J. Coat. Technol. Res.* **2017**, *14*, 1107–1113. [CrossRef]
25. Palomino Agurto, E.M.; Vega Gutierrez, M.S.; Chen, H.-L.; Robinson, C.S. Wood-Rotting Fungal Pigments as Colorant Coatings on Oil-Based Textile Dyes. *Coatings* **2017**, *7*, 152. [CrossRef]
26. Hussain, T.; Ali, R. Comparison of properties of cotton fabric dyed with pigment and reactive dye. *J. Text. Inst.* **2009**, *100*, 95–98. [CrossRef]
27. Punrattanasin, N.; Nakpathom, M.; Somboon, B.; Narumol, N.; Rungruangkitkrai, N.; Mongkholrattanasit, R. Silk fabric dyeing with natural dye from mangrove bark (Rhizophora apiculata Blume) extract. *Ind. Crops Prod.* **2013**, *49*, 122–129. [CrossRef]
28. Rajendran, R.; Selvi, B.T. Natural Dyeing of Cotton Fabrics with Pigment Extracted from Roseomonas Fauriae. *Univers. J. Environ. Res. Technol.* **2014**, *4*, 54–59.
29. Hinsch, E.; Robinson, S. Comparing Colorfastness to Light of Wood-Staining Fungal Pigments and Commercial Dyes: An Alternative Light Test Method for Color Fastness. *Coatings* **2018**, *8*, 189. [CrossRef]
30. Hernández, V.A.; Machuca, Á.; Saavedra, I.; Chavez, D.; Astuya, A.; Barriga, C. Talaromyces australis and Penicillium murcianum pigment production in optimized liquid cultures and evaluation of their cytotoxicity in textile applications. *World J. Microbiol. Biotechnol.* **2019**, *35*, 160. [CrossRef] [PubMed]
31. Bajpai, D. Laundry detergents: An overview. *J. Oleo Sci.* **2007**, *56*, 327–340. [CrossRef] [PubMed]
32. Golinski, P.; Krick, T.P.; Blanchette, R.A.; Mirocha, C.J. Chemical characterization of a red pigment (5, 8-dihydroxy-2, 7-dimethoxy-1, 4-naphthalenedione) produced by Arthrographis cuboidea in pink stained wood. *Holzforsch.-Int. J. Biol. Chem. Phys. Technol. Wood* **1995**, *49*, 407–410.
33. Tudor, D.; Robinson, S.C.; Cooper, P.A. The influence of pH on pigment formation by lignicolous fungi. *Int. Biodeterior. Biodegrad.* **2013**, *80*, 22–28. [CrossRef]

34. Dlugogorski, B.Z.; Kennedy, E.M.; Mackie, J.C. Low temperature oxidation of linseed oil: A review. *Fire Sci. Rev.* **2012**, *1*, 3. [CrossRef]
35. Mills, J.; White, R. *The Organic Chemistry of Museum Objects*; Oddy, A., Ed.; Butterworth-Heinmann Ltd.: Oxford, UK, 1994.
36. Shashoua, Y.; Skals, I. Development of a conservation strategy for a collection of waterproofed military uniforms. *Conservator* **2004**, *28*, 57–65. [CrossRef]
37. Gast, L.; Schneider, W.J.; Cowan, J. Polyester amides from linseed oil for protective coatings. *J. Am. Oil Chem. Soc.* **1966**, *43*, 418–421. [CrossRef]
38. Gast, L.; Schneider, W.J.; Cowan, J. Polyesteramides from linseed oil for protective coatings low acid-value polymers. *J. Am. Oil Chem. Soc.* **1968**, *45*, 534–536. [CrossRef]
39. Rattee, I. Bonds between dyes and fibres. *Sci. Prog. (1933-)* **1964**, *52*, 581–592.
40. Casselman, K.D. *Lichen Dyes: The New Source Book;* Courier Corporation: North Chelmsford, MA, USA, 2001.
41. Lewis, D.M.; Vo, L.T. Dyeing cotton with reactive dyes under neutral conditions. *Color. Technol.* **2007**, *123*, 306–311. [CrossRef]
42. Samanta, A.K.; Agarwal, P. *Application of Natural Dyes on Textiles;* Scientific Research Publishing: Wuhan, China, 2009.
43. Vega Gutierrez, M.S.; He, Y.; Cao, Y.; Stone, D.; Walsh, Z.; Malhotra, R.; Chen, H.-L.; Chang, C.-H.; Robinson, C.S. Feasibility and Surface Evaluation of the Pigment from Scytalidium cuboideum for Inkjet Printing on Textiles. *Coatings* **2019**, *9*, 266. [CrossRef]

Journal of
Fungi

MDPI

Review

Use of the Versatility of Fungal Metabolism to Meet Modern Demands for Healthy Aging, Functional Foods, and Sustainability

Jacqueline A. Takahashi [1,*], Bianca V. R. Barbosa [1], Bruna de A. Martins [1], Christiano P. Guirlanda [2] and Marília A. F. Moura [2]

1 Department of Chemistry, Exact Sciences Institute, Universidade Federal de Minas Gerais, Pres. Antônio Carlos Avenue, 6627, Pampulha, Belo Horizonte 31270-901, MG, Brazil; biancavrb99@gmail.com (B.V.R.B.); bruna.almeidamartins@gmail.com (B.d.A.M.)
2 Department of Food Science, Faculty of Pharmacy, Universidade Federal de Minas Gerais, Pres. Antônio Carlos Avenue, 6627, Pampulha, Belo Horizonte 31270-901, MG, Brazil; cpguirlanda@gmail.com (C.P.G.); mourafmari@gmail.com (M.A.F.M.)
* Correspondence: jat@qui.ufmg.br

Received: 31 August 2020; Accepted: 27 September 2020; Published: 15 October 2020

check for updates

Abstract: Aging-associated, non-transmissible chronic diseases (NTCD) such as cancer, dyslipidemia, and neurodegenerative disorders have been challenged through several strategies including the consumption of healthy foods and the development of new drugs for existing diseases. Consumer health consciousness is guiding market trends toward the development of additives and nutraceutical products of natural origin. Fungi produce several metabolites with bioactivity against NTCD as well as pigments, dyes, antioxidants, polysaccharides, and enzymes that can be explored as substitutes for synthetic food additives. Research in this area has increased the yields of metabolites for industrial applications through improving fermentation conditions, application of metabolic engineering techniques, and fungal genetic manipulation. Several modern hyphenated techniques have impressively increased the rate of research in this area, enabling the analysis of a large number of species and fermentative conditions. This review thus focuses on summarizing the nutritional, pharmacological, and economic importance of fungi and their metabolites resulting from applications in the aforementioned areas, examples of modern techniques for optimizing the production of fungi and their metabolites, and methodologies for the identification and analysis of these compounds.

Keywords: fungi; secondary metabolites; metabolomics; NTCD; additives; functional foods; nutraceuticals; sustainability; healthy aging

---

1. Health and Modern Food Demands

Health concerns have always existed among humans. Although some conditions and diseases cannot be avoided yet, the manifestation of several non-transmissible chronic diseases (NTCD) with high prevalence in patients over 60 years of age such as diabetes, cardiovascular, and neurodegenerative diseases can be delayed by adhering to a healthy lifestyle, which among other factors, is directly correlated to eating habits. The physiological effects associated with the consumption of certain foods are thus becoming very popular. Several types of diets and foods such as the fat-free diet, low-carb diet [1], Mediterranean diet [2], and soy-based diet [3] have been adopted in the quest for healthy aging. Several effects of NTCD have also been postponed through calorie restriction diets in animal models [4]. However, prolonged caloric restriction in humans generates undesirable effects; thus, alternative ways of preventing NTCD have been sought through the development of drugs, foods, and/or nutraceuticals that have both health-promoting and anti-aging effects, without causing adverse effects [5]. There is an increasing trend to combine the use of nutraceuticals with pharmacotherapy, even among individuals with non-aging-related diseases.

As oxidative stress is among the metabolic factors and pathways most related to cell aging, the consumption of nutraceuticals and functional foods with antioxidant activity has increased. Antioxidants can benefit the human body by directly or indirectly neutralizing reactive oxygen species (ROS), modulating metabolic pathways and gene expression, and activating mechanisms of cellular stress and autophagy that delay aging through pathways unrelated to ROS [5]. The nutraceutical market has already reached market values up to USD 117 billion [6]. Nutraceuticals can be classified into several categories based on the level of innovation and area of application (Table 1) [7,8]. Another trend associated with health improvement and NTCD prevention is that of consuming natural foods or foods containing natural, rather than synthetic, additives such as natural flavoring agents, acidulants, and colorants. Several people also prefer vegetarian and vegan diets, which involve restrictions in food and additives of animal origin to different extents; such diets are mainly motivated by the environmental impacts of livestock farming and animal welfare and have prompted studies on the possible effects of these diets on health [9] and increased the market demand for new vegetarian and vegan food products.

Filamentous fungi are capable of responding to different demands for the development of functional foods, nutraceuticals, and bioactive substances that can be used as medicines or in the food industry, either through the use of their biomass or the metabolites produced by them. A comprehensive review of the various aspects of fungal biotechnology and industrial applications was recently published by Meyer et al. [10].

The consumption of fungi as food mainly involves the consumption of mushrooms (Ascomycota and Basidiomycota phyla), which are enjoyed worldwide, sometimes as delicatessen or gourmet products. However, the role of fungi in human and animal health extends much further than the recognized health benefits of mushrooms [11]. Many fungal species are commercially available as supplements or nutraceuticals, and the fungal metabolites produced by these organisms including many non-Basidiomycota species as functional foods have multiple pharmacological activities. Examples of fungal species used as functional foods include many species of microscopic filamentous fungi that are easily cultivated under ex situ scalable conditions such as some well-known *Penicillium*, *Aspergillus*, and *Fusarium* species. However, lesser known species such as *Ashbya gossypii* also play an important role in the production of food additives such as riboflavin (vitamin B2) **(1)** (Figure 1) [12]. Other important fungal products associated with health improvement are enzymes such as β-galactosidase, which hydrolyzes lactose from dairy foods, is produced by filamentous fungi such as *Trichoderma* sp., and is helpful for lactose-intolerant individuals [13].

Table 1. Examples of food additives and nutraceuticals from fungi.

Category	Active Component	Bioactivity	Fungal Source	References
Nutrient	Vitamin C **(55)**	Antioxidant	*Dictyophora indusiata*	[14]
Secondary Metabolites	Resveratrol	Antioxidant	*Pleurotusostreatus*	[15]
	Agmatine **(28)**	Neurological benefits	*Aspergillus oryzae*	[16]
ω-6 Polyunsaturated fatty acid	γ-linolenic acid	Anti-inflammatory	*Mucor circinelloides*	[17]
	Arachidonic acid	Development of the nervous central system and enhancement of immune response	*Mortierella alpina*	[18]
Probiotic	Whole cell	Increase of beneficial bacteria population in gastrointestinal tract	*Saccharomyces boulardii*	[19]
Nutraceutical Enzymes	Fibrino(geno)lytic enzymes	Antithrombotic	*Penicillium* sp. *Rhizopus microsporus*	[20] [21]
	Lipase (Lipopan F)	Decrease glycemic response	*Rhizopus oryzae*	[22]
Fortified Nutraceuticals	Folate in fermented maize-based porridge		*Saccharomyces cerevisiae*	[23]

This review discusses the usefulness of metabolites produced by fungi for food and pharmaceutical purposes closely linked to health improvement and the prevention of NTCD, respectively, as lead compounds, additives, nutraceuticals, supplements, and functional ingredients. The health benefits of fungal metabolites are highlighted. Cutting-edge tools for yield improvement and thus the scaling up of fungal metabolite production as well as the means for ensuring a successful circular economy in this area are also discussed.

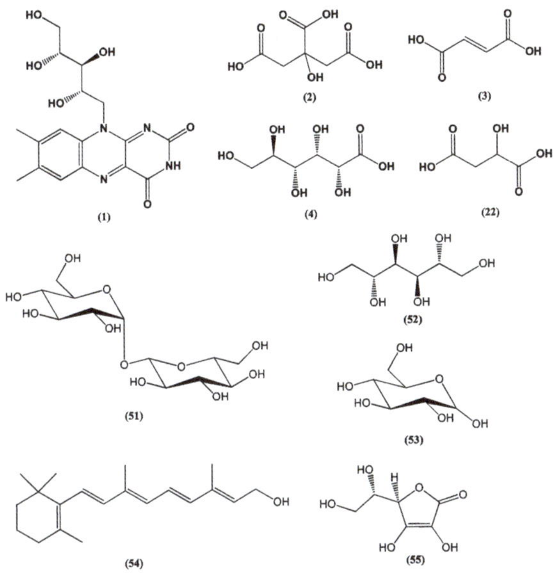

Figure 1. Chemical structures of some fungal-derived food additives **(1–4, 22, 51–55)**.

2. Natural Food Additives from Fungi

Prompted by the growing evidence for the association between natural compound intake and health, there is a demand for the replacement of synthetic food additives with natural products. Fungal metabolites feature many properties that have been explored for such replacement. This section presents the potential of fungal metabolites as food additives, focusing on their versatility as coloring agents, producing species, and recent achievements and challenges in this area.

Fungi provide several food additives and technology adjuvants such as organic acids, colorants, and fatty acids including some ω-3 and ω-6 class fatty acids, which are essential for human metabolism. Citric acid **(2)** and fumaric acid **(3)** (Figure 1), metabolites of *Aspergillus niger* and *Rhizopus oryzae*, respectively, are outstanding food additives that are industrially produced. Citric acid **(2)** and gluconic acid **(4)** (Figure 1) are fungal products with the highest commercial-scale production [24]. Citric acid **(2)** has a number of applications as an antioxidant, preservative, acidulant, pH control agent, and flavor regulation agent, with market numbers predicted to reach USD 3.6 billion in 2020 [24].

The harmful effects of synthetic food colorants on human health such as attention-deficit/hyperactivity disorder, asthma and allergies, cancer, and neurological disorders have accelerated the search for natural substitutes. The use of the yellow coloring compounds tartrazine, quinoline, and sunset as well as the red coloring compound amaranth [25] has been regulated and supervised by the World Health Organization. In addition, sustainability issues have contributed to the decreased acceptability of non-biodegradable synthetic colorants that are difficult to remove from effluents, causing toxic effects on plants, bacteria, algae, fishes, and crustaceans [26]. Fungal pigments and dyes have emerged as alternatives to these synthetic food additives [27].

Food dyes are coloring materials soluble in food substrates, while food pigments are insoluble in food and need to be carried by vehicles that bind to the food instead [28] (ACS, 2020). Food colorants are pigments or dyes approved for use as food additives [29] (FDA, 2017). A wide spectrum of natural colors can be obtained from metabolites of fungi of different genera such as *Eurotium*, *Fusarium*,

Monascus, and *Penicillium*, isolated from marine and terrestrial environments and extracted by techniques considered environmentally friendly including those employing ultrasound, pressurized liquid, microwaves, and pulsed electric field [30]. The fungal metabolites have a wide color range, which can be represented by carotenoids such as lycopene (red) **(5)**, β-carotene (yellow-orange) **(6)**, and astaxanthin (pink-red) **(7)**; azafilones such as monascorubrin (orange) **(8)** and ankaflavin (yellow) **(9)**; and the quinine derivatives alizarin (purple-red) **(10)** and bikaverin (red) **(11)** (Figure 2). Arpink Red **(12)** (Figure 2), an anthraquinone produced by *Penicillium oxalicum*, has been approved by the Codex Alimentarius for use as a food colorant in meat products, dairy, confectionery, ice cream, and alcoholic and non-alcoholic beverages [31,32]. Naftoquinone hydrosoluble metabolites (purple color) were produced by a soil-originated strain of *Fusarium oxysporum* using a simple culture medium containing glucose, ammonium sulfate, and salts [33]. Crystalline neoechinulin A **(13)** (ivory color), neoechinulin B **(14)**, and cladosporin **(15)** (all having a yellow color) (Figure 2) are described as metabolites of *Eurotium amstelodami*, *Eurotium herbariorum*, and *Eurotium rubrum* isolated from outdoor and indoor samples in Canada and grown in medium containing sucrose, yeast extract, and salts [34].

The filamentous fungus *Monascus purpureus* is one of the first species that was used to produce natural colorants. It is traditionally consumed in Asia in fermented red rice, which is important in Chinese cuisine, and used in folk medicine as a regulator of digestive and circulatory functions [35,36]. Some species of the genus *Monascus* produce the secondary metabolite monacolin K **(16)** (Figure 2), which consists of a lactone with a free hydroxyl-acid moiety. Aside fromits coloring property, monacolin K **(16)** has a high antioxidant activity and is marketed as a hypocholesterolemic drug known as lovastatin. The pigments of yellow, orange, and red color produced by the *Monascus* genus are considered safe for human consumption [37].

The development of mutant strains of *Monascus* has enabled an increase in available pigments including monashin **(17)** (Figure 2), which is obtained through the mutation of the enzyme polyketide synthase and has an antioxidant activity [35]. The monascin **(18)** (Figure 2) (yellow) produced by *M. purpureus* after activation by the transcription factors DAF-16/FOXO increased the production of superoxide dismutase and thermal shock protein HSP16.2, improving survival in a worm model [38]. Monascin **(18)** also reduced non-alcoholic fatty liver disease and increased AMPK levels and γ 1α co-activator of the receptor activated by peroxisome proliferator in mice [39]. In another study, a new azaphylone, monapurpureusone **(19)**, and a new brownish natural product, monapurpureusin **(20)** (Figure 2), were obtained from a mutant strain of *M. purpureus* cultivated in fermented rice extract. Both compounds presented superoxide radical scavenging activity (EC_{50} = 176.2 and 271.2 μM, respectively), with monapurpureusone **(19)** superior to the control gallic acid (237.1 μM), and anti-inflammatory activity (IC_{50} = 27.5 and 24.9 μM, respectively), with both compounds superior to the control quercetin (35.9 μM) [36].

Red and orange pigments were produced by a strain of *Talaromyces albobiverticillius* isolated from a marine environment at pH 6.5, and their colors were shown to be dependent on the fermentation period (198.6 and 229.0 h, respectively) [40]. Optimal conditions for the production of pigments were also described for *F. oxysporum* (red color, rate C:N = 9, blue LED light, and absence of co-culture) and *Aspergillus chevalieri* (yellow color, rate C:N = 20, glucose as the carbon source, UV and red light, and co-cultivation with *Kluyveromyces marxianus*) [41]. The production of natural food colorants such as melanins, azaphilones, flavins, phenazines, and quinines by filamentous fungi has also been reported for Basidiomycetes [28].

The replacement of synthetic colorants with fungal pigments and dyes enables the production of safer and healthier foods. Compared with plant sources of natural pigments, fungal sources are more economically attractive, considering the relative ease of yield maximization by manipulating fungal fermentative parameters instead of relying on seasonal factors, as may occur in the production of pigments and dyes of plant origin.

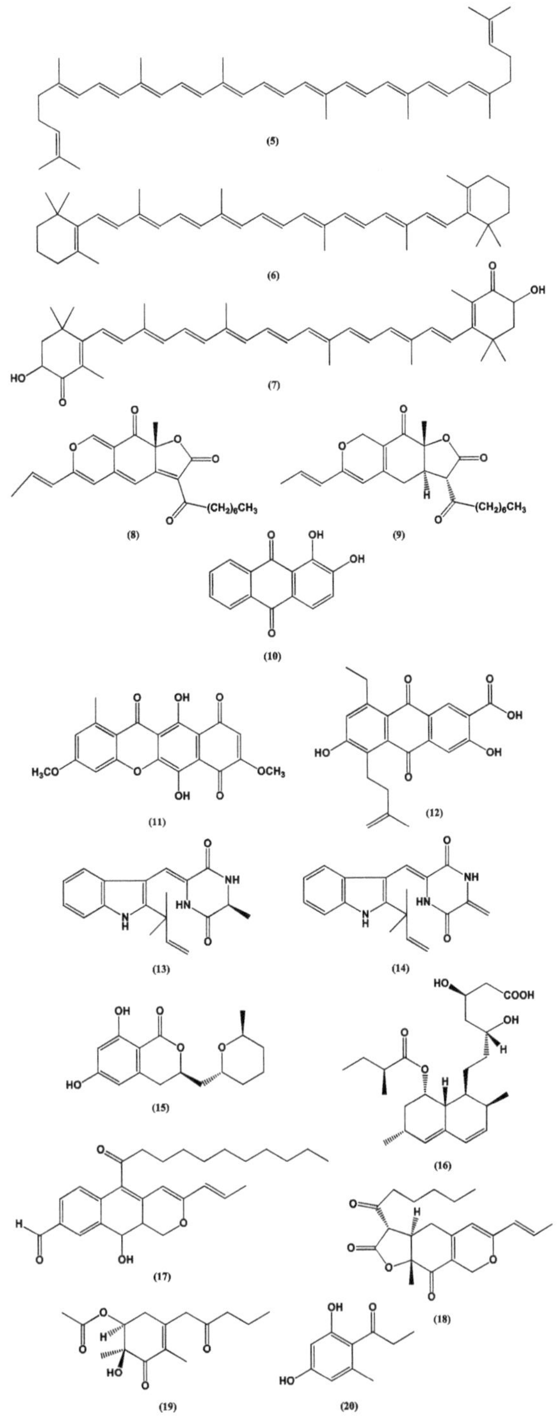

Figure 2. Chemical structures of some colored fungal secondary metabolites **(5–20)**.

Nevertheless, some challenges need to be considered for regulating new food-related colorants such as the possibility of interactions with the food matrix, which causes undesirable sensorial changes, loss of color stability, and contamination by toxic substances such as mycotoxins. Some alternatives proposed for these problems are based on the controlled release of the colorant using microencapsulation as well as the use of nanoformulations to eliminate undesirable aromas and flavors [30]. Spray-drying microencapsulation was successfully used in broth fermented by three species producing yellow dyes, *Aspergillus keveii*, *Penicillium flavigenum*, and *Epicoccum nigrum*. Encapsulation with three adjuvants (maltodextrin, modified starch, and gumarabic) provided pigment retention above 70% [41]. With respect to problems linked to the production of mycotoxins such as citrinin **(21)** (Figure 3) produced by *Monascus*, strategies vary from changes in cultivation and fermentation conditions to disruption of genes encoding the production of mycotoxins in question to create non-mycotoxin-producing mutant strains [30].

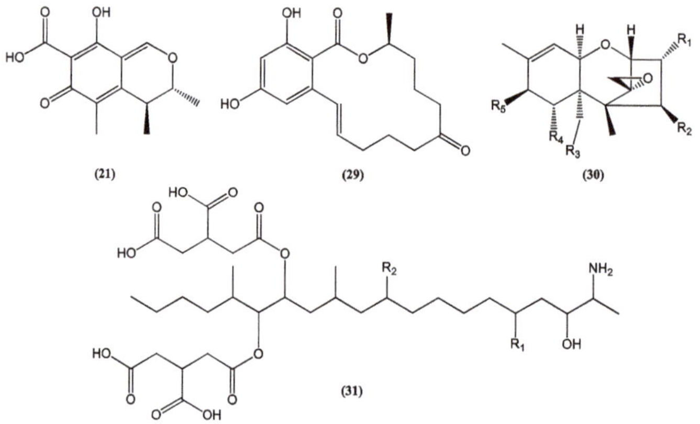

Figure 3. Chemical structures of some mycotoxins **(21, 29)** and classes of mycotoxins **(30, 31)** frequently produced by fungal species.

The use of fungal metabolites as natural colorants contributes greatly to the development of healthier foods. Nonetheless, several other additives of fungal origin such as substances related to taste and pH (acidulants, flavorings, and sweeteners), texture (thickeners and emulsifiers), and increased shelf life (antioxidants and preservatives) have been studied. Some examples can be found in Table 2. The application of some of these additives including exopolysaccharides is not restricted to the food industry. Fungal exopolysaccharides produced by species such as *Phellinus linteus*, *Ganoderma lucidum*, *Fusarium* sp., *Pleurotus* spp., *Inonotus obliquus*, and *Aureobasidium pullulans* have been employed to aid moisture retention in confectionery, increase the viscosity and crystallization of sugar, and as stabilizers, emulsifiers, and thickening agents [42,43]. These compounds are also of great interest to textiles, food, cosmetics, and pharmaceutical industries and are also important in agriculture as preservatives, bioherbicides, and microbicides.

Table 2. Examples of non-coloring natural additives from fungal origin.

Function	Compound Name and Structure	Fungus	Main Uses	References
Acidulant	Malic acid 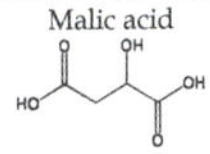	*A. niger*	Candies, soft drinks, baked goods	[44]
Antioxidant	Gallic acid 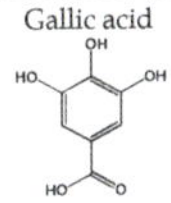	*A. niger*	Any food susceptible to oxidation	[45]
Thickener agent	Galactomannans and β-1,3-glucans 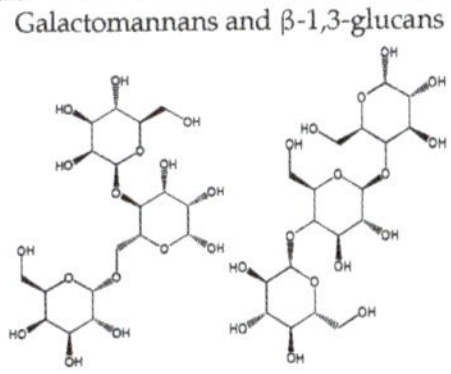	*Aspergillus terreus*	Sauces, soups, puddings, fillings	[46]
Emulsifier	Sorbitol 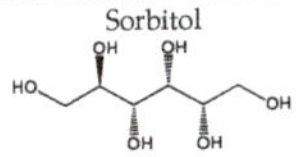	*Neopestalotiopsis* sp.	Baked goods and desserts	[47]
Flavoring agent	Dihydro-β-ionone 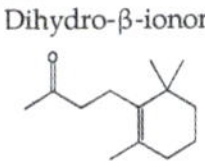	*Volvariellavolvacea*	Beverages, ice creams, candies	[48]
Preservative	Natamycin 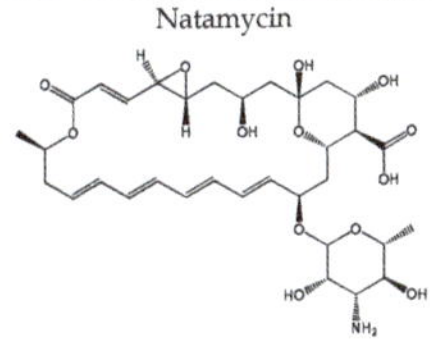	*Streptomyces gilvosporeus*	Fungicide used as biopreservative in dairy and meat products	[49]
Sweetener	Mogroside V 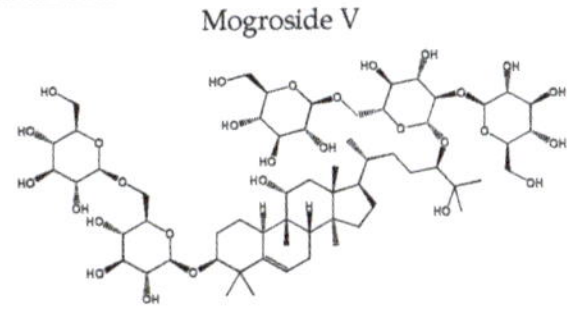	*Diaporthe angelicae* and *Fusarium solani*	Juices, soft drinks, cereals, confectionary, baked goods	[50]

Metabolic engineering of microorganisms has been successfully employed for producing 42 out of the 316 food additives from numerous species of fungi and bacteria currently approved by the European Union [44]. Using this technique, the production of glutamic acid, a metabolite capable of providing "umami" flavor to food, though fermentation with *Corynebacterium glutamicum* was optimized. The production of malic acid **(22)** (Figure 1), an acidulant used in food and beverages, was increased after overexpression of the genes encoding its precursors in the species *S. cerevisiae*, *Aspergillus flavus*, *Aspergillus oryzae*, and *A. niger* [51,52]. Metabolic engineering can also be used to introduce heterologous routes for enzyme production in microorganisms (e.g., to produce enzymes of plant origin) [51]. Changes in the molecular structures and colors of microbial pigments were made using the engineering system CRISPR-Cas9, which cleaves specific microbial DNA sites and introduces changes in one or more target genes [30]. For instance, *Escherichia coli* was used for the production of β-carotene **(6)** through genomic editing to introduce insertions, deletions, and substitutions in regions of the lacZ, galK, and ldhA genes using the CRISPR-Cas9 system [53].

A yeast species, *Yarrowia lipolytica*, was engineered by introducing genes for the production of β-ketolase and β-hydroxylase of seaweed and bacteria origin, increasing the yield of astaxanthin

(7) (285 ± 19 mg/L; 47% of total carotenoids) [54]. This was a great achievement, given the lower productivity of astaxanthin **(7)** by yeasts like *Xanthophyllomyces dendrorhous* (10.2 mg/L) [55] and microalgae like *Haematococcus pluvialis* (84.8 mg/L) [56] and *Chlorella zofingiensis* (12.5 mg/L) [57]. Astaxanthin **(7)** has a high market value (USD 2500–7000/kg), 5–20 times higher than that of β-carotene **(6)** produced by the microalga *Dunaliella salina* (USD 300–500/kg) [54,58].

Overall, tools such as genetic manipulation based on transcriptomic analysis, induction of mutations, cloning, and insertion of heterologous plasmids into species with well-known genomes have enabled the production of large quantities of metabolites by previously non-producing species [59]. The elucidation and manipulation of the different stages in the transcription and secretion of amylases, xylanases, and cellulases in filamentous fungi has enabled their overexpression [60]; these enzymes are widely used in bakery products to improve the quality of dough through hydrolysis of long-chain carbohydrates and non-starch polysaccharides (cellulose and arabinoxylans).

It was hypothesized that enzymes produced by *A. oryzae* and *A. niger* could cleave the anti-nutritional factors present in flour, in addition to hydrolyzing proteins and carbohydrates into smaller molecules and thus making them more accessible for digestion and releasing phenolic compounds from the matrix. However, the fungi used the amino acids present in the flour for their own metabolism, which negatively affected the protein quality of the final product [61]. A new fermented food product was thus developed using stale bread as a substrate for the fungus *Neurospora intermedia*, which converted 65% of the starch into 21% of protein, in addition to supplementing the final product with minerals and vitamins; although there was a reduction in the amounts of proline, glutamic acid, and phenylalanine, the overall amino acid composition was improved [62].

3. Benefits, Research, and Industrial Applications of Fungal Metabolites

The scope of the pharmacological activity of fungal metabolites seems to be as endless as the structural diversity. However, several issues in this area must be addressed such as the extent of in vivo effects. The wide range of beneficial biological effects of fungal metabolites can be related to the prevention and treatment of NTCD, and some such effects have identified several potential compounds for developing new drugs.

Terrein **(23)** (Figure 4), for instance, is a secondary metabolite biosynthesized in high concentrations (537.26 ± 23.42 g/kg crude extract) by *A.terreus* [63]. The anti-inflammatory and antioxidant properties of terrein **(23)** were reported in in vitro studies [64]. These important medicinal properties and the high initial yield of this metabolite allow the large-scale production and technological development of the *A. terreus* crude extract for the prevention of some age-related NTCD.

Compounds presenting cytotoxicity against tumor cells form one of the most important classes of fungal metabolites. The world incidence of cancers is high, with cancer affecting one in five people at some point in life. The number of cancer cases registered in 2018 was 18.1 million people, a number that could double in 2040 [65]. Cancer is a multifactorial disease that affects people regardless of age, gender, or origin. As an NTCD, cancer is worrying in several ways, since working-age people can be affected, treatment is costly and long, mortality rates are high, and individuals often have health issues even after cure.

Many reports on fungal metabolites with cytotoxic activity show the vast arsenal of molecules that fungi provide to combat cancer. Table 3 provides examples of antitumor metabolites produced by fungi. Species from different genera or secondary metabolites produced by them sometimes present inhibitory power more pronounced than or relatively close to that presented by standard compounds. This was observed for hypocriol A **(24)** and F **(25)** (Figure 4), isolated from the strain *Hypocrea* sp. [66], and for sesquiterpene strichocaranes E **(26)** and F **(27)** (Figure 4), isolated from the entomopathogenic fungus *Isaria fumosorosea* [67], in comparison to the standard cisplatin. Many human tumor cell lines can be inhibited by fungal metabolites such as breast (MDA and MCF-7) [67], cervical and lung [66], colorectal [63], gastric and liver [68], and pancreatic [69] cell lines. The majority of compounds presented in Figure 4 are terpenes with different degrees of hydroxylation as well as linear and cyclic nitrogen-bearing compounds with free NH groups.

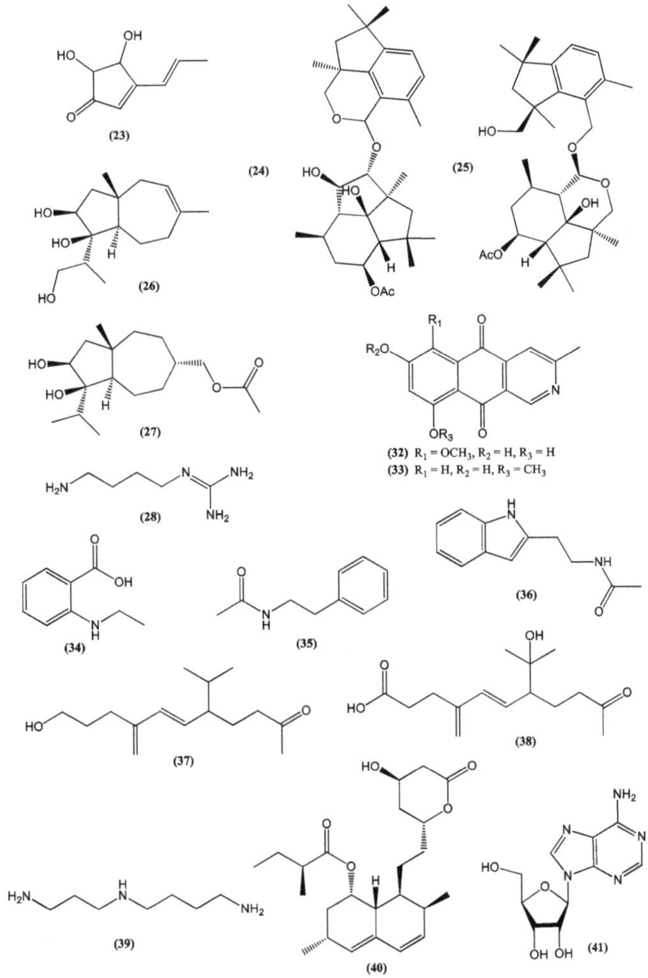

Figure 4. *Cont.*

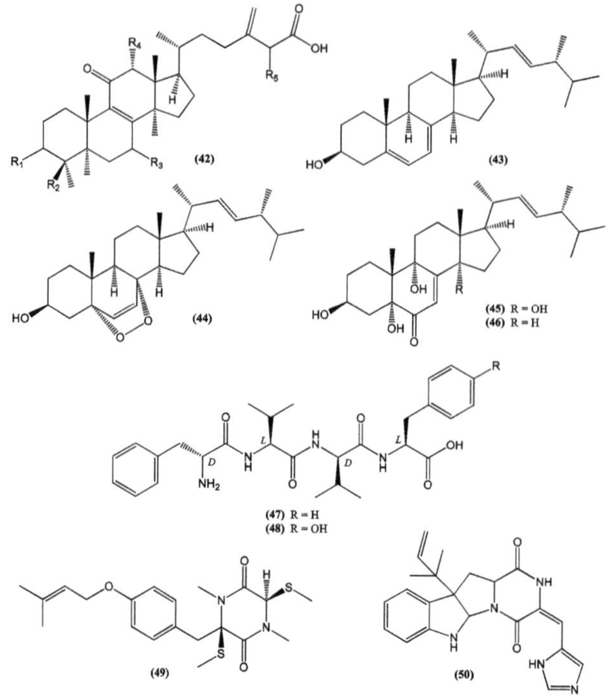

Figure 4. Chemical structures of metabolites **(23–28, 32–41, 43–50)** and class of bioactive compounds **(42)** produced by fungal species.

Table 3. Cytotoxic activity of some fungal secondary metabolites against several human tumor cell lines.

Metabolite	Fungal Origin	Yield	Cytotoxic Activity				Reference
			Human Tumor Cell	IC_{50}	Control	IC_{50}	
7-Desmethyl-6-methylbostrycoidin (**32**)	*F. solani*	7 mg/4.3 g extract	Breast (MDA MB 231)	0.73 µM	Doxorubicin	0.07 µM	[69]
			Pancreatic (MIA PaCa2)	0.64 µM		0.04 µM	
			Cervical (HeLa)	0.71 µM		0.05 µM	
			Non small-cell lung (NCI HI975)	0.34 µM		0.03 µM	
			Lung fibroblast (WI38)	6.42 µM		0.35 µM	
7-Desmethylscorpinone (**33**)	*F. solani*	5 mg/4.3 g extract	Breast (MDA MB 231)	1.51 µM	Doxorubicin	0.07 µM	[69]
			Pancreatic (MIA PaCa2)	0.98 µM		0.04 µM	
			Cervical (HeLa)	0.96 µM		0.05 µM	
			Non small-cell lung (NCI HI975)	0.61 µM		0.03 µM	
			Lung fibroblast (WI38)	5.84 µM		0.35 µM	
Ergosterol (**43**)	*Penicillium chrysogenum*	ni	Breast (MCF-7)	0.10 mM	ni	ni	[70]
	V.volvacea	500 mg/725 g extract	Prostate (PC-3M)	27.98 ± 0.97 µM	5-Fluorouracil	64.35 µM	[68]
Ergosterol peroxide (**44**)	*V. volvacea*	15 mg/725 g extract	Prostate (PC-3M)	23.15 ± 1.54 µM	5-Fluorouracil	64.35 µM	[68]
3β, 5α, 9α, 14 α-Tetrahydroxy-ergosta-7,22-dien-6-one (**45**)	*V. volvacea*	5.9 mg/725 g extract	Liver (HepG2)	20.72 ± 0.76 µM	5-Fluorouracil	54.74 µM	[68]
3β, 5α, 9α-Trihydroxy-ergosta-7,22-dien-6-one (**46**)	*V. volvacea*	12.5 mg/725 g extract	Gastric (SGC-7901)	12.03 ± 0.77 µM	5-Fluorouracil	75.05 µM	[68]
			Liver (HepG2)	5.90 ± 0.44 µM		54.74 µM	
Hypocriol A (**24**)	*Hypocrea* sp.	50.3 mg/63.2 g extract	Colorectal (HCT116)	18.6 ± 0.7 µM	Cisplatin	18.8 ± 1.9 µM	[66]
			Cervical (HeLa)	7.7 ± 0.4 µM		14.7 ± 0.8 µM	
			Lung (A549)	25.3 ± 2.5 µM		13.8 ± 1.2 µM	
			Breast (MCF-7)	19.7 ± 0.4 µM		17.6 ± 2.4 µM	
Hypocriol F (**25**)	*Hypocrea* sp.	11.2 mg/63.2 g extract	Colorectal (HCT116)	2.7 ± 0.6 µM	Cisplatin	18.8 ± 1.9 µM	[66]
			Cervical (HeLa)	4.6 ± 0.1 µM		14.7 ± 0.8 µM	
			Lung (A549)	15.3 ± 1.6 µM		13.8 ± 1.2 µM	
			Breast (MCF-7)	23.6 ± 1.3 µM		17.6 ± 2.4 µM	
Terrein (**23**)	*A. terreus*	537.26 ± 23.42 g/kg extract	Colorectal (HCT-116)	12.13 µM	Doxorubicin	0.11 µM	[63]
			Hepatocellular (HepG2)	22.53 µM		0.85 µM	
Trichocarane E (**26**)	*I. fumosorosea*	30 mg/200 g extract	Breast (MDA)	0.13 µg/mL	Cisplatin	2.90 µg/mL	[67]
			Breast (MCF-7)	2.46 µg/mL		1.14 µg/mL	
			Ovary (SKOV-3)	1.01 µg/mL		3.80 µg/mL	
			Cervical (Hela)	2.32 µg/mL		2.24 µg/mL	
			Lung (A549)	1.40 µg/mL		2.13 µg/mL	
			Liver (HepG2)	1.87 µg/mL		0.62 µg/mL	
Trichocarane F (**27**)	*I. fumosorosea*	41 mg/200 g extract	Breast (MDA)	0.89 µg/mL	Cisplatin	2.90 µg/mL	[67]
			Breast (MCF-7)	4.38 µg/mL		1.14 µg/mL	
			Ovary (SKOV-3)	1.46 µg/mL		3.80 µg/mL	
			Cervical (Hela)	4.57 µg/mL		2.24 µg/mL	
			Lung (A549)	1.66 µg/mL		2.13 µg/mL	
			Liver (HepG2)	3.66 µg/mL		0.62 µg/mL	

Note: ni = not informed.

Despite the huge number of fungal bioactive metabolites, there is controversy regarding the consumption or development of new drugs from metabolites biosynthesized by toxin-producing fungi. Terrein **(23)** and agmatine **(28)** (Figure 4)are examples of biologically active metabolites produced by *Aspergillus*, a genus often associated with the production of mycotoxins [71,72]. However, species from *Aspergillus* and *Fusarium* genera, known sources of toxins such as zearalenone **(29)**, trichothecenes **(30)**, and fumonisins **(31)** (Figure 3), have shown a biotechnological potential beyond the production of mycotoxins. For example, azaanthraquinone derivatives, 7-desmethyl-6-methylbostrycoidin **(32)**, and 7-desmethylscorpinone **(33)** (Figure 4), isolated from *F. solani* cultures, showed significant activity against tumor cell lines (Table 3) [69].The mycoprotein "Quorn", a popular fungal food for human consumption, is produced using mycelia of *Fusarium* sp. [72]. Therefore, disregarding the potential of species from mycotoxin-producing genera is unnecessary in the research and industrial development of food products and medicines.

Cancer and neurodegenerative diseases are pathologies targeted by antioxidant therapies, not only for treatment but also for prevention, as proven by pre-clinical, clinical, and epidemiological studies [73]. Although compounds and foods with antioxidant activity have been targeted by many studies [5,74], some brief considerations are worthwhile. Despite encouraging data from several screenings pointing to a very significant number of fungal products with antioxidant activity, it is important to note that most of the experiments in these studies were not applied to tissues and organ systems, and the pharmacokinetic aspects of the absorption of these substances by the human organism were not evaluated [74]. Thus, the straightforward extrapolation of the results of these screenings to in vivo human applications is not possible [75].

Conversely, the in vitro antioxidant profile of a natural product can be improved. This was observed in a study in which three strains of *M. circinelloides* (CBS 277.49, WJ11, and CBS 108.16) were grown for different time periods (three, five, and seven days) in different culture media (standard Kendrick medium and Ratledge and modified Kendrick and Ratledge media (MKR), with nitrogen deficiency) and evaluated for phenolic compounds and antioxidant capacity. The total phenolics (TPC) and flavonoids (TFC) were improved for CBS 277.49, while CBS 108.16 produced a higher amount of condensed tannins (TCT). The ethanol extract obtained from CBS 277.49 (five days of growth in MKR medium) presented the best results regarding neutralization of the2,2′-Azino-bis(3-ethylbenzothiazoline-6-sulfonic acid)($ABTS^+$) radical, copper reducing power, and ferric reducing power. The latter had a higher ferric reducing power than those of the standards BHT and α-tocopherol [76].

Despite the relevance of cancer as an NTCD, several other diseases linked to aging limit a healthy life as people age. Accumulated dietary deficiencies throughout life and loss of immunity increase mortality rates associated with non-fatal diseases such as infections. Chronic inflammation is another worrisome condition that increases the risk of developing various pathologies such as hypertension, diabetes, and neurodegenerative diseases [77]. Although early diagnosis and measures to avoid aging-related pathologies and their consequences are essential, drugs for existing diseases continue to be required. Table 4 presents the structures of fungal metabolites associated with the treatment or control of infections, inflammatory and neuroinflammatory conditions, and Alzheimer's disease and examples and extents of the biological activities of these compounds.

Table 4. Examples of non-cytotoxic biologically active fungal secondary metabolites.

Fungal Species	Bioactive Compound	Bioactivity			Health Benefit	References
		Target of Inhibitory Activity	Value	Control		
Cladosporium sphaerospermum	Cladosin L 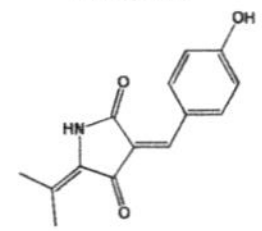	*Staphylococcus aureus*	MIC = 25–50 μM	ni	Antibacterial	[78]
Fusarium chlamydosporum	Chlamydosterol A 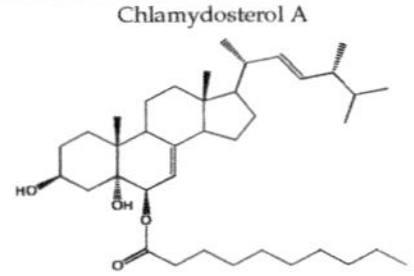	5-lipoxygenase (5-LOX)	IC_{50} = 3.06 μM	Indomethacin IC_{50} =1.13 μM	Anti-inflammatory	[79]
Hypoxylon sp.	Hypoxylon xanthone A 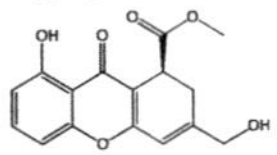	LPS-induced NO production	>70% at 1 μM	Minocycline >60% at 1 μM	Anti-neuroinflammatory	[80]
Rhizopycnis vagum	Rhizovagine A 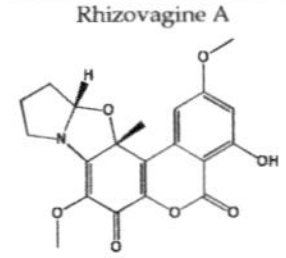	Acetylcholinesterase enzyme	IC_{50} = 43.1 μM	Tacrine hydrochloride IC_{50} = 6.1 μM	Treatment for Alzheimer's disease	[81]
Saccharicola sp.	Speciosin U 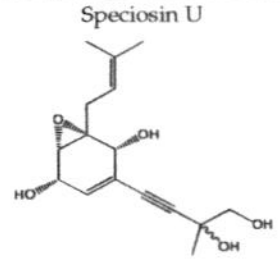	Acetylcholinesterase (huAChE-ICER)	IC_{50} = 0.037 ± 0.01 mg.mL^{-1}	Galantamine IC_{50} = 0.076 ± 0.01 mg.mL^{-1}	Treatment for Alzheimer's disease	[82]
Trichoderma sp.	Coniothyrinone A 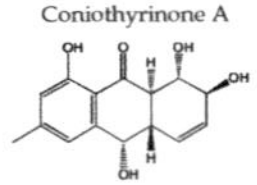	*Vibrio anguillarum*	MIC = 1.56 μM	Ciprofloxacin MIC = 0.625 μM	Antibacterial	[83]

ni = not informed.

The use of epigenetics for the production of new secondary metabolites through the activation of biosynthetic routes involving transcriptional genes previously silenced is an approach to increase and diversify the production of bioactive secondary metabolites by fungi. The use of a mutant strain of *Fusarium graminearum* lacking H3K27 methyltransferase after removal of the secondary metabolism repressor *kmt6* gene was reported as being advantageous for obtaining 22 metabolites, three of which had not been previously identified to be produced by this species: N-ethyl anthranilic acid **(34)**, N-phenetylacetamide **(35)**, and N-acetyltryptamine **(36)** (Figure 4). The production of the double mutant *kmt6fus 1*, with elimination of the production of compounds of the class fusarins, enabled the discovery of two new sesquiterpenes, tricinolone **(37)** and tricinolonic acid **(38)** (Figure 4) [84].

4. Mushrooms as Functional Foods for Preventing Aging-Related Non-Transmissible Chronic Diseases (NTCD)

The intake of edible mushrooms as functional foods associated with health improvement has been widely described [85]. Some metabolites obtained from mushrooms have been introduced into the market as they have antitumor properties or immunostimulants such as the polysaccharide lentinan produced by the edible mushrooms *Lentinula edodes* (Shiitake) and *G. lucidum*. *G. lucidum* latter is consumed in traditional East Asian medicine as bitter tea (hot aqueous extraction) and can also be obtained as a dry powder. The pharmacological properties of several mushroom metabolites have been demonstrated such as those of polysaccharides produced by *G. lucidum*, which were able to increase the survival of worms by activating the transcription factors DAF-16/FOXO [86].

Hirsutella sinensis produces polysaccharides with prebiotic properties related to insulin resistance and diabetes control [87,88]. Spermidine **(39)** (Figure 4) produced by several species [89], proved capable of prolonging the life of mice by reducing histone acetyltransferase EP300 [90], while epidemiological evidence indicates that spermidine **(39)** intake may also contribute to the reduction of human mortality [91]. Considerable amounts of the metabolite lovastatin **(40)** (Figure 4), a drug registered as low-density lipoprotein (LDL) cholesterol lowering, were detected in the widely consumed mushrooms *Agaricus bisporus* (30.79 mg/100 g dry weight), *Cantharellus cibarius* (67.89 mg/100 g dry weight), *Imleri abadia* (6.21 mg/100 g dry weight), and *L. edodes* (0.95 mg/100 g dry weight) [92].

However, the most prominent example of a mushroom with nutraceutical properties may be *Cordyceps militaris*, a rare and naturally occurring entomopathogenic medicinal mushroom in the Himalayan Mountains, Tibet, Nepal, and India. Studies have reported that consuming *C. militaris* extracts significantly increases glucose metabolism, thereby decreasing the glucose level in the blood. In addition, consumption of this mushroom provides protection against diabetic nephropathy [93]. *C. militaris* produces polysaccharides in the fructification body that are active toward α-glucosidase [94] and have immunomodulatory activity; this suggests their incorporation into functional foods and dietary supplements [95]. Anti-adipogenic activity was reported for a fermentation mix containing strawberry, silkworm pupae, and *C. militaris* [96]. Several nutraceutical products containing *Cordyceps* are available in the global market. Among the benefits claimed by their manufacturers are the promotion of mental health and benefits to the vascular system ("*Cordyceps* active"), cognition support ("Mushroom Plus"), anticancer and antioxidant activity ("Bhutan *Cordyceps* Tea"), strengthening of the cardiovascular system ("MRM *Cordyceps*CS-4 Strain"), and immune system support ("MycoNutri *Cordyceps* Organic") [97]. Cordycepin **(41)** (Figure 4) is the main metabolite produced by *C. militaris* and is very effective in reducing the accumulation of LDL, total cholesterol, triglycerides, and hyperlipidemia caused by high-fat diets [98]. Cordycepin **(41)** has several pharmacological properties such as anti-inflammatory, immunomodulatory, antioxidant, anti-aging, anticancer, antiviral, cardio, and hepatoprotective properties, among others [97]. This range of activities results in health effects that may help in postponing aging-linked NTCD. *C. militaris* also has high nutritional value and contains proteins, phenolic compounds, steroids, and lectins [99].

Another mushroom species with high functional and nutraceutical potential of high market value is *Agaricus subrufescens* (synonymy *Agaricus blazei* and *Agaricus brasiliensis*) [100]. It is commercialized in several countries such as Brazil (brand name "Sun mushroom"), China (Ji Song Rong), and Japan (Himematsutake) [101]. Sun mushroom contains polyphenols and polysaccharides and is known to decrease oxidative stress and prevent NTCD; it is indicated to have antioxidant, antitumor, anti-inflammatory, and immunomodulatory properties [102].

Another class of bioactive compounds of increasing prominence found in edible mushrooms includes antcins **(42)** (Figure 4), steroids that contain an ergostane-type skeleton and are produced by *Antrodia* species such as *Antrodia cinnamomea* and *Antrodia salmomea*. Studies suggest that these compounds are promising agents in the treatment of cancer, inflammation, diabetes, and diseases resulting from oxidative stress, among others. The aforementioned species have been historically used in communities of Taiwan for treating various diseases such as diarrhea, abdominal pain, hypertension, dermatological irritation, and intoxication by food, alcohol, and drugs [103]. A study tracking 36,499 middle-aged and elderly Japanese men over an average of 13.2 years found a positive relationship between regular mushroom consumption and decreased incidence of prostate cancer [104]. Table 5 summarizes the classes of compounds and health benefits from some mushroom species cited in this section.

Table 5. Mushroom-originated compounds and health benefits related to non-transmissible chronic diseases (NTCD).

Mushroom Species	Popular Name	Origin	Related Compounds	Health Benefits	References
G. lucidum	Reishi	China and Eastern Asia	Polysaccharides	Antioxidant activity related to DAF-16/FOXO activation	[85]
H.sinensis	Caterpillar	Tibet	Polysaccharides	Prebiotic properties related to insulin resistance and diabetes control	[87,88]
L. edodes	Shitake	Eastern Asia	Spermidine **(39)**	Reduction on age-dependent memory impairment	[91]
A. bisporus	Champignon	Eastern Europe	Lovastatin **(40)**	LDL-cholesterol lowering	[94]
C. cibarius	Chanterelle				[92]
I. badia	Bay bolete				[92]
L. edodes	Shitake				[92]
C. militaris	Caterpillar	China, Tibet	Polysaccharides	Immunomodulation improvement	[94,95]
			Cordycepin **(41)**	Total and LDL-cholesterol lowering, reduction of hyperlipidemia caused by high-fat diets	[96]
A. subrufescens	Sun Mushroom	Eastern North America	Polyphenols, polysaccharides	Decrease of oxidative stress, preventing diseases like cancer and inflammation	[100–102]
A. cinnamomea	Niu-Chang-Chih	Taiwan	Antcins **(42)**		[103]

5. Toward a Sustainable Production of Fungal Metabolites

As the demand for preventive medicines, nutraceuticals, new drugs, food additives, and other health-related products of natural origin grows, the need for scaling up is also increasing. For bioactive metabolites of plant origin, efforts to increase production may be slow because production sometimes dependent on the seasonality of plants and a long plant growth period. Thus, efforts to increase the production of bioactive compounds have been directed toward microorganism-based options, such as metabolic engineering. Modifications in the culturing of fungi have been successful in the yield improvement of biomass and bioactive compounds. Moreover, special consideration is being given to endophytic fungi, especially those able to produce metabolites biosynthesized by their host plants. Table 6 shows some interesting examples of bioactive compounds produced by endophytic fungi after optimization of fermentation conditions. The good outcomes in this area were exemplified by Torres-Mendoza et al. [105] who reported, from 2001 to 2019, 224 patents related to metabolites from endophytic fungi applied to agricultural, biotechnology, pharmaceutical, and food industries, most of which used species from the *Aspergillus*, *Fusarium*, *Trichoderma*, *Penicillium*, and *Phomopsis* genera.

Table 6. Bioactive compounds produced by endophytic fungi under different fermentation conditions.

Fungal Species	Host Plant	Target Compound	Health Benefit	Methodology	Target Parameters	Reference
A.terreus	Coconut tree	L-Asparaginase	Treatment of acute lymphocytic leukemia	Factorial experimental design. Increase of scale(5-l bioreactor system).	pH, temperature, inoculum concentration.	[106]
F. solani	*Chonemorpha fragrans*	Camptothecin	Anticancer	Box–Behnken design using one factor at a time method.	Carbon and nitrogen sources, ethanol concentration, pH, temperature, stirring speed, incubation period, precursors and elicitors.	[107]
Penicillium bilaiae	*Phoenix dactylifera*	Acidic protease	Increasing in food digestibility	Response surface methodology. Plackett-Burman design. Box-Behnken design.	Temperature, initial pH, carbon and nitrogen sources, metal ions, detergents and enzyme inhibitors.	[108]
P. ostreatus	ni	Lovastatin **(40)**	Anti-hypercholesterolemic	Response surface methodology.	Nutrients, particle size of the solid substrate, temperature, incubation time.	[109]
Meyerozymaguilliermondii	leaves of *Hibiscus rosa-sinensis*			One parameter at time approach.	Nutrients, pH, inoculum size, temperature, addition of metallic ions, modulators, precursors.	[110]
A.niger	ni	Urease	Diuretic	Response surface methodology.	Strains, incubation time, temperature, pH, biomass, inoculum size, nitrogen content and moisture.	[111]
Spissiomycesendophytica	*Balanophorafungosa*	Melanin	Radioprotective, thermoregulator, antitumor, and antiviral	One parameter at time approach.	Inhibitors, culture medium, temperature, pH.	[112]
E. nigrum	*Taxus baccata*	Taxol	Anticancer	One parameter at a time approach. Mutant strains.	Culture medium, stirring speed, temperature, incubation period, pH, medium volume, inoculum age, inoculum size, carbon source, nitrogen source, phosphorus source, gamma radiation dose.	[113]
Alternaria brassicae	*Huperzia serrata*	Huperzine A	Acetylcholinesterase inhibitor	Multifactorial statistical approaches. Plackett–Burman. Central composite designs.	Culture medium composition, medium volume, inoculum age, inoculum size, incubation period, ethanol addition, pH, temperature.	[114]
F. oxysporum	*Dioscoreazingiberensis*	Diosgenin	Anti-cancer, anti-thrombic, anti-diabetic, cardioprotective, osteoarthritis protective activity	One parameter at time approach.	Culture medium, antibiotics, temperature.	[115]
E. nigrum	*Terminalia arjuna*	Digoxin	Regulating the heart rhythm and strengthening heart diffusion	One parameter at time approach.	Culture medium, temperature, elicitors, incubation time, pH, medium volume, inoculum age, inoculum size, gamma irradiation mutagenesis.	[113]
Penicillium mallochii	A beech tree bark from Balikesir, Turkey	Orange-red pigment	Decreasing in allergic responses to synthetic pigments	One parameter at time approach.	Culture medium, pH, temperature.	[116]
Tausonia pullulans	*Vinca minor*	Vincamine	Improvement of cerebrovascular and cognitive disorders and reduction the effects of certain types of stroke	Protoplasts optimization.	Incubation time, temperature, modulators, protoplast inactivation method (heat, ultraviolet, microwave, sodium nitrite, and diethyl sulfate).	[117]

ni = not informed.

Yield enhancement in microbial production is one of the most challenging issues, but good progress has been reported. The production of fungal metabolites can also be optimized by co-cultivation with other fungal or bacterial species and variation of chemical parameters, such as composition of the culture medium, and physical parameters, such as temperature, pH, stirring speed, intensity and color of light, and oxygenation. The metabolic modulation resulting from these approaches depends on the fungal species; thus different fungi have been explored to produce fungal metabolites with new industrial applications around the world [38].

Biosynthesis of the natural polyamide agmatine **(28)** by *A. oryzae*, a fungus generally recognized as safe, was described during the fermentative process of sake production (Japanese rice wine) [118]. Optimization of the initial yield of agmatine **(28)** produced by *A. oryzae* in the presence of *S. cerevisiae* in a solid state (3.5 mM agmatine) was achieved by varying some fermentative parameters. At pH 5.3, the production of agmatine **(28)** increased to 6.3 mM. An increase of over 100% in the initial productivity was obtained by adding L-lactic (pH 3.0, 8.2 mM agmatine), succinic acid (pH 3.5, 8.7 mM agmatine), and citric acid **(2)** (pH 3.2, 8.3 mM agmatine) in the fermentation medium. Therapeutic evidences indicate that agmatine **(28)** is a promising lead compound against several NTCD that affect the central nervous system, such as Alzheimer's disease [119]. In 2019 only, the number of people affected by dementia was estimated at 50 million, with the worrying forecast of this number to be trebled by 2050 [120]. Therefore, new molecules for the treatment of Alzheimer's disease and other types of dementia are highly desired. The pharmacological properties and yield of agmatine **(28)** in fermentation by *A. oryzae* suggest the possibility of its incorporation into nutraceuticals [119]. In excess, agmatine **(28)** may present toxicity and can enhance the toxic action of other biogenic amines, such as histamine and tyramine, produced by the decarboxylation of amino acids by food-fermenting microorganisms. Therefore, determining a proper intake for agmatine **(28)** is necessary to mitigate risks and allows for the numerous aforementioned health benefits [16].

The preference for faster, less expensive, and green approaches has sped up research on bioactive fungal metabolites. Some promising analytical tools, such as matrix-assisted laser desorption ionization time-of-flight mass spectrometry (MALDI-ToF MS), have been useful for addressing some bottleneck problems in this area, such as fungal identification. Genetic and morphological methods, although very efficient, require expertise, time, and resources that may be saved by the use of analytical instrumentation [121]. Such improvements start during screening steps, in the initial laboratory prospection of relevant species and promising fermentative conditions, and continue to the production steps, with the use of alternative substrates for fungal growth, such as agro industrial residues, towards a circular economy. Thus, extracts intended for screening are prepared on a reduced scale, using lower amounts of reagents, such as solvents, during extraction. In addition, modern analytical tools coupled with hyphenated techniques have been recently used for the analysis of a greater number of extracts, without the need for prior isolation of fungal metabolites from extracts. Biochemometrically derived fingerprints obtained by gas chromatography, high-performance liquid chromatography, MS, and nuclear magnetic resonance spectroscopy (NMR), together with statistical analysis also enable the establishment of straightforward associations between metabolite content and biological activities detected for a large number of crude extracts [70,122].

Although MS is the most widely used technique for this kind of analysis [123], NMR, in particular ^{1}H NMR metabolomic analysis, has gained ground by enabling the nondestructive identification of molecules in complex mixtures originating from various biological materials, such as plants and fungi extracts, foods, biological fluids, and tissues [124,125]. NMR is comparable to MS in criteria, such as efficiency, speed, reproducibility, and ease of sample preparation; however, it has advantages in terms of sample recovery and isotope detection. It is also independent of ionization potential and does not require complex internal standards [126]. Bi-dimensional biochemometric NMR evaluation of the crude extract of a marine-derived strain of *P. chrysogenum* enabled the attribution of the anti-proliferative activity of the extract against breast cancer cell lines to ergosterol **(43)** (Figure 4; Table 3), a structural metabolite common in fungi [70]. In addition, other metabolites of the ergostane class, ergosterol peroxide **(44)**, 3β, 5α, 9α, 14 α-tetrahydroxy-ergosta-7,22-dien-6-one **(45)**, and 3β, 5α, 9α-trihydroxy-ergosta-7,22-dien-6-one **(46)** (Figure 4), were detected as metabolites of the edible mushroom *V. volvacea* and proved to be active against human cancer cell lines (Table 3) [68].

An interesting application of NMR-based metabolomics for the comparison and origin identification of edible mushrooms was reported. Multivariate analyses, such as PCA, were used to compare the chemical profiles of species, enabling the differentiation of *Kuehneromyces mutabilis* and *Hypholom acapnoides,* and the direct identification of 17 secondary metabolites in their extracts. Statistically significant differences were observed inthe^{1}H NMR data in the variation of composition based on the collection site and restrictedness of metabolites in *K. mutabilis*. Upon comparing *H. capnoides* and *K. mutabilis*, a higher diversity of sugars was observed in *H. capnoides*, while *K. mutabilis* presented significantly higher amounts of some organic compounds, such as fumaric acid **(3)**, showing the versatility of the ^{1}H NMR technique for comparing fungi in terms of metabolites [127].

Other sophisticated approaches combining UPLC–ESI-TOF/MS, differential off-line LC–NMR, and quantitative^{1}H NMR (qHNMR) analysis were used to identify D-Phe-L-Val-D-Val-L-Tyr **(47)**, D-Phe-L-Val-D-Val-L-Phe **(48)**, and cis-bis(methylthio)silvatin **(49)** (Figure 4) in the aerobic fermentation of *Penicillium roqueforti* with and without L-Tryptophan enrichment. Another metabolite, roquefortine C **(50)** (Figure 4), was identified as an antimicrobial agent against *Bacillus subtilis* and *E. coli* [126]. ^{1}H NMR-based metabolomics, applied in a study on extracts of 11 species of edible mushrooms, identified dry *Pleurotus geesteranum* and *Hericium erinaceus* and fresh *Pleurotus sapidus* as the most prominent species in terms of metabolite production. In addition, the use of ^{1}H NMR led to the detection of more than 100 different metabolites of interest to the food industry including trehalose **(51)**, mannitol **(52)**, and glucose **(53)** (Figure 1). The analysis also showed high levels of carbohydrates and proteins, in addition to considerable amounts of vitamins A **(54)** and C **(55)** (Figure 1) and amino acids as nutritional compounds [14]. Therefore, NMR-based metabolomics has been successful in selecting fungi with outstanding potential for further technological development. Metabolomic tools have also been applied for the evaluation of organoleptic properties and interference of external factors with food quality [128] and nutrient quantification such as in the determination of the nutritional contents of 11 *Capsicum annuum* cultivars in terms of ascorbic acid (vitamin C) **(55)** [129]. In the near future, once there is sufficient data in this area, these useful metabolomic tools could be applied in several meta-studies such as for the determination of vitamin C **(55)** content in edible mushrooms. One study [130] reported vitamin C contents between 31.16 ± 0.93 (*Calvatia gigantea*) and 108.11 ± 3.22 (*Lepis tagilva*) mg/kg dry matter. In another study [14], the vitamin C contents ranged from 0.5 to 111.4 mg/100g dry weight. The highest levels of vitamin C were produced by the species *D. indusiata* and *A. subrufescens* at 111.4 and 69.7 mg/100 g dry weight, respectively [14], values close to the daily levels of vitamin C **(55)** recommended in some countries (75 to 110 mg) [131].

Metabolomic approaches have also been applied to the quantification of aflatoxins in industrialized baby food [132], to determine adulteration and its effect on the safety of nutraceuticals [133], and for differentiating extracts of various biomaterials such as crops [134] and fungi [14].

While metabolomic tools provide speed for research, technological development for future industrial applications of fungal metabolites must consider that the modern economy is increasingly challenged to transform traditional processes into sustainable production chains, minimizing waste, and reusing biomass for applications in the food and drug industries, even for well-established processes such as citric acid **(2)** production. The global market value of citric acid **(2)** was estimated to increase to USD 3.6 billion in 2020 and maintain an annual growth of 5% until 2025. Despite the successful industrial experience for citric acid **(2)** production, research to increase sustainability in bio-refineries is ongoing, opening new possibilities such as clustering-related fermentation production including olive oil and wine to reach models for circular bioeconomy [135].

Fortunately, the production of nutraceuticals and functional foods from biomass, extracts, and fungal metabolites greatly addresses the challenge of sustainability improvement, allowing the incorporation of low-cost, underutilized, and abundant materials into the industrial fermentation process. Many agro-industrial residues available all over the world are sources of peptides, prebiotic dietary fibers, and hydrolyzed or smaller organic molecules such as phenolics, carotenoids, and tocopherols, among other classes. These features make them good materials for fungal cultivation as well as interesting substrates for the production of bioactive components from macromolecules [59,136]. For example, fermentation of pomegranate bark residue by *A. niger* resulted in the production of citric acid **(2)** with reduced production costs, while adding value to a residue usually directed for disposal [137]. High yields (7.8 to 14.4 g/L) of dry biomass, with protein contents ranging from 33.8 to 50.8%, and ethanol were recovered from the fermentation of effluents from the wheat starch industry by the fungi *A. oryzae* and *R. oryzae*. This kind of process contributes to the development of a circular economy, because the large volume and high chemical demand of oxygen in effluents from the starch industry represent an environmental burden and an additional cost as they create a need for effluents to be treated before being discarded [138]. The filamentous fungi *Actinomucor elegans* and *Umbelopsis isabelin* were simultaneously used for the enrichment of white grape-producing bagasse, a residue with little nutritional attractiveness. The fungi were able to increase the γ-linolenic acid and carotenoid contents, improving the nutritional content of bagasse, and thus creating the possibility of reintroducing bagasse into the production chain as a low-cost functional food for humans [139].

6. Conclusions

Fungi and their metabolites have important industrial applications in high-value-added products and have potential for the development of nutraceuticals that can contribute to the prevention of NTCD and improve health, especially in terms of human aging. In addition, fungi are suitable in the production of natural food additives such as colorants and stabilizers that have lower health risks than synthetic food additives, and bioactive metabolites for pharmacological use such as enzymes, statins, and antitumor agents. Fungal antioxidants have applications in both food preservation and the combat of oxidative stress in the human body, with positive outcomes for several diseases such as cancer. The use of metabolic engineering techniques has facilitated the overcoming of some obstacles to explore the pharmacological potential of fungi, even those producing toxic substances such as some species of the genera *Monascus*, *Aspergillus*, and *Fusarium*. Modern approaches have been successfully utilized to evaluate the interference of additives derived from fungi with the organoleptic properties and quality of food. Strategies currently available for scaling up metabolite production include direct genetic alteration with tools such as CRISPR-Cas9 and gene recombination. Research on the use of agro-industrial byproducts for sustainable fungal fermentation has shed light on its remarkable economic importance to the production of natural additives, food, drugs, and nutraceuticals. Further in vivo antioxidant activity studies of fungal metabolites are still scarce; however, new insights are required to expand the use of metabolites from filamentous fungi to improve human health in the 21st century.

Author Contributions: All authors participated in the layout of this review. J.A.T., B.V.R.B., B.d.A.M., C.P.G., and M.A.F.M. identified the literature for review and drafted the manuscript. In addition, J.A.T. conceived and critically revised the manuscript, B.V.R.B. and B.d.A.M. provided the tables and figures, and C.P.G. and M.A.F.M. formatted the document. All authors have read and approved the final version of the manuscript.

Funding: The authors gratefully acknowledge financial support from the Brazilian Agencies Fundação de Amparo à Pesquisa do Estado de Minas Gerais (FAPEMIG PPM-00255-18); Conselho Nacional de Desenvolvimento Científico e Tecnológico (CNPq Grants 304922/2018-8, 141601/2018-3, 142517/2018-6, 121994/2019-8); the National Institute of Science and Technology–INCT BioNat/CNPq (Grant 465637/2014-0); and the Coordenação de Aperfeiçoamento de Pessoal de Nível Superior (CAPES)–Finance Code 001.

Acknowledgments: We thank Michel Almeida for kindly preparing the graphical abstract illustration.

Conflicts of Interest: The authors declare no conflict of interest.

References

1. Piffieri, F.; Aujard, F. Caloric restriction, longevity and aging: Recent contributions from human and non-human primate studies. *Prog. Neuropsychopharmacol. Biol. Psychiatry* **2019**, *95*, 109702. [CrossRef] [PubMed]
2. Aiello, A.; Accardi, G.; Caruso, C.; Candore, G. Effects of nutraceuticals of Mediterranean diet on aging and longevity. In *The Mediterranean Diet: An Evidence-Based Approach*, 2nd ed.; Preedy, V.R., Wastson, R.R., Eds.; Academic Press: Cambridge, MA, USA, 2020; pp. 547–553. [CrossRef]
3. Zhang, Z.; He, S.; Cao, X.; Ye, Y.; Yang, L.; Wang, J.; Liu, H.; Sun, H. Potential prebiotic activities of soybean peptides Maillard reaction products on modulating gut microbiota to alleviate aging-related disorders in D-galactose-induced ICR mice. *J. Funct. Foods* **2020**, *65*, 103729. [CrossRef]
4. Cava, E.; Fontana, L. Will calorie restriction work in humans? *Aging* **2013**, *5*, 507–514. [CrossRef] [PubMed]
5. Martel, J.; Ojcius, D.M.; Ko, Y.F.; Chang, C.J.; Young, J.D. Antiaging effects of bioactive molecules isolated from plants and fungi. *Med. Res. Rev.* **2019**, *39*, 1515–1552. [CrossRef] [PubMed]
6. Sachdeva, V.; Roy, A.; Bharadvaja, N. Current prospects of nutraceuticals: A review. *Curr. Pharm. Biotechnol.* **2020**, *21*, 884–896. [CrossRef]
7. Ruchi, S. Role of nutraceuticals in health care: A review. *Int. J. Green Pharm.* **2017**, *11*, S385–S394. [CrossRef]
8. Padmavathi, D. A general review on "nutraceuticals": Its golden health impact over human community. *Int. J. Food. Sci. Nut.* **2018**, *3*, 214–217.
9. Tomova, A.; Bukovsky, I.; Rembert, E.; Yonas, W.; Alwarith, J.; Barnard, N.D.; Kahleova, H. The effects of vegetarian and vegan diets on gut microbiota. *Front. Nutr.* **2019**, *6*, 47. [CrossRef]
10. Meyer, V.; Basenko, E.Y.; Philipp Benz, J.; Braus, G.H.; Caddick, M.X.; Csukai, M.; Vries, R.P.; Endy, D.; Frisvad, J.C.; Gunde-Cimerman, N.; et al. Growing a circular economy with fungal biotechnology: A white paper. *Fungal Biol. Biotechnol.* **2020**, *7*, 1–23. [CrossRef]
11. Kumari, K. Mushrooms as source of dietary fiber and its medicinal value: A review article. *J. Pharmacogn. Phytochem.* **2020**, *9*, 2075–2078.
12. Kato, T.; Azegami, J.; Yokomori, A.; Dohra, H.; Enshasy, H.A.; Park, E.Y. Genomic analysis of a riboflavin—Overproducing *Ashbya gossypii* mutant isolated by disparity mutagenesis. *BMC Genom.* **2020**, *21*, 1–17. [CrossRef] [PubMed]
13. Jesus, L.F.M.C. Produção de β-Galactosidase Por Fungos Filamentosos: Screening, Purificação e Caracterização Bioquímica. Master's Thesis, Universidade Estadual Paulista "Julio Mesquita Filho", UNESP, São Paulo, Brazil, 2020.
14. Liu, D.; Chen, Y.Q.; Xiao, X.W.; Zhong, R.T.; Yang, C.F.; Liu, B.; Zhao, C. Nutrient properties and nuclear magnetic resonance-based metabonomic analysis of macrofungi. *Foods* **2019**, *8*, 397. [CrossRef] [PubMed]
15. Koutrotsios, G.; Kalogeropoulos, N.; Stathopoulos, P.; Kaliora, A.C.; Zervakis, G.I. Bioactive compounds and antioxidant activity exhibit high intraspecific variability in *Pleurotusostreatus* mushrooms and correlate well with cultivation performance parameters. *World J. Microb. Biotechnol.* **2017**, *33*, 98. [CrossRef] [PubMed]
16. Akasaka, N.; Fujiwara, S. The therapeutic and nutraceutical potential of agmatine, and its enhanced production using *Aspergillus oryzae*. *Amino Acids* **2020**, *52*, 181–197. [CrossRef]
17. Chan, L.G.; Dias, F.F.; Saarni, A.; Cohen, J.; Block, D.; Taha, A.Y.; de Moura Bell, J.M. Scaling up the bioconversion of cheese whey permeate into fungal oil by *Mucor circinelloides*. *J. Am. Oil Chem. Soc.* **2020**, *97*, 703–716. [CrossRef]
18. Mamani, L.D.G.; Magalhães, A.I., Jr.; Ruan, Z.; de Carvalho, J.C.; Soccol, C.R. Industrial production, patent landscape, and market trends of arachidonic acid-rich oil of *Mortierella alpina*. *Biotech. Res. Innov.* **2019**, *3*, 103–119. [CrossRef]
19. Anand, S.; Singh, K.S.; Aggarwal, D. Expanding Avenues for Probiotic Yeast. In *Microbial Cell Factories*, 1st ed.; Sharma, D., Saharan, B.S., Eds.; CRC Press: Boca Raton, FL, USA, 2018; pp. 125–141.
20. Baggio, L.M.; Panagio, L.A.; Gasparin, F.G.M.; Sartori, D.; Celligoi, M.A.P.C.; Baldo, C. Production of fibrinogenolytic and fibrinolytic enzymes by a strain of *Penicillium* sp. isolated from contaminated soil with industrial effluent. *Acta Sci. Health Sci.* **2019**, *41*, e40606. [CrossRef]
21. Zhang, S.; Wang, Y.; Zhang, N.; Sun, Z.; Shi, Y.; Cao, X.; Wang, H. Purification and characterization of a fibrinolytic enzyme from *Rhizopus microsporus* var. tuberosus. *Food Technol. Biotech.* **2015**, *53*, 243–248. [CrossRef]

22. Huang, Z.; Brennan, C.S.; Zheng, H.; Mohan, M.S.; Stipkovits, L.; Liu, W.; Kulasiri, D.; Guan, W.; Zhao, H.; Liu, J. The effects of fungal lipase-treated milk lipids on bread making. *LWT* **2020**, 109455. [CrossRef]
23. Hjortmo, S.B.; Hellström, A.M.; Andlid, T.A. Production of folates by yeasts in Tanzanian fermented togwa. *FEMS Yeast Res.* **2008**, *8*, 781–787. [CrossRef]
24. Copetti, M.V. Fungi as industrial producers of food ingredients. *Curr. Opin. Food Sci.* **2019**, *25*, 52–56. [CrossRef]
25. Dikshit, R.; Tallapragada, P. Comparative Study of Natural and Artificial Flavoring Agents and Dyes. In *Natural and Artificial Flavoring Agents and Food Dyes*; Grumezescu, A.M., Holban, A.M., Eds.; Elsevier: London, UK, 2018; Volume 7, pp. 83–111.
26. Tkaczyk, A.; Mitrowska, K.; Posyniak, A. Synthetic organic dyes as contaminants of the aquatic environment and their implications for ecosystems: A review. *Sci. Total Environ.* **2020**, *717*, 137222. [CrossRef] [PubMed]
27. Venil, C.K.; Velmurugan, P.; Dufossé, L.; Devi, P.R.; Ravi, A.V. Fungal Pigments: Potential coloring compounds for wide ranging applications in textile dyeing. *J. Fungi* **2020**, *6*, 68. [CrossRef]
28. [ACS] American Chemical Society. Dyes, Pigments and Inks. 2020. Available online: https://www.acs.org/content/acs/en/careers/college-to-career/chemistry-careers/dyes-pigments-ink.html (accessed on 8 September 2020).
29. [FDA] U.S. Food and Drug Administration. Color Additives History. 2017. Available online: https://www.fda.gov/industry/color-additives/color-additives-history#:~{}:text=A%20color%20additive%2C%20as%20defined,or%20to%20the%20human%20body.&text=One%20of%20the%20U.S.%20Food,are%20safely%20and%20appropriately%20used (accessed on 8 September 2020).
30. Kalra, R.; Conlan, X.A.; Goel, M. Fungi as a potential source of pigments: Harnessing filamentous fungi. *Front. Chem.* **2020**, *8*. [CrossRef] [PubMed]
31. Dufossé, L. Microbial production of food grade pigments. *Food Technol. Biotech.* **2006**, *44*, 313–321.
32. Sen, T.; Barrow, C.J.; Deshmukh, S.K. Microbial pigments in the food industry—Challenges and the way forward. *Front. Nutr.* **2019**, *6*. [CrossRef]
33. Lebeau, J.; Petit, T.; Clerc, P.; Dufossé, L.; Caro, Y. Isolation of two novel purple naphtoquinone pigments concomitant with the bioactive red bikaverin and derivates thereof produced by *Fusarium oxysporum*. *Biotech. Prog.* **2019**, *35*. [CrossRef]
34. Slack, G.J.; Puniani, E.; Frisvad, J.C.; Samson, R.A.; Miller, J.D. Secondary metabolites from *Eurotium* species, *Aspergillus calidoustus* and *A. insuetus* common in Canadian homes with a review of their chemistry and biological activities. *Mycol. Res.* **2009**, *113*, 480–490. [CrossRef]
35. Mondal, S.; Pandit, S.G.; Puttananjaiah, M.H.; Harohally, N.V.; Dhale, M.A. Structural and functional characterization of new pigment molecule monashin from *Monascuspurpureus* CFR410-11. *Process. Biochem.* **2019**, *82*, 173–178. [CrossRef]
36. Wu, H.C.; Cheng, M.J.; Wu, M.D.; Chen, J.J.; Chen, Y.L.; Chang, H.S.; Chen, K.P. Secondary metabolites from the fermented rice of the fungus *Monascus purpureus* and their bioactivities. *Nat. Prod. Res.* **2019**, *33*, 3541–3550. [CrossRef]
37. Kraboun, K.; Kongbangkerd, T.; Rojsuntornkitti, K.; Phanumong, P. Factors and advances on fermentation of *Monascus* sp. for pigments and monacolin K production: A review. *Int. Food. Res. J.* **2019**, *26*, 751–761.
38. Shi, Y.C.; Pan, T.M.; Liao, V.H.C. Monascin from *Monascus*-fermented products reduces oxidative stress and amyloid-beta toxicity via DAF-16/FOXO in *Caenorhabditis elegans*. *J. Agric. Food. Chem.* **2016**, *64*, 7114–7120. [CrossRef] [PubMed]
39. Hsu, W.H.; Chen, T.H.; Lee, B.H.; Hsu, Y.W.; Pan, T.M. Monascin and ankaflavin act as natural AMPK activators with PPARα-agonist activity to down-regulate nonalcoholic steatohepatitis in high-fat diet-fed C57BL/6 mice. *Food Chem. Toxicol.* **2014**, *64*, 94–103. [CrossRef] [PubMed]
40. Venkatachalam, M.; Shum-Chéong-Sing, A.; Dufossé, L.; Fouillaud, M. Statistical optimization of the physico-chemical parameters for pigment production in submerged fermentation of *Talaromyces albobiverticillius* 30548. *Microorganisms* **2020**, *8*, 711. [CrossRef]
41. Palacio-Barrera, A.M.; Areiza, D.; Zapata, P.; Atehortúa, L.; Correa, C.; Peñuela-Vásquez, M. Induction of pigment production through media composition, abiotic and biotic factors in two filamentous fungi. *Biotechnol. Rep.* **2018**, *20*, e00308. [CrossRef]

42. Osemwegie, O.O.; Adetunji, C.O.; Ayeni, E.A.; Adejobi, O.I.; Arise, R.O.; Nwonuma, C.O.; Oghenekaro, A.O. Exopolysaccharides from bacteria and fungi: Current status and perspectives in Africa. *Heliyon* **2020**, *6*, e04205. [CrossRef]
43. Mahapatra, S.; Banerjee, D. Fungal exopolysaccharide: Production, composition and applications. *Microbiol. Insights* **2013**, *6*, 1–16. [CrossRef]
44. Iyyappan, J.; Bharathiraja, B.; Baskar, G.; Kamalanaban, E. Process optimization and kinetic analysis of malic acid production from crude glycerol using *Aspergillus niger*. *Bioresour. Technol.* **2019**, *281*, 18–25. [CrossRef]
45. Saeed, S.; Aslam, S.; Mehmood, T.; Naseer, R.; Nawaz, S.; Mujahid, H.; Firyal, S.; Anjum, A.A.; Sultan, A. Production of gallic acid under solid-state fermentation by utilizing waste from food processing industries. *Waste Biomass Valoriz.* **2020**, 1–9. [CrossRef]
46. Costa, C.R.L.M.; Menolli, R.A.; Osaku, E.F.; Tramontina, R.; Melo, R.H.; Amaral, A.E.; Duarte, P.A.D.; Carvalho, M.M.; Smiderle, F.R.; Silva, J.L.C.; et al. Exopolysaccharides from *Aspergillus terreus*: Production, chemical elucidation and immunoactivity. *Int. J. Biol. Macromol.* **2019**, *139*, 654–664. [CrossRef]
47. Fooladi, T.; Soudi, M.R.; Alimadadi, N.; Savedoroudi, P.; Heravi, M.M. Bioactive exopolysaccharide from *Neopestalotiopsis* sp. strain SKE15: Production, characterization and optimization. *Int. J. Biol. Macromol.* **2019**, *129*, 127–139. [CrossRef]
48. Xu, X.; Xu, R.; Jia, Q.; Feng, T.; Huang, Q.; Ho, C.T.; Song, S. Identification of dihydro-β-ionone as a key aroma compound in addition to C8 ketones and alcohols in *Volvariella volvacea* mushroom. *Food Chem.* **2019**, *293*, 333–339. [CrossRef] [PubMed]
49. Zeng, X.; Zeng, H.; Meng, Y.; Xu, D.; Zhang, B.; Xin, B.; Liu, S.; Zhai, B.; Yu, F.; Zhu, M.; et al. Continuous natamycin production by using immobilized *Streptomyces gilvosporeus* Z8 via repeated batch culture. *J. Chem. Technol. Biotechnol.* **2020**, *95*, 73–77. [CrossRef]
50. Bin, C.; Fangming, Y.; Zhi, J. Mogroside V-producing endophytic fungi isolated from *Siraitia grosvenorii*. *Planta Med.* **2020**, *86*, 983–987. [CrossRef] [PubMed]
51. Kallscheuer, N. Engineered microorganisms for the production of food additives approved by the European Union—A systematic analysis. *Front. Microbiol.* **2018**, *9*, 1746. [CrossRef]
52. Kovilein, A.; Kubisch, C.; Cai, L.; Ochsenreither, K. Malic acid production from renewables: A review. *J. Chem. Technol. Biotechnol.* **2019**, *95*, 513–526. [CrossRef]
53. Li, Y.; Lin, Z.; Huang, C.; Zhang, Y.; Wang, Z.; Tang, Y.; Chen, T.; Zhao, X. Metabolic engineering of *Escherichia coli* using CRISPR-Cas9 mediated genome editing. *Metab. Eng.* **2015**, *31*, 13–21. [CrossRef] [PubMed]
54. Tramontin, L.R.R.; Kildegaard, K.R.; Sudarsan, S.; Borodina, I. Enhancement of astaxanthin biosynthesis in oleaginous yeast *Yarrowialipolytica* via *microalgal pathway*. *Microorganisms* **2019**, *7*, 472. [CrossRef] [PubMed]
55. Harith, Z.T.; Charalampopoulos, D.; Chatzifragkou, A. Rapeseed meal hydrolysate as substrate for microbial astaxanthin production. *Biochem. Eng. J.* **2019**, *151*, 107330. [CrossRef]
56. Hwang, S.W.; Choi, H.I.; Sim, S.J. Acidic cultivation of *Haematococcuspluvialis* for improved astaxanthin production in the presence of a lethal fungus. *Bioresour. Technol.* **2019**, *278*, 138–144. [CrossRef] [PubMed]
57. Ip, P.F.; Wong, K.H.; Chen, F. Enhanced production of astaxanthin by the green microalga *Chlorella zofingiensis* in mixotrophic culture. *Process. Biochem.* **2004**, *39*, 1761–1766. [CrossRef]
58. Tang, D.Y.Y.; Khoo, K.S.; Chew, K.W.; Tao, Y.; Ho, S.H.; Show, P.L. Potential utilization of bioproducts from microalgae for the quality enhancement of natural products. *Bioresour. Technol.* **2020**, *304*, 122997. [CrossRef] [PubMed]
59. Villena, G.K.; Kitazono, A.A.; Hernández-Macedo, M.L. Bioengineering Fungi and Yeast for the Production of Enzymes, Metabolites, and Value-Added Compounds. In *Fungal Biotechnology and Bioengineering*; Hesham, A.E.L., Upadhyay, R.S., Sharma, G.D., Manoharachary, C., Gupta, V.K., Eds.; Springer: Cham, Switzerland, 2020; pp. 209–237. [CrossRef]
60. Sun, X.; Su, X. Harnessing the knowledge of protein secretion for enhanced protein production in filamentous fungi. *World J. Microbiol. Biotech.* **2019**, *35*, 54. [CrossRef] [PubMed]
61. Kumitch, H.M.; Stone, A.; Nosworthy, M.G.; Nickerson, M.T.; House, J.D.; Korber, D.R.; Tanaka, T. Effect of fermentation time on the nutritional properties of pea protein-enriched flour fermented by *Aspergillus oryzae* and *Aspergillus niger*. *Cereal Chem.* **2019**, *97*, 104–113. [CrossRef]

62. Gmoser, R.; Fristedt, R.; Larsson, K.; Undeland, I.; Taherzadeh, M.J.; Lennartsson, P.R. From stale bread and brewers spent grain to a new food source using edible filamentous fungi. *Bioengineered* **2020**, *11*, 582–598. [CrossRef]
63. Asfour, H.Z.; Awan, Z.A.; Bagalagel, A.A.; Elfaky, M.A.; Abdelhameed, R.F.A.; Elhady, S.S. Large-scale production of bioactive terrein by *Aspergillus terreus* strain s020 isolated from the Saudi Coast of the Red Sea. *Biomolecules* **2019**, *9*, 480. [CrossRef]
64. Lee, Y.H.; Lee, S.J.; Jung, J.E.; Kim, J.S.; Lee, N.H.; Yi, H.K. Terrein reduces age-related inflammation induced by oxidative stress through Nrf2/ERK1/2/HO-1 signalling in aged HDF cells. *Cell Biochem. Funct.* **2015**, *33*, 479–486. [CrossRef]
65. [WHO]. *WHO Report on Cancer: Setting Priorities, Investing Wisely and Providing Care for All*; World Health Organization: Geneva, Switzerland, 2020; Available online: https://www.who.int/publications-detail-redirect/who-report-on-cancer-setting-priorities-investing-wisely-and-providing-care-for-all (accessed on 28 August 2020).
66. Ren, F.; Zhu, S.; Wang, B.; Li, L.; Liu, X.; Su, R.; Che, Y. Hypocriols AF heterodimeric botryane ethers from *Hypocrea* sp., an insect-associated fungus. *J. Nat. Prod.* **2016**, *79*, 1848–1856. [CrossRef] [PubMed]
67. Zhang, J.; Liu, S.S.; Yuan, W.Y.; Wei, J.J.; Zhao, Y.X.; Luo, D.Q. Carotane-type sesquiterpenes from cultures of the insect pathogenic fungus *Isaria fumosorosea*. *J. Asian Nat. Prod. Res.* **2017**, *21*, 1–7. [CrossRef]
68. Chen, P.; Qin, H.J.; Li, Y.W.; Ma, G.X.; Yang, J.S.; Wang, Q. Study on chemical constituents of an edible mushroom *Volvariella volvacea* and their antitumor activity *in vitro*. *Nat. Prod. Res.* **2018**, *34*, 1–6. [CrossRef]
69. Chowdhury, N.S.; Sohrab, M.H.; Rana, M.S.; Hasan, C.M.; Jamshidi, S.; Rahman, K.M. Cytotoxic naphthoquinone and azaanthraquinone derivatives from an endophytic *Fusarium solani*. *J. Nat. Prod.* **2017**, *80*, 1173–1177. [CrossRef] [PubMed]
70. Ory, L.; Nazih, E.H.; Daoud, S.; Mocquard, J.; Bourjot, M.; Margueritte, L.; Delsuc, M.A.; Bard, J.M.; Pouchus, Y.F.; Bertrand, S.; et al. Targeting bioactive compounds in natural extracts—Development of a comprehensive workflow combining chemical and biological data. *Anal. Chim. Acta* **2019**, *1070*, 29–42. [CrossRef] [PubMed]
71. Varga, J.; Baranyi, N.; Chandrasekaran, M.; Vágvölgyi, C.; Kocsubé, S. Mycotoxin producers in the Aspergillus genus: An update. *Acta. Biol. Szeged.* **2015**, *59*, 151–167.
72. Abdel-Azeem, A.M.; Abdel-Azeem, M.A.; Darwish, A.G.; Nafady, N.A.; Ibrahim, N.A. Fusarium: Biodiversity, Ecological Significances, and Industrial Applications. In *Recent Advancement in White Biotechnology through Fungi, Fungal Biology*; Yadav, A.N., Singh, S., Mishra, S., Gupta, A., Eds.; Springer Nature: Cham, Switzerland, 2019; Volume 3, pp. 201–261.
73. Neha, K.; Haider, M.R.; Pathak, A.; Yar, M.S. Medicinal prospects of antioxidants: A review. *Eur. J. Med. Chem.* **2019**, *178*, 687–704. [CrossRef]
74. Wolfe, K.L.; Liu, R.H. Cellular antioxidant activity (CAA) assay for assessing antioxidants, foods, and dietary supplements. *J. Agric. Food Chem.* **2007**, *55*, 8896–8907. [CrossRef]
75. Granato, D.; Shahidi, F.; Wrolstad, R.; Kilmartin, P.; Melton, L.D.; Hidalgo, F.J.; Miyashita, K.; Camp, J.; Alasalvar, C.; Ismail, A.B.; et al. Antioxidant activity, total phenolics and flavonoids contents: Should we ban *in vitro* screening methods? *Food Chem.* **2018**, *264*, 471–475. [CrossRef]
76. Hameed, A.; Hussain, S.A.; Yang, J.; Ijaz, U.M.; Liu, Q.; Suleria, H.A.R.; Song, Y. Antioxidants potential of the filamentous fungi (*Mucor circinelloides*). *Nutrients* **2017**, *9*, 1101. [CrossRef]
77. Furman, D.; Campisi, J.; Verdin, E.; Carrera-Bastos, P.; Targ, S.; Franceschi, C.; Ferrucci, L.; Gilroy, D.W.; Fasano, A.; Miller, G.W.; et al. Chronic inflammation in the etiology of disease across the life span. *Nat. Med.* **2019**, *25*, 1822–1832. [CrossRef]
78. Pan, F.; El-Kashef, D.H.; Kalscheuer, R.; Müller, W.E.G.; Lee, J.; Feldbrügge, M.; Mándi, A.; Kurtán, T.; Liu, Z.; Wu, W.; et al. Cladosins L-O, new hybrid polyketides from the endophytic fungus Cladosporium sphaerospermum WBS017. *Eur. J. Med. Chem.* **2020**, *191*, 112159. [CrossRef]
79. Al-Rabia, M.W.; Mohamed, G.A.; Ibrahim, S.R.M.; Asfoura, H.Z. Anti-inflammatory ergosterol derivatives from the endophytic fungus *Fusarium chlamydosporum*. *Nat. Prod. Res.* **2020**. [CrossRef]
80. Xiao, Y.; Xu, B.; Kang, Y.; Li, Y.; Cui, Y.; Liu, W.; Xiang, Z. A neuroinflammation inhibitor, hypoxylon xanthone A, from soil fungus *Hypoxylon* sp. *Lett. Org. Chem.* **2020**, *17*, 116–120. [CrossRef]
81. Wang, A.; Zhao, S.; Gu, G.; Xu, D.; Zhang, X.; Lai, D.; Zhou, L. Rhizovagine A, an unusual dibenzo-a-pyrone alkaloid from the endophytic fungus *Rhizopycnisvagum nitaf* 22. *RSC Adv.* **2020**, *10*, 27894–27898. [CrossRef]

82. Chapla, V.M.; Honório, A.E.; Gubiani, J.R.; Vilela, A.F.L.; Young, M.C.M.; Cardoso, C.L.; Pavan, F.R.; Cicarelli, R.M.; Ferreira, P.M.P.; Bolzani, V.S.; et al. Acetylcholinesterase inhibition and antifungal activity of cyclohexanoids from the endophytic fungus *Saccharicola* sp. *Phytochem. Lett.* **2020**, *39*, 116–123. [CrossRef]
83. Qi, J.; Zhao, P.; Zhao, L.; Jia, A.; Liu, C.; Zhang, L.; Xia, X. Anthraquinone derivatives from a sea cucumber-derived *Trichoderma* sp. fungus with antibacterial activities. *Chem. Nat. Compd.* **2020**, *56*, 112–114. [CrossRef]
84. Adpressa, D.A.; Connolly, L.R.; Konkel, Z.M.; Neuhaus, G.F.; Chang, X.L.; Pierce, B.R.; Smith, K.M.; Freitag, M.; Loesgen, S. A metabolomics-guided approach to discover *Fusarium graminearum* metabolites after removal of a repressive histone modification. *Fungal Genet. Biol.* **2019**, *132*, 103256. [CrossRef] [PubMed]
85. Giavasis, I. Bioactive fungal polysaccharides as potential functional ingredients in food and nutraceuticals. *Curr. Opin. Biotechnol.* **2014**, *26*, 162–173. [CrossRef] [PubMed]
86. Chuang, M.H.; Chiou, S.H.; Huang, C.H.; Yang, W.B.; Wong, C.H. The lifespan-promoting effect of acetic acid and Rishi polysaccharide. *Bioorg. Med. Chem.* **2009**, *17*, 7831–7840. [CrossRef] [PubMed]
87. Wu, T.R.; Lin, C.S.; Chang, C.J.; Lin, T.L.; Martel, J.; Ko, Y.F.; Ojcius, D.M.; Lu, C.C.; Young, J.D.; Lai, H.C. Gut commensal *Parabacteroides goldsteinii* plays a predominant role in the anti-obesity effects of polysaccharides isolated from *Hirsutella sinensis*. *Gut* **2019**, *68*, 248–262. [CrossRef]
88. Chang, C.J.; Lin, C.S.; Lu, C.C.; Martel, J.; Ko, Y.F.; Ojcius, D.M.; Tseng, S.F.; Wu, T.R.; Chen, Y.Y.M.; Young, J.D.; et al. *Ganoderma lucidum* reduces obesity in mice by modulating the composition of the gut microbiota. *Nat. Commun.* **2015**, *6*, 7489. [CrossRef]
89. Rajtilak, M.; Matt, L.; Brian, M.; Rakesh, M.; Subhash, M.; Carol, C.W.; Christine, S.; Kanniah, R.; Jeffrey, C. The *Aspergillus flavus* spermidine synthase (spds) gene, is required for normal development, aflatoxin production, and pathogenesis during infection of Maize Kernels. *Front. Plant Sci.* **2018**, *9*, 317. [CrossRef]
90. Madeo, F.; Eisenberg, T.; Pietrocola, F.; Kroemer, G. Spermidine in health and disease. *Science* **2018**, *359*. [CrossRef] [PubMed]
91. Kiechl, S.; Pechlaner, R.; Willeit, P.; Notdurfter, M.; Paulweber, B.; Willeit, K.; Werner, P.; Ruckenstuhl, C.; Iglseder, B.; Weger, S.; et al. Higher spermidine intake is linked to lower mortality: A prospective population-based study. *Am. J. Clin. Nutr.* **2018**, *108*, 371–380. [CrossRef] [PubMed]
92. Kała, K.; Kryczyk-Poprawa, A.; Rzewińska, A.; Muszyńska, B. Fruiting bodies of selected edible mushrooms as a potential source of lovastatin. *Eur. Food. Res. Technol.* **2020**, *246*, 713–722. [CrossRef]
93. Dong, Y.; Jing, T.; Meng, Q.; Liu, C.; Hu, S.; Ma, Y.; Liu, Y.; Lu, J.; Cheng, Y.; Wang, D.; et al. Studies on the antidiabetic activities of *Cordyceps militaris* extract in diet-streptozotocin-induced diabetic Sprague-Dawley rats. *Biomed Res. Int.* **2014**, *2014*, 160980. [CrossRef] [PubMed]
94. Wu, L.; Sun, H.; Hao, Y.; Zheng, X.; Song, Q.; Dai, S.; Zhu, Z. Chemical structure and inhibition on α-glucosidase of the polysaccharides from *Cordyceps militaris* with different developmental stages. *Int. J. Biol. Macromol.* **2020**, *148*, 722–736. [CrossRef]
95. He, S.; Wang, Y.; Xie, J.; Gao, H.; Li, X.; Huang, Z. 1H NMR-based metabolomic study of the effects of flavonoids on citrinin production by *Monascus*. *Food Res. Int.* **2020**, *137*, 109532. [CrossRef]
96. Guo, L.; Li, K.; Kang, J.S.; Kang, N.J.; Son, B.G.; Choi, Y.W. Strawberry fermentation with *Cordyceps militaris* has anti-adipogenesis activity. *Food Biosci.* **2020**, *35*, 100576. [CrossRef]
97. Ashraf, S.A.; Elkhalifa, A.E.O.; Siddiqui, A.J.; Patel, M.; Awadelkareem, A.M.; Snoussi, M.; Ashraf, M.S.; Adnan, M.; Hadi, S. Cordycepin for health and wellbeing: A potent bioactive metabolite of an entomopathogenic *Cordyceps* medicinal fungus and its nutraceutical and therapeutic potential. *Molecules* **2020**, *25*, 2735. [CrossRef]
98. Guo, P.; Kai, Q.; Gao, J.; Lian, Z.Q.; Wu, C.M.; Wu, C.A.; Zhu, H.B. Cordycepin prevents hyperlipidemia in hamsters fed a high-fat diet via activation of AMP-activated protein kinase. *J. Pharmacol. Sci.* **2010**, *113*, 395–403. [CrossRef]
99. Kopalli, S.R.; Cha, K.M.; Lee, S.H.; Hwang, S.Y.; Lee, Y.J.; Koppula, S.; Kim, S.K. Cordycepin, an active constituent of nutrient powerhouse and potential medicinal mushroom *Cordyceps militaris* Linn., ameliorates age-related testicular dysfunction in rats. *Nutrients* **2019**, *11*, 906. [CrossRef]
100. Kerrigan, R.W. *Agaricus subrufescens*, a cultivated edible and medicinal mushroom, and its synonyms. *Mycologia* **2005**, *97*, 12–24. [CrossRef]

101. Taofiq, O.; Rodrigues, F.; Barros, L.; Peralta, R.M.; Barreiro, M.F.; Ferreira, I.C.F.R.; Oliveira, M.B.P.P. *Agaricus blazei* Murrill from Brazil: An ingredient for nutraceutical and cosmeceutical applications. *Food Funct.* **2019**, *10*, 565–572. [CrossRef] [PubMed]
102. Navegantes-Lima, K.C.; Monteiro, V.V.S.; Gaspar, S.L.F.; Oliveira, A.L.B.; de Oliveira, J.P.; Reis, J.F.; Gomes, R.S.; Rodrigues, C.A.; Stutz, H.; Sovrani, V.; et al. *Agaricus brasiliensis* mushroom protects against sepsis by alleviating oxidative and inflammatory response. *Front. Immunol.* **2020**, *11*, 1238. [CrossRef] [PubMed]
103. Kumar, K.J.S.; Vani, M.G.; Chen, C.Y.; Hsiao, W.W.; Li, J.; Lin, Z.X.; Chu, F.H.; Yen, G.C.; Wang, S.Y. A mechanistic and empirical review of antcins, a new class of phytosterols of formosan fungi origin. *J. Food Drug Anal.* **2020**, *28*, 38–59. [CrossRef] [PubMed]
104. Zhang, S.; Sugawara, Y.; Chen, S.; Beelman, R.B.; Tsuduki, T.; Tomata, Y.; Matsuyama, S.; Tsuji, I. Mushroom consumption and incident risk of prostate cancer in Japan: A pooled analysis of the Miyagi Cohort Study and the Ohsaki Cohort Study. *Int. J. Cancer* **2020**, *146*, 2712–2720. [CrossRef] [PubMed]
105. Torres-Mendoza, D.; Ortega, H.E.; Cubilla-Rios, L. Patents on endophytic fungi related to secondary metabolites and biotransformation applications. *J. Fungi* **2020**, *6*, 58. [CrossRef]
106. Da Rocha, W.R.V.; Costa-Silva, T.A.; Agamez-Montalvo, G.S.; Feitosa, V.A.; Machado, S.E.F.; de Souza Lima, G.M.; Pessoa, A.J.; Alves, H.S. Screening and optimizing fermentation production of l-asparaginase by *Aspergillus terreus* strain S-18 isolated from the Brazilian Caatinga Biome. *J. Appl. Microbiol.* **2019**, *126*, 1426–1437. [CrossRef]
107. Clarance, P.; Khusro, A.; Lalitha, J.; Sales, J.; Paul, A. Optimization of camptothecin production and biomass yield from endophytic fungus *Fusarium solani* strain ATLOY-8. *J. Appl. Pharm. Sci.* **2019**, *9*, 35–46. [CrossRef]
108. Mefteh, F.B.; Frikha, F.; Daoud, A.; Bouket, A.C.; Luptakova, L.; Alenezi, F.N.; Al-Anzi, B.S.; Oszako, T.; Gharsallah, N.; Belbahri, L. Response surface methodology optimization of an acidic protease produced by *Penicillium bilaiae* isolate TDPEF30, a newly recovered endophytic fungus from healthy roots of date palm trees (*Phoenix dactylifera* L.). *Microorganisms* **2019**, *7*, 74. [CrossRef]
109. Atlı, B.; Yamaç, M.; Yıldız, Z.; Şőlener, M. Solid state fermentation optimization of *Pleurotus ostreatus* for lovastatin production. *Pharm. Chem. J.* **2019**, *53*, 858–864. [CrossRef]
110. Ravuri, M.; Shivakumar, S. Optimization of conditions for production of lovastatin, a cholesterol lowering agent, from a novel endophytic producer *Meyerozyma guilliermondii*. *J. Biol. Product. Nat.* **2020**, *10*, 192–203. [CrossRef]
111. Khan, Y.M.; Munir, H.; Anwar, Z. Optimization of process variables for enhanced production of urease by indigenous *Aspergillus niger* strains through response surface methodology. *Biocatal. Agric. Biotechnol.* **2019**, *20*, 101202. [CrossRef]
112. Suwannarach, N.; Kumla, J.; Watanabe, B.; Matsui, K.; Lumyong, S. Characterization of melanin and optimal conditions for pigment production by an endophytic fungus, *Spissiomyces endophytica* SDBR-CMU319. *PLoS ONE* **2019**, *14*, e0222187. [CrossRef]
113. El-Sayed, E.S.R.; Zaki, A.G.; Ahmed, A.S.; Ismaiel, A.A. Production of the anticancer drug taxol by the endophytic fungus *Epicoccum nigrum* TXB502: Enhanced production by gamma irradiation mutagenesis and immobilization technique. *Appl. Microbiol. Biotechnol.* **2020**, 1–13. [CrossRef] [PubMed]
114. Zaki, A.G.; El-Shatoury, E.H.; Ahmed, A.S.; Al-Hagar, O.E. Production and enhancement of the acetylcholinesterase inhibitor, huperzine A, from an endophytic *Alternaria brassicae* AGF041. *Appl. Microbiol. Biotechnol.* **2019**, *103*, 5867–5878. [CrossRef]
115. Biswas, D.; Nazir, R.; Biswas, P.; Kumar, V.; Nandy, S.; Mukherjee, A.; Mukherjee, A.; Dey, A.; Pandey, D.K. Endophytic sources of diosgenin, a natural steroid with multiple therapeutic values. *S. Afr. J. Bot.* **2020**. [CrossRef]
116. Bouhri, Y.; Askun, T.; Tunca, B.; Deniz, G.; Aksoy, S.A.; Mutlu, M. The orange-red pigment from *Penicillium mallochii*: Pigment production, optimization, and pigment efficacy against Glioblastoma cell lines. *Biocatal. Agric. Biotechnol.* **2020**, *23*, 101451. [CrossRef]
117. Xu, S.; Ren, N.; Liu, J.; Wu, Y.; Yuan, G. Improvement of vincamine production of endophytic fungus through inactivated protoplast fusion. *Int. Microbiol.* **2020**, 1–11. [CrossRef] [PubMed]
118. Akasaka, N.; Kato, S.; Kato, S.; Hidese, R.; Wagu, Y.; Sakoda, H.; Fujiwara, S. Agmatine production by *Aspergillus oryzae* is elevated by low pH during solid-state cultivation. *Appl. Environ. Microbiol.* **2018**, *84*, e00722-18. [CrossRef] [PubMed]

119. Barua, S.; Kim, J.Y.; Kim, J.Y.; Kim, J.H.; Lee, J.E. Therapeutic effect of agmatine on neurological disease: Focus on ion channels and receptors. *Neurochem. Res.* **2019**, *44*, 735–750. [CrossRef]
120. [ADI] Alzheimer's Disease International. World Alzheimer Report. In *Attitudes to Dementia*; Alzheimer's Disease International: London, UK, 2019; Available online: https://www.alz.co.uk/research/world-report-2019 (accessed on 29 August 2020).
121. Patel, R. A moldy application of MALDI: MALDI-ToF mass spectrometry for fungal identification. *J. Fungi* **2019**, *5*, 4. [CrossRef] [PubMed]
122. Esteki, M.; Shahsavari, Z.; Simal-Gandara, J. Gas Chromatographic fingerprinting coupled to chemometrics for food authentication. *Food Rev. Int.* **2019**, *36*. [CrossRef]
123. Li, Z.; Shi, Y.; Zhang, X.; Xu, J.; Wang, H.; Zhao, L.; Wang, Y. Screening immunoactive compounds of *Ganoderma lucidum* spores by mass spectrometry molecular networking combined with in vivo Zebrafish assays. *Front. Pharmacol.* **2020**, *11*, 287. [CrossRef] [PubMed]
124. Markley, J.L.; Brüschweiler, R.; Edison, A.S.; Eghbalnia, H.R.; Powers, R.; Raftery, D.; Wishart, D.S. The future of NMR-based metabolomics. *Curr. Opin. Biotech.* **2017**, *43*, 34–40. [CrossRef]
125. Li, S.; Tian, Y.; Jiang, P.; Lin, Y.; Liu, X.; Yang, H. Recent advances in the application of metabolomics for food safety control and food quality analyses. *Crit. Rev. Food Sci. Nutr.* **2020**, 1–22. [CrossRef]
126. Hammerl, R.; Frank, O.; Schmittnägel, T.; Ehrmann, M.A.; Hofmann, T. Functional metabolome analysis of *Penicillium roqueforti* by means of differential off-line LC−NMR. *J. Agric. Food Chem.* **2019**, *67*, 5135–5146. [CrossRef]
127. Alanne, A.L.; Issakainen, J.; Pihlaja, K.; Jokioja, J.; Sinkkonen, J. Metabolomic discrimination of the edible mushrooms *Kuehneromyce smutabilis* and *Hypholom acapnoides* (Strophariaceae, Agaricales) by NMR spectroscopy. *Z. Naturforsch.* **2019**, *74*, 201–210. [CrossRef]
128. González-Domínguez, R.; Sayago, A.; Fernández-Recamales, A. Metabolomics: An emerging tool for wine characterization and the investigation of health benefits. In *Engineering Tools in the Beverage Industry: The Science of Beverages*; Grumezescu, A.M., Holban, A.M., Eds.; Wood Head Publishing: Sawston, UK, 2019; Volume 3, pp. 315–350. [CrossRef]
129. Florentino-Ramos, E.; Villa-Ruano, N.; Hidalgo-Martínez, D.; Ramírez-Meraz, M.; Méndez-Aguilar, R.; Velásquez-Valle, R.; Zepeda-Vallejo, L.G.; Pérez-Hernández, N.; Becerra-Martínez, E. ^{1}H NMR-based fingerprinting of eleven Mexican *Capsicum annuum* cultivars. *Food Res. Int.* **2019**, *121*, 12–19. [CrossRef]
130. Gąsecka, M.; Siwulski, M.; Mleczek, M. Evaluation of bioactive compounds content and antioxidant properties of soil-growing and wood-growing edible mushrooms. *J. Food Process. Preserv.* **2017**, e13386. [CrossRef]
131. Carr, A.C.; Lykkesfeldt, J. Discrepancies in global vitamin C recommendations: A review of RDA criteria and underlying health perspectives. *Crit. Rev. Food Sci. Nutr.* **2020**, 1–14. [CrossRef]
132. Da Silva, L.P.; Vargas, E.A.; Madureira, F.D.; Faria, A.F.; Augusti, R. Development and validation of a novel analytical method to quantify aflatoxins in baby food samples by employing dispersive solid phase extraction with multi-walled carbon nanotubes. *Food Anal. Methods* **2020**, *13*, 1530–1537. [CrossRef]
133. Orhan, I.E.; Senol, F.S.; Skalicka-Wozniak, K.; Georgiev, M.; Sener, B. Adulteration and safety issues in nutraceuticals and dietary supplements: Innocent or risky? In *Nutraceuticals: Nanotechnology in the Agri-Food Industry*; Grumezescu, A.M., Ed.; Academic Press: Cambridge, UK, 2016; Volume 4, pp. 153–182.
134. Colosimo, R.; Gabriele, M.; Cifelli, M.; Longo, V.; Domenici, V.; Pucci, L. The effect of sourdough fermentation on *Triticum dicoccum* from Garfagnana: ^{1}H NMR characterization and analysis of the antioxidant activity. *Food Chem.* **2020**, *305*, 125510. [CrossRef] [PubMed]
135. Papadaki, E.; Mantzouridou, F.T. Citric acid production from the integration of Spanish-style green olive processing wastewaters with white grape pomace by *Aspergillus niger*. *Bioresour. Technol.* **2019**, *280*, 59–69. [CrossRef] [PubMed]
136. Faustino, M.; Veiga, M.; Sousa, P.; Costa, E.M.; Silva, S.; Pintado, M. Agro-food byproducts as a new source of natural food additives. *Molecules* **2019**, *24*, 1056. [CrossRef]
137. Roukas, T.; Kotzekidou, P. Pomegranate peel waste: A new substrate for citric acid production by *Aspergillus niger* in solid-state fermentation under non-aseptic conditions. *Environ. Sci. Pollut. Res.* **2020**, *27*, 13105–13113. [CrossRef]

138. Souza Filho, P.F.; Zamani, A.; Taherzadeh, M.J. Edible protein production by filamentous fungi using starch plant wastewater. *Waste Biomass Valoriz.* **2019**, *10*, 2487–2496. [CrossRef]
139. Dulf, F.V.; Vodnar, D.C.; Toşa, M.I.; Dulf, E.H. Simultaneous enrichment of grape pomace with γ-linolenic acid and carotenoids by solid-state fermentation with Zygomycetes fungi and antioxidant potential of the bioprocessed substrates. *Food Chem.* **2020**, *310*, 125927. [CrossRef] [PubMed]

Publisher's Note: MDPI stays neutral with regard to jurisdictional claims in published maps and institutional affiliations.

Journal of *Fungi*

MDPI

Review

Selenium Biofortification of Crop Food by Beneficial Microorganisms

Yuanming Ye, Jingwang Qu, Yao Pu, Shen Rao, Feng Xu and Chu Wu *

College of Horticulture & Gardening, Yangtze University, Jingzhou 434025, China; y.m.ye2020@gmail.com (Y.Y.); qujw@yangtzeu.edu.cn (J.Q.); puyao456@163.com (Y.P.); raoshen1989@163.com (S.R.); xufeng198@126.com (F.X.)

* Correspondence: wuchu08@yangtzeu.edu.cn; Tel.: +86-716-806-6262

Received: 30 March 2020; Accepted: 26 April 2020; Published: 3 May 2020

Abstract: Selenium (Se) is essential for human health, however, Se is deficient in soil in many places all around the world, resulting in human diseases, such as notorious Keshan disease and Keshin–Beck disease. Therefore, Se biofortification is a popular approach to improve Se uptake and maintain human health. Beneficial microorganisms, including mycorrhizal and root endophytic fungi, dark septate fungi, and plant growth-promoting rhizobacteria (PGPRs), show multiple functions, especially increased plant nutrition uptake, growth and yield, and resistance to abiotic stresses. Such functions can be used for Se biofortification and increased growth and yield under drought and salt stress. The present review summarizes the use of mycorrhizal fungi and PGPRs in Se biofortification, aiming to improving their practical use.

Keywords: selenium; biofortification; transporters; mycorrhizal fungi; plant growth-promoting rhizobacteria (PGPRs)

1. Introduction

At present, it is widely accepted that selenium (Se) possesses multiple physiological functions in various biological systems as an integral part of a range of proteins containing Se. Therefore Se is important for human health. However, Se distribution in the earth's crust is greatly uneven, ranging from 0.005 mg·kg^{-1} in Finland to 8000 mg·kg^{-1} in Tuva-Russia [1]. Se deficiency has been reported in many places all around the world including China, North America, New Zealand, Australia, Sweden, and Finland [2–5]. Some notorious diseases are directly related to Se deficiency, such as Keshan disease and Keshin–Beck disease, two endemic diseases related to Se deficiency. Keshan disease was first prevalent at alarge scale in 1935 in Keshan county, Heilongjiang province, China. Keshan disease generally occurs in children and women of childbearing age and its symptoms are related to impairment of cardiac function, cardiac enlargement, and arrhythmia [6]. Although the main factor was not determined for the disease in etiology, it was closely related to Se because it was found that there was an obvious Se deficiency in local soil, and Se supplementation could partly control the disease. An investigation analyzed some physiological parameters, including blood Se level, glutathione peroxidase-1 (GPx-1) activity, and variance at codon 198 in *GPx-1* gene, and found that the main risk factors for the disease were low GPx-1 activity, Keshan disease family history, and living in an endemic area [7], suggesting that Keshan disease is closely related to low GPx-1 activity. Kaschin–Beck disease is an osteoarthropathy, which manifests as severe dysarthrosis of joints, shortened fingers and toes, and in severe cases dwarfism. In China, the disease is prevalent in the Tibetan Plateau [8,9]. An investigation carried out by Zhang et al. [8] showed that the levels of environmental Se were very low, and Kaschin–Beck disease in the Tibetan Plateau was much severe with decreasing environmental Se under the Se-deficient condition, suggesting the relationship between Kaschin–Beck disease and Se deficiency in the Tibetan Plateau. In addition, Se is related to other human diseases and health, such as cancer [10–13], muscle disease [14], and healthy aging and longevity [15–17]. Therefore, it is essential to maintain Se homeostasis in human body [18,19]. It was estimated that Se intake of >900 μg·day^{-1} is harmful, and intake of <30 μg·day^{-1} is not enough [20]. Some data have shown that over 800million people all around the world might suffer from Se deficiency [21–26]. Therefore, sufficient dietary Se uptake is important for human health.

Acquired Se is converted into some proteins thatcontain at least one of the two amino acids (i.e., selenocysteine (SeCys) and selenomethionine (SeMet)) as a key component (i.e., selenoproteins). Human health and diseases are related to selenoproteins, and selenocysteine is regarded as the 21st proteinogenic amino acid. The human genome encodes about 30 selenoproteins. In the article written by Reeves and Hoffmann [27], they described functions of selenoproteins in detail. Among the selenoproteins in human, glutathione peroxidases (GPxs) seem to be more important, because they include eight proteins (GPx1–GPx8) having antioxidant properties with multidimensional roles in living cells, ranging from H_2O_2 homeostasis to regulation of apoptosis [28]. Therefore, enough Se uptake is essential for functional maintenance of these selenoproteins. Since Se is deficient in many places all around the world, Se fortification in food is necessary. In view of high toxicity of selenite and selenate, Se biofortification is relatively bio-safe. Organic Seleno-compounds act as potential therapeutic and chemo-preventive agents that function as antioxidants, enzyme modulators, antitumor, antimicrobials, antihypertensive agents, antivirals, and cytokine inducers [29]. Organic seleno-compounds are provided with crop food [30–35], vegetables [36–42], fruits [43–45], and even nuts [46–48]. Therefore, how to increase concentrations of organic seleno-compounds in these plants is of significance for improvement of dietary Se acquisition by human being.

Se biofortification may be carried out by multiple ways, such as application of Se fertilizers on leaves [49–53] and in soil [52,54,55]. Se-enriched organic fertilizers are also applied. For example, Bañuelos et al. [56] used Se-enriched *Stanleya pinnata* to cultivate Se-enriched broccoli and carrots, and found that more than 90% of organic Se was converted to inorganic selenate and selenite. Se foliar application seems to be most effective way to fortify Se uptake in most arable crops [52,57]. However, a contrary result was observed by Lyons et al. [58]. They found foliar application was less efficient than application to soil at planting (at application rates of 40 and 120 g·ha^{-1}, respectively) in Australian trials. The agronomic application of Se fertilizers are more expensive and short-term solutions, especially in large-scale fields. Relatively, agronomic Se biofortification with beneficial microorganisms (BMOs) is a more inexpensive and long-term solution, especially in poor places and Se-rich places, such as Enshi, Hubei province, China [59] and Pineridge Natural Area, a seleniferous site west of Fort Collins, CO, USA [60].

In the present article, we focus on the roles of beneficial microorganisms in Se biofortification and our aim is to improve use of beneficial microorganisms in practice.

2. Improvement of Se Biofortification by BMOs

Symbiosis of plants with BMOs is helpful for plant growth and to increase in micronutrition uptake and resistance to abiotic and biotic stresses. Based on the characteristics of BMOs, BMOs can be used for Se biofortification. BMOs, including mycorrhizal fungi (endo- and ectomycorrhizal fungi), root endophytic fungi (REFs), and PGPRs, are popular in biofilmed biofertilizers. Arbuscular mycorrhizal fungi (AMFs) are preferential to colonize in roots of angiosperms, and ectomycorrhizal fungi are popular in gymnosperms. Most REFs possess a wide range of plant hosts.

2.1. *Arbuscular Mycorrhizal Fungi*

Arbuscular mycorrhizal fungi are used for Se biofortification because of their ability to enhance nutrition uptake of their host plants (Table 1). Functions of mycorrhizal fungi have been the primary focus of research, especially those involved in phosphate uptake. The genomes of these fungi encode some high-affinity inorganic phosphate transporters and some of them have been isolated and identified [61–65]. On the other hand, *in planta*, some symbiosis-specific phosphate transporters can be induced by symbiosis with mycorrhizal fungi [61,66–72]. Thus, the interaction between plants and mycorrhizal fungi strengthens phosphate uptake and transportation to host plants [73–77]. Similarly, there are some sulfate transporters encoded by genomes of mycorrhizal fungi, such as sulfate transporters GBC38160.1 and GBC25943.1 and sulfate permeases PKY50973.1 in arbuscular mycorrhizal fungus *Rhizophagus irregularis*, sulfate transporters EDR02618.1 and EDR02177.1, and sulfate permeases EDR11271.1 and EDR00466.1 in ectomycorrhizal fungus *Laccaria bicolor*. Since Se and sulfur (S) belong to the same element family (VI-A), the chemical properties of Se are very similar to S. Se is absorbed as selenate or selenite, which is metabolized via the sulfur assimilation pathway in plants, leading to biosynthesis of SeCys, SeMet, and other Se isologs of various S metabolites [78–82]. Se can be transported by sulfur transporters to host plants, just like phosphate transported between mycorrhizal fungi and their host plants, such as the high-affinity sulfate permease [83] and the high-affinity sulfate transporters Sultr1:1 and Sultr1:2 [84–86]. The two sulfate transporters are proton-sulfate symporters, such that for every molecule of selenate entry into root cells, three protons are taken up. Sulfate transporters function in Se accumulation in food crops. Wheat genotype 'Puelche' is the most Se-tolerant and has the greatest Se accumulation among the three wheat genotypes studied (i.e., 'Puelche', 'Tinto', and 'Kumpa'), such that its Se accumulation was related to the strongest transcript level of the sulfate transporter TaeSultr4.1 in roots [87]. In addition, other transporters also take part in Se transport, such as silicon transporters in rice [88] and tomato [89], phosphate/orthophosphate transporters in wheat [90], rice [91–93], tomato [89], and yeast (*Saccharomyces cerevisiae*) [94,95], and monocarboxylates transporters in yeast (*S. cerevisiae*) [96]. Thus, it is reasonable to explain the experimental results that plant availability of selenate and selenite was influenced by the competing ions phosphate and sulfate [97,98]. Competition between phosphate and Se uptake led to decrease in Se accumulation translocation coefficients, and Se concentrations in wheat roots, stems, leaves, and spikes when phosphate fertilizers were applied to selenite fertilized soil [99]. However, a different case occurred. An investigation was carried out on sulfate and selenate uptake in *Astragalus* species (two Se hyperaccumulators *A. racemosus* and *A. bisulacatus* and two closely related non-accumulators *A. glycyphyllos* and *A. drummondii*), and results showed that sulfur deficiency increased Se accumulation, and increased Se supply increased sulfate accumulation in both root and shoot tissues [100]. In certain *Astragalus* species, the high expression of sulfate transporters led to enhanced ability of Se uptake and translocation, and therefore contributed to the Se hyperaccumulation trait. At present, except for sulfate and phosphate transporters, it is not clear whether other transporters have their homologous proteins in mycorrhizal fungi. If these homologous proteins occur in mycorrhizal fungi, they could mediate Se transport to host plants. On the other hand, decreases in sulfate bioavailability and mycorrhizal symbiosis enhanced expression of sulfate transporters, resulting in increase in ability to absorb sulfate and consequent uptake of Se [101–104]. Similarly, Se deficiency also enhances expression of sulfate transporters, resulting in an increase in Se uptake, and mycorrhizal symbiosis also enhances Se uptake.

Table 1. Arbuscular mycorrhizal fungi (AMFs) and root endophytic fungi (REFs) often used for Se biofortification.

Microbes	Microbial Types	Host Plants	References
Funneliformis mosseae	AMF	*Triticum aestivum, Lactuca sativa, Asparagus officinalis,*	[105–108]
Glomus claroideum	AMF	*Triticum aestivum*	[109]
Glomus fasciculatum	AMF	*Allium sativum*	[110]
Glomus irtraradices	AMF	*Allium sativum*	[55]
Glomus mosseae	AMF	*Lolium perenne, Allium sativum, Medicago sativa, Glycine max, Zea mays*	[110–112]
Glomus versiform	AMF	*Triticum aestivum*	[105]

Table 1. *Cont.*

Microbes	Microbial Types	Host Plants	References
Rhizophagus intraradices	AMF	*Lactuca sativa, Asparagus officinalis, Lactuca sativa, Allium cepa*	[106–108,113]
Alternaria seleniiphila	REF	*Stanleya pinnata*	[114]
Alternaria astragali	REF	*Astragalus bisulcatus*	[114]
Aspergillus leporis	REF	*Stanleya pinnata*	[114]
Fusarium acuminatum	REF	*Astragalus racemosus*	[114]
Trichoderma harzianum	REF	*Allium cepa*	[106]

Some evidence supports the role of mycorrhizal fungi in enhancing Se uptake in plants. Wheat seedlings were inoculated with *Glomusversiform* or *Funneliformis mosseaein* hydroponic culture medium for eight weeks, the two arbuscular mycorrhizal fungi significantly increased selenate and selenite uptake by wheat root, but they did not show effect on uptake of SeMet [105]. Meanwhile, compared to non-mycorrhizal roots, mycorrhizal roots showed significantly higher V_{max} for selenate and selenite uptake (179.6 vs. 55.93 nmol·g^{-1}DW·h^{-1} for selenate and 1688.0 vs. 860.3 nmol·g^{-1}DW·h^{-1} for selenite). Higher Se accumulation was carried out through up-regulating the expression of three genes encoding sulfate transporters, i.e., TaSultr1:1, TaSultr1:3, and TaSultr2:1, in the mycorrhizal roots, especially TaSultr1:1. In mycorrhizal roots with G. *versiform* and *F. mosseae*, the relative expressions of *TaSultr1:1* gene was significantly up-regulated by 2.18-fold and 2.12-fold, respectively. Garlic (*Allium sativum* L.) is an important condimental species. This species ispopular all around the world because of its diallyl disulfide, a component of garlic, which can inhibit proliferation of various cancer cells (e.g., colon, lung, and skin cancer cells) and WEHI-3 leukemia cells [115–117]. Garlic is used for Se biofortification with mycorrhizal fungi. A survey of applying selenate fertilizer and mycorrhizal fungus *Glomus irtraradices* to soil was conducted, and the results showed mycorrhizal addition increased the Se uptake of garlic by10-fold to 15 μg·g^{-1}DW, and fertilization with selenate and amendment of mycorrhizal fungi strongly increased the Se concentrations in garlic to around 1% [55]. Further analyses showed that the amendment of soil with the mycorrhizal fungus and/or selenate increased selenate concentrations in garlic, but did not affect distribution of detected Se species in garlic. In Se-contaminated soil, mycorrhizal fungi inoculation increased Se accumulation of plants. Alfalfa, maize, and soybean seedlings were cultivated in the soil contaminated with different levels of Se, and results showed that mycorrhizal fungi inoculation decreased Se accumulation in roots and shoots of all the plants at low Se levels (0 or 2 mg·kg^{-1}), but increased Se accumulation in alfalfa shoots and maize roots and shoots at Se level of 20 mg·kg^{-1} [112]. Contrary results were observed on ryegrass (*Lolium perenne* cv. 'Barclay') [111]. Their results showed that Se concentrations in roots of ryegrass were not affected by mycorrhizal inoculation with the AMF *G. mosseae*, but mycorrhizal inoculation significantly reduced Se concentrations in shoots [111], further decreasing Se uptake in whole plants. Lettuce (*Lactuca sativa* L.) is one of the most consumed leaf vegetables in some places around the world because of its good properties, such as high levels of antioxidants (such as carotenoids, polyphenols, ascorbate, α-tocopherol) and dietary fiber [118–120], thus it is suitable for Se biofortification to enhance dietary Se consumption. When two lettuce cultivars 'Batavia Rubia Munguia' (BRM) and 'Maravilla de Verano' (MV) were treated with Se compounds (selenite, organic Se compounds SeU and $SeCH_3$) and AMFs (a mixture of *Rhizophagus intraradices* and *Funneliformis mosseae*), their growths were continuously improved by AMFs, except for BRM under treatment of $SeCH_3$ [106]. The positive effect of AMFs on plant biomass was different among lettuce cultivars and forms of seleno-compounds, and BRM lettuce plants showed the highest mycorrhizal efficiency index (MEI) under treatment of SeU, MV lettuce plants with the highest MEI under $SeCH_3$, suggesting that the two lettuce cultivars possessed preference for different seleno-compounds when they were inoculated with AMFs. Meanwhile, AMFs inoculation significantly affected mineral accumulation in the leaves of BRM lettuce. In general, mycorrhizal inoculation significantly increased levels of macro and micronutrients, but significantly reduced Se levels in leaves of BRM lettuce. Significant interaction occurred about Se levels in shoots of BRM lettuce between seleno-compounds and AMFs inoculation. Similar status occurred on MV lettuce. Under treatment of selenite, AMFs inoculation reduced Se concentrations in leaves of MV lettuce. In contrast, under treatment of organic seleno-compound SeCH3, MV lettuce never accumulated detectable levels of Se in leaves, regardless of whether they were inoculated with AMFsor not. Treatment of organic seleno-compound SeU slightly increased Se concentrations in leaves of MV lettuce without AMFs inoculation [106]. Other research showed similar results [108]. All the results suggest that combination of seleno-compounds and AMFs inoculation does not increase Se levels in lettuce leaves, although it increases levels of some macro- and micronutrients and antioxidants. Therefore, some AMFs are not suitable for Se biofortification in lettuce. Of course, other AMFs should be chosen to examine their role in Se biofortification in lettuce under treatment of seleno-compounds. At present, it is not clear whether lettuce symbioses with some ectomycorrhizal fungi. Thus, more research is necessary for Se biofortification in lettuce.

Consversa et al. [107] investigated the effect of Se fern application and AMFs (*Rhizophagus intraradices* and *Funneliformis mosseae*) inoculation on Se biofortification for two years, such that Se fern application was carried out on green asparagus (*Asparagus officinalis* L.). Their experimental results showed that Se levels in non-mycorrhizal *A. officinaliscv.* 'Grande' plants increased in trial A1 as exogenous selenate levels increased. Under selenate treatment of 75 and 125 g·ha^{-1}, Se concentrations in spears increased 4.7 and 6.4-fold on a dry weight basis compared to control, respectively. Similar results occurred in trail B1. In trail B1, Se concentrations in spears were significantly affected by the interaction between Se amendment and AMFs inoculation. In spears of plants without Se amendment, Se levels were similar in mycorrhizal and non-mycorrhizal plants. All the results suggest a combination of Se amendment and mycorrhizal fungi greatly improve Se biofortification in *A. officinalis* and the combination should be recommended in field by large scale. However, contrary results have also been observed. When the AMF *Glomus mosseae* was used for inoculation with alfalfa (*Medicago sativa* L. cv. 'Chuangxin'), maize (*Zea mays* cv. 'ND108'), and soybean (*Glycine max* cv. 'Zhonghuang No. 17'), mycorrhizal inoculation significantly decreased Se concentrations in roots with the highest reduction for alfalfa (50–70%), while it was less than 40% for maize and soybean, Se concentrations in shoots decreased by 7–38% for mycorrhizal treatment, and the difference caused by inoculation influence was insignificant among the plant species [112]. When Se was added at the levels of 0 and 2 mg·kg^{-1}, the total Se accumulation in roots and shoots of all the three plant species were lower in mycorrhizal than in non-mycorrhizal treatment, while the opposite pattern was observed in roots of maize and shoots of alfalfa and maize when Se was applied at 20 mg·kg^{-1} [112]. These results show negative effects on Se accumulation in these plant species when low levels of exogenous Se were added.

In addition, some ectomycorrhizal fungi can accumulate Se in their fruit bodies [121–123], suggesting their ability to acquire Se. Some of these ectomycorrhizal fungi are edible, thus they are used for biofortification of Se in fruit bodies. Few investigations on the role of REFs in Se biofortification have beencarried out (Table 1). In general, REFs, especially members of the genus *Trichoderma*, can colonize roots of some host plants, thus they can be widely used for Se biofortification. At present, there are not reports on roles of dark septate fungi in Se biofortification of food crops.

Taken together, mycorrhizal inoculation might increase Se accumulation in some crop species, leading to Se biofortification of crops. For some crop species, more investigations are needed, especially for interactions between mycorrhizal fungi and crop species. For the abovementioned negative effects of G. *mosseae* on Se accumulation in alfalfa, maize, and soybean, more mycorrhizal fungi and root endophytic fungi should be used to investigation.

2.2. Se Biofortification by PGPRs

Plant growth-promoting rhizobacteria (PGPRs) are popular in improving nutrition uptake, plant growth, and resistance to abiotic stresses [124–127]. Some of them possess the ability to solubilize phosphate in soil. Such ability could be used for Se biofortification (Table 2), because in some soil agrotypes, such as volcanic Andisols in southern Chile, Se bioavailability is very low. On the one hand, Se can form stable complexes with clays and/or can be strongly absorbed onto oxy-hydroxides of aluminum, iron, or manganese, and remain low in terms of bioavailability to plants [128–130]. On the other hand, oxyanions of Se, i.e., selenite and selenate, are bioavailable to plants. When selenate and selenite are supplied to soil, they are rapidly reduced to insoluble forms (e.g., Se–metal ion complex), leading to their low bioavailability (less than 10% only). The Se fertilizers that are not acquired by plant roots readily after application are not bioavailable to plants in the next season or the next year [131]. Thus, Se re-solubility in soil is very important. Although there are no report concernsregarding Se-solubilizing PGPRs at present, some seleno-bacteria have been studied [30,35,109,132,133]. Trivedi et al. [35] isolated and identified some endophytic seleno-bacteria from the various tissues of *Ricinus communis* plants and molecular identification analyses showed that they were *Paraburkholderia megapolitana*, *Alcaligenes faecalis*, and *Stenotrophomonas maltophilia*. Among the three bacteria, *P. megapolitana* was most effective in improving the growth of *Glycine max* plants under drought and enhancing Se biofortification which was 7.4-fold higher compared to control. The synergistic effect on Se biofortification and increased drought tolerance is important for plants grown in arid and semi-arid places with Se deficiency. A great number of people all around the world are dependent on wheat as their main component of diet, thus it is important to fortify Se in wheat grains [30]. Many studies have been carried out on Se biofortification in wheat. Durán et al. [109] evaluated the effects of Se acquisition by wheat plants through the co-inoculation of native seleno-bacteria strains *Stenotrophomonas* sp. B19, *Enterobacter* sp. B16, *Bacillus* sp. R12, and *Pseudomnas* sp. R8, both individually and in mixture, as a seleno-nanosphere source with AMF *Glomus claroideum*. They found that Se concentrations in plant tissues in inoculated plants were significantly higher than those of un-inoculated controls.Meantime, regardless of presence of AMF *G. claroideum*, Se concentrations in grains of wheat plants inoculated with *Enterobacter* sp. B16 were higher than those of plants inoculated with the rest of the microbial strains. In addition, PGPRs showed their synergistic role in improving Se concentrations with AMFs. When plants were inoculated with the seleno-bacteria strains and *G. claroideum*, Se concentrations in grains were 23.5% higher than those in non-mycorrhizal plants. The synergisms might be related to the relationship between seleno-bacteria strains and AMFs, because the seleno-bacteria could acquire more nutrition from the hyphae of their neighboring AMFs or ectomycorrhizal fungi [134–137]. Moreover, Durán et al. [132] isolated two Se-tolerant endophytic bacteria *Acinetobacters* sp. E6.2 and *Bacillus* sp. E5. They studied production of seleno-compounds (SeMet and seleno-methyl-selenocysteins (MeSeCys)) by the two bacteria, but they did not study the effects of the two bacteria on Se biofortification. Co-application of Se fertilizers and seleno-bacteria sometimes leads to changes in bacterial population. When Se-tolerant bacteria and Se amendment were supplied to wheat in Andisols, Se amendment stimulated population growth of two bacterial groups (*Paenibacillaceae* and *Brucellaceae*), but inhibited other bacterial groups (Clostridia, *Burkholderiales*, *Chitinophagaceae*, and *Oxalobacteraceae*) [133]. Meanwhile, Se concentrations in roots and leaves of wheat plants inoculated with Se-tolerant bacterial strains *Pseudomonas* sp. R8 and *Stenotrophomonas* sp. B19 were significantly higher than those of the un-inoculated controls. Higher Se biofortification is related to the Se tolerance of the two bacteria, because higher Se concentrations in roots and leaves were also observed when wheat plants inoculated with *Stenotrophomonas* sp. B19 were grown at concentrations of 5 and 10 mM of selenite, compared to those grown at 2 mM [133]. The results suggested that Se in seleno-bacteria could be transferred into their host plants. Effects of other Se-tolerant bacteria on Se biofortification were also investigated. When wheat plants were inoculated with two Se-tolerant bacterial strains *Bacillus cereus* YAP6 and *Bacillus licheniformis* YAP7, Se concentrations in the stems of the Se-treated wheat plants were increased up to 375%, and Se concentrations in kernels increased up to 154% of those in un-inoculated Se-treated wheat plants [34]. Meanwhile, the *Bacillus* strains can produce auxin, leading to increased number of leaves and greater biomass and shoot length [34]. When wheat plants were inoculated with *Bacillus pichinotyi* in the presence of selenate, they posed significantly higher biomass, shoot length, and spike length compared to un-inoculated plants [33]. Meanwhile wheat plants inoculated with *B. pichinotyi* had significantly higher Se concentrations in wheat kernels (167%) and stems (252%), compared to un-inoculated plants. Overall, greater biomass means higher Se biofortification, which is important for crops cultivated by large scale in field. Rhizobia not only fixes nitrogen, but also helps Se accumulation. Data from Alford et al. [138] showed rhizobia significantly increased shoot biomass and Se accumulation in shoots of the Se-hyperaccumulator *Astragalus bisulcatus* and the nonhyperaccumulator *A. drummondii*. The dual roles of rhizobia are of significance for organic Se production.

Table 2. Plant growth-promoting rhizobacteria (PGPRs) often used for Se biofortification.

Microbes	Host Plants	References
Acinetobacters sp. *E6.2*	-	[132]
Acinetobater sp.	*Triticum aestivum*	[139]
Alcaligenes faecalis	*Ricinus communis, Glycine max*	[35]
Anabaena sp.	*Triticum aestivum*	[30,140]
Bacillus amyloliquefaciens	*Arabidopsis thaliana*	[141]
Bacillus axarquiens	*Triticum aestivum*	[139]
Bacillus cereus	*Triticum aestivum*	[34]
Bacillus licheniformis	*Triticum aestivum*	[34]
Bacillus mycoides	*Brassica juncea*	[142]
Bacillus pichinotyi	*Triticum aestivum*	[33]
Bacillus sp. *E5*	-	[132]
Bacillus sp. *E6.1*	*Triticum aestivum*	[139]
Bacillus sp. *R12*	*Triticum aestivum*	[109]
Bacillus subtilis	*Allium cepa*	[113]
Calothrix sp.	*Triticum aestivum*	[30,140]
Enterobacter ludwigii	*Triticum aestivum*	[139]
Enterobacter sp. *B16*	*Triticum aestivum*	[109]
Klebsiella oxytoca	*Triticum aestivum*	[139]
Paraburkholderia megapolitana	*Ricinus communis, Glycine max*	[35]
Providencia sp.	*Triticum aestivum*	[30,140]
Pseudomnas sp. *R8*	*Triticum aestivum*	[109,133]
Rhizobium sp.	*Astragalus bisulcatus, A. drummondii*	[138]
Rhizosphere bacteria	*Scirpus robustus, Polypogon monspeliensis*	[143]
Se-tolerant bacteria	*Brassica juncea*	[144]
Stenotrophomonas maltophilia	*Ricinus communis, Glycine max, Brassica juncea*	[35,142]
Stenotrophomonas sp. *B19*	*Triticum aestivum*	[109,133]

Interestingly, volatile organic compounds (VOCs) released by PGPRs improve Se biofortification of plants. VOCs from *Bacillus amyloliquefaciens* BF06 significantly increased photosynthesis and growth of *Arabidopsis* plants and these VOCs led to an obvious increase in expressions of some genes encoding sulfate transporters and Se concentrations in plants [141]. VOCs released by *B. amyloliquefaciens* could not increase Se biofortification of *Arabidopsis Sultr1:2* mutants. All the results suggested sulfate transporters with high expression mediate Se uptake, as shown above. Meanwhile, the results indicate an unknown mechanism that PGPRs improves Se biofortification. The question is inevitable, how do the VOCs improve expression of sulfate transporters?

Taken together, Se amendment could improve population growth of some Se-tolerant bacteria; if these bacteria show synergistic effect on Se biofortification, they could be mixed in some biofilmed biofertilizers specific to certain crops and vegetables, thus, their combinative amendment along with Se fertilizers become a Se biofortification tool in sustainable agriculture [52,145].

3. Concluding Remarksand Perspectives

Since Se is essential for human health, Se biofortification must be carried out in Se-deficient places by various ways on food crops, vegetables, fruits, and nuts. Foliar and soil fertilization are effective for enhancing Se accumulation in crops. However, the two ways are expensive for large-scaled food crops, especially in poor places. Moreover, the effect of the two ways is short-term and they easily cause area source pollution. BMOs improve Se uptake and accumulation in food crops. Therefore, combination of BMOs and soil fertilization is a good approach to Se biofortification of crop food. At present, for Se biofortification by BMOs, there remain some questions to resolve. The first relates to the synergisms among these beneficial microorganisms. Biofilmed biofertilizers often include many BMOs. As they can compete for nutrition from their common host plants, some of them are possibly antagonistic. Therefore, before they are mixed in biofilmed biofertilizers, the synergism should be examined in detail. The second relates to Se biofortification and phytoremediation. Phytoremediation of Se is popular in Se-rich places. The plants harvested in phytoremediation could be used as the organic source of Se. However, attention must be paid to the fact that there are possibly other heavy metals in the harvested plants. In addition, transgenic plant technology has been used for phytoremediation and Se biofortification. Since some people are very sensitive to genetically modification of crop plants, application of transgenic plants for Se biofortification should be careful. The third relates to theuse of Se hyperaccumulators. Some Se hyperaccumulators, such as *Stanleya pinnata* and *Astragalus bisulcatus* and *Cardamine enshiensis*, should be paid more attentions, especially *C. enshiensis*, because it is edible and can be directly used in food. The forth relates to the use of Se nanoparticles. Relatively, Se nanoparticles are less toxic and more eco-friendly for both humans and the environment. More researches are necessary for use of Se nanoparticles, especially for the production of Se nanoparticles using plants and fungi. The fifth relates to increased plant resistance to abiotic stresses. Exogenous Se compounds and seleno-bacteria synergistically improve plant resistance to abiotic stresses, thus the synergism should be well used for plant resistance to abiotic stresses, especially drought and salt stress. Finally, the sixth relates to functions of root endophytic fungi and dark septate fungi. The two types of fungi possess ecological functions similar to mycorrhizal fungi, and they often colonize many plant species. However, very little attention has been paid to them.

Author Contributions: C.W. and F.X. planned the whole article and outline, and provided some references. C.W. and J.Q. revised the second manuscript. Y.Y. wrote the rough manuscript; Y.P. and S.R. reviewed the rough manuscript. All authors have read and agreed to the published version of the manuscript.

Funding: This research was funded by National Natural Science Foundation of China (grant numbers 31870378); Fujian University Key Laboratory for Plant-Microbe Interaction (grant numberPMI2018KF2).

Conflicts of Interest: The authors declare no conflict of interest. The funders had no role in the design of the study; in the collection, analyses, or interpretation of data; in the writing of the manuscript, or in the decision to publish the results.

References

1. Chasteen, T.G.; Bentley, R. Biomethylation of selenium and tellurium: Microorganisms and plants. *Chem. Rev.* **2003**, *103*, 1–25. [CrossRef] [PubMed]
2. Gissel-Nielsen, G.; Gupta, U.C.; Lamand, M.; Westermarck, T. Selenium in soil and plants and its importance in livestock and human nutrition. *Adv. Agron.* **1984**, *37*, 397–460.
3. Gupta, U.C.; Gupta, S.C. Selenium in soils and crops, its deficiencies in livestock and humans: Implications for management. *Commun. Soil Sci. Plant Anal.* **2000**, *31*, 1791–1807. [CrossRef]
4. Ylaranta, T. Sorption of selenite and selenate added in the soil. *Ann. Agr. Fenn.* **1983**, *22*, 9–39.

5. Blazina, T.; Sun, Y.; Voegelin, A.; Lenz, M.; Berg, M.; Winkel, L.H.E. Terrestrial selenium distribution in China is potentially linked to monsoonal climate. *Nat. Commun.* **2014**, *5*, 4717. [CrossRef]
6. Xu, G.; Wang, S.; Gu, B.; Yang, Y.; Song, H.; Xue, W.; Liang, W.; Zhang, P. Further investigation on the role of selenium deficiency in the aetiology and pathogenesis of Keshan disease. *Biomed. Environ. Sci.* **1997**, *10*, 316–326.
7. Lei, C.; Niu, X.; Ma, X.; Wei, J. Is selenium deficiency really the cause of Keshan disease? *Environ. Geochem. Health* **2011**, *33*, 183–188. [CrossRef]
8. Zhang, B.; Yang, L.; Wang, W.; Li, Y.; Li, H. Environmental selenium in the Kaschin-Beck disease area, Tibetan Plateau, China. *Environ. Geochem. Health* **2011**, *33*, 495–501. [CrossRef]
9. Sun, L.Y.; Yuan, L.J.; Fu, Y.; Deng, J.Y.; Wang, L.H. Prevalence of Kaschin-Beck disease among Tibetan children in Aba Tibetan and Qiang Autonomous Prefecture: A 3-year epidemiological survey. *J. Pediatrics* **2012**, *8*, 140–144. [CrossRef]
10. Encio, I.; Sanmartin, C.; Palop, J.A.; Font, M.; Moreno, E.; Plano, D.; Lamberto, I. Bisacylimidoselenocarbamates cause G2/M arrest associated with the modulation of CDK1 and Chk2 in human breast cancer MCF-7 Cells. *Curr. Med. Chem.* **2013**, *20*, 1609–1619.
11. Hatfield, D.L.; Tsuji, P.A.; Carlson, B.A.; Gladyshev, V.N. Selenium and selenocysteine: Roles in cancer, health, and development. *Trends Biochem. Sci.* **2014**, *39*, 112–120. [CrossRef] [PubMed]
12. Yoo, M.H.; Carlson, B.A.; Tsuji, P.A.; Tobe, R.; Naranjo-Suarez, S.; Lee, B.J.; Davis, C.D.; Gladyshev, V.N.; Hatfield, D.L. Selenoproteins harboring a split personality in both preventing and promoting cancer. In *Selenium: Its Molecular Biology and Role in Human Health*; Hatfield, D.L., Berry, M.J., Gladyshev, V.N., Eds.; Springer: New York, NY, USA, 2011; pp. 325–333.
13. Patra, A.R.; Hajra, S.; Baral, R.; Bhattacharya, S. Use of selenium as micronutrients and for future anticancer drug: A review. *Nucleus* **2019**. [CrossRef]
14. Michalke, B. Uncovering the importance of selenium in muscle disease. In *Selenium*; Michalke, B., Ed.; Molecular and Integrative, Toxicology; Springer: Cham, Switzerland, 2018; pp. 345–362.
15. Mocchegiani, E.; Malavolta, M. Role of zinc and selenium in oxidative stress and immunosenescence: Implications for healthy aging and longevity. In *Handbook of Immunosenescence*; Fulop, T., Franceschi, C., Hirokawa, K., Pawelec, G., Eds.; Springer: Cham, Switzerland, 2019; pp. 2539–2573.
16. Zhang, Y. Trace elements and healthcare: A bioinformatics perspective. *Adv. Exp. Med. Biol.* **2017**, *1005*, 63. [PubMed]
17. Varlamova, E.G.; Maltseva, V.N. Micronutrient selenium: Uniqueness and vital functions. *Biophysics* **2019**, *64*, 510–521. [CrossRef]
18. Rayman, M.P. The importance of selenium to human health. *Lancet* **2000**, *356*, 233. [CrossRef]
19. Combs, G.F. Selenium in global food systems. *Br. J. Nutr.* **2001**, *85*, 517–547. [CrossRef]
20. Fairweather-Tait, S.J.; Bao, Y.; Broadley, M.R.; Collings, R.; Ford, D.; Hesketh, J.E.; Hurst, R. Selenium in human health and disease. *Antioxid. Redox Signal.* **2011**, *14*, 1337–1383. [CrossRef]
21. Fordyce, F.M. Selenium deficiency and toxicity in the environment. In *Essentials of Medical Geology*; Springer: Dordrecht, The Netherlands, 2013; pp. 375–416.
22. Khalid, S.; Asghar, H.N.; Akhtar, M.J.; Aslam, A.; Zahir, Z.A. Biofortification of iron in chickpea by plant growth promoting rhizobacteria. *Pak. J. Bot.* **2015**, *47*, 1191–1194.
23. Kumar, A.; Patel, J.S.; Bahadur, I.; Meena, V.S. *The Molecular Mechanisms of KSMs for Enhancement of Crop Production Under Organic Farming*; Springer: New Delhi, India, 2016.
24. Jat, L.K.; Singh, Y.V.; Meena, S.K.; Meena, S.K.; Parihar, M.; Jatav, H.S.; Meena, R.K.; Meena, V.S. Does integrated nutrient management enhance agricultural productivity? *J. Pur. Appl. Microbiol.* **2015**, *9*, 1211–1221.
25. Malagoli, M.; Schiavon, M.; Dall'Acqua, S.; Pilon-Smits, E.A.H. Effects of selenium biofortification on crop nutritional quality. *Front. Plant Sci.* **2015**, *6*, 280. [CrossRef]
26. Ahmad, M.; Nadeem, S.M.; Naveed, M.; Zahir, Z.A. Potassium-solubilizing bacteria and their application in agriculture. In *Potassium Solubilizing Microorganisms for Sustainable Agriculture*; Meena, V.S., Maurya, B.R., Verma, J.P., Meena, R.S., Eds.; Springer: New Delhi, India, 2016; pp. 293–313.
27. Reeves, M.A.; Hoffmann, P.R. The human selenoproteome: Recent insights into functions and regulation. *Cell. Mol. Life Sci.* **2009**, *66*, 2457. [CrossRef] [PubMed]

28. Brigelius-Flohé, R.; Maiorino, M. Glutathione peroxidases. *Biochim. Biophys. Acta* **2013**, *1830*, 3289–3303. [CrossRef]
29. Soriano-Garcia, M. Organoselenium compounds as potential therapeutic and chemopreventive agents: A review. *Curr. Med. Chem.* **2004**, *11*, 1657–1669. [CrossRef] [PubMed]
30. Abadin, Z.U.; Yasin, M.; Faisal, M. Bacterial-mediated selenium biofortification of *Triticum aestivum*: Strategy for improvement in selenium phytoremediation and biofortification. In *Agriculturally Important Microbes for Sustainable Agriculture*; Meena, V., Mishra, P., Bisht, J., Pattanayak, A., Eds.; Springer: Singapore, 2017; pp. 299–315.
31. Chomchan, R.; Siripongvutikorn, S.; Puttarak, P.; Rattanapon, R. Influence of selenium bio-fortification on nutritional compositions, bioactive compounds content and anti-oxidative properties of young ricegrass (*Oryza sativa* L.). *Funct. Foods Heal. Dis.* **2017**, *7*, 195–209. [CrossRef]
32. D'Amato, R.; Fontanella, M.C.; Falcinelli, B.; Beone, G.M.; Bravi, E.; Marconi, O.; Benincasa, P.; Businelli, D. Selenium Biofortification in Rice (*Oryza sativa* L.) Sprouting: Effects on Se yield and nutritional traits with focus on phenolic acid profile. *J. Agric. Food Chem.* **2018**, *66*, 4082–4090. [CrossRef]
33. Yasin, M.; El-Mehdawi, A.F.; Anwar, A.; Pilon-Smits, E.A.H.; Faisal, M. Microbial-enhanced selenium and iron biofortification of wheat (*Triticum aestivum* L.)—Applications in phytoremediation and biofortification. *Int. J. Phytoremediation* **2015**, *17*, 341–347. [CrossRef] [PubMed]
34. Yasin, M.; El-Mehdawi, A.F.; Pilon-Smits, E.A.H.; Faisal, M. Selenium-fortified wheat: Potential of microbes for biofortification of selenium and other essential nutrients. *Int. J. Phytoremediation* **2015**, *17*, 777–786. [CrossRef] [PubMed]
35. Trivedi, G.; Patel, P.; Saraf, M. Synergistic effect of endophytic selenobacteria on biofortification and growth of *Glycine max* under drought stress. *South Afr. J. Bot.* **2019**, *2019*, 1–9. [CrossRef]
36. Businelli, D.; D'Amato, R.; Onofri, A.; Tedeschini, E.; Tei, F. Se-enrichment of cucumber (*Cucumis sativus* L.), lettuce (*Lactuca sativa* L.) and tomato (*Solanum lycopersicum* L. Karst) through fortification in pre-transplanting. *Sci. Hortic.* **2015**, *197*, 697–704. [CrossRef]
37. Dhillon, K.S.; Dhillon, S.K. Accumulation and distribution of selenium in some vegetable crops grown in selenate-Se treated clay loam soil. *Front. Agric. China* **2009**, *3*, 366–373. [CrossRef]
38. Dil, T.; Alex, A.; Indika, M.; Clarice, C.; Pushparajah, T.; Shiv, K. Selecting lentil accessions for global selenium biofortification. *Plants* **2017**, *6*, 34.
39. Bachiega, P.; Salgado, J.M.; de Carvalho, A.L.T.G.; Schwarz, K.; Tezotto, T.; Morzelle, M.C. Antioxidant and antiproliferative activities in different maturation stages of broccoli (*Brassica oleracea* Italica) biofortified with selenium. *Food Chem.* **2016**, *190*, 771–776. [CrossRef]
40. Peng, Q.; Guo, L.; Ali, F.; Li, J.; Qin, S.; Feng, P.; Liang, D. Influence of Pak choi plant cultivation on Se distribution, speciation and bioavailability in soil. *Plant Soil* **2016**, *403*, 331–342. [CrossRef]
41. Slekovec, M.; Goessier, W. Accumulation of selenium in natural plants and selenium supplemented vegetable and selenium speciation by HPLC-ICPMS. *Chem. Speciat. Bioavailab.* **2015**, *17*, 63–73. [CrossRef]
42. Susana, G.M.; Fabián, P.L.; Ema, G.E.; Paola, L.M.; Julia, M.M.; Irma, D.R.; Antonio, J.M.; Erika, R.M.; Adalberto, B.M. Selenium and sulfur to produce Allium functional crops. *Molecules* **2017**, *22*, 558.
43. Jing, D.-W.; Du, Z.-Y.; Ma, H.-L.; Ma, B.-Y.; Liu, F.-C.; Song, Y.-G.; Xu, Y.-F.; Li, L. Selenium enrichment, fruit quality and yield of winter jujube as affected by addition of sodium selenite. *Sci. Hortic.* **2017**, *225*, 1–5. [CrossRef]
44. Nie, J.; Kuang, L.; Li, Z.; Pang, R.; Yang, L.; Chen, Q.; Li, A.; Zhao, X.; Xu, W. Selenium content of main deciduous fruits from China and its dietary exposure assessment. *Sci. Agric. Sinica* **2015**, *48*, 3015–3026.
45. Sun, X.; Yi, H.; Chen, Y.; Luo, Y.; Ping, T. Effects of different concentrations of Se6+ on selenium absorption, transportation, and distribution of citrus seedlings (*C. junos* cv. Ziyang xiangcheng). *J. Plant Nutr.* **2018**, *41*, 168–177.
46. Cominetti, C.; de Bortoli, M.C.; Garrido, A.B., Jr.; Cozzolino, S.M.F. Brazilian nut consumption improves selenium status and glutathione peroxidase activity and reduces atherogenic risk in obese women. *Nutr. Res.* **2012**, *32*, 403–407. [CrossRef]
47. Ip, C.; Lisk, D.J. Bioactivity of selenium from Brazil nut for cancer prevention and selenoenzyme maintenance. *Nutr. Cancer* **1994**, *21*, 203–212. [CrossRef]

48. Lima, L.W.; Stonehouse, G.C.; Walters, C.; El Mehdawi, A.F.; Fakra, S.C.; Pilon-Smits, E.A. Selenium accumulation, speciation and localization in Brazil nuts (Bertholletia excelsa H.B.K.). *Plants* **2019**, *8*, 289. [CrossRef] [PubMed]
49. Kápolna, E.; Gergely, V.; Dernovics, M.; Illés, A.; Fodor, P. Fate of selenium species in sesame seeds during simulated bakery process. *J. Food Eng.* **2007**, *79*, 494–501. [CrossRef]
50. Graham, L. Biofortification of cereals with foliar selenium and iodine could reduce hypothyroidism. *Front. Plant Sci.* **2018**, *9*, 730.
51. Kapolna, E.; Hillestrom, P.R.; Laursen, K.H.; Husted, S.; Larsen, E.H. Effect of foliar application of selenium on its uptake and speciation in carrot. *Food Chem.* **2009**, *115*, 1357–1363. [CrossRef]
52. Ros, G.H.; van Rotterdam, A.M.D.; Bussink, D.W.; Bindraban, P.S. Selenium fertilization strategies for bio-fortification of food: An agro-ecosystem approach. *Plant Soil* **2016**, *404*, 99–112. [CrossRef]
53. Xiong, L.; Li, B.; Yang, Y. Effects of foliar selenite on the nutrient components of turnip (*Brassica rapa* var. rapa Linn.). *Front. Chem.* **2018**, *6*, 42.
54. Broadley, M.R.; Alcock, J.; Alford, J.; Cartwright, P.; Foot, I.; Fairweather-Tait, S.J.; Hart, D.J.; Hurst, R.; Knott, P.; Mcgrath, S.P. Selenium biofortification of high-yielding winter wheat (*Triticum aestivum* L.) by liquid or granular Se fertilisation. *Plant Soil* **2010**, *332*, 5–18. [CrossRef]
55. Larsen, E.H.; Łobiński, R.; Burger-Meÿer, K.; Hansen, M.; Ruzik, L.; Mazurowska, L.; Rasmussen, P.H.; Sloth, J.J.; Scholten, O.; Kik, C. Uptake and speciation of selenium in garlic cultivated in soil amended with symbiotic fungi (mycorrhiza) and selenate. *Anal. Bioanal. Chem.* **2006**, *385*, 1098–1108. [CrossRef]
56. Bañuelos, G.S.; Arroyo, I.; Pickering, I.J.; Yang, S.I.; Freeman, J.L. Selenium biofortification of broccoli and carrots grown in soil amended with Se-enriched hyperaccumulator *Stanleya pinnata*. *Food Chem.* **2015**, *166*, 603–608. [CrossRef]
57. Aspila, P. History of selenium supplemented fertilization in Finland. In Proceedings of the Twenty Years of Selenium Fertilization, Helsinki, Finland, 8–9 September 2005; pp. 8–13.
58. Lyons, G.H.; Lewis, J.; Lorimer, M.F.; Holloway, R.E.; Graham, R.D. High-selenium wheat: Agronomic biofortification strategies to improve human nutrition. *J. Food Agric. Environ.* **2004**, *22*, 171–178.
59. Deng, X.; Zhao, Z.; Zhou, J.; Chen, J.; Lv, C.; Liu, X. Compositional analysis of typical selenium ore from Enshi and its effect on selenium enrichment in wetland and dryland crops. *Plant Soil* **2018**, *433*, 55–64. [CrossRef]
60. Sura-de Jong, M.; Reynolds, R.J.B.; Richterova, K.; Musilova, L.; Staicu, L.C.; Chocholata, I.; Cappa, J.J.; Taghavi, S.; van der Lelie, D.; Frantik, T.; et al. Selenium hyperaccumulators harbor a diverse endophytic bacterial community characterized by high selenium resistance and plant growth promoting properties. *Front. Plant Sci.* **2015**, *6*, 113. [CrossRef] [PubMed]
61. Carrino-Kyker, S.R.; Kluber, L.A.; Coyle, K.P.; Burke, D.J. Detection of phosphate transporter genes from arbuscular mycorrhizal fungi in mature tree roots under experimental soil pH manipulation. *Symbiosis* **2017**, *72*, 123–133. [CrossRef]
62. Hawkesford, M.J.; Buchner, P.; Hopkins, L.; Howarth, J.R. Phosphate transporters in arbuscular myccorhizal symbiosis. In *Arbuscular mycorrhzias: Physiology and Functions*; Koltai, H., Kapulnik, Y., Eds.; Springer: Dordrencht, The Netherlands, 2010; pp. 117–135.
63. Tatry, M.-V.; El Kassis, E.; Lambilliotte, R.; Corratgé, C.; van Aarle, I.; Amenc, L.K.; Alary, R.; Zimmermann, S.; Sentenac, H.; Plassard, C. Two differentially regulated phosphate transporters from the symbiotic fungus Hebeloma cylindrosporum and phosphorus acquisition by ectomycorrhizal Pinus pinaste. *Plant J.* **2009**, *57*, 1092–1102. [CrossRef] [PubMed]
64. Fiorilli, V.; Lanfranco, L.; Bonfante, P. The expression of GintPT, the phosphate transporter of Rhizophagus irregularis, depends on the symbiotic status and phosphate availability. *Planta* **2013**, *237*, 1267–1277. [CrossRef] [PubMed]
65. Xie, X.; Lin, H.; Peng, X.; Xu, C.; Sun, Z.; Jian, K.; Huang, A.; Wu, X.; Tang, N. Arbuscular mycorrhizal symbiosis requires a phosphate transporter in the *Gigaspora margarita* fungal symbiont. *Mol. Plant* **2016**, *9*, 1583–1608. [CrossRef] [PubMed]
66. Glassop, D.; Smith, S.E.; Smith, F.W. Cereal phosphate transporters associated with the mycorrhizal pathway of phosphate uptake into roots. *Planta* **2005**, *222*, 688–698. [CrossRef]
67. Loth-Pereda, V.; Orsini, E.; Courty, P.E.; Lota, F.; Martin, F. Structure and expression profile of the phosphate Pht1 transporter gene family in Mycorrhizal *Populus trichocarpa*. *Plant Physiol.* **2011**, *156*, 2141–2154. [CrossRef]

68. Pumplin, N.; Zhang, X.; Noar, R.D.; Harrison, M.J. Polar localization of a symbiosis-specific phosphate transporter is mediated by a transient reorientation of secretion. *PNAS* **2012**, *109*, E665–E672. [CrossRef]
69. Nagy, R.; Karandashov, V.; Chagué, V.; Kalinkevich, K.; Tamasloukht, M.; Xu, G.; Jakobsen, I.; Levy, A.A.; Amrhein, N.; Bucher, M. The characterization of novel mycorrhiza-specific phosphate transporters from *Lycopersicon esculentum* and *Solanum tuberosum* uncovers functional redundancy in symbiotic phosphate transport in solanaceous species. *Plant J.* **2005**, *42*, 236–250. [CrossRef]
70. Paszkowski, U.; Kroken, S.; Roux, C.; Briggs, S.P. Rice phosphate transporters include an evolutionarily divergent gene specifically activated in arbuscular mycorrhizal symbiosis. *Proc. Natl. Acad. Sci. USA* **2002**, *99*, 13324–13329. [CrossRef] [PubMed]
71. Tian, H.; Drijber, R.A.; Li, X.; Miller, D.N.; Wienhold, B.J. Arbuscular mycorrhizal fungi differ in their ability to regulate the expression of phosphate transporters in maize (*Zea mays* L.). *Mycorrhiza* **2013**, *23*, 507–514. [CrossRef] [PubMed]
72. Zhang, T.; Wen, X.; Ding, G. Ectomycorrhizal symbiosis enhances tolerance to low phosphorous through expression of phosphate transporter genes in masson pine *(Pinus massoniana)*. *Acta Physiol. Plant* **2017**, *39*, 101. [CrossRef]
73. Smith, S.E.; Read, D.J. Arbuscular mycorrhizaes. In *Mycorrhizal Symbiosis*, 3rd ed.; Smith, S.E., Read, D.J., Eds.; Academic Press: London, UK, 2008; pp. 13–187.
74. Smith, S.E.; Jakobsen, I.; Grønlund, M.; Smith, F.A. Roles of arbuscular mycorrhizas in plant phosphorus nutrition: Interactions between pathways of phosphorus uptake in arbuscular mycorrhizal roots have important implications for understanding and manipulating plant phosphorus acquisition. *Plant Physiol.* **2011**, *156*, 1050–1057. [CrossRef] [PubMed]
75. Higo, M.; Takahashi, Y.; Gunji, K.; Isobe, K. How are arbuscular mycorrhizal associations related to maize growth performance during short-term cover crop rotation? *Sci. Food Agric.* **2018**, *98*, 1388–1396. [CrossRef]
76. Higo, M.; Tatewaki, Y.; Gunji, K.; Kaseda, A.; Isobe, K. Cover cropping can be a stronger determinant than host crop identity for arbuscular mycorrhizal fungal communities colonizing maize and soybean. *Peer J.* **2019**, *7*, e6403. [CrossRef]
77. Higo, M.; Tatewaki, Y.; Iida, K.; Yokota, K.; Isobe, K. Amplicon sequencing analysis of arbuscular mycorrhizal fungal communities colonizing maize roots in different cover cropping and tillage systems. *Sci. Rep.* **2020**, *10*, 6093. [CrossRef]
78. Eah, P.-S. Se in plants. In *Progress in Botany*; Lüttge, U., Beyschlag, W., Eds.; Springer: Cham, Switzerland, 2015; Volume 76, pp. 93–107.
79. Ellis, D.R.; Salt, D.E. Plants, selenium and human health. *Curr. Opin. Plant Biol.* **2003**, *6*, 273–279. [CrossRef]
80. Sors, T.G.; Ellis, D.R.; Na, G.N.; Lahner, B.; Salt, D.E. Analysis of sulfur and selenium assimilation in Astragalus plants with varying capacities to accumulate selenium. *Plant J.* **2005**, *42*, 785–797. [CrossRef]
81. Wessjohann, L.A.; Schneider, A.; Abbas, M.; Brandt, W. Selenium in chemistry and biochemistry in comparison to sulfur. *Biol. Chem.* **2007**, *388*, 997–1006. [CrossRef]
82. Chauhan, R.; Awasthi, S.; Srivastava, S.; Dwivedi, S.; Pilon-Smits, E.A.H.; Dhankher, O.P.; Tripathi, R.D. Understanding selenium metabolism in plants and its role as a beneficial element. *Crit. Rev. Environ. Sci. Technol.* **2019**, *49*, 1937–1958. [CrossRef]
83. Lauchli, A. Selenium in plants: Uptake, functions, and environmental toxicity. *Bot. Acta* **1993**, *106*, 455–468. [CrossRef]
84. Lass, B.; Ullrich-Eberius, C.I. Evidence for proton/sulfate cotransport and its kinetics inLemna gibba G1. *Planta* **1984**, *161*, 53–60. [CrossRef] [PubMed]
85. Hawkesford, M.; Davidian, J.C.; Grignon, C. Sulphate/proton cotransport in plasma-membrane vesicles isolated from roots of *Brassica napus* L.: Increased transport in membranes isolated from sulphur-starved plants. *Planta* **1993**, *190*, 297–304. [CrossRef]
86. Shibagaki, N.; Rose, A.; McDermott, J.P.; Fujiwara, T.; Hayashi, H.; Yoneyama, T.; Davies, J.P. Selenate-resistant mutants of *Arabidopsis thaliana* identify *Sultr1:2*, a sulfate transporter required for efficient transport of sulfate into roots. *Plant J.* **2002**, *29*, 475–486. [CrossRef] [PubMed]
87. Inostroza-Blancheteau, C.; Reyes-Díaz, M.; Alberdi, M.; Godoy, K.; Rojas-Lillo, Y.; Cartes, P.; de la Luz Mora, M. Influence of selenite on selenium uptake, differential antioxidant performance and gene expression of sulfate transporters in wheat genotypes. *Plant Soil* **2013**, *369*, 47–59. [CrossRef]

88. Zhao, X.Q.; Mitani, N.; Yamaji, N.; Shen, R.F.; Ma, J.F. Involvement of silicon influx transporter OsNIP2;1 in selenite uptake in rice. *Plant Physiol.* **2010**, *153*, 1871–1877. [CrossRef]
89. Wang, M.; Yang, W.; Zhou, F.; Du, Z.; Xue, M.; Chen, T.; Liang, D. Effect of phosphate and silicate on selenite uptake and phloem-mediated transport in tomato (*Solanum lycopersicum* L.). *Environ. Sci. Pollut. Res.* **2019**, *26*, 20475–20484. [CrossRef]
90. Li, H.F.; Mcgrath, S.P.; Zhao, F.J. Selenium uptake, translocation and speciation in wheat supplied with selenate or selenite. *New Phytol.* **2008**, *178*, 92–102. [CrossRef]
91. Chen, G.H.; Yan, W.; Yang, S.P.; Wang, A.; Zhu, Y.L. Overexpression of rice phosphate transporter gene OsPT2 enhances tolerance to low phosphorus stress in soybean. *J. Agric. Sci. Technol.* **2015**, *17*, 469–482.
92. Zhang, L.; Hu, B.; Li, W.; Chen, R.; Deng, K.; Li, H.; Yu, F.; Ling, H.; Li, Y.; Chu, C. OsPT2, a phosphate transporter, is involved in the active uptake of selenite in rice. *New Phytol.* **2014**, *201*, 1183–1191. [CrossRef]
93. Zhang, M.; Wilson, L.; Xing, G.; Jiang, L.; Tang, S. Optimizing root architecture and increasing transporter gene expression are strategies to promote selenium uptake by high-se accumulating rice cultivar. *Plant Soil* **2020**, *447*, 319–332. [CrossRef]
94. Lazard, M.; Blanquet, S.; Fisicaro, P.; Labarraque, G.; Plateau, P. Uptake of selenite by *Saccharomyces cerevisiae* involves the high and low affinity orthophosphate transporters. *J. Biol. Chem.* **2010**, *285*, 32029–32037. [CrossRef]
95. Pérez-Sampietro, M.; Serra-Cardona, A.; Canadell, D.; Casas, C.; Ario, J.; Herrero, E. The yeast Aft2 transcription factor determines selenite toxicity by controlling the low affinity phosphate transport system. *Sci. Rep.* **2016**, *6*, 32836. [CrossRef] [PubMed]
96. McDermott, J.R.; Rosen, B.P.; Liu, Z. Jen1p: A high affinity selenite transporter in yeast. *Mol. Biol. Cell* **2010**, *21*, 3934–3941. [CrossRef] [PubMed]
97. Hopper, J.L.; Parker, D.R. Plant availability of selenite and selenate as influenced by the competing ions phosphate and sulfate. *Plant Soil* **1999**, *210*, 199–207. [CrossRef]
98. Yang, X.; Lu, Y.; Zhong, J.; Qian, Y.; Zhao, Z.; Liu, X. The Positive effect of sulfur on selenium detoxification under selenite condition in wheat. *Commun. Soil Sci. Plant Anal.* **2017**, *48*, 1564–1573. [CrossRef]
99. Zhang, D.; Dong, T.; Ye, J.; Hou, Z. Selenium accumulation in wheat (*Triticum aestivum* L.) as affected by coapplication of either selenite or selenate with phosphorus. *Soil Sci. Plant Nutr.* **2017**, *63*, 37–44. [CrossRef]
100. Cabannes, E.; Buchner, P.; Broadley, M.R.; Hawkesford, M.J. A Comparison of sulfate and selenium accumulation in relation to the expression of sulfate transporter genes in *Astragalus* Species. *Plant Physiol.* **2011**, *157*, 2227–2239. [CrossRef]
101. Hawkesford, J.M. Plant responses to sulphur deficiency and the genetic manipulation of sulphate transporters to improve S-utilization efficiency. *J. Exp. Bot.* **2000**, *51*, 131–138. [CrossRef]
102. Hawkesford, M.J.; Buchner, P.; Hopkins, L.; Howarth, J.R. The plant sulfate transporter family: Specialized functions and integration with whole plant nutrition. In *Sulfur Transport and Assimilation in Plants: Regulation, Interaction and Signaling*; Davidian, J.C., Grill, D., DeKok, L.J., Stulen, I., Hawkesford, M.J., Schnug, E., Rennenberg, H., Eds.; Backhuys Publishers: Leiden, The Netherlands, 2003; pp. 1–10.
103. Shinmachi, F.; Buchner, P.; Stroud, J.L.; Parmar, S.; Zhao, F.J.; McGrath, S.P.; Hawkesford, M.J. Influence of sulfur deficiency on the expression of specific sulfate transporters and the distribution of sulfur, selenium, and molybdenum in wheat. *Plant Physiol.* **2010**, *153*, 327–336. [CrossRef] [PubMed]
104. Giovannetti, M.; Tolosano, M.; Volpe, V.; Kopriva, S.; Bonfante, P. Identification and functional characterization of a sulfate transporter induced by both sulfur starvation and mycorrhiza formation in Lotus japonicus. *New Phytol.* **2014**, *204*, 609–619. [CrossRef]
105. Luo, W.; Li, J.; Ma, X.; Niu, H.; Hou, S.; Wu, F. Effect of arbuscular mycorrhizal fungi on uptake of selenate, selenite, and selenomethionine by roots of winter wheat. *Plant Soil* **2019**, *438*, 71–83. [CrossRef]
106. Sanmartín, C.; Garmendia, I.; Romano, B.; Díaz, M.; Palop, J.A.; Goicoechea, N. Mycorrhizal inoculation affected growth, mineral composition, proteins and sugars in lettuces biofortified with organic or inorganic selenocompounds. *Sci. Hortic.* **2014**, *180*, 40–51. [CrossRef]
107. Conversa, G.; Lazzizera, C.; Chiaravalle, A.E.; Miedico, O.; Bonasia, A.; La Rotonda, P.; Elia, A. Selenium fern application and arbuscular mycorrhizal fungi soil inoculation enhance Se content and antioxidant properties of green asparagus (*Asparagus officinalis* L.) spears. *Sci. Hortic.* **2019**, *252*, 176–191. [CrossRef]

108. Goicoechea, N.; Garmendia, I.; Fabbrin, E.G.; Bettoni, M.M.; Palop, J.A.; Sanmartín, C. Selenium fertilization and mycorrhizal technology may interfere in enhancing bioactive compounds in edible tissues of lettuces. *Sci. Hortic.* **2015**, *195*, 163–172. [CrossRef]
109. Durán, P.; Acuña, J.J.; Jorquera, M.A.; Azcón, R.; Mora, M.L. Enhanced selenium content in wheat grain by co-inoculation of selenobacteria and arbuscular mycorrhizal fungi: A preliminary study as a potential Se biofortification strategy. *J. Cereal Sci.* **2013**, *57*, 275–280. [CrossRef]
110. Patharajan, S.; Raaman, N. Influence of arbuscular mycorrhizal fungi on growth and selenium uptake by garlic plants. *Arch. Phytopathol. Plant Prot.* **2012**, *45*, 138–151. [CrossRef]
111. Munier-Lamy, C.; Deneux-Mustin, S.; Mustin, C.; Merlet, D.; Berthelin, J.; Leyval, C. Selenium bioavailability and uptake as affected by four different plants in a loamy clay soil with particular attention to mycorrhizae inoculated ryegrass. *J. Environ. Radioact.* **2007**, *97*, 148–158. [CrossRef]
112. Yu, Y.; Zhang, S.; Wen, B.; Huang, H.; Luo, L. Accumulation and Speciation of Selenium in Plants as Affected by Arbuscular Mycorrhizal Fungus Glomus mosseae. *Biol. Trace Elem. Res.* **2011**, *143*, 1789–1798. [CrossRef]
113. Golubkina, N.A.; Zamana, S.; Seredin, T.; Poluboyarinov, P.A.; Sokolov, S.; Baranova, H.; Krivenkov, L.; Pietrantonio, L.; Caruso, G. Effect of selenium biofortification and beneficial microorganism inoculation on yield, quality and antioxidant properties of shallot bulbs. *Plants* **2019**, *8*, 102. [CrossRef] [PubMed]
114. Lindblom, S.D.; Valdez-Barillas, J.R.; Fakra, S.C.; Marcus, M.A.; Pilon-Smits, E.A.H. Influence of microbial associations on selenium localization and speciation in roots of *Astragalus* and *Stanleya* hyperaccumulators. *Environ. Exp. Bot.* **2013**, *88*, 33–42. [CrossRef]
115. Lei, X.; Yao, S.; Zu, X.; Huang, Z.; Liu, L.; Zhong, M.; Zhu, B.; Tang, S.; Liao, D. Apoptosis induced by diallyl disulfide in human breast cancer cell line MCF-7. *Acta Pharm. Sin.* **2008**, *29*, 1233–1239. [CrossRef] [PubMed]
116. Yang, J.S.; Kok, L.F.; Lin, Y.H.; Kuo, T.C.; Chung, J.G. Diallyl disulfide inhibits WEHI-3 leukemia cells in vivo. *Anticancer Res.* **2006**, *26*, 219–225. [PubMed]
117. Suangtamai, T.; Tanyong, D.I. Diallyl disulfide induces apoptosis and autophagy via mTOR pathway in myeloid leukemic cell line. *Tumor Biol.* **2016**, *37*, 10993–10999. [CrossRef]
118. Nicolle, C.; Cardinault, N.; Gueux, E.; Jaffrelo, L.; Rock, E.; Mazur, A.; Amouroux, P.; Rémésy, C. Health effect of vegetable-based diet: Lettuce consumption improves cholesterol metabolism and antioxidant status in the rat. *Clin. Nutr.* **2004**, *23*, 605–614. [CrossRef]
119. Serafini, M.; Bugianesi, R.; Salucci, M.; Azzini, E.; Maiani, G. Effect of acute ingestion of fresh and stored lettuce (Latuca sativa) on plasma total antioxidant capacity and antioxidant levels in human subjects. *Br. J. Nutr.* **2003**, *88*, 615–623. [CrossRef]
120. Llorach, R.; Martínez-Sánchez, A.; Tomás-Barberán, F.A.; Gil, M.I.; Ferreres, F. Characterisation of polyphenols and antioxidant properties of five lettuce varieties and escarole. *Food Chem.* **2008**, *108*, 1028–1038. [CrossRef]
121. Slejkovec, Z.; Elteren, J.T.V.; Woroniecka, U.D.; Kroon, K.J.; Byrne, A.R. Preliminary study on the determination of selenium compounds in some selenium-accumulating mushrooms. *Biol. Trace Elem. Res.* **2000**, *75*, 139–155. [CrossRef]
122. Stijve, T.; Noorloos, T.; Byrne, A.R.; Šlejkovec, Z.; Goessler, W.T. High selenium levels in edible Albatrellus mushroom. *Deut Lebens-Rundsch* **1998**, *94*, 275–279.
123. Borovička, J.; Řanda, Z. Distribution of iron, cobalt, zinc and selenium in macrofungi. *Mycol. Prog.* **2007**, *6*, 249–259. [CrossRef]
124. Etesami, H.; Maheshwari, D.K. Use of plant growth promoting rhizobacteria (PGPRs) with multiple plant growth promoting traits in stress agriculture: Action mechanisms and future prospects. *Ecotoxicol. Environ. Saf.* **2018**, *156*, 225–246. [CrossRef] [PubMed]
125. Sayyed, R.Z.; Reddy, M.S.; Antonius, S. *Plant Growth Promoting Rhizobacteria (PGPR): Prospects for Sustainable Agriculture*; Springer: Singapore, 2019.
126. Sapre, S.; Gontia-Mishra, I.; Tiwari, S. *ACC Deaminase-Producing Bacteria: A Key Player in Alleviating Abiotic Stresses in Plants*; Springer: Singapore, 2019.
127. Etesami, H.; Adl, S.M. Plant growth-promoting rhizobacteria (PGPR) and their action mechanisms in availability of nutrients to plants. In *Phyto-Microbiome in Stress Regulation*; Kumar, M., Kumar, V., Prasad, R., Eds.; Environmental and Microbial Biotechnology; Springer: Singapore, 2020; pp. 147–203.
128. Cartes, P.; Gianfreda, L.; Mora, M.L. Uptake of Selenium and its Antioxidant Activity in Ryegrass When Applied as Selenate and Selenite Forms. *Plant Soil* **2005**, *276*, 359–367. [CrossRef]

129. Mora, M.D.L.L.; Pinilla, L.; Rosas, A.; Cartes, P. Selenium uptake and its influence on the antioxidative system of white clover as affected by lime and phosphorus fertilization. *Plant Soil* **2008**, *303*, 139–149. [CrossRef]
130. Nakamaru, Y.M.; Altansuvd, J. Speciation and bioavailability of selenium and antimony in non-flooded and wetland soils: A review. *Chemosphere* **2014**, *111*, 366–371. [CrossRef]
131. Yli-Halla, M. Influence of Se fertilization on soil Se status. In Proceedings of the Twenty Years of Se Fertilization, Helsinki, Finland, 8–9 September 2005; pp. 25–32.
132. Durán, P.; Acuña, J.J.; Gianfreda, L.; Azcón, R.; Funes-Collado, V.; Mora, M.L. Endophytic selenobacteria as new inocula for selenium biofortification. *Appl. Soil Ecol.* **2015**, *96*, 319–326. [CrossRef]
133. Acuña, J.J.; Jorquera, M.A.; Barra, P.J.; Crowley, D.E.; María, D.L.L.M. Selenobacteria selected from the rhizosphere as a potential tool for Se biofortification of wheat crops. *Biol. Fertil. Soils* **2013**, *49*, 175–185. [CrossRef]
134. Artursson, V.; Finlay, R.D.; Jansson, J.K. Interactions between arbuscular mycorrhizal fungi and bacteria and their potential for stimulating plant growth. *Environ. Microbiol.* **2006**, *8*, 1–10. [CrossRef]
135. Barea, J.; Pozo, M.; Azcón, R.; Azcón-Aguilar, C. Microbial co-operation in the rhizosphere. *J. Exp. Bot.* **2005**, *56*, 1761–1778. [CrossRef]
136. Marschner, P.; Baumann, K. Changes in bacterial community structure induced by mycorrhizal colonisation in split-root maize. *Plant Soil* **2003**, *251*, 279–289. [CrossRef]
137. Hrynkiewicz, K.; Ciesielska, A.; Haug, I.; Baum, C. Ectomycorrhiza formation and willow growth promotion as affected by associated bacteria: Role of microbial metabolites and use of C sources. *Biol. Fertil. Soils* **2010**, *46*, 139–150. [CrossRef]
138. Alford, É.; Lindblom, S.; Pittarello, M.; Freeman, J.; Fakra, S.; Marcus, M.; Broeckling, C.; Pilon-Smits, E.; Paschke, M. Roles of rhizobial symbionts in selenium hyperaccumulation in Astragalus (Fabaceae). *Am. J. Bot.* **2014**, *101*, 1895–1905. [CrossRef] [PubMed]
139. Durán, P.; Acuña, J.J.; Jorquera, M.A.; Azcón, R.; Paredes, C.; Rengel, Z.; Mora, M.D.L.L. Endophytic bacteria from selenium-supplemented wheat plants could be useful for plant-growth promotion, biofortification andGaeumannomyces graminisbiocontrol in wheat production. *Biol. Fertil. Soils* **2014**, *50*, 983–990. [CrossRef]
140. Rana, A.; Joshi, M.; Prasanna, R.; Shivay, Y.S.; Nain, L. Biofortification of wheat through inoculation of plant growth promoting rhizobacteria and cyanobacteria. *Eur. J. Soil Biol.* **2012**, *50*, 118–126. [CrossRef]
141. Jianfei, W.; Cheng, Z.; Xin, X.; Yue, X.; Lin, Z.; Zhongyou, M. Enhanced iron and selenium uptake in plants by volatile emissions of *Bacillus amyloliquefaciens* (BF06). *Appl. Sci.* **2017**, *7*, 85.
142. Lampis, S.; Ferrari, A.; Cunhaqueda, A.C.; Alvarenga, P.; Di, G.S.; Vallini, G. Selenite resistant rhizobacteria stimulate SeO(3) (2−) phytoextraction by *Brassica juncea* in bioaugmented water-filtering artificial beds. *Environ. Sci. Pollut. Res. Int.* **2009**, *16*, 663. [CrossRef]
143. Souza, M.P.D.; Huang, C.P.A.; Chee, N.; Terry, N. Rhizosphere bacteria enhance the accumulation of selenium and mercury in wetland plants. *Planta* **1999**, *209*, 259–263. [CrossRef]
144. Souza, M.P.D.; Chu, D.; Zhao, M.; Zayed, A.M.; Ruzin, S.E.; Terry, S.N. Rhizosphere bacteria enhance selenium accumulation and volatilization by Indian mustard. *Plant Physiol.* **1999**, *119*, 565–573. [CrossRef]
145. Patel, P.J.; Trivedi, G.R.; Shah, R.K.; Saraf, M. Selenorhizobacteria: As biofortification tool in sustainable agriculture. *Biocatal. Agric. Biotechnol.* **2018**, *14*, 198–203. [CrossRef]

Journal of *Fungi*

MDPI

Review

Antagonistic Yeasts: A Promising Alternative to Chemical Fungicides for Controlling Postharvest Decay of Fruit

Xiaokang Zhang [1,2,†], Boqiang Li [1,†], Zhanquan Zhang [1], Yong Chen [1] and Shiping Tian [1,2,*]

1 Key Laboratory of Plant Resources, Institute of Botany, Innovative Academy of Seed Design, Chinese Academy of Sciences, Beijing 100093, China; zhangxk@ibcas.ac.cn (X.Z.); bqli@ibcas.ac.cn (B.L.); zhangzhanquan82@ibcas.ac.cn (Z.Z.); chenyong@ibcas.ac.cn (Y.C.)
2 College of Life Sciences, University of Chinese Academy of Sciences, Beijing 100049, China
* Correspondence: tsp@ibcas.ac.cn; Tel.: +86-10-62836559
† These authors contributed equally to this work.

Received: 12 July 2020; Accepted: 28 August 2020; Published: 31 August 2020

check for updates

Abstract: Fruit plays an important role in human diet. Whereas, fungal pathogens cause huge losses of fruit during storage and transportation, abuse of chemical fungicides leads to serious environmental pollution and endangers human health. Antagonistic yeasts (also known as biocontrol yeasts) are promising substitutes for chemical fungicides in the control of postharvest decay owing to their widespread distribution, antagonistic ability, environmentally friendly nature, and safety for humans. Over the past few decades, the biocontrol mechanisms of antagonistic yeasts have been extensively studied, such as nutrition and space competition, mycoparasitism, and induction of host resistance. Moreover, combination of antagonistic yeasts with other agents or treatments were developed to improve the biocontrol efficacy. Several antagonistic yeasts are used commercially. In this review, the application of antagonistic yeasts for postharvest decay control is summarized, including the antagonistic yeast species and sources, antagonistic mechanisms, commercial applications, and efficacy improvement. Issues requiring further study are also discussed.

Keywords: yeast; biological control; postharvest decay; fruit

1. Introduction

As an important part of the human diet, fruit provides the body with beneficial vitamins, minerals, organic acids, and antioxidants. Fruits have been shown to have many health-related effects, such as anti-cancer effects, skin protecting effects, and postponing of senescence [1–4]. As orchards are usually far away from urban areas, and the fruit maturity occurs in a relatively short period, leading to a disparity between supply and demand in the market, which necessitates a certain period of storage and transportation to adjust for this disparity. However, postharvest spoilage, which involves rot, nutrient loss, and water content loss, occurs most often during storage and transportation, which leads to considerable economic losses. It has been reported that about 25% of total fruit production is wasted after harvest in developed countries, and the postharvest losses in developing countries account for >50% of total fruit production because of lack of efficient transportation and refrigeration facilities [5].

Fungi are the main cause of postharvest spoilage. Fruit rot can be induced by wound generated during harvesting, packaging, storage, and transportation, as well as the favorable growth conditions for pathogens (e.g., high water and nutrient content, low pH, and decreased resistance after harvest) [6]. During the process of infection, many fungi produce mycotoxins, which may enter the food chain via fresh and processed fruit products and then endanger human health. For example, *Penicillium expansum*, which causes blue mold in many fruits, leads to not only fruit decay but also the contamination of patulin, a teratogenic, carcinogenic, and immunotoxic mycotoxin [7]. Chemical fungicides have long been used to control postharvest decay. However, overdependence on traditional chemical fungicides has resulted in a variety of problems, such as fungicide residues, environmental pollution, and increased pathogen resistance to fungicides. Therefore, identifying safe and effective approaches to control postharvest fungal disease is urgent.

Since Gutter and Littauer first reported the use of *Bacillus subtilis* to combat citrus fruit pathogens in 1953, the biocontrol capability of microorganisms against postharvest decay has attracted widespread attention [8,9]. Among the various microbial antagonists, yeast and yeast-like fungi occupy an important position as they are environmentally friendly, exhibit good biocontrol efficacy against pathogens, possess adequate stress tolerance, and can potentially be genetically improved; additionally, there is a well-developed system for culturing, fermentation, storage, and handling of these antagonistic yeasts [10]. Moreover, yeasts have been used in food and beverage production for thousands of years and currently play an important role in the food industry. Thus, the utilization of yeasts is generally considered safe, and easily acceptable by market. With the great properties and application superiority, antagonistic yeasts are considered as a promising alternate to synthetic chemical fungicides [5,9]. Over the past few decades, great progresses have been made in biological control based on antagonistic yeasts, including strain isolation and screening, mode of action, improvement of biocontrol efficacy, and formulation. Particularly, several antagonistic yeasts with excellent biocontrol performance have been developed and registered as commercial products. Nonetheless, the widespread use of yeast antagonists to manage postharvest diseases still faces many challenges. A deeper understanding of the mode of action of antagonistic yeasts in postharvest biocontrol system is still needed; the inconsistency of performance of antagonistic yeasts under commercial conditions need be overcome; the market penetration of products is difficult.

Here, a comprehensive overview of the applications of antagonistic yeasts in postharvest decay control is presented, including the features of antagonistic yeasts, antagonistic mechanisms, efficacy improvement, and commercial applications. The latest research results are highlighted, and issues requiring further study are also discussed.

2. Features of Antagonistic Yeasts

Yeasts are a group of eukaryotic fungi, most of which are unicellular and reproduce by budding [11]. There are also a variety of phylogenetically different groups of yeast-like fungi, such as *Aureobasidium pullulans*. Antagonistic yeasts (also known as biocontrol yeasts) refers to yeast or yeast-like fungi that can inhibit or interfere with the growth, development, reproduction, or activity of phytopathogens. Wilson and Wisniewski summarized the criteria for the selection of ideal biocontrol agents in 1989 [12]. With the extensive research on antagonistic yeasts, the screening criteria for antagonistic yeasts have gradually improved [13]. An ideal antagonistic yeast should be genetically stable, have simple nutrient requirements, be effective in adverse environmental conditions and at low concentrations, and be effective against multiple fungal pathogens on various fruits [6,9]. Moreover, an antagonistic yeast should have favorable commercial potential: It should be able to grow on an inexpensive growth medium, be easy to store and dispense, and be compatible with other physical and chemical treatments (e.g., controlled atmosphere, low/high temperature, chemical fungicides/pesticides, and phytohormones [5]. As for biosafety, a desirable antagonistic yeast would be environmentally friendly, have no pathogenicity regarding the host fruits, produce no metabolites that are harmful to humans, and be unable to cause infection in humans [5,9].

The isolation and screening process is the first step in the development of a biocontrol agent. Most antagonistic yeasts were isolated directly from fruit surfaces [14,15], but they have a wider distribution in nature, such as on leaves and roots and in seawater and soil (even Antarctic soil) [16–20]. So far, a large number of antagonistic yeasts have been isolated and screened. Some of them have been widely studied, such as *Candida* spp., *Cryptococcus* spp., *Metschnikowia* spp., *Pichia* spp., *Rhodotorula* spp., and yeast-like fungus *A. pullulans*, and several species, such as *Candida oleophila*, *Candida sake*, *Metschnikowia fructicola*, *A. pullulans*, *Saccharomyces cerevisiae*, and *Cryptococcus albidus*, have been developed as commercial products [21–28]. They have been demonstrated to antagonize common postharvest pathogens, including *Botrytis cinerea*, *Penicillium* spp., *Rhizopus stolonifer*, *Colletotrichum* spp., *Monilinia fructicola*, *Alternaria alternata*, and *Aspergillus niger*. Representative antagonistic yeasts that were isolated from various sources and are used for the management of postharvest diseases are shown in Figure 1.

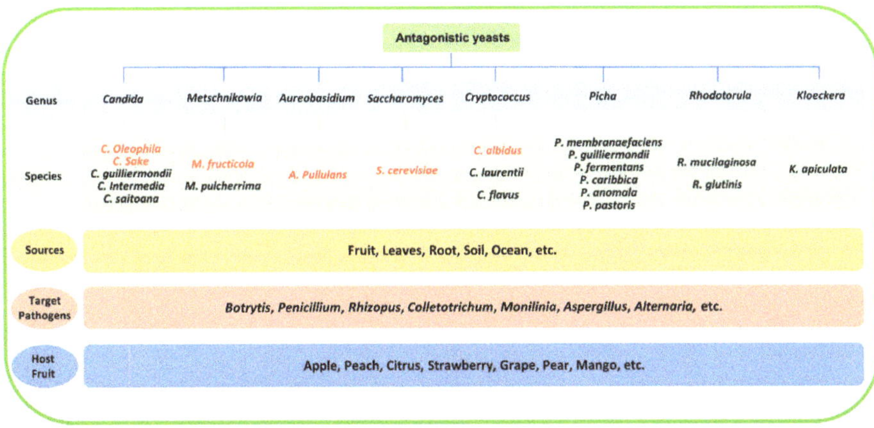

Figure 1. Representative antagonistic yeasts from various sources used for the management of postharvest decay. Species that have been already in commercial use are highlighted in red.

3. Mechanisms of Action

Elucidating the mechanisms of action is the foundation for the development and application of antagonistic yeasts [29]. Compared with the impressive results achieved regarding the identification of antagonistic yeasts, the study of their mechanisms of action is relatively slow due to the complexity of the postharvest biocontrol system. In this system, the antagonistic yeasts, pathogenic fungi, and fruit hosts interact with each other under the influence of the environment, and the influence of the epiphytic microbiome should also be taken into consideration (Figure 2) [12,29–31].

The antagonistic yeasts are likely to function via multiple mechanisms, including competition for nutrients and space, mycoparasitism, induction of host resistance, production of volatile organic compounds (VOCs), and toxins [9,10,31]. With the increase in the number of annotated yeast genomes and the development of "omics" technologies and transformation technologies, the modes of action of antagonistic yeasts will be further deciphered in the near future [32–34].

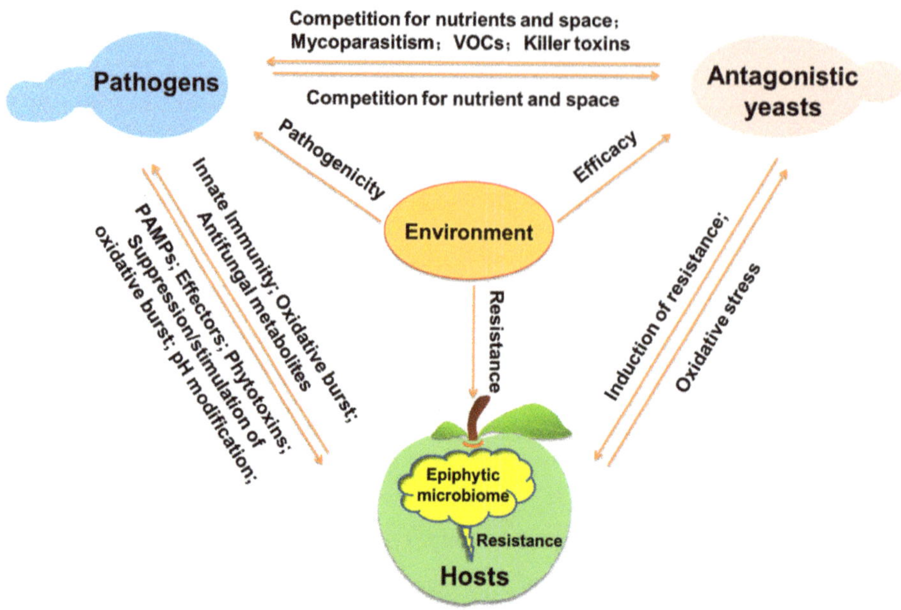

Figure 2. Schematic diagram of the possible interactions among components of the biocontrol system, including the pathogens, antagonistic yeasts, host, epiphytic microbiome, and environment. Antagonistic yeasts can inhibit pathogens through competition for nutrient and space, mycoparasitism, VOCs, and killer toxins. Conversely, pathogens also compete with antagonistic yeasts for nutrient and space to affect their colonization and growth. In addition, antagonistic yeasts can induce the resistance of hosts to inhibit infection, while reactive oxygen species (ROS) produced by hosts may be an oxidative stress to yeasts. During the interaction between fruit hosts and pathogens, hosts can resist the pathogen attack through oxidative burst, innate immune system, and antifungal metabolites, while pathogens can suppress host resistance through pathogen-associated molecular patterns (PAMPs), effectors, phytotoxins, pH modification, and suppression or stimulation of the oxidative burst. The epiphytic microbiome on hosts is also associated with the host resistance. Moreover, environmental conditions have a wide influence on the pathogenicity of pathogens, the efficacy of antagonistic yeasts, and the resistance of hosts.

3.1. Competition for Nutrients and Space

Both postharvest pathogens and antagonistic yeasts require nutrients (e.g., carbohydrates and nitrogen) and space to colonize and develop. Therefore, the competition for nutrients and space has been considered the primary mode by which antagonistic yeasts suppress postharvest fungal pathogens [5,29]. Once the antagonistic yeasts come into contact with the surface of the injured fruit, they will occupy the wounds and rapidly deplete the nutrients, which limits the germination of fungal spores [35,36]. After that, other mechanisms of action (besides the competition for nutrition and space) cooperatively come into operation to control the postharvest pathogens [8].

Carbon, nitrogen, and iron ions are the main nutrients needed for the growth of microbes. Compared with carbohydrates, nitrogen is considered to be a key factor limiting the growth of postharvest fruit pathogens, because most fruits are rich in sugar but limited in nitrogen sources such as amino acids. The application of exogenous amino acids reduced the antagonistic effect of the yeast *A. pullulans* against *Penicillium expansum* on apple fruit, indicating the importance of nitrogen competition to biocontrol efficacy [37]. Moreover, iron plays a crucial role in the growth and virulence of pathogens. Iron is a component of cytochromes, other heme proteins, and non-heme proteins; it is also a cofactor of various enzymes in fungal cells [9,38]. The yeast *Metschnikowia pulcherrima* can produce iron chelators to compete for the iron required by pathogens, thus strongly inhibiting the growth of the pathogens [39]. Parafati et al. also proposed that the consumption of iron ions plays an important role in the biocontrol effect of *M. pulcherrima* [40]. Some antagonistic yeasts can also produce siderophores to compete for iron in a low-iron microenvironment, thus inhibiting the germination and growth of pathogens. For example, rhodotorulic acid is a dihydroxamate siderophore produced by *Rhodotorula glutinis* that improves the biocontrol against *P. expansum* [41]. Siderophores produced by *A. pullulans* plays an important role in yeast growth and pathogen inhibition under iron deficiency environment [13,42].

Biofilms are dense microbial communities attached on fruit surfaces, and encapsulated by polymeric extracellular matrix (ECM) [43]. Formation of biofilm is considered as another strategy utilized by antagonistic yeasts to compete for space and nutrient [10,44]. Scherm et al. found that the biofilm formation of *S. cerevisiae* M25 was directly related to its biocontrol effect, with only the *S. cerevisiae* cells collected during the biofilm formation phase effectively controlling *P. expansum* on apples [45]. Biofilm formation has also been hypothesized to be a key mechanism of action of *Metschnikowia citriensis* against *Penicillium digitatum* and *Penicillium italicum* on citrus fruit [46]. Notably, it was reported that *Pichia fermentans* formed biofilms and inhibited postharvest decay in apple fruits but caused rapid decay in peach fruits in the absence of a plant pathogen [47], indicating the potential risk of dimorphic antagonistic yeast becoming pathogens.

3.2. Mycoparasitism

Mycoparasitism refers to the phenomenon of antagonistic yeasts feeding on fungal pathogens via attaching to the fungal pathogen hyphae and then secreting cell wall-degrading enzymes to destroy or lyse the fungal structures. Especially in the case of nutritional deficiencies, antagonistic yeasts tend to absorb nutrients from pathogenic cells, leading to the death of these "prey" cells. During mycoparasitism, a variety of enzymes are involved in the degradation of the fungal pathogen cell wall, especially β-1,3-glucanase (GLU), chitinase (CHT), and proteases [29], and these secreted enzymes are thought to play an important role in biocontrol [48]. Wisniewski et al. first reported the mycoparasitism of *Pichia guilliermondii* [49]. They observed that the yeast strongly adhered to the *B. cinerea* mycelium and caused hyphal collapse, which was presumably due to a lectin-like interaction. It has also been reported that both *Pichia membranefaciens* and *C. albidus* can attach to and degrade the hyphae of *P. expansum*, *M. fructicola*, and *R. stolonifer* [50]. Banani et al. found that the chitinase gene, *MfChi*, of the yeast *M. fructicola* was significantly induced by cell wall of the postharvest pathogen *M. fructicola*, and MfChi-overexpressing *Pichia pastoris* inhibited the brown rot of peach fruits [51]. In *C. oleophila*, GLU was demonstrated to be associated with inhibiting conidial germination and hyphal growth of *P. expansum* [52].

3.3. Induction of Host Resistance

Induction of host resistance, as one of the major mechanism of antagonistic yeasts for postharvest decay control in fruits, has also been extensively studied [53,54]. Antagonistic yeasts have been reported to act as biological elicitors in the interactions with fruit hosts [29,48]. Treatment with antagonistic yeasts can increase the expression of defense-related genes and enhance the activities of defense-related enzymes. Strongly induced activities of defense-related enzymes, such as CHT, GLU, phenylalanine ammonia-lyase (PAL), and peroxidase (POD), have been reported to be responsible for the biocontrol efficacy of *Cryptococcus laurentii* on postharvest decay caused by *A. alternata*, *M. fructicola*, and *P. expansum* [55–57]. Chan et al. found that the antagonistic yeast *P. membranaefaciens* could induce the activities of three pathogenesis-related (PR) proteins, which may contribute to the resistance improvement of peach fruit to *P. expansum* [33]. Similarly, induced expression of defense-related genes and the activities of defense-related enzymes by *W. anomalus* were considered as one of the possible mechanisms in inhibiting blue mold decay caused by *P. expansum* in pears [58].

Moreover, application of antagonistic yeasts can enhance activity of antioxidant enzymes, which may alleviate oxidative damage cause by reactive oxygen species (ROS) produced by hosts in response to pathogen infection. *P. membranaefaciens* has been reported to affect the activities of antioxidant enzymes, including POD, catalase (CAT), glutathione peroxidase (GPX), superoxide dismutase (SOD), and polyphenol oxidase (PPO), in peaches and sweet cherry fruits after inoculation with *P. expansum* [33,59]. Additionally, four antagonistic yeasts (*P. membranaefaciens*, *C. laurentii*, *Candida guilliermondii*, and *R. glutinis*) have been reported to increase the activities of POD and CAT, upregulate the expression of the corresponding genes, and reduce the levels of protein carbonylation in peach fruits caused by *M. fructicola* [60].

Antagonistic yeasts can also induce changes in secondary metabolites and cell structure related to disease resistance. Droby et al. found that the application of *C. oleophila* increased the levels of the phytoalexins umbelliferone, scoparone, and scopoletin in grape fruit peels [61]. El-Ghaouth et al. found that the antagonistic yeast *Candida saitoana* could induce host cell deformation, generate mastoid structures, and consequently inhibit *B. cinerea* infection [62].

Multiple mechanisms may be simultaneously involved in the resistance induction by antagonistic yeasts. For example, several antagonistic yeasts, such as *C. laurentii* [63], *P. membranaefaciens* [33], *P. guilliermondii* [64], *R. glutinis* [60], and *R. paludigenum* [25], induced changes in activities of both defense-related enzymes and antioxidant enzymes in fruit. Induction of disease resistance by antagonistic yeasts is also affected by pathogens and environmental conditions. As shown in Figure 2, there are complex interactions between the hosts, pathogens, antagonistic yeasts, and environment, which remains to be elucidated.

3.4. Production of VOCs and Killer Toxins

Compared to filamentous fungi, yeasts have a lower secretory capacity and produce only few secondary metabolites. Nevertheless, VOCs and killer toxins are metabolites that have been reported to exhibit antifungal activity.

VOCs are volatile compounds with low molecular weight (<300 Da), low polarity, and high vapor pressure. Some antagonistic yeasts can produce VOCs, and the mixture of VOCs has been proposed to play an important role in the control of postharvest pathogens under airtight conditions [48,65]. It was reported that *Candida intermedia* 410 inhibited the growth of *B. cinerea* on strawberries by releasing VOCs without direct contact; the absorption of VOCs by activated carbon abolished the biocontrol activity of *C. intermedia* 410 [16]. Two strains of *A. pullulans* (L1 and L8) have been reported to produce VOCs to inhibit the growth and infection of postharvest pathogens, including *B. cinerea*, *Colletotrichum acutatum*, and *Penicillium* spp. [66]. Moreover, VOCs have been reported to suppress the mycelial growth, sporulation, and ochratoxin A biosynthesis of *Aspergillus carbonarius* and *Aspergillus ochraceus* [67,68]. VOCs are considered to be potential biological fumigants because of their volatility, which allows them to control postharvest decay without direct contact with the edible commodities. Contarino et al. found that the main VOCs emitted by common antagonistic yeasts include ethyl alcohol, phenylethyl alcohol, 3-methyl-1-butanol, ethyl acetate, and isoamyl acetate [69]. However, VOCs produced by *Muscodor albus* have been reported to cause DNA damage and cytotoxicity in bacterial cells, indicating that some VOCs may be toxic [70]. Therefore, the safety of VOCs should be thoroughly evaluated in future studies.

Several toxins have been reported to be able to control postharvest pathogens, and proteinaceous killer toxins are the most prominent antifungal toxins produced by yeasts [10]. Killer toxins provide a competitive advantage to yeasts, and they can kill fungi (including other yeasts) by a variety of mechanisms, including hydrolyzation of the cell wall, destruction of the cell structure, and inhibition of DNA synthesis [71]. Yeast strains with a particular killer phenotype are immune to their own killer toxins and those in the same class while being lethal to other yeast strains [71]. Owing to this characteristic, killer toxins have long been used in the wine industry to control spoilage yeasts. As natural antifungal proteins, killer toxins are environmentally friendly, nontoxic to mammals, have a good acid tolerance, and have a low probability of inducing resistance. Therefore, killer toxins have been proposed as potential biocontrol agents. Killer toxins produced by *Wickerhamomyces anomalus* BS91 are encoded by the genes *WaEXG1* and *WaEXG2*, showed exoglucanase activity, and associated with biocontrol capabilities against *B. cinerea*, *P. digitatum*, *P. italicum*, *Monilinia fructigena*, and *M. fructicola* [40,72–75]. *P. membranaefaciens* was found to produce killer toxins PMKT and PMKT2 that target (1→6)-β-D-glucans and mannoproteins in pathogen cell walls and thereby inhibit the growth of postharvest pathogens [76]. Moreover, killer toxins produced by *Debaryomyces hansenii* have been reported to suppress human pathogenic *Candida* yeasts, but only within a certain temperature and pH range, indicating the influence of environmental factors on the antifungal activity of killer toxins [77]. Furthermore, the effects of yeast killer toxins on beneficial microorganisms need to be further evaluated, especially regarding microorganisms in the phyllosphere, on edible commodities, and in the human gut.

4. Constraints on the Application of Antagonistic Yeasts, Improvement of Their Biocontrol Efficacy, and Commercial Application

4.1. Constraints on the Application of Antagonistic Yeasts

Over the past few decades, numerous yeasts with antifungal properties have been identified, but only a few have been developed as commercial antifungal products. This has mainly been due to the fact that besides having excellent biocontrol efficacy, for commercial application, an antagonistic yeast needs to meet additional requirements. Many commercial factors restrict the development and commercialization of antagonistic yeasts, including the immature commercialization technology, high development costs, small postharvest market, and low market acceptance [8,78]. Furthermore, as the utilization of antagonistic yeasts to control postharvest decay is an emerging industry, the research on antagonistic yeasts remains insufficient. In particular, although many studies have reported on the biocontrol mechanisms of antagonistic yeasts, the specific mechanisms require further clarification.

Biosafety is one of the main reasons for using antagonistic yeasts instead of chemical fungicides. Most of the identified antagonistic yeasts have been directly isolated from the surface of fruits, and humans are already exposed to these yeasts when they eat fresh fruits and vegetables in their daily lives, so there is often less concern about the biosafety of antagonistic yeasts. However, some yeasts may be the origin of human infection under rare circumstances [79–81]. Therefore, the biosafety of antagonistic yeasts, including their safety related to skin irritation and ingestion, needs to be fully evaluated. Registration is also an obstacle to the commercialization of many antagonistic yeasts. Biocontrol agents must be approved by relevant regulatory agencies before commercial application. Compared with synthetic chemical fungicides, the registration of an antagonistic yeast is less costly and time-consuming, but it is still a factor to be considered in the development process. The registration of an antagonistic yeast requires an accredited safety assessment report and biocontrol efficacy data. Furthermore, the difficulty of registration varies in different regions. For example, the registration of biocontrol agents in the United States takes an average of 2 years, while in Europe, it takes about 7 years [6]. In China, with government incentives, the registration of biocontrol agents takes about 2–3 years.

Compared with chemical fungicides, antagonistic yeasts still need to be improved in many respects, which also limits their commercialization and market acceptance. Antagonistic yeasts are more expensive than chemical fungicides and are inconvenient to use. Moreover, an ideal biocontrol agent for controlling postharvest decay of fruits and vegetables must be highly effective (>95%) [31]. However, according to the reported researches so far, the biocontrol efficacy using antagonistic yeast alone cannot reach the level demonstrated by chemical fungicides. In addition, the biocontrol efficacy of antagonistic yeasts regarding postharvest decay depends on the high activity and reproductive capacity of the yeasts. In addition, there are issues associated with the use of many antagonistic yeasts, such as their unstable antifungal activity, short shelf life, and strict required storage conditions.

4.2. Improvement of the Biocontrol Efficacy

As mentioned above, the use of antagonistic yeast alone to prevent postharvest decay is generally inferior to the use of chemical fungicides. Therefore, while identifying new high-efficacy yeast strains, researchers are also constantly searching for effective ways to strengthen the biocontrol efficacy of existing antagonistic yeasts. The combined use of biological control and physical or chemical methods is an effective way to improve the biocontrol efficacy. For example, hot water treatment (HWT) by immersing fruit in a circulating water bath at 42 °C for 40 min improved the biocontrol efficacy of the antagonistic yeasts *C. guilliermondii* and *P. membranaefaciens* without affecting their growth [82].

Salicylic acid (SA) is an important hormone in plants that is related to the induction of the plant response against pathogens [54]. Qin et al. found that SA treatment increased the antagonism of *R. glutinis* against *P. expansum* and *A. alternata* in sweet cherry fruits [83]. SA at low concentrations increased the activities of defense-related enzymes but had little effect on the growth of the yeast and the two pathogens. This indicated that the biocontrol efficacy enhanced by SA may be related to the triggering of host resistance. The ability of SA to enhance the biocontrol efficacy of biocontrol microbes has been demonstrated in many yeast species [84–86]. Methyl jasmonate (MeJA) is another phytohormone that can induce host defense responses [74]. MeJA has also been reported to improve the biocontrol effects of antagonistic yeasts [87,88].

Moreover, it has been reported that exogenous application of brassinosteroids or nitric oxide can induce plant host resistance [89,90], but their synergistic effects when used with antagonistic yeasts remain to be studied. Many natural plant extracts can inhibit the growth and development of pathogenic fungi, such as methyl thujate [91,92], hinokitiol [93], and cinnamic acid [94]. Li et al. reported that cinnamic acid improved the biocontrol efficacy of *C. laurentii* [95], which indicates the potential of combined application of natural plant extracts with antagonistic yeasts for controlling postharvest pathogens. Several other microbial metabolites, such as epsilon-polylysine, natamycin, and rapamycin, have been reported to control postharvest pathogens [96–98], and the combined application of microbial metabolites with antagonistic yeasts is worth exploring.

The use of certain chemical reagents or other antifungal methods can also enhance the biocontrol efficacy of antagonistic yeasts. For example, $CaCl_2$ has been reported to enhance the efficacy of antagonistic yeasts [99–101]. Additionally, chitosan has antifungal properties and can induce host defense responses, and multiple studies have shown that chitosan can enhance the biocontrol efficacy of antagonistic yeasts such as *C. saitoana* [102], *C. laurentii* [103], and *P. membranaefaciens* [104]. Furthermore, inorganic salts (e.g., ammonium molybdate, sodium bicarbonate, and trisodium phosphate) [105–107], minerals (e.g., silicon and boron) [108,109], and sugar protectants (e.g., maltose and lactose) [110] have been reported to enhance the biocontrol efficacy of antagonistic yeasts. It has also been reported that the use of a combination of an antagonistic yeast and a low-dose chemical fungicide can achieve a similar biocontrol efficacy to the use of the fungicide alone at a commercial dosage, which is considered to be an effective method to reduce fungicide use [14].

The mixed application of various antagonistic yeasts is also considered to be an effective way to broaden the antifungal spectrum of biocontrol reagents and to enhance the biocontrol efficacy. Calvo et al. found that the combined application of *R. glutinis* and *C. laurentii* improved their ability to control gray mold on apples [111]. However, it should be noted that compatibility between mixed antagonistic yeasts is necessary to ensure that their normal growth and function are maintained. Moreover, Zhao et al. reported that the heterologous expression of flagellin in *S. cerevisiae* significantly induced resistance in the host plant and improved the biocontrol efficacy of the yeast against *B. cinerea*, which suggests that the heterologous expression of elicitors in yeasts may be an effective strategy to improve the biocontrol efficacy [112].

4.3. Commercial Application

The commercialization of an antagonistic yeast is a long and costly process requiring extensive testing of toxicology and biocontrol efficacy under commercial conditions. Encouragingly, over the past few decades, a few antagonistic yeasts have been developed and commercialized (Table 1). Aspire (based on *C. oleophila*) and YieldPlus (based on *C. albidus*) are the first-generation commercial antagonistic yeasts [27]. They were available on the market for several years, but they have now been withdrawn due to reasons such as difficulties in market development, low profitability, and inconsistent and low efficacy under commercial conditions [29]. After that, Nexy (another product based on *C. oleophila*) was developed for controlling decay on pome, citrus, and banana, and it was approved for registration throughout the European Union in 2013. Shemer (based on *M. fructicola*) was originally registered in Israel and was successfully used for managing pre- and postharvest diseases on various fruits and vegetables [113]. It was subsequently acquired by Bayer CropScience (Germany) and then sublicensed to Koppert Biological Systems (the Netherlands) to expand its sales [114]. Moreover, Bio-ferm, an Austrian company, developed two products based on *A. pullulans* strains DSM 14940 and DSM 14941, Blossom Protect (Boni-Protect) and Botector. With the mode of action of competition for nutrients and space, Blossom Protect is used to control postharvest decay caused by several fungal pathogens in pome fruit, while Botector is mainly used against gray mold in grape, strawberry, and tomato.

Table 1. Antagonistic yeast-based commercial products developed for the management of postharvest pathogens (adapted from [9] and [115] with modification).

Product	Yeast	Fruit	Target Pathogens	Manufacturer	In Use
Aspire	*Candida oleophila*	Stone fruit, pome, citrus, strawberry	*Botrytis, Penicillium, Monilinia*	Ecogen, USA	No
Blossom Protect	*Aureobasidium pullulans*	Pome	*Penicillium, Botrytis, Monilinia*	Bio-ferm, Austria	Yes
Botector	*Aureobasidium pullulans*	Grape, strawberry and tomato	*Botrytis cinerea*	Bio-ferm, Austria	Yes
Candifruit	*Candida sake*	Pome	*Penicillium, Botrytis, Rhizopus*	IRTA/Sipcam-Inagra, Spain	No
Nexy	*Candida oleophila*	Pome, banana, citrus	*Botrytis, Penicillium*	Lesaffre, Belgium	Yes
Noli	*Metschnikowia fructicola*	Strawberry, blueberry, grape, stone fruit	*Botrytis, Monilinia*	Koppert, The Netherlands	Yes
Remeo	*Saccharomyces cerevisiae*	Grape	*Botrytis, Erysiphe, Plasmopara*	BASF/Agrauxine, France	Yes
Shemer	*Metschnikowia fructicola*	Pome, strawberry, grape, stone fruit	*Botrytis, Penicillium, Rhizopus, Aspergillus*	Bayer/Koppert, The Netherlands	Yes
YieldPlus	*Cryptococcus albidus*	Pome, citrus	*Botrytis, Penicillium, Mucor*	Lallem, South Africa	No

5. Conclusions and Perspectives

The environmental pollution and health hazards caused by chemical fungicides have attracted increasing attention from regulatory agencies and consumers, and there is now global interest in reducing or eliminating the use of chemical fungicides. As a potential substitute for chemical fungicides, antagonistic yeasts have been extensively studied over the past few decades, and considerable progress has been made regarding the identification and development of antagonistic yeasts. However, so far, the use of antagonistic yeast alone is still insufficient to completely replace chemical fungicides. There remain many aspects of antagonistic yeasts that could be improved, even for the few commercially available antagonistic yeasts.

Although the application of antagonistic yeasts is limited by many obstacles, there is still great potential for their improvement and development. Due to the regulatory restrictions on chemical fungicides and the declining consumer acceptance of them, it is foreseeable that the use of chemical fungicides will be gradually decreased or even discontinued. The reduction in available products on the market and the increasing demand for safe and effective antifungal products provide opportunities for the development of antagonistic yeast products. The biocontrol efficacy of antagonistic yeasts could be further improved in the future through a variety of strategies, such as combining an antagonistic yeast with a chemical or physical treatment, using multiple antagonistic yeasts, and genetically altering antagonistic yeasts. Moreover, the advancement of molecular biotechnologies and the emergence of "omics" technologies are providing powerful tools for the development and application of antagonistic yeasts.

To promote the commercial application of antagonistic yeasts, efforts can be made in the following aspects: (a) the full verification of biosafety; (b) the in-depth exploration of the involved mechanisms of action; (c) the enhancement and maintenance of biocontrol efficacy under commercial conditions; (d) the development of broad-spectrum antifungal products; (e) the extension of shelf-life; (f) the control of cost and the development of the market; and (g) the understanding of the complex interactions between the components of the biocontrol system, including the antagonistic yeast, pathogen, host, natural microbiome, and environment. Furthermore, gene editing has been considered to be a potentially effective strategy to improve the performance of antagonistic yeasts, though genetically modified microorganisms (GMOs) are restricted due to government policies and low consumer acceptance at present.

Author Contributions: S.T. designed the manuscript content; X.Z. and B.L. wrote the manuscript under the coordination of S.T.; Z.Z. and Y.C. provided some data. All authors have read and agreed to the published version of the manuscript.

Funding: This work is supported by National Natural Science Foundation of China (31930086; 31530057; 31722043) and The National Key Research and Development Program of China (2016YFD0400902).

Conflicts of Interest: The authors declare no conflict of interest. The funders had no role in the design of the study; in the collection, analyses, or interpretation of data; in the writing of the manuscript, or in the decision to publish the results.

References

1. Diane, F.; Ziegler, R.G.; Michaud, D.S.; Giovannucci, E.L.; Speizer, F.E.; Willett, W.C.; Colditz, G.A. Prospective study of fruit and vegetable consumption and risk of lung cancer among men and women. *J. Natl. Cancer Inst.* **2000**, *92*, 1812–1823.
2. Zakrevskii, V.V. Fruit in The Prevention of Cancer. *Biomed. J. Sci. Tech. Res.* **2018**, *9*, 7099–7101. [CrossRef]
3. Choi, M.; Jo, H.; Kim, M.; Kang, M.; Shin, H. Fruit juice supplementation alters human skin antioxidant levels in vivo: Case study of Korean adults by resonance Raman spectroscopy. *Biotechnol. Bioprocess Eng.* **2018**, *23*, 116–121. [CrossRef]
4. Shin, S.; Son, D.; Kim, M.; Lee, S.; Roh, K.B.; Ryu, D.; Lee, J.; Jung, E.; Park, D. Ameliorating effect of Akebia quinata fruit extracts on skin aging induced by advanced glycation end products. *Nutrients* **2015**, *7*, 9337–9352. [CrossRef]
5. Liu, J.; Sui, Y.; Wisniewski, M.; Droby, S.; Liu, Y. Review: Utilization of antagonistic yeasts to manage postharvest fungal diseases of fruit. *Int. J. Food Microbiol.* **2013**, *167*, 153–160. [CrossRef]
6. Nunes, C.A. Biological control of postharvest diseases of fruit. *Eur. J. Plant Pathol.* **2012**, *133*, 181–196. [CrossRef]
7. Chen, Y.; Peng, H.; Wang, X.; Li, B.; Long, M.; Tian, S. Biodegradation mechanisms of patulin in *Candida guilliermondii*: An iTRAQ-based proteomic analysis. *Toxins* **2017**, *9*, 48. [CrossRef]
8. Droby, S.; Wisniewski, M.; Macarisin, D.; Wilson, C. Twenty years of postharvest biocontrol research: Is it time for a new paradigm? *Postharvest Biol. Technol.* **2009**, *52*, 137–145. [CrossRef]
9. Dukare, A.S.; Paul, S.; Nambi, V.E.; Gupta, R.K.; Singh, R.; Sharma, K.; Vishwakarma, R.K. Exploitation of microbial antagonists for the control of postharvest diseases of fruits: A review. *Crit. Rev. Food Sci. Nutr.* **2018**, *59*, 1498–1513. [CrossRef]
10. Freimoser, F.M.; Rueda-Mejia, M.P.; Tilocca, B.; Migheli, Q. Biocontrol yeasts: Mechanisms and applications. *World J. Microbiol. Biotechnol.* **2019**, *35*, 154. [CrossRef]
11. Barnett, J.A.; Payne, R.W.; Yarrow, D. (Eds.) *Yeasts: Characteristics and Identification*, 3rd ed.; Cambridge University Press: Cambridge, UK, 2000.
12. Wilson, C.L.; Wisniewski, M.E. Biological control of postharvest diseases of fruits and vegetables: An emerging technology. *Annu. Rev. Phytopathol.* **1989**, *27*, 425–441. [CrossRef]

13. Zajc, J.; Černoša, A.; Di Francesco, A.; Casteria, R.; De Curtis, F.; Lima, G.; Badri, H.; Jijakli, H.; Ippolito, A.; GostinČar, C.; et al. Characterization of *Aureobasidium pullulans* isolates selected as biocontrol agents against fruit decay pathogens. *Fungal Genomics Biol.* **2020**, *10*, 163.
14. Fan, Q.; Tian, S. Postharvest biological control of Rhizopus rot of nectarine fruits by *Pichia membranefaciens*. *Plant Dis.* **2000**, *84*, 1212–1216.
15. Liu, J.; Wisniewski, M.; Droby, S.; Tian, S.; Hershkovitz, V.; Tworkoski, T. Effect of heat shock treatment on stress tolerance and biocontrol efficacy of *Metschnikowia fructicola*. *FEMS Microbiol. Ecol.* **2011**, *76*, 145–155. [CrossRef]
16. Huang, R.; Li, G.; Zhang, J.; Yang, L.; Che, H.; Jiang, D.; Huang, H. Control of postharvest Botrytis fruit rot of strawberry by volatile organic compounds of *Candida intermedia*. *Phytopathology* **2011**, *101*, 859–869. [CrossRef] [PubMed]
17. Long, C.; Wu, Z.; Deng, B. Biological control of *Penicillium italicum* of citrus and *Botrytis cinerea* of grape by strain 34–9 of *Kloeckera apiculata*. *Eur. Food Res. Technol.* **2005**, *221*, 197–201. [CrossRef]
18. Wang, Y.; Bao, Y.; Shen, D.; Feng, W.; Yu, T.; Zhang, J.; Zheng, X.D. Biocontrol of *Alternaria alternata* on cherry tomato fruit by use of marine yeast *Rhodosporidium paludigenum* Fell & Tallman. *Int. J. Food Microbiol.* **2008**, *123*, 234–239.
19. Hu, H.; Yan, F.; Wilson, C.; Shen, Q.; Zheng, X. The ability of a cold-adapted *Rhodotorula mucilaginosa* strain from Tibet to control blue mold in pear fruit. *Antonie Van Leeuwenhoek* **2015**, *108*, 1391–1404. [CrossRef]
20. Vero, S.; Garmendia, G.; González, M.B.; Bentancur, O.; Wisniewski, M. Evaluation of yeasts obtained from Antarctic soil samples as biocontrol agents for the management of postharvest diseases of apple (*Malus× domestica*). *FEMS Yeast Res.* **2013**, *13*, 189–199. [CrossRef]
21. Qin, G.; Tian, S.; Xu, Y. Biocontrol of postharvest diseases on sweet cherries by four antagonistic yeasts in different storage conditions. *Postharvest Biol. Technol.* **2004**, *31*, 51–58. [CrossRef]
22. Qin, G.; Tian, S. Biocontrol of postharvest diseases of jujube fruit by *Cryptococcus laurentii* combined with a low dosage of fungicides under different storage conditions. *Plant Dis.* **2004**, *88*, 497–501. [CrossRef]
23. Lahlali, R.; Serrhini, M.; Jijakli, H. Efficacy assessment of *Candida oleophila* (strain O) and *Pichia anomala* (strain K) against major postharvest diseases of citrus fruits in Morocco. *Commun. Appl. Biol. Sci. Ghent Univ.* **2004**, *69*, 601–609.
24. Spadaro, D.; Vola, R.; Piano, S.; Gullino, M.L. Mechanisms of action and efficacy of four isolates of the yeast *Metschnikowia pulcherrima* active against postharvest pathogens on apples. *Postharvest Biol. Technol.* **2002**, *24*, 123–134. [CrossRef]
25. Lu, L.; Ye, C.; Guo, S.; Sheng, K.; Shao, L.; Zhou, T.; Yu, T.; Zheng, X. Preharvest application of antagonistic yeast *Rhodosporidium paludigenum* induced resistance against postharvest diseases in mandarin orange. *Biol. Control* **2013**, *67*, 130–136. [CrossRef]
26. Ippolito, A.; Ghaouth, A.E.; Wilson, C.L.; Wisniewski, M. Technology Control of postharvest decay of apple fruit by *Aureobasidium pullulans* and induction of defense responses. *Postharvest Biol. Technol.* **2000**, *19*, 265–272. [CrossRef]
27. Janisiewicz, W.J.; Korsten, L. Biological control of postharvest diseases of fruits. *Annu. Rev. Phytopathol.* **2002**, *40*, 411–441. [CrossRef]
28. Mari, M.; Martini, C.; Guidarelli, M.; Neri, F. Postharvest biocontrol of *Monilinia laxa*, *Monilinia fructicola* and *Monilinia fructigena* on stone fruit by two *Aureobasidium pullulans* strains. *Biol. Control* **2012**, *60*, 132–140. [CrossRef]
29. Spadaro, D.; Droby, S. Development of biocontrol products for postharvest diseases of fruit: The importance of elucidating the mechanisms of action of yeast antagonists. *Trends Food Sci. Technol.* **2016**, *47*, 39–49. [CrossRef]
30. Massart, S.; Martinez-Medina, M.; Jijakli, M.H. Biological control in the microbiome era: Challenges and opportunities. *Biol. Control* **2015**, *89*, 98–108. [CrossRef]
31. Sharma, R.; Singh, D.; Singh, R. Biological control of postharvest diseases of fruits and vegetables by microbial antagonists: A review. *Biol. Control* **2009**, *50*, 205–221. [CrossRef]
32. Massart, S.; Jijakli, H.M. Use of molecular techniques to elucidate the mechanisms of action of fungal biocontrol agents: A review. *J. Microbiol. Methods* **2007**, *69*, 229–241. [CrossRef] [PubMed]
33. Chan, Z.; Qin, G.; Xu, X.; Li, B.; Tian, S. Proteome approach to characterize proteins induced by antagonist yeast and salicylic acid in peach fruit. *J. Proteome Res.* **2007**, *6*, 1677–1688. [CrossRef] [PubMed]

34. Tian, S.; Torres, R.; Ballester, A.R.; Li, B.; Vilanova, L.; González-Candelas, L. Molecular aspects in pathogen-fruit interactions: Virulence and resistance. *Postharvest Biol. Technol.* **2016**, *122*, 11–21. [CrossRef]
35. Li, B.; Zhou, Z.; Tian, S. Combined effects of endo-and exogenous trehalose on stress tolerance and biocontrol efficacy of two antagonistic yeasts. *Biol. Control* **2008**, *46*, 187–193. [CrossRef]
36. Liu, J.; Wisniewski, M.; Droby, S.; Norelli, J.; Hershkovitz, V.; Tian, S.; Farrell, R. Increase in antioxidant gene transcripts, stress tolerance and biocontrol efficacy of *Candida oleophila* following sublethal oxidative stress exposure. *FEMS Microbiol. Ecol.* **2012**, *80*, 578–590. [CrossRef]
37. Bencheqroun, S.K.; Bajji, M.; Massart, S.; Labhilili, M.; El Jaafari, S.; Jijakli, M.H. In vitro and in situ study of postharvest apple blue mold biocontrol by *Aureobasidium pullulans*: Evidence for the involvement of competition for nutrients. *Postharvest Biol. Technol.* **2007**, *46*, 128–135. [CrossRef]
38. Talibi, I.; Boubaker, H.; Boudyach, E.; Ait Ben Aoumar, A. Alternative methods for the control of postharvest citrus diseases. *J. Appl. Microbiol.* **2014**, *117*, 1–17. [CrossRef]
39. Gore-Lloyd, D.; Sumann, I.; Brachmann, A.O.; Schneeberger, K.; Ortiz-Merino, R.A.; Moreno-Beltrán, M.; Schläfli, M.; Kirner, P.; Santos Kron, A.; Rueda-Mejia, M.P. Snf2 controls pulcherriminic acid biosynthesis and antifungal activity of the biocontrol yeast *Metschnikowia pulcherrima*. *Mol. Microbiol.* **2019**, *112*, 317–332. [CrossRef]
40. Parafati, L.; Vitale, A.; Restuccia, C.; Cirvilleri, G. Biocontrol ability and action mechanism of food-isolated yeast strains against *Botrytis cinerea* causing post-harvest bunch rot of table grape. *Food Microbiol.* **2015**, *47*, 85–92. [CrossRef]
41. Calvente, V.; Benuzzi, D.; de Tosetti, M.S. Antagonistic action of siderophores from *Rhodotorula glutinis* upon the postharvest pathogen *Penicillium expansum*. *Int. Biodeterior. Biodegrad.* **1999**, *43*, 167–172. [CrossRef]
42. Chi, Z.; Wang, X.X.; Ma, Z.C.; Buzdar, M.A.; Chi, Z.M. The unique role of siderophore in marine-derived *Aureobasidium pullulans* HN6.2. *Biometals* **2012**, *25*, 219–230. [CrossRef] [PubMed]
43. Costa-Orlandi, C.B.; Sardi, J.C.O.; Pitangui, N.S.; De Oliveira, H.C.; Scorzoni, L.; Galeane, M.C.; Medina-Alarcón, K.P.; Melo, W.C.M.A.; Marcelino, M.Y.; Braz, J.D.; et al. Fungal biofilms and polymicrobial diseases. *J. Fungi* **2017**, *3*, 22. [CrossRef] [PubMed]
44. Rendueles, O.; Ghigo, J.M. Mechanisms of competition in biofilm communities. *Microbiol Spectr.* **2015**, *3*. [CrossRef]
45. Scherm, B.; Ortu, G.; Muzzu, A.; Budroni, M.; Arras, G.; Migheli, Q. Biocontrol activity of antagonistic yeasts against *Penicillium expansum* on apple. *J. Plant Pathol.* **2003**, *85*, 205–213.
46. Liu, Y.; Yao, S.; Deng, L.; Ming, J.; Zeng, K. Different mechanisms of action of isolated epiphytic yeasts against *Penicillium digitatum* and *Penicillium italicum* on citrus fruit. *Postharvest Biol. Technol.* **2019**, *152*, 100–110. [CrossRef]
47. Giobbe, S.; Marceddu, S.; Scherm, B.; Zara, G.; Mazzarello, V.L.; Budroni, M.; Migheli, Q. The strange case of a biofilm-forming strain of *Pichia fermentans*, which controls Monilinia brown rot on apple but is pathogenic on peach fruit. *FEMS Yeast Res.* **2007**, *7*, 1389–1398. [CrossRef]
48. Di Francesco, A.; Martini, C.; Mari, M. Biological control of postharvest diseases by microbial antagonists: How many mechanisms of action? *Eur. J. Plant Pathol.* **2016**, *145*, 711–717. [CrossRef]
49. Wisniewski, M.; Biles, C.; Droby, S.; McLaughlin, R.; Wilson, C.; Chalutz, E. Mode of action of the postharvest biocontrol yeast, *Pichia guilliermondii*. I. Characterization of attachment to *Botrytis cinerea*. *Physiol. Mol. Plant Pathol.* **1991**, *39*, 245–258. [CrossRef]
50. Chan, Z.; Tian, S. Interaction of antagonistic yeasts against postharvest pathogens of apple fruit and possible mode of action. *Postharvest Biol. Technol.* **2005**, *36*, 215–223. [CrossRef]
51. Banani, H.; Spadaro, D.; Zhang, D.; Matic, S.; Garibaldi, A.; Gullino, M.L. Postharvest application of a novel chitinase cloned from *Metschnikowia fructicola* and overexpressed in *Pichia pastoris* to control brown rot of peaches. *Int. J. Food Microbiol.* **2015**, *199*, 54–61. [CrossRef]
52. Tamayo-Urbina, C.; Guerrero-Prieto, V.; Guigon-Lopez, C.; Vargas-Albores, F.; Berlanga-Reyes, D.; Acosta-Muniz, C.; Ojeda-Barrios, D. Purification and characterization of β-1, 3-glucanase from *Candida oleophila* for the biocontrol of *Penicillium expansum*. *Res. Rev. J. Bot. Sci.* **2016**, *5*, 38–45.
53. Yao, H.; Tian, S. Effects of pre- and post-harvest application of salicylic acid or methyl jasmonate on inducing disease resistance of sweet cherry fruit in storage. *Postharvest Biol. Technol.* **2005**, *35*, 253–262. [CrossRef]
54. Romanazzi, G.; Sanzani, S.M.; Bi, Y.; Tian, S.; Gutiérrez Martínez, P.; Alkan, N. Induced resistance to control postharvest decay of fruit and vegetables. *Postharvest Biol. Technol.* **2016**, *122*, 82–94. [CrossRef]

55. Tian, S.; Wan, Y.; Qin, G.; Xu, Y. Induction of defense responses against Alternaria rot by different elicitors in harvested pear fruit. *Appl. Microbiol. Biotechnol.* **2006**, *70*, 729–734. [CrossRef]
56. Tian, S.; Yao, H.; Deng, X.; Xu, X.; Qin, G.; Chan, Z. Characterization and expression of β-1,3-glucanase genes in jujube fruit induced by the microbial biocontrol agent *Cryptococcus laurentii*. *Phytopathology* **2007**, *97*, 260–268. [CrossRef]
57. Meng, X.; Tian, S. Effects of preharvest application of antagonistic yeast combined with chitosan on decay and quality of harvested table grape fruit. *J. Sci. Food Agric.* **2009**, *89*, 1838–1842. [CrossRef]
58. Zhang, Q.; Zhao, L.; Li, Z.; Li, C.; Li, B.; Gu, X.; Zhang, X.; Zhang, H. Screening and identification of an antagonistic yeast controlling postharvest blue mold decay of pears and the possible mechanisms involved. *Biol. Control* **2019**, *133*, 26–33. [CrossRef]
59. Chan, Z.; Tian, S. Induction of H_2O_2-metabolizing enzymes and total protein synthesis by antagonistic yeast and salicylic acid in harvested sweet cherry fruit. *Postharvest Biol. Technol.* **2006**, *39*, 314–320. [CrossRef]
60. Xu, X.; Qin, G.; Tian, S. Effect of microbial biocontrol agents on alleviating oxidative damage of peach fruit subjected to fungal pathogen. *Int. J. Food Microbiol.* **2008**, *126*, 153–158. [CrossRef]
61. Droby, S.; Vinokur, V.; Weiss, B.; Cohen, L.; Daus, A.; Goldschmidt, E.; Porat, R. Induction of resistance to *Penicillium digitatum* in grapefruit by the yeast biocontrol agent *Candida oleophila*. *Phytopathology* **2002**, *92*, 393–399. [CrossRef]
62. El-Ghaouth, A.; Wilson, C.L.; Wisniewski, M. Ultrastructural and cytochemical aspects of the biological control of *Botrytis cinerea* by *Candida saitoana* in apple fruit. *Phytopathology* **1998**, *88*, 282–291. [CrossRef] [PubMed]
63. Lai, J.; Cao, X.; Yu, T.; Wang, Q.; Zhang, Y.; Zheng, X.; Lu, H. Effect of *Cryptococcus laurentii* on inducing disease resistance in cherry tomato fruit with focus on the expression of defense-related genes. *Food Chem.* **2018**, *254*, 208–216. [CrossRef] [PubMed]
64. Zhao, Y.; Tu, K.; Shao, X.; Jing, W.; Su, Z. Effects of the yeast *Pichia guilliermondii* against *Rhizopus nigricans* on tomato fruit. *Postharvest Biol. Technol.* **2008**, *49*, 113–120. [CrossRef]
65. Mari, M.; Bautista-Baños, S.; Sivakumar, D. Decay control in the postharvest system: Role of microbial and plant volatile organic compounds. *Postharvest Biol. Technol.* **2016**, *122*, 70–81. [CrossRef]
66. Di Francesco, A.; Ugolini, L.; Lazzeri, L.; Mari, M. Production of volatile organic compounds by *Aureobasidium pullulans* as a potential mechanism of action against postharvest fruit pathogens. *Biol. Control* **2015**, *81*, 8–14. [CrossRef]
67. Farbo, M.G.; Urgeghe, P.P.; Fiori, S.; Marcello, A.; Oggiano, S.; Balmas, V.; Hassan, Z.U.; Jaoua, S.; Migheli, Q. Effect of yeast volatile organic compounds on ochratoxin A-producing *Aspergillus carbonarius* and A. ochraceus. *Int. J. Food Microbiol.* **2018**, *284*, 1–10. [CrossRef]
68. Tilocca, B.; Balmas, V.; Hassan, Z.U.; Jaoua, S.; Migheli, Q. A proteomic investigation of *Aspergillus carbonarius* exposed to yeast volatilome or to its major component 2-phenylethanol reveals major shifts in fungal metabolism. *Int. J. Food Microbiol.* **2019**, *306*, 108265. [CrossRef]
69. Contarino, R.; Brighina, S.; Fallico, B.; Cirvilleri, G.; Parafati, L.; Restuccia, C. Volatile organic compounds (VOCs) produced by biocontrol yeasts. *Food Microbiol* **2019**, *82*, 70–74. [CrossRef]
70. Alpha, C.J.; Campos, M.; Jacobs-Wagner, C.; Strobel, S.A. Mycofumigation by the volatile organic compound-producing fungus *Muscodor albus* induces bacterial cell death through DNA damage. *Appl. Environ. Microbiol.* **2015**, *81*, 1147–1156. [CrossRef]
71. Mannazzu, I.; Domizio, P.; Carboni, G.; Zara, S.; Zara, G.; Comitini, F.; Budroni, M.; Ciani, M. Yeast killer toxins: From ecological significance to application. *Crit. Rev. Biotechnol.* **2019**, *39*, 603–617. [CrossRef]
72. Platania, C.; Restuccia, C.; Muccilli, S.; Cirvilleri, G. Efficacy of killer yeasts in the biological control of *Penicillium digitatum* on Tarocco orange fruits (*Citrus sinensis*). *Food Microbiol.* **2012**, *30*, 219–225. [CrossRef] [PubMed]
73. Parafati, L.; Cirvilleri, G.; Restuccia, C.; Wisniewski, M. Potential role of exoglucanase genes (*WaEXG1* and *WaEXG2*) in the biocontrol activity of *Wickerhamomyces anomalus*. *Microb. Ecol.* **2017**, *73*, 876–884. [CrossRef]
74. Aloui, H.; Licciardello, F.; Khwaldia, K.; Hamdi, M.; Restuccia, C. Physical properties and antifungal activity of bioactive films containing *Wickerhamomyces anomalus* killer yeast and their application for preservation of oranges and control of postharvest green mold caused by *Penicillium digitatum*. *Int. J. Food Microbiol.* **2015**, *200*, 22–30. [CrossRef]

75. Grzegorczyk, M.; Zarowska, B.; Restuccia, C.; Cirvilleri, G. Postharvest biocontrol ability of killer yeasts against *Monilinia fructigena* and *Monilinia fructicola* on stone fruit. *Food Microbiol.* **2017**, *61*, 93–101. [CrossRef]
76. Belda, I.; Ruiz, J.; Alonso, A.; Marquina, D.; Santos, A. The biology of *Pichia membranifaciens* killer toxins. *Toxins* **2017**, *9*, 112. [CrossRef] [PubMed]
77. Banjara, N.; Nickerson, K.W.; Suhr, M.J.; Hallen-Adams, H.E. Killer toxin from several food-derived *Debaryomyces hansenii* strains effective against pathogenic Candida yeasts. *Int. J. Food Microbiol.* **2016**, *222*, 23–29. [CrossRef] [PubMed]
78. Droby, S. Biological control of postharvest diseases of fruits and vegetables: Difficulties and challenges. *Phytopathol. Pol.* **2006**, *39*, 105–117.
79. Opulente, D.A.; Langdon, Q.K.; Buh, K.V.; Haase, M.A.; Sylvester, K.; Moriarty, R.V.; Jarzyna, M.; Considine, S.L.; Schneider, R.M.; Hittinger, C.T. Pathogenic budding yeasts isolated outside of clinical settings. *FEMS Yeast Res.* **2019**, *19*, foz032. [CrossRef] [PubMed]
80. Enache-Angoulvant, A.; Hennequin, C. Invasive *Saccharomyces* infection: A comprehensive review. *Clin. Infect. Dis.* **2005**, *41*, 1559–1568. [CrossRef] [PubMed]
81. De Llanos, R.; Querol, A.; Pemán, J.; Gobernado, M.; Fernández-Espinar, M.T. Food and probiotic strains from the *Saccharomyces cerevisiae* species as a possible origin of human systemic infections. *Int. J. Food Microbiol.* **2006**, *110*, 286–290. [CrossRef]
82. Zong, Y.; Liu, J.; Li, B.; Qin, G.; Tian, S. Effects of yeast antagonists in combination with hot water treatment on postharvest diseases of tomato fruit. *Biol. Control* **2010**, *54*, 316–321. [CrossRef]
83. Qin, G.; Tian, S.; Xu, Y.; Wan, Y. Enhancement of biocontrol efficacy of antagonistic yeasts by salicylic acid in sweet cherry fruit. *Physiol. Mol. Plant Pathol.* **2003**, *62*, 147–154. [CrossRef]
84. Farahani, L.; Etebarian, H.R. Enhancement of the efficacy of two antagonistic yeasts with salicylic acid against *Penicillium expansum*. *Arch. Phytopathol. Plant Prot.* **2012**, *45*, 260–267. [CrossRef]
85. Xu, X.; Chan, Z.; Xu, Y.; Tian, S. Effect of *Pichia membranaefaciens* combined with salicylic acid on controlling brown rot in peach fruit and the mechanisms involved. *J. Sci. Food Agric.* **2008**, *88*, 1786–1793. [CrossRef]
86. Shao, Y.; Zeng, J.; Tang, H.; Zhou, Y.; Li, W. The chemical treatments combined with antagonistic yeast control anthracnose and maintain the quality of postharvest mango fruit. *J. Integr. Agric.* **2019**, *18*, 1159–1169. [CrossRef]
87. Yao, H.; Tian, S. Effects of a biocontrol agent and methyl jasmonate on postharvest diseases of peach fruit and the possible mechanisms involved. *J. Appl. Microbiol.* **2005**, *98*, 941–950. [CrossRef]
88. Guo, J.; Fang, W.; Lu, H.; Zhu, R.; Lu, L.; Zheng, X.; Yu, T. Inhibition of green mold disease in mandarins by preventive applications of methyl jasmonate and antagonistic yeast *Cryptococcus laurentii*. *Postharvest Biol. Technol.* **2014**, *88*, 72–78. [CrossRef]
89. Zhu, Z.; Zhang, Z.; Qin, G.; Tian, S. Effects of brassinosteroids on postharvest disease and senescence of jujube fruit in storage. *Postharvest Biol. Technol.* **2010**, *56*, 50–55. [CrossRef]
90. Lai, T.; Wang, Y.; Li, B.; Qin, G.; Tian, S. Defense responses of tomato fruit to exogenous nitric oxide during postharvest storage. *Postharvest Biol. Technol.* **2011**, *62*, 127–132. [CrossRef]
91. Ji, D.; Chen, T.; Ma, D.; Liu, J.; Xu, Y.; Tian, S. Inhibitory effects of methyl thujate on mycelial growth of *Botrytis cinerea* and possible mechanisms. *Postharvest Biol. Technol.* **2018**, *142*, 46–54. [CrossRef]
92. Ma, D.; Ji, D.; Liu, J.; Xu, Y.; Chen, T.; Tian, S. Efficacy of methyl thujate in inhibiting *Penicillium expansum* growth and possible mechanism involved. *Postharvest Biol. Technol.* **2020**, *161*, 111070. [CrossRef]
93. Wang, Y.; Liu, X.; Chen, T.; Xu, Y.; Tian, S. Antifungal effects of hinokitiol on development of *Botrytis cinerea* in vitro and in vivo. *Postharvest Biol. Technol.* **2020**, *159*, 111038. [CrossRef]
94. Zhang, Z.; Qin, G.; Li, B.; Tian, S. Effect of cinnamic acid for controlling gray mold on table grape and its possible mechanisms of action. *Curr. Microbiol.* **2015**, *71*, 396–402. [CrossRef] [PubMed]
95. Li, J.; Li, H.; Ji, S.; Chen, T.; Tian, S.; Qin, G. Enhancement of biocontrol efficacy of *Cryptococcus laurentii* by cinnamic acid against *Penicillium italicum* in citrus fruit. *Postharvest Biol. Technol.* **2019**, *149*, 42–49. [CrossRef]
96. Li, H.; He, C.; Li, G.; Zhang, Z.; Li, B.; Tian, S. The modes of action of epsilon-polylysine (ε-PL) against *Botrytis cinerea* in jujube fruit. *Postharvest Biol. Technol.* **2019**, *147*, 1–9. [CrossRef]
97. He, C.; Zhang, Z.; Li, B.; Xu, Y.; Tian, S. Effect of natamycin on *Botrytis cinerea* and *Penicillium expansum*—Postharvest pathogens of grape berries and jujube fruit. *Postharvest Biol. Technol.* **2019**, *151*, 134–141. [CrossRef]

98. Ma, D.; Ji, D.; Zhang, Z.; Li, B.; Qin, G.; Xu, Y.; Chen, T.; Tian, S. Efficacy of rapamycin in modulating autophagic activity of *Botrytis cinerea* for controlling gray mold. *Postharvest Biol. Technol.* **2019**, *150*, 158–165. [CrossRef]
99. Tian, S.; Fan, Q.; Xu, Y.; Jiang, A. Effects of calcium on biocontrol activity of yeast antagonists against the postharvest fungal pathogen *Rhizopus stolonifer*. *Plant Pathol.* **2002**, *51*, 352–358. [CrossRef]
100. Gramisci, B.R.; Lutz, M.C.; Lopes, C.A.; Sangorrín, M.P. Enhancing the efficacy of yeast biocontrol agents against postharvest pathogens through nutrient profiling and the use of other additives. *Biol. Control* **2018**, *121*, 151–158. [CrossRef]
101. Tournas, V.H.; Katsoudas, E.J. Effect of $CaCl_2$ and various wild yeasts from plant origin on controlling *Penicillium expansum* postharvest decays in Golden Delicious apples. *Microbiol. Insights* **2019**, *12*, 1–6. [CrossRef]
102. Elghaouth, A.; Smilanick, J.L.; Wilson, C.L. Enhancement of the performance of *Candida saitoana* by the addition of glycolchitosan for the control of postharvest decay of apple and citrus fruit. *Postharvest Biol. Technol.* **2000**, *19*, 103–110. [CrossRef]
103. Meng, X.; Qin, G.; Tian, S. Influences of preharvest spraying *Cryptococcus laurentii* combined with postharvest chitosan coating on postharvest diseases and quality of table grapes in storage. *LWT Food Sci. Technol.* **2010**, *43*, 596–601. [CrossRef]
104. Zhou, Y.; Zhang, L.; Zeng, K. Efficacy of *Pichia membranaefaciens* combined with chitosan against *Colletotrichum gloeosporioides* in citrus fruits and possible modes of action. *Biol. Control* **2016**, *96*, 39–47. [CrossRef]
105. Janisiewicz, W.J.; Saftner, R.A.; Conway, W.S.; Yoder, K.S. Control of blue mold decay of apple during commercial controlled atmosphere storage with yeast antagonists and sodium bicarbonate. *Postharvest Biol. Technol.* **2008**, *49*, 374–378. [CrossRef]
106. Wan, Y.; Tian, S.; Qin, G. Enhancement of biocontrol activity of yeasts by adding sodium bicarbonate or ammonium molybdate to control postharvest disease of jujube fruits. *Lett. Appl. Microbiol.* **2003**, *37*, 249–253. [CrossRef]
107. Cai, J.; Chen, J.; Lu, G.; Zhao, Y.; Tian, S.; Qin, G. Control of brown rot on jujube and peach fruits by trisodium phosphate. *Postharvest Biol. Technol.* **2015**, *99*, 93–98. [CrossRef]
108. Qin, G.; Tian, S. Enhancement of biocontrol activity of *Cryptococcus laurentii* by silicon and the possible mechanisms involved. *Phytopathology* **2005**, *95*, 69–75. [CrossRef]
109. Cao, B.; Hua, L.; Tian, S.; Qin, G. Boron improves the biocontrol activity of *Cryptococcus laurentii* against *Penicillium expansum* in jujube fruit. *Postharvest Biol. Technol.* **2012**, *68*, 16–21. [CrossRef]
110. Zheng, F.; Zhang, W.; Sui, Y.; Ding, R.; Yi, W.; Hu, Y.; Liu, H.; Zhu, C. Sugar protectants improve the thermotolerance and biocontrol efficacy of the biocontrol Yeast, *Candida oleophila*. *Front. Microbiol.* **2019**, *10*, 187. [CrossRef]
111. Calvo, J.; Calvente, V.; Orellano, M.E.D.; Benuzzi, D.; Tosetti, M.I.S.D. Improvement in the biocontrol of postharvest diseases of apples with the use of yeast mixtures. *Biocontrol* **2003**, *48*, 579–593. [CrossRef]
112. Zhao, S.; Guo, Y.; Wang, Q.; Luo, H.; He, C.; An, B. Expression of flagellin at yeast surface increases biocontrol efficiency of yeast cells against postharvest disease of tomato caused by *Botrytis cinerea*. *Postharvest Biol. Technol.* **2020**, *162*, 111112. [CrossRef]
113. Blachinsky, D.; Antonov, J.; Bercovitz, A.; El-ad, B.; Feldman, K.; Husid, A.; Lazare, M.; Marcov, N.; Shamai, I.; Droby, S. Commercial applications of "Shemer" for the control of pre- and postharvest diseases. *IOBC WPRS Bull.* **2007**, *30*, 75–78.
114. Droby, S.; Wisniewski, M.; Teixidó, N.; Spadaro, D.; Jijakli, M.H. The science, development, and commercialization of postharvest biocontrol products. *Postharvest Biol. Technol.* **2016**, *122*, 22–29. [CrossRef]
115. Wisniewski, M.; Droby, S.; Norelli, J.; Liu, J.; Schena, L. Alternative management technologies for postharvest disease control: The journey from simplicity to complexity. *Postharvest Biol. Technol.* **2016**, *122*, 3–10. [CrossRef]

Journal of
Fungi

MDPI

Review

Saccharomyces boulardii: What Makes It Tick as Successful Probiotic?

Pedro Pais [1,2,†], **Vanda Almeida** [1,2,†], **Melike Yılmaz** [1,2] **and Miguel C. Teixeira** [1,2,*]

[1] Department of Bioengineering, Instituto Superior Técnico, Universidade de Lisboa, 1049-001 Lisboa, Portugal; pedrohpais@tecnico.ulisboa.pt (P.P.); vanda_pintalmeida@hotmail.com (V.A.); melikeyilmaz@tecnico.ulisboa.pt (M.Y.)

[2] Biological Sciences Research Group, IBB-Institute for Bioengineering and Biosciences, Instituto Superior Técnico, 1049-001 Lisboa, Portugal

* Correspondence: mnpct@tecnico.ulisboa.pt; Tel.: +351-218417772

† These authors contributed equally for this work.

Received: 9 May 2020; Accepted: 2 June 2020; Published: 4 June 2020

Abstract: *Saccharomyces boulardii* is a probiotic yeast often used for the treatment of GI tract disorders such as diarrhea symptoms. It is genetically close to the model yeast *Saccharomyces cerevisiae* and its classification as a distinct species or a *S. cerevisiae* variant has long been discussed. Here, we review the main genetic divergencies between *S. boulardii* and *S. cerevisiae* as a strategy to uncover the ability to adapt to the host physiological conditions by the probiotic. *S. boulardii* does possess discernible phenotypic traits and physiological properties that underlie its success as probiotic, such as optimal growth temperature, resistance to the gastric environment and viability at low pH. Its probiotic activity has been elucidated as a conjunction of multiple pathways, ranging from improvement of gut barrier function, pathogen competitive exclusion, production of antimicrobial peptides, immune modulation, and trophic effects. This review summarizes the participation of *S. boulardii* in these mechanisms and the multifactorial nature by which this yeast modulates the host microbiome and intestinal function.

Keywords: *Saccharomyces boulardii; Saccharomyces cerevisiae*; probiotics; gastrointestinal tract

1. Introduction

Probiotics are defined as live organisms which, when administered in adequate amounts, confer a health benefit to the host, independently of where the action takes place and of the type of administration. They are normally recommended to help strengthen host systems, for example the gastrointestinal (GI) tract, and assist in the recovery of certain diseases. According to this definition, probiotics in food must contain at least 10^6 CFU/g of viable and active microorganisms, while freeze-dried supplements have shown good results with 10^7 to 10^{11} viable microorganisms per day [1–5]. It is also preferable that these are of human origin and that they cannot transfer any antibiotic resistance, pathogenicity or toxicity factors [4].

The most commonly used probiotics comprise lactic acid producing bacteria (*Lactobacillus* spp., *Bacillus* spp., *Bifidobacterium* spp., *Streptococcus* spp., and *Enterococcus* spp.) that are found in the human gastrointestinal tract, usually ingested in fermented foods [4]. These probiotics can be used by themselves or combined with each other, although it should be noted that not all combinations are stable and different strains of the same probiotic bacteria can have different capabilities or enzymatic activities, even if they belong to the same species [4,6]. Probiotic properties widely differ between species, strains or even between strain variants, which means these properties can be strain/variant-specific [4].

The ability of a given organism to display probiotic activity is also dependent on its ability to compete for a host niche. Probiotics must compete with pathogens that adhere specifically to host cells, such as those of the GI tract, including *Helicobacter pylori* or *Clostridium difficile*, but also *Borellia* spp., *Treponema* spp. or *Spirilium* spp. [2]. This means that the competition between probiotic microorganisms and pathogens is dependent on habitat-related idiosyncrasies [2]. Host factors can also influence the effectiveness of a probiotic. Genetic factors, baseline immune functions or microbiome diversity vary among individuals, which together with environmental factors (e.g., diet or stress) account for unique backgrounds where the same probiotic will have distinct outcomes [4].

Several bacteria have been identified as probiotics and their modes of action scrutinized to some extent, but yeasts may also exhibit probiotic properties. The baker's yeast *Saccharomyces cerevisiae* does not seem to present significant advantageous attributes for human health [1]. On the other hand, the closely related *Saccharomyces boulardii* is effective in complementing the treatment of acute gastrointestinal diseases such as diarrhea or chronic diseases such as inflammatory bowel disease (IBD) [7,8]. To date, this is the only yeast used as a probiotic [4] and its probiotic properties are supported by scientific evidence from the *S. boulardii* CNCM I-745 (or *S. boulardii* Hansen CBS 5926) strain produced by Laboratoires Biocodex, highlighted by more than 80 randomized clinical trials [1]. Nevertheless, the efficacy of this strain cannot be extrapolated to other strains, like *S. boulardii* CNCM 1079 [1].

In this review, current knowledge on *S. boulardii* traits that support its probiotic nature and the correlation with distinctive features when compared with the non-probiotic *S. cerevisiae* will be explored. Focus will be given on reviewing the biology, genetics, ability to colonize the human gut and compete with gastrointestinal pathogens as features that may underlie the probiotic activity of *S. boulardii*. Unanswered questions, mostly related to the genetic basis underlying the probiotic phenotype, are discussed.

2. *S. boulardii* and *S. cerevisiae*: Similar but Different

The budding yeast known as *S. boulardii* is usually referred to as a distinct species within the *Saccharomyces* genus, despite being genetically close and sharing a similar karyotype to the model yeast *S. cerevisiae* [9–11]. Molecular typing studies resorting to pulsed-field gel electrophoresis (PFGE), randomly amplified polymorphic DNA-polymerase chain reaction (RAPD-PCR), and restriction fragment length polymorphisms (RFLP) of non-transcribed spacer (NTS) or internal transcribed spacer (ITS) reveal that *S. boulardii* strains from distinct origins all belong to a clearly delimited cluster within the *S. cerevisiae* species, arguing that they should be considered different strains of the same species [10,12]. Likewise, a DNA/RNA hybridization spotted microarrays study also concluded that *S. boulardii* is a strain of *S. cerevisiae* that has lost all intact Ty1/2 elements rather than a different species [13], while another study identified Ty1/3/4 as absent elements, but not Ty2/5 [11]. Phylogenetic analysis also shows that *S. boulardii* clusters are closely related to *S. cerevisiae* wine strains [11]. In spite of such similarities, microsatellite polymorphisms may provide a way to differentiate both species and identify *S. boulardii* properly [14,15].

Despite the striking relatedness in molecular phylogeny and typing, *S. boulardii* does possess identifiable distinct traits and is physiologically and metabolically distinct from *S. cerevisiae* (Table 1). Namely, *S. boulardii* is incapable of producing ascospores, switching to haploid form, or using galactose as carbon source [11,16–19]. It is more resistant to temperature and acidic stresses, but less resistant to bile salts [12,18].

Table 1. Metabolic, physiological and genetic features of *S. cerevisiae* and *S. boulardii*. The data shown was collected from several studies [11–13,16–21].

<table>
<tr><th colspan="2">Features</th><th>S. Cerevisiae</th><th>S. Boulardii</th></tr>
<tr><td colspan="2">Optimal growth temperature [12]</td><td>30 °C</td><td>37 °C</td></tr>
<tr><td colspan="2">High temperature resistance (52 °C) [12]</td><td>45% viability</td><td>65% viability</td></tr>
<tr><td colspan="2">Acid pH resistance (pH = 2 for one hour) [12,18]</td><td>No—30% viability</td><td>Yes—75% viability</td></tr>
<tr><td colspan="2">Tolerance to bile acids (>0.3%(w/v)) [12]</td><td>No—Survival up to 0.15% (w/v)</td><td>No—Survival up to 0.10% (w/v)</td></tr>
<tr><td colspan="2">Basic pH resistance (pH = 8) [12,18]</td><td>Yes</td><td>Yes</td></tr>
<tr><td colspan="2">Assimilation of galactose [16,17,19]</td><td>Yes</td><td>No</td></tr>
<tr><td colspan="2">Ploidy [18]</td><td>Diploid or haploid</td><td>Always diploid</td></tr>
<tr><td colspan="2">Homo or heterothallic [11]</td><td>Homothallic</td><td>Homothallic</td></tr>
<tr><td colspan="2">Mating type [13]</td><td>Both</td><td>Both</td></tr>
<tr><td colspan="2">Sporulation [16,18]</td><td>Sporogenous</td><td>Asporogenous, but produces fertile hybrids with S. cerevisiae</td></tr>
<tr><td colspan="2">Pseudo-hyphal switching [18]</td><td>Normal</td><td>Increased</td></tr>
<tr><td colspan="2">Retrotransposon (Ty elements) [11]</td><td>Intact Ty elements</td><td>No intact Ty1, 3 or 4 elements</td></tr>
<tr><td rowspan="3">Adhesion to epithelial cells</td><td>Normal microbiome (mice and human) [18,20]</td><td>No</td><td>No</td></tr>
<tr><td>Gnotobiotic mice [21]</td><td>Unknown</td><td>Yes</td></tr>
<tr><td>Humans treated with ampicillin [20]</td><td>Unknown</td><td>Yes</td></tr>
</table>

3. *S. boulardii* Genomic Variations Provide Hints for Its Physiological Properties

S. boulardii and *S. cerevisiae* genomes were found to differ in internal regions of lower copy number in three chromosomes: chromosome I (*PRM9, MST28, YAR047C, YAR050W, CUP1, YAR060W* and *YAR061W*); chromosome VII (*YGL052W* and *MST27*) and chromosome XII (*ASP3* and *YLR156W*). *PRM9, MST27* and *MST28* genes encode nonessential membrane proteins specific to the *Saccharomyces sensu stricto* species [18]. *YAR050W* encodes a lectin-like protein that participates in flocculation; Asp3 is a nitrogen catabolite-regulated cell wall L-asparaginase II. *CUP1* had a two times lower number of copies than the average for *S. cerevisiae* species, possibly causing the increased sensitivity to copper in *S. boulardii* when compared to other *S. cerevisiae* strains [18].

Within genes with higher copy number, two functions are well represented: protein synthesis (*RPL31A, RPL41A, RPS24B, RPL2B* and *RSA3*) and stress response (*HSP26, SSA3, SED1, HSP42, HSP78* and *PBS2*). It is possible that these genes aid in increased growth rate and pseudo-hyphal switching and in higher resistance to high pH [18]. Duplicated and triplicated genes mostly encode stress response proteins, elongation factors, ribosomal proteins, kinases, transporters and fluoride export, which might aid in *S. boulardii* adaptation to stress conditions [11]. Altered gene copy number and mutations when compared to *S. cerevisiae* in the *SDH1* and *WHI2* genes was associated with increased acetic acid production by *S. boulardii*, correlated with antimicrobial activity [22].

S. boulardii was shown to display enhanced ability for pseudo-hyphal switching during nitrogen starvation compared to other *S. cerevisiae* strains [18]. Several genes related to pseudo-hyphal growth have considerably different number of copies: *CDC42*, *DFG16*, *RGS2*, *CYR1* and *CDC25* have higher copy number; *STE11*, *SKM1* and *RAS1* have lower copy numbers [18]. Some of these genes are involved in cyclic adenosine monophosphate (cAMP) pathways, suggesting its hyperactivation can lead to increased pseudo-hyphal growth. As a possible consequence, *S. boulardii* ability to create pseudo-hyphae was observed to be faster and more extensive than *S. cerevisiae* [18].

Variations in the number of repetitive sequences within flocculation genes was also identified in *S. boulardii*, namely in *FLO1*. The encoded flocculin was found to harbor additional copies of residue repeats when compared with most *S. cerevisiae* strains [11]. The Flo8 protein was also found to differ between *S. boulardii* and some *S. cerevisiae* strains where a point mutation results in a truncated protein (including in the reference strain S288c), resulting in defective flocculation and adhesion [23]. Other flocculation genes (*FLO10* and *FLO11*) detected in *S. boulardii* were not found to harbor differences in the copy number and period length of the repeats [11]. The higher maximum number of repeats (e.g., *FLO1*) in *S. boulardii* may affect its adhesion and flocculation ability, as well as sensitivity to stress [11].

Several studies have shown that *S. boulardii* is unable to use galactose as a carbon source, despite harboring all galactose uptake and fermentation genes [16,17,19]. Some studies have proposed that it is able to assimilate, but not ferment, galactose, possibly due to energy requirements [24,25]. More recently, a mutation in the gene *PGM2* was also associated with the inefficient use of galactose [17]. *S. boulardii* is also unable to use palatinose, possibly related with the absence of 3 isomaltase encoding genes (*IMA2*, *IMA3* and *IMA4*), which is involved in palatinose uptake and metabolism [11,19].

4. Adaptation to Host Environment

Probiotics must be able to endure in adverse conditions. *S. boulardii* optimal growth temperature corresponds to the human host temperature (37 °C), while *S. cerevisiae* grows optimally at 30 °C. *S. boulardii* is also more resistant to very high temperatures keeping 65% viability after one hour at 52 °C, while *S. cerevisiae* loses viability down to 45% [12].

The main obstacles in the stomach are the very acidic pH (2 to 3) and the presence of proteases such as pepsin that kill most microorganisms, including probiotics that enter the organism [26]. Diseases like hypochlorhydria decrease the bactericide properties of the stomach and make the patient more susceptible to infections by *H. pylori* and *Salmonella* spp. and to migrations of pathogenic microorganisms to the small intestine where they establish themselves [26]. In the case of the small intestine, main stressors include the high concentrations of bile salts, pancreatic enzymes, hydrolytic enzymes, pancreatin, organic acids, the integrity of the epithelial and brush border, the immune defense and the native microbiome and its secondary metabolism products (H_2S, bacteriocins, organic acids) [27]. Bile salts are detergents produced in the liver from cholesterol and secreted to the intestine to improve nutrient absorption. As detergent like molecules, bile salts can be toxic to GI tract microorganisms by disrupting their cellular membrane lipid bilayer structures [12]. However, some probiotics are able to resist degradation by hydrolytic enzymes and bile salts [6]. For example, *S. boulardii* and *Bacillus coagulans* remain viable after exposure to simulated gastric juice containing pepsin and hydrochloric acid. These probiotics were also seen to be stable to the impact of bile salts [6]. *Bacillus clausii* was partially resistant to these conditions [6]. On the other hand, most *Lactobacillus* and *Bifidobacterium* spp. have reduced viability under exposure to gastrointestinal agents such as pepsin, hydrochloric acid and bile [6].

In vitro testing of probiotic formulations consisting of *S. boulardii* and bacterial probiotics (*Lactobacillus* spp. and *Bifidobacterium* spp.) highlighted the ability of *S. boulardii* to survive GI tract conditions. *S. boulardii* was able to survive after incubation in a gastric-like environment and in an intestinal environment (bile salts, pancreatin, pH 7.0) for 3 h, whereas the viability of the bacterial probiotics was severely impaired [6]. *S. boulardii* is also more resistant to a gastric environment than *S. cerevisiae*, while the viability of both species in an intestinal environment (sodium chloride, pepsin, pancreatin, pH 8.0) is not affected after 1 h [12]. Accordingly, 1 h was enough to show that *S. boulardii* is more resistant to low pH than *S. cerevisiae*, particularly at pH 2.0 [18]. Interestingly, although *S. boulardii* can survive the GI environments, its viability is significantly increased for 2 h if encapsulated by a double layer with sodium alginate and gelatin [28]. Tolerance displayed by *S. boulardii* to bile salts has also been tested. Surprisingly, *S. cerevisiae* is more tolerant to bile salts than *S. boulardii*. However, after 1 h, both species show a tolerance threshold bellow what would be considered as resistance to bile salts [12].

Dynamic modelling of the stomach and small intestine conditions also showed *S. boulardii* to be resilient to gastric and lower intestinal conditions, while modelling of the colon environment revealed the yeast is not able to colonize the colon, but had an individual-dependent effect in the microbiotic profile [29]. Other studies also point to the inability of *S. boulardii* to colonize the gut, suggesting that this yeast does not strongly adhere to intestinal epithelial cells and is quickly removed from the gastrointestinal system in healthy individuals [18]. However, it has been shown to colonize the intestine of gnotobiotic mice after a single administration [21]. This may mean that, although *S. boulardii* can colonize the gut, competition with intestinal microbiome is limiting [18]. Indeed, both *S. boulardii* and other *Saccharomyces* strains were shown to be unable to remain attached to human and mouse epithelial cells, in vitro and in vivo, respectively [18]. However, they do adhere to Caco-2 cells through an extracellular factor, probably secreted mucus [18]. Colonization of the gut was observed to be dependent, both in mice and human, on repeated administration over several days [20,21,30]. Moreover, administration of ampicillin increased *S. boulardii* cell concentration [20], reinforcing the notion that competition with intestinal microbiome plays a relevant role in the establishment of this yeast.

5. Mechanisms of Action

The gut microbiome is responsible for a multitude of roles, including protection against pathogen colonization, epithelial barrier maintenance or modulation of immune activity [31]. The mechanisms by which gut microbiome homeostasis is maintained are not yet fully understood. Probiotics are believed to display a variety of mechanisms: antitoxin effects, physiological protection, modulation of the normal microbiome, metabolic regulation and signaling pathway modification, nutritional and trophic effects, immune system regulation, pathogen competition, interactions with the brain-gut axis, cellular adhesion, cellular antagonism and mucin production [1,4,31]. *S. boulardii* has been described as participating in a number of these effects as part of its probiotic activity (Figure 1). The genetic basis and mechanistic details that underlie these observations are not fully understood and their clarification could be key to better exploit this yeast and how to potentiate general probiotic activity.

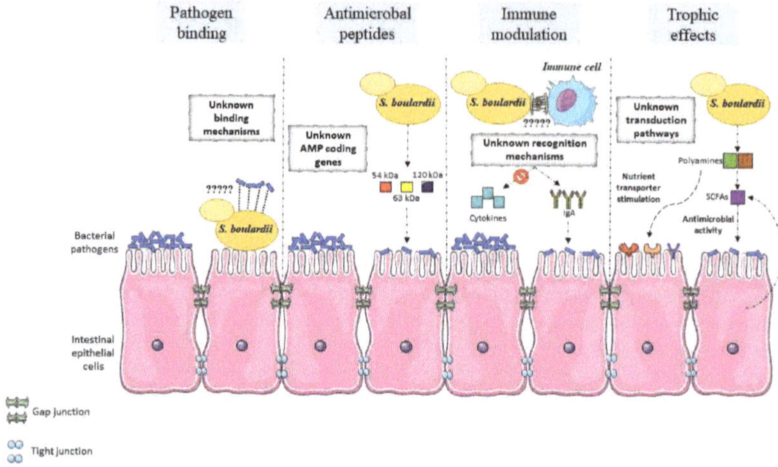

Figure 1. Overview of the main modes of action that support S. boulardii probiotic activity in the intestinal epithelium. Studies have described the outcome of S. boulardii administration in pathogen exclusion, antimicrobial properties, immune modulation, and trophic effects. The genetic basis and mechanistic details that underlie these observations are not fully understood and their clarification could be key to better exploit this yeast and how to potentiate general probiotic activity. Pathogen exclusion is mainly achieved by pathogen binding to the yeast cells, rather than competition for epithelial binding sites with the pathogens. The yeast cell wall components responsible for the binding, the correspondent pathogenic receptors and the binding dynamics have not been fully investigated. Antimicrobial action is achieved, at least partially, by the secretion of still unknown proteins with antimicrobial effects. The genes that code for these proteins have not been identified and could provide further clues on the mode of action of S. boulardii. Immune modulation and the effect of S. boulardii on inflammatory pathways has been uncovered to some extent. The mechanistic insights and dynamics of S. boulardii interaction with immune cells still need to be ascertained to better understand the yeast action in the immunological function. Multiple trophic effects have been described to be stimulated by S. boulardii on intestinal epithelial cells. Some pathways have been elucidated, although the multitude of trophic effects suggests concerted action and crosstalk between yeast and host cell sensory pathways.

5.1. Modulation of The Normal Microbiome

Modulation of the normal microbiome may be favored directly by transiting probiotics which produce antimicrobial substances, or indirectly contribute to immune modulation or gut barrier function [31]. The use of probiotics has typically been applied to reestablish the normal gut microbiome upon dysbiosis. Gut dysbiosis refers to changes in the microbiome's quantitative and qualitative composition. These changes may contribute to a disease state frequently associated to inflammation and can be a result of antibiotic-associated diarrhea, acute infectious diarrhea or IBD [31,32]. Probiotics treatment helps to stabilize the gut microbial community and lead to an improved disease outcome [32]. While some probiotics may become a part of the microbiome, others simply pass through the GI tract and modulate or influence the existing microbiome before exiting the body [31].

Several factors can have deleterious effects on the gut microbiome and hinder its protective role to the host epithelial lining, such as antibiotic use or surgery [1]. This may result in host susceptibility to colonization by pathogens until the normal microbiome is reestablished, which can take several weeks [33]. *S. boulardii* helps to restore the normal microflora in this type of patient and the use of probiotics as modulators of the normal microbiome through colonization during the susceptibility period may work as a surrogate until the normal microbiome is reestablished [34].

5.2. Antimicrobial Activity

Antagonism against pathogens can be achieved by colonization and exclusion of pathogens, modulation of metabolic and signaling pathways, production of inhibitory compounds or immune modulation [31]. Competition is one of the main mechanisms associated with probiotic activity against gut pathogens: consumption of nutrients by probiotics results in nutrient limitation for pathogenic organisms [31,35]. On the other hand, the ability of probiotics to grow and colonize the gut can lead to a decrease of the gut pH due to the production of metabolites, leading to stressful conditions for pathogens [35]. A possible role for *S. boulardii* in managing pathogenic activity was associated with a protective effect of *S. boulardii* against pathogenic bacteria in yeast-treated mice, although the mode of action is not associated with a reduction of the pathogenic population [21], as well as with another study which observed a protective effect against *Candida albicans* in a murine model [36].

The production of compounds with antimicrobial activity is yet another major mode of action of probiotics. Several components of the probiotic metabolome, such as organic acids, bacteriocins, hydrogen peroxide, diacetyl, or amines limit the growth of pathogenic bacteria [31]. In particular, bacteriocins play a crucial role in the antimicrobial action of probiotic bacteria, especially *Lactobacillus* spp. [35,37–40]. As for the production of antimicrobial substances by other probiotics, *S. boulardii* possibly secretes proteins that reduce *Citrobacter rodentium* adhesion to host epithelial cells by modulating virulence factors [41]. It also displays antimicrobial activity by secreting 54-kDa, 63-kDa and 120-kDa proteins that cleave microbial toxins or reduce cAMP levels. *S. boulardii* can block toxin receptors or function as a decoy receptor for toxins. The 54-kDa serine protease produced by *S. boulardii* cleaves toxins A and B from *C. difficile* and the enterocytic receptor to which the toxins bind, which causes inflammation, fluid secretion, mucosal permeability and injury in the intestines [42,43]. Other mechanisms that *S. boulardii* uses against *C. difficile* infection are growth inhibition and decreased toxin production due to secreted factors and stimulation of host mucosal disaccharidase activity [44,45]. Another study refers to the ability of *S. boulardii* to inhibit *Escherichia coli* surface endotoxins by dephosphorylation. A 63-kDa alkaline phosphatase targets the lipopolysaccharide (LPS) and contributes to decreased tumor necrosis factor α (TNF-α) cytokine levels [46]. *S. boulardii* also produces a 120-kDa protein that decreases the chloride secretions stimulated by cholera toxin by reducing cAMP levels [47]. *S. boulardii* is also able to adhere to cholera toxin via its cell wall, thus blocking its toxic effects [48]. Despite these observations, the sequencing of *S. boulardii* genomes did not provide a clear identification of the genes encoding these 54-kDa, 63-kDa and 120-kDa proteins [25].

S. boulardii also confers protection against the lethal toxin produced by *Bacillus anthracis*. The bacterium causes ulcerative lesions from the jejunum to cecum and uses its toxin to disrupt intestinal epithelium integrity, causing mucosal erosion, ulceration and bleeding [49]. The protective effect of *S. boulardii* is associated with maintenance of barrier function and reduction of harmful physiological responses elicited by the toxin, such as formation of stress fibers [50]. The protective effect is achieved by release of proteases and cleavage of the lethal toxin [50].

Some *S. boulardii* strains are able to produce high concentrations of acetic acid, which was found to exert an inhibitory effect in *E. coli* [22]. In turn, the decrease in pH due to acetic acid production is essential for the antimicrobial activity of short-chain organic acids. The combined effect of high acetic acid concentration and lower pH may be an additional mechanism that makes *S. boulardii* an effective probiotic. Moreover, acetic acid is produced under aerobic conditions by *S. boulardii*. Due to the radial oxygen gradient between the epithelial surface (high oxygen levels) and the center of the gut lumen (low oxygen levels), microorganisms colonizing the epithelial surface have greater availability of oxygen [51]. Since acetic acid is produced under aerobic conditions by *S. boulardii*, its production should be higher near the epithelial surface. During antibiotic treatment and pathogen infection, oxygen concentration also increases in the gastrointestinal tract [52,53], which could support the antimicrobial action of *S. boulardii*.

5.3. Adhesion

In order for the host not to mechanically eliminate the gut microbiome, it is crucial that its components adhere to host surfaces [1,4]. Some probiotics express surface adhesins that mediate the attachment to the mucous layer by recognizing host molecules such as transmembrane proteins (integrins or cadherins) and extracellular matrix components (collagen, fibronectin, laminin or elastin) [1,4]. Probiotics can also influence the production of mucin and the barrier function of the intestine, thus hindering adhesion and consequent invasion of pathogenic microorganisms [54].

Mucin is produced by epithelial cells to avert adhesion by pathogenic bacteria, whereas successful probiotics should be able to adhere to the intestinal mucous, as is the case of *S. boulardii* [13,55]. The adhesion of *S. boulardii* to the mucus membrane contributes to reducing the availability of binding sites for pathogens [13]. Five *S. cerevisiae* cell wall proteins (encoded by *CIS3*, *CWP2*, *FKS3*, *PIR3* and *SCW4*) were found to mediate adhesion of the yeast cells to the pathogenic bacteria *E. coli*, *Salmonella enterica* serovar *typhimurium* (*S. typhimurium*) and *Salmonella enterica* serovar *typhi* (*S. typhi*) [55]. Other studies have shown that these bacteria are also bound to *S. boulardii* [55–58]. Additionally, *S. boulardii* also inhibits *C. difficile* adhesion to epithelial cells and displays inhibitory activity on *Entamoeba histolytica* adhesion to erythrocytes [59,60]. This interaction limits the ability of pathogens to bind directly to the intestinal receptors and proceed with host invasion. In fact, *S. boulardii* hinders epithelium invasion by *S. typhimurium* due to steric hindrance caused by its larger size as compared to bacteria [61]. As *S. boulardii* does not significantly bind to epithelial cells of healthy individuals and is quickly flushed out, pathogens bound to *S. boulardii* are possibly flushed together with the yeast cells [13,31,55]. However, *S. boulardii* does have several flocculation genes required for protection against environmental stress and biofilm formation [11]. The characterization of this gene family in the context of host adhesion and colonization could provide further insight on the probiotic features of *S. boulardii*.

The ability of *S. boulardii* to bind bacterial pathogens has been associated with the presence of mannose residues in the yeast cell wall [56]. This is a similar mechanism to the previously characterized adhesion of bacterial pathogens to the epithelial surface via mannose residues [62], which is the basis for the addition of exogenous sugars as a strategy to inhibit pathogen adhesion [63]. Cell wall mannan oligosaccharides are a common feature in yeast, but the affinity between *E. coli* and *S. boulardii* is higher than with *S. cerevisiae* [56]. Further investigations revealed that bile salts decrease adhesion of bacteria to yeast cells [55], which can have relevant implications for yeasts as successful probiotics. Accordingly, bile salts also decrease the adhesion of probiotic bacteria to intestine epithelia due to diminished surface hydrophobicity and higher surface potential [64], bolstering how important it is for probiotic microorganisms to evolve adaptation strategies within the host.

5.4. Immune Modulation

Metabolites produced by the gut microbiome can perform immunomodulatory and anti-inflammatory functions that stimulate immune cells. This ability arises from the interaction between the probiotics and the epithelial cells, dendritic cell monocytes, macrophages and/or lymphocytes [1,31]. Probiotics also promote enhanced phagocytic activity, cell proliferation and production of secretory immunoglobulins IgA and IgM [65].

S. boulardii can modulate immunological function by acting as a stimulant or a pro-inflammatory inhibitor. It is capable of modulating the inflammatory process upon *S. typhimurium* infection by decreasing the levels of the pro-inflammatory molecules such as cytokine interleukin 8 (IL-8), mitogen activated protein (MAP) kinases and the (nuclear factor kappa B) NF-κB signaling pathway [58]. An inhibitory effect of *S. boulardii* over MAP kinases and IL-8 levels upon *C. difficile* infection was also observed [66]. Likewise, *S. boulardii* contributes to increasing the levels of anti-inflammatory cytokines (IL-4 and IL-10) and decreasing pro-inflammatory cytokines (IL-1β) upon infection with *E. coli* and *C. albicans* [67]. On the other hand, *S. boulardii* was associated with increased IgA and IgG levels in serum in response to *C. difficile* toxins A and B [68,69]. *S. boulardii* was also found to attach to the surface of dendritic cells [70] and modulate the expression of toll-like receptors (TLRs) and cytokines [70–72]. Moreover, *S. boulardii* also causes the imprisonment of T helper cells in mesenteric lymphatic nodes, reducing inflammation [73].

Another study found that in the early phase of *S. typhimurium* infection, *S. boulardii* induces pro-inflammatory cytokine production (interferon-γ—IFN-γ) and represses the production of anti-inflammatory cytokines (IL-10) in the small intestine, but increases the levels of both cytokines in the cecum [57]. This suggests that *S. boulardii* can differentially modulate immune activity through the GI tract [57]. Overall, probiotics may be able to persistently modulate both the innate and adaptive immune responses either locally or systemically [1,31]. The data from several studies indicates that *S. boulardii* plays a pivotal role in immune modulation against the most common GI tract pathogens.

5.5. Trophic Effects

S. boulardii is a modulator of enzyme activity required to maintain a healthy gastrointestinal tract. It exerts trophic effects such as stimulation of brush border membrane digestive enzymes and nutrient transporter activity [74]. Several studies have shown a wide array of trophic effects stimulated by *S. boulardii*: brush border sucrase, lactase, and maltase activities [44,75–78]; iso-maltase activity [78]; glucoamylase and *N*-aminopeptidase activity [76]; leucine-aminopeptidase activity [79]; α,α-trehalase activities in the endoluminal fluid and intestinal mucosa; brush border α-glucosidase [80]; spermine, spermidine and putrescine levels in rat jejunal mucosa [75,77]; adenosine triphosphatase, γ-glutamyl transpeptidase, lipase, and trypsin activities and TNF-α, IL-10, transforming growth factor beta (TGF-β), and secretory IgA [5]; diamine oxidase activities, brush border sodium/glucose cotransporter 1 expression and sodium-dependent glucose uptake [74,77]; GRB2-SHC-CrkII-Ras-GAP-Raf-ERK1,2 transduction pathway in rats and decreased p38 MAPK and NF-κB [81–83].

Probiotics can modulate short chain fatty acids (SCFA: acetate, propionate, or butyrate) and/or branched-chain fatty acid (BCFA: iso-butyrate, 2-methylbutyrate, or isovalerate) synthesis. SCFAs have a complex role in the physiological and biochemical functions in different tissues (intestine, liver, adipose, muscle and brain). *S. boulardii* assists in reestablishing SCFA levels, which are depressed during disease [84,85]. Acetate and butyrate are major SCFAs in intestinal epithelial cells, playing a role in barrier function, anti-inflammatory and immune modulation pathways [86,87]. A study reported that a short-term treatment (6 days) with *S. boulardii* diminishes the incidence of diarrhea in patients receiving enteral nutrition by increasing SCFA levels, particularly butyrate [84]. SCFAs can also present antimicrobial activity, and a study probing several *S. boulardii* and *S. cerevisiae* strains for inhibitory effects in *E. coli* described the production of acetic acid exclusively by *S. boulardii* as an antimicrobial mechanism [22]. Moreover, acetate also stimulates T regulatory cells, induces mucus secretion gene expression, inhibits proinflammatory cytokine CXCL8 and serves as a substrate for the production of butyrate by the microbiome [22].

The activity of many digestive enzymes (sucrase-iso-maltase, maltase-glucoamylase, lactase-phlorizin hydrolase, alanine aminopeptidase and alkaline phosphatase) and nutrient transporters (sodium-glucose transport proteins) may be induced by polyamines secreted by *S. boulardii* [74]. *S. boulardii* secretes polyamines that promote RNA binding and stabilization and, hence, growth and protein (lactase, maltase, sucrase, among others) synthesis [74]. These molecules are also able to shield lipids from oxidation and boost SCFA activity. Polyamines may also affect kinase activities and external signal transduction pathways, therefore modulating the GRB2-SHC-CrkII-Ras-GAP-Raf-ERK1,2 and the PI3K pathways [74]. They can also aid the generation of specific transcripts by interacting with DNA [74]. All of these polyamine functions lead to a general polyamine-triggered metabolic activation in order to regenerate brush border damage and maturation of enterocytes [74,75,80]. Not only does *S. boulardii* induce the enzymatic activities of lactase-phlorizin hydrolase, α-glucosidases, alkaline phosphatases and aminopeptidases, but it also increases glucose intestinal absorption, one of the products of lactose degradation [74]. Production of lactase by the host, partially stimulated by *S. boulardii*, mediates lactose degradation thus alleviating lactose intolerance.

6. *S. boulardii* Safety and Clinical Efficacy

Although many probiotics are documented as safe, common safety issues regarding the use of probiotics include: transfer of antibiotic resistance genes, translocation of live organisms from the intestine to other sites of the body, persistence in the intestine and development of adverse reactions [1]. Most of these concerns have been dismissed when evaluating *S. boulardii* safety. *S. boulardii* is not known to acquire resistance genes, unlike bacterial probiotics such as *Lactobacillus* spp. [88,89]. Animal studies show that there is reduced translocation in the treatment with *S. boulardii* when compared with *S. cerevisiae* [90–92]. *S. boulardii* does not persist in the intestine after three to five days after discontinuation of the ingestion, according to pharmacokinetic studies [20]. The data available from 90 clinical trials assessing the efficacy and safety of *S. boulardii* has been thoroughly assessed elsewhere [1]. Randomized and controlled trials clearly show the absence of any serious adverse reactions, while only some presented moderate adverse reactions, such as constipation in patients with *C. difficile* infection [93]. Although fungemia is viewed as a potential problem, there were no fungemia cases reported in clinical trials [1]. *S. boulardii*-associated fungemia was observed in patients with serious co-morbidity factors and central venous catheters, which responded well to fluconazole or amphotericin B therapy [91,94–96]. Importantly, *S. cerevisiae*-associated fungemia has a worse prognosis than that caused by *S. boulardii* [97].

Clinical trials have investigated the efficacy of *S. boulardii* in the improvement of several GI conditions' outcome. This yeast was seen to improve the outcome of several diarrhea diseases, including pediatric diarrhea, antibiotic-associated diarrhea, acute diarrhea, traveler's diarrhea caused by bacterial, viral or parasites, and enteral nutrition-related diarrhea [1,15,98]. *S. boulardii* also improves the outcome in patients suffering from *H. pylori* or *C. difficile* infections by helping bacteria eradication, preventing relapses, reducing adverse reactions, and reducing treatment-associated diarrhea [1,15,98]. IBD is a prevalent GI tract disorder associated with inflammatory diarrheal diseases such as ulcerative colitis, pouchitis and Crohn's disease [1]. Clinical trial data points to a possible role of *S. boulardii* in reducing treatment relapses [1,15,98], which are frequent in these conditions, although further studies are required to reach compelling conclusions. Irritable Bowel Syndrome (IBS) symptoms also improve with *S. boulardii* administration. It is a condition frequently characterized by abdominal bloating, abdominal pain, and disturbed intestinal transit. These symptoms were shown to be alleviated in 50% of patients upon *S. boulardii* use [99].

7. Conclusions

S. boulardii is a probiotic yeast with proven efficacy in the treatment of GI conditions, especially when used as an adjuvant to antibiotic treatment. Present data indicate that the benefits of *S. boulardii* appear to be transient and independent of host gut colonization, differentiating its mode of action from other widely used bacterial probiotics. The absence of colonization appears to correlate with pathogen binding as a mechanism to halt pathogen colonization, rather than competitive exclusion due to yeast adhesion. Genomics studies have contributed to pinpoint distinct genome features that mainly confer on *S. boulardii* the ability to resist host stresses, conferring higher viability through GI passage than observed for other common probiotics. *S. boulardii* also elicits a complex immunomodulatory effect with roles in fine-tuning immunological pathways during pathogen infection or chronic diseases. This yeast also contributes to the homeostasis of the normal microbiome and plays a relevant role in modulating secretory functions by intestinal epithelial cells, thus benefitting nutritional requirements of the host. Overall, *S. boulardii* displays a multifactorial role as a probiotic, with proven efficacy and safety in alleviating the symptomology of a number of GI conditions. However, there is a significant knowledge gap between *S. boulardii* phenotypic effects and the underlying genetic basis, especially when compared to *S. cerevisiae*. What are the 54-kDa, 63-kDA and 120-kDA proteins secreted by *S. boulardii* that cleave microbial toxins or reduce cAMP levels? What are the proteins responsible for higher adhesion of *S. boulardii* to pathogenic bacteria, when compared to *S. cerevisiae*, especially considering the differences in flocculin encoding genes? What are the mechanisms that allow *S. boulardii* to overcome the negative impact of bile salts during host adaptation? What are the proteins or cellular components that mediate immune recognition of *S. boulardii* and modulation of the immune response? These questions remain unanswered. Further research on the genetic basis of *S. boulardii* probiotic activity will certainly increase our understanding of this fascinating yeast, while providing important clues for the selection and optimization of even more powerful probiotic fungi.

Author Contributions: P.P. and M.C.T. delineated manuscript organization. P.P., with contributions from V.A. and M.Y., wrote the manuscript, under the coordination of M.C.T. All authors have read and agreed to the published version of the manuscript.

Funding: This research was funded by "Fundação para a Ciência e a Tecnologia" (FCT) [Contract PTDC/BII-BIO/28216/2017], as well as by Programa Operacional Regional de Lisboa 2020 [LISBOA-01-0145-FEDER-022231—The BioData.pt Research Infrastructure]. Funding received by iBB from FCT (UIDB/04565/2020), and from Programa Operacional Regional de Lisboa 2020 (LISBOA-01-0145-FEDER-007317).

Conflicts of Interest: The authors declare no conflict of interest. The funders had no role in the design of the study; in the collection, analyses, or interpretation of data; in the writing of the manuscript, or in the decision to publish the results.

References

1. McFarland, L.V. Common Organisms and Probiotics: Saccharomyces boulardii. In *The Microbiota in Gastrointestinal Pathophysiology*; Academic Press: Cambridge, MA, USA, 2017; pp. 145–164.
2. Schrezenmeir, J.; de Vrese, M. Probiotics, prebiotics, and synbiotics—approaching a definition. *Am. J. Clin. Nutr.* **2001**, *73*, 361s–364s. [CrossRef] [PubMed]
3. Moradi, R.; Nosrati, R.; Zare, H.; Tahmasebi, T.; Saderi, H.; Owlia, P. Screening and characterization of in-vitro probiotic criteria of saccharomyces and kluyveromyces strains. *Iran. J. Microbiol.* **2018**, *10*, 123–131. [PubMed]
4. Plaza-Diaz, J.; Ruiz-Ojeda, F.J.; Gil-Campos, M.; Gil, A. Mechanisms of Action of Probiotics. *Adv. Nutr.* **2019**, *10*, S49–S66. [CrossRef]
5. Sun, Y.; Rajput, I.R.; Arain, M.A.; Li, Y.; Baloch, D.M. Oral administration of Saccharomyces boulardii alters duodenal morphology, enzymatic activity and cytokine production response in broiler chickens. *Anim. Sci. J.* **2017**, *88*, 1204–1211. [CrossRef] [PubMed]
6. Kabluchko, T.V.; Bomko, T.V.; Nosalskaya, T.N.; Martynov, A.V.; Osolodchenko, T.P. In the gastrointestinal tract exist the protective mechanisms which prevent overgrowth of pathogenic bacterial and its incorporation. *Ann. Mechnikov Inst.* **2017**, *1*, 28–33.
7. Kelesidis, T.; Pothoulakis, C. Efficacy and safety of the probiotic Saccharomyces boulardii for the prevention and therapy of gastrointestinal disorders. *Therap. Adv. Gastroenterol.* **2012**, *5*, 111–125. [CrossRef] [PubMed]
8. Sen, S.; Mansell, T.J. Yeasts as probiotics: Mechanisms, outcomes, and future potential. *Fungal Genet. Biol.* **2020**, *137*, 103333. [CrossRef]
9. Van Der Aa Kühle, A.; Jespersen, L. The Taxonomic Position of Saccharomyces boulardii as Evaluated by Sequence Analysis of the D1/D2 Domain of 26S rDNA, the ITS1-5.8S rDNA-ITS2 Region and the Mitochondrial Cytochrome-c Oxidase II Gene. *Syst. Appl. Microbiol.* **2003**, *26*, 564–571. [CrossRef]
10. Mitterdorfer, G.; Mayer, H.K.; Kneifel, W.; Viernstein, H. Clustering of Saccharomyces boulardii strains within the species S. cerevisiae using molecular typing techniques. *J. Appl. Microbiol.* **2002**, *93*, 521–530. [CrossRef]
11. Khatri, I.; Tomar, R.; Ganesan, K.; Prasad, G.S.; Subramanian, S. Complete genome sequence and comparative genomics of the probiotic yeast Saccharomyces boulardii. *Sci. Rep.* **2017**, *7*, 1–13. [CrossRef]
12. Fietto, J.L.; Araújo, R.S.; Valadão, F.N.; Fietto, L.G.; Brandão, R.L.; Neves, M.J.; Gomes, F.C.; Nicoli, J.R.; Castro, I.M. Molecular and physiological comparisons between Saccharomyces cerevisiae and Saccharomyces boulardii. *Can. J. Microbiol.* **2004**, *50*, 615–621. [CrossRef] [PubMed]
13. Edwards-ingram, L.C.; Gent, M.E.; Hoyle, D.C.; Hayes, A.; Stateva, L.I.; Oliver, S.G. Comparative Genomic Hybridization Provides New Insights Into the Molecular Taxonomy of the Saccharomyces Sensu Stricto Complex. *Genome Res.* **2004**, *14*, 1043–1051. [CrossRef] [PubMed]

14. Hennequin, C.; Thierry, A.; Richard, G.F.; Lecointre, G.; Nguyen, H.V.; Gaillardin, C.; Dujon, B. Microsatellite Typing as a New Tool for Identification of Saccharomyces cerevisiae Strains. *J. Clin. Microbiol.* **2001**, *39*, 551–559. [CrossRef] [PubMed]
15. McFarland, L.V. Systematic review and meta-analysis of saccharomyces boulardii in adult patients. *World J. Gastroenterol.* **2010**, *16*, 2202–2222. [CrossRef]
16. McFarland, L.V. Saccharomyces boulardii Is Not Saccharomyces cerevisiae. *Clin. Infect. Dis.* **1996**, *22*, 200–201. [CrossRef]
17. Liu, J.J.; Zhang, G.C.; Kong, I.I.; Yun, E.J.; Zheng, J.Q.; Kweon, D.H.; Jin, Y.S. A mutation in PGM2 causing inefficient galactose metabolism in the probiotic yeast Saccharomyces boulardii. *Appl. Environ. Microbiol.* **2018**, *84*, e02858-17. [CrossRef]
18. Edwards-ingram, L.; Gitsham, P.; Burton, N.; Warhurst, G.; Clarke, I.; Hoyle, D.; Oliver, S.G.; Stateva, L. Genotypic and Physiological Characterization of Saccharomyces boulardii, the Probiotic Strain of Saccharomyces cerevisiae. *Appl. Environ. Microbiol.* **2007**, *73*, 2458–2467. [CrossRef]
19. GMitterdorfer; WKneifel; HViernstein Utilization of prebiotic carbohydrates by yeasts of therapeutic relevance. *Lett. Appl. Microbiol.* **2001**, *34*, 251–255.
20. Klein, S.M.; Elmer, G.W.; McFarland, L.V.; Surawicz, C.M.; Levy, R.H. Recovery and Elimination of the Biotherapeutic Agent, Saccharomyces boulardii, in Healthy Human Volunteers. *Pharm. Res. An Off. J. Am. Assoc. Pharm. Sci.* **1993**, *10*, 1615–1619.
21. Rodrigues, A.C.P.; Mardi, R.M.; Bambirra, E.A.; Vieira, E.G.; Nicoli, U.R. Effect of Saccharomyces boulardii against experimental oral infection with Salmonella typhimurium and Shigella flexnerim conventional and gnotobiotic mice. *J. Appl. Bacteriol.* **1996**, *81*, 251–256. [CrossRef]
22. Offei, B.; Vandecruys, P.; De Graeve, S.; Foulquié-moreno, M.R.; Thevelein, J.M. Unique genetic basis of the distinct antibiotic potency of high acetic acid production in the probiotic yeast Saccharomyces cerevisiae var. boulardii. *Genome Res.* **2019**, 1478–1494. [CrossRef] [PubMed]
23. Liu, H.; Styles, C.A.; Fink, G.R. Saccharomyces cerevisiae S288C has a mutation in FLO8, a gene required for filamentous growth. *Genetics* **1996**, *144*, 967–978. [PubMed]
24. van den Brink, J.; Akeroyd, M.; van der Hoeven, R.; Ponk, J.T.; de Winde, J.H.; Daran-Lapujade, P.A.S. Energetic limits to metabolic flexibility: Responses of Saccharomyces cerevisiae to glucose-galactose transitions. *Microbiology* **2009**, *155*, 1340–1350. [CrossRef] [PubMed]
25. Khatri, I.; Akhtar, A.; Kaur, K.; Tomar, R.; Prasad, G.S. Gleaning evolutionary insights from the genome sequence of a probiotic yeast Saccharomyces boulardii. *Gut Pathog.* **2013**, *5*, 30. [CrossRef] [PubMed]
26. Martinsen, T.C.; Bergh, K.; Waldum, H.L. Gastric Juice: A Barrier Against Infectious Diseases. *Basic Clin. Pharmacol. Toxicol.* **2005**, *96*, 94–102. [CrossRef] [PubMed]
27. Holzapfel, W.H.; Haberer, P.; Snel, J.; Schillinger, U.; Huis In'T Veld, J.H.J. Overview of gut flora and probiotics. *Int. J. Food Microbiol.* **1998**, *41*, 85–101. [CrossRef]
28. Du Le, H.; Trinh, K.S. Survivability of Lactobacillus acidophilus, Bacillus clausii and Saccharomyces boulardii encapsulated in alginate gel microbeads. *Carpathian J. Food Sci. Technol.* **2018**, *10*, 95–103.
29. Cordonnier, C.; Thévenot, J.; Etienne-Mesmin, L.; Denis, S.; Alric, M.; Livrelli, V.; Blanquet-Diot, S. Dynamic In Vitro Models of the Human Gastrointestinal Tract as Relevant Tools to Assess the Survival of Probiotic Strains and Their Interactions with Gut Microbiota. *Microorganisms* **2015**, *3*, 725–745. [CrossRef]
30. Filho-Lima, J.V.M.; Vieira, E.C.; Nicoli, J.R. Antagonistic effect of Lactobacillus acidophilus, Saccharomyces boulardii and Escherichia coli combinations against experimental infections with Shigella flexneri and Salmonella enteritidis subsp. typhimurium in gnotobiotic mice. *J. Appl. Microbiol.* **2000**, *88*, 365–370. [CrossRef]
31. Bajaj, B.K.; Claes, I.J.J.; Lebeer, S. Functional mechanisms of probiotics. *J. Microbiol. Biotechnol. Food Sci.* **2015**, *4*, 321–327. [CrossRef]
32. Ceapa, C.; Wopereis, H.; Rezaïki, L.; Kleerebezem, M.; Knol, J.; Oozeer, R. Influence of fermented milk products, prebiotics and probiotics on microbiota composition and health. *Best Pract. Res. Clin. Gastroenterol.* **2013**, *27*, 139–155. [CrossRef] [PubMed]
33. Dethlefsen, L.; Huse, S.; Sogin, M.L.; Relman, D.A. The Pervasive Effects of an Antibiotic on the Human Gut Microbiota, as Revealed by Deep 16S rRNA Sequencing. *PLoS Biol.* **2008**, *6*, e280. [CrossRef]
34. McFarland, L.V. Use of probiotics to correct dysbiosis of normal microbiota following disease or disruptive events: A systematic review. *BMJ Open* **2014**, *4*, e005047. [CrossRef] [PubMed]

35. Bermudez-Brito, M.; Plaza-Díaz, J.; Muñoz-Quezada, S.; Gómez-Llorente, C.; Gil, A. Probiotic Mechanisms of Action. *Ann. Nutr. Metab.* **2012**, *61*, 160–174. [CrossRef] [PubMed]
36. Ducluzeau, R.; Bensaada, M. Comparative effect of a single or continuous administration of "Saccharomyces boulardii" on the establishment of various strains of "candida" in the digestive tract of gnotobiotic mice. *Ann. Microbiol. (Paris)* **1982**, *133*, 491–501. [PubMed]
37. de Arauz, L.J.; Jozala, A.F.; Mazzola, P.G.; Vessoni Penna, T.C. Nisin biotechnological production and application: A review. *Trends Food Sci. Technol.* **2009**, *20*, 146–154. [CrossRef]
38. Vilagravel, B.; Esteve-Garcia, E.; Brufau, J. Probiotic micro-organisms: 100 years of innovation and efficacy; Modes of action. *Worlds. Poult. Sci. J.* **2010**, *66*, 369–380.
39. Corr, S.C.; Li, Y.; Riedel, C.U.; O'Toole, P.W.; Hill, C.; Gahan, C.G.M. Bacteriocin production as a mechanism for the antiinfective activity of Lactobacillus salivarius UCC118. *Proc. Natl. Acad. Sci. USA* **2007**, *104*, 7617–7621. [CrossRef] [PubMed]
40. Nielsen, D.S.; Cho, G.S.; Hanak, A.; Huch, M.; Franz, C.M.A.P.; Arneborg, N. The effect of bacteriocin-producing Lactobacillus plantarum strains on the intracellular pH of sessile and planktonic Listeria monocytogenes single cells. *Int. J. Food Microbiol.* **2010**, *141*, S53–S59. [CrossRef] [PubMed]
41. Wu, X.; Vallance, B.A.; Boyer, L.; Bergstrom, K.S.B.; Walker, J.; Madsen, K.; O'Kusky, J.R.; Buchan, A.M.; Jacobson, K. Saccharomyces boulardii ameliorates Citrobacter rodentium-induced colitis through actions on bacterial virulence factors. *Am. J. Physiol. Gastrointest. Liver Physiol.* **2007**, *294*, G295–G306. [CrossRef] [PubMed]
42. Castagliuolo, I.; Riegler, M.F.; Valenick, L.; LaMont, J.T.; Pothoulakis, C. Saccharomyces boulardii protease inhibits the effects of Clostridium difficile toxins A and B in human colonic mucosa. *Infect. Immun.* **1999**, *67*, 302–307. [CrossRef]
43. Castagliuolo, I.; Thomas Lamont, J.; Nikulasson, S.T.; Pothoulakis, C. Saccharomyces boulardii protease inhibits Clostridium difficile toxin A effects in the rat ileum. *Infect. Immun.* **1996**, *64*, 5225–5232. [CrossRef]
44. Buts, J.P.; Bernasconi, P.; Van Craynest, M.P.; Maldague, P.; Meyer, R. Response of human and rat small intestinal mucosa to oral administration of saccharomyces boulardii. *Pediatr. Res.* **1986**, *20*, 192–196. [CrossRef] [PubMed]
45. Buts, J.P.; Bernasconi, P.; Vaerman, J.P.; Dive, C. Stimulation of secretory IgA and secretory component of immunoglobulins in small intestine of rats treated with Saccharomyces boulardii. *Dig. Dis. Sci.* **1990**, *35*, 251–256. [CrossRef] [PubMed]
46. Buts, J.P.; Dekeyser, N.; Stilmant, C.; Delem, E.; Smets, F.; Sokal, E. Saccharomyces boulardii produces in rat small intestine a novel protein phosphatase that inhibits Escherichia coli endotoxin by dephosphorylation. *Pediatr. Res.* **2006**, *60*, 24–29. [CrossRef] [PubMed]
47. Czerucka, D.; Roux, I.; Rampal, P. Saccharomyces boulardii inhibits secretagogue-mediated adenosine 3′,5′-cyclic monophosphate induction in intestinal cells. *Gastroenterology* **1994**, *106*, 65–72. [CrossRef]
48. Brandão, R.L.; Castro, I.M.; Bambirra, E.A.; Amaral, S.C.; Fietto, L.G.; Tropia, M.J.M.; Neves, M.J.; Dos Santos, R.G.; Gomes, N.C.M.; Nicoli, J.R. Intracellular signal triggered by cholera toxin in Saccharomyces boulardii and Saccharomyces cerevisiae. *Appl. Environ. Microbiol.* **1998**, *64*, 564–568. [CrossRef]
49. Sweeney, D.A.; Hicks, C.W.; Cui, X.; Li, Y.; Eichacker, P.Q. Anthrax infection. *Am. J. Respir. Crit. Care Med.* **2011**, *184*, 1333–1341. [CrossRef]
50. Pontier-bres, R.; Rampal, P.; Peyron, J.; Munro, P.; Lemichez, E.; Czerucka, D. The Saccharomyces boulardii CNCM I-745 Strain Shows Protective Effects against the B. anthracis LT Toxin. *Toxin* **2015**, *7*, 4455–4467. [CrossRef]
51. Albenberg, L.; Esipova, T.V.; Judge, C.P.; Bittinger, K.; Chen, J.; Laughlin, A.; Grunberg, S.; Baldassano, R.N.; Lewis, J.D.; Li, H.; et al. Correlation between intraluminal oxygen gradient and radial partitioning of intestinal microbiota. *Gastroenterology* **2014**, *147*, 1055–1063. [CrossRef]
52. Vacca, I. Microbiome: The microbiota maintains oxygen balance in the gut. *Nat. Rev. Microbiol.* **2017**, *15*, 574. [CrossRef] [PubMed]
53. Rivera-Chávez, F.; Lopez, C.A.; Bäumler, A.J. Oxygen as a driver of gut dysbiosis. *Free Radic. Biol. Med.* **2017**, *105*, 93–101. [CrossRef] [PubMed]
54. Gogineni, V.K.; Morrow, L.E. Probiotics: Mechanisms of Action and Clinical Applications. *J. Probiotics Heal.* **2013**, *1*, 2. [CrossRef]

55. Tiago, F.C.P.; Martins, F.S.; Souza, E.L.S.; Pimenta, P.F.P.; Araujo, H.R.C.; Castro, I.M.; Branda, R.L.; Nicoli, J.R.; Nicoli, J.R. Adhesion to the yeast cell surface as a mechanism for trapping pathogenic bacteria by Saccharomyces probiotics. *J. Med Microbiol.* **2012**, *61*, 1194–1207. [CrossRef] [PubMed]
56. Gedek, B.R. Adherence of Escherichia coli serogroup 0 157 and the Salmonella Typhimurium mutant DT 104 to the surface of Saccharomyces boulardii. *Mycoses* **1999**, *42*, 261–264. [CrossRef] [PubMed]
57. Pontier-bres, R.; Munro, P.; Boyer, L.; Anty, R.; Rampal, P.; Lemichez, E. Saccharomyces boulardii Modifies Salmonella Typhimurium Traffic and Host Immune Responses along the Intestinal Tract. *PLoS ONE* **2014**, *9*, e103069. [CrossRef] [PubMed]
58. Martins, F.S.; Dalmasso, G.; Arantes, R.M.E.; Doye, A.; Lemichez, E.; Lagadec, P.; Imbert, V.; Peyron, J.F.; Rampal, P.; Nicoli, J.R.; et al. Interaction of Saccharomyces boulardii with Salmonella enterica serovar typhimurium protects mice and modifies T84 cell response to the infection. *PLoS ONE* **2010**, *5*, e8925. [CrossRef]
59. Tasteyre, A.; Barc, M.C.; Karjalainen, T.; Bourlioux, P.; Collignon, A. Inhibition of in vitro cell adherence of Clostridium difficile by Saccharomyces boulardii. *Microb. Pathog.* **2002**, *32*, 219–225. [CrossRef]
60. Rigothier, M.C.; Maccario, J.; Gayral, P. Inhibitory activity of saccharomyces yeasts on the adhesion of Entamoeba histolytica trophozoites to human erythrocytes in vitro. *Parasitol. Res.* **1994**, *80*, 10–15. [CrossRef]
61. Pontier-Bres, R.; Prodon, F.; Munro, P.; Rampal, P.; Lemichez, E.; Peyron, J.F.; Czerucka, D. Modification of salmonella typhimurium motility by the probiotic yeast strain saccharomyces boulardii. *PLoS ONE* **2012**, *7*, e33796. [CrossRef]
62. Kline, K.A.; Fälker, S.; Dahlberg, S.; Normark, S.; Henriques-Normark, B. Bacterial Adhesins in Host-Microbe Interactions. *Cell Host Microbe* **2009**, *5*, 580–592. [CrossRef] [PubMed]
63. Cusumano, C.K.; Hultgren, S.J. Bacterial adhesion—A source of alternate antibiotic targets. *IDrugs* **2009**, *12*, 699–705. [PubMed]
64. Gómez Zavaglia, A.; Kociubinski, G.; Pérez, P.; Disalvo, E.; De Antoni, G. Effect of bile on the lipid composition and surface properties of bifidobacteria. *J. Appl. Microbiol.* **2002**, *93*, 794–799. [CrossRef]
65. Kaur, I.P.; Kuhad, A.; Garg, A.; Chopra, K. Probiotics: Delineation of Prophylactic and Therapeutic Benefits. *J. Med. Food* **2009**, *12*, 219–235. [CrossRef]
66. Chen, K.-H.; Miyazaki, T.; Tsai, H.-F.; Bennett, J.E. The bZip transcription factor Cgap1p is involved in multidrug resistance and required for activation of multidrug transporter gene CgFLR1 in Candida glabrata. *Gene* **2007**, *386*, 63–72. [CrossRef]
67. Fidan, I.; Kalkanci, A.; Yesilyurt, E.; Yalcin, B.; Erdal, B.; Kustimur, S.; Imir, T. Effects of Saccharomyces boulardii on cytokine secretion from intraepithelial lymphocytes infected by Escherichia coli and Candida albicans. *Mycoses* **2009**, *52*, 29–34. [CrossRef]
68. Qamar, A.; Aboudola, S.; Warny, M.; Michetti, P.; Kelly, N.P.; Division, G.; Israel, B.; Medical, D. Saccharomyces boulardii Stimulates Intestinal Immunoglobulin A Immune Response to Clostridium difficile Toxin A in Mice. *Infect. Immun.* **2001**, *69*, 2762–2765. [CrossRef] [PubMed]
69. Kyne, L.; Warny, M.; Qamar, A.; Kelly, C.P. Association between antibody response to toxin A and protection against recurrent Clostridium difficile diarrhoea. *Lancet* **2001**, *357*, 189–193. [CrossRef]
70. Rajput, I.R.; Hussain, A.; Li, Y.L.; Zhang, X.; Xu, X.; Long, M.Y.; You, D.Y.; Li, W.F. *Saccharomyces boulardii* and *Bacillus subtilis* B10 Modulate TLRs Mediated Signaling to Induce Immunity by Chicken BMDCs. *J. Cell. Biochem.* **2014**, *115*, 189–198. [CrossRef]
71. Badia, R.; Zanello, G.; Chevaleyre, C.; Lizardo, R.; Meurens, F.; Martínez, P.; Brufau, J.; Salmon, H. Effect of Saccharomyces cerevisiae var. Boulardii and β-galactomannan oligosaccharide on porcine intestinal epithelial and dendritic cells challenged in vitro with Escherichia coli F4 (K88). *Vet. Res.* **2012**, *43*, 4. [CrossRef]
72. Badia, R.; Brufau, M.T.; Guerrero-Zamora, A.M.; Lizardo, R.; Dobrescu, I.; Martin-Venegas, R.; Ferrer, R.; Salmon, H.; Martínez, P.; Brufau, J. β-galactomannan and Saccharomyces cerevisiae var. boulardii modulate the immune response against Salmonella enterica serovar typhimurium in porcine intestinal epithelial and dendritic cells. *Clin. Vaccine Immunol.* **2012**, *19*, 368–376. [CrossRef]
73. Dalmasso, G.; Cottrez, F.; Imbert, V.; Lagadec, P.; Peyron, J.F.; Rampal, P.; Czerucka, D.; Groux, H. Saccharomyces boulardii Inhibits Inflammatory Bowel Disease by Trapping T Cells in Mesenteric Lymph Nodes. *Gastroenterology* **2006**, *131*, 1812–1825. [CrossRef] [PubMed]

74. Moré, M.I.; Vandenplas, Y. Saccharomyces boulardii CNCM I-745 Improves Intestinal Enzyme Function: A Trophic Effects Review. *Clin. Med. Insights Gastroenterol.* **2018**, *11*, 1179552217752679. [CrossRef] [PubMed]
75. Buts, J.P.; De Keyser, N.; Raedemaeker, L. De Saccharomyces boulardii enhances rat intestinal enzyme expression by endoluminal release of polyamines. *Pediatr. Res.* **1994**, *36*, 522–527. [CrossRef] [PubMed]
76. Zaouche, A.; Loukil, C.; De Lagausie, P.; Peuchmaur, M.; Macry, J.; Fitoussi, F.; Bernasconi, P.; Bingen, E.; Cezard, J.P. Effects of oral Saccharomyces boulardii on bacterial overgrowth, translocation, and intestinal adaptation after small-bowel resection in rats. *Scand. J. Gastroenterol.* **2000**, *35*, 160–165. [CrossRef]
77. Buts, J.P.; De Keyser, N.; Marandi, S.; Hermans, D.; Sokal, E.M.; Chae, Y.H.E.; Lambotte, L.; Chanteux, H.; Tulkens, P.M. Saccharomyces boulardii upgrades cellular adaptation after proximal enterectomy in rats. *Gut* **1999**, *45*, 89–96. [CrossRef]
78. Harms, H.-K.; Bertele-Harms, R.-M.; Bruer-Kleis, D. Enzyme-Substitution Therapy with the Yeast *Saccharomyces cerevisiae* in Congenital Sucrase-Isomaltase Deficiency. *N. Engl. J. Med.* **1987**, *316*, 1306–1309. [CrossRef]
79. Buts, J.-P.; De Keyser, N.; Stilmant, C.; Sokal, E.; Marandi, S. Saccharomyces boulardii Enhances N-Terminal Peptide Hydrolysis in Suckling Rat Small Intestine by Endoluminal Release of a Zinc-Binding Metalloprotease. *Pediatr. Res.* **2002**, *51*, 528–534. [CrossRef]
80. Jahn, H.U.; Ullrich, R.; Schneider, T.; Liehr, R.M.; Schieferdecker, H.L.; Holst, H.; Zeitz, M. Immunological and trophical effects of saccharomyces boulardii on the small intestine in healthy human volunteers. *Digestion* **1996**, *57*, 95–104. [CrossRef]
81. Buts, J.P.; Keyser, N. De Transduction pathways regulating the trophic effects of Saccharomyces boulardii in rat intestinal mucosa. *Scand. J. Gastroenterol.* **2010**, *45*, 175–185. [CrossRef]
82. Buts, J.P.; Dekeyser, N. Raf: A key regulatory kinase for transduction of mitogenic and metabolic signals of the probiotic Saccharomyces boulardii. *Clin. Res. Hepatol. Gastroenterol.* **2011**, *35*, 596–597. [CrossRef]
83. Chang, C.; Wang, K.; Zhou, S.N.; Wang, X.D.; Wu, J.E. Protective Effect of Saccharomyces boulardii on Deoxynivalenol-Induced Injury of Porcine Macrophage via Attenuating p38 MAPK Signal Pathway. *Appl. Biochem. Biotechnol.* **2017**, *182*, 411–427. [CrossRef]
84. Schneider, S.M.; Girard-Pipau, F.; Filippi, J.; Hébuterne, X.; Moyse, D.; Hinojosa, G.C.; Pompei, A.; Rampal, P. Effects of Saccharomyces boulardii on fecal short-chain fatty acids and microflora in patients on long-term total enteral nutrition. *World J. Gastroenterol.* **2005**, *11*, 6165–6169. [CrossRef]
85. Girard-Pipau, F.; Pompei, A.; Schneider, S.; Nano, J.L.; Hebuterne, X.; Boquet, P.; Rampal, P. Intestinal microflora, short chain and cellular fatty acids, influence of a probiotic Saccharomyces boulardii. *Microb. Ecol. Health Dis.* **2002**, *14*, 221–228. [CrossRef]
86. Macfarlane, G.T.; Macfarlane, S. Bacteria, colonic fermentation, and gastrointestinal health. *J. AOAC Int.* **2012**, *95*, 50–60. [CrossRef]
87. Fukuda, S.; Toh, H.; Hase, K.; Oshima, K.; Nakanishi, Y.; Yoshimura, K.; Tobe, T.; Clarke, J.M.; Topping, D.L.; Suzuki, T.; et al. Bifidobacteria can protect from enteropathogenic infection through production of acetate. *Nature* **2011**, *469*, 543–549. [CrossRef]
88. Wannaprasat, W.; Koowatananukul, C.; Ekkapobyotin, C.; Chuanchuen, R. Quality analysis of commercial probiotic products for food animals. *Southeast Asian J. Trop. Med. Public Health* **2009**, *40*, 1103–1112.
89. Salminen, M.K.; Rautelin, H.; Tynkkynen, S.; Poussa, T.; Saxelin, M.; Valtonen, V.; Jarvinen, A. Lactobacillus Bacteremia, Species Identification, and Antimicrobial Susceptibility of 85 Blood Isolates. *Clin. Infect. Dis.* **2006**, *42*, e35–e44. [CrossRef] [PubMed]
90. Karen, M.; Yuksel, O.; Akyürek, N.; Ofluoğlu, E.; Çağlar, K.; Şahin, T.T.; Paşaoğlu, H.; Memiş, L.; Akyürek, N.; Bostanci, H. Probiotic Agent Saccharomyces boulardii Reduces the Incidenceof Lung Injury in Acute Necrotizing Pancreatitis Induced Rats. *J. Surg. Res.* **2010**, *160*, 139–144. [CrossRef] [PubMed]
91. Lessard, M.; Dupuis, M.; Gagnon, N.; Nadeau, É.; Matte, J.J.; Goulet, J.; Fairbrother, J.M. Administration of Pediococcus acidilactici or Saccharomyces cerevisiae boulardii modulates development of porcine mucosal immunity and reduces intestinal bacterial translocation after Escherichia coli challenge. *J. Anim. Sci.* **2009**, *87*, 922–934. [CrossRef] [PubMed]
92. Byron, J.K.; Clemons, K.V.; McCusker, J.H.; Davis, R.W.; Stevens, D.A. Pathogenicity of Saccharomyces cerevisiae in complement factor five-deficient mice. *Infect. Immun.* **1995**, *63*, 478–485. [CrossRef] [PubMed]
93. Mcfarland, L.V.; Surawicz, C.M.; Elmer, G.W.; Moyer, K.A.; Melcher, S.A.; Fekety, R.; Bowen, K.E.; Cox, J.L.; Noorani, Z.; Harrington, G.; et al. A Randomized Placebo-Controlled Trial of Saccharomyces boulardii in

Combination With Standard Antibiotics for Clostridium difficile Disease. *JAMA J. Am. Med. Assoc.* **1994**, *271*, 1913–1918. [CrossRef]

94. Thygesen, J.B.; Glerup, H.; Tarp, B. Saccharomyces boulardii fungemia caused by treatment with a probioticum. *BMJ Case Rep.* **2012**, *2012*, bcr0620114412. [CrossRef] [PubMed]
95. Appel-da-Silva, M.C.; Narvaez, G.A.; Perez, L.R.R.; Drehmer, L.; Lewgoy, J. Saccharomyces cerevisiae var. boulardii fungemia following probiotic treatment. *Med. Mycol. Case Rep.* **2017**, *18*, 15–17. [CrossRef] [PubMed]
96. Cassone, M.; Serra, P.; Mondello, F.; Girolamo, A.; Scafetti, S.; Pistella, E.; Venditti, M. Outbreak of Saccharomyces cerevisiae Subtype boulardii Fungemia in Patients Neighboring Those Treated with a Probiotic Preparation of the Organism. *J. Clin. Microbiol.* **2003**, *41*, 5340–5343. [CrossRef] [PubMed]
97. Enache-Angoulvant, A.; Hennequin, C. Invasive Saccharomyces Infection: A Comprehensive Review. *Clin. Infect. Dis.* **2005**, *41*, 1559–1568. [CrossRef]
98. Dinleyici, E.C.; Kara, A.; Ozen, M.; Vandenplas, Y. Saccharomyces boulardii CNCM I-745 in different clinical conditions. *Expert Opin. Biol. Ther.* **2014**, *14*, 1593–1609. [CrossRef]
99. Pineton de Chambrun, G.; Neut, C.; Chau, A.; Cazaubiel, M.; Pelerin, F.; Justen, P.; Desreumaux, P. A randomized clinical trial of Saccharomyces cerevisiae versus placebo in the irritable bowel syndrome. *Dig. Liver Dis.* **2015**, *47*, 119–124. [CrossRef]

Journal of *Fungi*

Review

Application of Probiotic Yeasts on *Candida* Species Associated Infection

Lohith Kunyeit [1,2,3], Anu-Appaiah K A [1,2] and Reeta P. Rao [3,*]

1 Department of Microbiology and Fermentation Technology, CSIR- Central Food Technological Research Institute (CFTRI), Mysuru 570020, India; lohichanthala@yahoo.in (L.K.); anuappaiah@cftri.res.in (A.K.A.)
2 Academy of Scientific and Innovative Research (AcSIR), Ghaziabad 201002, India
3 Department of Biology and Biotechnology, Worcester Polytechnic Institute, Worcester, MA 01609, USA
* Correspondence: rpr@wpi.edu; Tel.: +1-508-831-5000

Received: 8 August 2020; Accepted: 19 September 2020; Published: 25 September 2020

Abstract: Superficial and life-threatening invasive *Candida* infections are a major clinical challenge in hospitalized and immuno-compromised patients. Emerging drug-resistance among *Candida* species is exacerbated by the limited availability of antifungals and their associated side-effects. In the current review, we discuss the application of probiotic yeasts as a potential alternative/ combination therapy against *Candida* infections. Preclinical studies have identified several probiotic yeasts that effectively inhibit virulence of *Candida* species, including *Candida albicans*, *Candida tropicalis*, *Candida glabrata*, *Candida parapsilosis*, *Candida krusei* and *Candida auris*. However, *Saccharomyces cerevisiae* var. *boulardii* is the only probiotic yeast commercially available. In addition, clinical studies have further confirmed the in vitro and in vivo activity of the probiotic yeasts against *Candida* species. Probiotics use a variety of protective mechanisms, including posing a physical barrier, the ability to aggregate pathogens and render them avirulent. Secreted metabolites such as short-chain fatty acids effectively inhibit the adhesion and morphological transition of *Candida* species. Overall, the probiotic yeasts could be a promising effective alternative or combination therapy for *Candida* infections. Additional studies would bolster the application of probiotic yeasts.

Keywords: *Candida albicans*; non-albicans *Candida* species; *Candida auris*; *Saccharomyces boulardii*; *Saccharomyces cerevisiae*; aromatic alcohols

1. Introduction

The fermented foods are a rich source of beneficial microorganisms, and they have a long history of exhibiting health benefits, particularly *S. cerevisiae* and lactic acid bacteria (LAB). Their safety is evidenced by consumption of fermented foods and beverages over centuries. Today, it is well accepted that the rich microbial profile of fermented food provides more than just nutrition. For example, functional activity of microorganisms in food helps enhance the bio-availability of micronutrients, improving the sensory quality and shelf life of the food, degrading anti-nutritive factors (such as trypsin inhibitors and phytate degradation), enriching antioxidant and antimicrobial compounds, and fortifying health-promoting bioactive compounds [1,2]. These attractive microbial activities in the fermented foods have been a draw in the field of probiotics.

Characteristics of bacterial strains such as *Lactobacillus* and *Bifidobacterium* species have been extensively studied and commercially available as probiotic supplements. Yeasts, which are also common in fermented foods, remain largely unexplored for probiotic potential. We and other researchers have observed that yeasts that originate from fermented sources such as apple cider, wine, fermented coconut palm, and fermented dairy products survive the harsh condition of the gastrointestinal (GI) tract and retain the ability to attach to intestinal epithelium [3–5]. More recently, live bacteria have been used in fecal transplants to prevent and/or treat several GI complications [6]. The probiotic bacteria, such as lactic acid bacteria (LAB) and *Bifidobacterium* species, have effectively treated several GI complications, including candidiasis [7,8]. However, other than *Saccharomyces boulardii*, potential probiotic yeasts such as *S. cerevisiae* and several other non-*Saccharomyces* yeasts are largely unexplored use as biotherapeutics, specifically for *Candida* infections. In reviewing the current literature here, we focus on the biotherapeutic potential and mechanism(s) of action of beneficial yeasts against *Candida* infections.

The vast majority of fungal infections are caused by *Candida albicans*, a polymorphic commensal yeast as well as some non-albicans *Candida* species. Disease range from superficial infections, such as cutaneous and mucosal, to life-threatening bloodstream infections (BSI), or invasive deep tissue infections. Superficial infections usually affect the nails, skin, and mucosal membrane of the host and are recalcitrant to treatment. For example, vulvovaginal candidiasis (VVC) has infected 75% of women population at least once in their lifetime. Furthermore, a small population (5–8%) suffers from at least four recurrent VVC per year [9].

Bloodstream infection (BSI) and other invasive *Candida* infections cause high morbidity and mortality especially among immune-compromised patients [10]. *Candida* species are the fourth-leading cause of nosocomial infections in the world, and *Candida* BSI attributes to 35% mortality rate in all the *Candida* associated infections [11]. Furthermore, the National Nosocomial Infection Surveillance System (NNIS), USA, has revealed total 27,200 nosocomial infections between January 1980 through April 1990, among these *C. albicans* and non-albicans *Candida* species were involved total 19,621 (72%) of the overall infections [12].

Though *C. albicans* is a major commensal yeast flora of the GI tract, non-albicans *Candida* species such as *Candida glabrata*, *Candida tropicalis*, *Candida parapsilosis*, and *Candida krusei* have been frequently identified in a healthy individual's gut. On the other hand, among 15–20 pathogenic non-albicans *Candida* species, *Candida glabrata*, *Candida tropicalis*, *Candida parapsilosis*, and *Candida krusei* are predominant constituting 35–65% of the overall infections [13]. As an opportunistic pathogen, certain groups of immune-compromised individuals have a higher susceptibility towards *Candida* infection. Invasive *Candida* infections are also closely associated with advanced medical techniques such as medical implants and stents [14]. For instance, the patients who are on antibiotic therapy and chemotherapy, central venous catheters, total parenteral nutrition, extensive surgery, burns, renal failure and hemodialysis, or mechanical ventilation are at a major risk for superficial and invasive *Candida* infections [14].

2. Morphological Transition and Metabolic Flexibility Promote Virulence of *Candida* In Vivo

As a polymorphic yeast, *C. albicans* and few non-albicans *Candida* strains, such as *C. tropicalis* and *C. glabrata*, exhibit multiple morphological structures such as yeast form, germ tubes, pseudo-hyphae, and/or hyphae that play a key role in the infection. For example, filamentous morphology is well-known for epithelial invasion and is primarily involved in biofilm formation [15]. Yeast form cells are planktonic and are important for dissemination. Once they attach, they initiate germ tubes, pseudo-hyphae, and/or hyphae that enhance adhesion to surfaces. Attachment to abiotic surfaces initiates biofilm formation. Biofilms on implanted medical devises may lead to invasive fungal infections—a major risk factor for *Candida* infection-associated mortality [16]. Attachment to live cells (such as epithelium) causes damage, evokes an immune response and ultimately gains access to deeper tissues. Therefore, the polymorphism of *Candida* is an important consideration in its infectious outcomes.

The host's innate immunity is a major factor in fungal clearance, normally through a process called phagocytosis where immune cells ingest and biochemically eliminate the pathogens [17]. However, the switch from yeast to filamentous form is a common escape mechanism of *Candida* species [18]. Therefore, *C. albicans* filament has less susceptibility for phagocytosis by innate immune cells than the yeast form [19]. In addition, metabolic flexibility of *C. albicans* facilitates colonization by adapting to varying nutritional availability [20]. For instance, in case of *Candida* meningoencephalitis (*Candida* infection in brain tissue), glucose and vitamins are the major nutrient sources for the pathogen, while in liver, it utilizes glycogen as a nutrient source [9]. A study revealed that adaptation to alternative carbon sources such as lactate and other nutrient sources increased environment stress response and virulence [21]. All of these attributes make *C. albicans* and non-albicans *Candida* species a unique pathogen among the microbial community.

3. Drug Resistance Is a Major Hurdle to Antifungal Therapy

Antifungal drugs used to treat *Candida* associated infections, work either by killing or inhibiting the growth of *Candida* species. A sparse number of antifungal classes such as polyenes, azoles, and echinocandins are used depending on conditions of invasive *Candida* infections [22]. Multiple *Candida* strains have already developed resistance to these drugs making this a public health threat [23]. For example, surveillance data from health-care facilities revealed widespread fluconazole resistance among clinical isolates of both *C. albicans* and non-albicans *Candida* strains [24–26]. Azoles such as fluconazole is a first-line antifungal drug that is used extensively for therapy and prophylaxis against *Candida* infections. Several resistant mechanisms have been connected to drug-resistant *Candida* species including overexpression of drug efflux pumps, alteration in drug targets, and changes in membrane sterol composition [22]. The structural heterogeneity of *Candida* biofilm has a major significance in clinical context due to higher resistance against most antifungal agents. Furthermore, these drugs can be toxic for the patients with several side effects that include GI disturbances, hepatotoxicity, and neurotoxicity due to their target resemblance to its host cell, antifungal metabolism in liver and cross drug interaction in the host [27,28].

More recently, multi-drug resistant *Candida auris* has emerged as a "super bug" posing significant clinical challenges and a major threat to public health. *Candida auris*, is often involved in the nosocomial bloodstream infection world-wide [29]. *C. auris* has been shown to last in the hospital settings and spread from person-to-person by direct contact or contaminated surfaces [23]. In addition, *C. auris* is closely related with *Candida haemulonii* and is often misidentified as such. This requires a specialized laboratory method for identification [23], further delaying implementation of infection control. Therefore, now more than ever, there is an urgent need for effective alternatives to conventional modes to treat *Candida* infections.

Some attempts have been made to using specific diets that avoid high sugar-containing food such as bakery products, milk, and dairy product. The claim is that it reduces *Candida* colonization of the GI tract [30]. Intestinal overgrowth of *C. albicans* contributes to Crohn's disease that affects 1.6 million Americans [31,32]. *C. albicans* overgrowth is caused by an imbalance in the intestinal microbiota and host immune status. To restore the balance and modulate host immunity, foods rich in antioxidants and other nutritional supplements have been suggested [30,33]. More recently, studies on the human microbiome have opened new insight into the role of the resident gut microbiota in physical health and mental wellbeing. Applications of beneficial microbes as fecal transplants [34] or fermented milk products [35] for the treatment of irritable bowel syndrome (IBS) and irritable bowel disease (IBD) has gained traction. Here we discuss the potential of probiotic yeasts against *Candida* virulence and pathogenesis.

4. Use of Probiotics as Biotherapeutics

As stated by Hippocrates, "let food be thy medicine and medicine be thy food". Today, the idea of food and/or diet is not just extended towards mere survival or hunger satisfaction. The health-conscious population deeply cares about additional aspects including health improvement and prevention of diseases. In this context, functional food plays a significant role where, the concept of food has not only intended to provide humans with necessary nutrients, but also to prevent diseases and increase physical and mental well-being. Probiotic, considered as a functional food, is mostly consumed in the form of traditional fermented food products such as milk products, fermented vegetables, and meats [36].

Probiotics are defined as "live microorganisms which, when consumed in adequate amounts, confer health benefits on the host" [37]. The archived scientific documents have explained the diverse positive effects of probiotics on a wide range of diseases and disorders including lactose indigestion, diarrhea, immune modulation, inflammatory bowel syndrome, constipation, infection, allergy, serum cholesterol, blood pressure, and reduction of urinary tract infections [38]. In addition, the Human Microbiome Project by National Institute of Health (NIH), USA, changed the views on beneficial microbial research; it exposed the influence of gut microbiome and human health during various infections and disease conditions including, mental health.

5. Interaction of Probiotics Yeast and *Candida* Species

Several reports suggest that probiotic bacteria are effective against GI complications such as diarrhea, leaky gut syndrome, as well as *Helicobacter pylori* and *Clostridium difficile* infections [39,40]. However, *Saccharomyces cerevisiae* var. *boulardii* is the only yeast currently available for human use as probiotics. Its efficacy against *Candida* has been explored previously. Specifically, pathogen-free mice that were infected with *C. albicans* and subsequently treated with *S. boulardii* prevented the translocation of *Candida* to internal organs [41–43]. These groups further confirmed that *S. boulardii* effectively reduced *C. albicans* translocation colonization and inflammation in in vivo models.

Clinical reports around the use of probiotic yeasts are limited. One study, reports that oral administration of *S. boulardii* to infants reduced the fungal colonization and invasive fungal infections [44]. Another study conducted in preteen children focused on the effects of probiotics against *Candida* infection. They used a probiotic cocktail of yeast and bacteria in combination with prebiotics and demonstrated a reduction in colonization of *C. albicans* [45].

6. Probiotic Yeasts Exhibit Multiple Inhibitory Mechanisms against *Candida* Species

Pre-clinical and/or clinical studies indicate that *S. boulardii* and other potential probiotic yeasts ameliorate complications associated with *Candida* infection by mechanisms outlined in Table 1. However, there was a lack of specific mechanistic insights on how these probiotic yeasts interact with *Candida* species especially in the context of a live host. Pathogens in GI tract induce necrosis and apoptosis of intestinal epithelia by reducing the production of mucin or its degradation. Pathogens also downregulate IgA and other proteins of the tight junction thereby increasing intestinal permeability [46,47]. *S. boulardii* has been shown to increase IgA production in *Clostridium difficile* colitis and antibiotic-associated diarrhea in mice model [48]. *S. boulardii* also decreases epithelial necrosis, apoptosis, and increases the production of antioxidant enzymes such as superoxide dismutase, catalase, and glutathione peroxidase in mouse models in necrotizing enterocolitis in mice [49]. In addition, *S. boulardii* activates the intestinal epithelial restoration in GI tract [50]. Together these cellular responses may contribute to its beneficial properties and prevent *Candida* infection.

Table 1. List of probiotic yeasts and its mechanisms against virulence and pathogenesis of *Candida* species.

Probiotic Yeast Strains	Mechanisms of Probiotic Yeasts against *Candida* Species Virulence and Pathogenesis
S. boulardii	• Inhibits *C. albicans* and non-albicans *Candida* species include *C. tropicalis, C. krusei, C. parapsilosis, C. glabrata,* and *C. auris* adhesion, biofilm formation and/or filamentation, in in vitro, ex vivo, in vivo, and clinical settings [51,52] • Secrets small bioactive molecules [52,53] • Reduces inflammatory cytokines TNFα and INF γ in colon epithelial [41] • Prevents the *C. albicans* translocation in GI tract [43]

Table 1. *Cont.*

Probiotic Yeast Strains	Mechanisms of Probiotic Yeasts against *Candida* Species Virulence and Pathogenesis
S. cerevisiae *	• Inhibits *C. albicans* and non-albicans *Candida* species adhesion, colonization, biofilm formation, and filamentation in in vitro, ex vivo, and in vivo models [53,54] • Inhibits the *Candida* adhesion to epithelial cells by initiation of co-aggregation [54] • *S. cerevisiae* form a barrier over the biotic surfaces and inhibits the *Candida* adhesion [54] • Reduces virulence gene expressions of *C. albicans* during infection [54] • Secrets bioactive molecules [53] • Decreases the pro-inflammatory cytokine TNF-α and enhances IL-10 expressions in the host [55] • Decreases the colorization and host cell damage during the infection [53,54] • β-glucan decreased intestinal inflammation [55]
I. occidentalis *	• Inhibits *C. albicans* and non-albicans *Candida* species such as *C. tropicalis, C. krusei, C. parapsilosis, C. glabrata,* and *C. auris* adhesion, colonization, biofilm formation and/ or filamentation in vitro, ex vivo, in vivo models [53] • An unidentified metabolite (s) inhibits virulence of non-albicans *Candida* species [53]

* Potential probiotic yeast, not commercialized.

6.1. *Immunogenic Response and Anti-Virulence Ability of Probiotic Yeasts*

Since resistance to antifungal drugs has emerged as a significant problem, researchers have explored alternative means of treating recalcitrant fungal infections. Modulation of host immunity is one avenue that is being considered as an alternative [56,57]. For example, *S. boulardii* has been shown to reduce pro-inflammatory cytokines such as IL-1β and TNF, and increase anti-inflammatory cytokines IL-4 and IL-10 during *Candida* infection [42,58]. Other alternative therapies target virulence strategies such as adhesion and filamentation of *C. albicans* [59]. These maybe used to treat abiotic surfaces to deter microbes from binding. Probiotics also have the ability to inhibit virulence factors of the pathogen. We and others have demonstrated that cells, as well as the cell-free secretome of probiotic yeasts such as *S. boulardii, S. cerevisiae,* and a non-*Saccharomyces* yeast *Issatchenkia occidentalis* inhibit adhesion, filamentation, and biofilm development of *C. albicans* [52] and other non-albicans *Candida* species such as *C. tropicalis, C. krusei, C. glabrata,* and *Candida parapsilopsis.* Biofilms are complex multispecies structures that include *C. albicans* among other microbes [60,61]. Probiotics yeasts have been shown to be effective against fungal biofilms composed of *C. albicans* and non-albicans *Candida* species [53]; however, no studies have been focused on their efficacy on cross-kingdom biofilms. These studies implicated the involvement of yeast metabolite(s) in inhibiting adhesion and/ or morphological transition in vitro [53]. These studies also indicate that probiotic yeast affect a broad spectrum and not limited to *C. albicans;* rather, it can inhibit virulence across the *Candida* genus.

Cultured intestinal epithelial models such as Caco-2, Intestin 407 and HT-29 have been extensively used to study microbial interactions or host-microbe interactions. These cell lines recapitulate various features of the intestinal epithelial surface including the formation of villi, production of mucus, and antibodies such as IgA [62]. We and others have demonstrated that probiotic yeasts effectively reduce adhesion of *C. albicans* and non-albicans *Candida* species to these cultured epithelial cell lines [52,53]. In addition, yeast *S. boulardii* has been shown to pose a barrier and preserve the integrity of the epithelium by the reduction of pro-inflammatory cytokines in the intestine [40,52].

Even though live probiotic cells are known to play a significant role in preventing virulence of *C. albicans*, the role of exact cellular components involved are less investigated. For example, administration of cell wall components of *S. cerevisiae* reduced the *Candida* associated inflammation and colonization in animal models [55]. Interestingly, one of the four *S. cerevisiae* strains used in this study (strain Sc-4) increased the mortality and inflammation in the host, suggesting strain-specific effects of the probiotic yeasts against *Candida* species [55]. Such strain specificity as also been reported in the interaction of *Lactobacillus* strains with *C. albicans* [63]. Furthermore, heat-killed *S. cerevisiae* reduced the vaginal colonization of *C. albicans* when applied against vaginal candidiasis in a murine model [54]. These effects could either mediated by yeast cell wall components such as β-glucan or simply that the biomass of heat-inactivated probiotic cells form a physical barrier that occludes host factors that facilitate *C. albicans* attachment (Figure 1A) [54].

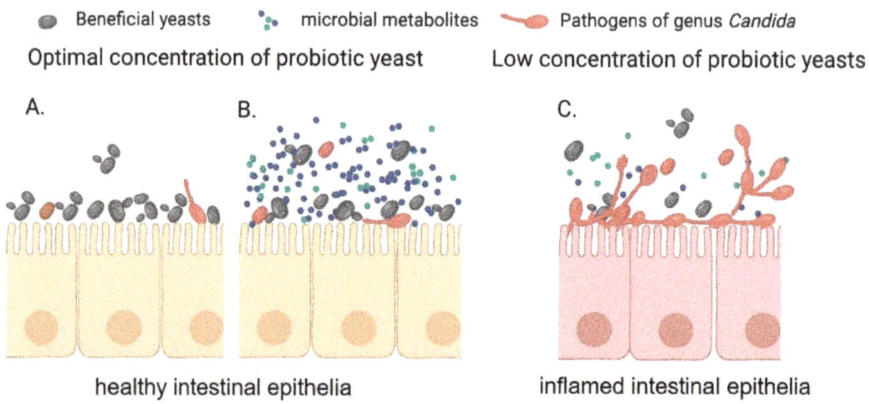

Figure 1. Probiotic yeast either form a physical barrier on epithelial surfaces (**A**) or secretes bioactive metabolite (**B**) to inhibit the adhesion and morphological transition of *Candida* species on epithelial cells. Further, suitable probiotic yeasts cell number is required for the effective inhibition of *Candida* virulence in the host GI tract (**C**).

6.2. Role of Small Bioactive Metabolites in Probiotic Action

Beneficial microbes or probiotics in the intestine are thought to control pathogen overgrowth by competing for limited nutrients. There is a growing body of literature that supports the notion that inhibitory function is primarily mediated by secreted small molecules with suitable probiotic cell number (Figure 1B,C) [53,64]. Microorganisms produce metabolites that have been shown to alter the course of an infection by synergistic or antagonistic interactions with infectious agents. Such metabolites include hydrogen peroxide, bacteriocins, and organic acids that effectively inhibit the virulence and growth of various *Candida* species [64,65] (Table 2). On the other hand, few interesting microbial metabolites, such as tyrosol and indole-3 acetic acid, trigger the filamentation in *C. albicans* [66,67]. Small molecules derived from bacteria have been evaluated for activity against *Candida* virulence and pathogenesis. For example, lectins of lactobacilli and bifidobacterial strains isolated from humans have been shown to inhibit the growth of drug-resistant *C. albicans* [68]. The Gram-positive pathogenic bacteria, *Enterococcus faecalis*, produces a peptide called EntV which has been shown to reduce *C. albicans* virulence [69]. Furthermore, organic acids such as acetic acid and lactic acid have been shown to enhance antifungal treatment of *C. albicans* and *C. glabrata* [70]. Many *Lactobacillus*, *Bifidobacterium*, and yeasts strains produce these organic acids. *S. boulardii* produces several bioactive compounds such as *Saccharomyces* anti-inflammatory factor (SAIF), anti-toxin factors, short-chain fatty acids, bioactive proteins of 54 kDa, and 120 kDa which play a major role in preventing bacterial infections [38,71]. However, there has been very limited knowledge on probiotic yeast metabolites on *Candida* species. Recently a group showed that yeast *S. boulardii* metabolite capric acid (Decanoic acid)—a saturated fatty acid, inhibits the filamentation of *C. albicans* interaction [52].

In natural habitats, potential interaction of microbial communities has been a key element for the ecological dynamics. Bacteria and eukaryotic microorganisms exhibit both symbiotic and/or antagonistic interaction in the natural environment. In fact, *C. albicans* co-exists with other non-albicans *Candida* species or bacteria in the biofilm as well as the human GI tract. These inter-species interactions between C. albicans and other microbes typically affect filamentation of *C. albicans*. For instance, certain secretory molecules of *Salmonella typhimurium* and *Streptococcus mutants* inhibit cell growth and filamentation of *C. albicans* in the co-culture conditions [72,73]. Another well studied bacterium is *Pseudomonas aeruginosa*, where bacterial toxin phenazine inhibits the filamentation of *C. albicans* [74,75].

The morphological transition of yeast has been controlled by cell density and/or quorum sensing molecules. Apart from bacteria, the quorum sensing mechanism is also well studied in yeast such as *C. albicans* and *S. cerevisiae*. Farnesol and tyrosol are known cell density molecules in *C. albicans* which controls the morphological transition. Similarly, yeasts such as *S. cerevisiae* and other many non-*Saccharomyces* yeast produce alcoholic signaling molecules called phenylethanol and tryptophol. An abundant usage and availability of well-curated genetic database indicate that *S. cerevisiae* has gained more attention on quorum sensing mechanisms than the non-*Saccharomyces* yeast strains. There are few studies claiming that factors such as low nitrogen content and cell density play a significant role in the production of phenylethanol and tryptophol in *S. cerevisiae* and regulates its morphological transition mechanism [76]. Furthermore, these signal molecules are controlled by the expression of *ARO8*, *ARO9*, and *ARO10*, where *ARO8* and *ARO9* encode the aromatic aminotransferases and *ARO10* encodes the aromatic decarboxylase reaction [77,78].

Table 2. Microbial metabolites and its functions against *Candida* species.

Microbial Strains	Bioactive Metabolite	Functions
S. boulardii [52]	Short-chain fatty acids (capric acid)	Filamentation inhibition, and antifungal activity against *C. albicans*
S. cerevisiae [53,54]	Unknown	Adhesion and filamentation inhibition
I. occidentalis [53]	Unknown	Adhesion and filamentation inhibition
Lactobacillus acidophilus, L. crispatus, L. vaginalis [65,68]	Lectins, hydrogen peroxide, lactic acid	Inhibit cell growth of *C. albicans*
Bifidobacterium adolescentis, B. bifidum, B. gallinarum [68]	Lectins	Inhibit cell growth of *C. albicans*
Enterococcus faecalis [69]	Peptide EntV	Filamentation inhibition
Pseudomonas aeruginosa [75,79]	Phenazine, 3-oxo-C12 homoserine lactone	Filamentation inhibition
Salmonella typhimurium [72]	Unknown	Inhibit cell growth and filamentation
Streptococcus mutants [73]	Unknown	Inhibit cell growth and filamentation

Several research groups have predicted and/or observed an antagonistic nature of aromatic alcohols, phenylethanol, and tryptophol against fungi. Winters et al., (2019) reported that high concentrations of *S. cerevisiae* inhibited non-*Saccharomyces* strains in mixed cultures and under fermentation conditions [78]. Although there were direct evidence of inhibition due to these secondary metabolites, commercially procured phenylethanol and tryptophol have been shown to inhibit filamentation of *C. albicans* [77]. This result is bolstered by the observation that administration of tryptophol enhances survival of *Galleria mellonella* larval that are infected with *Candida* [80]. Furthermore, a cocktail of phenylethanol, isoamyl alcohol, E-nerolidol, and farnesol provides protection against *Candida* infection in a murine model of infection [81]. Together these studies establish a paradigm for inhibition of fungal virulence that is mediated by aromatic alcohols.

7. Gaps in our Understanding of Biotherapeutic Application of Probiotics for *Candida* Infection

Probiotic yeasts yield several positive outcomes in in vitro, ex vivo, and in vivo readouts during colonization of *Candida* species. Information about their effect during systemic infection is an area that needs further investigation. Numerous animal and handful of clinical experiments have revealed that probiotics and metabolites such as short-chain fatty acids, tryptophol and phenylethanol play an abundant role in human health and diseases. However, the origin of these metabolites is ill-defined and their effects on clinical manifestations of *Candida* infection need further investigation. These studies would provide substantive information to improve biotherapeutic properties of beneficial microbes against *Candida* infections.

Emergence of drug resistance and complications associated with side effects have sparked interest in alternative therapies. Applications of food-derived yeasts have been shown to have positive outcomes against *C. albicans* and non-albicans *Candida* species virulence and infection in pre-clinical and clinical settings. Food-derived beneficial yeasts are also generally safe and pose an effective alternative to traditional antifungals. They may also be used in combination therapy with conventional antifungal drugs since the synergistic effect of probiotics and antifungal agents would prevent emergence of drug resistance.

Funding: This research was funded by NIH-NCCIH, grant number NIH-NCCIH 1R15AT009926-01, DST-INSPIRE program, Department of Science and Technology, Government of India, award number DST/INSPIRE Fellowship/2013/553 and Fulbright-Nehru doctoral fellowship, United States–India Education Foundation (USIEF), India, award number 2310/DR/2018-2019.

Acknowledgments: We thank the Director, CSIR-Central Food Technological Research Institute (CFTRI) for encouragement and research support. This work is partially supported by NIH-NCCIH 1R15AT009926-01 grant to RPR. LK is grateful to the INSPIRE program, Department of Science and Technology, Government of India and Fulbright-Nehru doctoral fellowship, United States–India Education Foundation (USIEF), India for the financial support for his doctoral research.

Conflicts of Interest: The authors declare no conflict of interest.

References

1. Tamang, J.P.; Shin, D.H.; Jung, S.J.; Chae, S.W. Functional Properties of Microorganisms in Fermented Foods. *Front. Microbiol.* **2016**, *7*, 578. [CrossRef]
2. Nkhata, S.G.; Ayua, E.; Kamau, E.H.; Shingiro, J.B. Fermentation and germination improve nutritional value of cereals and legumes through activation of endogenous enzymes. *Food Sci. Nutr.* **2018**, *6*, 2446–2458. [CrossRef]
3. Lohith, K.A.; Anu-Appaiah, K.A. In vitro probiotic characterization of yeasts of food and environmental origin. *Int. J. Probiotics Prebiotics* **2014**, *9*, 1–6.
4. Lohith, K.A.; Anu-Appaiah, K.A. Antagonistic effect of *Saccharomyces cerevisiae* KTP and *Issatchenkia occidentalis* ApC on hyphal development and adhesion of *Candida albicans*. *Med. Mycol.* **2018**, *56*, 1023–1032. [CrossRef]
5. Kumura, H.; Tanoue, Y.; Tsukahara, M.; Tanaka, T.; Shimazaki, K. Screening of dairy yeast strains for probiotic applications. *J. Dairy Sci.* **2004**, *87*, 4050–4056. [CrossRef]
6. Gupta, S.; Allen-Vercoe, E.; Petrof, E.O. Fecal microbiota transplantation: In perspective. *Ther. Adv. Gastroenterol.* **2016**, *9*, 229–239. [CrossRef]
7. Verna, E.C.; Lucak, S. Use of probiotics in gastrointestinal disorders: What to recommend? *Ther. Adv. Gastroenterol.* **2010**, *3*, 307–319. [CrossRef] [PubMed]
8. Wagner, R.D.; Pierson, C.; Warner, T.; Dohnalek, M.; Farmer, J.; Roberts, L.; Hilty, M.; Balish, E. Biotherapeutic effects of probiotic bacteria on candidiasis in immunodeficient mice. *Infect. Immun.* **1997**, *65*, 4165–4172. [CrossRef]
9. Mayer, F.L.; Wilson, D.; Hube, B. *Candida albicans* pathogenicity mechanisms. *Virulence* **2013**, *4*, 119–128. [CrossRef]
10. Delaloye, J.; Calandra, T. Invasive candidiasis as a cause of sepsis in the critically ill patient. *Virulence* **2014**, *5*, 161–169. [CrossRef]
11. Calderone, R.A.; Fonzi, W.A. Virulence factors of *Candida albicans*. *Trends Microbiol.* **2001**, *9*, 327–335. [CrossRef]
12. Jarvis, W.R. Epidemiology of nosocomial fungal infections, with emphasis on Candida species. *Clin. Infect. Dis Off. Publ. Infect. Dis. Soc. Am.* **1995**, *20*, 1526–1530. [CrossRef] [PubMed]
13. Pappas, P.G.; Lionakis, M.S.; Arendrup, M.C.; Ostrosky-Zeichner, L.; Kullberg, B.J. Invasive candidiasis. *Nat. Rev. Dis. Prim.* **2018**, *4*, 18026. [CrossRef] [PubMed]

14. Spampinato, C.; Leonardi, D. Candida infections, causes, targets, and resistance mechanisms: Traditional and alternative antifungal agents. *BioMed Res. Int.* **2013**, *2013*, 204237. [CrossRef]
15. Wilson, D. Candida albicans. *Trends Microbiol.* **2019**, *27*, 188–189. [CrossRef]
16. Taff, H.T.; Mitchell, K.F.; Edward, J.A.; Andes, D.R. Mechanisms of Candida biofilm drug resistance. *Future Microbiol.* **2013**, *8*, 1325–1337. [CrossRef] [PubMed]
17. Munoz, J.F.; Delorey, T.; Ford, C.B.; Li, B.Y.; Thompson, D.A.; Rao, R.P.; Cuomo, C.A. Coordinated host-pathogen transcriptional dynamics revealed using sorted subpopulations and single macrophages infected with *Candida albicans*. *Nat. Commun.* **2019**, *10*, 1607. [CrossRef] [PubMed]
18. Sudbery, P.; Gow, N.; Berman, J. The distinct morphogenic states of *Candida albicans*. *Trends Microbiol.* **2004**, *12*, 317–324. [CrossRef]
19. Wartenberg, A.; Linde, J.; Martin, R.; Schreiner, M.; Horn, F.; Jacobsen, I.D.; Jenull, S.; Wolf, T.; Kuchler, K.; Guthke, R.; et al. Microevolution of *Candida albicans* in macrophages restores filamentation in a nonfilamentous mutant. *PLoS Genet.* **2014**, *10*, e1004824. [CrossRef]
20. Lorenz, M.C.; Fink, G.R. The glyoxylate cycle is required for fungal virulence. *Nature* **2001**, *412*, 83–86. [CrossRef]
21. Ene, I.V.; Adya, A.K.; Wehmeier, S.; Brand, A.C.; MacCallum, D.M.; Gow, N.A.; Brown, A.J. Host carbon sources modulate cell wall architecture, drug resistance and virulence in a fungal pathogen. *Cell. Microbiol.* **2012**, *14*, 1319–1335. [CrossRef] [PubMed]
22. Berman, J.; Krysan, D.J. Drug resistance and tolerance in fungi. *Nat. Rev. Microbiol.* **2020**, *18*, 319–331. [CrossRef] [PubMed]
23. Centers for Disease Control and Prevention (CDC). Candida auris. Available online: https://www.cdc.gov/fungal/candida-auris/candida-auris-qanda.html (accessed on 22 July 2020).
24. Shi, C.; Liu, J.; Li, W.; Zhao, Y.; Meng, L.; Xiang, M. Expression of fluconazole resistance-associated genes in biofilm from 23 clinical isolates of *Candida albicans*. *Braz. J. Microbiol.* **2019**, *50*, 157–163. [CrossRef]
25. Rybak, J.M.; Muñoz, J.F.; Barker, K.S.; Parker, J.E.; Esquivel, B.D.; Berkow, E.L.; Lockhart, S.R.; Gade, L.; Palmer, G.E.; White, T.C.; et al. Mutations in tac1b: A novel genetic determinant of clinical fluconazole resistance in Candida auris. *mBio* **2020**, *11*, e00365-20. [CrossRef]
26. Castanheira, M.; Deshpande, L.M.; Messer, S.A.; Rhomberg, P.R.; Pfaller, M.A. Analysis of global antifungal surveillance results reveals predominance of Erg11 Y132F alteration among azole-resistant *Candida parapsilosis* and *Candida tropicalis* and country-specific isolate dissemination. *Int. J. Antimicrob. Agents* **2020**, *55*, 105799. [CrossRef]
27. Lewis, R.E. Current concepts in antifungal pharmacology. *Mayo Clin. Proc.* **2011**, *86*, 805–817. [CrossRef]
28. Wang, J.L.; Chang, C.H.; Young-Xu, Y.; Chan, K.A. Systematic review and meta-analysis of the tolerability and hepatotoxicity of antifungals in empirical and definitive therapy for invasive fungal infection. *Antimicrob. Agents Chemother.* **2010**, *54*, 2409–2419. [CrossRef]
29. Ben-Ami, R.; Berman, J.; Novikov, A.; Bash, E.; Shachor-Meyouhas, Y.; Zakin, S.; Maor, Y.; Tarabia, J.; Schechner, V.; Adler, A.; et al. Multidrug-Resistant *Candida haemulonii* and *Candida auris*, Tel Aviv, Israel. *Emerg. Infect. Dis.* **2017**, *23*, 195. [CrossRef]
30. Martins, N.; Ferreira, I.C.; Barros, L.; Silva, S.; Henriques, M. Candidiasis: Predisposing factors, prevention, diagnosis and alternative treatment. *Mycopathologia* **2014**, *177*, 223–240. [CrossRef]
31. Loftus, E.V., Jr. Clinical epidemiology of inflammatory bowel disease: Incidence, prevalence, and environmental influences. *Gastroenterology* **2004**, *126*, 1504–1517. [CrossRef]
32. Molodecky, N.A.; Soon, S.; Rabi, D.M.; Ghali, W.A.; Ferris, M.; Chernoff, G.; Benchimol, E.I.; Panaccione, R.; Ghosh, S.; Barkema, H.W.; et al. Increasing incidence and prevalence of the inflammatory bowel diseases with time, based on systematic review. *Gastroenterology* **2012**, *142*, 46–54 e42. [CrossRef]
33. Rehaume, L.M.; Jouault, T.; Chamaillard, M. Lessons from the inflammasome: A molecular sentry linking *Candida* and Crohn's disease. *Trends Immunol.* **2010**, *31*, 171–175. [CrossRef]
34. Xu, D.; Chen, V.L.; Steiner, C.A.; Berinstein, J.A.; Eswaran, S.; Waljee, A.K.; Higgins, P.D.; Owyang, C. Efficacy of fecal microbiota transplantation in irritable bowel syndrome: A systematic review and meta-analysis. *Am. J. Gastroenterol.* **2019**, *114*, 1043–1050. [CrossRef]
35. Simren, M.; Ohman, L.; Olsson, J.; Svensson, U.; Ohlson, K.; Posserud, I.; Strid, H. Clinical trial: The effects of a fermented milk containing three probiotic bacteria in patients with irritable bowel syndrome—A randomized, double-blind, controlled study. *Aliment. Pharmacol. Ther.* **2010**, *31*, 218–227. [CrossRef]

36. Suryabhan, P.; Lohith, K.; Anu-Appaiah, K.A. Sucrose and sorbitol supplementation on maltodextrin encapsulation enhance the potential probiotic yeast survival by spray drying. *LWT* **2019**, *107*, 243–248. [CrossRef]
37. Morelli, L.; Capurso, L. FAO/WHO guidelines on probiotics: 10 years later. *J. Clin. Gastroenterol.* **2012**, *46*, S1–S2. [CrossRef]
38. Vandenplas, Y.; Brunser, O.; Szajewska, H. Saccharomyces boulardii in childhood. *Eur. J. Pediatr.* **2009**, *168*, 253–265. [CrossRef]
39. Tung, J.M.; Dolovich, L.R.; Lee, C.H. Prevention of *Clostridium difficile* infection with *Saccharomyces boulardii*: A systematic review. *Can. J. Gastroenterol. J. Can. Gastroenterol.* **2009**, *23*, 817–821. [CrossRef]
40. Terciolo, C.; Dapoigny, M.; Andre, F. Beneficial effects of *Saccharomyces boulardii* CNCM I-745 on clinical disorders associated with intestinal barrier disruption. *Clin. Exp. Gastroenterol.* **2019**, *12*, 67–82. [CrossRef]
41. Jawhara, S.; Poulain, D. *Saccharomyces boulardii* decreases inflammation and intestinal colonization by *Candida albicans* in a mouse model of chemically-induced colitis. *Med. Mycol.* **2007**, *45*, 691–700. [CrossRef]
42. Algin, C.; Sahin, A.; Kiraz, N.; Sahinturk, V.; Ihtiyar, E. Effectiveness of bombesin and *Saccharomyces boulardii* against the translocation of *Candida albicans* in the digestive tract in immunosuppressed rats. *Surg. Today* **2005**, *35*, 869–873. [CrossRef]
43. Berg, R.; Bernasconi, P.; Fowler, D.; Gautreaux, M. Inhibition of *Candida albicans* translocation from the gastrointestinal tract of mice by oral administration of *Saccharomyces boulardii*. *J. Infect. Dis.* **1993**, *168*, 1314–1318. [CrossRef]
44. Demirel, G.; Celik, I.H.; Erdeve, O.; Saygan, S.; Dilmen, U.; Canpolat, F.E. Prophylactic *Saccharomyces boulardii* versus nystatin for the prevention of fungal colonization and invasive fungal infection in premature infants. *Eur. J. Pediatr.* **2013**, *172*, 1321–1326. [CrossRef]
45. Kumar, S.B.A.; Chakrabarti, A.; Singhi, S. Evaluation of efficacy of probiotics in prevention of Candida colonization in a PICU—A randomized controlled trial. *Crit. Care Med.* **2013**, *41*, 565–572. [CrossRef]
46. Zeng, Q.; He, X.; Puthiyakunnon, S.; Xiao, H.; Gong, Z.; Boddu, S.; Chen, L.; Tian, H.; Huang, S.H.; Cao, H. Probiotic mixture golden bifido prevents neonatal Escherichia coli K1 translocation via enhancing intestinal defense. *Front. Microbiol.* **2017**, *8*, 1798. [CrossRef]
47. Allert, S.; Förster, T.M.; Svensson, C.M.; Richardson, J.P.; Pawlik, T.; Hebecker, B.; Rudolphi, S.; Juraschitz, M.; Schaller, M.; Blagojevic, M.; et al. *Candida albicans*-Induced Epithelial Damage Mediates Translocation through Intestinal Barriers. *mBio* **2018**, *9*, e00915-18. [CrossRef]
48. Qamar, A.; Aboudola, S.; Warny, M.; Michetti, P.; Pothoulakis, C.; LaMont, J.T.; Kelly, C.P. Saccharomyces boulardii stimulates intestinal immunoglobulin A immune response to *Clostridium difficile* toxin A in mice. *Infect. Immun.* **2001**, *69*, 2762–2765. [CrossRef]
49. Wang, Y.; Li, Y.; Chen, C.; Zhang, Y.; Li, J.; Fan, Y.; Chen, X.; Wang, S.; Wang, J. Saccharomyces boulardii exerts anti-apoptosis and anti-necroptosis effects on neonatal mice necrotizing enterocolitis by increasing reactive oxygen species consumption. *Int. J. Clin. Exp. Med.* **2019**, *12*, 10019–10028.
50. Canonici, A.; Siret, C.; Pellegrino, E.; Pontier-Bres, R.; Pouyet, L.; Montero, M.P.; Colin, C.; Czerucka, D.; Rigot, V.; André, F. *Saccharomyces boulardii* improves intestinal cell restitution through activation of the alpha2beta1 integrin collagen receptor. *PLoS ONE* **2011**, *6*, e18427. [CrossRef]
51. Murzyn, A.; Krasowska, A.; Augustyniak, D.; Majkowska-Skrobek, G.; Łukaszewicz, M.; Dziadkowiec, D. The effect of *Saccharomyces boulardii* on *Candida albicans*-infected human intestinal cell lines Caco-2 and Intestin 407. *FEMS Microbiol. Lett.* **2010**, *310*, 17–23. [CrossRef]
52. Murzyn, A.; Krasowska, A.; Stefanowicz, P.; Dziadkowiec, D.; Lukaszewicz, M. Capric acid secreted by *S. boulardii* inhibits *C. albicans* filamentous growth, adhesion and biofilm formation. *PLoS ONE* **2010**, *58*, e12050. [CrossRef] [PubMed]
53. Kunyeit, L.; Kurrey, N.K.; Anu-Appaiah, K.A.; Rao, R.P. Probiotic yeasts inhibit virulence of non-albicans *Candida* species. *mBio* **2019**, *10*, e02307-19. [CrossRef]
54. Pericolini, E.; Gabrielli, E.; Ballet, N.; Sabbatini, S.; Roselletti, E.; Cayzeele Decherf, A.; Pélerin, F.; Luciano, E.; Perito, S.; Jüsten, P.; et al. Therapeutic activity of a *Saccharomyces cerevisiae*-based probiotic and inactivated whole yeast on vaginal candidiasis. *Virulence* **2017**, *8*, 74–90. [CrossRef]
55. Jawhara, S.; Habib, K.; Maggiotto, F.; Pignede, G.; Vandekerckove, P.; Maes, E.; Dubuquoy, L.; Fontaine, T.; Guerardel, Y.; Poulain, D. Modulation of intestinal inflammation by yeasts and cell wall extracts: Strain dependence and unexpected anti-inflammatory role of glucan fractions. *PLoS ONE* **2012**, *7*, e40648. [CrossRef]

56. Vylkova, S.; Nayyar, N.; Li, W.; Edgerton, M. Human beta-defensins kill *Candida albicans* in an energy-dependent and salt-sensitive manner without causing membrane disruption. *Antimicrob. Agents Chemother.* **2007**, *51*, 154–161. [CrossRef]
57. Edgerton, M.; Koshlukova, S.E.; Araujo, M.W.; Patel, R.C.; Dong, J.; Bruenn, J.A. Salivary histatin 5 and human neutrophil defensin 1 kill *Candida albicans* via shared pathways. *Antimicrob. Agents Chemother.* **2000**, *44*, 3310–3316. [CrossRef] [PubMed]
58. Fidan, I.; Kalkanci, A.; Yesilyurt, E.; Yalcin, B.; Erdal, B.; Kustimur, S.; Imir, T. Effects of *Saccharomyces boulardii* on cytokine secretion from intraepithelial lymphocytes infected by *Escherichia coli* and *Candida albicans*. *Mycoses* **2009**, *52*, 29–34. [CrossRef]
59. Fazly, A.; Jain, C.; Dehner, A.C.; Issi, L.; Lilly, E.A.; Ali, A.; Cao, H.; Fidel, P.L.; Rao, R.P.; Kaufman, P.D. Chemical screening identifies filastatin, a small molecule inhibitor of Candida albicans adhesion, morphogenesis, and pathogenesis. *Proc. Natl. Acad. Sci. USA* **2013**, *110*, 13594–13599. [CrossRef]
60. Harriott, M.M.; Noverr, M.C. Importance of Candida-bacterial polymicrobial biofilms in disease. *Trends Microbiol.* **2011**, *19*, 557–563. [CrossRef]
61. Coco, B.J.; Bagg, J.; Cross, L.J.; Jose, A.; Cross, J.; Ramage, G. Mixed *Candida albicans* and *Candida glabrata* populations associated with the pathogenesis of denture stomatitis. *Oral Microbiol. Immunol.* **2008**, *23*, 377–383. [CrossRef]
62. Lea, T. Epithelial Cell Models; General Introduction. In *The Impact of Food Bioactives on Health: In Vitro and Ex Vivo Models*; Verhoeckx, K., Cotter, P., Lopez-Exposito, I., Kleiveland, C., Lea, T., Mackie, A., Requena, T., Swiatecka, D., Wichers, H., Cham, C.H., Eds.; Springer International Publishing: Cham, Switzerland, 2015; pp. 95–102.
63. Strus, M.; Kucharska, A.; Kukla, G.; Brzychczy-Włoch, M.; Maresz, K.; Heczko, P.B. The in vitro activity of vaginal Lactobacillus with probiotic properties against Candida. *Infect. Dis. Obstet. Gynecol.* **2005**, *13*, 69–75. [CrossRef] [PubMed]
64. Ohshima, T.; Kojima, Y.; Seneviratne, C.J.; Maeda, N. Therapeutic application of synbiotics, a fusion of probiotics and prebiotics, and biogenics as a new concept for oral candida infections: A Mini Review. *Front. Microbiol.* **2016**, *7*, 10. [CrossRef] [PubMed]
65. Parolin, C.; Marangoni, A.; Laghi, L.; Foschi, C.; Ñahui Palomino, R.A.; Calonghi, N.; Cevenini, R.; Vitali, B. Isolation of vaginal lactobacilli and characterization of anti-Candida activity. *PLoS ONE* **2015**, *10*, e0131220. [CrossRef]
66. Rao, R.P.; Hunter, A.; Kashpur, O.; Normanly, J. Aberrant synthesis of indole-3-acetic acid in *Saccharomyces cerevisiae* triggers morphogenic transition, a virulence trait of pathogenic fungi. *Genetics* **2010**, *185*, 211–220. [CrossRef] [PubMed]
67. Chen, H.; Fujita, M.; Feng, Q.; Clardy, J.; Fink, G.R. Tyrosol is a quorum-sensing molecule in *Candida albicans*. *Proc. Natl. Acad. Sci. USA* **2004**, *101*, 5048–5052. [CrossRef]
68. Lakhtin, M.; Alyoshkin, V.; Lakhtin, V.; Afanasyev, S.; Pozhalostina, L.; Pospelova, V. Probiotic Lactobacillus and Bifidobacterial lectins against *Candida albicans* and *Staphylococcus aureus* clinical strains: New Class of the Pathogen Biofilm Destructors. *Probiot. Antimicrob. Proteins* **2010**, *2*, 186–196. [CrossRef]
69. Graham, C.E.; Cruz, M.R.; Garsin, D.A.; Lorenz, M.C. Enterococcus faecalis bacteriocin EntV inhibits hyphal morphogenesis, biofilm formation, and virulence of *Candida albicans*. *Proc. Natl. Acad. Sci. USA* **2017**, *114*, 4507–4512. [CrossRef]
70. Lourenco, A.; Pedro, N.A.; Salazar, S.B.; Mira, N.P. Effect of Acetic Acid and Lactic Acid at Low pH in Growth and Azole Resistance of *Candida albicans* and *Candida glabrata*. *Front. Microbiol.* **2018**, *9*, 3265. [CrossRef]
71. Czerucka, D.; Piche, T.; Rampal, P. Review article: Yeast as probiotics–*Saccharomyces boulardii*. *Aliment. Pharmacol. Ther.* **2007**, *26*, 767–778. [CrossRef]
72. Tampakakis, E.; Peleg, A.Y.; Mylonakis, E. Interaction of *Candida albicans* with an intestinal pathogen, *Salmonella enterica* serovar *Typhimurium*. *Eukaryot. Cell* **2009**, *8*, 732–737. [CrossRef]
73. Shareck, J.; Belhumeur, P. Modulation of Morphogenesis in *Candida albicans* by Various Small Molecules. *Eukaryot. Cell.* **2011**, *10*, 1004–1012. [CrossRef] [PubMed]
74. Morales, D.K.; Grahl, N.; Okegbe, C.; Dietrich, L.E.; Jacobs, N.J.; Hogan, D.A. Control of *Candida albicans* metabolism and biofilm formation by *Pseudomonas aeruginosa* phenazines. *mBio* **2013**, *4*, e00526-12. [CrossRef] [PubMed]

75. Hogan, D.A.; Kolter, R. Pseudomonas-Candida interactions: An ecological role for virulence factors. *Science* **2002**, *296*, 2229–2232. [CrossRef] [PubMed]
76. Wuster, A.; Babu, M.M. Transcriptional control of the quorum sensing response in yeast. *Mol. bioSyst.* **2010**, *6*, 134–141. [CrossRef] [PubMed]
77. Chen, H.; Fink, G.R. Feedback control of morphogenesis in fungi by aromatic alcohols. *Genes Dev.* **2006**, *20*, 1150–1161. [CrossRef]
78. Winters, M.; Arneborg, N.; Appels, R.; Howell, K. Can community-based signalling behaviour in *Saccharomyces cerevisiae* be called quorum sensing? A critical review of the literature. *FEMS Yeast Res.* **2019**, *19*. [CrossRef]
79. Hogan, D.A.; Vik, A.; Kolter, R. A *Pseudomonas aeruginosa* quorum-sensing molecule influences *Candida albicans* morphology. *Mol. Microbiol.* **2004**, *54*, 1212–1223. [CrossRef]
80. Singkum, P.; Muangkaew, W.; Suwanmanee, S.; Pumeesat, P.; Wongsuk, T.; Luplertlop, N. Suppression of the pathogenicity of *Candida albicans* by the quorum-sensing molecules farnesol and tryptophol. *J. Gen. Appl. Microbiol.* **2019**, *65*, 277–283. [CrossRef]
81. Martins, M.; Lazzell, A.L.; Lopez-Ribot, J.L.; Henriques, M.; Oliveira, R. Effect of exogenous administration of *Candida albicans* autoregulatory alcohols in a murine model of hematogenously disseminated candidiasis. *J. Basic Microbiol.* **2012**, *52*, 487–491. [CrossRef]

Journal of *Fungi*

MDPI

Review

Immunomodulatory Effects of Edible and Medicinal Mushrooms and Their Bioactive Immunoregulatory Products

Shuang Zhao [1], Qi Gao [1], Chengbo Rong [1], Shouxian Wang [1], Zhekun Zhao [1,2], Yu Liu [1] and Jianping Xu [3,*]

1 Institute of Plant and Environment Protection, Beijing Academy of Agriculture and Forestry Sciences, Beijing 100097, China; zhaoshuang@ipepbaafs.cn (S.Z.); gaoqi@ipepbaafs.cn (Q.G.); rongchengbo@ipepbaafs.cn (C.R.); wangshouxian@ipepbaafs.cn (S.W.); zhaozhekun@ipepbaafs.cn (Z.Z.); liuyu@ipepbaafs.cn (Y.L.)

2 College of Life Sciences and Food Engineering, Hebei University of Engineering, Handan 056038, China

3 Department of Biology, McMaster University, Hamilton, ON L8S 4K1, Canada

* Correspondence: jpxu@mcmaster.ca

Received: 10 October 2020; Accepted: 2 November 2020; Published: 8 November 2020

check for updates

Abstract: Mushrooms have been valued as food and health supplements by humans for centuries. They are rich in dietary fiber, essential amino acids, minerals, and many bioactive compounds, especially those related to human immune system functions. Mushrooms contain diverse immunoregulatory compounds such as terpenes and terpenoids, lectins, fungal immunomodulatory proteins (FIPs) and polysaccharides. The distributions of these compounds differ among mushroom species and their potent immune modulation activities vary depending on their core structures and fraction composition chemical modifications. Here we review the current status of clinical studies on immunomodulatory activities of mushrooms and mushroom products. The potential mechanisms for their activities both in vitro and in vivo were summarized. We describe the approaches that have been used in the development and application of bioactive compounds extracted from mushrooms. These developments have led to the commercialization of a large number of mushroom products. Finally, we discuss the problems in pharmacological applications of mushrooms and mushroom products and highlight a few areas that should be improved before immunomodulatory compounds from mushrooms can be widely used as therapeutic agents.

Keywords: bioactive compounds; FIP; human health; immunomodulation; induced apoptosis; lectin; medicinal mushrooms; polysaccharide; terpenes and terpenoids

1. Introduction

In clinical practice, immunomodulators are usually classified into three categories: immunosuppressants, immunostimulants, and immunoadjuvants [1]. Their market share has increased rapidly over the past few years due to wide-ranging medical applications for patients that require human immune system modulations. Immune system modulations are also commonly used as prophylactic medicine for an increasing number of healthy people [2,3]. While most immunomodulators are synthetic or semi-synthetic compounds, there has been a growing interest in natural immunomodulators. Many natural compounds have shown significant immunomodulatory and overall health-benefiting effects to humans, with no or minimal toxicity. These natural-based products with potential pharmacological and beneficial effects are increasingly perceived as safer than synthetic compounds by the general public [4,5]. Indeed, many of the currently used chemical drugs have negative side effects and the market share of natural immunomodulators is increasing rapidly with an annual growth rate of 8.6% [1,6].

Medicinal mushrooms (MMs) are an important source of natural immunomodulators. Representing a subset of all mushrooms, MMs are broadly defined as macroscopic fungi that are used in the form of extracts or whole mushroom powder for human health benefits [7]. The health benefits may come in the form of helping to prevent and/or treat diseases in humans, and to create a dietary balance of a healthy diet. Dating back to thousands of years, MMs have been historically used as herbal medicines for human health, such as for the treatment of infectious diseases, gastrointestinal disorders and asthmatic conditions [8]. The biomass or specific extracts from all developmental stages of MMs, including the fruiting body, sclerotium, mycelium, and spores, have been used as health foods or dietary supplements [9,10]. Some of the extracted nutrients from mushrooms known as mushroom nutraceuticals have been made into capsules or tablets as dietary supplements. Regular intake of these nutraceuticals has been associated with enhancement of the human immune response, leading to increased resistance to infectious diseases and faster recovery from a diversity of diseases [11].

At present, thousands of branded MM products are sold all over the world. The health benefits of MM products include anticancer, immune-stimulation, antioxidant, antihyperglycemic, antihypertensive, neuroprotective, hepatoprotective, antidiabetic, antifungal, antibacterial, and antiviral activities [7,12]. Their effects have been attributed to many components, such as minerals, essential amino acids, dietary fiber, proteins, polysaccharides, lipopolysaccharides, glycoproteins, and secondary metabolites. Among these, some of the complex organic compounds have shown immunomodulatory effects [7]. For example, polysaccharides from MMs can activate natural killer cells, macrophages, and neutrophils, as well as induce innate immune cytokines and interleukins. In addition, secondary metabolites from MMs, such as sterols, terpenes, and phenols can enhance the survival of hosts by stabilizing their important metabolic functions [8].

Different MMs contain different functional components that may impact the same or different immunomodulatory pathways at varying efficacies. In the following sections, we first provide a brief description of the known MMs. We then summarize the diversity, structure, function, and molecular mechanism of action of functional ingredients from MMs that have shown to be involved in immunomodulation. We finish by briefly describing how genomics can accelerate research on medicinal mushrooms.

2. Medicinal Mushrooms

As mentioned above, medicinal mushrooms refer to all macroscopic fungi whose extracts or powder form from any stages of the mushroom development have shown documented beneficial effects on health [13]. These beneficial effects may have been shown in the forms of in vitro, ex vivo, or in vivo activities. Their effects may cover different groups of organisms such as antagonistic effects against human pathogens and parasites, and/or beneficial effects for human and animal cell lines, or animal and human individuals [11]. Since many edible mushrooms and their products have shown to be a beneficial component of the human diet, some of these edible mushrooms are also commonly included as medicinal mushrooms [7]. In our literature search, a large number of MMs have been documented. For example, terpenes and terpenoids from *Ganoderma lucidum* could stimulate the expressions of genes coding for proteins in the nuclear factor (NF)–kB pathway and modulate immune system functions [14]. Heteroglycan and heteroglycan-peptide from the mushroom of *Hericium erinaceus* can modulate the immuno-effects by inducing nitric oxide production and increasing expression of tumor necrosis factor (TNF)-α, interleukin (IL)-1β, IL-12 [15]. These mushrooms belong to two fungal phyla, Ascomycota and Basidiomycota. Most of the MMs are in phylum Basidiomycota. Table 1 shows the major medicinal mushrooms, including their taxonomy and geographic/ecological distributions. As can be seen, some of these mushrooms are broadly distributed (e.g., the button mushroom *Agaricus bisporus*) while others are geographically more restricted (e.g., the Himalayan caterpillar fungus *Ophiocordyceps sinensis*). Some of the mushrooms included in the table e.g., *Amanita phalloides* are highly poisonous when consumed by humans. However, the dilutions of an *A. phalloides* extract that contains the toxin amanitin have shown to be effective as an anti-tumor therapy [16].

Table 1. Major medicinal mushrooms and their main distributions.

MM Species	Common Name	Taxonomy	Geographic/Ecological Distribution
Agrocybe aegerita	Black Poplar mushroom	Basidiomycota Agaricomycetes Agaricales Bolbitiaceae	North temperate and subtropical zone
Agaricus bisporus	Button mushroom, Portobello mushroom, Common mushroom	Basidiomycota Agaricomycetes Agaricales Agaricaceae	USA, China, France, Netherlands, United Kingdom, Italy, Poland, Spain, Germany, Canada, Ireland, Belgium, Indonesia, Hungary and Mexico
Agaricus blazei (syn. *Agaricus brasiliensis*)	Royal Sun Agaricus, Almond Portobello	Basidiomycota Agaricomycetes Agaricales Agaricaceae	America, Brasil, Japan, China
Amanita phalloides	Death Cap	Basidiomycota Agaricomycetes Agaricales Amanitaceae	Europe, North American, Asia
Boletus edulis	Cep, Porcini, Penny Bun Bolete	Basidiomycota Agaricomycetes Boletales Boletaceae	China, Italy, France, Swiss, Germany
Boletus speciosus	Red-Capped Butter Bolete	Basidiomycota Agaricomycetes Boletales Boletaceae	Eastern North America, Southwest of China and Europe
Chroogomphus rutilus	Copper Spike	Basidiomycota Agaricomycetes Boletales Gomphidiaceae	China
Clitocybe nebularis	Clouded Funnel	Basidiomycota Agaricomycetes Agaricales Tricholomataceae	China, Japan, Taiwan, Europe, North America, North Africa
Cryptoporus volvatus	Veiled Polypore	Basidiomycota Agaricomycetes Polyporales Polyporaceae	Trunks of pine, fir and spruce
Dichomitus squalens	Common White-Rot fungus	Basidiomycota Agaricomycetes Polyporales Polyporaceae	Trunks of conifers such as pine and larch
Flammulina velutipes	Golden Needle mushroom	Basidiomycota Agaricomycetes Agaricales Physalacriaceae	Subtropical zone such as Japan, Russia, Australia and other countries as well as Europe, North America
Floccularia luteovirens (syn. *Armillaria luteovirens*)	Scaly Yellow mushroom	Basidiomycota Agaricomycetes Agaricales Tricholomataceae	Meadow at altitudes of 3000–4000 m above sea level

Table 1. *Cont.*

MM Species	Common Name	Taxonomy	Geographic/Ecological Distribution
Ganoderma atrum	Black Ling-zhi	Basidiomycota Agaricomycetes Polyporales Polyporaceae	Tropical regions
Ganoderma capense	Dark Ling-zhi	Basidiomycota Agaricomycetes Polyporales Polyporaceae	Tropical regions
Ganoderma japonicum	Bloody Ling-zhi	Basidiomycota Agaricomycetes Polyporales Polyporaceae	Majority in tropical and subtropical regions of Asia, Australia, Africa and America, minority in temperate zone
Ganoderma lucidum	Reitake, Ling-zhi, Spirit Plant Reishi	Basidiomycota Agaricomycetes Polyporales Polyporaceae	Majority in tropical and subtropical regions of Asia, Australia, Africa and America, minority in temperate zone
Ganoderma microsporum	Small-Spored Ling-zhi	Basidiomycota Agaricomycetes Polyporales Polyporaceae	Subtropics zone
Ganoderma lingzhi	Ling-zhi	Basidiomycota Agaricomycetes Polyporales Polyporaceae	China, North Korea, Japan
Ganoderma sinensis	Zi-zhi	Basidiomycota Agaricomycetes Polyporales Polyporaceae	China, North Korea, Japan
Ganoderma tsugae	Hemlock Varnish Shelf	Basidiomycota Agaricomycetes Polyporales Polyporaceae	Northern and Montaine zone
Grifola frondosa	Maitake Hen of the Woods	Basidiomycota Agaricomycetes Polyporales Grifolaceae	Japan, China
Hericium erinaceus	Lion's Mane mushroom, Bearded Tooth mushroom, Monkey-Head mushroom	Basidiomycota Agaricomycetes Russulales Hericiaceae	Broad-leaved forest or coniferous and broad-leaved mixed forest in northern temperate zone such as Western Europe, North America, China, Japan, Russia
Inonotus obliquus	Clinker Polypore, Birch Conk, Chaga	Basidiomycota Agaricomycetes Hymenochaetacles Hymenochaetaceae	Russia, China
Lentinula edodes	Shiitake, Black Forest mushroom, Golden Oak mushroom	Basidiomycota Agaricomycetes Agaricales Omphalotaceae	Distributed in an arc area on the west side of the Pacific Ocean, Japan, Papua New Guinea, Nepal, the Mediterranean coast and northern Africa

Table 1. *Cont.*

MM Species	Common Name	Taxonomy	Geographic/Ecological Distribution
Lignosus rhinocerotis	Tiger Milk mushroom	Basidiomycota Agaricomycetes Polyporales Polyporaceae	China, Indonesia, Philippines, Sri Lanka, Australia, Thailand, Malaysia, Papua New Guinea and rainforests of East Africa
Leucocalocybe mongolica (syn. *Tricholoma mongolicum*)	Mongolia mushroom	Basidiomycota Basidiomycetes Agaricales Agaricales incertae sedis	Inner Mongolia in China
Marasmius oreades	Fairy Ring mushroom	Basidiomycota Agaricomycetes Agaricales Marasmiaceae	North America and Asia
Morchella esculenta	Common Morel, Yellow Morel, Sponge Morel	Ascomycota Pezizomycetes Pezizales Morohellaceae	Widely cultured over the world such as France, Germany, America, India, China, Russia, Sweden, Mexico, Spain, Czechoslovakia and Pakistan
Morchella conica	Black Morel, Sponge mushroom	Ascomycota Pezizomycetes Pezizales Morohellaceae	Distributed under broad-leaved forest, coniferous broad-leaved mixed forest, forest edge open space and weeds
Naematelia aurantialba (syn. *Tremella aurantialba*)	Golden Tremella	Basidiomycota Tremellomycetes Tremellales Naemateliaceae	Mountain forest of quercus, mutualism with *Stereum* spp.
Ophiocordyceps sinensis	Caterpillar fungus, Himalaya Viagra	Ascomycota Sordariomycetes Hypocreales Ophiocordycipitaceae	Southwest China, Nepal
Pholiota adiposa	Chestnut mushroom	Basidiomycota Agaricomycetes Agaricales Strophariaceae	Distributed on the dead willows in the forest in China
Pleurotus citrinopileatus	Golden Oyster mushroom, Tamogitake	Basidiomycota Agaricomycetes Agaricales Pleurotaceae	Widely cultured all over the world
Pleurotus ostreatus	Oyster mushroom	Basidiomycota Agaricomycetes Agaricales Pleurotaceae	Widely cultured all over the world
Cerioporus squamosus (syn. *Polyporus squamosus*)	Dryad's Saddle, Pheasant's Back mushroom	Basidiomycota Agaricomycetes Polyporales Polyporaceae	Widely distributed in hardwood forest of North America, Australia, Asia and Europe
Poria cocos	Fuling, China Root	Basidiomycota Agaricomycetes Polyporales Laetiporaceae	Parasitic on the roots of Pinaceae plants, mainly distributed in China
Rhodonia placenta (syn. *Postia placenta*)	Rosy Crust	Basidiomycota Agaricomycetes Polyporales Dacryobolaceae	Widely distributed all over the world

Table 1. *Cont.*

MM Species	Common Name	Taxonomy	Geographic/Ecological Distribution
Pseudosperma umbrinellum (syn. *Inocybe umbrinella*)	Fibrous Hat	Basidiomycota Agaricomycetes Agaricales Inocybaceae	France
Russula delica	Milk-White Brittlegill	Basidiomycota Agaricomycetes Russulales Russulaceae	Taiga forest and mixed forests
Russula lepida	Rosy Russula	Basidiomycota Agaricomycetes Russulales Russulaceae	Widely distributed all over the world
Sarcodon aspratus	Black Tiger Paw	Basidiomycota Agaricomycetes Thelephorales Thelephoraceae	Southwest of China
Schizophyllum commune	Split Gill	Basidiomycota Agaricomycetes Agaricales Schizophyllaceae	Widely distributed all over the world
Stropharia rugosoannulata	Wine Cap Stropharia, Garden Giant, Burgundy mushroom, King Stropharia	Basidiomycota Agaricomycetes Agaricales Strophariaceae	Europe, North America, Asia
Taiwanofungus camphoratus (syn. *Antrodia camphorate*)	Poroid Brown-rot fungus, Stout Camphor fungus	Basidiomycota Agaricomycetes Polyporales incertae sedis	Mountain forest in Taiwan with altitudes of 450–2000 m
Trametes versicolor (syn. *Polystictus versicolor*)	Turkey Tail fungus	Basidiomycota Agaricomycetes Polyporales Polyporaceae	Global distribution; Broad-leaf woods
Tropicoporus linteus (syn. *Phelllinus linteus*)	Mesima, Black Hoof fungus	Basidiomycota Agaricomycetes Hymenochaetales Hymenochaetaceae	Distributed on the dead trees and trunks in China
Xerocomellus chrysenteron (syn. *Xerocomus Chrysenteron*)	Red Cracking Bolete	Basidiomycota Agaricomycetes Agaricales Agaricales incertae sedis	China
Xylaria hypoxylon	Candlestick fungus, Candlesnuff fungus, Carbon Antlers, Stag's Horn fungus	Ascomycota Sordariomycetes Xylariales Xylariaceae	Northern Europe
Xylaria nigripes	Dead Moll's Fingers	Ascomycota Sordariomycetes Xylariales Xylariaceae	China, mutualism with white ant
Volvariella volvacea	Straw mushroom	Basidiomycota Basidiomycetes Agaricales Pluteaceae	China, East Asia, Southeast Asia

Some of these MM species have been used as herbal medicine for centuries, including *Ganoderma lucidum*, *Ganoderma lingzhi*, *Lentinula edodes*, *Inonotus obliquus*, *Fomitopsis officinalis*, *Piptoporus betulinus*, and *Fomes fomentarius* [7,17]. While these mushrooms have attracted most of the medical attention among the MMs, other species in multiple genera have also shown immunomodulatory and anti-tumor effects, such as those in genera *Agaricus*, *Albatrellus*, *Antrodia*, *Calvatia*, *Clitocybe*, *Cordyceps*, *Flammulina*, *Fomes*, *Funlia*, *Ganoderma*, *Inocybe*, *Inonotus*, *Lactarius*, *Phellinus*, *Pleurotus*, *Russula*, *Schizophyllum*, *Suillus*, *Trametes*, and *Xerocomus* [12].

Figure 1 shows a few representative medicinal mushroom species in their natural habitats. Some medicinal mushrooms are only found in the wild, e.g., the ectomycorrhizal mushrooms *Boletus edulis* and *Russula lepida*. However, a large number of medicinal mushrooms are also commercially cultivated, including Shiitake, Ling-zhi, and Lion's Mane. Figure 2 shows a few representative medicinal mushrooms under cultivation.

Figure 1. A few representative medicinal mushrooms from the wild.

Figure 2. A few representative cultivated medicinal mushrooms.

3. Immunomodulatory Compounds and Commercial Products of Medicinal Mushrooms

As shown above, there is a large number of medicinal mushrooms in diverse taxonomic groups. Some of these medicinal mushrooms are commercially cultivated for food but there is an increasing trend for developing the immune-active compounds from these cultivated mushrooms into nutraceuticals. Tables 2–5 summarize the major groups of bioactive compounds in medicinal mushrooms and their demonstrated immunomodulatory effects to specific pathologies, including the relevant references.

The main classes of compounds from medicinal mushrooms with immunomodulatory properties are terpenes and terpenoids, lectins, fungal immunomodulatory proteins (FIPs), and polysaccharides (particularly β-D-glucans, but also include polysaccharopeptides and polysaccharide proteins) [1]. Below we describe specific examples in each of these groups.

3.1. Polysaccharides

Among the bioactive compounds derived from mushrooms with immunomodulatory activity, those based on polysaccharides, with or without side chain modifications (including polysaccharopeptides and polysaccharide proteins) are the most reported during the last several decades [18]. Table 2 presents a list of polysaccharides from medicinal mushrooms that have shown immunomodulatory activities. Among the reported polysaccharides with immunomodulatory and antitumor activities, the best-known is lentinan, isolated from shiitake (*L. edodes*), as well as schizophyllan from *Schizophyllum commune*. Both lectinan and schizophyllan contain β-1,3-D-glucans with β-1,6 branches. Specifically, lentinan showed immunomodulatory properties against gastric cancer while schizophyllan was effective against head and neck cancer. Both products have been licensed and approved in Japan since 1986 for clinical use, in combination with chemotherapy against the two respective cancers [19,20].

Other polysaccharide-based compounds showing immunomodulatory properties have a similar core polysaccharide chemical structure but contain different branching linkages and/or branches with different conjugates. These polysaccharide-conjugate complexes are called heteroglucans, with α(1–4)- and β(1–3) glycosidic linkages to protein components. For example, in the presence of fucose (a hexose deoxy sugar with the chemical formula C6H12O5), the turkey tail mushroom *Trametes versicolor* produces a Krestin bound β-glucan polysaccharide K (PSK). PSK is commercially produced from this mushroom in Japan and has been approved for clinical use since 1977 [21]. Several subsequent reports confirmed the effectiveness of PSK as an adjuvant to conventional cancer therapies through inhibition of cancer metastasis [22], induction of cancer cell apoptosis [23], improvement of inflammatory cytokines gene expression [24,25].

Another compound isolated from *T. versicolor* is a polysaccharide peptide (PSP). PSP contains rhamnose and arabinose, two monosaccharides not found in PSK. In addition, the conjugated protein was also different. PSP has been commercially available in the Chinese market since 1987 [21]. It has been documented to improve the quality of life in cancer patients by providing substantial pain relief and enhancing immune status in 70–97% of patients with stomach, esophagus, lung, ovary and cervical cancers. Specifically, PSP has been shown to be capable of boosting immune cell production, ameliorating chemotherapy symptoms, and enhancing tumor infiltration by dendritic and cytotoxic T-cells [26].

Two other well-known polysaccharide–protein complexes produced by *Macrocybe gigantea* (syn. *Tricholoma giganteum*) and *Agaricus brazei*, respectively, have also shown immunomodulatory effects. A polysaccharide-protein complex (PSPC) isolated from *T. giganteum* showed that it could help restore and improve the phagocytic function of macrophages in tumor-bearing mice [13,27]. Similarly, AbM isolated from *A. brazei* contains diverse polysaccharide–protein complexes with different chemical linkages such as β-1,6-glucan, α-1,6- and α-1,4-glucan, glucomannan and β-1,3-glucan. AbM has been shown to have immunomodulatory and antineoplastic properties [28]. Polysaccharides and polysaccharide–protein complexes from other medicinal mushrooms that have shown immunomodulatory effects are listed in Table 2.

Table 2. Major immunomodulatory polysaccharides from medicinal mushrooms and their immunomodulatory effects.

Source	Active Compound	Immunomodulatory Effect	Refs
Agaricus blazei (syn. *Agaricus brasiliensis*)	Heteroglycan, Glycoprotein, Glucomannan-protein complex, β-1,3-D-glucan, with β-1,6-D-glucan branch	Stimulates Natural Killer (NK) cells, macrophages, dendritic cells, and granulocytes; induction of Tumor Necrosis Factor (TNF), Interferon (IFN)-γ, and Interleukin (IL)-8 production	[29]
Auricularia auricula-judae	AF1 β-1,3-D-glucan main chain with two β-1,6-D-glucosyl residues	Induces apoptosis of cancer cell	[30]
Gymnopus dryophilus (syn. *Collybia dryophila*)	β-D-glucan	Inhibits NO production in activated macrophages	[31]
Ophiocordyceps sinensis	β-D-glucan, heteroglycan, cordyglucan	Increase in IL-5 induction with decrease in IL-4 and IL-17	[32]
Cryptoporus volvatus	β-1,3-D-Glucan	Decreases in TLR2 and activate NF-κB	[33]
Flammulina velutipes	Glycoprotein, *Flammulina velutipes* peptidoglycan (FVP), β-1,3-D-glucan	Increases NO, IL-1 production, and TNF-α secretion	[34]
Ganoderma lucidum	Ganoderan, Heteroglycan, mannoglucan, glycopeptide	Stimulates TNF-α, IL-1, IFN-γ production, activate NF-κB.	[35]
Grifola frondosa	Grifolan (1–6-monoglucosyl-branched β-1,3-D-glucan), proteoglycan, heteroglycan, galactomannan	Macrophage activation, induction of IL-1, IL-6, and TNF-α secretion	[36]
Hericium erinaceus	Heteroglycan, heteroglycan-peptide, β-1,3 branched-β-1,2-mannan	Induces NO production, increase expression of TNF-α, IL-1β, IL-12	[15]
Inonotus obliquus	β-D-glucan	Enhance expression of IL-1β, IL-6, TNF-α, and inducible nitric oxide synthase (iNOS) in macrophages	[37]
Lentinula edodes (syn. *Lentinus edodes*)	Lentinan, glucan, mannoglucan, proteoglycan, β-(1-6)-D-glucan, α-(1-3)-D-glucan	Induces non-specific cytotoxicity in macrophage and enhance cytokine production Induces cytotoxic effect on a breast cancer cell line	[38–40]
Lentinus squarrosulus	Glucan	Activation of macrophages, splenocytes and thymocytes	[41]
Morchella esculenta	Galactomannan, β-1,3-D-glucan	Macrophage activation, activate NF-κB	[42]
Morchella conica	Galactomannan	Induces NO, IL-1β, IL-6 production	[43]
Naematelia aurantialba (syn. *Tremella aurantialba*)	Heteroglycan	Enhances mouse spleen lymphocyte proliferation	[44]
Pleurotus sp. 'Florida' (syn. *Pleurotus florida*)	α-1,6-glucan and α-1, 3-, β-1,6-D-glucan	Stimulates macrophages, splenocytes and thymocytes	[45,46]
Pleurotus ostreatus	Pleuran, heterogalactan, proteoglycan	Induces IL-4 and IFN-γ production	[47]
Poria cocos	β-pachyman, β-Glucan, β-1,3-D-glucan, α-1, 3-D-glucan	Promotes the immune reaction; increases the expression of cytokines	[48,49]
Sarcodon aspratus	Fucogalactan, 1,6-α-D-glucopyranosyl residue	Increases the release of TNF-α and NO in macrophage	[50]

Table 2. *Cont.*

Source	Active Compound	Immunomodulatory Effect	Refs
Schizophyllum commune	Schizophyllan, 1,6-monoglucosyl branched β-1, 3-D-glucan	Activation of T cell, increases interleukin, and TNF-α production	[51]
Sparassis crispa	β-Glucan	Enhances IL-6 and INF-γ production	[52]
Taiwanofungus camphoratus (syn. *Antrodia camphorate)*	β-1,3-D-Gluco-pyranans with β-1,6-D-glucosyl branches, proteoglycan	Induction of INF-γ, TNF-α	[53]
Tropicoporus linteus (*syn.* Phellinus linteus*)*	Acidic polysaccharides	Activation of murine B cells, Induces IL-12 and IFN-γ production, Blocks NF-κB, TNF-α, IL-1α, IL-1β, and IL-4 production	[54]
Trametes versicolor	Polysaccharide peptide Krestin (PSK), β-1,3-glycosidic bond with β-1,6-glycosidic branches	Increases the expression of cytokines; stimulates the macrophage phagocytes	[1,55]
Tremella fuciformis	Acidic glucuronoxylomannan α-1,3-D-mannan backbone with β-linked D-glucuronic acid	Induces human monocytes to express interleukins	[56,57]
Macrocybe gigantea (syn. *Tricholoma giganteum*)	Polysaccharide-protein complex (PSPC)	Increases phagocytic function of macrophages by activating macrophages to release mediators such as NO and TNF-α and inhibits S180 and HL-60 cells	[13,27]
Xylaria nigripes	β-Glucan	Inhibits NO, IL-1β, IL-6, TNF-α, and IFN-γ production	[58]

3.2. Mushroom Proteins and Protein–Conjugate Complexes

Mushroom proteins and protein–conjugate complexes are also well-known as immunomodulatory compounds. Similar to the polysaccharide-based compounds, these protein-based immunomodulatory compounds in medicinal mushrooms can also be grouped into different categories. Here, these compounds are grouped into two major categories: fungal immunomodulatory proteins (FIPs) and lectins. FIPs differ from lectins by having no conjugate while each lectin contains specific carbohydrates conjugated to a polypeptide.

Table 3 lists all the lectins from medicinal mushrooms isolated so far that have shown immunomodulatory effects. These lectins have been shown to be capable of stimulating nitrite production, upregulating the expressions of tumor necrosis factor (TNF)-α and interleukins, activating lymphocytes, and promoting the production of macrophage-activating factors etc. The medicinal mushroom species containing such lectins are very diverse, including *Floccularia luteovirens* (syn. *Armillaria luteovirens*), *Ganoderma capense*, *Grifola frondosa*, *Pseudosperma umbrinellum* (syn. *Inocybe umbrinella*), *Pholiota adipose*, *Pleurotus citrinopileatus*, *Russula delica*, *S. commune*, *Leucocalocybe mongolica* (syn. *Tricholoma mongolicum*), *Volvariella volvacea*, and *Xerocomus spadiceus* [59–64]. In addition, several mushroom lectins have also shown potent antiviral, mitogenic, antimicrobial and antioxidant activities [59,63,65–69].

Similarly, a large number FIPs have been identified. The FIP names, the medicinal mushrooms that produce them, and evidence for their specific immunomodulatory effects are presented in Table 4. Among these, the best known is probably Ling-Zhi-8 from *G. lucidum* which acts as an immunosuppressive agent [1]. In addition, aside from immunomodulation, many FIPs have also shown antitumor activities in pharmacological tests, including the inhibition of cell growth and proliferation, the induction of apoptosis and autophagy, and the reduction of invasion and migration of tumor cells. At present, most of these tests are conducted using tissue cultures. Further tests using animal models and clinical trials are needed in order to confirm their safety and efficacy in humans. If confirmed, these FIPs could be more efficiently produced and commercialized through genetic engineering for clinical use.

Table 3. Major immunomodulatory lectins from medicinal mushrooms and their immunomodulatory effects.

Source	Lectin name	Immunomodulatory effect	Refs
Agaricus bisporus	*Agaricus bisporus* lectin (ABL)	Stimulate mice splenocytes mitogenicity and inhibit proliferation of L1210 and HT-29 cells	[70,71]
Agrocybe aegerita	*Agrocybe aegerita* lectin (AAL)	Inhibit proliferation of 4T1, HeLa, SW480 SGC7901, MGC803, BGC823, HL-60 and S180 cells	[72,73]
Amanita phalloides	-	Inhibit proliferation of L1210 cells	[74]
Floccularia luteovirens (syn. *Armillaria luteovirens*)	*Armillaria luteovirens* lectin (ALL)	Stimulate mice splenocytes mitogenicity and inhibit proliferation of L1210, Mouse myeloma MBL2 and HeLa cells	[75]
Boletus edulis	*Boletus edulis* lectin (BEL)	Stimulate mice splenocytes mitogenicity and inhibit proliferation of human hepatocyte carcinoma G2 (HepG2) and HT-29 cells	[76]
Boletus speciosus	*Boletus speciosus* hemagglutinin (BSH)	Inhibit proliferation of HepG2 and L1210 cells	[77]
Clitocybe nebularis	*Clitocybe nebularis* lectin (CNL)	Inhibit proliferation of human leukemic T cells	[78]
Flammulina velutipes	*Flammulina velutipes* agglutinin (FVA)	Stimulate mice splenocytes mitogenicity and inhibit proliferation of L1210 cells	[79]
Ganoderma capense	-	Stimulate mice splenocytes mitogenicity and inhibit proliferation of L1210, M1, HepG2 cells	[62]
Grifola frondosa	*Grifola frondosa* lectin (GFL)	Inhibit proliferation of HeLa	[80]
Hericium erinaceus (Syn.*Hericium erinaceum*)	*Hericium erinaceus* agglutinin (HEA)	Inhibit proliferation of HepG2 and human breast cancer MCF7 cells	[81]
Kurokawa leucomelas	*Kurokawa leucomelas* KL-15	Inhibit proliferation of U937 cells	[82]
Lactarius flavidulus	*Lactarius flavidulus* lectin (LFL)	Inhibit proliferation of HepG2 and L1210 cells	[83]
Lignosus rhinocerotis	*Lignosus rhinocerotis* lectin (LRL)	Inhibit proliferation of HeLa, MCF7 and A549 cells	[84]
Marasmius oreades	*Marasmius oreades* agglutinin (MOA)	Inhibit proliferation of SW480, HepG2 and NIH-3T3 cells	[85]
Pholiota adiposa	*Pholiota adiposa* lectin (PAL)	Inhibit proliferation of HepG2 and MCF7 cells	[61]
Pleurotus citrinopileatus	-	Stimulate mice splenocytes mitogenicity and inhibit proliferation of S180 cells	[59]
Pleurotus eous	*Pleurotus eous* lectin (PEL)	Inhibit proliferation of MCF7, K562 and HepG2	[86]
Cerioporus squamosus (syn. *Polyporus squamosus*)	*Polyporus squamosus* lectin 1a (PSL1a)	Inhibit proliferation of HeLa cells	[87]
Pseudosperma umbrinellum (syn. *Inocybe umbrinella*)	*Inocybe umbrinella* lectin (IUL)	Inhibit proliferation of HepG2 and MCF7 cells	[60]
Russula delica	-	Inhibit proliferation of HepG2 and MCF7 cells	[64]

Table 3. *Cont.*

Source	Lectin name	Immunomodulatory effect	Refs
Russula lepida	*Russula lepida* lectin (RLL)	Inhibit proliferation of HepG2 and MCF7 cells	[88]
Schizophyllum commune	*Schizophyllum commune* lectin (SCL)	Stimulate mice splenocytes mitogenicity and inhibit proliferation of KB, HepG2 and S180 cells	[63,89]
Stropharia rugosoannulata	*Stropharia rugosoannulata* lectin (SRL)	Inhibit proliferation of HepG2 and L1210 cells	[90]
Leucocalocybe mongolica (syn. *Tricholoma mongolicum*)	*Tricholoma mongolicum* lectin 1 (TML-1), *Tricholoma mongolicum* lectin 2 (TML-2)	Inhibit proliferation of S180 cells	[91]
Volvariella volvacea	*Volvariella volvacea* lectin (VVL)	Inhibit proliferation of S180 cells and enhance IL-2 and IFN-γ transcriptions	[92,93]
Xerocomellus chrysenteron (syn. *Xerocomus chrysenteron*)	*Xerocomus chrysenteron* lectin (XCL)	Inhibit proliferation of NIH-3T3 and HeLa cells	[94]
Xylaria hypoxylon	*Xylaria hypoxylon* lectin (XHL)	Inhibit proliferation of HepG2 cells	[95]

Table 4. Major fungal immunomodulatory proteins (FIPs) from medicinal mushrooms and their immunomodulatory effects.

FIP Name	Source	Immunomodulatory Effect	Refs
FIP-aca	*Taiwanofungus camphoratus* (Syn. *Antrodia camphorate*)	Induce expression of different cytokines (IL-1b, IL-6, IL-12, TNF-α) and chemokines (CCL3, CCL4, CCL5, CCL10)	[96]
FIP-cru1	*Chroogomphis rutilus*	Stimulate the proliferation of murine splenocytes and enhanced the secretion of IL-2	[97]
FIP-dsq (FIP-dsq2)	*Dichomitus squalens*	Induce apoptosis and interrupt migration of A549 cells	[98]
FIP-fve	*Flammulina velutipes*	Stimulate mitogenesis in human peripheral lymphocytes, suppress systemic anaphylaxis reaction, enhance transcription of IL-3, INF-g	[99]
FIP-gja	*Ganoderma japonicum*	-	GenBank: AAX98241
FIP-gat	*Ganoderma atrum*	-	[100]
FIP-glu1 (LZ-8)	*Ganoderma lucidum*	Enhance transcription of IL-2, IL-3, IL-4, IFN-g, TNF-α	[101]
FIP-gmi	*Ganoderma microsporum*	Down regulation of TNF-α	[102]
FIP-gsi	*Ganoderma sinensis*	Enhance production of IL-2, IL-3, IL-4, INF-g, TNF-a	[103]
FIP-gts	*Ganoderma tsugae*	Induce cytokine secretion, cellular proliferation of human peripheral mononuclear cells (HPBMCs), enhance IFN-g expression	[104]
FIP-glu2 (LZ-9)	*Ganoderma lucidum*	Activate THP-1 macrophages and induce pro-inflammatory cytokine transcription	[105]
FIP-SN15	Intrageneric shuffled library	Induce U-251 MG cells apoptosis	[106]
FIP-Irh	*Lignosus rhinocerotis*	Inhibit the proliferation of MCF7, HeLa and A549 cancer cell lines	[84]
FIP-pcp	*Poria cocos*	Enhance production of IL-1b, IL-6, IL-18, TNF-a, NO	[107]
FIP-ppl	*Postia placenta*	Stimulate mouse splenocyte cell proliferation and enhance interleukin-2 (IL-2) release, inhibit proliferation and induce apoptotic effects on gastric tumor cells (MGC823)	[108]
FIP-tve2 (FIP-tvc)	*Trametes versicolor*	Increase human peripheral blood lymphocytes, enhanced production of TNF-a, NO	[109]
FIP-vvo	*Volvariella volvacea*	Enhance expression of IL-2, IL-4, IFN-g, TNF-a	[110]

"-" not tested.

3.3. Terpenes and Terpenoids

Terpenes are a large and diverse class of hydrocarbon compounds derived biosynthetically from units of isopentenyl pyrophosphate. They are widespread in nature, produced by a variety of plants, particularly conifers, some insects, and fungi, including mushrooms. The addition of functional groups (usually oxygen-containing) to terpenes produce terpenoids. Both terpenes and terpenoids from a number of medicinal mushrooms have shown immunoregulatory activities with medical significance. Table 5 shows the types of terpenes and terpenoids that have been isolated from medicinal mushrooms, including evidence for their specific immunomodulatory activities. For example, *Ganoderma* sp. are known for their high content of triterpenoids and these triterpenoids have shown high immunomodulating and anti-infective activities [111–113]. A study showed that terpenes and terpenoids modulate immune system functions by stimulating the expressions of genes coding for proteins in the nuclear factor (NF)–kB pathway and for mitogen-activated protein kinases [14].

Table 5. Major immunomodulatory terpenes and terpenoids from medicinal mushrooms and their immunomodulatory effects.

Type of Terpenes	Source	Compound	Immunomodulatory Effect	Refs
Monoterpenoids	*Pleurotus cornucopiae*	-	Inhibit the proliferation of HeLa and HepG2 cells	[114]
Sesquiterpenoids	*Stereum hirsutum*	-	Inhibit the proliferation of HepG2 and A549 cells	[111]
	Inonotus rickii	3α,6α-Hydroxycinnamolide	Inhibit the proliferation of SW480 cells	[112]
	Pleurotus cornucopiae	Pleurospiroketals A, B, C	Inhibit the proliferation of HeLa cells	[113]
	Anthracophyllum sp. *BCC18695*	Anthracophyllone	Inhibit the proliferation of MCF7, NCI-H187, KB and Vero cells	[115]
	Flammulina velutipes	Enokipodins B, D, J 2,5-cuparadiene-1,4-dione Flammulinolides A, B, C, F	Inhibit the proliferation of HepG2, MCF7, SGC7901, KB, HeLa and A549 cells	[116,117]
	Neonothopanus nambi	Nambinones C	Inhibit the proliferation of NCI-H187 cells	[118]
Diterpenoids	*Cyathus africanus*	Neosarcodonin O, 11-O-acetylcyathatriol, Cyathins H	Inhibit the proliferation of K562 and Hela cells	[119]
	Pleurotus eryngii	Eryngiolide A	Inhibit the proliferation of Hela and HepG2 cells	[120]
	Sarcodon scabrosus	Sarcodonin G	Inhibit the proliferation of HOC-21, HEC-1, U251-SP, MM-1CB and HMV-1 cells	[121]
	Tricholoma sp.	Tricholomalide A, B, C	Inhibit the proliferation of HeLa cells	[122]

Table 5. *Cont.*

Type of Terpenes	Source	Compound	Immunomodulatory Effect	Refs
Triterpenoids	*Ganoderma boninense*	Ganoboninketals A, B, C	Inhibit the proliferation of A549 and HeLa cells	[123]
	Ganoderma orbiforme BCC 22324	Ganoderic acid T and its C-3 epimer compound	Inhibit the proliferation of NCIH187, MCF7 and KB cells	[124]
	Ganoderma lucidum	lucialdehydes B, C, ganodermanondiol, ganodermanonol, ganoderic acid DM, ganoderic acid X	Inhibit the proliferation of T-47D, LLC, Meth-A, and Sarcoma 180 cells; Decrease the protein levels of CDK2, CDK6, p-Rb, cycle D1 and c-Myc in MCF7 cells; inhibit activity against topoisomerases I and II α and promote apoptosis	[125–127]
	Ganoderma concinna	5α-lanosta-7,9(11),24-triene-3β-hydroxy-26-al, 5α-lanosta-7,9(11),24-triene-15α-26-dihydroxy-3-one, 8α,9α-epoxy-4,4,14α-trimethyl-3,7,11,15,20-pentaoxo-5α-pregrane	Induce apoptosis in promyelocyticleukemia HL-60 cells	[128]
	Ganoderma tsugae	Tsugaric acid A, 3β-hydroxy-5α-lanosta-8, 24-dien-21-oic acid	Inhibit the proliferation of HT-3, T-24, and CaSKi cells	[129]
	Hypholoma fasciculare (syn. *Naematoloma fasciculare*)	Fusciculol C, L, M, G	Inhibit the proliferation of HCT-15, SK-OV-3, SK-MEL-2 and A549 cells	[130]
	Astraeus odoratus	Astraodoric acids A, B, D	Inhibit the proliferation of KB, NCI-H187, and MCF7 cells	[131]
	Russula lepida Russula amarissima	Cucurbitane hydroxyl acid	Inhibit the proliferation of WISH, CAKI 1 and A549 cells	[132]
	Leucopaxillus gentianeus	Cucurbitacin B Leucopaxillone A	Inhibit the proliferation of MCF7, HepG2, kidney carcinoma CAKI-1 and A549 cells	[133]
	Hebeloma versipelle	24(E)-3β-hydroxylanosta-8,24-dien-26-al-21-oic acid	Inhibit the proliferation of HL60, Bel-7402, SGC-7901 and A549 cells	[134]
	Tricholoma saponaceum	Saponaceol A	Inhibit the proliferation of HL-60 cells	[135]
	Elfvingia applanata	The methyl ester of elfvingic acid H	Inhibit the proliferation of Ehrlich and Kato III cells	[136]

4. Immunomodulation and Other Human Health Effects of Medicinal Mushrooms

The human immune system is tightly linked to tumor development. With the increasing impact of tumor on human health, a large number of studies have been undertaken to identify mushroom extracts/fractions/compounds with antitumor activities. Indeed, some of the observed antitumor activities by medicinal mushroom extracts were based on the activation of the immune system (Tables 2–5). Davis et al. (2020) recently suggested that 17 medicinal mushroom species (*A. brazei*, *Cordyceps militaris*, *Flammulina velutipe*, *F. fomentarius*, *F. officinalis*, *Ganoderma applanatum*, *G. lucidum*, *Ganoderma oregonense*, *G. frondosa*, *Hericium erinaceus* (syn. *Hericium erinaceum*), *I. obliquus*, *L. edodes*, *Tropicoporus linteus* (syn. *Phellinus linteus*), *P. betulinus*, *Pleurotus ostreatus*, *S. commune*) could support both immune-activation for cancer treatments and help resolve host defense-induced inflammatory reactions and facilitate a post-response return to homeostasis for cancer patients. Furthermore, a medicinal mushroom formulation consisting of *G. lucidum*, *L. edodes* and *G. frondosa* showed synergistic antitumor and immuno-modulatory activity in human macrophages [137].

As shown in Tables 2–5, many medicinal mushrooms each can produce different categories of compounds with immunomodulatory effects. Furthermore, different extractions of the same mushroom may show non-overlapping but complementary activities. For example, in *L. edodes*, its heterogalactan (fucomannogalactan) has anti-inflammatory properties [138], lentinan has anti-tumor effect [139], crude water-soluble polysaccharides can activate macrophages and increase the productions of nitric oxide (NO), cytokines, and proteins related to phagocytosis [140], and polysaccharides with both antioxidant effects [141] and antiviral activities [142].

The significance of functional components from medicinal mushrooms has been shown not only from clinical perspectives but also from foods. Because many medicinal mushrooms are commercially cultivated for food, there has been an increasing trend of including mushrooms and their components into other foods to develop functional foods, including adding new flavors or promoting certain types of functions. For example, Ulziijargal et al. used mushroom mycelia of *T. camphoratus*, *A. blazei*, *H. erinaceus*, and *P. linteus* to substitute 5% of wheat flour to make bread. The final product contains substantial amounts of the amino acids gaminobutyric acid (GABA) and ergothioneine and showed beneficial health effects [143]. Kim et al. developed noodles that contained *L. edodes* paste, resulting in a higher quality, fibre-rich functional food with antioxidant and hypocholesterolemic properties [144]. Components of other mushrooms, e.g., *Pleurotus sajor-caju* dry powder, *A. bisporus* extracts, the freeze-dried powder from *A. aegerita*, *Suillus luteus*, and *Coprinopsis atramentaria* have been used to develop snacks and cheese-related products with much commercial success [145–148]. The benefits include increased proteins, minerals, crude fiber or ingredients with antioxidant potentials and free radical scavenging capacities. The diversified flavors and tastes as well as the enhanced nutritional values of food due to the addition of mushroom components represent an exciting direction for the edible and medicinal mushroom industries.

5. Mechanisms for the Immunomodulation Effects of Medicinal Mushroom Compounds

The immune system consists of a network of cells, tissues and organs that work together to defend the body against attacks by "foreign" invaders [2]. The network is connected by lymphatic vessels from organ to organ. The network includes protective barriers that constantly communicate with lymphatic fluid rich in white blood cells and leukocytes. When pathogens break our physical barriers (i.e., skin and mucosal membranes of the mouth, nose, the gastrointestinal tract, and the urogenital tract), the next line of the body's defense response is activated. This line of defense includes granulocytes and monocytes that also function as antigen-presenting cells (APCs) for helper T lymphocytes. These cells synthetize and secrete lipid mediators such as prostaglandins as well as cytokines which act as messengers in regulating immune response and stimulating adaptive immunity. For example, natural killer (NK) cells can recognize infected and abnormal cells, such as cancer cells and kill these cells by inducing them to undergo apoptosis or by producing cytokines, such as interferon-gamma (IFN-γ). They also activate macrophages and kill phagocytosed microbes. Figure 3 presents the overview of the key immune response against microbial pathogens.

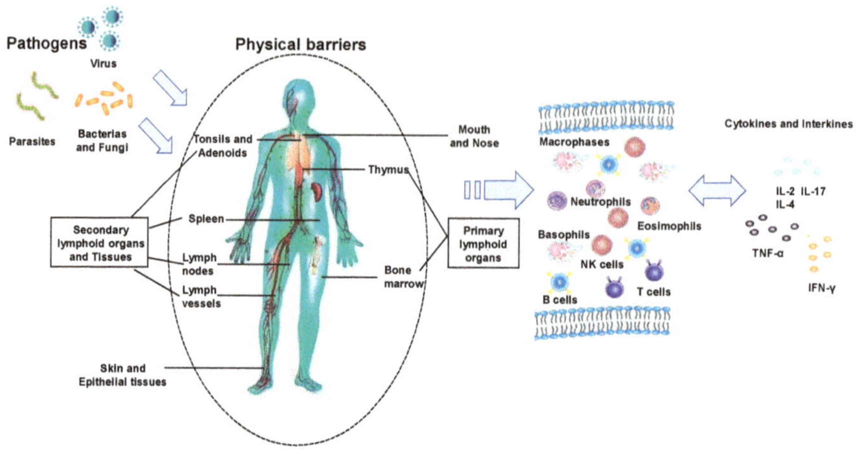

Figure 3. Key components and pathways in host immune response against pathogen infections.

It is well known that the human immune system can be modulated by foods, supplements or endogenous bioactive agents [2]. Different types of immunoregulatory compounds have been isolated from medicinal mushrooms, including mushroom fruiting bodies and fermented mycelia. Tables 2–5 present the specific components isolated from different mushroom species that have shown significant immunomodulatory activities, including their (potential) mechanisms of action. For example, two polysaccharides from two different mushroom species have shown significant immunoenhancing activities. In the first, a glucuronoxylomannan TAP-3 obtained from *Naematelia aurantialba* (syn. *Tremella aurantialba*) showed marked immune enhancement activity and promoted NO, IL-1β and TNF-α secretions from macrophages [149]. Similarly, another study showed that at a concentration of 40 μg/mL, polysaccharide CCP from *Craterellus cornucopioides* strengthened the phagocytic function of macrophages, increased the expression of cytokines, upregulated the expression of cell membrane receptor TLR4 and downstream protein kinase products through activation of the TLR4–NFκB pathway [150].

Some of these bioactive compounds can also directly attack cancer cells while showing immunoregulatory effects. For example, Li et al. reported that polysaccharide LRP-1 purified from *Leccinum rugosiceps* inhibited the growth of human hepatoma cells HepG2 and human breast carcinoma MCF-7 cells, and induced the secretion of NO, IL-6 and TNF-α in vitro [151]. Similarly, a recent report showed that an aqueous extract of *Sarcodon imbricatus* (SIE) effectively inhibited the growth, migration, and invasion properties of breast cancer cells in vitro and reduced tumor growth in vivo, while showing increased expression of PD-L1 and increased NK cell viability [152]. Furthermore, Xue et al. reported that a triterpenoid EAe from *Pleurotus eryngii* inhibited MCF-7 cell lines proliferation with an EC50 of 298.26 μg/mL, and significantly inhibited the growth of CD-1 tumors (inhibition rate of 65%) in mice in a dose-dependent manner without toxicity [153].

6. Relationship between Structure and Activity of Immunomodulatory Compounds from Medicinal Mushrooms

Immunomodulators from medicinal mushrooms have been shown to be capable of stimulating both innate and adaptive immune responses. They activate innate immune system components such as natural killer (NK) cells, neutrophils, and macrophages, and stimulate the expression and secretion of cytokines. These cytokines in turn activate adaptive immunity by promoting B cell proliferation and differentiation for antibody production and by stimulating T cell differentiation to T helper (Th) 1 and Th2 cells, which mediate cellular and humoral immunities, respectively [154].

As shown in Tables 2–5, a large number of immunoregulatory compounds from medicinal mushrooms have been reported. These compounds differ greatly in their molecular weight and structure. Below we describe the relationships between their molecular structures and immunoregulatory activities.

6.1. Polysaccharides

Polysaccharides are the most commonly reported natural immunomodulators from mushrooms. The immunomodulating polysaccharides are highly diversified in their sugar compositions, main chain polymer structures, degrees of branching, conformations, molecular weights, and other physical properties, which together have significant effects on the bioactivity and mode of action of each polysaccharide [18]. These polysaccharides are either homoglycans (polysaccharides that contain residues of only one type of monosaccharide molecule) or heteroglycans (polysaccharides that contain residues of two or more types of monosaccharide molecules), and are able to combine with other molecules such as oligo- or poly-peptides to make peptidoglycan or polysaccharide–protein complexes. In general, higher molecular weight polysaccharides exhibit greater bioactivity [155]. These large polysaccharides are not able to penetrate the immune cells, but instead act to bind cell receptors. For example, the highest immunomodulatory activity of PSK was associated with the highest molecular weight fraction of this compound, at >200 kDa [156]. Similarly, the highest activity of a polysaccharide fraction of *G. frondosa* extract was ascribed to one with a molecule weight of over 800 kDa [157]. In contrast, low molecular weight polysaccharides can penetrate immune cells and exert stimulatory effects from within.

The number and lengths of short branched chains in mushroom polysaccharides can significantly influence their bioactivity [155]. In most cases, the bioactive immunomodulator polysaccharides are characterized by a main chain of 1,3-β-D-glucan with a small number of short branched chains with 1,6-β-linkage (Figure 4). Studies have shown that immunologically active polysaccharides generally have a degree of branching number (DB) between 20% and 40%. For example, lentinan has a DB number of 40%, schizophyllan of 33%, and PSK of 20%. While a high DB number is generally correlated with a high activity, in some cases, debranching of polysaccharides can also increase their bioactivity. For example, the partially debranched form of pachymaran from *Poria cocos* showed greater activity than the original native form [158]. Even in the well-studied lentinan, its maximal immunomodulating and antitumor activities were achieved when the molecule had a DB of 32% [159], and there was a negative correlation between their biological activity and DB number between 32 and 40% [160].

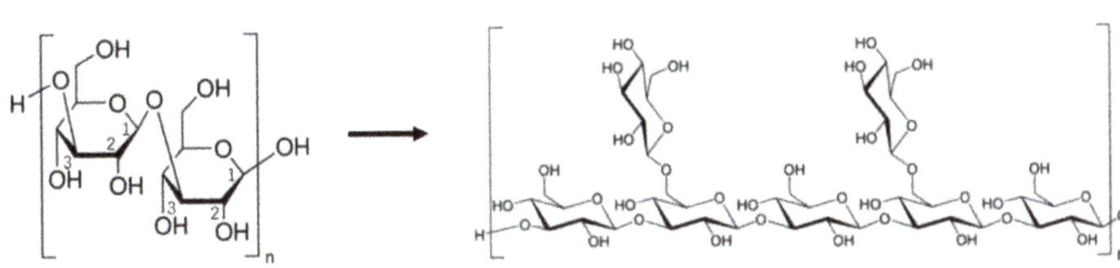

Figure 4. The basic structures of bioactive polysaccharides 1-3,-β-D-glucan and short branched chains with 1-6,-β-linkage.

Aside from the main-chain structure and branching pattern, the conformation of polysaccharides can also impact their bioactivity, e.g., by influencing the stability of the structure. In polysaccharides, the triple helix conformation is usually more stable than other conformations and bears the cytokine stimulating activity of the β-D-glucan. Lentinan, schizophyllan, scleroglucan, and PSK all have a triple helix structure [161]. However, not all polysaccharide immunomodulators from mushrooms have the triple helix structure. For example, mushroom polysaccharides with a random coil conformation can also have potent immunomodulating and anticancer activity [56].

Chemical modification is an effective and common approach to increase biological activities of polysaccharides. This approach has been applied to develop a number of effective immunomodulators from mushroom polysaccharides. Those modifications include carboxymethylation, hydroxylation, formyl-methylation, amino-ethylation, or sulfation. The introduction of such chemical groups may increase the possible contacts between the modified polysaccharides and the immune cell receptors through hydrogen bonding and/or electrostatic attraction, and thus increase the immunological response. For example, the sulfated cell wall glucan from *L. edodes* exhibited higher immunomodulatory and anticancer activities compared to the native polysaccharides [162]. The increased effect from sulfation may be related to the increased solubility, as shown in the hyper branched β-glucan TM3b. Taken together, molecular weight, branching, chemical configuration, and chemical modification can all have strong influence on the bioactivity of polysaccharides from mushrooms.

6.2. Lectins

Lectins belong to a unique group of proteins that can recognize and interact with various cell surface carbohydrates/glycoproteins. Mushroom lectins have shown specific immunomodulatory, antiproliferative, and antitumor activities [163]. The diverse sources of mushroom lectins have different immunomodulatory mechanisms: some mediate their actions by activating the immune system while others produce potent cytotoxic effects towards cells [91]. For example, two lectins extracted from *L. mongolica* (syn.*T. mongolicum*), TML-1 and TML-2, show immunomodulatory and antitumor activities. These two lectins stimulate the production of nitrite and tumor necrosis factor (TNF)-α and inhibit the growth of mouse lymphoblast-like (p815) mastocytoma cells by the production of macrophage-activating factors. These factors include interferon (IFN)-γ and other cytokines, activated through upregulation of inducible nitric oxide synthase (NOS), interleukin (IL)-1β, and transforming growth factor-β [91]. *G. frondosa* lectin is reported as having a potent cytotoxic effect against HeLa cells in vitro, even at very low concentrations. A 15.9-kDa homodimeric, lactose-binding, ricin-B-like lectin (CNL) from *Clitocybe nebularis* exhibited antiproliferative activity against human leukemic T cells [78], which induces the maturation and activation of dendritic cells (DCs) and stimulates several proinflammatory cytokines such as IL-6, IL-8, and TNF-α [164]. The encoding gene of CNL from *C. nebularis* has been cloned and successfully expressed in *Escherichia coli* [165].

6.3. FIPs

The fungal immunomodulatory proteins are a group of proteins with highly similar amino acid sequences. They exist as dimers in a dumbbell-shaped structure similar to that of the variable region of immunoglobulin heavy chains [166]. The FIPs have shown diverse functions. Through binding to Toll-like receptors (TLRs), FIPs stimulate antigen presenting cells and release cytokines such as NO and IL-12. By activating phosphorylation of p38/MAPK and increasing the production of NF-κB, FIPs can promote the proliferation and differentiation of helper T cells (Th0) to form Th1 cells and Th2 cells, activate macrophages and B cells, produce a variety of cell factors (Figure 5). For example, FIP-fve from *Flammulina velutipes* can upregulate the expression of intercellular adhesion molecules on the T cell surface by phosphorylation of p38/MAPK, and activate Th1 cells to produce IL-2, IFN-γ, to exert its immunomodulatory effect [99]. FIP-vvo can not only activate Th1 cells and enhance IL-2, TNF-α and IFN-γ transformations, but also induce Th2 cells to produce IL-4, B cell differentiation, and the transformation of immunoglobulin and production of antibody IgE. Several studies have also shown that by interacting with TLRs, FIP can activate other signaling pathways besides the p38/MAPK and NF-κ B. For example, FIP from *Ganoderma tsugae* (FIP-gts) can stimulate human peripheral blood monocyte to produce IFN-γ and activates the PI3K/Akt signaling pathway [104].

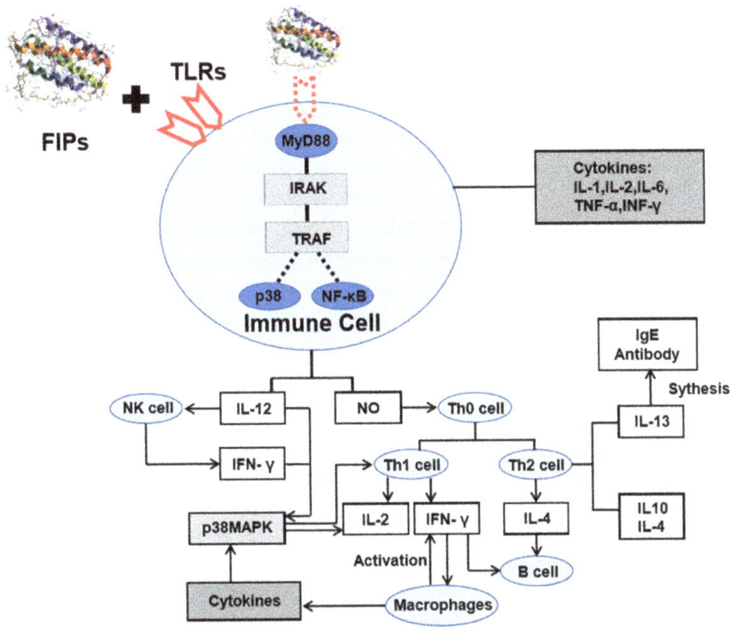

Figure 5. FIPs immunomodulatory mechanism by toll-like receptors (TLRs) signaling pathway.

FIPs typically exist in low quantities in their native mushrooms. The low yield/production has been a major limitation of their research and application. Therefore, techniques are being rapidly developed to enhance the production of recombinant FIPs in other organisms such as the yeast *Pichia pastoris* and the bacterium *E. coli*. For example, the LZ-8 gene of *G. lucidum* has been expressed in *P. pastoris* to produce a recombinant LZ-8 protein (rLZ-8). While the recombinant protein lacks the carbohydrate moiety of the native protein, it shows similar bioactivity for IL-2 induction as the native protein. The FIP-fve protein has also been successfully expressed in *E. coli* [1]. Interestingly, the recombinant FIPs showed higher immunomodulatory activity and induced greater expression of specific cytokines than that extracted from the mushrooms [167].

6.4. Terpenes and Terpenoids

Terpenes and terpenoids are widely distributed in mushrooms. They are a large and diversified group of organic compounds but share the core of isoprene five-carbon atom units of molecular formula $(C_5H_6)n$ as the main building block (Figure 6) [1,13]. Among this group of compounds, the best-known is probably the triterpenoids from *G. lucidum* and *G. lingzhi*. These triterpenoids can help reduce drug nephrotoxicity and minimize inflammation. Figure 6 shows a diversity of terpene derivatives in *G. lucidum* and *G. lingzhi*, including ganodermic and ganoderic acids, ganoderals, ganoderols, ganodermanontriol, lucidone, and ganodermanondiol. All these compounds have shown immunomodulating, antitumor, and/or anti-infective activities [168]. At present, their mechanisms of action and structure–activity relationships are little understood. However, their broad activities suggest significant potential for research and for clinical therapeutic applications.

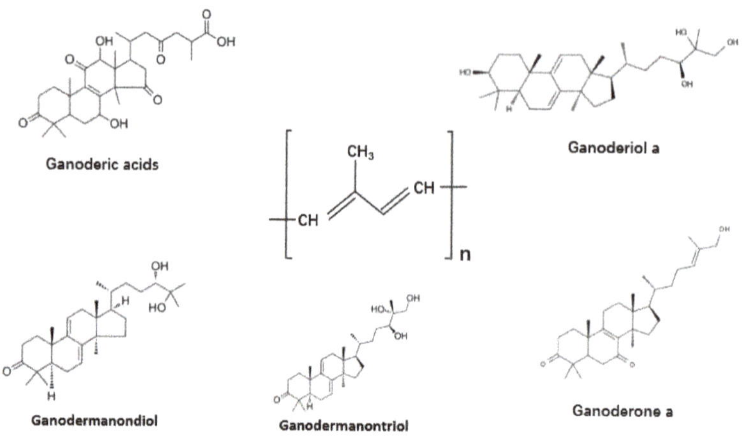

Figure 6. The structures of representative terpenes and terpenoids in the fungus *Ganoderma lucidum* and *Ganoderma lingzhi*.

7. Genomes and Molecular Techniques in the Study of Immunomodulatory Compounds in Medicinal Mushrooms

Due to their environmental, agricultural, commercial, and/or medical interests, the genomes of a number of medicinal mushrooms have been sequenced and annotated. Table 6 lists the genomic features of 12 representative medicinal mushrooms. These species differ in genome size and/or gene content. Analyses of these genomes have revealed some of the genes related to the synthesis and production of immunomodulatory compounds in medicinal mushrooms. Not surprisingly, the most commonly identified genes related to immunomodulatory effects are those coding for FIPs (Table 6). However, in *G. lucidum*, genes involved in the synthesis of several other immunomodulators have also been identified. The identification and confirmation of those genes require genetic manipulation systems which are not available at present for most medicinal mushrooms. In *G. lucidum*, such a system is available.

Ganoderic acids (GAs) are among the main active ingredients of *G. lucidum* with immunomodulatory effects. GAs belong to the triterpenoid secondary metabolites. Genome sequence analyses and functional studies showed that the terpenoids in *G. lucidum* are synthesized through the Mevalonate (MVA) pathway. Several genes in this pathway in *G. lucidum* have been cloned and their functional roles confirmed, including those encoding 3-Hydroxy-3-methylglutaryl-CoA reductase (HMGR) and Farnesyl diphosphate synthase (FPPs). Genome sequence data mining also identified the putative genes involved in the modification of the triterpene backbone, such as those involved in cyclization and glycosylation, which are very important for the synthesis of the diversity of GAs in *G. lucidum*.

At present, most of the genes and metabolic pathways involved in the synthesis of immunomodulators in the categories of polysaccharides, lectins, and terpenoids in medicinal mushrooms have not been identified or confirmed. However, the availability of increasing genomic resources coupled with the broad pharmacological activities and therapeutic effects of medicinal mushrooms should help facilitate the identification of genes and metabolic pathways involved in their biosynthesis. Such understandings could help future productions of those compounds through biotechnology using surrogate hosts.

Table 6. Major genomic features of representative medicinal mushrooms.

Medicinal Mushroom	Genome Size Mb	Number of Genes	GC Content (%)	Known Genes Related to Immunomodulatory Effects	Genetic Manipulations (Transformation Method)	Refs
Agrocybe aegerita	44.7908	14110	49.2		Polyethylene glycol–mediated transformation (PEG)	[169]
Agaricus bisporus	30.78	10863	46.5		PEG, Electroporation, Particle bombardment, *Agrobacterium tumefaciens*-mediated transformation (ATMT)	[170–173]
Flammulina velutipes	35.64		49.76	*Fip-fve*	PEG, Electroporation, Electro-injection, Restriction enzyme-mediated integration (REMI), ATMT	[173–177]
Ganoderma atrum				*fip-gat*		[178]
Ganoderma lucidum	43.68		55.4	*fip-glu*, Mevalonate (MVA) pathway genes: *AACT* (acetyl-CoA acetyltransferase); *HMGS* (3-hydroxy-3-methylglutaryl-CoA synthase), *HMGR* (3-hydroxy-3-methylglutaryl-CoA reductase), *MVK* (mevalonate kinase), *MPK* (phosphomevalonate kinase), *MVD* (pyrophosphomevalonate decarboxylase), *IDI* (isopentenyl-diphosphate isomerase), *GPPs* (geranyl diphosphate synthase), *FPPs* (farnesyl diphosphate synthase), *SQS* (squalene synthase), *SE* (squalene monooxygenase), *OSC* (2,3-oxidosqualene-lanosterol cyclase), *P450* (cytochrome P450), *UGTs* (uridine diphosphate glycosyltransferases)	PEG, Electroporation, REMI	[173,179–181]
Ganoderma sinensis	48.96	15478	55.6	*fip-gsi*		[103,182]
Ganoderma tsugae	45.5			*fip-gts*		[183]
Hericium erinaceus (syn. *Hericium erinaceum*)	41.21		52.43		ATMT	[184,185]
Lentinula edodes	39.92	12051	46		PEG, Electro-injection, REMI, ATMT	[173,186]
Pleurotus ostreatus	34.36	12296	50.76		PEG, Electroporation, REMI, Particle bombardment	[173,187–189]
Postia placenta	66.6724	12716	47.2	*fip-ppl*		[108,190]
Trametes versicolor	44.794	14572	57.3	*fip-tvc*		[191,192]
Volvariella volvacea	35.72		48.8	*fip-vvo*	PEG, Particle bombardment, ATMT	[193,194]

8. Conclusions and Perspectives

Many edible and medicinal mushrooms contain compounds with significant immunoregulatory activities. This paper attempts to provide a comprehensive review on the types of these compounds; their distributions, structures and functions; and their potential mechanisms of actions. These compounds have shown their activities through in vitro, ex vivo, tissue cultures, and/or in vivo studies. Some of these compounds have been commercialized and licensed for clinical use.

Aside from the above described compounds, other compounds from edible and medicinal mushrooms may also have great potentials. One such group is chitin and chitin-related compounds. Chitin is the most common aminopolysaccharide polymer in nature and the main material that gives strength to the fungal cell walls (as well as to the exoskeletons of crustaceans and insects). Through deacetylation, either chemically or enzymatically, chitin can be converted to chitosan, a well-known derivative. Through hydrolysis, both chitin and chitosan can be converted to chito-oligosaccharides. Several fungal chitin, chitosan, and chito-oligosaccharides have shown promising benefits to humans and human health [195,196]. For example, chitins from filamentous molds such as *Aspergillus niger* and *Mucor rouxii* and other organisms have been used in plant protection and food processing; chitosan in diagnosis, drug delivery, infection control, molecular imaging, and wound healing; and chito-oligosaccharides in antimicrobial and antitumor activities [195,196]. At present, none of those tested chitin, chitosan, and chito-oligosaccharides for human effects have come from edible or medicinal mushrooms yet. However, due to the similar chemical structures of chitin from different groups of organisms, it's highly likely that this group of natural products from edible and medicinal mushrooms will have similar effects and they represent a promising area of future development for edible and medicinal mushrooms.

While the future looks bright, significant issues remain before the full potential of medicinal mushrooms can be reached. Specifically, during our review of the literature, we identified several significant gaps and areas for future research and development. In the first, there is an urgent need to identify the structures and mechanisms of action for active ingredients in many extracts and formulations from medicinal mushrooms. More rigorous chemical analyses as well as understanding the in vivo pharmacokinetics and pharmacodynamics of individual compounds are needed to fill this gap of knowledge. The second promising area of study is to identify the genes and metabolic pathways involved in producing these immunomodulators in medicinal mushrooms. As shown above, aside from the few FIPs where the specific encoding genes have been identified, we have little information about the genes and how they are regulated in producing the other types of mushroom immunomodulators. While the availability of high through-put technologies and genome sequences are facilitating the discoveries, experimental investigations are needed in order to confirm and identify the conditions for increased productions of these compounds. Fortunately, gene editing technologies and -omics tools are becoming increasingly accessible to the broader life sciences communities. In the third, most immunomodulators described above exist in low quantities in medicinal mushrooms and their extractions can take a long time and be costly. For efficient production, it is very important to develop alternative approaches, e.g., by cloning and expressing the relevant biosynthesis genes in alternative hosts, using industrial fermentations, and developing efficient extraction and purification protocols from such commercial cell cultures. Lastly, the potential interactions between immunomodulators from mushrooms and other medicines, foods, and food supplements need to be critically analyzed in order to establish guidelines for safe and effective use of immunomodulators from medicinal mushrooms [197]. Indeed, at present, safety data about many medicinal mushroom products are not available from controlled clinical trials and associated negative side effects have been reported in several cases for certain types of usages of medicinal mushroom products [198,199]. There is also a cultural difference between Oriental and Western cultures about the use of medicinal mushrooms [197], presenting both a challenge and an opportunity for researchers and policy makers on the broad implications of these products on human health.

Author Contributions: S.Z. and J.X. conceived and prepared the manuscript, Q.G., C.R. and Z.Z. participated in the manuscript editing; S.W. and Y.L. provided the photos of mushrooms. All authors have read and agreed to the published vision of manuscript.

Funding: This work was supported by National Natural Science Foundation of China (NSFC31701975), by the Collaborative Innovation Center of Beijing Academy of Agricultural and Forestry Sciences (KJCX201915), and by Beijing Academy of Agriculture and Forestry Sciences (KJCX20200208).

Acknowledgments: We thank Junliang Zhou from Kunming University for providing the photos of mushroom species.

Conflicts of Interest: The authors declare no conflict of interest.

References

1. El Enshasy, H.A.; Hatti-Kaul, R. Mushroom Immunomodulators: Unique Molecules with Unlimited Applications. *Trends Biotechnol.* **2013**, *31*, 668–677. [CrossRef] [PubMed]
2. Reis, F.S.; Martins, A.; Vasconcelos, M.H.; Morales, P.; Ferreira, I.C.F.R. Functional Foods Based on Extracts or Compounds Derived from Mushrooms. *Trends. Food Sci. Technol.* **2017**, *66*, 48–62. [CrossRef]
3. Himanshi, R.; Shalinee, P.; Satyawati, S. Mushroom nutraceuticals for improved nutrition and better human health: A review. *PharmaNutrition* **2017**, *5*, 35–46. [CrossRef]
4. Mannino, G.; Stefano, V.D.; Lauria, A.; Pitonzo, R.; Gentile, C. Vaccinium Macrocarpon (Cranberry)-Based Dietary Supplements: Variation in Mass Uniformity, Proanthocyanidin Dosage and Anthocyanin Profile Demonstrates Quality Control Standard Needed. *Nutrients* **2020**, *12*, 992. [CrossRef]
5. Brya, P. Who Reads Food Labels? Selected Predictors of Consumer Interest in Front-of-Package and Back-of-Package Labels during and after the Purchase. *Nutrients* **2020**, *12*, 2605. [CrossRef]
6. Shukla, S.; Bajpai, V.K.; Kim, M. Plants as Potential Sources of Natural Immunomodulators. *Rev. Environ. Sci. Biol.* **2014**, *13*, 17–33. [CrossRef]
7. Wasser, S.P. Medicinal Mushroom Science: History, Current Status, Future Trends, and Unsolved Problems. *Int. J. Med. Mushrooms* **2010**, *12*, 1–16. [CrossRef]
8. Davis, R.; Taylor, A.; Nally, R.; Benson, K.F.; Stamets, P.; Jensen, G.S. Differential Immune Activating, Anti-Inflammatory, and Regenerative Properties of the Aqueous, Ethanol, and Solid Fractions of a Medicinal Mushroom Blend. *J. Inflamm. Res.* **2020**, *13*, 117–131. [CrossRef]
9. Chilton, J. A New Analytical Fingerprinting Method for Quality Control of Medicinal Mushroom Products. In Proceedings of the 2016 International Society for Mushroom Science, Amsterdam, The Netherlands, 30 May–2 June 2016; Sonnenberg, B., Ed.; International Society for Mushroom Science (ISMS): Amsterdam, The Netherlands, 2016; pp. 267–270.
10. Dubey, S.K.; Chaturvedi, V.K.; Mishra, D.; Bajpeyee, A.; Tiwari, A.; Singh, M.P. Role of Edible Mushroom as A Potent Therapeutics for the Diabetes and Obesity. *3 Biotech* **2019**, *9*. [CrossRef]
11. Chaturvedi, V.K.; Agarwal, S.; Gupta, K.K.; Ramteke, P.W.; Singh, M.P. Medicinal Mushroom: Boon for Therapeutic Applications. *3 Biotech* **2018**, *8*. [CrossRef]
12. Wong, J.H.; Ng, T.B.; Chan, H.H.L.; Liu, Q.; Man, G.C.W.; Zhang, C.Z.; Guan, S.; Ng, C.C.W.; Fang, E.F.; Wang, H.; et al. Mushroom Extracts and Compounds with Suppressive Action on Breast Cancer: Evidence from Studies Using Cultured Cancer Cells, Tumor-Bearing Animals, and Clinical Trials. *Appl. Microbiol. Biotechnol.* **2020**, *104*, 4675–4703. [CrossRef]
13. Moradali, M.F.; Mostafavi, H.; Ghods, S.; Hedjaroude, G.A. Immunomodulating and Anticancer Agents in the Realm of Macromycetes Fungi (Macrofungi). *Int. Immunopharmacol.* **2007**, *7*, 701–724. [CrossRef]
14. Gao, Y.H.; Zhou, S.F.; Chen, G.L.; Dai, X.H.; Ye, J.X. A Phase I/II Study of a *Ganoderma lucidum* (Curt.: Fr.) P. Karst. Extract (Ganopofy) in Patients with Advanced Cancer. *Int. J. Med. Mushrooms* **2002**, *4*, 207–214. [CrossRef]
15. Lee, J.S.; Cho, J.Y.; Hong, E.K. Study on Macrophage Activation and Structural Characteristics of Purified Polysaccharides from the Liquid Culture Broth of *Hericium erinaceus*. *Carbohydr. Polym.* **2009**, *78*, 162–168. [CrossRef]
16. Riede, I. Tumor Therapy with *Amanita phalloides* (Death Cap): Stabilization of B-Cell Chronic Lymphatic Leukemia. *J. Altern. Complement. Med.* **2010**, *16*, 1129–1132. [CrossRef]

17. Van, G.; Leo, J.L.D. Culinary-Medicinal Mushrooms: Must Action Be Taken? *Int. J. Med. Mushrooms* **2009**, *11*, 281–286. [CrossRef]
18. Chakraborty, I.; Sen, I.K. Bioactive Polysaccharides from Natural Sources: A review on the Antitumor and Immunomodulating Activities. *Biocatal. Agric. Biotechnol.* **2019**, *22*, 101425. [CrossRef]
19. Ina, K.; Kataoka, T.; Ando, T. The Use of Lentinan for Treating Gastric Cancer. *Anti-Cancer Agent Med. Chem.* **2013**, *13*, 681–688. [CrossRef]
20. Ngwuluka, N.C.; Ochekpe, N.A.; Aruoma, O.I. Functions of Bioactive and Intelligent Natural Polymers in the Optimization of Drug Delivery. In *Industrial Applications for Intelligent Polymers and Coatings*; Springer International Publishing: New York, NY, USA, 2016; pp. 165–184. [CrossRef]
21. Cui, J.; Chisti, Y. Polysaccharopeptides of *Coriolus versicolor*: Physiological Activity, Uses, and Production. *J. Shanxi Med. Univ.* **2003**, *21*, 109–122. [CrossRef]
22. Oba, K.; Teramukai, S.; Kobayashi, M.; Matsui, T.; Kodera, Y.; Sakamoto, J. Efficacy of Adjuvant Immunochemotherapy with Polysaccharide K for Patients with Curative Resections of Gastric Cancer. *Cancer Immunol. Immun.* **2007**, *56*, 905–911. [CrossRef] [PubMed]
23. Hattori, T.S.; Komatsu, N.; Shichijo, S.; Itoh, K. Protein-Bound Polysaccharide K Induced Apoptosis of the Human Burkitt Lymphoma Cell Line, Namalwa. *Biomed. Pharm.* **2004**, *58*, 226–230. [CrossRef] [PubMed]
24. Kato, M.; Hirose, K.; Hakozaki, M.; Ohno, M.; Saito, Y.; Izutani, R.; Noguchi, J.; Hori, Y.; Okumoto, S.; Kuroda, D. Induction of Gene Expression for Immunomodulating Cytokines in Peripheral Blood Mononuclear Cells in Response to Orally Administered PSK, An Immunomodulating Protein-bound Polysaccharide. *Cancer Immunol. Immun.* **1995**, *40*, 152–156. [CrossRef] [PubMed]
25. Price, L.A.; Wenner, C.A.; Sloper, D.T.; Slaton, J.W.; Novack, J.P. Role for Toll-like Receptor 4 in TNF-alpha Secretion by Murine Macrophages in Response to Polysaccharide Krestin, a *Trametes versicolor* mushroom extract. *Fitoterapia* **2010**, *81*, 914–919. [CrossRef]
26. Kidd, P.M. The Use of Mushroom Glucans and Proteoglycans in Cancer Treatment. *Altern. Med. Rev.* **2000**, *5*, 4–27.
27. Ooi, V.E.C. Pharmacological Studies on Certain Mushrooms from China. *Int. J. Med. Mushrooms* **2001**, *3*, 1. [CrossRef]
28. Biedron, R.; Tangen, J.M.; Maresz, K.; Hetland, G. *Agaricus blazei* Murill—Immunomodulatory Properties and Health Benefits. *Funct. Foods Health Dis.* **2012**, *2*, 428–447. [CrossRef]
29. Firenzuoli, F.; Gori, L.; Lombardo, G. The Medicinal Mushroom *Agaricus blazei* Murrill: Review of Literature and Pharmaco-Toxicological Problems. *Evid.-Based Complement. Altern. Med.* **2008**, *5*, 3–15. [CrossRef]
30. Ma, Z.; Wang, J.; Zhang, L.; Zhang, Y.; Ding, K. Evaluation of Water Soluble β-d-glucan from *Auricularia auricular-judae* as Potential Anti-tumor Agent. *Carbohydr. Polym.* **2010**, *80*, 977–983. [CrossRef]
31. Pacheco-Sanchez, M.; Boutin, Y.; Angers, P.; Gosselin, A.; Tweddell, R.J. A Bioactive (1 -> 3)-, (1 -> 4)-beta-D-glucan from *Collybia dryophila* and Other Mushrooms. *Mycologia* **2006**, *98*, 180–185. [CrossRef]
32. Chen, P.X. Properties of *Cordyceps sinensis*: A Review. *J. Funct. Foods* **2013**, *5*, 550–569. [CrossRef]
33. Yao, H.Y.; Zhang, L.H.; Shen, J.; Shen, H.J.; Jia, Y.L.; Yan, X.F.; Xie, Q.M. Cyptoporus Polysaccharide Prevents Lipopolysaccharide-Induced Acute Lung Injury Associated with Down-regulating Toll-like Receptor 2 Expression. *J. Ethnopharmacol.* **2011**, *137*, 1267–1274. [CrossRef] [PubMed]
34. Yin, H.; Wang, Y.; Wang, Y.; Chen, T.; Tang, H.; Wang, M. Purification, Characterization and Immuno-Modulating Properties of Polysaccharides Isolated from *Flammulina velutipes* Mycelium. *Am. J. Chin. Med.* **2010**, *38*, 191–204. [CrossRef]
35. Zhu, X.L.; Chen, A.F.; Lin, Z.B. *Ganoderma lucidum* Polysaccharides Enhance the Function of Immunological Effector Cells in Immunosuppressed Mice. *J. Ethnopharmacol.* **2007**, *111*, 219–226. [CrossRef]
36. Matsui, K.; Kodama, N.; Nanba, H. Effects of Maitake (*Grifola frondosa*) D-Fraction on the Carcinoma Angiogenesis. *Cancer Lett.* **2001**, *172*, 193–198. [CrossRef]
37. Dong, P.W.; Lee, J.S.; Kwon, D.S.; Lee, K.E.; Shin, W.C.; Hong, E.K. Immunostimulating Activity by Polysaccharides Isolated from Fruiting Body of *Inonotus obliquus*. *Mol. Cells* **2011**, *31*, 165–173. [CrossRef]
38. Mizuno, T. The Extraction and Development of Antitumor-Active Polysaccharides from Medicinal Mushrooms in Japan (Review). *Int. J. Med. Mushrooms* **1999**, *1*, 9–29. [CrossRef]
39. Bisen, P.S.; Baghel, R.K.; Sanodiya, B.S.; Thakur, G.S.; Prasad, G.B.K.S. *Lentinus edodes*: A Macrofungus with Pharmacological Activities. *Curr. Med. Chem.* **2010**, *17*, 2419–2430. [CrossRef]

40. Morales, D.; Rutckeviski, R.; Villalva, M.; Abreu, H.; Smiderle, F.R. Isolation and Comparison of α- and β-D-glucans from Shiitake Mushrooms (*Lentinula edodes*) with Different Biological Activities. *Carbohydr. Polym.* **2019**, *229*, 115521. [CrossRef] [PubMed]
41. Bhunia, S.K.; Dey, B.; Maity, K.K.; Patra, S.; Mandal, S.; Maiti, S.; Maiti, T.K.; Sikdar, S.R.; Islam, S.S. Isolation and Characterization of an Immunoenhancing Glucan from Alkaline Extract of an Edible mushroom, *Lentinus squarrosulus* (Mont.) Singer. *Carbohyd. Res.* **2011**, *346*, 2039–2044. [CrossRef]
42. Cui, H.L.; Chen, Y.; Wang, S.S.; Kai, G.Q.; Fang, Y.M. Isolation, Partial Characterisation and Immunomodulatory Activities of Polysaccharide from *Morchella esculenta*. *J. Sci. Food Agric.* **2011**, *91*, 2180–2185. [CrossRef]
43. Su, C.A.; Xu, X.Y.; Liu, D.Y.; Ming, W. Isolation and Characterization of Exopolysaccharide with Immunomodulatory Activity from Fermentation Broth of *Morchella conica*. *DARU J. Pharm. Sci.* **2013**, *21*, 5. [CrossRef]
44. Du, X.; Zhang, J.; Yang, Y.; Tang, Q.; Jia, W.; Pan, Y. Purification, Chemical Modification and Immunostimulating Activity of Polysaccharides from *Tremella aurantialba* Fruit Bodies. *J. Zhejiang Univ. Sci. B* **2010**. [CrossRef]
45. Roy, S.K.; Das, D.; Mondal, S.; Maiti, D.; Bhunia, B.; Maiti, T.K.; Islam, S.S. Structural Studies of an Immunoenhancing Water-soluble Glucan Isolated from Hot Water Extract of an Edible Mushroom, *Pleurotus florida*, Cultivar Assam Florida. *Carbohyd. Res.* **2009**, *344*, 2596–2601. [CrossRef] [PubMed]
46. Dey, B.; Bhunia, S.K.; Maity, K.K.; Patra, S.; Mandal, S.; Maiti, S.; Maiti, T.K.; Sikdar, S.R.; Islam, S.S. Glucans of *Pleurotus florida* Blue variant: Isolation, Purification, Characterization and Immunological Studies. *Int. J. Biol. Macromol.* **2012**, *50*, 591–597. [CrossRef] [PubMed]
47. Gern, R.M.M.; Wisbeck, E.; Rampinelli, J.R.; Ninow, J.L.; Furlan, S.A. Alternative Medium for Production of *Pleurotus ostreatus* Biomass and Potential Antitumor Polysaccharides. *Bioresour. Technol.* **2008**, *99*, 76–82. [CrossRef]
48. Chen, X.; Xu, X.; Zhang, L.; Kennedy, J.F. Flexible Chain Conformation of (1→3)-β-d-glucan from *Poria cocos* sclerotium in NaOH/urea Aqueous Solution. *Carbohyd. Polym.* **2009**, *75*, 586–591. [CrossRef]
49. Lin, Y.; Zhang, L.; Chen, L.; Jin, Y.; Zeng, F.; Jin, J.; Wan, B.; Cheung, P.C.K. Molecular Mass and Mntitumor Activities of Sulfated Derivatives of α-glucan from *Poria cocos* Mycelia. *Int. J. Biol. Macromol.* **2004**, *34*, 231–236. [CrossRef]
50. Han, X.Q.; Wu, X.M.; Chai, X.Y.; Chen, D.; Dai, H.; Dong, H.L.; Ma, Z.Z.; Gao, X.M.; Tu, P.F. Isolation, Characterization and Immunological Activity of a Polysaccharide from the Fruit Bodies of an Edible Mushroom, *Sarcodon aspratus* (Berk.) S. Ito. *Food Res. Int.* **2011**, *44*, 489–493. [CrossRef]
51. Hobbs, C. The Chemistry, Nutritional Value, Immunopharmacology, and Safety of the Traditional Food of Medicinal Split-Gill Fugus *Schizophyllum commune* Fr.:Fr. (Schizophyllaceae). A Literature Review. *Int. J. Med. Mushrooms* **2005**, *7*, 127–140. [CrossRef]
52. Bimczok, D.; Wrenger, J.; Schirrmann, T.; Rothkotter, H.J.; Wray, V.; Rau, U. Short Chain Regioselectively Hydrolyzed Scleroglucans Induce Maturation of Porcine Dendritic Cells. *Appl. Microbiol. Biotechnol.* **2009**, *82*, 321–331. [CrossRef]
53. Liu, J.J.; Huang, T.S.; Hsu, M.L.; Chen, C.C.; Lin, W.S.; Lu, F.J.; Chang, W.H. Antitumor Effects of the Partially Purified Polysaccharides from *Antrodia camphorata* and the Mechanism of Its Action. *Toxicol. Appl. Pharm.* **2004**, *201*, 186–193. [CrossRef]
54. Wu, S.J.; Liaw, C.C.; Pan, S.Z.; Yang, H.C.; Ng, L.T. *Phellinus linteus* Polysaccharides and Their Immunomodulatory Properties in Human Monocytic Cells. *J. Funct. Foods* **2013**, *5*, 679–688. [CrossRef]
55. Rau, U.; Kuenz, A.; Wray, V.; Nimtz, M.; Wrenger, J.; Cicek, H. Production and Structural Analysis of the Polysaccharide Secreted by *Trametes (Coriolus) versicolor* ATCC 200801. *Appl. Microbiol. Biotechnol.* **2009**, *81*, 827–837. [CrossRef]
56. Wasser, S. Medicinal Mushrooms as a source of Antitumor and Immunomodulating Polysaccharides. *Appl. Microbiol. Biotechnol.* **2002**, *60*, 258–274. [CrossRef]
57. Baets, S.D.; Vandamme, E.J. Extracellular *Tremella* polysaccharides: Structure, properties and applications. *Biotechnol. Lett.* **2001**, *23*, 1361–1366. [CrossRef]
58. Ko, H.J.; Song, A.; Lai, M.N.; Ng, L.T. Immunomodulatory Properties of *Xylaria nigripes* in Peritoneal Macrophage Cells of Balb/c Mice. *J. Ethnopharmacol.* **2011**, *138*, 762–768. [CrossRef] [PubMed]

59. Li, Y.R.; Liu, Q.H.; Wang, H.X.; Ng, T.B. A Novel Lectin with Potent Antitumor, Mitogenic and HIV-1 Reverse Transcriptase Inhibitory Activities from the Edible Mushroom *Pleurotus citrinopileatus*. *Biochim. Biophys. Acta (BBA)-Gen. Subj.* **2008**, *1780*, 51–57. [CrossRef]
60. Zhao, J.K.; Wang, H.X.; Ng, T.B. Purification and Characterization of a Novel Lectin from the Toxic Wild Mushroom *Inocybe umbrinella*. *Toxicon* **2009**, *53*, 360–366. [CrossRef]
61. Zhang, G.Q.; Sun, J.; Wang, H.X.; Ng, T.B. A Novel Lectin with Antiproliferative Activity from the Medicinal Mushroom *Pholiota adiposa*. *Acta. Biochim. Pol.* **2009**, *56*, 415–421. [CrossRef]
62. Ngai, P.H.K.; Ng, T.B. A Mushroom (*Ganoderma capense*) Lectin with Spectacular Thermostability, Potent Mitogenic Activity on Splenocytes, and Antiproliferative Activity toward Tumor Cells. *Biochem. Biophys. Res. Commun.* **2004**, *314*, 988–993. [CrossRef]
63. Han, C.H.; Liu, Q.H.; Ng, T.B.; Wang, H.X. A Novel Homodimeric Lactose-binding Lectin from the Edible Split Gill Medicinal Mushroom *Schizophyllum commune*. *Biochem. Biophys. Res. Commun.* **2005**, *336*, 252–257. [CrossRef]
64. Zhao, S.; Zhao, Y.; Li, S.; Zhao, J.; Zhang, G.; Wang, H.; Ng, T.B. A Novel Lectin with Highly Potent Antiproliferative and HIV-1 Reverse Transcriptase Inhibitory Activities from the Edible Wild Mushroom *Russula delica*. *Glycoconj. J.* **2010**, *27*, 259–265. [CrossRef]
65. Licastro, F.; Morini, M.C.; Kretz, O.; Dirheimer, G.; Creppy, E.E.; Stirpe, F. Mitogenic Activity and Immunological Properties of Bolesatine, A Lectin Isolated from the Mushroom *Boletus satanas* Lenz. *Int. J. Biochem.* **1993**, *25*, 789. [CrossRef]
66. Ho, J.C.K.; Sze, S.C.W.; Shen, W.Z.; Liu, W.K. Mitogenic Activity of Edible Mushroom Lectins. *Biochim. Biophys. Acta (BBA)-Gen. Subj.* **2004**, *1671*, 9–17. [CrossRef] [PubMed]
67. Silvana, A.; Paola, M.; Cerdeiras, M.; Fraguas, L.F. Screening for Lectins from Basidiomycetes and Isolation of *Punctularia atropurpurascens* Lectin. *J. Basic Microbiol.* **2014**, *54*, 89–96. [CrossRef]
68. Amano, K.; Katayama, H.; Saito, A.; Ando, A.; Nagata, Y.; Yoshiho, N. *Aleuria aurantia* Lectin Exhibits Antifungal Activity Against *Mucor racemosus*. *Biosci. Biotechnol. Biochem.* **2012**, *76*, 967–970. [CrossRef]
69. Rana, T.; Bera, A.K.; Das, S.; Bhattacharya, D.; Pan, D.; Bandyopadhyay, S.; Mondal, D.K.; Samanta, S.; Bandyopadhyay, S.; Das, S.K. *Pleurotus florida* Lectin Normalizes Duration Dependent Hepatic oxidative Stress Responses Caused by Arsenic in Rat. *Exp. Toxicol. Pathol.* **2012**. [CrossRef] [PubMed]
70. Zhang, G.Q.; Chen, Q.J.; Hua, J.; Liu, Z.L.; Sun, Y.; Xu, X.; Han, P.; Wang, H.X. An Inulin-Specific Lectin with Anti-HIV-1 Reverse Transcriptase, Antiproliferative, and Mitogenic Activities from the Edible Mushroom *Agaricus bitorquis*. *BioMed Res. Int.* **2019**, *2019*, 1341370. [CrossRef]
71. Yu, L.G.; Fernig, D.G.; Smith, J.A.; Milton, J.D.; Rhodes, J.M. Reversible Inhibition of Proliferation of Epithelial Cell Lines by *Agaricus bisporus* (Edible Mushroom) Lectin. *Cancer Res.* **1993**, *53*, 4627–4632. [CrossRef]
72. Yang, Q.; Yin, Y.; Pan, Y.; Ye, X.; Xu, B.; Yu, W.; Zeng, H.; Sun, H. Anti-metastatic Activity of *Agrocybe aegerita* Galectin (AAL) in a Mouse Model of Breast Cancer Lung Metastasis. *J. Func. Foods* **2018**, *41*, 163–170. [CrossRef]
73. Zhao, C.; Sun, H.; Tong, X.; Qi, Y. An Antitumour Lectin from the Edible Mushroom *Agrocybe aegerita*. *Biochem. J.* **2003**, *374*, 321–327. [CrossRef]
74. Lutsik-Kordovsky, M.D.; Stasyk, T.V.; Stoika, R.S. Analysis of Cytotoxicity of Lectin and Non-lectin Proteins from *Amanita* Mushrooms. *Exp. Oncol.* **2001**, *23*, 43–45.
75. Feng, K.; Liu, Q.H.; Ng, T.B.; Liu, H.Z.; Li, J.Q.; Chen, G.; Sheng, H.Y.; Xie, Z.L.; Wang, H.X. Isolation and Characterization of a Novel Lectin from the Mushroom *Armillaria luteo-virens*. *Biochem. Biolphys. Res. Commun.* **2006**, *345*, 1573–1578. [CrossRef] [PubMed]
76. Zheng, S.; Li, C.; Ng, T.B.; Wang, H.X. A Lectin with Mitogenic Activity from the Edible wild Mushroom *Boletus edulis*. *Process Biochem.* **2007**, *42*, 1620–1624. [CrossRef]
77. Sun, J.; Ng, T.B.; Wang, H.X.; Zhang, G.Q. A Novel Hemagglutinin with Antiproliferative Activity against Tumor Cells from the Hallucinogenic Mushroom *Boletus speciosus*. *BioMed Res. Int.* **2014**, *2014*, 340467. [CrossRef]
78. Pohleven, J.; Obermajer, N.A.; Saboti, J.A.; Lovar, S.; Sep, I.K.; Kos, J.; Kralj, B.; Trukelj, B.; Brzin, J.E. Purification, Characterization and Cloning of a Ricin B-like Lectin from Mushroom *Clitocybe nebularis* with Antiproliferative Activity Against Human Leukemic T cells. *Biochim. Biophys. Acta (BBA)-Gen. Subj.* **2009**, *1790*, 173–181. [CrossRef] [PubMed]

79. Ng, T.B.; Ngai, P.H.K.; Xia, L. An Agglutinin with Mitogenic and Antiproliferative Activities from the Mushroom *Flammulina velutipes*. *Mycologia* **2006**, *98*, 167–171. [CrossRef]
80. Kawagishi, H.; Nomura, A.; Mizuno, T.; Kimura, A.; Chiba, S. Isolation and Characterization of a Lectin from *Grifola frondosa* Fruiting Bodies. *Biochim. Biophys. Acta (BBA)-Gen. Subj.* **1990**, *1034*, 247–252. [CrossRef]
81. Li, Y.; Zhang, G.; Ng, T.B.; Wang, H. A Novel Lectin with Antiproliferative and HIV-1 Reverse Transcriptase Inhibitory Activities from Dried Fruiting Bodies of the Monkey Head Mushroom *Hericium erinaceum*. *J. Biomed. Biotechnol.* **2010**, *2010*, 716515. [CrossRef]
82. Koyama, Y.; Katsuno, Y.; Miyoshi, N.; Hayakawa, S.; Mita, T.; Muto, H.; Isemura, S.; Aoyagi, Y.; Isemura, M. Apoptosis Induction by Lectin Isolated from the Mushroom *Boletopsis leucomelas* in U937 Cells. *Biosci. Biotechnol. Biochem.* **2002**. [CrossRef]
83. Wu, Y.; Wang, H.; Ng, T.B. Purification and Characterization of a Lectin with Antiproliferative Activity toward Cancer Cells from the Dried Fruit Bodies of *Lactarius flavidulus*. *Carbohyd. Res.* **2011**, *346*, 2576–2581. [CrossRef]
84. Pushparajah, V.; Fatima, A.; Chong, C.H.; Gambule, T.Z.; Chan, C.J.; Ng, S.T.; Tan, C.S.; Fung, S.Y.; Lee, S.S.; Tan, N.H. Characterisation of a New Fungal Immunomodulatory Protein from Tiger Milk mushroom, *Lignosus rhinocerotis*. *Sci. Rep.* **2016**, *6*, 30010. [CrossRef] [PubMed]
85. Cordara, G.; Winter, H.C.; Goldstein, I.J.; Krengel, U.; Sandvig, K. The Fungal Chimerolectin MOA Inhibits Protein and DNA Synthesis in NIH/3T3 Cells and May Induce BAX-mediated Apoptosis. *Biochem. Biophys. Res. Commun.* **2014**, *447*, 586–589. [CrossRef]
86. Mahajan, R.G.; Patil, S.I.; Mohan, D.R.; Shastry, P. *Pleurotus eous* Mushroom Lectin (PEL) with Mixed Carbohydrate Inhibition and Antiproliferative Activity on Tumor Cell Lines. *J. Biochem. Mol. Biol. Biophys.* **2002**, *6*, 341–345. [CrossRef]
87. Manna, D.; Pust, S.; Torgersen, M.L.; Cordara, G.; Sandvig, K. *Polyporus squamosus* Lectin La (PSL1a) Exhibits Cytotoxicity in Mammalian Cells by Disruption of Focal Adhesions, Inhibition of Protein Synthesis and Induction of Apoptosis. *PLoS ONE* **2017**, *12*, e0170716. [CrossRef]
88. Zhang, G.; Sun, J.; Wang, H.; Ng, T.B. First Isolation and Characterization of a Novel Lectin with Potent Antitumor Activity from a *Russula* Mushroom. *Phytomedicine* **2010**, *17*, 775–781. [CrossRef]
89. Chumkhunthod, P.; Rodtong, S.; Lambert, S.J.; Fordham-Skelton, A.P.; Reynolds, C.D. Purification and Characterization of an N-acetyl-D-galactosamine-specific Lectin from the Edible Mushroom *Schizophyllum commune*. *Biochim. Biophys. Acta (BBA)-Gen. Subj.* **2006**, *1760*, 326–332. [CrossRef]
90. Zhang, W.; Tian, G.; Geng, X.; Zhao, Y.; Tzi, N.; Zhao, L.; Wang, H. Isolation and Characterization of a Novel Lectin from the Edible Mushroom *Stropharia rugosoannulata*. *Molecules* **2014**, *19*, 19880–19891. [CrossRef]
91. Wang, H.X.; Ng, T.B.; Ooi, V.E.C.; Liu, W.K.; Chang, S.T. Actions of Lectins from the Mushroom *Tricholoma mongolicum* on Macrophages, Splenocytes and Life-span in Sarcoma-bearing Mice. *Anticancer Res.* **1997**, *17*, 419–424.
92. Liua, W.K.; Ho, J.C.K.; Ng, T.B. Suppression of Cell Cycle Progression by a Fungal Lectin: Activation of Cyclin-dependent Kinase Inhibitors. *Biochem. Pharmacol.* **2001**, *61*, 33–37. [CrossRef]
93. She, Q.B.; Ng, T.B.; Liu, W.K. A Novel Lectin with Potent Immunomodulatory Activity Isolated from Both Fruiting Bodies and Cultured Mycelia of the Edible Mushroom *Volvariella volvacea*. *Biochem. Biophys. Res. Commun.* **1998**. [CrossRef]
94. Marty-Detraves, C.; Francis, F.; Baricault, L.; Fournier, D.; Paquereau, L. Inhibitory Action of a New lectin from *Xerocomus chrysenteron* on Cell-substrate Adhesion. *Mol. Cell. Biochem.* **2004**, *258*, 49–55. [CrossRef]
95. Liu, Q.; Wang, H.; Ng, T.B. First Report of a Xylose-specific Lectin with Potent Hemagglutinating, Antiproliferative and Anti-mitogenic Activities from a Wild Ascomycete Mushroom. *Biochim. Biophys. Acta (BBA)-Gen. Subj.* **2006**, *1760*, 1914–1919. [CrossRef]
96. Sheu, F.; Chien, P.J.; Hsieh, K.Y.; Chin, K.L.; Huang, W.T.; Tsao, C.Y.; Chen, Y.F.; Cheng, H.C.; Chang, H.H. Purification, Cloning, and Functional Characterization of a Novel Immunomodulatory Protein from *Antrodia camphorata* (Bitter Mushroom) That Exhibits TLR2-Dependent NF-κB Activation and M1 Polarization within Murine Macrophages. *J. Agric. Food Chem.* **2009**, *57*, 4130–4141. [CrossRef]
97. Lin, J.W.; Guan, S.Y.; Duan, Z.W.; Shen, Y.H.; Li, T.L. Gene Coning of a Novel Fungal Immunomodulatory Protein from *Chroogomphis rutilus* and Its Expression in Pichia Pastoris. *J. Chem. Technol. Biotechnol.* **2016**, *91*, 2761–2768. [CrossRef]

98. Li, S.; Jiang, Z.; Sun, L.; Liu, X.; Huang, Y.; Wang, F.; Xin, F. Characterization of a new fungal immunomodulatory protein, FIP-dsq2 from *Dichomitus squalens*. *J. Biol.* **2017**, *246*, 45–51. [CrossRef]
99. Wang, P.H.; Hsu, C.I.; Tang, S.C.; Huang, Y.L.; Lin, J.Y. Fungal Immunomodulatory Protein from *Flammulina velutipes* Induces Interferon-γ Production Through p38 Mitogen-Activated Protein Kinase Signaling Pathway. *J. Agric. Food Chem.* **2004**. [CrossRef]
100. Su, K.Q.; Wang, X.F.; Zhou, X.W. Cloning and Bioinformatics Analysis of Fugal Immunomodulatory Protein Gene from *Ganoderma astum*. *J. Shanghai Jiaotong Univ.* **2012**, *30*, 65–71. (In Chinese)
101. Li, Q.Z.; Wang, X.F.; Bao, T.W.; Liang, R.; Lin, J.; Zhou, X. In Vitro Synthesis of a Recombinant Fungal Immunomodulatory Protein from Lingzhi or Reishi Medicinal Mushroom, *Ganoderma lucidum* (W.Curt.:Fr.) P.Karst. (Aphyllophoromycetideae) and Analysis of Its Immunomodulatory Activity. *Int. J. Med. Mushrooms* **2010**, *12*, 347–358. [CrossRef]
102. Lin, C.H.; Hsiao, Y.M.; Ou, C.C.; Lin, Y.W.; Chiu, Y.L.; Lue, K.H.; Chang, J.G.; Ko, J.L. GMI, a *Ganoderma* Immunomodulatory Protein, Down-regulates Tumor Necrosis Factor α-Induced Expression of Matrix Metalloproteinase 9 via NF-κB Pathway in Human Alveolar Epithelial A549 Cells. *J. Agric. Food Chem.* **2010**, *58*, 12014–12021. [CrossRef]
103. Li, Q.; Wang, X.; Chen, Y.; Lin, J.; Zhou, X. Cytokines Expression Induced by *Ganoderma sinensis* Fungal Immunomodulatory Proteins (FIP-gsi) in Mouse Spleen Cells. *Appl. Biochem. Biotechnol.* **2010**, *162*, 1403–1413. [CrossRef]
104. Hsiao, Y.M.; Huang, Y.L.; Tang, S.C.; Shieh, G.J.; Lai, J.Y.; Wang, P.H.; Ying, T.H.; Ko, J.L. Effect of a Fungal Immunomodulatory Protein from *Ganoderma tsugae* on Cell Cycle and Interferon-gamma Production Through Phosphatidylinositol 3-kinase Signal Pathway. *Process Biochem.* **2008**, *43*, 423–430. [CrossRef]
105. Bastiaan-Net, S.; Chanput, W.; Hertz, A.; Zwittink, R.D.; Mes, J.J.; Wichers, H.J. Biochemical and Functional Characterization of Recombinant Fungal Immunomodulatory Proteins (rFIPs). *Int. Immunopharmacol.* **2013**, *15*, 167–175. [CrossRef]
106. Cong, W.R.; Xu, H.; Liu, Y.; Li, Q.Z.; Li, W.; Zhou, X.W. Production and Functional Characterization of a Novel Fungal Immunomodulatory Protein FIP-SN15 Shuffled from Two Genes of *Ganoderma* species. *Appl. Microbiol. Biotechnol.* **2014**, *98*, 5967–5975. [CrossRef]
107. Chang, H.H.; Sheu, F. A Novel Fungal Immunomodulatory Protein (PCP) Isolated from *Poria cocos* Activates Mouse Peritoneal Macrophage Involved in Toll-like Receptor 4. *FASEB J.* **2007**, *21*, 702–715. [CrossRef]
108. Li, S.Y.; Shi, L.J.; Ding, Y.; Nie, Y.; Tang, X.M. Identification and Functional Characterization of a Novel Fungal immunomodulatory Protein from *Postia placenta*. *Food Chem. Toxicol.* **2015**, *78*, 64–70. [CrossRef]
109. Feng, L.; Wen, H.A.; Zhang, Y.J.; Min, A.; Liu, X.Z. Purification and Characterization of a Novel Immunomodulatory Protein from the Medicinal Mushroom *Trametes versicolor*. *Sci. China Life Sci.* **2011**, *54*, 91–97. [CrossRef]
110. Hsu, H.Y.; Hua, K.F.; Wu, W.C.; Hsu, J.; Weng, S.T.; Lin, T.L.; Liu, C.Y.; Hseu, R.S.; Huang, C.T. Reishi Immuno-modulation Protein Induces Interleukin-2 Expression via Protein Kinase-dependent Signaling Pathways Within Human T Cells. *J. Cell. Physiol.* **2008**, *215*, 15–26. [CrossRef]
111. Ma, K.; Bao, L.; Han, J.; Jin, T.; Yang, X.; Zhao, F.; Li, S.; Song, F.; Liu, M.; Liu, H. New Benzoate Derivatives and Hirsutane Type Sesquiterpenoids with Antimicrobial Activity and Cytotoxicity from the Solid-state Fermented Rice by the Medicinal Mushroom *Stereum hirsutum*. *Food Chem.* **2014**, *143*, 239–245. [CrossRef]
112. Chen, H.P.; Dong, W.B.; Feng, T.; Yin, X.; Li, Z.H.; Dong, Z.J.; Li, Y.; Liu, J.K. Four New Sesquiterpenoids from Fruiting Bodies of the Fungus *Inonotus rickii*. *J. Asian Nat. Prod. Res.* **2014**, *16*, 1–6. [CrossRef]
113. Wang, S.J.; Bao, L.; Han, J.J.; Wang, Q.X.; Yang, X.L.; Wen, H.A.; Guo, L.-D.; Li, S.J.; Zhao, F.; Liu, H.W. Pleurospiroketals A–E, Perhydrobenzannulated 5,5-Spiroketal Sesquiterpenes from the Edible Mushroom *Pleurotus cornucopiae*. *J. Nat. Prod.* **2013**. [CrossRef] [PubMed]
114. Wang, S.; Bao, L.; Zhao, F.; Wang, Q.; Li, S.; Ren, J.; Li, L.; Wen, H.; Guo, L.; Liu, H. Isolation, Identification, and Bioactivity of Monoterpenoids and Sesquiterpenoids from the Mycelia of Edible Mushroom *Pleurotus cornucopiae*. *J. Agric. Food Chem.* **2013**, *61*, 5122–5129. [CrossRef]
115. Intaraudom, C.; Boonyuen, N.; Supothina, S.; Tobwor, P.; Prabpai, S.; Kongsaeree, P.; Pittayakhajonwut, P. Novel Spiro-sesquiterpene from the Mushroom *Anthracophyllum* sp. BCC18695. *Phytochem. Lett.* **2013**, *6*, 345–349. [CrossRef]
116. Li, L.; Yang, X.; Li, S.; Gao, H.; Yao, X.S.; Wen, H.; Liu, H.W. Bioactive Sesquiterpenoids from the Solid Culture of the Edible Mushroom *Flammulina velutipes* Growing on Cooked Rice. *Food Chem.* **2012**. [CrossRef]

117. Wang, Y.; Li, B.; Liu, D.; Yang, X.; Li, S.; Hao, G.; Yao, X.; Wen, H.; Liu, H. Two New Sesquiterpenes and Six Norsesquiterpenes from the Solid Culture of the Edible Mushroom *Flammulina velutipes*. *Tetrahedron* **2012**, *68*, 3012–3018. [CrossRef]
118. Kanokmedhakul, S.; Lekphrom, R.; Kanokmedhakul, K.; Hahnvajanawong, C.; Bua-Art, S.; Saksirirat, W.; Prabpai, S.; Kongsaeree, P. Cytotoxic Sesquiterpenes from Luminescent Mushroom *Neonothopanus nambi*. *Tetrahedron* **2012**, *68*, 8261–8266. [CrossRef]
119. Han, J.J.; Chen, Y.H.; Bao, L.; Yang, X.L.; Liu, D.; Li, S.J.; Zhao, F.; Liu, H. Anti-inflammatory and Cytotoxic Cyathane Diterpenoids from the Medicinal Fungus *Cyathus africanus*. *Fitoterapia* **2013**, *84*, 22–31. [CrossRef]
120. Wang, S.J.; Li, Y.X.; Bao, L.; Han, J.J.; Yang, X.L.; Li, H.R.; Wang, Y.Q.; Li, S.J.; Liu, H.W. Eryngiolide A, a Cytotoxic Macrocyclic Diterpenoid with an Unusual Cyclododecane Core Skeleton Produced by the Edible Mushroom *Pleurotus eryngii*. *Org. Lett.* **2012**, *14*, 3672–3675. [CrossRef] [PubMed]
121. Suzuki. Anti-proliferative and Apoptosis-inducible Activity of Sarcodonin G from *Sarcodon scabrosus* in HeLa Cells. *Int. J. Oncol.* **1992**. [CrossRef]
122. Tsukamoto, S.; Macabalang, A.D.; Nakatani, K.; Obara, Y.; Ohta, T. Tricholomalides A-C, New Neurotrophic Diterpenes from the Mushroom *Tricholoma* sp. *J. Nat. Prod.* **2004**, *66*, 1578–1581. [CrossRef]
123. Ma, K.; Ren, J.; Han, J.; Bao, L.; Liu, H. Ganoboninketals A-C, Antiplasmodial 3,4-seco-27-Norlanostane Triterpenes from *Ganoderma boninense* Pat. *J. Nat. Prod.* **2014**, *77*, 1847–1852. [CrossRef] [PubMed]
124. Isaka, M.; Chinthanom, P.; Kongthong, S.; Srichomthong, K.; Choeyklin, R. Lanostane Triterpenes from Cultures of the Basidiomycete *Ganoderma orbiforme* BCC 22324. *Phytochemistry* **2013**. [CrossRef] [PubMed]
125. Gao, J.J.; Min, B.S.; Ahn, E.M.; Nakamura, N.; Lee, H.K.; Hattori, M. New Triterpene Aldehydes, Lucialdehydes A–C, from *Ganoderma lucidum* and Their Cytotoxicity against Murine and Human Tumor Cells. *J. ChemInform.* **2002**. [CrossRef]
126. Wu, G.S.; Lu, J.J.; Guo, J.J.; Li, Y.B.; Tan, W.; Dang, Y.Y.; Zhong, Z.F.; Xu, Z.T.; Chen, X.P.; Wang, Y.T. Ganoderic Acid DM, a Natural Triterpenoid, Induces DNA Damage, G1 Cell Cycle Arrest and Apoptosis in Human Breast Cancer Cells. *Fitoterapia* **2012**, *83*, 408–414. [CrossRef] [PubMed]
127. Li, C.H.; Chen, P.Y.; Chang, U.M.; Kan, L.S.; Fang, W.H.; Tsai, K.S.; Lin, S.B. Ganoderic Acid X, a lanostanoid Triterpene, Inhibits Topoisomerases and Induces Apoptosis of Cancer Cells. *Life Sci.* **2005**, *77*, 252–265. [CrossRef]
128. Lez, A.G.G.; León, F.; Rivera, A.; Lez-Plata, J.G.; Padron, J.I. New Lanostanoids from the Fungus *Ganoderma concinna*. *J. Nat. Prod.* **2002**, *65*, 417–421. [CrossRef]
129. Su, H.J.; Fann, Y.F.; Chung, M.I.; Won, S.J.; Lin, C.N. New Lanostanoids of *Ganoderma tsugae*. *J. Nat. Prod.* **2000**, *63*, 514–516. [CrossRef]
130. Kim, K.H.; Moon, E.; Sang, U.C.; Sun, Y.K.; Kang, R.L. Lanostane Triterpenoids from the Mushroom *Naematoloma fasciculare*. *J. Nat. Prod.* **2013**, *76*, 845. [CrossRef]
131. Arpha, K.; Phosri, C.; Suwannasai, N.; Mongkolthanaruk, W.; Sodngam, S. Astraodoric acids A–D: New Lanostane Triterpenes from Edible Mushroom *Astraeus odoratus* and Their Anti-Mycobacterium Tuberculosis H37Ra and Cytotoxic Activity. *J. Agric. Food Chem.* **2012**, *60*, 9834–9841. [CrossRef]
132. Clericuzio, M.; Cassino, C.; Corana, F.; Vidari, G. Terpenoids from *Russula lepida* and *R. amarissima* (Basidiomycota, Russulaceae). *Phytochemistry* **2012**, *84*, 154–159. [CrossRef]
133. Clericuzio, M.; Tabasso, S.; Bianco, M.A.; Pratesi, G.; Beretta, G.; Tinelli, S.; Zunino, F.; Vidari, G. Cucurbitane Triterpenes from the Fruiting Bodies and Cultivated Mycelia of *Leucopaxillus gentianeus*. *J. Nat. Prod.* **2006**, *69*, 1796–1799. [CrossRef]
134. Shao, H.J.; Chen, Q.; Fei, W.; Zhang, Y.L.; Luo, D.Q.; Liu, J.K. A New Cytotoxic Lanostane Triterpenoid from the Basidiomycete *Hebeloma versipelle*. *Cheminform* **2010**, *37*, 828–831. [CrossRef]
135. Yoshikawa, K.; Kuroboshi, M.; Ahagon, S.; Arihara, S. Three Novel Crustulinol Esters, Saponaceols A-C, from *Tricholoma saponaceum*. *Chem. Pharm. Bull.* **2004**. [CrossRef]
136. Yoshikawa, K.; Nishimura, N.; Bando, S.; Arihara, S.; Matsumura, E. New Lanostanoids, Elfvingic Acids A–H, from the Fruit Body of *Elfvingia applanata*. *J. Nat. Prod.* **2002**. [CrossRef]
137. Mallard, B.; Leach, D.N.; Wohlmuth, H.; Tiralongo, J. Synergistic Immuno-modulatory Activity in Human Macrophages of a Medicinal Mushroom Formulation Consisting of Reishi, Shiitake and Maitake. *PLoS ONE* **2019**, 14. [CrossRef]

138. Carbonero, E.R.; Gracher, A.H.P.; Komura, D.L.; Marcon, R.; Freitas, C.S.; Baggio, C.H.; Santos, A.R.S.; Torri, G.; Gorin, P.A.J.; Iacomini, M. *Lentinus edodes* heterogalactan: Antinociceptive and anti-inflammatory effects. *Food Chem.* **2008**, *111*, 531–537. [CrossRef]
139. Finimundy, T.C.; Dillon, A.J.P.; Henriques, J.A.P.; Ely, M.R. A Review on General Nutritional Compounds and Pharmacological Properties of the *Lentinula edodes*. Mushroom. *Int. J. Food Sci. Nutr.* **2014**, *5*, 1095–1105. [CrossRef]
140. Lee, H.H.; Lee, J.S.; Cho, J.Y.; Kim, Y.E.; Hong, E.K. Study on Immunostimulating Activity of Macrophage Treated with Purified Polysaccharides from Liquid Culture and Fruiting Body of *Lentinus edodes*. *J. Microbiol. Biotechnol.* **2009**, *19*, 566–572. [CrossRef]
141. Chen, H.; Ju, Y.; Li, J.; Yu, M. Antioxidant activities of polysaccharides from *Lentinus edodes* and Their Significance for Disease Prevention. *Int. J. Biol. Macromol.* **2012**, *50*, 214–218. [CrossRef]
142. Rincao, V.P.; Yamamoto, K.A.; Ricardo, N.M.P.S.; Soares, S.A.; Meirelles, L.D.P.; Nozawa, C.; Linhares, R.E.C. Polysaccharide and Extracts from *Lentinula edodes*: Structural Features and Antiviral activity. *J. Virol.* **2012**, *9*. [CrossRef]
143. Ulziijargal, E.; Yang, J.H.; Lin, L.Y.; Chen, C.P.; Mau, J.L. Quality of Bread Supplemented with Mushroom Mycelia. *Food Chem.* **2013**, *138*, 70–76. [CrossRef]
144. Kim, S.Y.; Kang, M.Y.; Kim, M.Y. Quality Characteristics of Noodle Added with Browned Oak Mushroom (*Lentinus edodes*). *Korean J. Food Cook. Sci.* **2008**, *24*, 665–671. (In Korean)
145. Parab, D.N.; Dhalagade, J.R.; Sahoo, A.K.; Ranveer, R.C. Effect of Incorporation of Mushroom (*Pleurotus sajor-caju*) Powder on Quality Characteristics of Papad (Indian snack food). *Int. J. Food Sci. Nutr.* **2012**, *63*, 866–870. [CrossRef]
146. Singla, R.; Ghosh, M.; Ganguli, A. Phenolics and Antioxidant Activity of a Ready-to-eat Snack Food Prepared from the Edible Mushroom (*Agaricus bisporous*). *Food Sci. Nutr.* **2009**, *39*, 227–234. [CrossRef]
147. Margaret, B.; Emma, D.; Tiwari, B.K.; Brennan, C.S. Enrichment of Extruded Snack Products with Coproducts from Chestnut Mushroom (*Agrocybe aegerita*) Production: Interactions between Dietary Fiber, Physicochemical Characteristics, and Glycemic Load. *J. Agric. Food Chem.* **2012**, *60*, 4396–4401.
148. Ribeiro, A.; Ruphuy, G.; Lopes, J.C.; Dias, M.M.; Barros, L.; Barreiro, F.; Ferreira, I.C.F.R. Spray-drying Microencapsulation of Synergistic Antioxidant Mushroom Extracts and Their use as Functional Food Ingredients. *Food Chem.* **2015**, *188*, 612–618. [CrossRef]
149. Yuan, Q.; Zhang, X.; Ma, M.; Long, T.; Xiao, C.; Zhang, J.; Liu, J.; Zhao, L. Immunoenhancing Glucuronoxylomannan from *Tremella Aurantialba* Bandoni et Zang and Its Low-Molecular-Weight Fractions by Radical Depolymerization: Properties, Structures and Effects on Macrophages. *Carbohydr. Polym.* **2020**, *238*, 116–184. [CrossRef] [PubMed]
150. Guo, M.; Meng, M.; Zhao, J.; Wang, X.; Wang, C. Immunomodulatory Effects of the Polysaccharide from *Craterellus cornucopioides* Via Activating the TLR4-NFkappaB Signaling Pathway in Peritoneal Macrophages of BALB/c Mice. *Int. J. Biol. Macromol.* **2020**, *160*, 871–879. [CrossRef]
151. Li, Y.; You, L.; Dong, F.; Yao, W.; Chen, J. Structural Characterization, Antiproliferative and Immunoregulatory Activities of a Polysaccharide from *Boletus leccinum* Rugosiceps. *Int. J. Biol. Macromol.* **2020**, *157*, 106–118. [CrossRef]
152. Tan, X.; Chen, W.; Jiao, C.; Liang, H.; Yun, H.; He, C.; Chen, J.; Ma, X.; Xie, Y. Anti-tumor and Immunomodulatory Activity of the Aqueous Extract of *Sarcodon imbricatus* in Vitro and in Vivo. *Food Funct.* **2020**, *11*, 1110–1121. [CrossRef] [PubMed]
153. Zhao, H.X.; Li, J.Z.; Wan, C.Y.; Hua, W.; Xiao, H.K.; Lv, Y.P.; Dong, M.H. Antitumor and Immunomodulatory Activity of *Pleurotus Eryngii* Extract. *J. Food Biochem.* **2014**, *39*, 19–27. [CrossRef]
154. Borchers, A.T.; Krishnamurthy, A.; Keen, C.L.; Meyers, F.J.; Gershwin, M.E. The Immunobiology of Mushrooms. *Exp. Biol. Med.* **2008**, *233*, 259–276. [CrossRef] [PubMed]
155. Xin, M.; He, B.L.; Li, X.L. Antitumor Polysaccharides from Mushrooms: A Review on the Structural Characteristics, Antitumor Mechanisms and Immunomodulating Activities. *Carbohydr. Res.* **2016**, 30–41. [CrossRef]
156. Kim, F.; Sakagami, H.; Tanuma, S.I.; Konno, K. Stimulation of Interferon-γ-induced Human Myelogenous Leukemic Cell Differentiation by High Molecular Weight PSK Subfraction. *Anticancer Res.* **1990**, *10*, 55–58. [CrossRef]

157. Adachi, Y.; Ohno, N.; Ohsawa, M.; Oikawa, S.; Yadomae, T. Change of Biological Activities of (1.RAR.3)-.BETA.-D-glucan from *Grifola frondosa* upon Molecular Weight Reduction by Heat Treatment. *Chem. Pharm. Bull.* **1990**, *38*, 477–481. [CrossRef] [PubMed]
158. Chihara, G.; Hamuro, J.; Maeda, Y.; Arai, Y.; Fukuoka, F. Antitumour Polysaccharide Derived Chemically from Natural Glucan (Pachyman). *Nature* **1970**, *225*, 943–944. [CrossRef]
159. Bae, I.Y.; Kim, H.W.; Yoo, H.J.; Kim, E.S.; Lee, S.; Dong, Y.P.; Lee, H.G. Correlation of Branching Structure of Mushroom β-glucan with its Physiological Activities. *Food Res. Int.* **2013**, *51*, 195–200. [CrossRef]
160. Ren, L.; Perera, C.; Hemar, Y. Antitumor Activity of Mushroom Polysaccharides: A Review. *Food Funct.* **2012**, *3*, 1118–1130. [CrossRef]
161. Palleschi, A.; Bocchinfuso, G.; Coviello, T.; Alhaique, F. Molecular Dynamics Investigations of the Polysaccharide Scleroglucan: First Study on the Triple Helix Structure. *Carbohydr. Res.* **2005**, *340*, 2154–2162. [CrossRef]
162. Zhang, P.; Cheung, P.C.K. Evaluation of Sulfated *Lentinus edodes* α-(1 → 3)-D-glucan as a Potential Antitumor Agent. *J. Agric. Chem. Soc. Jpn.* **2002**, *66*, 1052–1056. [CrossRef]
163. Singh, R.S.; Walia, A.K.; Kennedy, J.F. Mushroom Lectins in Biomedical Research and Development. *Int. J. Boil. Macromol.* **2019**, *151*, 1340–1350. [CrossRef]
164. Švajger, U.; Pohleven, J.; Kos, J.; Štrukelj, B.; Jeras, M. CNL, a Ricin B-like Lectin from Mushroom *Clitocybe nebularis*, Induces Maturation and Activation of Dendritic Cells via the Toll-like Receptor 4 Pathway. *Immunology* **2011**, *134*, 409–418. [CrossRef] [PubMed]
165. Pohleven, J.; Renko, J.; Magister, S.; Smith, S.; Magister, D.F.; Kuenzler, M.; Strukelj, B.; Turk, D.; Kos, J.; Sabotic, J. Bivalent Carbohydrate Binding Is Required for Biological Activity of *Clitocybe nebularis* Lectin (CNL), the N,N'-Diacetyllactosediamine (GalNAcβ1-4GlcNAc, LacdiNAc)-Specific Lectin from Basidiomycete C. nebularis. *J. Biol. Chem.* **2012**, *287*, 10602–10612. [CrossRef]
166. Zhang, L.Q.; Zhang, Z.Y.; Wei, Z.X. Fungal Immunomodulatory Proteins: Characteristic, Potential Antitumor Activities and Their Molecular Mechanisms. *Drug Discov. Today* **2018**. [CrossRef]
167. Wang, X.F.; Li, Q.Z.; Bao, T.W.; Cong, W.R.; Song, W.X.; Zhou, X.W. In Vitro Rapid Evolution of Fungal Immunomodulatory Proteins by DNA Family Shuffling. *Appl. Microbiol. Biotechnol.* **2013**, *97*, 2455–2465. [CrossRef]
168. Jeong, Y.T.; Yang, B.K.; Jeong, S.C.; Kim, S.M.; Song, C.H. *Ganoderma applanatum*: A Promising Mushroom for Antitumor and Immunomodulating Activity. *Phytother. Res. Ptr.* **2008**, *22*, 614–619. [CrossRef]
169. Tao, N.; Cui, X.H.; Cai, H.M.; Ma, Y.H.; Zhao, Y.C.; Chen, W.M. Establishment of Genetic Transformation System of *Agrocybe aegerita* Using PEG-Mediated Method. *Mycosystema* **2020**, *39*, 1100–1108. [CrossRef]
170. Sonnenberg, A.S.M.; Gao, W.; Lavrijssen, B.; Hendrickx, P.; Visser, R.G.F. A Detailed Analysis of the Recombination Landscape of the Button Mushroom *Agaricus bisporus* Var. *bisporus*. *Fungal Genet. Biol.* **2016**, *93*, 35–45. [CrossRef]
171. Morin, E.; Kohler, A.; Baker, A.R.; Foulongne-Oriol, M.; Lombard, V.; Nagy, L.G.; Ohm, R.A.; Patyshakuliyeva, A.; Brun, A.; Aerts, A.L.; et al. Genome Sequence of the Button Mushroom *Agaricus bisporus* Reveals Mechanisms Governing Adaptation to a Humic-rich Ecological Niche (vol 109, pg 17501, 2012). *Proc. Natl. Acad. Sci. USA* **2013**, *110*, 4146. [CrossRef]
172. Chen, M.Y.; Liao, J.H.; Guo, Z.J.; Li, H.R.; Lu, Z.H.; Cai, D.F.; Wang, Z.S. The Expression Vector Construction and Transformation of Thermotolerance-related Gene of *Agaricus bisporus*. *Mycosystema* **2009**, *28*, 797–801. (In Chinese)
173. Cheng, K.; Zhu, Z.; Wang, J.; Chen, J.; Bao, D.; Chen, M.; Zhang, J.; Tan, Q. Progress in the Genetic Transformation of Edible Fungi. *Acta Edulis Fungi* **2012**, *19*, 92–99. (In Chinese)
174. Kurata, A.; Fukuta, Y.; Mori, M.; Kishimoto, N.; Shirasaka, N. Draft Genome Sequence of the Basidiomycetous Fungus *Flammulina velutipes* TR19. *Genome Announc.* **2016**, *4*, e00505–e00516. [CrossRef]
175. Young-Jin, P.; Hun, B.K.; Seonwook, L.; Changhoon, K.; Hwanseok, R.; Hyungtae, K.; Jeong-Sun, S.; Hae-Ran, P.; Dae-Eun, Y.; Jae-Young, N.; et al. Whole Genome and Global Gene Expression Analyses of the Model Mushroom *Flammulina velutipes* Reveal a High Capacity for Lignocellulose Degradation. *PLoS ONE* **2014**, *9*, e93560. [CrossRef]
176. Hyeokjun, Y.; You, Y.H.; Ju-Ri, W.; Young-Jin, P.; Kong, W.S.; Byoung-Moo, L.; Jong-Guk, K. The Mitochondrial Genome of the White-Rot Fungus *Flammulina velutipes*. *J. Gen. Appl. Microbiol.* **2012**, *58*, 331–337. [CrossRef]

177. Ko, J.L.; Lin, S.J.; Hsu, C.I.; Kao, C.L.; Lin, J.Y. Molecular Cloning and Expression of a Fungal Immunomodulatory Protein, FIP-fve, from *Flammulina velutipes*. *J. Med. Assoc.* **1997**, *96*, 517–524. [CrossRef]
178. Xu, H.; Kong, Y.Y.; Chen, X.; Guo, M.Y.; Bai, X.H.; Lu, Y.J.; Li, W.; Zhou, X.W. Recombinant FIP-gat, a Fungal Immunomodulatory Protein from *Ganoderma atrum*, Induces Growth Inhibition and Cell Death in Breast Cancer Cells. *J. Agric. Food Chem.* **2016**, *64*, 2690–2698. [CrossRef]
179. Li, Q.Z.; Chang, Y.Z.; He, Z.M.; Chen, L.; Zhou, X.W. Immunomodulatory Activity of *Ganoderma lucidum* Immunomodulatory Protein Via PI3K/Akt and MAPK Signaling Pathways in RAW264.7 Cells. *J. Cell. Physiol.* **2019**, *234*, 23337–23348. [CrossRef]
180. Chen, S.L. Genome Sequence of the Model Medicinal Mushroom *Ganoderma lucidum*. *Nat. Commun.* **2012**, *3*. [CrossRef]
181. Liu, D.; Jing, G.; Dai, W.; Kang, X.; Zhuo, H.; Zhang, H.M.; Wei, L.; Le, L.; Ma, J.; Xia, Z. The Genome of *Ganderma lucidum* Provide Insights into Triterpense Biosynthesis and Wood Degradation. *PLoS ONE* **2012**. [CrossRef]
182. Zhu, Y.; Xu, J.; Sun, C.; Zhou, S.; Xu, H.; Nelson, D.R.; Qian, J.; Song, J.; Luo, H.; Xiang, L. Chromosome-level Genome Map Provides Insights into Diverse Defense Mechanisms in the Medicinal Fungus *Ganoderma sinense*. *Sci. Rep.* **2015**, *5*, 11087. [CrossRef]
183. Wang, P.H.; Yang, S.F.; Chen, G.D.; Han, C.P.; Chen, S.C.; Lin, L.Y.; Ko, J.L. Human Nonmetastatic Clone 23 Type 1 Gene Suppresses Migration of Cervical Cancer Cells and Enhances the Migration Inhibition of Fungal Immunomodulatory Protein from *Ganoderma tsugae*. *Reprod. Sci.* **2007**, *14*, 475–485. [CrossRef]
184. Gong, W.; Wang, Y.; Xie, C.; Zhou, Y.; Peng, Y. Whole Genome Sequence of an Edible and Medicinal mushroom, *Hericium erinaceus* (Basidiomycota, Fungi). *Genomics* **2020**, *112*. [CrossRef]
185. Liu, L.; Xiao, Z.; Guo, L.; Lin, J.; You, L.; Liao, J. Establishment of Genetic Transformation System of *Hericium erinaceus* Using PEG Mediated Method. *Mycosystema* **2014**, *33*, 121–128. (In Chinese)
186. Gong, W.B.; Li, L.; Zhou, Y.; Bian, Y.B.; Kwan, H.S.; Cheung, M.K.; Xiao, Y. Genetic Dissection of Fruiting Body-related Traits Using Quantitative Trait loci Mapping in *Lentinula edodes*. *Appl. Microbiol. Biotechnol.* **2016**, *100*, 5437–5452. [CrossRef]
187. Qu, J.; Zhao, M.; Tom, H.; Feng, X.; Zhang, J.; Huang, C. Identification and Characterization of Small Noncoding RNAs in Genome Sequences of the Edible Fungus *Pleurotus ostreatus*. *BioMed Res. Int.* **2016**, *2016*, 2503023. [CrossRef]
188. Riley, R.; Salamov, A.A.; Brown, D.W.; Nagy, L.G.; Floudas, D.; Held, B.W.; Levasseur, A.; Lombard, V.; Morin, E.; Otillar, R. Extensive Sampling of Basidiomycete Genomes Demonstrates Inadequacy of the White-Rot/Brown-Rot Paradigm for Wood Decay Fungi. *Proc. Natl. Acad. Sci. USA* **2014**, *111*, 9923–9928. [CrossRef]
189. Yong, W.; Zeng, F.; Chau, H.C.; Zhang, Y.; Ching, L.F.C. The Mitochondrial Genome of the Basidiomycete Fungus *Pleurotus ostreatus* (oyster mushroom). *FEMS Microbiol. Lett.* **2010**. [CrossRef]
190. Martinez, D.; Challacombe, J.; Morgenstern, I.; Hibbett, D.; Schmoll, M.; Kubicek, C.P.; Ferreira, P.; Ruiz-Duenas, F.J.; Martinez, A.T.; Kersten, P.; et al. Genome, Transcriptome, and Secretome Analysis of Wood Decay Fungus *Postia placenta* Supports Unique Mechanisms of Lignocellulose Conversion. *Proc. Natl. Acad. Sci. USA* **2009**, *106*, 1954–1959. [CrossRef]
191. Li, F.; Wen, H.; Liu, X.; Zhou, F.; Chen, G. Gene Cloning and Recombinant Expression of a Novel Fungal Immunomodulatory Protein from *Trametes versicolor*. *Protein Expr. Purif.* **2012**, *82*, 339–344. [CrossRef]
192. Floudas, D.; Binder, M.; Riley, R.; Barry, K.; Blanchette, R.A.; Henrissat, B.T.; Martinez, A. The Paleozoic Origin of Enzymatic Lignin Decomposition Reconstructed from 31 Fungal Genomes. *Science* **2012**, *336*, 1715–1719. [CrossRef]
193. Sun, X.; Huang, W.; Xiao, S.; Liang, C.; Zhang, S.; Liu, Z.; Sun, F. Extracellular Expression and Efficient Purification of a Functional Recombinant *Volvariella volvacea* Immunomodulatory Protein (FIP-vvo) Using Pichia Pastoris System. *Protein Expr. Purif.* **2014**, *94*, 95–100. [CrossRef] [PubMed]
194. Bao, D.; Gong, M.; Zheng, H.; Chen, M.; Zhang, L.; Wang, H.; Jiang, J.; Wu, L.; Zhu, Y.; Zhu, G. Sequencing and Comparative Analysis of the Straw Mushroom (*Volvariella volvacea*) Genome. *PLoS ONE* **2013**, *8*, e58294. [CrossRef] [PubMed]

195. Muzzarelli, R.A.A.; Boudrant, J.; Meyer, D.; Manno, N.; Demarchis, M.; Paoletti, M.G. Current views on fungal chitin/chitosan, human chitinases, food preservation, glucans, pectins and inulin: A tribute to Henri Braconnot, precursor of the carbohydrate polymers science, on the chitin bicentennial. *Carbohydr. Polym.* **2012**, *87*, 995–1012. [CrossRef]
196. Yuan, X.; Zheng, J.; Jiao, S.; Cheng, G.; Feng, C.; Du, Y.; Liu, H. A review on the preparation of chitosan oligosaccharides and application to human health, animal husbandry and agricultural production. *Carbohydr. Polym.* **2019**, *220*, 60–70. [CrossRef]
197. Grundemann, C.; Reinhardt, J.K.; Lindequist, U. European medicinal mushrooms: Do they have potential for modern medicine? An update. *Phytomedicine* **2020**, *66*, 153131. [CrossRef] [PubMed]
198. Wanmuang, F.; Leopairut, J.; Kositchaiwat, C.; Wananukul, W.; Bunyaratvej, S. Fatal fulminant hepatitis associated with Ganoderma lucidum (Lingzhi) mushroom powder. *J. Med. Assoc. Thail.* **2007**, *90*, 179–181.
199. Hisamochi, A.; Kage, M.; Arinaga, T.; Ide, T.; Miyajima, I.; Ogata, K.; Kuwahara, R.; Koga, Y.; Kumashiro, R.; Sata, M. Drug-induced liver injury associated with *Agaricus blazei* Murill which is very similar to autoimmune hepatitis. *Clin. J. Gastroenterol.* **2013**, *6*, 139–144. [CrossRef]

Journal of *Fungi*

Review

Sphinganine-Analog Mycotoxins (SAMs): Chemical Structures, Bioactivities, and Genetic Controls

Jia Chen [1], Zhimin Li [1], Yi Cheng [1], Chunsheng Gao [1], Litao Guo [1], Tuhong Wang [1] and Jianping Xu [1,2,*]

1 Institute of Bast Fiber Crops and Center of Southern Economic Crops, Chinese Academy of Agricultural Sciences, Changsha 410205, China; chenjia01@caas.cn (J.C.); lizhimin@caas.cn (Z.L.); chengyi@caas.cn (Y.C.); gaochunsheng@caas.cn (C.G.); guolitao@caas.cn (L.G.); wangtuhong@caas.cn (T.W.)

2 Department of Biology, McMaster University, Hamilton, ON L8S 4K1, Canada

* Correspondence: jpxu@mcmaster.ca

Received: 31 October 2020; Accepted: 22 November 2020; Published: 24 November 2020

Abstract: Sphinganine-analog mycotoxins (SAMs) including fumonisins and *A. alternata* f. sp. *Lycopersici* (AAL) toxins are a group of related mycotoxins produced by plant pathogenic fungi in the *Fusarium* genus and in *Alternaria alternata* f. sp. Lycopersici, respectively. SAMs have shown diverse cytotoxicity and phytotoxicity, causing adverse impacts on plants, animals, and humans, and are a destructive force to crop production worldwide. This review summarizes the structural diversity of SAMs and encapsulates the relationships between their structures and biological activities. The toxicity of SAMs on plants and animals is mainly attributed to their inhibitory activity against the ceramide biosynthesis enzyme, influencing the sphingolipid metabolism and causing programmed cell death. We also reviewed the detoxification methods against SAMs and how plants develop resistance to SAMs. Genetic and evolutionary analyses revealed that the *FUM* (fumonisins biosynthetic) gene cluster was responsible for fumonisin biosynthesis in *Fusarium* spp. Sequence comparisons among species within the genus *Fusarium* suggested that mutations and multiple horizontal gene transfers involving the *FUM* gene cluster were responsible for the interspecific difference in fumonisin synthesis. We finish by describing methods for monitoring and quantifying SAMs in food and agricultural products.

Keywords: sphinganine-analog mycotoxins; fumonisins; AAL-toxin; chemical structure; toxicity; genetics and evolution; biosynthesis

1. Introduction

Mycotoxins are secondary metabolites produced by various fungi. These metabolites have important ecological functions on living systems in their natural habitats. As secondary metabolites, mycotoxins are regarded as not essential for fungal growth or reproduction. However, their toxic effects to plants, animals, as well as humans are attracting increasing attention from chemists, biologists, food scientists, and healthcare professionals. Many fungi are capable of synthesizing mycotoxins, including certain saprophytic molds, poisonous mushrooms, human fungal pathogens, and plant fungal pathogens. Mycotoxins produced by plant pathogenic fungi can be divided into two groups: (i) host-selective (or host-specific) toxins (HSTs) and (ii) non-host-specific toxins (nHSTs), depending on whether they are specifically toxic to host plant (HSTs) or to a wide range of species (nHSTs). The known mycotoxins are typically low molecular-weight chemicals but with diverse structures and modes of actions. One group of mycotoxins are structurally analogous to sphingosine, the backbone precursor of sphingolipids that play essential structural and cellular roles in eukaryotic cells. These toxins are called sphinganine-analog mycotoxins (SAMs), with fumonisins and the *Alternaria alternata* f. sp. *Lycopersici* (AAL) toxins as the two most widely studied groups of SAMs. SAMs are toxic to plants and animals. They act by inhibiting the ceramide synthase (CerS), thereby influencing the sphingolipid metabolism and initiating apoptosis in animals and programmed cell death (PCD) in plants [1–3]. The objective of this paper is to provide an updated review on the structural diversity, syntheses, modes of action, and health impacts of SAMs.

The discovery of fumonisin was first reported in 1988 and the organism producing it was *Fusarium verticillioides* (syn. *Gibberella fujikuroi* mating population A, syn. *G. moniliformis* Wineland, syn. *F. moniliforme* Sheldon) [4]. Fumonisins have since been found to be produced by at least 18 species of the *Fusarium* genus, with *F. verticillioides* and *F. proliferatum* being the most prominent, and by three unrelated fungal genera, *Aspergillus* section *Nigri* (such as *Asp. niger*, *Asp. Welwitschiae* (syn. *Asp. awamori*) and so on, known as black aspergilli), *Tolypocladium* (*T. inflatum*, *T. cylindrosporum*, and *T. geodes*), and *Alternaria* (the tomato pathotype of *A. alternata,* formerly known as *A. arborescens*) [5–10]. Species of the *Fusarium* genus can be found as saprophytes in soil and as endophytes and pathogens of many plants worldwide. A common group of diseases caused by *Fusarium* pathogens is rotting that can happen to all tissues during all stages of plant development [11,12]. In addition, the *Fusarium* species can infect crops at the post-harvest period during storage [13]. The fungal propagules surviving in the soil can also infect new crop plants and can be carried to new fields by wind or by anthropogenic activities, such as when seedlings are transplanted [14]. *Fusarium* strains can synthesize fumonisins during all stages of their growth, including the saprophytic stage in the soil, during their pathogenesis, and as endophytes in different parts of plants, as well as during crop storage after harvest [15].

Fumonisins, as a nHST, are major contaminants of cereals and grains, including corn, rice, wheat, barley, rye, oat, millet, and products made based on these crops [16]. The consumption of food contaminated by fumonisins significantly increases health problems for humans, leading to a variety of cancers such as esophageal cancer and neurological defects [17,18]. For example, the International Agency for Research on Cancer (IARC) characterized fumonisin FB_1 as a group 2B carcinogens for humans [16]. Fumonisins can also cause diseases and adverse effects in other species, especially in livestock when the feeds are contaminated [19]. Well-known diseases in livestock caused by fumonisins include leukoencephalomalacia in horses and pulmonary edema syndrome in pigs [20,21].

Similar to fuminisins, the AAL-toxins include a family of structurally analogous metabolites. AAL-toxins are a group of HST produced by the ascomycete fungal pathogen *A. alternata* f. sp. *Lycopersici*, the causal agent of tomato stem canker disease [22]. It should be noted that several other pathotypes of *A. alternata* could also produce other HSTs responsible for fungal pathogenesis on their specific host plants, respectively [23]. Unlike other HSTs produced by *A. alternata*, besides the susceptible tomato host, AAL-toxins can also affect many other weeds and crops of dicotyledonous species and at least 25 species of *Solanaceae* [24,25]. Furthermore, the tomato pathotype of *A. alternata* was also reported to produce fumonisins B (FBs) [8,26]. AAL-toxin and FBs were not only detected in the necrosis plant tissues and culture media inoculated by *A. alternata* but also in spores and mycelia of this pathogen [27]. However, AAL-toxin remains the only toxin as a pathogenicity factor for stem canker disease of sensitive tomato varieties, while fumonisins are toxigenic virulence factors [28].

Because of the adverse impacts of SAMs on animal and human health, these toxins are also attracting increasing attention from food inspectors and public health agencies. Over the last three decades, significant progress has been made in our understanding of SAMs. Our objectives of this review are to capture these developments on SAMs with regard to their chemical structural diversity, the relationship between structure and activity, PCD induction, detoxification, genetics and evolution of SAMs biosynthesis, and laboratory detections.

2. Chemical and Structural Properties

2.1. Chemical and Structural Properties of Sphingolipids

SAMs have a distinct structural similarity to sphinganine (Figure 1). Sphinganine (dihydrosphingosine, DHS) is the simplest class of sphingolipids and has a backbone that consists of a linear aliphatic group with 18-carbon, an amino at C-2, and two hydroxyls (-OH) at C-1 and C-3, respectively. Phytosphingosine is obtained if a hydroxyl is introduced at C-4. Sphingosine consists of the sphinganine backbone but with a double bond at C-4. Ceramides are synthesized by linking an amide fatty acid at C-2 of sphingosine. Ceramides is a waxy lipid molecule, which is found in high concentrations in the membrane of eukaryotic cells. More complex sphingolipids can be formed by linking different chemical groups to hydroxyl (C1) of ceramides. Sphingolipids are one type of lipids widely found in their membranes in eukaryotes and a few prokaryotes, and they form complex and diverse interactions with other molecules [29]. Sphingolipids play important structural and functional roles, they are involved in a variety of signal transductions and crucial cellular processes [30,31]. For example, in humans, ceramides help form the skin's barrier and regulate immune response, protecting the skin against environmental irritants, pollutants, and water loss. Without the proper ratio of ceramides on our epidermal cells, the barrier of the skin will be damaged, resulting in dryness, itching, and irritation [32].

sphingolipid

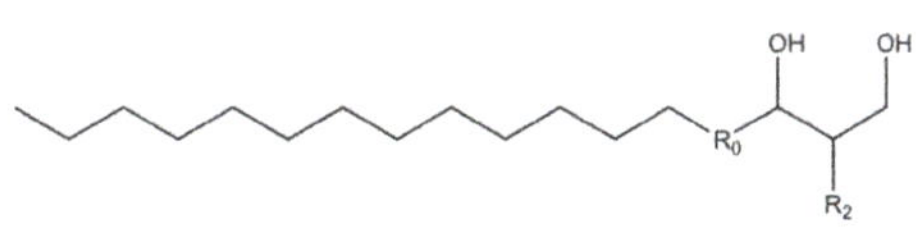

sphingolipid	R_0	R_2
sphingosine	CH=CH	NH_2
sphinganine	CH_2-CH_2	NH_2
phytosphingosine	CH_2-CHOH	NH_2
ceramide	CH=CH	Y-$(CH_2)_n$-CH_3

Figure 1. Chemical structure of sphingolipids. The table shows the different substituents in the chemical scaffold of the most essential sphingolipid.

2.2. Chemical and Structural Properties of Fumonisins

SAMs consist of two main types of toxins, fuminisins and AAL-toxins. Fumonisins can be divided into seven groups (FA, FB, FC, FD, FP, FP_y, and FL_a). These groups differ in the nitrogen functional group and the length of the carbon backbone [5]. Most fumonisins contain a 19–20 (FD contain 17 or 18 carbon) linear backbone similar to sphinganine with one nitrogen functional group (except for FP_ys and FL_as), two to four hydroxyl, two methyl, and two propane-1,2,3-tricarboxylic acid (PTCA) side chains esterified to the backbones [26,33]. The structural features of the seven groups of fumonisins are shown in Figure 2. Among them, the B group is the dominant one. For example, FB_1 accounts for 70–80% of the total fumonisins produced by *F. verticillioides* and is the predominant toxic form [5]. FB_2 and FB_3 are isomers of each other but with one less hydroxyl group than FB_1. The B series of fumonisins (FBs) are also the main food contaminants. Group A fumonisins (FA) are acetylated derivates of group B toxins, with lower toxicity and bioactivity than their FB counterparts [34]. Group C fumonisins (FC) have the same nitrogen functional group as FB_1 but lack the terminal methyl group at C-1 [35]. Three forms of acetylated FC_1 have been discovered in *F. oxysporum* [36]. Group P fumonisins (FP) have a nitrogen functional group of 3-hydroxypyridinium instead of the amino group in FB at the R_2 position [37]. The FC and FP groups have similar phytotoxic and cytotoxic effects to those caused by FB_1 or AAL-toxin [38]. Aside from these four main groups, there are several other lesser-known fumonisin analogs, with one or two PTCA replaced by a hydroxyl or carbonyl or other carboxylic acids group at C-13 and/or C-14 of the backbone (for example, HFB_1, as show in

Figure 2). Rheeder et al. summarized the 28 fumonisin analogs that have been characterized between 1988 and 2002 [5]. By reversed-phase high-performance liquid chromatography/electrospray ionization ion trap multistage mass spectrometry (RP-HPLC/ESI-IT-MS^n), Bartok et al. detected 58 fumonisins (including FD) or fumonisin-like compounds from *F. verticillioides* in rice cultures, and 28 isomers of FB_1 [33,39]. Indeed, the recent application of a semi-targeted method revealed over 100 structurally related compounds from SAMs-producing fungi, including a hydroxyl-FB_1, and two new classes of non-aminated fumonisins (FP_ys and FL_as) [26].

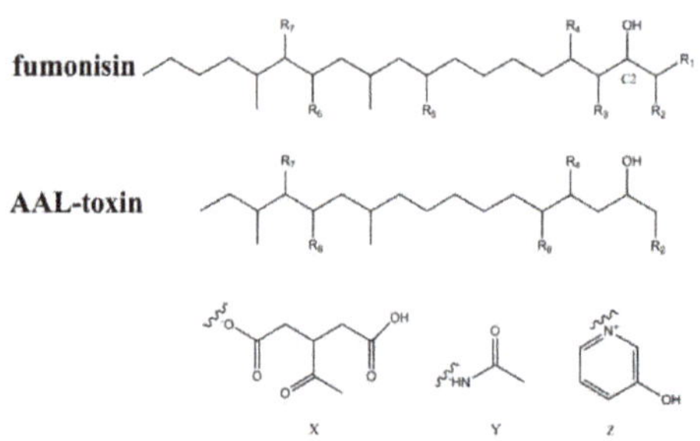

Panel A

AAL-toxin	R_2	R_4	R_8	R_6	R_7
TA_1	NH_2	OH	OH	X	H
TA_2	NH_2	OH	OH	H	X
TB_1	NH_2	OH	H	X	H
TB_2	NH_2	OH	H	H	X
TC_1	NH_2	H	H	X	H
TC_2	NH_2	H	H	H	X
TD_1	Y	OH	H	X	H
TD_2	Y	OH	H	H	X
TE_1	Y	H	H	X	H
TE_2	Y	H	H	H	X

Panel B

fumonisin	R_1	R_2	R_3	R_4	R_5	R_6=R_7
FA_1	CH_3	Y	H	OH	OH	X
FA_2	CH_3	Y	H	OH	H	X
FA_3	CH_3	Y	H	H	OH	X
FA_4	CH_3	Y	H	H	H	X
$isoFA^{a}_{1a,b}$	CH_3	Y	H	H	H	X
FC_1	H	NH_2	H	OH	OH	X
FC_2	H	NH_2	H	OH	H	X
FC_3	H	NH_2	H	H	OH	X
FC_4	H	NH_2	H	H	H	X
$isoFC_1$	H	NH_2	OH	H	OH	X
$isoFC^{b}_{2,3}$	H	NH_2	H	H	H	X
FC_1OH	H	NH_2	OH	OH	OH	X
FD^{c}	H	NH_2	H	H	OH	X
FP_1	CH_3	Z	H	OH	OH	X
FP_2	CH_3	Z	H	OH	H	X
FP_3	CH_3	Z	H	H	OH	X
FP_{y2}	CH_3	=O	H	OH	H	X
FP_{y4}	CH_3	=O	H	H	H	X
FP_{y6}	CH_3	=O	OH	OH	H	X
FL_{a2}	CH_3	OH	H	OH	H	X
FL_{a4}	CH_3	OH	H	H	H	X

fumonisin B	R_1	R_2	R_3	R_4	R_5	R_6=R_7
FB_1	CH_3	NH_2	H	OH	OH	X
FB_2	CH_3	NH_2	H	OH	H	X
FB_3	CH_3	NH_2	H	H	OH	X
FB_4	CH_3	NH_2	H	H	H	X
FB_6	CH_3	NH_2	OH	OH	H	X
$isoFB_1$	CH_3	NH_2	OH	H	OH	X
$isoFB^{b}_{1a-d}$	CH_3	NH_2	H	H	OH	X
$isoFB^{b}_{2,3a-e}$	CH_3	NH_2	H	H	H	X
FB^{a}_5/ $isoFB^{a}_{5a-d}$	CH_3	NH_2	H	H	OH	X
FB_1OH	CH_3	NH_2	OH	OH	OH	X
FBK_1 2TCA[d]	CH_3	NH_2	H	OH	OH	X
FBK_4 2TCA[d]	CH_3	NH_2	H	OH	H	X
HFB_1	CH_3	NH_2	H	OH	OH	OH
$PHFB^{e}_{1a,b}$	CH_3	NH_2	H	OH	OH	X/OH

[a] There are two other OH groups on the backbone
[b] There is another OH group on the backbone
[c] C18 compound or C17 compound (C2:without OH, R_3:ketone, R_4:OH
[d] C2 ketone
[e] A PTCA (X) was removed by hydroxylation at C13 or C14

Figure 2. Chemical structure of sphinganine-analog mycotoxins (SAMs). Panel A shows the AAL-toxins, Panel B shows fumonisin. In the table of each panel, the different substituents present in the chemical scaffolds of individual compounds are shown.

2.3. Chemical and Structural Properties of AAL-Toxin

The AAL-toxins have a structural similarity to fumonisins (Figure 2). The main difference between fumonisins and AAL-toxins is that AAL-toxins have one fewer PTCA side chain than fumonisins. The AAL-toxins have been divided into five pairs based on their side chain structures: A, B, C, D, and E pairs (TA, TB, TC, TD and TE). These pairs differ in their nitrogen functional group and hydroxylation at C-4 or C-5 positions of the backbone [40–42]. Each pair of AAL-toxins is composed of two regioisomers with PTCA esterified to C-13 or C-14 of the backbone, respectively. The TA pair is the major pair of toxins, with the TB and TC pairs formed by removing hydroxyl groups one by one from C-5 and C-4 of the TA pair. The TD and TE pairs were acetylated derivatives of TB and TC respectively, while the acetylated form of TA and keto derivatives of AAL-toxins (2-keto or 14-keto analogues predicted) were also found in 2015 [26]. These four regioisomeric pairs (TB, TC, TD, and TE) of AAL-toxins can all induce genotype-specific necrosis characteristics in tomato leaflets in the same pattern as that of the TA pair, but they differ as much as 1000-fold in their relative toxicity [42].

2.4. Chemical and Structural Properties of Analogs of SAMs

In addition to fumonisins and AAL-toxins, several fungal secondary metabolites have also been identified as structural analogs of sphinganine and CerS inhibitors (summarized in Figure 3 and Table 1). These metabolites include myriocins, sphingofungins, viridiofungins, 2-amino-14,16-dimethyl-octadecan-3-ol (2-AOD-3-ol), and a new C17-SAM identified from mussels contaminated by marine fungi including *Aspergillus*, *Fusarium*, and *Trichoderma.* Australifungin, a structurally unrelated mycotoxin produced by *Sporormiella australis*, was also shown to inhibit sphingolipid synthesis in plants, similar to those of SAMs.

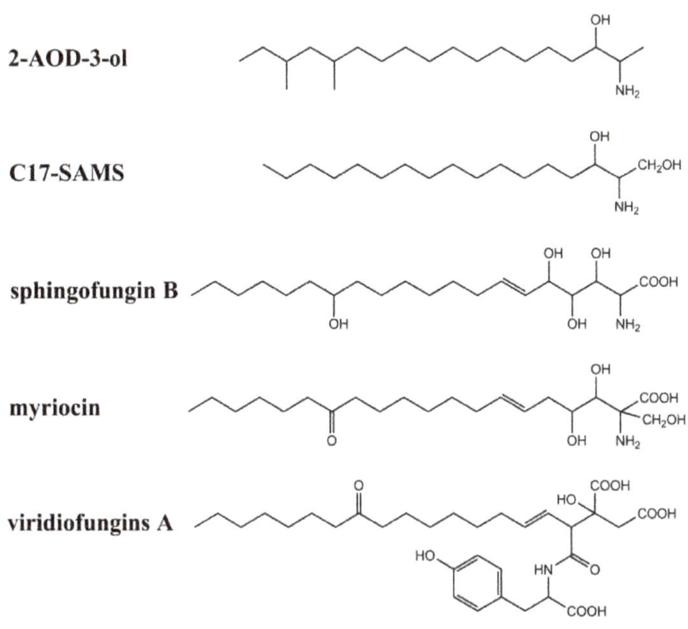

Figure 3. Chemical structures of other sphinganine-analog metabolites.

Myriocins, sphingofungins, and viridiofungins inhibit serine palmitoyltransferase (SPT), while fumonisins, AAL-toxin, and australifungin inhibit sphinganine-N acyltransferase. Serine palmitoyl transferase, one of the key enzymes in the synthesis of sphingolipids, was also reported to play a positive role in PCD regulation. The increase of SPT activity promoted PCD in plants. In contrast, by inhibiting SPT activity, the excessive accumulation of sphingosine can be alleviated, leading to reduced PCD [43]. Therefore, myriocin are usually used as a SPT inhibitor to pretreat *Arabidopsis thaliana* and tomato plants to induce their resistance to FB_1 and AAL-toxin, respectively [44,45].

Table 1. Analogs of sphinganine-analog mycotoxins (SAMs), their fungal producer(s), and their activities.

Analog of SAMs		Fungi/Origin	Activities	Scopus Citation (Review)	Reference
Myriocins (thermozymocidin, ISP-I)		*Myriococcum albomyces* *Melanconis flavovirens* *Isaria sinclairii*	Antifungal activity Inhibitor of serine palmitoyltransferase (SPT) Immunosuppressive activity Protective effect on hepatotoxicity Relieve fumonisin B_1 (FB_1)-induced toxicity and cell death Multi-pharmacological function on human	421(34)	[46–53]
Sphingofungins	E/F A/B/C/D/I G/H	*Paecilomyces variotii* *Asp. fumigatus* *Asp. penicilliodes*	Inhibitor of SPT Antifungal activity	65(15)	[54–58]
Viridiofungins	A/B/C	*Trichoderma viride Pers* *Tri. harzianum*	Inhibitors of SPT and squalene synthase Antifungal but lack antibacterial activity	21(5)	[59–61]
Australifungin		*Sporormiella australis*	Inhibitors of sphinganine-N acyl transferase Antifungal activity, phytotoxicity	26(7)	[62,63]
2-AOD-3-ol		*F. avenaceum*	Animal cell toxicity as fumonisin B	5	[64]
C17-sphinganine analog mycotoxin		Contaminated mussels	Blocking skeletal muscle contraction	1	[65]

3. Relationships between SAMs' Structure and Biological Activities

The biological effects of SAMs, such as their toxicity, are similar among different SMAs. Many SAMs have a similar spectrum of susceptible plant species [34]. Tomato tissues and cells are similarly sensitive to AAL-toxins and to FB_1 and FB_2 toxins. In some other plants, AAL-toxins can cause necrotic cell death, similar to that of fumonisins [66]. For animal tissue cultures, the TA toxins can induce cytotoxicity in both rat liver and dog kidney cells as FB_1 toxin [67,68]. Besides, AAL-toxin and *F. verticillioides* could also inhibit larval growth and reduced pupal weights of tobacco budworrn *Heliothis virescens* [69]. Such similarities have been attributed to the structural similarities between the SAMs and sphinganine. However, there are differences among SAMs in their biological effects and those differences are related to their structural differences. Below, we summarize the main findings in this area.

The amino functional group of SAMs is essential for their toxic activity. The peracetylated derivatives of AAL-toxins and FB_1 are biologically inactive or have significantly reduced toxicity in both the plant bioassay and the animal tissue culture systems [66,70,71]. These results were consistent with initial reports on these toxins showing that blocking the free primary amines of AAL-toxins by specific reagents could abolish the biological activities of these toxins in plants [72]. In an in vitro test of rat primary hepatocytes, it was noted that the N-acetyl analogue of FB_1, FA_1, also showed CerS inhibition [68]. Later, FA was found to spontaneously undergo isomerization, rearranging its O-acetylation group to form different analogs. The impact of these rearrangement products on inhibition of CerS in rat liver slices also supported the important role of a primary amino for both CerS inhibition and toxicity [73]. Derivatization of the amino group with fluorogenic reagents also makes the FBs' detection possible by the high-performance liquid chromatography (HPLC) assay [74]. FBs can bind covalently to proteins by reacting with amino groups in abiogenic conversions, which may increase the toxicity of those conversion products [75]. Similarly, the terminal amino group of FB_1 can conjugate to bovine serum albumin (BSA) and work as an immunogen to produce monoclonal antibodies for enzyme-linked immunosorbent assay (ELISA) detection [76]. Amino group of fumonisins can also work as an electron donor and react with the electrophilic carbon within the isothiocyanate (ITC) group. Consequently, FBs can be degraded by fumigation treatment with ITC-containing compounds [77].

The hydrolysis product of FB_1 (HFB_1) was shown as less toxic than both FB_1 and TA to plants [78]. Neither HFB_1 nor the yeast sphingolipids (completely acetylated) contain PTCA. While both had adverse effects on duckweed growth, they showed lower phytotoxicity than TA and FB_1 that contained one and two PTCA, respectively [79]. In contrast, the hydrolysis products of AAL-toxins largely maintain the toxicities of their parental compounds to the susceptible tomato lines [66]. These results indicate that PTCA is important to phytotoxicity of FBs and there is specificity of interaction between AAL-toxins and tomatoes.

Different from those in plants, an in vitro test using primary hepatocytes of rat showed that the HFBs had greater cytotoxicity than FBs. However, the HFBs could not initiate cancer development due to the lack of PTCA moiety, which was proposed to play an active role in the fumonisins absorption from the gut [70]. In the pregnant LM/Bc mouse model, HFB_1 did not cause neural tube defects. In contrast, 10 mg of FB_1/kg body weight of mice disrupted maternal sphingolipid metabolism, caused hepatic apoptosis in the female mice, increased fetus mortality, and reduced fetus weight [80]. In the SAMs-sensitive pig model, HFB_1 was shown to have limited intestinal or hepatic toxicity but only slightly disrupted sphingolipids metabolism [81]. The toxic effects of FB_1 and HFB_1 exposure on intestinal barrier function and immunity in a pig intestinal porcine epithelial cells and porcine peripheral blood mononuclear cells co-culture model was also investigated. FB_1 aggravated lipopolysaccharide (LPS)/deoxynivalenol (DON)-induced intestinal inflammation, while HFB_1 showed less toxicity to the immune system [82]. In addition, when HFB_1 and HFB_2 were acylated by CerS, the N-acyl-metabolites were toxic in vitro to the human colonic cell line and in vivo to the intraperitoneal rat tissues [83].

Fumonisins are capable of binding to polysaccharides and proteins via their two PTCA side chains in thermal-treated food and form fumonisin artifacts [84]. The activities of SAMs vary depending on where hydroxylation occurs along the carbon backbone. For example, FB_2 had a greater cytotoxic effect than FB_3 and FB_1 in primary rat hepatocytes [70]. However, different from most other side groups, the C-1 terminal methyl group, which differed between FC and AAL-toxin from other fumonisins, seemed not required for the biological activity in SAMs.

Similar symptoms but less phytotoxicities of SAMs were observed when long-chain sphingoid bases or simple sphingolipids were applied to duckweed, which indicated that the phytotoxicity of SAMs might be resulted from the accumulation of phytotoxic sphingolipid intermediates [71,85]. This result was consistent with the induction of PCD through ceramide-based signaling pathways (described below).

Although AAL-toxins and fumonisins are structurally related chemicals with similar phytotoxicity, the latter are 10 times less efficient. AAL-toxins have been considered to serve as an herbicide at a very low dosage against a wide variety of broadleaf weeds (e.g., jimsonweed, prickly sida, and black nightshade). However, monocotyledonous crops (e.g., maize, wheat, and resistant varieties of tomato) are tolerant to AAL-toxins [24,86,87]. Until 2013, the mode of action through CerS inhibition was not among the 21 molecular target sites of the commonly used herbicides. Using AAL-toxin as a lead compound has the potential to develop novel and safe bioherbicide, which has phytotoxicity but reduced or no mammalian toxicity [88,89].

4. Detoxification of SAMs

Using agricultural and manufacturing practices for preventing the spread and growth of toxin-producing fungi and limiting mycotoxin production is the preferred method to eliminate food contamination by fungal toxins at the pre-harvest period [90]. However, it is extremely difficult to completely prevent fungal pathogen growth and mycotoxin contamination in agricultural practices and in food storage and processing. Since the toxicity of SAMs is structurally dependent, our knowledge on the relationships between SAMs' structure and biological activity provides clues for developing effective management strategies to minimize the impact of SAMs in food and feed products. Indeed, structural modifications such as hydrolysis have been demonstrated as effective at reducing the toxicity of SAMs [91].

Over the last three decades, chemical, biological, and physical strategies have been developed to degrade mycotoxins in food and feed products [92]. For example, nixtamalization was applied to reduce FBs by cooking fumonisin-contaminated maize with lime, as well as by using atmospheric ammoniation treatment [93–95]. Chlorine dioxide also showed the ability to degrade FB_1 [96]. Two common cooking methods include extrusion and nixtamalization were shown to reduce the toxicity of FB_1-contaminated corn [97]. Cold atmospheric pressure plasma was used as a physical treatment to successfully degrade pure FB_1 and AAL-toxins within 60 s, while the presence of the matrix slowed down the degradation [98,99]. Ozone was applied to disrupt fungal cells of *Fusarium* and *Aspergillus* by oxidizing sulfhydryl and amino acid groups of enzymes or attacking the polyunsaturated fatty acids of the cell wall [100]. However, not all SAMs are susceptible to physical and chemical treatments. In addition, some of these treatments may also result in derivatives with unknown toxicity and be detrimental for the treated commodities, as shown in some cases [101,102].

Another method to reduce SAM toxicity is through microbial actions. Microorganisms can carry out biotransformation reactions to detoxify SAMs [91]. Such methods include deamination, acetylation, hydrolysis, glucosylation, and decarboxylation. For example, Benedetti et al. isolated a Gram-negative rod bacterial strain from soil capable of degrading fumonisin to four metabolites when fumonisin was supplied as the sole carbon source [103]. The bacterium *Sphingopyxis* sp. could detoxify fumonisin B1 by at least two enzymatic steps, including an initial de-esterification reaction followed by de-amination of hydrolyzed product [104]. Chlebicz and Śliżewska found that 12 strains of *Lactobacillus* sp. bacteria and 6 strains of *Saccharomyces cerevisiae* yeast could reduce the concentration of FB_1 and FB_2 by 40% [105]. Similarly, Burgess demonstrated that fumonisin-producing *Asp. welwitschiae* have the ability to produce enzymes to synthesize non-aminated fumonisins that are less toxic than FB, and that those enzymes could be used for fumonisin detoxification [106]. Indeed, using enzymes to detoxify by modification of chemical structures has become a promising method for mycotoxins control after grains harvest [107,108]. For example, fumD (carboxylesterase) from *Sphingopyxis* catalyze detoxification of FB_1 to the hydrolyzed form by hydrolysis of both PTCA side chains. Then, the aminotransferase FumI could degrade FB_1 by catalyzing the deamination of HFB_1. FumD has also been tested for interference of fumonisins adsorption in turkey, swine, and pig [109–111]. Finally, several other enzymes such as manganese peroxidase from lignocellulose-degrading fungi and laccase from *Pleurotus eryngii* were all capable of degrading fumonisins [112,113].

Another potential method to reduce SAMs from food and feed products is to use adsorbent materials to soak up and remove the toxins. Many materials have shown the capacity to adsorb mycotoxins in vitro, thus the use of adsorbents in livestock diet as feed additives can potentially decrease the bioavailability of mycotoxins to humans and animals. As feed additives, cholestyramine, nanosilicate clay platelets, and refined calcium montmorillonite clay all reduced FB_1 toxicosis [114–116]. Moreover, natural products such as the phenolic compound chlorophorin, honey, and cinnamon oil have all shown promise as fumonisin-reducing agents [117–119].

5. Programmed Cell Death and Sphingolipids

Almost all cells die eventually. There are four main types of cell death: necroptosis, pyroptosis, ferroptosis, and apoptosis, classified based on their distinct molecular and cellular processes and different outcomes. Apoptosis or programmed cell death (PCD) is a kind of cell suicide that strictly regulates cells that are no longer needed or are a threat to the wellbeing of multicellular organisms. Both plants and animals have PCD and they are functionally analogous to each other [120–123]. PCD plays essential roles to maintain normal physiological activities in multicellular organisms such as plants and animals and is an active self-regulating process to selectively eliminate redundant, aged, and damaged cells. PCD can be predicted for specific cells at defined developmental stages. However, PCD can also be induced by membrane-bound and cytosolic proteins stimulated by stress-induced signals. Such signals can trigger cell death via intricate cascades of transcriptional changes and post-translational protein modifications [122,124]. The characteristics of PCD include reduced cell volume, chromatin marginalization and condensation, nuclear lamina disassembly, DNA fragmentation, and apoptotic body formation, etc. [1].

PCD triggers and propagation involve many factors, including the expressions of certain cell surface receptors, transmembrane domains of several membrane proteins, intracellular proteins related to the propagation of death signals, secondary messengers including inositol triphosphate and ceramides, calcium (Ca^{2+}) fluxes, reactive oxygen species (ROS), regulatory factors of cell cycle, and other suppressors or activators proteins. Many of these subcellular components, genes, and signal transduction pathways involved in PCD are functionally conserved across all domains of cellular organisms, from bacteria to fungi to plants and animals. However, there are differences in the actual mechanisms among organisms, as summarized in References [125,126].

Sphingolipids have been implicated to play an important role in cell growth, development, response to external environment, and PCD. As the main component of the cell membrane system, sphingolipids help to maintain the structural stability and transport of molecules across cell membranes [127,128]. In mammals, sphingolipids are especially abundant in the nervous system cells, with important functions in cell contact, growth, differentiation, communication, response to stress signals, and apoptosis [30,129]. In plants, sphingolipids are involved in response to both biotic and abiotic stresses, such as to pathogen infection, drought, and low temperature [31,130,131]. Indeed, the linkage of ceramide signaling to apoptosis has been widely reported in both plants and animals. Consequently, actions by SAMs to disrupt the functions of sphingolipids could have significant negative consequences. However, our knowledge about the roles of sphingolipids on apoptosis have also led to increasing interests on potential novel therapies using sphingolipids as treatment targets against degenerative and proliferative diseases in humans and animals, such as cancer and Parkinson's disease [132,133].

A large number of studies have shown that sphingolipids could serve as critical secondary messengers in signal transduction to regulate PCD [134,135]. For example, in neutrophils, sphingolipids have been linked to increased superoxide formation and Ca^{2+} influx, which are universal signaling molecules involved in many cellular functions [136]. The induction of PCD in *Arabidopsis* by ceramides was also verified to be partly dependent on ROS in mitochondria or regulated by the release of Ca^{2+} [137,138]. In addition, increase of sphingosine level can activate the mitogen-activated protein kinase (MAPK) pathway in which MPK6 participates, promote the accumulation of sialic acid (SA), and then induce PCD [43]. Lachaud et al. found that the activated calcium-dependent kinase (CPK3) can regulate the process of sphingosine-induced PCD by dissociating CPK3 from the 14-3-3 protein-complex under increased calcium concentration induced by sphingosine. The activated CPK3 is then degraded followed by PCD induction [139].

Mutational studies of *Arabidopsis* and in vitro experiments have shown that ceramides, and free sphingoid bases such as sphingosine, sphinganine, and phytosphingosine, can all induce PCD. In contrast, phosphorylated products of these compounds can inhibit or alleviate PCD in plants [44,138,140–142]. However, phosphorylated sphingosine can inhibit the growth of yeast cells, which suggests that there are different mechanisms of action between plants and yeasts [143]. It is worth mentioning that not all ceramides can induce PCD. Nagano et al. found that when the C2 position of fatty acid in the side chain of ceramides was hydroxylated, PCD was inhibited rather than induced [144]. These results indicate that the occurrence of PCD in plants depends not only on the absolute content of sphingolipids, but also the relative ratios of various modified forms. In the next section, we will describe how SAMs are involved in PCD.

6. SAMs Trigger PCD through Ceramide-Based Signaling Pathways

In plants, pathogen invasions can lead to disruptions in host cellular homeostasis, and trigger cell death in susceptible varieties or even in resistant varieties with a hypersensitive response (HR). Because of its many similarities with PCD, HR is often considered as a form of PCD in plants. AAL-toxins, as the pathogenic factor of tomato stem canker disease, can induce PCD in sensitive tomato varieties resulting in fragmentation of chromosomal DNA and formation of apoptotic bodies in cells [145]. Similarly, when treated by FB_1 produced by pathogenic *Fusarium*, *Arabidopsis* protoplasts showed symptoms similar to PCD in animal cells [146]. At the tissue and organ levels, *Arabidopsis* leaves treated with the FB_1 toxin showed characteristic disease symptoms. Cells of the diseased leaves had overall phenotypes similar to HR, including callose accumulation, ROS production, and pathogenesis-related (PR) gene induction [147]. The damages caused by SAMs on host plants can further increase pathogen infection and colonization. Similar to those found in plants, SAMs can induce neuro-/renal-responses, heptatoxicosis, and neoplasms, as well as cell death in animals. The relationship between apoptosis and ceramide signaling has been established in both plants and animals in their response to SAMs [120]. For example, the induction of cell death in both tomato and African green monkey kidney (CV-1) cells occurred under similar toxin concentrations and time frames. For both types of cells, morphological markers characteristic of apoptosis were observed, including cells with positive terminal deoxynucleotidyl transferase end labeling (TUNEL), DNA fragmentations, and the formation of apoptotic-like bodies [145,148].

SAMs are structurally analogous to sphinganine and are thus effective inducers of PCD. The emerging mechanism of their actions is that SAMs can competitively bind to CerS in cells. Such binding leads to the accumulation of free sphingoid bases, the substrates of CerS, while ceramides as products of CerS were consumed and reduced, activating PCD in plant and animal cell lines [3,85]. For example, it was reported that FB_1 is a potent competitive inhibitor of CerS from liver and brain microsomes in several mammalian cell lines [149,150]. An increase in sphinganine was observed in an in vivo test of CerS inhibition, as well as in FB-fed animals treated at high concentrations [2]. In addition, TA and FB_1 can inhibit CerS in rat hepatocytes and green tomato fruits [150,151]. Furthermore, it was found that FB_1 not only induced apoptosis in animal cells, but also altered cell morphology, cell–cell interactions, cell surface proteins behavior, protein kinase activity, and cell growth and viability in non-apoptotic cells [148,152]. In plant cells, after exposure to SAMs, sphingosine concentration increased significantly within a short time, followed by the accumulation of ROS in the cytoplasm and then apoptosis. These results suggested that the accumulation of sphingosine in cells was the upstream signal of ROS for cell death [43,44,85]. The induction of PCD by FB_1 is also related to the accumulation of ceramides. In *A. thaliana,* there are two types of CerSs that use different substrates. Class I CerSs use sphinganine and C-16 fatty acyl-CoA as substrates, while class II use phytosphingosine and very long-chain fatty acyl-CoA as substrates. FB_1 mainly inhibits the activity of class II CerSs. When treated with FB_1, phytosphingosine in cells increases significantly. At the same time, as the product of a previous step, sphinganine also increases, which provides more substrates for class I CerS. Consequently, the products of class I CerS in cells increase, leading to induced PCD [153,154].

7. Plant Resistance to SAMs

Phytohormones are also involved in the defense reaction induced by SAMs. Changes in ethylene (ET) were first discovered in AAL-induced necrosis of tomato [155]. Alteration in ethylene perception in "never ripe" mutants of tomatoes can markedly alleviate the tissue damage caused by SAMs, which indicated an ethylene-associated signal transduction during plant cell death [156]. Later, a transcription analysis of AAL-toxin-induced cell death was carried out in *Arabidopsis*. Genes responsive to ROS and ET were among the earliest upregulated genes [157]. Mase used VIGS (virus-induced gene silencing) analyses and verified that the ET signaling pathway and MAPK cascades were required for AAL-toxin-induced PCD in tobacco [158]. By SA-mediated ET suppression, glutathione (GSH) may be involved in resistance primarily against AAL-toxin-induced stress in Arabidopsis [159].

Unlike ethylene in host basal defense responses against the tomato pathotype of *A. alternata*, the jasmonate (JA)-dependent signaling pathway is not involved in host defense against the toxigenic *A. alternata* pathogen. JA affects pathogen acceptability via a toxin-independent mannerin in the interactions between plants and toxigenic necrotrophic fungal pathogens. It may act upstream of ethylene biosynthesis in AAL-toxin-triggered tomato cell death [160,161]. Later, a comparative proteomics analysis revealed that the COI1 (coronatine insensitive 1, JA receptor)-dependent JA pathway enhances AAL-toxin-induced PCD of tomato through regulating the redox status of the leaves, other phytohormone pathways, and/or important PCD components [162].

The sensitivity of tomato plants to the fungal pathogen *A. alternata* f. sp. lycopersici is controlled by the *Alternaria* stem canker resistance locus (*Asc*-locus) on chromosome 3 [163]. Mutations of tomato *Asc* locus gives resistance to the pathogen, while overexpression of the tomato *Asc-1* gene mediates high insensitivity to SAMs in tomato and confers resistance to pathogen infection in sensitive *Nicotiana* plants [164,165]. *Asc-1* is a homolog of the yeast longevity assurance gene *LAG1*, which encodes components of sphinganine N-acyltransferase. This resistance gene could prevent the disruption of sphingolipid metabolism during AAL-toxin-induced PCD. Both *Nicotiana* and *Lycopersicon* genera belong to Solanaceae. In tomato, insensitivity to SAMs and susceptibility to the pathogen is determined by *Asc-1* [166]. In contrast, the SAM-sensitive species in the *Nicotiana* (except for *N. umbratica*) still have *Asc-1* homologs and are resistant to *A. alternata* f. sp. lycopersici infection with HR, which indicates an additional (non-host) resistance mechanism between *Nicotiana* and this pathogen [167]. The multilayered defense systems also exist in *Arabidopsis* non-host resistance to *A. alternata* [168]. Similarly, although many *Fusarium* species produce fumonisin, they cannot infect AAL-sensitive tomato. This non-host resistance includes a multi-layer defense system involving both pre- and post-invasion, and help plants defend against various pathogens [169]. In addition, Zélicourt demonstrated that two of three Lag1 homologs in the *Orobanche cumana* genome were responsible for an enhanced sensitivity to AAL-toxin [170].

Aside from the above-mentioned genes, several other genes were identified from Arabidopsis and found to be involved in the AAL-induced PCD pathway, including Zinc *A. thaliana* 11 (a zinc finger protein ZAT11), fbr41 (FB_1 Resistant41), and baculovirus p35 gene (inhibitor of a class of cysteine proteases). All of them showed protective effects on AAL-toxin-induced cell death and pathogen infection in plants [171–173]. Discovery of resistant genes has provided a potential strategy for SAMs' control in crop production by plant transgenic modification.

Because of the high toxicity of fuminisins, especially FB_1, a large number of studies have focused on them. So far, the mechanism of FB_1 toxicity has been centered around its structural resemblance with sphinganine and consequent competitive inhibition of CerS and the disruption of lipidomic profiles. However, there is emerging evidence suggesting that FB_1 can disrupt mitochondrial function and generate excessive toxic ROS.

Table 2 shows a list of reviews summarizing the latest advances related to fumonisins, in their assessment, biosynthesis, detection, crop breeding of resistant varieties, and toxicity.

Table 2. Topical reviews on fumonisins over the last five years.

Subject	Content	Reference
Assessment	Biomarkers, metabolism, and biomonitoring of fumonisins in human biological fluids	[174]
Assessment	Impact on agriculture, food, and human health and their management strategies	[16]
Assessment	Risk assessment and intervention models for fumonisin of maize in South Africa	[175]
Assessment	Fumonisins and related *Fusarium* occurrence in wheat and its by-products	[176]
Assessment	Fumonisins and their modified forms	[177]
Assessment	Biological methods for fumonisins reduction and related *Fusarium* species control	[178]
Assessment	Fumonisins and *A. alternata* f. sp. *Lycopersici* (AAL) toxins in ruminants and their forages	[19]
Biosynthesis	Genetic regulation of fumonisins biosynthesis by specific genes and global regulators	[179]
Biosynthesis	Impact of environmental variables and genetics of maize resistance on fumonisin accumulation	[180]
Detection	Analytical methods for fumonisins detection in single corn kernels	[181]
Detection	Molecular methods for early detection of fumonisin-producing *F. verticillioides*	[182]
Plant resistant	Genomic, genes, and pathways in maize resistance to *Fusarium* ear rot and fumonisin accumulation	[183]
Plant resistant	Relationship between Bt maize hybrids and fumonisins contamination level	[184]
Toxicity	Mitochondrial toxicity induced by FB_1	[185]
Toxicity	Molecular mechanisms underlying FB_1-mediated toxicities and related interventions	[186]
Toxicity	CerS inhibition by fumonisins result in animal and plant disease	[187]
Toxicity	Dietary fumonisin and growth impairment in children and animals	[188]
Toxicity	Impact of fumonisin-contaminated feed on pig intestinal health	[189]
Toxicity	Oxidative stress-mediated toxicity and metabolism in vivo and in vitro	[190]

8. Genes Responsible for SAMs Production

In fungi, genes directly involved in the biosynthesis pathway of the same secondary metabolite are usually located at adjacent positions in gene clusters in the genome [191]. These genes are often co-expressed and co-regulated. The genes involved in fumonisin biosynthesis fit this general pattern. Specifically, *Fusarium* species capable of producing fumonisins typically contain one gene cluster involved in their synthesis, called the *FUM* gene cluster. At present, a total of 21 genes have been identified in *FUM* gene clusters of various species and verified to be involved in fumonisin biosynthesis and regulation or self-protection using a variety of approaches, such as gene knockouts, domain swapping, and heterologous expression (Table 4) [192–197]. In *F. verticillioides*, the *FUM* cluster responsible for fumonisin B biosynthesis includes 17 genes [193,197,198].

As described previously, SAMs are polyketide-derived compounds with structural similarity to sphinganine. Polyketides are synthesized by polyketide synthases (PKSs), which are large multifunctional enzymes. *FUM1* is a PKS gene previously designated as *FUM5* [199]. PKS encoded by *FUM1* catalyzes the synthesis of an octadecanoic acid precursor as the initial step for FB biosynthesis in *Fusarium* spp. [200,201]. The proposed biosynthetic pathway of FB is described in Figure 4 [197]. The precursor mentioned above undergoes condensation with L-alanine to synthesize the polyketide backbone, this reaction was catalyzed by the aminotransferase Fum8 [192,201]. In *F. oxysporum* strain O-1890, the orthologue of *FUM8* determines that *Fusarium* produces predominantly FCs [195]. The *fum8* deletion in some stains of *A. welwitschiae* is also considered to be associated with the loss of FB_2 production [202].

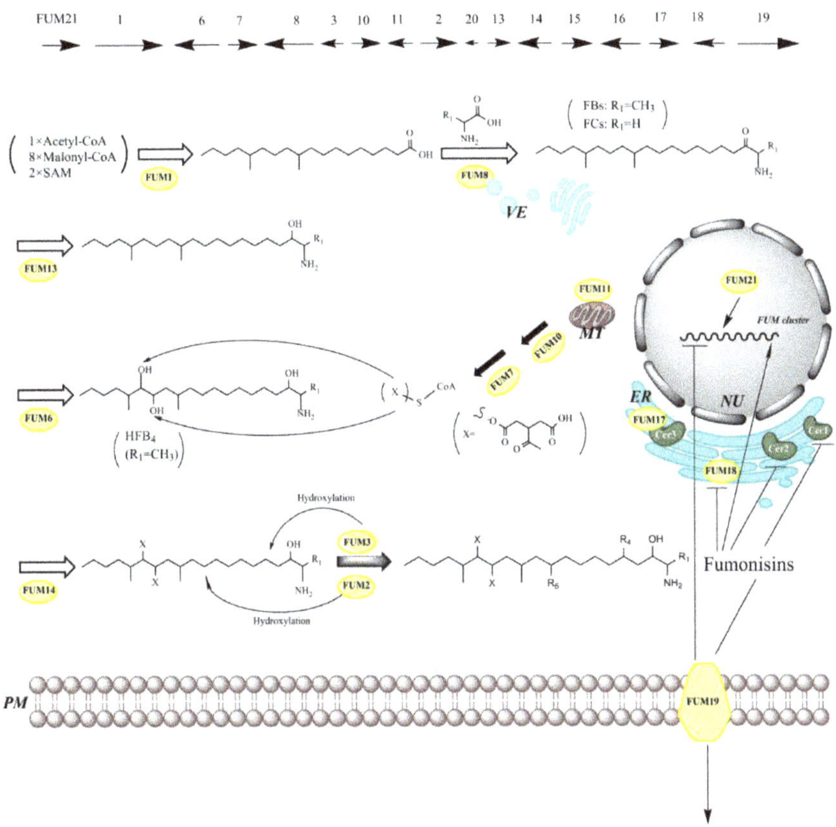

Figure 4. The *FUM* gene cluster, the proposed pathway of fumonisin biosynthesis, and the proposed mechanism for self-protection against fumonisins toxicity by the toxin-producing fungi (FBs and FCs). SAM, S-adenosyl methionine; VE, vesicles; MT, mitochondrion; ER, endoplasmic reticulum; NU, nucleus; PM, plasma membrane.

A likely mitochondrial carrier protein encoded by *FUM11* transport the substrate tricarboxylate for Fum7 (dehydrogenase) and Fum10 (acyl-CoA synthase) to produce CoA-activated tricarballylic acid, which are attached to the polyketide backbone by Fum14 (condensation-domain protein) [192,194]. The ensuring steps of fumonisin biosynthesis involving various modifications of the backbone (including primarily hydroxylation) were catalyzed by several enzymes, such as Fum6/Fum12/Fum15 (cytochrome P450 monooxygenase), Fum2 (hydroxylase of C10), Fum13 (short-chain dehydrogenase/3-ketoreduction), and finally, Fum3 (hydroxylase of C5, dioxygenase), catalyzed by *FUM9*-encoded protein (alleles of *FUM3*) [192,193,203–206].

In addition, *FUM21* encoding a GAL4-like Zn(II)2Cys6 transcription factor was verified to be involved in the regulation of fumonisin synthesis. However, it seemed that the deletion of *FUM16* had no apparent effect on fumonisin production in *F. verticillioides* [194,198,207–209]. Recently, *FUM17*–*FUM19* in *F. verticillioides* were found to help the fungus to avoid its own toxicity during fumonisin production. Fum19 is an ATP-binding cassette transporter (ABC transporter) and acts as a repressor of the *FUM* gene cluster. *FUM17* and *FUM18* are CerS homologs. *FUM18* could fully complement the yeast CerS null mutant *LAG1/LAC1*, while co-expression of *FUM17* and CER3 partially complemented. Both the Fum17 and Fum18 proteins enable *F. verticillioides* to increase its resistance of fumonisin by providing *FUM* cluster-encoded CerS activity as a first level of self-protection [197].

Aside from the *FUM* genes, other genes like *FST1* (transporter), *FUG1* (transcription or signal transduction factors), *CPP1* (protein phosphatase type 2A catalytic subunit), and *FvVEl* (regulator) in *F. verticillioides*, *PKS3* and *PKS11* in *F. proliferatum*, and GATA-type transcription factors *AreA* and *AreB* (known as the global nitrogen regulators) in *F. fujikuroi* have also been demonstrated to have an important role in fumonisin biosynthesis and regulation [209–214]. In addition, a degenerated, over-represented motif which is potentially involved in the cis-regulation of *FUM* genes and fumonisin biosynthesis was also identified from both *F. verticillioides* and *Asp. niger*, while it was not found in fumonisins non-producing fungi containing various *FUM* homologues [215].

Several abiotic and biotic factors have been found to affect the expression of *FUM* genes and regulate biosynthesis of fumonisin. These factors include water activity, temperature, carbon sources and other nutrients, host plant species and varieties or their extracts, and plant age [216–220]. Mature plants and extracts from those plants are often associated with higher concentrations of SAMs. It has been suggested that harvesting the crop at earlier stages other than full maturity could be one of the strategies to control fumonisin contamination [221,222].

The genome sequencing and analysis of *Asp. niger* revealed that its genome contained a gene cluster (*fum* cluster) homologous to the *FUM* cluster in *Fusarium* species (shown in Table 4). Specifically, 12 homologues of the fumonisin synthesis genes were found, including *fum1*, *fum3*, *fum6*, *fum7*, *fum8*, *fum10*, *fum13* to *fum16*, *fum19*, and *fum21* genes [208,223,224]. This gene cluster is also found in fumonisin-producing isolates of *Asp. welwitschiae* but is absent from the genomes of other sequenced *Aspergilli* that do not produce fumonisin, such as *Asp. fumigatus*, *Asp. oryzae*, and *Asp. nidulans* [7,208]. In addition, homologs of multiple *fum* genes have been found in several other *Aspergilli* spp. but where no fumonisin production has been detected (summarized in Table 3). Some of the *Aspergullus* spp. contain genes that are unique to them. For example, a dehydrogenase gene (*sdr1*) of a short-chain length was found in the *fum* cluster of *Asp. niger* but is absent in the *FUM* gene cluster of *Fusarium* spp. In contrast, the *Fusarium FUM2* gene with a function of hydroxylation at the C-10 backbone position of fumonisin is absent in the *Asp. niger fum* cluster [207,208]. This result is consistent with the study that shows that *Asp. niger* only produces fumonisins FB_2, FB_4, and FB_6, which lack a hydroxyl at C-10 [7,225,226]. However, isolates of several black aspergilli (including *A. niger*, *Asp. foetidus*, and *A. tubingensis*) isolated from peanuts and maize also produced FB_1 and FB_3, consistent with a complex biosynthesis pattern of the fumonisins in *Aspergilli* spp. [227].

Table 3. Difference in genomic context of fumonisin biosynthetic gene (*fum*) cluster between strains of *Aspergillus* spp.

Fungi	Stains	*fum* Cluster	Reference
Asp. niger	fumonisin-producing strains	*fum* cluster	[208]
	fumonisin-non-producing strains	Intact *fum* cluster	[7,235]
Asp. welwitschiae	fumonisin-producing strains	*fum* cluster	[7]
	fumonisin-non-producing strains	Three *fum* cluster types including an intact cluster	[7,235]
Asp. tubingensis	fumonisin-producing strains	Not tested	[227]
	fumonisin-non-producing strains	Multiple patterns of *fum* gene deletion	[7]
Asp. brasiliensis	fumonisin-non-producing strains	Multiple patterns of *fum* gene deletion	[7]
Asp. luchuensis	fumonisin-non-producing strains	Multiple patterns of *fum* gene deletion	[7,236]
Asp. fumigatus	fumonisin-non-producing strains	Not detected	[208]
Asp. oryzae	fumonisin-non-producing strains	Not detected	[208]
Asp. nidulans	fumonisin-non-producing strains	Not detected	[208]
Asp. foetidus	fumonisin-producing strains	Not tested	[227]

Table 4. Homologous genes and their functional roles in the biosynthesis of SAMs.

Homologue of Cluster Genes			Predict Gene Product and Function	Reference
fumonisin		AAL-toxin		
Fusarium spp.	*Aspergillus* spp.	*A. alternata*		[169,224]
FUM1 (fumonisin biosynthetic gene 1, previously designated as *FUM5*))	*fum1*	*ALT1* (AAL-toxin biosynthetic gene 1)	polyketide synthase	[12,199,200]
FUM2	absent		Dioxygenase for hydroxylation of C10	[228]
FUM3	*fum3*		Dioxygenase for hydroxylation of C5 (the same gene as *FUM9*)	[205,228,229]
FUM4			Not clear	[230]
FUM6	*fum6*	*ALT2*	Cytochrome P450 monooxygenase–reductase fusion proteins for hydroxylation of C14/C15	[192,231]
FUM7	*fum7*	*ALT3*	Type III alcohol dehydrogenases for PTCA (propane-1,2,3-tricarboxylic acid) side chain formation	[192,194,232]
FUM8	*fum8*		α-oxoamine synthase and homologous for amino transfer and FBs/FCs production	[192,195]
FUM10	*fum10*		Fatty acyl-CoA synthase for PTCA esterification	[193,194]
FUM11			mitochondrial transport protein for PTCA transport	[193,194]
FUM12			cytochrome P450 monooxygenases	[193]
FUM13	*fum13*	*ALT6*	Short-chain dehydrogenase/ketoreductase of C3	[193,204,206]
FUM14	*fum14*		Non-ribosomal peptide synthetase for PTCA esterification	[193,194,233]
FUM15	*fum15*		Cytochrome P450 monooxygenases	[193]
FUM16	*fum16*		Fatty acyl-CoA synthetase	[193,194,224]
FUM17			CerS for self-protection against fumonisins	[193,197]
FUM18			CerS for self-protection against fumonisins	[193,197]
FUM19	*fum19*		ABC transport protein as a repressor of FUM gene cluster	[193,197]
FUM20			Not clear	[234]
FUM21	*fum21*	*ALT13*	Zn(II)2Cys6 transcription factor	[198]
absent	*SDR1*		Short-chain dehydrogenase/reductase (SDR)	[7]

AAL-toxins are produced by *A. alternata* f. sp. *Lycopersici*, a specific pathotype of a common plant fungal pathogen in a genus different from *Aspergillus* and *Fusarium*. This pathotype can produce polyketide-derived compounds similar in structure to fumonisins produced by *Fusarium* species. The *ALT* (AAL-toxins synthesis) genes are also located as a cluster on a conditional disposable (CD) chromosome of ~1.0 Mb in all strains of the tomato pathotype of *A. alternata* from different countries [237]. Such CD chromosomes carrying a toxin biosynthesis gene cluster were also found in other pathotypes of *A. alternata* [169]. They control other HSTs production and pathogenicity to their host. They could maintain stably in a new genetic background to an expanded range of pathogenicity, which was verified by a protoplast fusion test [238]. The AAL-toxin gene cluster includes at least 13 genes in a 120 kb region, some of which showed significant similarity to the *FUM* gene cluster consisting of 17 genes in a 45.5 kb region. However, the arrangement of genes in the *ALT* and *FUM* clusters differs between these two groups of fungi. In addition, in one strain, As-27 of *A. alternata*, there were two sets of the AAL-toxin biosynthetic gene cluster on the CD chromosome [169]. The synthesis of AAL-toxins was found to be initiated by *ALT1*-encoded PKSs to produce the aminopentol backbone, which was then modified by other enzymes [239]. The functional similarity between *ALT1* and *FUM1* was confirmed when fumonisin biosynthesis in *FUM1*-disrupted *F. verticillioides* was restored when complemented by the *ALT1* from *A. alternata* [240]. Similarly, expression of *ALT1* and production of AAL-toxins were also found to be regulated by the global regulator *LaeA* [241]. AAL-toxin accumulation also benefits from high water activity (0.995 aw) and high temperature (above 30 °C) during the incubation period of the pathogen [242].

Interestingly, a PKS gene similar to *FUM1* and orthologs of the *FUM* gene cluster were found in the genome of *Cochliobolus* spp. by phylogenetic analysis of fungal polyketide. These fungi were also speculated to produce a fumonisin or other SAMs [243,244]. To predict the potential distribution of SAM production in fungi, Kim et al. proposed a hypothesis on SAM biosynthetic gene clusters based on fumonisin biosynthesis model. This putative gene cluster should include a PKS, an aminotransferase, and a dehydrogenase gene. Their model showed that sixty-nine species of the *Fusarium* genus and species of twenty-four other fungal genera were predicted to have at least one SAM cluster [245].

9. Evolution of SAMs Production

Horizontal gene transfer (HGT) has been proposed as a major mechanism responsible for the acquisition and evolution of fumonisins and AAL-toxins biosynthetic gene clusters among divergent fungi [7,9,169,224]. In *Fusarium*, genome sequence analyses revealed that the fumonisin biosynthetic genes (*FUM*) are clustered and show a consistent gene organization among most species. For example, the *FUM* clusters in *F. oxysporum*, *F. proliferatum*, and *F. verticillioides* exhibit relatively little variability, with the order and orientation of genes within the clusters all being the same as each other. In addition, their sequence variability among the orthologues of coding regions from *F. oxysporum* and *F. verticillioides* is relatively low [193,195]. The two different species of *Aspergillus* that produce fumonisins, *Asp. niger* and *Asp. welwitschiae*, are also similar to each other in their gene order but different from that of the *FUM* cluster in *Fusarium* [224]. At present, the tomato pathotype of *A. alternata* was the only species of genus *Alternaria* capable of producing SAMs and this pathotype has clustered genes (involved in AAL-toxin biosynthesis) similar to the *FUM* cluster in *Fusarium* [169,237]. Together, the gene structure and sequence analyses suggested that the SAMs biosynthetic gene cluster likely originated in *Fusarium* and transferred to *Asp. niger* and *A. alternata* by HGT. The similarities in chemical structure and cytotoxicity on plants and animals between fumonisins and AAL-toxins are also supportive of this hypothesis. However, the differences between *FUM* clusters and the AAL-toxins biosynthetic gene cluster also suggested that there has been significant divergence between them.

Analyses of the *FUM* gene cluster among *Fusarium* species also revealed evidence for gene gain, loss, and mutations of different genes. For example, not all *Fusarium* species can produce fumonisins. Even for species that can synthesize fumonisins, some strains produce more than others under the same experimental conditions, while other strains do not produce the toxins at all [9,196]. Indeed, for certain strains, while the *FUM* genes were detected, there was no detectable fumonisin. This was likely due to the accumulation of mutations leading to the *FUM* genes being nonfunctional. For example, several mutations have been found in *FUM7* and *FUM21* in *F. fujikuroi* [246]. Furthermore, several fumonisin-non-producing *Fusarium* species lack the fumonisin biosynthetic genes but retain homologs of several genes that flank the Fum cluster in *F. verticillioides* [247]. Interestingly, the flanking regions of the *FUM* cluster often differ between species, consistent with the independent origins of the *FUM* cluster, including independent acquisition and/or loss of the gene cluster by fumonisin-producing species [195]. For example, fumonisin-non-producing strains of *F. verticillioides* isolated from banana did not contain the functional fumonisin biosynthetic gene (*FUM*) cluster but did contain portions of *FUM21* and *FUM19* flanking the cluster, both of which are the terminal genes at each end of the *FUM* cluster. However, the banana strains are still pathogenic to banana, but they do not show the same pathology as the fumonisin-producing strains do on maize. When a banana strain was co-transformed with two overlapping cosmids containing the entire *FUM* gene cluster, fumonisin production and pathogenicity on maize seedlings were recovered [248]. Similar to *Fusarium*, FB-non-producing isolates of *Asp. niger* or other *Aspergillus* species also had an intact *fum* cluster or multiple patterns of *fum* gene deletion, respectively (shown in Table 3). Similarly, the AAL-toxin-production gene cluster in *A. alternata* was likely derived from *Fusarium* species by an HGT event. Evidence from the *ALT* gene cluster distributed in isolates of *A. alternata* also supports this hypothesis of HGTs within AAL-toxin-producing pathogens [237]. Together, these results suggest that there have been multiple HGTs of the cluster between species, as well as duplication and loss of the whole or part of the cluster after acquisition [9].

The hypothesis that multiple HGTs were involved in generating the current distribution of *FUM* gene clusters is further supported by phylogenetic studies. Specifically, phylogenetic trees based on genes from the *FUM* cluster often do not parallel that of the *Fusarium* species tree based on other genes [9]. In *Fusarium*, the translation elongation factor (tef-1α) gene is the most commonly used marker gene for taxonomic studies. However, sequences of the tef-1α are often insufficient for distinguishing fumonisin-producing isolates from different countries and/or host plants. In contrast, DNA sequence polymorphisms based on *FUM1* often provide better resolutions among pathotypes [196,249–253]. For example, phylogenetic analysis of 38 *F. proliferatum* isolates originating from different hosts showed that sequence variation among strains in the *FUM1* gene was correlated with that of the host plants. Specifically, phylogenetic analysis based on the partial *FUM1* sequences differentiated the host-related groups more clearly than that based on tef-1α sequences. The best distinguished group consists of garlic-derived isolates and formed a separate branch on a *FUM1*-based dendrogram [196,254]. Similarly, *FUM1* sequence divergence analysis on *F. proliferatum* and *F. verticillioides* strains isolated from pea also formed a distinct group when compared to strains derived from different host species [253].

Aside from gene differences among isolates from different host plants, variations of both toxigenic potential and growth patterns may also differ between isolates derived from the same host plants. While no difference was observed in FB levels measured among pea seeds, the FB productions differed between selected strains of *F. proliferatum* in rice cultures [253]. Among all these test isolates of *F. proliferatum*, the most varied group of isolates found so far were those isolated from maize [196,254]. Both the inter- and intra-specific variation in FBs synthesis level can at least partly be explained by the sequence differences inside the FUM cluster.

In summary, the analyses so far suggested that the *FUM* gene cluster was responsible for fumonisin biosynthesis. Mutation and deletion of some or all of the genes in the cluster could result in limited or no production of fumonisin, leading to a weaker disease development of the pathogen on the host plant. Sequence analyses showed that *A. alternata* has likely gained the ability for AAL-toxin production due to HGT of the SAMs gene cluster from fumonisin-producing *Fusarium* species followed by independent evolution in pathogen–host interaction. The divergent patterns of toxin biosynthesis gene sequence divergence may explain the differences between fumonisins and AAL-toxins in both their productions and their impacts on host–pathogen interactions.

10. Detection Method of SAMs

A variety of methods have been developed to detect SAMs, including HPLC with fluorescence/evaporative light scattering detection or mass spectrometry (MS), thin-layer chromatography (TLC), ELISA, Fourier transform near infrared (FT-NIR) spectroscopy, and so on [255–263]. While these traditional analytical methods were designed to detect and quantify known compounds for which standards are available, there is clear evidence that many unknown derivatives may exist in food and food products and some of these could be toxic to animals, including humans [39]. In 2015, a semi-targeted method combining product ion filtering and rapid polarity switching was designed for fast detection of all known fumonisins and AAL-toxins. Some new structurally related emerging toxins were also discovered by this method [26].

Aside from method development that targets potentially novel SAMs not reported before, there are also developments for efficient methods that target the detection of known SAMs. Indeed, several new methods based on immunoassay were developed for simple, rapid, and ultrasensitive on-site quantification of SAMs. For example, one method used chemiluminescent biosensors integrating a competitive lateral flow immunoassay and a charge-coupled device camera to detect FBs [264,265]. Another method uses gold nanoparticles or quantum dots nanobeads based on monoclonal antibodies against fumonisin and allows rapid detection of this mycotoxin in one step [266–268]. Furthermore, a loop-mediated isothermal amplification (LAMP) assay, based on the detection of *fum10*, could specifically detect the genes involved in FB_2 biosynthesis in *Aspergillus* species that could be evaluated using the naked eye in a short time [269]. This method was also applied to detect FB_1 targeting the *FUM1* gene in *Fusarium* [270]. The direct detection of genes involved in the biosynthesis of SAMs in agriculture production systems allows broad evaluations of the potential fumonisin-producing strains in food and feed products. To this end, multiplex PCR has shown great promise for detection of multiple fumonisin-producing *Fusarium* and *Aspergillus* species [271–273].

While the current focuses are on the known SAMs with known toxicities, there is increasing evidence that some of these toxins are masked and not easily detected or quantified. To ensure food safety, both the free forms and the masked forms of mycotoxins should be detected and quantified. The masked mycotoxins are usually modified forms of the mycotoxins by plant enzymes during infection and are not typically detectable during routine analysis. For example, the masked mycotoxins may conjugate with polar substances, store in the vacuole in the soluble form, or bind to macromolecules, and thus change their physiological properties. While the masked mycotoxins are often less toxic than the unmasked forms, they could be easily converted to the unmasked toxin forms, including during food digestion [274,275]. The most representative masked mycotoxins are the modified forms of Zearalenone (ZEN), DON, and fumonisins [276–279]. The so-called hidden fumonisins could form non-covalent bonds with food macro-constituents such as those in starch-based products. In certain situations, the masked fumonisins may be present in food at quantities much higher than the free forms. Many factors could influence the relative portions of the SAMs in masked forms, including crop growth conditions and food storage and processing techniques [279–282]. These hidden dangers require that novel method(s) be developed to allow the detection and quantification of the masked forms of SAMs as well as other mycotoxins.

11. Concluding Remarks

SAMs are highly toxic fungal compounds that have attracted significant attention from broad communities. They have toxicities to both plants and animals. These SAMs-producing pathogens are widely distributed in nature and closely related to agricultural production. Since its discovery in the mid-1980s, fumonisin has been among the mycotoxins with the greatest concern. As of now, Scopus database citations of fumonisin are above 5000, including 500+ reviews. SAMs are a series of compounds with structural similarity to sphingosine. As detection methods improve, additional new analogs have been continuously discovered. Research so far has shown that the toxicity and activity of SAMs are dependent on their structures. In this review, we summarized the detoxification method based on their structural properties using chemical, biological, and physical strategies. The toxicity of SAMs is mostly due to their inhibitory effects on CerS, disruption on sphingolipid metabolism, and initiation on PCD. Except for the adverse effect of SAMs on animals and humans, its phytotoxicity (e.g., AAL-toxin) could potentially be used for herbicide development and a model for studying the molecular mechanism of PCD in plants. Horizontal gene transfers on the SAMs biosynthesis gene cluster seemed widespread in these toxin-producing fungi, especially among *Fusarium* spp. Such phylogenetic distribution patterns suggest that there are potentially other fungi capable of producing SAMs, including their various modified forms. These and other issues require continued efforts from the scientific community on SAMs.

Author Contributions: J.X. conceived the review and finalized the manuscript, J.C. conducted the literature search and drafted the manuscript, Z.L. drafted all the figures and tables, Y.C. coordinated the format and submitted the manuscript to the journal, C.G., L.G., and T.W. helped to revise the final manuscript. All authors have read and agreed to the published version of the manuscript.

Funding: This research was supported by the National Key R&D Program of China (N0: 2018YFD0201106), Natural Science Foundation of Hunan (2020JJ5637) and Agricultural Science and Technology Innovation Program of the Chinese Academy of Agricultural Sciences (Grant CAAS-ASTIP-2015-IBFC).

Conflicts of Interest: The authors declare no conflict of interest.

References

1. Gilchrist, D.G. Programmed cell death in plant disease: The purpose and promise of cellular suicide. *Annu. Rev. Phytopathol.* **1998**, *36*, 393–414. [CrossRef]
2. Riley, R.T.; Wang, E.; Schroeder, J.J.; Smith, E.R.; Plattner, R.D.; Abbas, N.; Yoo, H.S.; Merrill, A.H., Jr. Evidence for disruption of sphingolipid metabolism as a contributing factor in the toxicity and carcinogenicity of fumonisins. *Nat. Toxins* **1996**, *4*, 3–15. [CrossRef] [PubMed]
3. Merrill, A.H., Jr.; Sullards, M.C.; Wang, E.; Voss, K.A.; Riley, R.T. Sphingolipid metabolism: Roles in signal transduction and disruption by fumonisins. *Environ. Health Persp.* **2001**, *109*, 283–289. [CrossRef]
4. Gelderblom, W.C.A.; Jaskiewicz, K.; Marasas, W.F.O.; Thiel, P.G.; Horak, R.M.; Vleggaar, R.; Kriek, N.P.J. Fumonisins-novel mycotoxins with cancer-promoting activity produced by *Fusarium moniliforme*. *Appl. Environ. Microb.* **1988**, *54*, 1806–1811. [CrossRef] [PubMed]
5. Rheeder, J.P.; Marasas, W.F.O.; Vismer, H.F. Production of fumonisin analogs by *Fusarium* species. *Appl. Environ. Microb.* **2002**, *68*, 2101–2105. [CrossRef]
6. Frisvad, J.C.; Smedsgaard, J.; Samson, R.A.; Larsen, T.O.; Thrane, U. Fumonisin B_2 production by *Aspergillus niger*. *J. Agric. Food Chem.* **2007**, *55*, 9727–9732. [CrossRef]
7. Susca, A.; Proctor, R.H.; Butchko, R.A.E.; Haidukowski, M.; Stea, G.; Logrieco, A.; Moretti, A. Variation in the fumonisin biosynthetic gene cluster in fumonisin-producing and nonproducing black aspergilli. *Fungal Genet. Biol.* **2014**, *73*, 39–52. [CrossRef]
8. Chen, J.; Mirocha, C.J.; Xie, W.; Hogge, L.; Olson, D. Production of the mycotoxin fumonisin B_1 by *Alternaria alternata* f. sp. *lycopersici*. *Appl. Environ. Microb.* **1992**, *58*, 3928–3931. [CrossRef]
9. Proctor, R.H.; Van Hove, F.; Susca, A.; Stea, G.; Busman, M.; van der Lee, T.; Waalwijk, C.; Moretti, A.; Ward, T.J. Birth, death and horizontal transfer of the fumonisin biosynthetic gene cluster during the evolutionary diversification of *Fusarium*. *Mol. Microbiol.* **2013**, *90*, 290–306. [CrossRef]
10. Mogensen, J.M.; Møller, K.A.; Von Freiesleben, P.; Labuda, R.; Varga, E.; Sulyok, M.; Kubátová, A.; Thrane, U.; Andersen, B.; Nielsen, K.F. Production of fumonisins B_2 and B_4 in *Tolypocladium* species. *J. Ind. Microbiol. Biot.* **2011**, *38*, 1329–1335. [CrossRef]
11. Alabouvette, C.; Lemanceau, P.; Steinberg, C. Recent advances in the biological control of fusarium wilts. *Pestic. Sci.* **1993**, *37*, 365–373. [CrossRef]
12. Desjardins, A.E.; Plattner, R.D.; Nelsen, T.C.; Leslie, J.F. Genetic analysis of fumonisin production and virulence of *Gibberella fujikuroi* mating population A (*Fusarium moniliforme*) on maize (Zea mays) seedlings. *Appl. Environ. Microb.* **1995**, *61*, 79–86. [CrossRef] [PubMed]
13. Chulze, S.N. Strategies to reduce mycotoxin levels in maize during storage: A review. *Food Addit. Contam. A* **2010**, *27*, 651–657. [CrossRef] [PubMed]

14. Munkvold, G.P.; McGee, D.C.; Carlton, W.M. Importance of different pathways for maize kernel infection by *Fusarium moniliforme*. *Phytopathology* **1997**, *87*, 209–217. [CrossRef]
15. Williams, L.D.; Glenn, A.E.; Bacon, C.W.; Smith, M.A.; Riley, R.T. Fumonisin production and bioavailability to maize seedlings grown from seeds inoculated with *Fusarium verticillioides* and grown in natural soils. *J. Agr. Food Chem.* **2006**, *54*, 5694–5700. [CrossRef]
16. Kamle, M.; Mahato, D.K.; Devi, S.; Lee, K.E.; Kang, S.G.; Kumar, P. Fumonisins: Impact on agriculture, food, and human health and their management strategies. *Toxins* **2019**, *11*, 328. [CrossRef]
17. Chu, F.S.; Li, G.Y. Simultaneous occurrence of fumonisin B_1 and other mycotoxins in moldy corn collected from the People's Republic of China in regions with high incidences of esophageal cancer. *Appl. Environ. Microb.* **1994**, *60*, 847–852. [CrossRef]
18. Marasas, W.F.O.; Riley, R.T.; Hendricks, K.A.; Stevens, V.L.; Sadler, T.W.; Gelineau-Van Waes, J.; Missmer, S.A.; Cabrera, J.; Torres, O.; Gelderblom, W.C.A.; et al. Fumonisins Disrupt Sphingolipid Metabolism, Folate Transport, and Neural Tube Development in Embryo Culture and In Vivo: A Potential Risk Factor for Human Neural Tube Defects among Populations Consuming Fumonisin-Contaminated Maize. *J. Nutr.* **2004**, *134*, 711–716. [CrossRef]
19. Gallo, A.; Giuberti, G.; Frisvad, J.C.; Bertuzzi, T.; Nielsen, K.F. Review on mycotoxin issues in ruminants: Occurrence in forages, effects of mycotoxin ingestion on health status and animal performance and practical strategies to counteract their negative effects. *Toxins* **2015**, *7*, 3057–3111. [CrossRef]
20. Ross, P.F.; Rice, L.G.; Osweiler, G.D.; Nelson, P.E.; Richard, J.L.; Wilson, T.M. A review and update of animal toxicoses associated with fumonisin-contaminated feeds and production of fumonisins by *Fusarium* isolates. *Mycopathologia* **1992**, *117*, 109–114. [CrossRef]
21. Harrison, L.R.; Colvin, B.M.; Greene, J.T.; Newman, L.E.; Cole, J.R., Jr. Pulmonary Edema and Hydrothorax in Swine Produced by Fumonisin B_1, a Toxic Metabolite of *Fusarium Moniliforme*. *J. Vet. Diagn. Investig.* **1990**, *2*, 217–221. [CrossRef] [PubMed]
22. Gilchrist, D.G.; Grogan, R.G. Production and nature of a host-specific toxin from *Alternaria alternata* f. sp. *lycopersici*. *Phytopathology* **1976**, *66*, 165–171. [CrossRef]
23. Meena, M.; Samal, S. *Alternaria* host-specific (HSTs) toxins: An overview of chemical characterization, target sites, regulation and their toxic effects. *Toxicol. Rep.* **2019**, *6*, 745–758. [CrossRef] [PubMed]
24. Abbas, H.K.; Duke, S.O.; Paul, R.N.; Riley, R.T.; Tanaka, T. AAL-toxin, a potent natural herbicide which disrupts sphingolipid metabolism of plants. *Pestic. Sci.* **1995**, *43*, 181–187. [CrossRef]
25. Mesbah, L.A.; Van Der Weerden, G.M.; Nijkamp, H.J.J.; Hille, J. Sensitivity among species of *Solanaceae* to AAL toxins produced by *Alternaria alternata* f.sp. *lycopersici*. *Plant Pathol.* **2000**, *49*, 734–741. [CrossRef]
26. Renaud, J.B.; Kelman, M.J.; Qi, T.F.; Seifert, K.A.; Sumarah, M.W. Product ion filtering with rapid polarity switching for the detection of all fumonisins and AAL-toxins. *Rapid Commun. Mass Sp.* **2015**, *29*, 2131–2139. [CrossRef]
27. Abbas, H.K.; Riley, R.T. The presence and phytotoxicity of fumonisins and AAL-toxin in *Alternaria alternata*. *Toxicon* **1996**, *34*, 133–136. [CrossRef]
28. Yamagishi, D.; Akamatsu, H.; Otani, H.; Kodama, M. Pathological evaluation of host-specific AAL-toxins and fumonisin mycotoxins produced by *Alternaria* and *Fusarium* species. *J. Gen. Plant Pathol.* **2006**, *72*, 323–327. [CrossRef]
29. Merrill, A.H. Sphingolipid Biosynthesis. In *Encyclopedia of Biological Chemistry*, 2nd ed.; Lennarz, W.J., Lane, M.D., Eds.; Elsevier Inc.: Amsterdam, The Netherlands, 2013; pp. 281–286. [CrossRef]
30. Hannun, Y.A.; Obeid, L.M. Sphingolipids and their metabolism in physiology and disease. *Nat. Rev. Mol. Cell Biol.* **2018**, *19*, 175–191. [CrossRef]
31. Huby, E.; Napier, J.A.; Baillieul, F.; Michaelson, L.V.; Dhondt-Cordelier, S. Sphingolipids: Towards an integrated view of metabolism during the plant stress response. *New Phytol.* **2020**, *225*, 659–670. [CrossRef]
32. Li, Q.; Fang, H.; Dang, E.; Wang, G. The role of ceramides in skin homeostasis and inflammatory skin diseases. *J. Dermatol. Sci.* **2020**, *97*, 2–8. [CrossRef] [PubMed]
33. Bartók, T.; Szécsi, Á.; Szekeres, A.; Mesterházy, Á.; Bartók, M. Detection of new fumonisin mycotoxins and fumonisin-like compounds by reversed-phase high-performance liquid chromatography/electrospray ionization ion trap mass spectrometry. *Rapid Commun. Mass Sp.* **2006**, *20*, 2447–2462. [CrossRef] [PubMed]
34. Abbas, H.K.; Duke, S.O.; Tanaka, T. Phytotoxicity of fumonisins and related compounds. *Toxin Rev.* **1993**, *12*, 225–251. [CrossRef]

35. Branham, B.E.; Plattner, R.D. Isolation and characterization of a new fumonisin from liquid cultures of *Fusarium moniliforme*. *J. Nat. Prod.* **1993**, *56*, 1630–1633. [CrossRef]
36. Seo, J.A.; Kim, J.C.; Lee, Y.W. N-acetyl derivatives of type C fumonisins produced by *Fusarium oxysporum*. *J. Nat. Prod.* **1999**, *62*, 355–357. [CrossRef]
37. Musser, S.M.; Gay, M.L.; Mazzola, E.P.; Plattner, R.D. Identification of a new series of fumonisins containing 3-hydroxypyridine. *J. Nat. Prod.* **1996**, *59*, 970–972. [CrossRef]
38. Abbas, H.K.; Shier, W.T.; Seo, J.A.; Lee, Y.W.; Musser, S.M. Phytotoxicity and cytotoxicity of the fumonisin C and P series of mycotoxins from *Fusarium* spp. fungi. *Toxicon* **1998**, *36*, 2033–2037. [CrossRef]
39. Bartók, T.; Tölgyesi, L.; Szekeres, A.; Varga, M.; Bartha, R.; Szécsi, A.; Bartók, M.; Mesterházy, A. Detection and characterization of twenty-eight isomers of fumonisin B_1 (FB_1) mycotoxin in a solid rice culture infected with *Fusarium verticillioides* by reversed-phase high-performance liquid chromatography/electrospray ionization time-of-flight and ion trap mass spectrometry. *Rapid Commun. Mass Sp.* **2010**, *24*, 35–42. [CrossRef]
40. Bottini, A.T.; Bowen, J.R.; Gilchrist, D.G. Phytotoxins. II. Characterization of a phytotoxic fraction from *Alternaria alternata* f. sp. *lycopersici*. *Tetrahedron Lett.* **1981**, *22*, 2723–2726. [CrossRef]
41. Bottini, A.T.; Gilchrist, D.G. Phytotoxins. I. A 1-amino dimethyl heptadecapentol from *alternaria alternata* f. sp. *lycopersici*. *Tetrahedron Lett.* **1981**, *22*, 2719–2722. [CrossRef]
42. Caldas, E.D.; Jones, A.D.; Ward, B.; Winter, C.K.; Gilchrist, D.G. Structural Characterization of Three New AAL Toxins Produced by *Alternaria alternata* f. sp. *lycopersici*. *J. Agric. Food Chem.* **1994**, *42*, 327–333. [CrossRef]
43. Saucedo-García, M.; Guevara-García, A.; González-Solís, A.; Cruz-García, F.; Vázquez-Santana, S.; Markham, J.E.; Lozano-Rosas, M.G.; Dietrich, C.R.; Ramos-Vega, M.; Cahoon, E.B.; et al. MPK6, sphinganine and the LCB2a gene from serine palmitoyltransferase are required in the signaling pathway that mediates cell death induced by long chain bases in *Arabidopsis*. *New Phytol.* **2011**, *191*, 943–957. [CrossRef] [PubMed]
44. Shi, L.; Bielawski, J.; Mu, J.; Dong, H.; Teng, C.; Zhang, J.; Yang, X.; Tomishige, N.; Hanada, K.; Hannun, Y.A.; et al. Involvement of sphingoid bases in mediating reactive oxygen intermediate production and programmed cell death in *Arabidopsis*. *Cell Res.* **2007**, *17*, 1030–1040. [CrossRef] [PubMed]
45. Spassieva, S.D.; Markham, J.E.; Hille, J. The plant disease resistance gene *Asc-1* prevents disruption of sphingolipid metabolism during AAL-toxin-induced programmed cell death. *Plant J.* **2002**, *32*, 561–572. [CrossRef]
46. Kluepfel, D.; Bagli, J.; Baker, H.; Charest, M.P.; Kudelski, A.; Sehgal, S.N.; Vézina, C. Myriocin, a new antifungal antibiotic from *Myriococcum albomyces*. *J. Antibiot.* **1972**, *25*, 109–115. [CrossRef]
47. Šašek, V.; Sailer, M.; Vokoun, J.; Musílek, V. Production of thermozymocidin (myriocin) by the pyrenomycete *Melanconis flavovirens*. *J. Basic Microb.* **1989**, *29*, 383–390. [CrossRef]
48. Fujita, T.; Inoue, K.; Yamamoto, S.; Ikumoto, T.; Sasaki, S.; Toyama, R.; Chiba, K.; Hoshino, Y.; Okumoto, T. Fungal metabolites. Part 11. A potent immunosuppressive activity found in *Isaria sinclairii* metabolite. *J. Antibiot.* **1994**, *47*, 208–215. [CrossRef]
49. Miyake, Y.; Kozutsumi, Y.; Nakamura, S.; Fujita, T.; Kawasaki, T. Serine palmitoyltransferase is the primary target of a sphingosine-like immunosuppressant, ISP-1/myriocin. *Biochem. Bioph. Res. Co.* **1995**, *211*, 396–403. [CrossRef]
50. Tatematsu, K.; Tanaka, Y.; Sugiyama, M.; Sudoh, M.; Mizokami, M. Host sphingolipid biosynthesis is a promising therapeutic target for the inhibition of hepatitis B virus replication. *J. Med. Virol.* **2011**, *83*, 587–593. [CrossRef]
51. Yu, S.; Jia, B.; Yang, Y.; Liu, N.; Wu, A. Involvement of PERK-CHOP pathway in fumonisin B_1- induced cytotoxicity in human gastric epithelial cells. *Food Chem. Toxicol.* **2020**, *136*. [CrossRef]
52. Liu, J.; Huang, X.; Withers, B.R.; Blalock, E.; Liu, K.; Dickson, R.C. Reducing sphingolipid synthesis orchestrates global changes to extend yeast lifespan. *Aging Cell* **2013**, *12*, 833–841. [CrossRef] [PubMed]
53. Reforgiato, M.R.; Milano, G.; Fabriàs, G.; Casas, J.; Gasco, P.; Paroni, R.; Samaja, M.; Ghidoni, R.; Caretti, A.; Signorelli, P. Inhibition of ceramide de novo synthesis as a postischemic strategy to reduce myocardial reperfusion injury. *Basic Res. Cardiol.* **2016**, *111*. [CrossRef] [PubMed]
54. Horn, W.S.; Smith, J.L.; Bills, G.F.; Raghoobar, S.L.; Helms, G.L.; Kurtz, M.B.; Marrinan, J.A.; Frommer, B.R.; Thornton, R.A.; Mandala, S.M. Sphingofungins E and F: Novel serinepalmitoyl trans-ferase inhibitors from *Paecilomyces variotii*. *J. Antibiot.* **1992**, *45*, 1692–1696. [CrossRef] [PubMed]

55. VanMiddlesworth, F.; Giacobbe, R.A.; Lopez, M.; Garrity, G.; Bland, J.A.; Bartizal, K.; Fromtling, R.A.; Polishook, J.; Zweerink, M.; Edison, A.M.; et al. Sphingofungins a, b, c, and d; a new family of antifungal agents: I. Fermentation, isolation, and biological activity. *J. Antibiot.* **1992**, *45*, 861–867. [CrossRef] [PubMed]
56. Song, Z.; Liu, Y.; Gao, J.; Hu, J.; He, H.; Dai, S.; Wang, L.; Dai, H.; Zhang, L.; Song, F. Antitubercular metabolites from the marine-derived fungus strain *Aspergillus fumigatus* MF029. *Nat. Prod. Res.* **2019**. [CrossRef]
57. Zhang, H.; Zhu, H.T.; Wang, D.; Yang, C.R.; Zhang, Y.J. Sphingofungins G and H: New five-membered lactones from *Aspergillus penicilliodes* Speg. *Nat. Prod. Res.* **2019**, *33*, 1284–1291. [CrossRef] [PubMed]
58. Hanada, K.; Nishijima, M.; Fujita, T.; Kobayashi, S. Specificity of inhibitors of serine palmitoyltransferase (SPT), a key enzyme in sphingolipid biosynthesis, in intact cells. A novel evaluation system using an SPT-defective mammalian cell mutant. *Biochem. Pharmacol.* **2000**, *59*, 1211–1216. [CrossRef]
59. Harris, G.H.; Turner Jones, E.T.; Meinz, M.S.; Nallin-Omstead, M.; Helms, G.L.; Bills, G.F.; Zink, D.; Wilson, K.E. Isolation and structure elucidation of viridiofungins A, B and C. *Tetrahedron Lett.* **1993**, *34*, 5235–5238. [CrossRef]
60. Mandala, S.M.; Thornton, R.A.; Frommer, B.R.; Dreikorn, S.; Kurtz, M.B. Viridiofungins, novel inhibitors of sphingolipid synthesis. *J. Antibiot* **1997**, *50*, 339–343. [CrossRef]
61. El-Hasan, A.; Walker, F.; Schöne, J.; Buchenauer, H. Detection of viridiofungin A and other antifungal metabolites excreted by Trichoderma harzianum active against different plant pathogens. *Eur. J. Plant Pathol.* **2009**, *124*, 457–470. [CrossRef]
62. Abbas, H.K.; Duke, S.O.; Merrill, A.H., Jr.; Wang, E.; Shier, W.T. Phytotoxicity of Australifungin, AAL-toxins and fumonisin B1 to *Lemna pausicostata*. *Phytochemistry* **1998**, *47*, 1509–1514. [CrossRef]
63. Mandala, S.M.; Thornton, R.A.; Frommer, B.R.; Curotto, J.E.; Rozdilsky, W.; Kurtz, M.B.; Giacobbe, R.A.; Bills, G.F.; Cabello, M.A. The Discovery of Australifungin, a Novel Inhibitor of Sphinganine N-Acyltransferase from *Sporormiella australis* Producing Organism, Fermentation, Isolation, and Biological Activity. *J. Antibiot.* **1995**, *48*, 349–356. [CrossRef]
64. Uhlig, S.; Petersen, D.; Flåøyen, A.; Wilkins, A. 2-Amino-14,16-dimethyloctadecan-3-ol, a new sphingosine analogue toxin in the fungal genus *Fusarium*. *Toxicon* **2005**, *46*, 513–522. [CrossRef] [PubMed]
65. Marrouchi, R.; Benoit, E.; Le Caer, J.P.; Belayouni, N.; Belghith, H.; Molgó, J.; Kharrat, R. Toxic C17-sphinganine analogue mycotoxin, contaminating Tunisian mussels, causes flaccid paralysis in rodents. *Mar. Drugs* **2013**, *11*, 4724–4740. [CrossRef] [PubMed]
66. Mirocha, C.J.; Gilchrist, D.G.; Shier, W.T.; Abbas, H.K.; Wen, Y.; Vesonder, R.F. AAL Toxins, funionisms (biology and chemistry) and host-specificity concepts. *Mycopathologia* **1992**, *117*, 47–56. [CrossRef] [PubMed]
67. Shier, W.T.; Abbas, H.K.; Mirocha, C.J. Toxicity of the mycotoxins fumonisins B_1 and B_2 and *Alternaria alternata* f. sp. *lycopersici* toxin (AAL) in cultured mammalian cells. *Mycopathologia* **1991**, *116*, 97–104. [CrossRef]
68. Van Der Westhuizen, L.; Shephard, G.S.; Snyman, S.D.; Abel, S.; Swanevelder, S.; Gelderblom, W.C.A. Inhibition of sphingolipid biosynthesis in rat primary hepatocyte cultures by fumonisin B_1 and other structurally related compounds. *Food Chem. Toxicol.* **1998**, *36*, 497–503. [CrossRef]
69. Abbas, H.K.; Mulrooney, J.E. Effect of Some Phytopathogenic Fungi and Their Metabolites on Growth of *Heliothis virescens* (F.) and Its Host Plants. *Biocontrol Sci. Techn.* **1994**, *4*, 77–87. [CrossRef]
70. Gelderblom, W.C.A.; Cawood, M.E.; Snyman, S.D.; Vleggaar, R.; Marasas, W.F.O. Structure-activity relationships of fumonisins in short-term carcinogenesis and cytotoxicity assays. *Food Chem. Toxicol.* **1993**, *31*, 407–414. [CrossRef]
71. Tanaka, T.; Abbas, H.K.; Duke, S.O. Structure-dependent phytotoxicity of fumonisins and related compounds in a duckweed bioassay. *Phytochemistry* **1993**, *33*, 779–785. [CrossRef]
72. Siler, D.J.; Gilchrist, D.G. Properties of host specific toxins produced by *Alternaria alternata* f. sp. *lycopersici* in culture and in tomato plants. *Physiol. Plant Pathol.* **1983**, *23*, 265–274. [CrossRef]
73. Norred, W.P.; Riley, R.T.; Meredith, F.I.; Poling, S.M.; Plattner, R.D. Instability of N-acetylated fumonisin B_1 (FA_1) and the impact on inhibition of ceramide synthase in rat liver slices. *Food Chem. Toxicol.* **2001**, *39*, 1071–1078. [CrossRef]
74. Shier, W.T.; Abbas, H.K. Current issues in research on fumonisins, mycotoxins which may cause nephropathy. *J. Toxicol. Toxin Rev.* **1999**, *18*, 323–335. [CrossRef]

75. Shier, W.T.; Abbas, H.K.; Abou-Karam, M.; Badria, F.A.; Resch, P.A. Fumonisins: Abiogenic Conversions of an Environmental Tumor Promoter and Common Food Contaminant. *J. Toxicol. Toxin Rev.* **2003**, *22*, 591–616. [CrossRef]
76. Savard, M.E.; Sinha, R.C.; Lau, R.; Séguin, C.; Buffam, S. Monoclonal antibodies for fumonisins B_1, B_2 and B_3. *Food Agric. Immunol.* **2003**, *15*, 127–134. [CrossRef]
77. Azaiez, I.; Meca, G.; Manyes, L.; Luciano, F.B.; Fernández-Franzón, M. Study of the chemical reduction of the fumonisins toxicity using allyl, benzyl and phenyl isothiocyanate in model solution and in food products. *Toxicon* **2013**, *63*, 137–146. [CrossRef]
78. Lamprecht, S.C.; Marasas, W.F.O.; Alberts, J.F.; Cawood, M.E.; Gelderblom, W.C.A.; Shephard, G.S.; Thiel, P.G.; Calitz, F.J. Phytotoxicity of fumonisins and TA-toxin to corn and tomato. *Phytopathology* **1994**, *84*, 383–391. [CrossRef]
79. Vesonder, R.F.; Peterson, R.E.; Labeda, D.; Abbas, H.K. Comparative phytotoxicity of the fumonisins, AAL-toxin and yeast sphingolipids in *Lemna minor* L. (duckweed). *Arch. Environ. Con. Tox.* **1992**, *23*, 464–467. [CrossRef]
80. Voss, K.A.; Riley, R.T.; Snook, M.E.; Gelineau-van Waes, J. Reproductive and sphingolipid metabolic effects of fumonisin B_1 and its alkaline hydrolysis product in LM/Bc mice: Hydrolyzed fumonisin B_1 did not cause neural tube defects. *Toxicol. Sci.* **2009**, *112*, 459–467. [CrossRef]
81. Grenier, B.; Bracarense, A.P.F.L.; Schwartz, H.E.; Trumel, C.; Cossalter, A.M.; Schatzmayr, G.; Kolf-Clauw, M.; Moll, W.D.; Oswald, I.P. The low intestinal and hepatic toxicity of hydrolyzed fumonisin B_1 correlates with its inability to alter the metabolism of sphingolipids. *Biochem. Pharmacol.* **2012**, *83*, 1465–1473. [CrossRef]
82. Gu, M.J.; Han, S.E.; Hwang, K.; Mayer, E.; Reisinger, N.; Schatzmayr, D.; Park, B.C.; Han, S.H.; Yun, C.H. Hydrolyzed fumonisin B_1 induces less inflammatory responses than fumonisin B_1 in the co-culture model of porcine intestinal epithelial and immune cells. *Toxicol. Lett.* **2019**, *305*, 110–116. [CrossRef] [PubMed]
83. Seiferlein, M.; Humpf, H.U.; Voss, K.A.; Sullards, M.C.; Allegood, J.C.; Wang, E.; Merrill, A.H., Jr. Hydrolyzed fumonisins HFB_1 and HFB_2 are acylated in vitro and in vivo by ceramide synthase to form cytotoxic N-acyl-metabolites. *Mol. Nutr. Food Res.* **2007**, *51*, 1120–1130. [CrossRef]
84. Seefelder, W.; Knecht, A.; Humpf, H.U. Bound fumonisin B_1: Analysis of fumonisin-B_1 glyco and amino acid conjugates by liquid chromatography-electrospray ionization-tandem mass spectrometry. *J. Agric. Food Chem.* **2003**, *51*, 5567–5573. [CrossRef] [PubMed]
85. Abbas, H.K.; Tanaka, T.; Duke, S.O.; Porter, J.K.; Wray, E.M.; Hodges, L.; Sessions, A.E.; Wang, E.; Merrill Jnr, A.H.; Riley, R.T. Fumonisin- and AAL-toxin-induced disruption of sphingolipid metabolism with accumulation of free sphingoid bases. *Plant Physiol.* **1994**, *106*, 1085–1093. [CrossRef] [PubMed]
86. Henriques, J.; Lima, M.; Rosa, S.; Dias, A.S.; Dias, L.S. Allelopathic plants. XVIII. *Solanum nigrum* L. *Allelopathy J.* **2006**, *17*, 1–15.
87. Duke, S.O.; Dayan, F.E. Modes of action of microbially-produced phytotoxins. *Toxins* **2011**, *3*, 1038–1064. [CrossRef]
88. Abbas, H.K.; Tanaka, T.; Shier, W.T. Biological activities of synthetic analogues of *Alternaria alternata* toxin (AAL-toxin) and fumonisin in plant and mammalian cell cultures. *Phytochemistry* **1995**, *40*, 1681–1689. [CrossRef]
89. Duke, S.O.; Dayan, F.E. Clues to new herbicide mechanisms of action from natural sources. *ACS Symposium Series* **2013**, *1141*, 203–215. [CrossRef]
90. Luo, Y.; Liu, X.; Li, J. Updating techniques on controlling mycotoxins—A review. *Food Control* **2018**, *89*, 123–132. [CrossRef]
91. McCormick, S.P. Microbial Detoxification of Mycotoxins. *J. Chem. Ecol.* **2013**, *39*, 907–918. [CrossRef]
92. Agriopoulou, S.; Stamatelopoulou, E.; Varzakas, T. Advances in occurrence, importance, and mycotoxin control strategies: Prevention and detoxification in foods. *Foods* **2020**, *9*, 137. [CrossRef] [PubMed]
93. Norred, W.P.; Voss, K.A.; Bacon, C.W.; Riley, R.T. Effectiveness of ammonia treatment in detoxification of fumonisin-contaminated corn. *Food Chem. Toxicol.* **1991**, *29*, 815–819. [CrossRef]
94. Chourasia, H.K. Efficacy of ammonia in detoxification of fumonisin contaminated corn. *Indian J. Exp. Biol.* **2001**, *39*, 493–495.
95. De Girolamo, A.; Lattanzio, V.M.T.; Schena, R.; Visconti, A.; Pascale, M. Effect of alkaline cooking of maize on the content of fumonisins B_1 and B_2 and their hydrolysed forms. *Food Chem.* **2016**, *192*, 1083–1089. [CrossRef] [PubMed]

96. Chang, X.; Wang, J.; Sun, C.; Liu, H.; Wu, S.; Sun, J.; Wu, Z. Research on degradation of chlorine dioxide in primary mycotoxins. *J. Chin. Cereals Oils Assoc.* **2016**, *31*, 113–118.
97. Voss, K.; Ryu, D.; Jackson, L.; Riley, R.; Gelineau-Van Waes, J. Reduction of Fumonisin Toxicity by Extrusion and Nixtamalization (Alkaline Cooking). *J. Agric. Food Chem.* **2017**, *65*, 7088–7096. [CrossRef] [PubMed]
98. ten Bosch, L.; Pfohl, K.; Avramidis, G.; Wieneke, S.; Viöl, W.; Karlovsky, P. Plasma-based degradation of mycotoxins produced by *Fusarium*, *Aspergillus* and *Alternaria* species. *Toxins* **2017**, *9*, 97. [CrossRef] [PubMed]
99. Wielogorska, E.; Ahmed, Y.; Meneely, J.; Graham, W.G.; Elliott, C.T.; Gilmore, B.F. A holistic study to understand the detoxification of mycotoxins in maize and impact on its molecular integrity using cold atmospheric plasma treatment. *Food Chem.* **2019**, *301*. [CrossRef] [PubMed]
100. Afsah-Hejri, L.; Hajeb, P.; Ehsani, R.J. Application of ozone for degradation of mycotoxins in food: A review. *Compr. Rev. Food Sci. Food Saf.* **2020**, *19*, 1777–1808. [CrossRef]
101. Milani, J.; Maleki, G. Effects of processing on mycotoxin stability in cereals. *J. Sci. Food Agric.* **2014**, *94*, 2372–2375. [CrossRef] [PubMed]
102. Jard, G.; Liboz, T.; Mathieu, F.; Guyonvarch, A.; Lebrihi, A. Review of mycotoxin reduction in food and feed: From prevention in the field to detoxification by adsorption or transformation. *Food Addit. Contam. A* **2011**, *28*, 1590–1609. [CrossRef] [PubMed]
103. Benedetti, R.; Nazzi, F.; Locci, R.; Firrao, G. Degradation of fumonisin B_1 by a bacterial strain isolated from soil. *Biodegradation* **2006**, *17*, 31–38. [CrossRef]
104. Heinl, S.; Hartinger, D.; Thamhesl, M.; Vekiru, E.; Krska, R.; Schatzmayr, G.; Moll, W.D.; Grabherr, R. Degradation of fumonisin B_1 by the consecutive action of two bacterial enzymes. *J. Biotechnol.* **2010**, *145*, 120–129. [CrossRef]
105. Chlebicz, A.; Śliżewska, K. In Vitro Detoxification of Aflatoxin B_1, Deoxynivalenol, Fumonisins, T-2 Toxin and Zearalenone by Probiotic Bacteria from Genus *Lactobacillus* and *Saccharomyces cerevisiae* Yeast. *Probiotics Antimicro.* **2020**, *12*, 289–301. [CrossRef] [PubMed]
106. Burgess, K.M.N.; Renaud, J.B.; McDowell, T.; Sumarah, M.W. Mechanistic Insight into the Biosynthesis and Detoxification of Fumonisin Mycotoxins. *ACS Chem. Biol.* **2016**, *11*, 2618–2625. [CrossRef] [PubMed]
107. Azam, M.S.; Yu, D.; Wu, A. Enzymes for degradation of *Fusarium* mycotoxins. In *Food Safety & Mycotoxins*; Wu, A., Ed.; Springer: Singapore, 2019; pp. 113–135. [CrossRef]
108. Alberts, J.; Schatzmayr, G.; Moll, W.D.; Davids, I.; Rheeder, J.; Burger, H.M.; Shephard, G.; Gelderblom, W. Detoxification of the fumonisin mycotoxins in maize: An enzymatic approach. *Toxins* **2019**, *11*, 523. [CrossRef] [PubMed]
109. Hartinger, D.; Schwartz, H.; Hametner, C.; Schatzmayr, G.; Haltrich, D.; Moll, W.D. Enzyme characteristics of aminotransferase FumI of *Sphingopyxis* sp. MTA144 for deamination of hydrolyzed fumonisin B_1. *Appl. Microbiol. Biot.* **2011**, *91*, 757–768. [CrossRef]
110. Masching, S.; Naehrer, K.; Schwartz-Zimmermann, H.E.; Sărăndan, M.; Schaumberger, S.; Dohnal, I.; Nagl, V.; Schatzmayr, D. Gastrointestinal degradation of fumonisin B_1 by carboxylesterase FumD prevents fumonisin induced alteration of sphingolipid metabolism in Turkey and swine. *Toxins* **2016**, *8*, 84. [CrossRef]
111. Schertz, H.; Kluess, J.; Frahm, J.; Schatzmayr, D.; Dohnal, I.; Bichl, G.; Schwartz-Zimmermann, H.; Breves, G.; Dänicke, S. Oral and intravenous fumonisin exposure in pigs—a single-dose treatment experiment evaluating toxicokinetics and detoxification. *Toxins* **2018**, *10*, 150. [CrossRef]
112. Loi, M.; Fanelli, F.; Cimmarusti, M.T.; Mirabelli, V.; Haidukowski, M.; Logrieco, A.F.; Caliandro, R.; Mule, G. In vitro single and combined mycotoxins degradation by Ery4 laccase from *Pleurotus eryngii* and redox mediators. *Food Control* **2018**, *90*, 401–406. [CrossRef]
113. Wang, X.; Qin, X.; Hao, Z.; Luo, H.; Yao, B.; Su, X. Degradation of four major mycotoxins by eight manganese peroxidases in presence of a dicarboxylic acid. *Toxins* **2019**, *11*, 566. [CrossRef] [PubMed]
114. Avantaggaito, G.; Solfrizzo, M.; Visconti, A. Recent advances on the use of adsorbent materials for detoxification of Fusarium mycotoxins. *Food Addit. Contam.* **2005**, *22*, 379–388. [CrossRef] [PubMed]
115. Brown, K.A.; Mays, T.; Romoser, A.; Marroquin-Cardona, A.; Mitchell, N.J.; Elmore, S.E.; Phillips, T.D. Modified hydra bioassay to evaluate the toxicity of multiple mycotoxins and predict the detoxification efficacy of a clay-based sorbent. *J. Appl. Toxicol.* **2014**, *34*, 40–48. [CrossRef] [PubMed]
116. Yuan, C.W.; Huang, J.T.; Chen, C.C.; Tang, P.C.; Huang, J.W.; Lin, J.J.; Huang, S.Y.; Chen, S.E. Evaluation of Efficacy and Toxicity of Exfoliated Silicate Nanoclays as a Feed Additive for Fumonisin Detoxification. *J. Agric. Food Chem.* **2017**, *65*, 6564–6571. [CrossRef] [PubMed]

117. Beekrum, S.; Govinden, R.; Padayachee, T.; Odhav, B. Naturally occurring phenols: A detoxification strategy for fumonisin B1. *Food Addit. Contam.* **2003**, *20*, 490–493. [CrossRef]
118. Siddoo-Atwal, C.; Atwal, A.S. A possible role for honey bee products in the detoxification of mycotoxins. *Acta Hortic.* **2012**, *963*, 237–245. [CrossRef]
119. Xing, F.; Hua, H.; Selvaraj, J.N.; Yuan, Y.; Zhao, Y.; Zhou, L.; Liu, Y. Degradation of fumonisin B_1 by cinnamon essential oil. *Food Control* **2014**, *38*, 37–40. [CrossRef]
120. Gilchrist, D.G. Mycotoxins reveal connections between plants and animals in apoptosis and ceramide signaling. *Cell Death Differ.* **1997**, *4*, 689–698. [CrossRef]
121. Jacobson, M.D.; Weil, M.; Raff, M.C. Programmed cell death in animal development. *Cell* **1997**, *88*, 347–354. [CrossRef]
122. Green, D.R. The Coming Decade of Cell Death Research: Five Riddles. *Cell* **2019**, *177*, 1094–1107. [CrossRef]
123. Das, A.; Kawai-Yamada, M.; Uchimiya, H. Programmed cell death in plants. In *Abiotic Stress Adaptation in Plants: Physiological, Molecular and Genomic Foundation*; Pareek, A., Sopory, S.K., Bohnert, H.J., Govindjee, Eds.; Springer: Amsterdam, The Netherlands, 2010; pp. 371–383. [CrossRef]
124. Bedoui, S.; Herold, M.J.; Strasser, A. Emerging connectivity of programmed cell death pathways and its physiological implications. *Nat. Rev. Mol. Cell Biol.* **2020**, *21*, 678–695. [CrossRef] [PubMed]
125. Rantong, G.; Gunawardena, A.H.L.A.N. Programmed cell death: Genes involved in signaling, regulation, and execution in plants and animals. *Botany* **2015**, *93*, 193–210. [CrossRef]
126. Valandro, F.; Menguer, P.K.; Cabreira-Cagliari, C.; Margis-Pinheiro, M.; Cagliari, A. Programmed cell death (PCD) control in plants: New insights from the *Arabidopsis thaliana* deathosome. *Plant Sci.* **2020**, *299*. [CrossRef] [PubMed]
127. Slotte, J.P. Biological functions of sphingomyelins. *Prog. Lipid Res.* **2013**, *52*, 424–437. [CrossRef]
128. Luttgeharm, K.D.; Kimberlin, A.N.; Cahoon, E.B. Plant sphingolipid metabolism and function. *Subcell. Biochem.* **2016**, *86*, 249–286. [CrossRef]
129. Chao, M.V. Ceramide: A potential second messenger in the nervous system. *Mol. Cell. Neurosci.* **1995**, *6*, 91–96. [CrossRef]
130. Ng, C.K.Y.; Carr, K.; McAinsh, M.R.; Powell, B.; Hetherington, A.M. Drought-induced guard cell signal transduction involves sphingosine-1-phosphate. *Nature* **2001**, *410*, 596–599. [CrossRef]
131. Lynch, D.V. Evidence that sphingolipid signaling is involved in responding to low temperature. *New Phytol.* **2012**, *194*, 7–9. [CrossRef]
132. Dixon, S.C.; Soriano, B.J.; Lush, R.M.; Borner, M.M.; Figg, W.D. Apoptosis: Its role in the development of malignancies and its potential as a novel therapeutic target. *Annu. Pharmacother.* **1997**, *31*, 76–82. [CrossRef]
133. Govindarajah, N.; Clifford, R.; Bowden, D.; Sutton, P.A.; Parsons, J.L.; Vimalachandran, D. Sphingolipids and acid ceramidase as therapeutic targets in cancer therapy. *Crit. Rev. Oncol. Hemat.* **2019**, *138*, 104–111. [CrossRef]
134. Magnin-Robert, M.; Le Bourse, D.; Markham, J.; Dorey, S.; Clément, C.; Baillieul, F.; Dhondt-Cordelier, S. Modifications of sphingolipid content affect tolerance to hemibiotrophic and necrotrophic pathogens by modulating plant defense responses in *Arabidopsis*. *Plant Physiol.* **2015**, *169*, 2255–2274. [CrossRef] [PubMed]
135. Zhou, D.R.; Eid, R.; Boucher, E.; Miller, K.A.; Mandato, C.A.; Greenwood, M.T. Stress is an agonist for the induction of programmed cell death: A review. *BBA-Mol. Cell Res.* **2019**, *1866*, 699–712. [CrossRef]
136. Wong, K.; Li, X.B.; Hunchuk, N. N-acetylsphingosine (C2-ceramide) inhibited neutrophil superoxide formation and calcium influx. *J. Biol. Chem.* **1995**, *270*, 3056–3062. [CrossRef]
137. Bi, F.C.; Liu, Z.; Wu, J.X.; Liang, H.; Xi, X.L.; Fang, C.; Sun, T.J.; Yin, J.; Dai, G.Y.; Rong, C.; et al. Loss of ceramide kinase in *Arabidopsis* impairs defenses and promotes ceramide accumulation and mitochondrial H_2O_2 burstsc w. *Plant Cell* **2014**, *26*, 3449–3469. [CrossRef]
138. Townley, H.E.; McDonald, K.; Jenkins, G.I.; Knight, M.R.; Leaver, C.J. Ceramides induce programmed cell death in *Arabidopsis* cells in a calcium-dependent manner. *Biol. Chem.* **2005**, *386*, 161–166. [CrossRef] [PubMed]
139. Lachaud, C.; Prigent, E.; Thuleau, P.; Grat, S.; Da Silva, D.; Brière, C.; Mazars, C.; Cotelle, V. 14-3-3-Regulated Ca 2+ -dependent protein kinase CPK3 is required for sphingolipid-induced cell death in *Arabidopsis*. *Cell Death Differ.* **2013**, *20*, 209–217. [CrossRef] [PubMed]
140. Greenberg, J.T.; Silverman, F.P.; Liang, H. Uncoupling salicylic acid-dependent cell death and defense-related responses from disease resistance in the *Arabidopsis* mutant acd5. *Genetics* **2000**, *156*, 341–350. [PubMed]

141. Liang, H.; Yao, N.; Song, J.T.; Luo, S.; Lu, H.; Greenberg, J.T. Ceramides modulate programmed cell death in plants. *Gene. Dev.* **2003**, *17*, 2636–2641. [CrossRef]
142. Simanshu, D.K.; Zhai, X.; Munch, D.; Hofius, D.; Markham, J.E.; Bielawski, J.; Bielawska, A.; Malinina, L.; Molotkovsky, J.G.; Mundy, J.W.; et al. *Arabidopsis* accelerated cell death 11, ACD11, Is a ceramide-1-phosphate transfer protein and intermediary regulator of phytoceramide levels. *Cell Rep.* **2014**, *6*, 388–399. [CrossRef]
143. Kim, S.; Fyrst, H.; Saba, J. Accumulation of phosphorylated sphingoid long chain bases results in cell growth inhibition in Saccharomyces cerevisiae. *Genetics* **2000**, *156*, 1519–1529. [PubMed]
144. Nagano, M.; Takahara, K.; Fujimoto, M.; Tsutsumi, N.; Uchimiya, H.; Kawai-Yamada, M. *Arabidopsis* sphingolipid fatty acid 2-hydroxylases (AtFAH1 and AtFAH2) are functionally differentiated in fatty acid 2-hydroxylation and stress responses. *Plant Physiol.* **2012**, *159*, 1138–1148. [CrossRef] [PubMed]
145. Wang, H.; Li, J.; Bostock, R.M.; Gilchrist, D.G. Apoptosis: A functional paradigm for programmed plant cell death induced by a host-selective phytotoxin and invoked during development. *Plant Cell* **1996**, *8*, 375–391. [CrossRef]
146. Asai, T.; Stone, J.M.; Heard, J.E.; Kovtun, Y.; Yorgey, P.; Sheen, J.; Ausubel, F.M. Fumonisin B1-induced cell death in *Arabidopsis* protoplasts requires jasmonate-, ethylene-, and salicylate-dependent signaling pathways. *Plant Cell* **2000**, *12*, 1823–1835. [CrossRef]
147. Stone, J.M.; Heard, J.E.; Asai, T.; Ausubel, F.M. Simulation of fungal-mediated cell death by fumonisin B1 and selection of fumonisin B1-resistant (fbr) *Arabidopsis* mutants. *Plant Cell* **2000**, *12*, 1811–1822. [CrossRef]
148. Wang, H.; Jones, C.; Ciacci-Zanella, J.; Holt, T.; Gilchrist, D.G.; Dickman, M.B. Fumonisins and *Alternaria alternata lycopersici* toxins: Sphinganine analog mycotoxins induce apoptosis in monkey kidney cells. *Proc. Natl. Acad. Sci. USA* **1996**, *93*, 3461–3465. [CrossRef] [PubMed]
149. Wang, E.; Norred, W.P.; Bacon, C.W.; Riley, R.T.; Merrill, A.H., Jr. Inhibition of sphingolipid biosynthesis by fumonisins. Implications for diseases associated with *Fusarium moniliforme*. *J. Biol. Chem.* **1991**, *266*, 14486–14490. [PubMed]
150. Merrill, A.H.; Wang, E.; Gilchrist, D.G.; Riley, R.T. Fumonisins and other inhibitors of de novo sphingolipid biosynthesis. *Adv. Lipid Res.* **1993**, *26*, 215–234. [PubMed]
151. Gilchrist, D.G.; Wang, H.; Bostock, R.M. Sphingosine-related mycotoxins in plant and animal diseases. *Can. J. Bot.* **1995**, *73*, s459–s467. [CrossRef]
152. Merrill, A.H., Jr.; Liotta, D.C.; Riley, R.I. Fumonisins: Fungal toxins that shed light on sphingolipid function. *Trends Cell Biol.* **1996**, *6*, 218–223. [CrossRef]
153. Markham, J.E.; Molino, D.; Gissot, L.; Bellec, Y.; Hématy, K.; Marion, J.; Belcram, K.; Palauqui, J.C.; Satiat-Jeunemaître, B.; Faure, J.D. Sphingolipids containing very-long-chain fatty acids define a secretory pathway for specific polar plasma membrane protein targeting in *Arabidopsis*. *Plant Cell* **2011**, *23*, 2362–2378. [CrossRef]
154. Ternes, P.; Feussner, K.; Werner, S.; Lerche, J.; Iven, T.; Heilmann, I.; Riezman, H.; Feussner, I. Disruption of the ceramide synthase LOH1 causes spontaneous cell death in *Arabidopsis thaliana*. *New Phytol.* **2011**, *192*, 841–854. [CrossRef]
155. Moussatos, V.V.; Yang, S.F.; Ward, B.; Gilchrist, D.G. AAL-toxin induced physiological changes in lycopersicon esculentum mill: Roles for ethylene and pyrimidine intermediates in necrosis. *Physiol. Mol. Plant Pathol.* **1994**, *44*, 455–468. [CrossRef]
156. Moore, T.; Martineau, B.; Bostock, R.M.; Lincoln, J.E.; Gilchrist, D.G. Molecular and genetic characterization of ethylene involvement in mycotoxin-induced plant cell death. *Physiol. Mol. Plant Pathol.* **1999**, *54*, 73–85. [CrossRef]
157. Gechev, T.S.; Gadjev, I.Z.; Hille, J. An extensive microarray analysis of AAL-toxin-induced cell death in *Arabidopsis thaliana* brings new insights into the complexity of programmed cell death in plants. *Cell. Mol. Life Sci.* **2004**, *61*, 1185–1197. [CrossRef]
158. Mase, K.I.; Mizuno, T.; Ishihama, N.; Fujii, T.; Mori, H.; Kodama, M.; Yoshioka, H. Ethylene signaling pathway and MAPK cascades are required for AAL toxin-induced Programmed cell death. *Mol. Plant Microbe Interact.* **2012**, *25*, 1015–1025. [CrossRef]
159. Sultana, A.; Boro, P.; Mandal, K.; Chattopadhyay, S. AAL-toxin induced stress in *Arabidopsis thaliana* is alleviated through GSH-mediated salicylic acid and ethylene pathways. *Plant Cell Tiss. Org.* **2020**, *141*, 299–314. [CrossRef]

160. Egusa, M.; Ozawa, R.; Takabayashi, J.; Otani, H.; Kodama, M. The jasmonate signaling pathway in tomato regulates susceptibility to a toxin-dependent necrotrophic pathogen. *Planta* **2009**, *229*, 965–976. [CrossRef] [PubMed]
161. Zhang, L.; Jia, C.; Liu, L.; Zhang, Z.; Li, C.; Wang, Q. The involvement of jasmonates and ethylene in *Alternaria alternata* f. sp. *lycopersici* toxin-induced tomato cell death. *J. Exp. Bot.* **2011**, *62*, 5405–5418. [CrossRef] [PubMed]
162. Zhang, M.; Koh, J.; Liu, L.; Shao, Z.; Liu, H.; Hu, S.; Zhu, N.; Dufresne, C.P.; Chen, S.; Wang, Q. Critical Role of COI1-Dependent Jasmonate Pathway in AAL toxin induced PCD in Tomato Revealed by Comparative Proteomics. *Sci. Rep.* **2016**, *6*. [CrossRef]
163. Witsenboer, H.M.A.; van de Griend, E.G.; Tiersma, J.B.; Nijkamp, H.J.J.; Hille, J. Tomato resistance to Alternaria stem canker: Localization in host genotypes and functional expression compared to non-host resistance. *Theor. Appl. Genet.* **1989**, *78*, 457–462. [CrossRef] [PubMed]
164. Van Der Biezen, E.A.; Nijkamp, H.J.J.; Hille, J. Mutations at the Asc locus of tomato confer resistance to the fungal pathogen *Alternaria alternata* f. sp. *lycopersici*. *Theor. Appl. Genet.* **1996**, *92*, 898–904. [CrossRef] [PubMed]
165. Brandwagt, B.F.; Kneppers, T.J.A.; John, J.H.; Nijkamp, H.J.J.; Hille, J. Overexpression of the tomato Asc-1 gene mediates high insensitivity to AAL toxins and fumonisin B1 in tomato hairy roots and confers resistance to *Alternaria alternata* f. sp. *lycopersici* in Nicotiana umbratica plants. *Mol. Plant Microbe Interact.* **2002**, *15*, 35–42. [CrossRef]
166. Brandwagt, B.F.; Mesbah, L.A.; Takken, F.L.W.; Laurent, P.L.; Kneppers, T.J.A.; Hille, J.; Nijkamp, H.J.J. A longevity assurance gene homolog of tomato mediates resistance to Altemaria alternata f. sp. *lycopersici* toxins and fumonisin B1. *Proc. Natl. Acad. Sci. USA* **2000**, *97*, 4961–4966. [CrossRef]
167. Brandwagt, B.F.; Kneppers, T.J.A.; Van der Weerden, G.M.; Nijkamp, H.J.J.; Hille, J. Most AAL toxin-sensitive Nicotiana species are resistant to the tomato fungal pathogen *Alternaria alternata* f. sp. *lycopersici*. *Mol. Plant Microbe Interact.* **2001**, *14*, 460–470. [CrossRef]
168. Egusa, M.; Miwa, T.; Kaminaka, H.; Takano, Y.; Kodama, M. Nonhost resistance of *Arabidopsis thaliana* against *Alternaria alternata* involves both pre- and postinvasive defenses but is collapsed by AAL-toxin in the absence of LOH2. *Phytopathology* **2013**, *103*, 733–740. [CrossRef] [PubMed]
169. Tsuge, T.; Harimoto, Y.; Akimitsu, K.; Ohtani, K.; Kodama, M.; Akagi, Y.; Egusa, M.; Yamamoto, M.; Otani, H. Host-selective toxins produced by the plant pathogenic fungus *Alternaria alternata*. *FEMS Microbiol. Rev.* **2012**, *37*, 44–66. [CrossRef] [PubMed]
170. de Zélicourt, A.; Montiel, G.; Pouvreau, J.B.; Thoiron, S.; Delgrange, S.; Simier, P.; Delavault, P. Susceptibility of *Phelipanche* and *Orobanche* species to AAL-toxin. *Planta* **2009**, *230*, 1047–1055. [CrossRef]
171. Lincoln, J.E.; Richael, C.; Overduin, B.; Smith, K.; Bostock, R.; Gilchrist, D.G. Expression of the antiapoptotic baculovirus p35 gene in tomato blocks programmed cell death and provides broad-spectrum resistance to disease. *Proc. Natl. Acad. Sci. USA* **2002**, *99*, 15217–15221. [CrossRef]
172. Qureshi, M.K.; Sujeeth, N.; Gechev, T.S.; Hille, J. The zinc finger protein ZAT11 modulates paraquat-induced programmed cell death in *Arabidopsis thaliana*. *Acta Physiol. Plant.* **2013**, *35*, 1863–1871. [CrossRef]
173. Shao, Z.; Zhao, Y.; Liu, L.; Chen, S.; Li, C.; Meng, F.; Liu, H.; Hu, S.; Wang, J.; Wang, Q. Overexpression of FBR41 enhances resistance to sphinganine analog mycotoxin-induced cell death and Alternaria stem canker in tomato. *Plant Biotechnol. J.* **2020**, *18*, 141–154. [CrossRef]
174. Al-Jaal, B.A.; Jaganjac, M.; Barcaru, A.; Horvatovich, P.; Latiff, A. Aflatoxin, fumonisin, ochratoxin, zearalenone and deoxynivalenol biomarkers in human biological fluids: A systematic literature review, 2001–2018. *Food Chem. Toxicol.* **2019**, *129*, 211–228. [CrossRef] [PubMed]
175. Alberts, J.; Rheeder, J.; Gelderblom, W.; Shephard, G.; Burger, H.M. Rural subsistence maize farming in South Africa: Risk assessment and intervention models for reduction of exposure to fumonisin mycotoxins. *Toxins* **2019**, *11*, 334. [CrossRef] [PubMed]
176. Cendoya, E.; Chiotta, M.L.; Zachetti, V.; Chulze, S.N.; Ramirez, M.L. Fumonisins and fumonisin-producing Fusarium occurrence in wheat and wheat by products: A review. *J. Cereal Sci.* **2018**, *80*, 158–166. [CrossRef]
177. Dall'Asta, C.; Battilani, P. Fumonisins and their modified forms, a matter of concern in future scenario? *World Mycotoxin J.* **2016**, *9*, 727–739. [CrossRef]

178. Alberts, J.F.; van Zyl, W.H.; Gelderblom, W.C.A. Biologically based methods for control of fumonisin-producing Fusarium species and reduction of the fumonisins. *Front. Microbiol.* **2016**, *7*. [CrossRef]
179. Gil-Serna, J.; Vázquez, C.; Patiño, B. Genetic regulation of aflatoxin, ochratoxin A, trichothecene, and fumonisin biosynthesis: A review. *Int. Microbiol.* **2020**, *23*, 89–96. [CrossRef]
180. Santiago, R.; Cao, A.; Butrón, A. Genetic factors involved in fumonisin accumulation in maize kernels and their implications in maize agronomic management and breeding. *Toxins* **2015**, *7*, 3267–3296. [CrossRef]
181. Chavez, R.A.; Cheng, X.; Stasiewicz, M.J. A review of the methodology of analyzing aflatoxin and fumonisin in single corn kernels and the potential impacts of these methods on food security. *Foods* **2020**, *9*, 297. [CrossRef]
182. Deepa, N.; Sreenivasa, M.Y. Molecular methods and key genes targeted for the detection of fumonisin producing *Fusarium verticillioides*–An updated review. *Food Biosci.* **2019**, *32*. [CrossRef]
183. Santiago, R.; Cao, A.; Malvar, R.A.; Butrón, A. Genomics of maize resistance to *Fusarium* ear rot and fumonisin contamination. *Toxins* **2020**, *12*, 431. [CrossRef]
184. Diaz-Gomez, J.; Marin, S.; Capell, T.; Sanchis, V.; Ramos, A.J. The impact of Bacillus thuringiensis technology on the occurrence of fumonisins and other mycotoxins in maize. *World Mycotoxin J.* **2016**, *9*, 475–486. [CrossRef]
185. Sheik Abdul, N.; Marnewick, J.L. Fumonisin B1-induced mitochondrial toxicity and hepatoprotective potential of rooibos: An update. *J. Appl. Toxicol.* **2020**. [CrossRef] [PubMed]
186. Liu, X.; Fan, L.; Yin, S.; Chen, H.; Hu, H. Molecular mechanisms of fumonisin B_1-induced toxicities and its applications in the mechanism-based interventions. *Toxicon* **2019**, *167*, 1–5. [CrossRef] [PubMed]
187. Riley, R.T.; Merrill, A.H., Jr. Ceramide synthase inhibition by fumonisins: A perfect storm of perturbed sphingolipid metabolism, signaling, and disease. *J. Lipid Res.* **2019**, *60*, 1183–1189. [CrossRef]
188. Chen, C.; Riley, R.T.; Wu, F. Dietary Fumonisin and Growth Impairment in Children and Animals: A Review. *Compr. Rev. Food Sci. Food Saf.* **2018**, *17*, 1448–1464. [CrossRef]
189. Pierron, A.; Alassane-Kpembi, I.; Oswald, I.P. Impact of two mycotoxins deoxynivalenol and fumonisin on pig intestinal health. *Porcine Health Manag.* **2016**, *2*. [CrossRef]
190. Wang, X.; Wu, Q.; Wan, D.; Liu, Q.; Chen, D.; Liu, Z.; Martinez-Larrañaga, M.R.; Martínez, M.A.; Anadón, A.; Yuan, Z. Fumonisins: Oxidative stress-mediated toxicity and metabolism in vivo and *in vitro*. *Arch. Toxicol.* **2015**, *90*, 81–101. [CrossRef]
191. Nützmann, H.W.; Scazzocchio, C.; Osbourn, A. Metabolic gene clusters in eukaryotes. *Annu. Rev. Genet.* **2018**, *52*, 159–183. [CrossRef]
192. Seo, J.A.; Proctor, R.H.; Plattner, R.D. Characterization of four clustered and coregulated genes associated with fumonisin biosynthesis in *Fusarium verticillioides*. *Fungal Genet. Biol.* **2001**, *34*, 155–165. [CrossRef]
193. Proctor, R.H.; Brown, D.W.; Plattner, R.D.; Desjardins, A.E. Co-expression of 15 contiguous genes delineates a fumonisin biosynthetic gene cluster in Gibberella moniliformis. *Fungal Genet. Biol.* **2003**, *38*, 237–249. [CrossRef]
194. Butchko, R.A.E.; Plattner, R.D.; Proctor, R.H. Deletion analysis of FUM genes involved in tricarballylic ester formation during fumonisin biosynthesis. *J. Agric. Food Chem.* **2006**, *54*, 9398–9404. [CrossRef] [PubMed]
195. Proctor, R.H.; Busman, M.; Seo, J.A.; Lee, Y.W.; Plattner, R.D. A fumonisin biosynthetic gene cluster in *Fusarium oxysporum* strain O-1890 and the genetic basis for B versus C fumonisin production. *Fungal Genet. Biol.* **2008**, *45*, 1016–1026. [CrossRef] [PubMed]
196. Stępień, Ł.; Koczyk, G.; Waśkiewicz, A. Genetic and phenotypic variation of *Fusarium proliferatum* isolates from different host species. *J. Appl. Genet.* **2011**, *52*, 487–496. [CrossRef] [PubMed]
197. Janevska, S.; Ferling, I.; Jojić, K.; Rautschek, J.; Hoefgen, S.; Proctor, R.H.; Hillmann, F.; Valiante, V. Self-Protection against the Sphingolipid Biosynthesis Inhibitor Fumonisin B1 Is Conferred by a FUM Cluster-Encoded Ceramide Synthase. *mBio* **2020**, *11*. [CrossRef]
198. Brown, D.W.; Butchko, R.A.E.; Busman, M.; Proctor, R.H. The *Fusarium verticillioides* FUM gene cluster encodes a Zn(II)2Cys6 protein that affects FUM gene expression and fumonisin production. *Eukaryot. Cell* **2007**, *6*, 1210–1218. [CrossRef]
199. Desjardins, A.E.; Munkvold, G.P.; Plattner, R.D.; Proctor, R.H. FUM1-A gene required for fumonisin biosynthesis but not for maize ear rot and ear infection by Gibberella moniliformis in field tests. *Mol. Plant Microbe Interact.* **2002**, *15*, 1157–1164. [CrossRef]

200. Proctor, R.H.; Desjardins, A.E.; Plattner, R.D.; Hohn, T.M. A polyketide synthase gene required for biosynthesis of fumonisin mycotoxins in *Gibberella fujikuroi* mating population A. *Fungal Genet. Biol.* **1999**, *27*, 100–112. [CrossRef]
201. Bojja, R.S.; Cerny, R.L.; Proctor, R.H.; Du, L. Determining the Biosynthetic Sequence in the Early Steps of the Fumonisin Pathway by Use of Tree Gene-Disruption Mutants of *Fusarium verticillioides*. *J. Agric. Food Chem.* **2004**, *52*, 2855–2860. [CrossRef]
202. Massi, F.P.; Sartori, D.; de Souza Ferranti, L.; Iamanaka, B.T.; Taniwaki, M.H.; Vieira, M.L.C.; Fungaro, M.H.P. Prospecting for the incidence of genes involved in ochratoxin and fumonisin biosynthesis in Brazilian strains of *Aspergillus niger* and *Aspergillus welwitschiae*. *Int. J. Food Microbiol.* **2016**, *221*, 19–28. [CrossRef]
203. Desjardins, A.E.; Plattner, R.D.; Proctor, R.H. Genetic and biochemical aspects of fumonisin production. *Adv. Exp. Med. Biol.* **1996**, *392*, 165–173. [CrossRef]
204. Butchko, R.A.E.; Plattner, R.D.; Proctor, R.H. *FUM13* encodes a short chain dehydrogenase/reductase required for C-3 carbonyl reduction during fumonisin biosynthesis in *Gibberella moniliformis*. *J. Agric. Food Chem.* **2003**, *51*, 3000–3006. [CrossRef] [PubMed]
205. Butchko, R.A.E.; Plattner, R.D.; Proctor, R.H. *FUM9* Is Required for C-5 Hydroxylation of Fumonisins and Complements the Meitotically Defined Fum3 Locus in *Gibberella moniliformis*. *Appl. Environ. Microb.* **2003**, *69*, 6935–6937. [CrossRef] [PubMed]
206. Yi, H.; Bojja, R.S.; Fu, J.; Du, L. Direct evidence for the function of FUM13 in 3-ketoreduction of mycotoxin fumonisins in *Fusarium verticillioides*. *J. Agric. Food Chem.* **2005**, *53*, 5456–5460. [CrossRef] [PubMed]
207. Proctor, R.H.; Plattner, R.D.; Desjardins, A.E.; Busman, M.; Butchko, R.A.E. Fumonisin production in the maize pathogen *Fusarium verticillioides*: Genetic basis of naturally occurring chemical variation. *J. Agric. Food Chem.* **2006**, *54*, 2424–2430. [CrossRef]
208. Pel, H.J.; De Winde, J.H.; Archer, D.B. Genome sequencing and analysis of the versatile cell factory *Aspergillus niger* CBS 513.88. *Nat. Biotechnol.* **2007**, *25*, 221–231. [CrossRef]
209. Rösler, S.M.; Sieber, C.M.K.; Humpf, H.U.; Tudzynski, B. Interplay between pathway-specific and global regulation of the fumonisin gene cluster in the rice pathogen *Fusarium fujikuroi*. *Appl. Microbiol. Biotechnol.* **2016**, *100*, 5869–5882. [CrossRef]
210. Myung, K.; Li, S.; Butchko, R.A.E.; Busman, M.; Proctor, R.H.; Abbas, H.K.; Calvo, A.M. FvVE1 regulates biosynthesis of the mycotoxins fumonisins and fusarins in *Fusarium verticiliioides*. *J. Agric. Food Chem.* **2009**, *57*, 5089–5094. [CrossRef]
211. Kim, H.; Woloshuk, C.P. Functional characterization of fst1 in *Fusarium verticillioides* during colonization of maize kernels. *Mol. Plant Microbe Interact.* **2011**, *24*, 18–24. [CrossRef]
212. Shin, J.H.; Kim, J.E.; Malapi-Wight, M.; Choi, Y.E.; Shaw, B.D.; Shim, W.B. Protein phosphatase 2A regulatory subunits perform distinct functional roles in the maize pathogen *Fusarium verticillioides*. *Mol. Plant Pathol.* **2013**, *14*, 518–529. [CrossRef]
213. Ridenour, J.B.; Bluhm, B.H. The novel fungal-specific gene *FUG1* has a role in pathogenicity and fumonisin biosynthesis in *Fusarium verticillioides*. *Mol. Plant Pathol.* **2017**, *18*, 513–528. [CrossRef]
214. Niehaus, E.M.; Münsterkötter, M.; Proctor, R.H.; Brown, D.W.; Sharon, A.; Idan, Y.; Oren-Young, L.; Sieber, C.M.; Novák, O.; Pěnčík, A.; et al. Comparative "omics" of the *Fusarium fujikuroi* species complex highlights differences in genetic potential and metabolite synthesis. *Genome Biol. Evol.* **2016**, *8*, 3574–3599. [CrossRef] [PubMed]
215. Montis, V.; Pasquali, M.; Visentin, I.; Karlovsky, P.; Cardinale, F. Identification of a cis-acting factor modulating the transcription of *FUM1*, a key fumonisin-biosynthetic gene in the fungal maize pathogen *Fusarium verticillioides*. *Fungal Genet. Biol.* **2013**, *51*, 42–49. [CrossRef] [PubMed]
216. Kohut, G.; Ádám, A.L.; Fazekas, B.; Hornok, L. N-starvation stress induced *FUM* gene expression and fumonisin production is mediated via the HOG-type MAPK pathway in *Fusarium proliferatum*. *Int. J. Food Microbiol.* **2009**, *130*, 65–69. [CrossRef] [PubMed]
217. Medina, A.; Schmidt-Heydt, M.; Cárdenas-Chávez, D.L.; Parra, R.; Geisen, R.; Magan, N. Integrating toxin gene expression, growth and fumonisin B_1 and B_2 production by a strain of *Fusarium verticillioides* under different environmental factors. *J. R. Soc. Interface* **2013**, *10*. [CrossRef]
218. Stepień, Ł.; Waśkiewicz, A.; Wilman, K. Host extract modulates metabolism and fumonisin biosynthesis by the plant-pathogenic fungus *Fusarium proliferatum*. *Int. J. Food Microbiol.* **2015**, *193*, 74–81. [CrossRef]

219. Wu, Y.; Li, T.; Gong, L.; Wang, Y.; Jiang, Y. Effects of different carbon sources on fumonisin production and fum gene expression by *Fusarium proliferatum*. *Toxins* **2019**, *11*, 289. [CrossRef] [PubMed]
220. Witaszak, N.; Lalak-Kańczugowska, J.; Waśkiewicz, A.; Stępień, Ł. The impacts of asparagus extract fractions on growth and fumonisins biosynthesis in *Fusarium proliferatum*. *Toxins* **2020**, *12*, 95. [CrossRef]
221. Johnson, L.; Harrison, J.H.; Hunt, C.; Shinners, K.; Doggett, C.G.; Sapienza, D. Nutritive value of corn silage as affected by maturity and mechanical processing: A contemporary review. *J. Dairy Sci.* **1999**, *82*, 2813–2825. [CrossRef]
222. Mansfield, M.A.; Archibald, D.D.; Jones, A.D.; Kuldau, G.A. Relationship of sphinganine analog mycotoxin contamination in maize silage to seasonal weather conditions and to agronomic and ensiling practices. *Phytopathology* **2007**, *97*, 504–511. [CrossRef]
223. Baker, S.E. *Aspergillus niger* genomics: Past, present and into the future. *Med. Mycol.* **2006**, *44*, 17–21. [CrossRef]
224. Brown, D.W.; Baker, S.E. Mycotoxins: A fungal genomics perspective. *Methods Mol. Biol.* **2017**, *1542*, 367–379. [CrossRef] [PubMed]
225. Noonim, P.; Mahakarnchanakul, W.; Nielsen, K.F.; Frisvad, J.C.; Samson, R.A. Fumonisin B2 production by *Aspergillus niger* in Thai coffee beans. *Food Addit. Contam. A* **2009**, *26*, 94–100. [CrossRef] [PubMed]
226. Månsson, M.; Klejnstrup, M.L.; Phipps, R.K.; Nielsen, K.F.; Frisvad, J.C.; Gotfredsen, C.H.; Larsen, T.O. Isolation and NMR characterization of fumonisin b_2 and a new fumonisin B6 from *aspergillus niger*. *J. Agric. Food Chem.* **2010**, *58*, 949–953. [CrossRef] [PubMed]
227. Palencia, E.R.; Mitchell, T.R.; Snook, M.E.; Glenn, A.E.; Gold, S.; Hinton, D.M.; Riley, R.T.; Bacon, C.W. Analyses of black *Aspergillus* species of peanut and maize for ochratoxins and fumonisins. *J. Food Prot.* **2014**, *77*, 805–813. [CrossRef]
228. Desjardins, A.E.; Plattner, R.D.; Proctor, R.H. Linkage among genes responsible for fumonisin biosynthesis in *Gibberella fujikuroi* mating population A. *Appl. Environ. Microb.* **1996**, *62*, 2571–2576. [CrossRef]
229. Ding, Y.; Bojja, R.S.; Du, L. Fum3p, a 2-Ketoglutarate-Dependent Dioxygenase Required for C-5 Hydroxylation of Fumonisins in *Fusarium verticillioides*. *Appl. Environ. Microb.* **2004**, *70*, 1931–1934. [CrossRef]
230. Plattner, R.D.; Desjardins, A.E.; Leslie, J.F.; Nelson, P.E. Identification and characterization of strains of *Gibberella fujikuroi* mating population A with rare fumonisin production phenotypes. *Mycologia* **1996**, *88*, 416–424. [CrossRef]
231. Uhlig, S.; Busman, M.; Shane, D.S.; Rønning, H.; Rise, F.; Proctor, R. Identification of early fumonisin biosynthetic intermediates by inactivation of the FUM6 gene in *Fusarium verticillioides*. *J. Agric. Food Chem.* **2012**, *60*, 10293–10301. [CrossRef]
232. Li, Y.; Lou, L.; Cerny, R.L.; Butchko, R.A.E.; Proctor, R.H.; Shen, Y.; Du, L. Tricarballylic ester formation during biosynthesis of fumonisin mycotoxins in *Fusarium verticillioides*. *Mycology* **2013**, *4*, 179–186. [CrossRef]
233. Zaleta-Rivera, K.; Xu, C.; Yu, F.; Butchko, R.A.E.; Proctor, R.H.; Hidalgo-Lara, M.E.; Raza, A.; Dussault, P.H.; Du, L. A bidomain nonribosomal peptide synthetase encoded by *FUM14* catalyzes the formation of tricarballylic esters in the biosynthesis of fumonisins. *Biochemistry* **2006**, *45*, 2561–2569. [CrossRef]
234. Brown, D.W.; Cheung, F.; Proctor, R.H.; Butchko, R.A.E.; Zheng, L.; Lee, Y.; Utterback, T.; Smith, S.; Feldblyum, T.; Glenn, A.E.; et al. Comparative analysis of 87,000 expressed sequence tags from the fumonisin-producing fungus *Fusarium verticillioides*. *Fungal Genet. Biol.* **2005**, *42*, 848–861. [CrossRef] [PubMed]
235. Susca, A.; Proctor, R.H.; Morelli, M.; Haidukowski, M.; Gallo, A.; Logrieco, A.F.; Moretti, A. Variation in fumonisin and ochratoxin production associated with differences in biosynthetic gene content in *Aspergillus niger* and *A. welwitschiae* isolates from multiple crop and geographic origins. *Front. Microbiol.* **2016**, *7*. [CrossRef] [PubMed]
236. Yamada, O.; Machida, M.; Hosoyama, A.; Goto, M.; Takahashi, T.; Futagami, T.; Yamagata, Y.; Takeuchi, M.; Kobayashi, T.; Koike, H.; et al. Genome sequence of Aspergillus luchuensis NBRC 4314. *DNA Res.* **2016**, *23*, 507–515. [CrossRef] [PubMed]
237. Akagi, Y.; Akamatsu, H.; Otani, H.; Kodama, M. Horizontal chromosome transfer, a mechanism for the evolution and differentiation of a plant-pathogenic fungus. *Eukaryot. Cell* **2009**, *8*, 1732–1738. [CrossRef]
238. Akagi, Y.; Taga, M.; Yamamoto, M.; Tsuge, T.; Fukumasa-Nakai, Y.; Otani, H.; Kodama, M. Chromosome constitution of hybrid strains constructed by protoplast fusion between the tomato and strawberry pathotypes of *Alternaria alternata*. *J. Gen. Plant Pathol.* **2009**, *75*, 101–109. [CrossRef]

239. Akamatsu, H.; Otani, H.; Kodama, M. Characterization of a gene cluster for host-specific AAL-toxin biosynthesis in the tomato pathotype of *Alternaria alternata*. *Fungal Genet. Rep.* **2003**, *50*, 355. [CrossRef]
240. Zhu, X.; Vogeler, C.; Du, L. Functional complementation of fumonisin biosynthesis in FUM1-disrupted *Fusarium verticillioides* by the AAL-toxin Polyketide synthase gene *ALT1* from *Alternaria alternata* f. sp. *Lycopersici*. *J. Nat. Prod.* **2008**, *71*, 957–960. [CrossRef]
241. Takao, K.; Akagi, Y.; Tsuge, T.; Harimoto, Y.; Yamamoto, M.; Kodama, M. The global regulator *LaeA* controls biosynthesis of host-specific toxins, pathogenicity and development of *Alternaria alternata* pathotypes. *J. Gen. Plant Pathol.* **2016**, *82*, 121–131. [CrossRef]
242. Vaquera, S.; Patriarca, A.; Cabrera, G.; Fernández Pinto, V. Temperature and water activity influence on simultaneous production of AAL toxins by *Alternaria arborescens* on tomato medium. *Eur. J. Plant Pathol.* **2017**, *148*, 1003–1009. [CrossRef]
243. Kroken, S.; Glass, N.L.; Taylor, J.W.; Yoder, O.C.; Turgeon, B.G. Phylogenomic analysis of type I polyketide synthase genes in pathogenic and saprobic ascomycetes. *Proc. Natl. Acad. Sci. USA* **2003**, *100*, 15670–15675. [CrossRef]
244. Condon, B.J.; Leng, Y.; Wu, D.; Bushley, K.E.; Ohm, R.A.; Otillar, R.; Martin, J.; Schackwitz, W.; Grimwood, J.; MohdZainudin, N.A.I.; et al. Comparative Genome Structure, Secondary Metabolite, and Effector Coding Capacity across *Cochliobolus* Pathogens. *PLoS Genet.* **2013**, *9*. [CrossRef] [PubMed]
245. Kim, H.S.; Lohmar, J.M.; Busman, M.; Brown, D.W.; Naumann, T.A.; Divon, H.H.; Uhlig, S.; Proctor, R.H. Identification and distribution of gene clusters required for synthesis of sphingolipid metabolism inhibitors in diverse species of the filamentous fungus *Fusarium*. *BMC Genom.* **2020**, *21*. [CrossRef]
246. Sultana, S.; Kitajima, M.; Kobayashi, H.; Nakagawa, H.; Shimizu, M.; Kageyama, K.; Suga, H. A natural variation of fumonisin gene cluster associated with fumonisin production difference in *Fusarium fujikuroi*. *Toxins* **2019**, *11*, 200. [CrossRef] [PubMed]
247. Fumero, M.V.; Villani, A.; Susca, A.; Haidukowski, M.; Cimmarusti, M.T.; Toomajian, C.; Leslie, J.F.; Chulze, S.N.; Moretti, A. Fumonisin and Beauvericin Chemotypes and Genotypes of the Sister Species *Fusarium subglutinans* and *Fusarium temperatum*. *Appl. Environ. Microb.* **2020**, *86*. [CrossRef] [PubMed]
248. Glenn, A.E.; Zitomer, N.C.; Zimeri, A.M.; Williams, L.D.; Riley, R.T.; Proctor, R.H. Transformation-mediated complementation of a *FUM* gene cluster deletion in *Fusarium verticillioides* restores both fumonisin production and pathogenicity on maize seedlings. *Mol. Plant Microbe Interact.* **2008**, *21*, 87–97. [CrossRef]
249. Geiser, D.M.; Jiménez-Gasco, M.D.M.; Kang, S.; Makalowska, I.; Veeraraghavan, N.; Ward, T.J.; Zhang, N.; Kuldau, G.A.; O'Donnell, K. FUSARIUM-ID v. 1.0: A DNA sequence database for identifying *Fusarium*. *Eur. J. Plant Pathol.* **2004**, *110*, 473–479. [CrossRef]
250. Kristensen, R.; Torp, M.; Kosiak, B.; Holst-Jensen, A. Phylogeny and toxigenic potential is correlated in *Fusarium* species as revealed by partial translation elongation factor 1 alpha gene sequences. *Mycol. Res.* **2005**, *109*, 173–186. [CrossRef]
251. Wang, J.; Wang, X.; Zhou, Y.; Du, L.; Wang, Q. Fumonisin detection and analysis of potential fumonisin-producing *Fusarium* spp. in *Asparagus* (*Asparagus officinalis* L.) in Zhejiang province of China. *J. Sci. Food Agric.* **2010**, *90*, 836–842. [CrossRef]
252. Jurado, M.; Marín, P.; Callejas, C.; Moretti, A.; Vázquez, C.; González-Jaén, M.T. Genetic variability and Fumonisin production by *Fusarium proliferatum*. *Food Microbiol.* **2010**, *27*, 50–57. [CrossRef]
253. Waśkiewicz, A.; Stepień, L.; Wilman, K.; Kachlicki, P. Diversity of pea-associated *F. proliferatum* and *F. verticillioides* populations revealed by *FUM1* sequence analysis and fumonisin biosynthesis. *Toxins* **2013**, *5*, 488–503. [CrossRef]
254. Stępień, Ł.; Koczyk, G.; Waśkiewicz, A. *FUM* cluster divergence in fumonisins-producing *Fusarium* species. *Fungal Biol.* **2011**, *115*, 112–123. [CrossRef] [PubMed]
255. Shephard, G.S.; Sydenham, E.W.; Thiel, P.G.; Gelderblom, W.C.A. Quantitative determination of fumonisins b_1 and b_2 by high-performance liquid chromatography with fluorescence detection. *J. Liq. Chromatogr.* **1990**, *13*, 2077–2087. [CrossRef]
256. Rottinghaus, G.E.; Coatney, C.E.; Minor, H.C. A rapid, sensitive thin layer chromatography procedure for the detection of fumonisin B_1 and B_2. *J. Vet. Diagn. Investig.* **1992**, *4*, 326–329. [CrossRef] [PubMed]
257. Schneider, E.; Usleber, E.; Märtlbauer, E. Rapid Detection of Fumonisin B_1 in Corn-Based Food by Competitive Direct Dipstick Enzyme Immunoassay/Enzyme-Linked Immunofiltration Assay with Integrated Negative Control Reaction. *J. Agric. Food Chem.* **1995**, *43*, 2548–2552. [CrossRef]

258. Plattner, R.D. Detection of fumonisins produced in *Fusarium moniliforme* cultures by HPLC with electrospray MS and evaporative light scattering detectors. *Nat. Toxins* **1995**, *3*, 294–298. [CrossRef]
259. Caldas, E.D.; Winter, C.K.; Daniel Jones, A.; Ward, B.; Gilchrist, D.G. Electrospray Ionization Mass Spectrometry of Sphinganine Analog Mycotoxins. *Anal. Chem.* **1995**, *67*, 196–207. [CrossRef]
260. Szurdoki, F.; Trousdale, E.; Ward, B.; Gee, S.J.; Hammock, B.D.; Gilchrist, D.G. Synthesis of protein conjugates and development of immunoassays for AAL toxins. *J. Agric. Food Chem.* **1996**, *44*, 1796–1803. [CrossRef]
261. Yu, W.; Yu, F.Y.; Undersander, D.J.; Chu, F.S. Immunoassays of selected mycotoxins in hay, silage and mixed feed. *Food Agric. Immunol.* **1999**, *11*, 307–319. [CrossRef]
262. Gaspardo, B.; Del Zotto, S.; Torelli, E.; Cividino, S.R.; Firrao, G.; Della Riccia, G.; Stefanon, B. A rapid method for detection of fumonisins B_1 and B_2 in corn meal using Fourier transform near infrared (FT-NIR) spectroscopy implemented with integrating sphere. *Food Chem.* **2012**, *135*, 1608–1612. [CrossRef]
263. Man, Y.; Liang, G.; Li, A.; Pan, L. Analytical Methods for the Determination of *Alternaria* Mycotoxins. *Chromatographia* **2017**, *80*, 9–22. [CrossRef]
264. Mirasoli, M.; Buragina, A.; Dolci, L.S.; Simoni, P.; Anfossi, L.; Giraudi, G.; Roda, A. Chemiluminescence-based biosensor for fumonisins quantitative detection in maize samples. *Biosens. Bioelectron.* **2012**, *32*, 283–287. [CrossRef] [PubMed]
265. Zangheri, M.; Di Nardo, F.; Anfossi, L.; Giovannoli, C.; Baggiani, C.; Roda, A.; Mirasoli, M. A multiplex chemiluminescent biosensor for type B-fumonisins and aflatoxin B_1 quantitative detection in maize flour. *Analyst* **2015**, *140*, 358–365. [CrossRef]
266. Shu, M.; Xu, Y.; Wang, D.; Liu, X.; Li, Y.; He, Q.; Tu, Z.; Qiu, Y.; Ji, Y.; Wang, X. Anti-idiotypic nanobody: A strategy for development of sensitive and green immunoassay for Fumonisin B_1. *Talanta* **2015**, *143*, 388–393. [CrossRef] [PubMed]
267. Di Nardo, F.; Alladio, E.; Baggiani, C.; Cavalera, S.; Giovannoli, C.; Spano, G.; Anfossi, L. Colour-encoded lateral flow immunoassay for the simultaneous detection of aflatoxin B_1 and type-B fumonisins in a single Test line. *Talanta* **2019**, *192*, 288–294. [CrossRef] [PubMed]
268. Hou, S.; Ma, J.; Cheng, Y.; Wang, H.; Sun, J.; Yan, Y. Quantum dot nanobead-based fluorescent immunochromatographic assay for simultaneous quantitative detection of fumonisin B1, dexyonivalenol, and zearalenone in grains. *Food Control* **2020**, *117*. [CrossRef]
269. Ferrara, M.; Logrieco, A.F.; Moretti, A.; Susca, A. A loop-mediated isothermal amplification (LAMP) assay for rapid detection of fumonisin producing *Aspergillus* species. *Food Microbiol.* **2020**, *90*. [CrossRef]
270. Wigmann, É.F.; Meyer, K.; Cendoya, E.; Maul, R.; Vogel, R.F.; Niessen, L. A loop-mediated isothermal amplification (LAMP) based assay for the rapid and sensitive group-specific detection of fumonisin producing *Fusarium* spp. *Int. J. Food Microbiol.* **2020**, *325*. [CrossRef] [PubMed]
271. Bluhm, B.H.; Flaherty, J.E.; Cousin, M.A.; Woloshuk, C.P. Multiplex polymerase chain reaction assay for the differential detection of trichothecene- and fumonisin-producing species of *Fusarium* in cornmeal. *J. Food Protect.* **2002**, *65*, 1955–1961. [CrossRef]
272. Kim, N.Y.; Lee, I.; Ji, G.E. Reliable and simple detection of ochratoxin and fumonisin production in black *Aspergillus*. *J. Food Protect.* **2014**, *77*, 653–658. [CrossRef] [PubMed]
273. Nagaraj, D.; Adkar-Purushothama, C.R.; Marikunte Yanjarappa, S. Multiplex PCR for the early detection of fumonisin producing *Fusarium verticillioides*. *Food Biosci.* **2016**, *13*, 84–88. [CrossRef]
274. Gareis, M.; Bauer, J.; Thiem, J.; Plank, G.; Grabley, S.; Gedek, B. Cleavage of Zearalenone-Glycoside, a "Masked" Mycotoxin, during Digestion in Swine. *J. Vet. Med. Ser. B* **1990**, *37*, 236–240. [CrossRef]
275. Berthiller, F.; Maragos, C.M.; Dall'Asta, C. Introduction to masked mycotoxin. In *Masked Mycotoxins in Food: Formation, Occurrence and Toxicological Relevance*; Dall'Asta, C., Berthiller, F., Eds.; Royal Society of Chemistry: Cambridge, UK, 2015; pp. 1–8.
276. Liu, J.; Applegate, T. Zearalenone (ZEN) in Livestock and Poultry: Dose, toxicokinetics, toxicity and estrogenicity. *Toxins* **2020**, *12*, 377. [CrossRef] [PubMed]
277. Ekwomadu, T.I.; Dada, T.A.; Nleya, N.; Gopane, R.; Sulyok, M.; Mwanza, M. Variation of *Fusarium* free, masked, and emerging mycotoxin metabolites in maize from Agriculture Regions of South Africa. *Toxins* **2020**, *12*, 149. [CrossRef]
278. Crews, C.; MacDonald, S.J. Natural occurrence of masked mycotoxins. In *Issues in Toxicology*; Berthiller, F., Dall'Asta, C., Eds.; Royal Society of Chemistry: Cambridge, UK, 2016; pp. 14–31. [CrossRef]

279. Bryła, M.; Roszko, M.; Szymczyk, K.; Jedrzejczak, R.; Obiedziński, M.W. Fumonisins and their masked forms in maize products. *Food Control* **2016**, *59*, 619–627. [CrossRef]
280. Falavigna, C.; Cirlini, M.; Galaverna, G.; Dall'Asta, C. Masked fumonisins in processed food: Co-occurrence of hidden and bound forms and their stability under digestive conditions. *World Mycotoxin J.* **2012**, *5*, 325–334. [CrossRef]
281. Latorre, A.; Dagnac, T.; Lorenzo, B.F.; Llompart, M. Occurrence and stability of masked fumonisins in corn silage samples. *Food Chem.* **2015**, *189*, 38–44. [CrossRef] [PubMed]
282. Generotti, S.; Cirlini, M.; Dall'Asta, C.; Suman, M. Influence of the industrial process from caryopsis to cornmeal semolina on levels of fumonisins and their masked forms. *Food Control* **2015**, *48*, 170–174. [CrossRef]

Publisher's Note: MDPI stays neutral with regard to jurisdictional claims in published maps and institutional affiliations.

Journal of *Fungi*

MDPI

Review

Fungal Pigments and Their Roles Associated with Human Health

Lan Lin [1,2] and Jianping Xu [2,*]

[1] School of Life Science and Technology, Department of Bioengineering, Key Laboratory of Developmental Genes and Human Diseases (MOE), Southeast University, Nanjing 210096, Jiangsu, China; linl04@seu.edu.cn

[2] Department of Biology, McMaster University, Hamilton, ON L8S 4K1, Canada

* Correspondence: jpxu@mcmaster.ca; Fax: +1-905-522-6066

Received: 15 October 2020; Accepted: 9 November 2020; Published: 12 November 2020

check for updates

Abstract: Fungi can produce myriad secondary metabolites, including pigments. Some of these pigments play a positive role in human welfare while others are detrimental. This paper reviews the types and biosynthesis of fungal pigments, their relevance to human health, including their interactions with host immunity, and recent progresses in their structure–activity relationships. Fungal pigments are grouped into carotenoids, melanin, polyketides, and azaphilones, etc. These pigments are phylogenetically broadly distributed. While the biosynthetic pathways for some fungal pigments are known, the majority remain to be elucidated. Understanding the genes and metabolic pathways involved in fungal pigment synthesis is essential to genetically manipulate the production of both the types and quantities of specific pigments. A variety of fungal pigments have shown wide-spectrum biological activities, including promising pharmacophores/lead molecules to be developed into health-promoting drugs to treat cancers, cardiovascular disorders, infectious diseases, Alzheimer's diseases, and so on. In addition, the mechanistic elucidation of the interaction of fungal pigments with the host immune system provides valuable clues for fighting fungal infections. The great potential of fungal pigments have opened the avenues for academia and industries ranging from fundamental biology to pharmaceutical development, shedding light on our endeavors for disease prevention and treatment.

Keywords: fungal pigments; melanin; carotenoids; polyketides; azaphilones; antitumor; medical roles

1. Introduction

Since prehistoric times, fungi have played critical roles in human daily activities, such as baking, brewing (wines and beers), and processing dairy products. With the advent of new technologies, especially biotechnology, over the past 60 years, the use of fungi has increased significantly for the production of commercially important products, such as edible mushrooms, drinks, alcohols, organic acids, enzymes, and colorants. After the breakthrough discovery of penicillin in the 1930s, fungi have become a very significant source of pharmaceutical products which give rise to life-saving medicines, including antibiotics, anticancer drugs, and cholesterol-lowering agents [1]. Apart from the benefits that fungi provide, some fungi can also cause significant diseases to a variety of crops, livestock, pet animals, and humans. Accumulating evidence has revealed that the incidence of human fungal diseases has been increasing rapidly since the 1980s, and is associated with excessive morbidity and mortality, particularly among immunosuppressed and immunocompromised patients [2]. Among the 625 fungal species known to be pathogenic to humans, those in the genera *Aspergillus*, *Candida*, *Cryptococcus*, and *Trichophyton* cause diseases in over 300 million people globally [2]. These diseases can be superficial, subcutaneous, or systemic. Some are acute infections, but others can be long-lasting chronic infections (e.g., athlete's foot). Accordingly, the detailed biological processes for generating useful products from beneficial fungi and for virulence in pathogenic fungi are very active areas of research.

To the general public, when we talk about pigments in living organisms, we often think of the colorful flowers, insects, and birds. Indeed, pigments in these groups of organisms are the subjects of curiosity for people from diverse walks of life, including painters, photographers, writers, gardeners, and biologists. The biological roles for many of those pigments have been extensively studied. In contrast, unless we are in the woods and see colorful mushrooms, we often do not associate pigments with fungi, with the possible exception of "black" molds in spoiled foods. However, fungi are prolific producers of a myriad of pigments of different chemical structures and a diverse range of colors. The majority of well-studied fungal pigments are from fungi of four genera: *Aspergillus*, *Penicillium*, *Paecilomyces*, and *Monascus* [3,4]. When chemically categorized, fungal pigments are grouped into carotenoids, melanins, polyketides, azaphilones (polyketide derivatives), etc. [5,6].

Carotenoids constitute an abundant group of pigments in nature. They are found in all plants where they play an essential role in photosynthesis. In fungi, over 200 fungal species have been documented as capable of producing carotenes [7]. Carotenoids play very important roles in human health. They not only serve as the precursors of vitamin A in humans, but can also alleviate and/or prevent human age-related diseases, such as cataracts and macular degeneration. Furthermore, they can reduce the incidence of coronary heart disease and carcinomas, including lung, breast, prostate, and colorectal cancers [8]. The health-promoting effects of carotenoids are believed to be associated with their bioactivity as antioxidants and free radical scavengers.

Another broadly distributed group of fungal pigments is melanin. Melanin contributes significantly to the fungal capacity to survive and thrive in unfavorable habitats. Fungal melanin molecules are polyphenolic and/or polyindolic compounds with high molecular masses and negative charges. Due to its physiochemical properties, melanin can mediate an array of cellular functions, allowing fungal adaptation to diverse environmental factors, including ultraviolet (UV) light, heat, ionizing radiation, and oxidative stressors [9]. These properties make melanin a potential artificial stress-protecting agent (e.g., radioprotective) in bioinspired applications. However, the relevance of melanin to human health is probably best exemplified in human fungal pathogens as a virulence factor. For example, in the opportunistic human pathogens of the *Cryptococcus* genus, such as *Cryptococcus neoformans* and *C. gattii*, strains unable to produce melanin are avirulent or have significantly reduced virulence [10,11]. Similarly, in *Aspergillus fumigatus* as well as *Paracoccidioides brasiliensis*, the melanized cells exhibited elevated resistance to phagocytosis [12].

In this paper, we provide a comprehensive and updated review on fungal pigments that have demonstrated relevance to human health either directly or indirectly. We first describe the categories and chemical structures of these fungal pigments, including representative fungi that produce them. This is followed by descriptions of their biosynthesis pathways, how they interact with the human immune system, and the relationships between their structure and activities. We finish by providing a perspective on future developments.

2. Types of Fungal Pigments and Their Relevance to Human Health

The majority of organisms living on this planet synthesize some pigments. These biological pigments confer on our world a wide array of colors via the absorption and refraction of specific wavelengths of light. In nature there are many types of biological pigments, ranging from monomeric (such as carotenoids, flavonoids, luciferin, and heme/porphyrin-based pigments) to polymeric (i.e., melanin and humic compounds). Some pigments possess conjugated moieties (i.e., aromatics rings) that facilitate electronic resonances and mediate energy transfers within and between cells. The energy captured and/or reflected by pigments has been shown to play multiple biological roles ranging from the maintenance of life, including the utilization of solar energy for metabolic functions and protection against radiation damage, to camouflage, as well as mate and pollinator attraction. Below, we describe the main types of fungal pigments relevant to human health.

2.1. Carotenoids

In addition to being produced by photosynthetic organisms ranging from cyanobacteria to flowering plants, carotenoids are also synthesized in a variety of heterotrophic microbes including fungi [13]. Furthermore, carotenoids are present in animals that are unable to synthesize them de novo, but instead acquire them from their diet.

Carotenoids contain an aliphatic polyene chain usually composed of eight isoprene units that include light-absorbing conjugated double bonds, giving rise to characteristic colors such as yellow, orange, or red [14,15]. Representative structures of carotenoids are shown in Figure 1. Some carotenoids are precursors of vitamin A, specifically those with β-ring end groups, such as β-carotene, zeaxanthin, and β-cryotoxanthin [8]. From the standpoint of human health, carotenoids are among the bioactive products with significant medical value. For example, carotenoids have been found to lower the risks of diseases, including cancer, cardiovascular diseases, and age-related eye disorders, such as macular degeneration and cataracts [16].

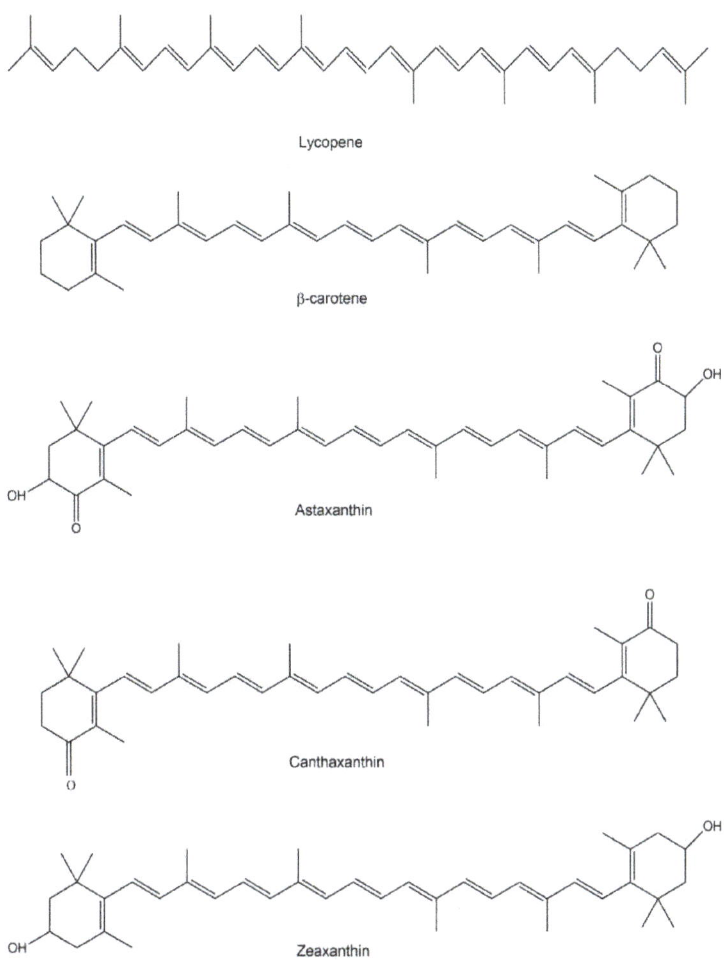

Figure 1. Structure of representative carotenoids.

Carotenoid-producing fungi are very diverse, including *Rhodotorula mucilaginosa* [17], *Blakeslea trispora, Phycomyces blakesleeanus, Mucor circinelloides, Fusarium sporotrichioides* [18], *Rhodosporium paludigenum*, and *Rhodotorula glutinis* [19]. Carotenoids protect fungi from oxidative stress and non-ionizing irradiation such as UV light. In addition, carotenoids are intermediates essential for synthesizing other biological molecules including bioactive apocarotenoids and related constituents. For instance, in many filamentous fungi, retinol, the precursor of vitamin A as well as an essential component of rhodopsins (membrane-bound photoreceptors), is produced via the oxidative cleavage of β-carotene [20]. In Mucorales, β-carotene is an intermediate during the synthesis of trisporoids, apocarotenoid derivatives that include the sexual hormones the trisporic acids. In addition, trisporoids have been proposed as substrates in the synthesis of sporopollenin, one of the most chemically inert biological polymers primarily found in the tough outer walls of plant spores and pollen grains [15]. In some fungi, such as the filamentous species *Blakeslea trispora* [21] and the yeast *Rhodotorula* spp. [22], the β-carotene biosynthetic pathway also produces lycopene, a carotenoid intermediate. Lycopene is a nutrient with antioxidant properties. It was originally identified in plants such as tomato and watermelon, giving their fruits the characteristic red and pink colours. Lycopene has been linked to several human health benefits ranging from improved cardiovascular health to protection against sunburns and certain types of cancers.

Aside from β-carotene and lycopene, another fungal carotenoid, astaxanthin, has attracted significant attention for its promise in the pharmaceutical, food and cosmetics industries. For example, astaxanthin has shown promise for treating Alzheimer's disease, Parkinson's disease, stroke, high cholesterol, age-related macular degeneration (age-related vision loss), and preventing cancer. *Xanthophyllomyces dendrorhous* (synonym *Phaffia rhodozyma*), a red- or orange-pigmented basidiomycetous yeast initially isolated from Japanese and Alaskan tree-exudates decades ago, is capable of producing astaxanthin [23,24]. Astaxanthin is synthesized via the mevalonate pathway starting with a condensation of two molecules of geranylgeranyl pyrophosphate, a four-step desaturation, cyclization, and a final 4-ketolation plus 3-hydroxylation [25,26]. In addition to astaxanthin, *X. dendrorhous* can also utilize the mevalonate pathway to produce several other types of carotenoids including β-carotene, canthaxanthin, and zeaxanthin. Canthaxanthin is used to reduce sensitivity to sunlight (photosensitivity) experienced by people who have a rare genetic disease called erythropoietic protoporphyria (EPP) [27]. In these people, sunlight can cause skin reactions such as rash, itch, and eczema. Similarly, zeaxanthin also protects human cells from the harmful effects of certain light sources such as the Sun. However, unlike canthaxanthin, that is preferentially deposited in the skin, zeaxanthin is an eye pigment that, once inside the body, accumulates in the eyes and helps to build a yellow-colored pigment shield to protect the eye cells. Indeed, the ability of *X. dendrorhous* to produce such a diversity of bioactive carotenoids makes it a potential model host to be engineered as a cell factory for the production of medically and industrially important carotenoids [28].

2.2. Melanin

Broadly speaking, melanins are heterogeneous polymers derived by the oxidation of phenols and subsequent polymerization of intermediate phenols and their resulting quinones. Fungal melanins are a subset of natural melanins and are polyphenolic and/or polyindolic compounds with high molecular masses and negative charges. Fungal melanins can appear dark green, brown, or black. In fungal cells, melanin serves as an antioxidant and a scavenger of reactive nitrogen species (RNS) and reactive oxygen species (ROS). Commonly found in pathogenic fungi, melanin pigments are thought to provide those pathogens protection from UV and other radiation and aid in evasion from the host's immune system attacks [29]. Fungal melanins belong to three major types: 1,8-dihydroxynaphthalene (DHN) melanin, 3,4-dihydroxyphenylalanine (DOPA)-melanin (also called eumelanin), and pyomelanin [30,31]. DHN melanin is produced from acetyl-coenzyme A via the polyketide synthase pathway. The fungus *A. fumigatus* is able to synthesize DHN melanin, which contributes to the gray-green appearance of its conidia. The deletion of its polyketide synthase PksP in *A. fumigatus* gave rise to colorless spores with weakened virulence [32]. Eumelanin is converted from *o*-diphenolic or *p*-diphenolic substrates by a polyphenol oxidase (laccase). In *Candida albicans*, eumelanin particles are observed in vitro and in infected mammalian tissues, such as murine kidney and human skin [33]. In this ascomycete fungus, melanin is externalized in the form of electron-dense melanosomes that are extracellularly secreted or loosely attached to the cell wall surface through binding with chitins [34]. Although *C. albicans* has laccase activity [33], the association of melanin with pathogenicity remains to be clarified for this fungus. Pyomelanin is an extracellular water-soluble pigment, in contrast to both DHN- and DOPA-melanins that are cell wall-immobilized and insoluble in water [31]. Pyomelanin is produced by the polymerization of homogentisic acid degraded from L-tyrosine/L-phenylalanine. In *A. fumigatus*, pyomelanin plays an important role in the germination of conidia and in the defensive responses toward external oxidative stresses [31].

Apart from the above-stated three major types of fungal melanins, there is another pathway of melanin biosynthesis in mushrooms such as the common cultivated mushroom *Agaricus bisporus* and others [35]. In these mushrooms, melanin is formed from a benzoquinone. Benzoquinone is converted from the precursor γ-glutaminyl-4-hydroxybenzene (GHB) by tyrosinase. GHB is the main phenolic compound in *A. bisporus* fruiting bodies and spores [36].

In addition to its involvement as a virulence factor in fungal pathogens, melanin enables fungi to survive and thrive in adverse and even extreme environments, providing protection against desiccation, non-ionizing (i.e., UV light) and ionizing radiation, as well as oxidative and nitrosative stresses [37,38]. Over the past decade, melanin pigments and their subunits have kindled great interest as soft biocompatible functional materials with antioxidant characteristics for engineering high-performance, low-impact biocompatible optoelectronic devices, such as light emitting diodes, memory devices, etc. [39]. Intriguingly, halophilic fungal isolates of two species, *Trimmatostroma salinum* and *Phaeotheca triangularis*, from the eastern coast of the Adriatic Sea, have been found to produce melanin in sodium chloride solutions at saturated concentrations [40]. Similarly, high-dose radiation resistant melanin was found in an Antarctic desert-dwelling (or desert-inhabiting) fungus, *Cryomyces antarcticus*. These fungal extremophiles could serve as promising sources of melanin with excellent physicochemical properties for a variety of industrial and medical applications.

2.3. Polyketides

Many fungal pigments are polyketide-based compounds. This group of fungal pigments is abundantly produced by most ascomycetous fungi, including the filamentous ascomycete genera *Neurospora* and *Monascus* [41].

Fungal polyketide pigments are made of tetraketides and octoketides which have eight C_2 units giving rise to the polyketide chain. Anthraquinones (including hydroxyanthraquinones), and naphthoquinones are representative classes which display an arsenal of various colors [42,43]. The basic structure of the anthraquinone class of pigments is a polycyclic aromatic hydrocarbon which is derived from anthracene or phthalic anhydride, i.e., the merger of three benzene rings. Anthraquinone itself is highly insoluble and is generally used in the manufacture of dyes for the textile and pulp industries [44,45]. However, other members of the anthraquinone class have been used in the pharmaceuticals industry.

Approximately 700 representatives of anthraquinones have been found in fungi, plants, and lichens, conferring a yellow, orange, or brown color to the mycelium of microscopic fungi, the fruiting bodies of macroscopic fungi (mushroom), as well as lichens [46]. The variety of anthraquinones results from the presence of different substituents, like -OH, -CH_3, -OCH_3, -CH_2OH, etc. as well as the reduction of carbonyl groups or double bonds in the benzene ring. Anthraquinones produce a yellow color, whereas the substituents produce various hues of the molecules ranging through yellow, orange, red, brown, and violet [47]. The most widely distributed variants found among fungi are 1, 8-dihydroxy and 1, 5, 8- or 1, 6, 8- trihydroxy anthraquinone derivatives [48]. However, substituents at other positions of the ring have been recently identified. For example, chrysophanol is a trihydroxyanthraquinone with a methyl substituent at C-3. It has shown antiviral and anti-inflammatory activities. Both chrysophanol and helminthosporin have shown inhibitory activity against cholinesterase with helminthosporin showing permeability across the blood-brain barrier [49]. Cholinesterase is a key enzyme involved in the development of Alzheimer's disease. Most anthraquinone-producing fungi produce a mixture of pigments of the quinone class. *Aspergillus crisfafus*, for instance, has been found to synthesize as many as 15 pigments related to anthraquinone. A strain of *Curvularia lunata* has been characterized that produces a mixture of three anthraquinone derivatives: chrysophanol, helminthosporin, and cynodontin, with cynodontin (1, 4, 5, 8-tetrahydroxy-3-methylanthraquinone) comprising more than 70% of the mixture and showing potential as a pharmaceutical with antifungal activity [44].

Chrysophanol, helminthosporin, and cynodontin are all hydroxyanthraquinones (HAQNs), derivatives of anthraquinone via the substitution of hydrogen atoms by hydroxyl groups plus the addition of other functional groups. Anthraquinones including HAQNs have been found to be produced by such fungi as *Aspergillus* spp., *Eurotium* spp., *Fusarium* spp., *Dreschlera* spp., *Penicillium* spp., *Emericella purpurea*, *Culvularia lunata*, *Mycosphaerella rubella*, *Microsporum* spp., etc. [48].

Filamentous fungi, such as *Penicillium* and *Aspergillus*, synthesize HAQNs through the polyketide pathway. This process has been employed for the production of natural food-grade colorants [50]. Apart from their usage as colorants in the food and cosmetics industries, they have been found to possess antiviral activities. For instance, emodin has been extracted from *Penicillium citrinum*, *P. islandicum*, *Aspergillus glaucus* and *A. variecolor* [51,52] and aloe-emodin from *Penicillium oxalicum* [53]. Emodin and its congener aloe-emodin are known antiviral agents [54]. Aspergilol H and I, HAQNs produced by *Aspergillus versicolor*, have been reported to have antiviral effects against HSV-1 [55]. Notably, HAQNs are able to absorb visible light and display colour, with the colours varying based on the position and number of the hydroxyl substituents in the different rings [56]. HAQNs are relatively stable and they exhibit superior brightness in comparison to azo pigments.

Naphthoquinone is a class of organic compounds structurally related to naphthalene. Naphthoquinones have very significant pharmacological activities against bacterial, fungal, viral, and insect pathogens and pests. In addition, they have anti-inflammatory, and antipyretic properties. *Fusarium* spp. have shown capable of producing a diversity of naphthoquinones with a broad spectrum of biological activities [57,58]. For instance, a compound produced by *Fusarium fujikuroi* and other fungi, bikaverin, a red heterotetracyclic pigment with the chemical structure of 6,11-dihydroxy-3,8-dimethoxy-1-methylbenzo[b]xanthene-7,10,12-trione, has shown anti-neoplastic activity, suggesting its potential as a pharmaceutical agent against lymphoma, carcinoma, and sarcoma, etc. [59–62].

2.4. Azaphilones

Azaphilones are a structurally variable group of fungal secondary metabolites (polyketide derivatives) [63], namely pigments with a highly oxygenated pyrone-quinone bicyclic core usually known as isochromene and a chiral quaternary centre [64,65]. Both a pyrone-quinone backbone and a chiral quaternary centre (Figure 2) are essential for compounds to be classified as azaphilones [66]. Azaphilone is so designated because of its ready insertion of nitrogen. For example, ammonia readily converts the pyran oxygen of monascorubrin into the violet-colored nitrogen analog monascorubramine (Figure 3) [67].

The azaphilone pigments are produced by a plethora of the ascomycetous and basidiomycetous fungi, particularly the former [68], including the genera *Aspergillus*, *Penicillium*, *Chaetomium*, *Talaromyces*, *Pestalotiopsis*, *Phomopsis*, *Emericella*, and *Epicoccum*, as well as *Monascus* and *Hypoxylon*, where they give rise to the yellow, red, or green hues of fruiting bodies and/or mycelia [69].

Fungal azaphilones, structurally bearing the isochromane-like ring, possess a broad-spectrum of biological activities including monoamine oxidase inhibition, phospholipase A2 inhibition, tumor promotion suppression, gp120-CD4 binding inhibition, and acyl-CoA: cholesterolacyltransferase (AGAT) inhibition.

However, the relationships between the structures of azaphilones, their spectrum of activities, and the corresponding mechanistic modes of action remain elusive [66].

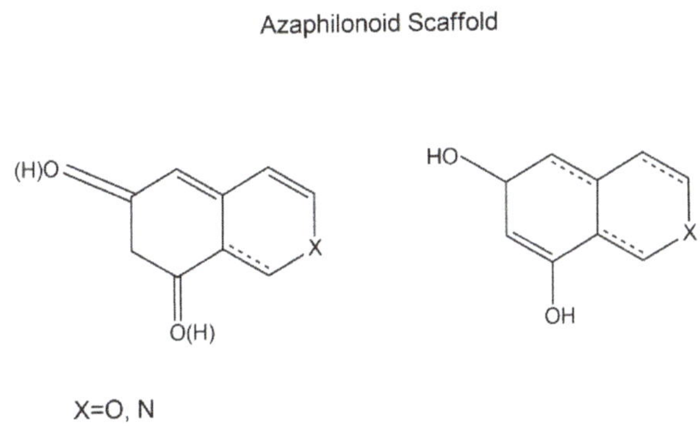

Figure 2. Azaphilonoid scaffold.

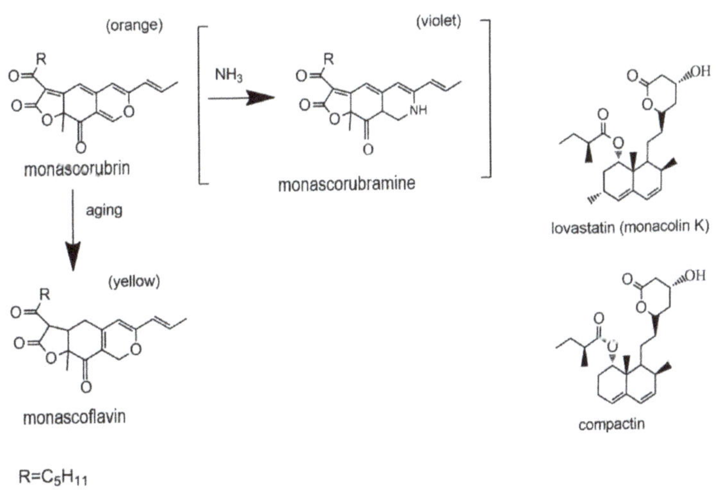

Figure 3. Azaphilone pigments derived from the yeast *Monascus*.

Monascin and ankaflavin (Figure 4B) are two yellow pigments of the azaphilonoid group. The fungal genus *Monascus*-derived monascin and ankaflavin have been found to potently inhibit preadipocyte differentiation, and promote lipolysis of mature adipocytes [70]. Furthermore, these two pigments are able to reduce blood lipids as well as lower the synthesis and accumulation of triglycerides (TG) [71]. For example, Lee et al. (2013) investigated the effects of eight-week administrations of monascin and ankaflavin on obesity factors using obese rats fed with a high-fat diet [72]. Their work illustrated the involvement of these two *Monascus* azaphilones in the modulation of preadipocyte differentiation, lipogenesis, and lipid absorption, highlighting their potential in the development of hypolipidemic drugs.

Monacolin K (viz. lovastatin) is a statin compound and previously thought to be a hypolipidemic component of *Monascus*-fermented products. While earlier studies suggested anti-obesity effects of monacolin K (Figure 4A), recent investigations indicated that monascin and ankaflavin (Figure 4B), rather than monacolin K, were likely the major players in the *Monascus*-mediated amelioration of diabetes and fatty liver [73].

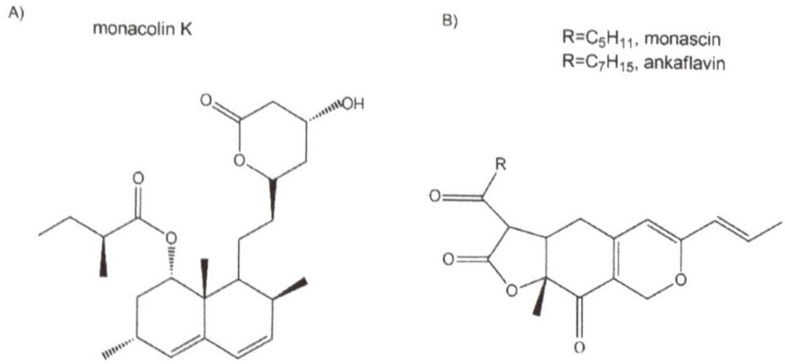

Figure 4. Chemical structure of monacolin K, monascin and ankaflavin. (**A**) Monacolin K; (**B**) Monascin and ankaflavin.

Indeed, there is mounting evidence that a long-term high-dosage administration of monacolin K might lead to numerous side effects such as rhabdomyolysis and reduction in the level of coenzyme Q_{10} (CoQ_{10}). Furthermore, although monacolin K reduces the serum total cholesterol (TC), triglycerides (TG) and low-density lipoprotein cholesterol (LDL-C) levels, it also substantially reduces high-density lipoprotein cholesterol (HDL-C) levels. On the other hand, Lee et al. (2010) showed that monascin and ankaflavin could increase HDL-C levels significantly while imposing no damage on the liver or kidneys [68]. In addition, monascin has been shown to protect the liver from chemical damage via its anti-inflammatory effect [74]. Other work with monascin has indicated that it significantly inhibits peroxynitrite ($ONOO^-$; PN), and ultraviolet light B (UVB)-induced skin carcinogenesis [75].

Together, the above results suggest that monascin and ankaflavin possess hypolipidemic, anti-atherosclerosis, antioxidative, and anti-inflammatory activities, thereby rendering them promising compounds that may be of value in the prevention and treatment of cardiovascular diseases [73].

Apart from *Monascus* species, the production of azaphilone pigments by other fungal species has also been documented [76]. For instance, chaetoviridins, a class of azaphilones synthesized by *Chaetomium globosum*, have been found to have strong antifungal activities. To date, four chaetoviridins, chaetoviridin A—D, have been characterized. Studies with mice have demonstrated that chaetoviridin A is able to suppress tumor progression via 12-O-tetradecanoylphorbal-13-acetate in two-phase carcinogenesis [77]. Independent work by Tomoda et al. revealed that chaetoviridin A and B might exert inhibitory effects on cholesteryl ester transfer protein [78].

2.5. Other Fungal Pigments

The pigment sclerotiorin has been isolated from cultures of *Penicillum sclerotiorum* 2AV2, which was recovered from Amazonian soil [6]. Sclerotiorin (Figure 5) has been found to have a variety of biological activities. Some of the biological activities most pertinent to human health include inhibitory effects on aldose reductase [79], endothelin receptor binding activity [80], antimicrobial effects including antifungal activity [17,81], apoptosis-triggering actions towards HCT-116 cancer cells [82], and the inhibition of integrase and protease of HIV-1 [83].

Studies on *Penicillum chrysogenum* have identified a polyketide synthase (PKS) gene required for the biosynthesis of sorbicillinoid, a yellow pigment [84]. The authors demonstrated that a highly reducing PKS enzyme, encoded by Pc21g05080 (*pks13*), is indispensable for the production of polyketide precursors such as sorbicillinol and dihydrosorbicillinol as well as their derivatives such as bisorbicillinoids. PKSs are also involved in the biosynthesis of sorbicillinoid in *Acremonium chrysogenum*, the fungal species used in the industrial production of cephalosporin C, an antibiotic in current clinical use [85].

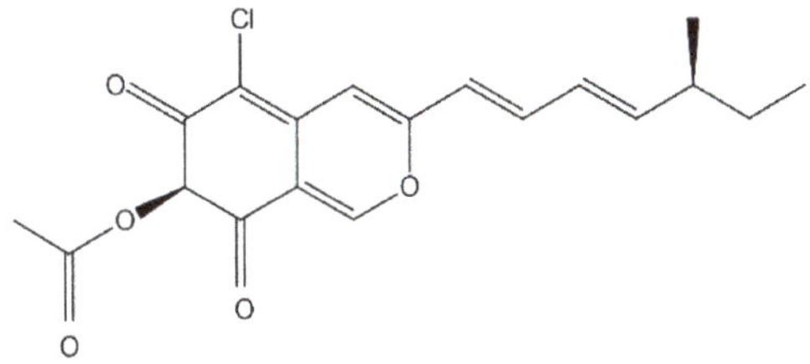

Figure 5. Chemical structure of the pigment sclerotiorin produced by *Penicillum sclerotiorum* 2AV2.

3. Biosynthesis of Fungal Pigments

So far, four major pathways have been identified as responsible for the biosynthesis of the main fungal pigments, including the polyketide synthetic pathway, the shikimate pathway, the terpenoid synthetic pathway, and the nitrogen-containing metabolite pathway.

3.1. Polyketide Synthetic Pathways

The polyketide synthetic pathway is involved in producing many fungal pigments with human health relevance. For example, there is increasing evidence that fungal pigments such as melanin, quinones, flavins, ankaflavin, and azaphilones, all relevant to human health [5,6], involve the polyketide synthesis pathway [86,87]. This pathway has been studied in diverse fungal species, including those in the genera *Monascus*, *Fusarium*, *Alternaria*, and *Epicoccum* [64]. The core polyketide synthetic pathway is shown in Figure 6. Fungal pigments are produced by this pathway via repetitive Claisen condensations of an acyl-coenzyme A (CoA) starting unit with malonyl-CoA elongation units in a manner analogous to fatty acid biosynthesis. The polyketide pathway generates either aromatic ketides or fatty acids. The extending chain of the aromatic ketides is stabilized by cyclization reactions and partial reduction. However, in the fatty acids, the carbonyl groups of the chain are reduced prior to the addition of the next C_2 group. One of the major differences between the two metabolic routes is that polyketides have varied degrees of β-keto processing. Specifically, some are not or only partially reduced, giving rise to the formation of (poly-)cyclic aromatic compounds. On the other hand, they may be significantly reduced, yielding linear or macrocyclic, non-aromatic carbon skeletons.

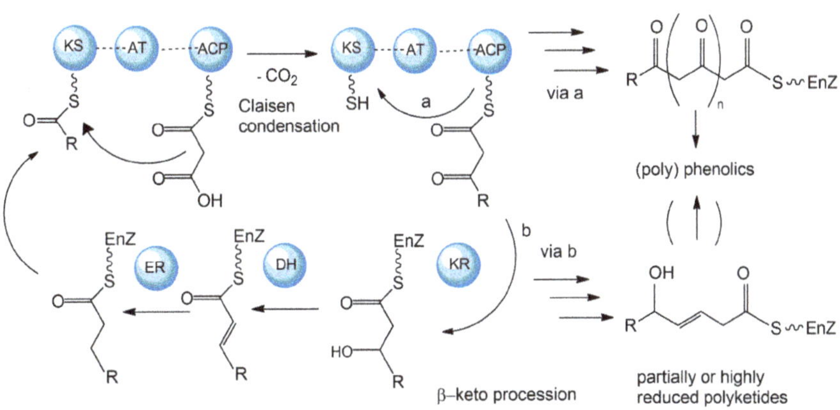

Figure 6. Mechanisms of fungal polyketide biosynthesis.

Aromatic ketides of this pathway include tetra-, hepta-, octa- and higher number ketides. Fungi possess a variety of pigments of the octaketide origin based on the anthra-9,10-quinone backbone with both rings being substituted. Anthraquinones are representatives of this type of pigment. In many cases anthraquinones are found in fungi in the corresponding colourless reductive states (anthranol, anthrone, anthrahydroquinone, and oxanthrone derivatives) that may be present in the forms of diverse glycoside conjugants. However, many naturally occurring, coloured anthraquinones are oligomers made by the coupling of two or more anthraquinone molecules. These oligomers differ in the number and positions from which monomers and amino acids are ligated to generate a diversity of fungal pigments [88,89].

A large number of enzymes have been identified in the fungal polyketide synthesis pathway. These include a core set of enzymes, including ketosynthase (KS), acyl transferase (AT), and acyl carrier protein (ACP) domains. Further, optional β-keto processing steps may be catalyzed by keto reductase (KR), dehydratase (DH) and enoyl reductase (ER) domains. Other optional ancillary domains involve cyclase (CYC) [90] and methyl transferase (MT) activities [91]. Based on their architecture and the presence or absence of additional β-keto processing domains, fungal PKSs are categorized into non-reducing or aromatic (NR-PKS), partially reducing (PR-PKS), and highly reducing PKS (HR-PKS).

The synthesized fungal polyketides can vary in their chain length, the degree of β-keto processing, and cyclization. Moreover, the tremendous structural diversity of polyketides is further obtained from derivatization of the polyketide carbon backbone by alkylation, acylation, and oxygenation, and by post-PKS modification or tailoring. Genome sequence analyses so far have revealed that all genes essential for fungal polyketide biosynthesis are clustered, including the PKS genes, genes encoding enzymes associated with tailoring, as well as regulatory genes [92].

While most fungal polyketide pigments are synthesized as aromatic ketides, the biosynthesis of azaphilones uses both the polyketide pathway and the fatty acid synthesis pathway. The polyketide pathway assembles the main polyketide chain of the azaphilone pigments from acetic acid (the starter unit) and five malonic acid molecules (the chain extender units) in a conventional way to generate the chromophore structure. The fatty acid synthesis pathway produces a medium-chain fatty acid (octanoic or hexanoic acid) that is then bound to the chromophore by a transesterification reaction [93,94].

3.2. *Shikimate Pathways*

Aromatic amino acids act as the precursor of various fungal secondary metabolites such as pigments and vitamins [95]. In the fungal phylum *Basidiomycota*, tyrosine is the precursor of a distinct class of pigments, the betalains, solely present in the genera of *Amanita* and *Hygrocybe* [96,97]. In the phylum *Ascomycota*, tyrosine-derived pigments, such as melanin and tyrosine betaine, are thought to contribute to stress tolerance (e.g., temperature, radiation) [98,99] and pathogenicity [100,101].

The shikimate pathway is present in the prokaryotes, microbial eukaryotes and plants studied. However, it is absent in those animal species investigated [102]. This pathway links carbohydrate metabolism to the biosynthesis of aromatic compounds. Metabolically, the shikimate pathway is a seven-step route utilized by fungi for the biosynthesis of aromatic amino acids like phenylalanine (Phe), tyrosine (Tyr), and tryptophan (Trp), as well as para-aminobenzoic acid, via the central intermediates shikimic and chorismic acids [86]. The first step involves the condensation of the glycolytic intermediate phosphoenol pyruvate and pentose phosphate pathway intermediate erythrose-4-phosphate to yield a seven-carbon heterocyclic compound, 3-deoxy-D-arabinose-heptulosonate-7-phosphate derivative (DAHP). The second step involves the generation of a highly substituted cyclohexane derivative, 3-dehydroquinate, by the replacement of the exocyclic C_7 of DAHP by the ring oxygen. The remaining five steps involve the introduction of a side chain and two of the three double bonds that convert this cyclohexane into a benzene ring, the core of aromatic amino acids [5]. The metabolic routes may vary for different classes of pigments.

The pigments produced via the shikimate pathway are generally water-soluble phenolic compounds, including terphenyls and pulvinic acids. Studies of *p*-terphenyls as a family of the mushroom pigments were initiated in 1877. Isolation of polyporic acid, atromentin and thelephoric acid represented the inception of the chemical investigation of fungal pigments. The elucidation of the structures of polyporic acid and atromentin by Kögl is recognized as a milestone in organic chemistry [103]. It has been demonstrated that some terphenyls display biological activities, such as immunosuppressive, antithrombotic, anticoagulant, neuroprotective, 5-lipoxygenase inhibitory (for the treatment of inflammatory bowel disease), and cytotoxic activities [104].

Many fungi have evolved pathways utilizing the aromatic products of shikimate metabolism (Figure 7). The initial steps in the biogenesis of *p*-terphenyls are the well-known reactions of primary metabolism flowing from shikimate to chorismic acids and subsequently to arylpyruvic acids. Experiments involving the feeding of ^{13}C- and ^{14}C-labeled precursors to fungal cultures revealed that *p*-terphenyls are assembled by initial condensation between two molecules of unbranched phenylpropanoid precursors, either phenylpyruvic acid or phenylalanine. Previous studies also revealed the involvement of 4-hydroxyphenylpyruvic acids or tyrosine in the initial condensation during the biosynthesis of terphenylquinones such as atromentin [105]. Atromentin is known as a key intermediate for further conversions, for instance, to more highly hydroxylated terphenylquinones and pulvinic acids [106,107].

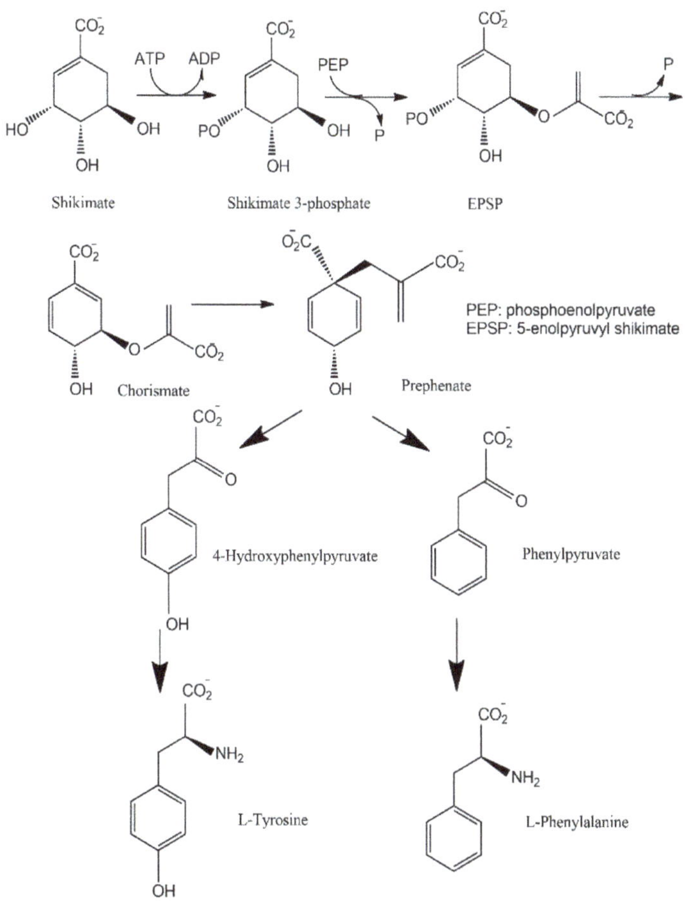

Figure 7. Shikimate pathway leading to biosynthesis of *p*-terphenyls.

3.3. *Terpenoid Synthetic Pathways*

Biochemically, the carotenoids are terpenoids [15]. Accordingly, the synthesis of carotenoids utilizes the terpenoid synthetic pathways deriving from the condensation of C_5 isoprene units [108]. The common terpenoid precursor is isopentenyl pyrophosphate (IPP), which is synthesized either via the mevalonate pathway (generated from hydroxymethylglutaryl coenzyme A, HMG-CoA), or via the non-mevalonate pathway (generated from the condensation of pyruvate and glyceraldehyde 3-phosphate [109]. So far, available research has demonstrated that fungal IPP is produced via the mevalonate pathway, whereas in bacteria and photosynthetic species IPP is produced via the non-mevalonate pathway. The early biosynthetic steps involve the sequential additions of IPP (isoprene, C_5) units to yield geranyl pyrophosphate (GPP, C_{10}), farnesyl pyrophosphate (FPP, C_{15}), and geranylgeranyl pyrophosphate (GGPP, C_{20}) [108,109]. The initial compound possessing the typical aliphatic carotenoid-like structure is 15-cis-phytoene (a colorless molecule), consisting of a symmetrical polyene chain generated via the condensation of two GGPP units catalyzed by phytoene synthase (Figure 8). The light-absorbing characteristics of carotenoids are attributed to the presence of a chromophore, which is comprised of an array of conjugated double bonds via the catalysis of desaturases. Different subsequent chemical modifications, usually the introduction of a cyclic end group (with β and ε rings being the most frequently introduced) into at least one of the ends of the molecule, and/or oxidative reactions (for instance, hydroxylation, carboxylation, epoxidation, esterification, etc.), may contribute to the huge arsenal of carotenoids [108].

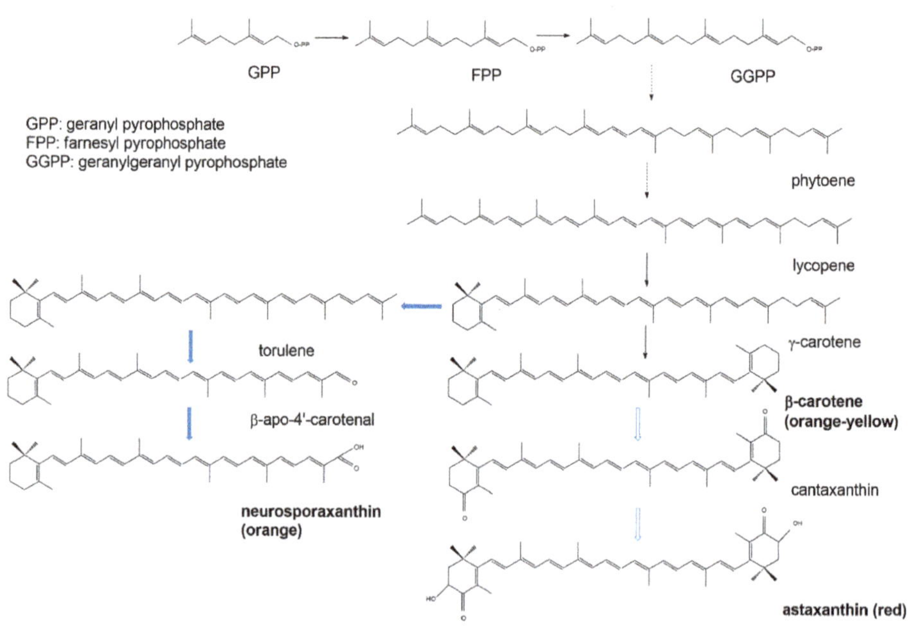

Figure 8. Simplified scheme of the biosynthetic pathways for β-carotene (black arrows) and the astaxanthin and neurosporaxanthin (blue arrows) from GPP.

3.4. *Nitrogen-Containing Metabolite Pathways*

Chalciporone (Figure 9) is a type of 2H-azepine alkaloid pigment produced by the mushroom *Chalciporus piperatus* (Basidiomycetes). This pigment is believed to act as a deterrent to insects and other predators and thereby can potentially protect the mushroom. The biosynthesis of chalciporone was illustrated in 2001 [110]. In their experimentations, the radioactive labelled sodium [U-$^{13}C_2$] acetate was not incorporated into chalciporone by *C. piperatus*. In contrast, the supplementation of a mixture of [U-^{13}C]-labelled fats led to chalciporone, in which seven intact acetate units were incorporated (*ca.* 10 atom % enrichment) between C_3 and C_{16}. However, the enrichment of the C_1 methyl group and the adjacent carbon (C_2) was not observed during this experiment, suggesting that the CH_3–CH–N moiety was derived from an α-amino acid. This was confirmed by the observation that administration of a mixture of [U-^{13}C]-labelled amino acids to *C. piperatus* gave rise to the enrichment of both C_1 and C_2 and of all the other carbon signals in the spectrum of chalciporone [107]. The work clearly demonstrated that the carbon backbone of chalciporone is generated from an amino acid plus seven acetate (=malonate) units.

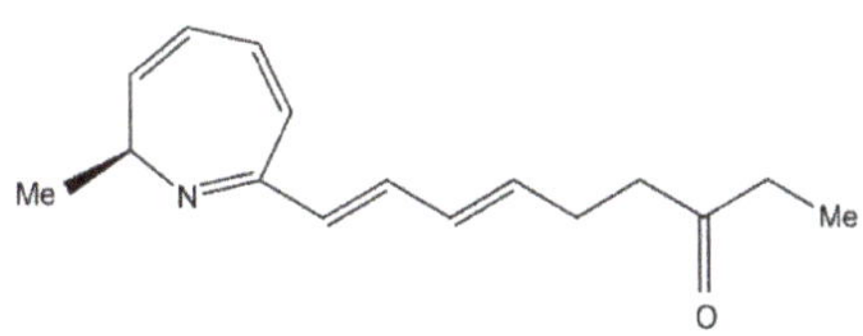

chalciporone (yellow)

Figure 9. Chalciporone.

The wood-rotting fungus *Pycnoporus cinnabarinus* (Basidiomycetes) can produce a red pigment, cinnabarinic acid (Figure 10), via oxidative dimerization of the precursor 3-hydroxyanthranilic acid in sporocarps as well as in culture broth [111]. This reaction is catalyzed by laccase and necessary for the production of antibacterial compounds by the fungus. Cinnabarinic acid shows inhibitory effects towards several Gram-positive bacteria of the *Streptococcus* genus. It is known that cinnabarinic acid shares structural homology with the antibiotic group of actinomycins (e.g., actinomycin D) produced by *Streptomyces* spp., having two cyclic pentapeptides linked to the phenoxazinone chromophore.

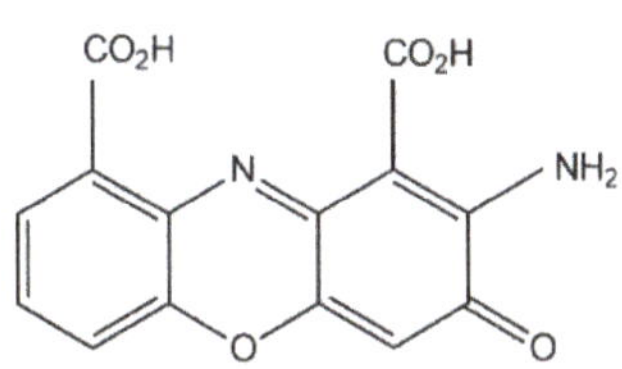

cinnabarinic acid (red)

Figure 10. Cinnabarinic acid.

Two indole pigments were isolated as free radical scavengers from the fruiting bodies of the mushroom *Agrocybe cylindracea* [112]. Based on spectroscopic data, they were identified as 6-hydroxy-1*H*-indole-3-carboxaldehyde and 6-hydroxy-1*H*-indole-3-acetamide (Figure 11). 6-hydroxy-1*H*-indole-3-acetamide is an amide derivative of 6-hydroxyindole-3-acetic acid. A previous study identified that 6-hydroxyindole-3-acetic acid could be produced through the transformation of indole-3-acetic acid by *Aspergillus niger* [113]. These two indolic compounds were found to possess potent inhibitory activity on lipid peroxidation in rat liver microsomes, with IC_{50} values of 4.1 and 3.9 μg per ml, respectively [109].

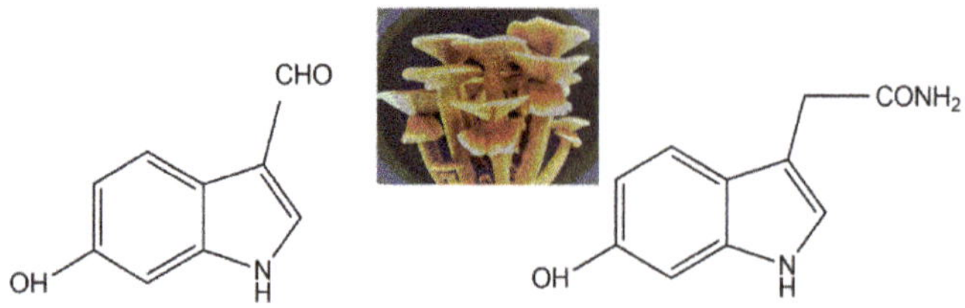

Figure 11. The indolic compounds 6-hydroxy-1H-indole-3-carboxaldehyde (**left**) and 6-hydroxy-1H-indole-3-acetamide (**right**) isolated from the fruiting bodies of *Agrocybe cylindracea* (middle, insert image).

4. Interaction of Fungal Pigments with the Host Immune System

One of the most interesting defense mechanisms of fish may involve non-specific immune responses that have evolved due to their long-term associations with fungi [114]. Specifically, the prevalent colonization of *Rhodotorula mucilaginosa* (a red yeast) in the fish gastrointestinal (GI) tract might have contributed to the stimulation of non-specific mechanisms of host immunoprotection as well as the maturation of the host GI tract [115]. Spectrophotometric analysis in wild fish (*Abramis brama, Rutilus rutilus, Perca fluviatilis*) indicated the presence of a set of three pigments—β-carotene, torularhodin, and torulene, which are produced by the commensal *Rh. mucilaginosa*. In addition to acting as scavengers of free radicals, these fungus-derived pigments are hypothesized to play a role in the immunostimulation of the host via enhanced activation of T lymphocytes, supporting the action of macrophages [116] or antimicrobial activity [115].

For decades, an accumulating body of evidence has revealed that carotenoids or their derivatives are involved in the activation of thymocites [117], the expression of immune-associated genes [118], and the increase in membrane fluidity [119], which represent vitally important functions in mounting immune responses in animals. Notably, the immunostimulatory effects of carotenoids are thought to be distinct from their antioxidant property [116]. Earlier investigations indicated that β-carotene and the non-provitamin A carotenoids are able to enhance cell-mediated and humoral immune responses in mammals [116]. It is not surprising that β-carotene-producing red yeast *Rh. mucilaginosa* might be beneficial to its hosts, in this case the wild fish, presumably through immuno-stimulation.

Given that (i) immunostimulatory effects of carotenoids in animals are thought to be independent of their antioxidant property [116], and (ii) innate immunity relies on effectors which produce cytotoxic molecules that may not only kill pathogens, but also harm host tissues, the mechanisms underlying the roles of carotenoid pigments in the well-being of animals need to be elucidated. Although the functionality of dietary antioxidants in invertebrate immunity is not fully understood, it has been found in vertebrates that carotenoids can scavenge cytotoxic radicals formed during the immune response. Carotenoids may consequently decrease the self-harming cost of immunity. A positive correlation between the effectiveness of the innate immune defense and the levels of circulating carotenoid might therefore be expected. In accordance with this hypothesis, a study using the amphipod species *Gammarus pulex* showed that the maintenance and use of the prophenoloxidase system were highly correlated with the carotenoid level in the haemolymph within the natural populations of this crustacean [120].

The enhanced immunity by carotenoids has also been demonstrated in mammals [121] where mice administered *Rhodotorula glutinis*-derived carotenoid could survive for 2 weeks post lethal-dose challenges of pathogenic *Pseudomonas aeruginosa* and *C. albicans*. This result points to the potential use of fungal carotenoids as food additives or dietary supplements to boost the human immune response against microbial infections.

Like carotenoids, melanins are ubiquitously produced by fungi, particularly pathogenic fungi such as *Cryptococcus neoformans*, *Aspergillus fumigatus*, and *Candida albicans*. As described previously, melanins in fungal pathogens contribute to fungal pathogenesis towards mammalian cells [10]. The melanin pigments contribute to antigen masking to circumvent recognition by host immune system and to the survival of fungal pathogens against phagocytosis [122]. For example, phagocytosed *C. neoformans* produces melanin to protect against the oxidative environment inside the phagolysosome [122]. For *A. fumigatus*, it is believed that melanins protect the conidia against reactive oxygen species, mask the recognition of various *A. fumigatus* pathogen-associated molecular patterns (PAMPs) by the host, inhibit macrophage apoptosis and phagolysosome fusion, and attenuate the host immune response. The protective effect of melanin might also be due to its role in modifying the surface properties of conidia and vegetative cells by altering the charge and hydrophobicity, thereby reducing the effectiveness of the immunological functions of the host [123]. For example, in contrast to the wild type conidia, melanin-deficient mutant conidia can activate human DCs (dendritic cells) and subsequent cytokine production, leading to their elimination by the host immune system.

The protective effects of melanin for fungal pathogens have also been demonstrated for autophagy. Autophagy plays an important role in host immunity to microbial pathogens. The autophagy system targets pathogens in phagosomes, promotes phagosome maturation and prevents pathogen escape into the cytosol. Autophagy protein LC3 is believed to be a key player during phagocytosis [124]. During *Aspergillus* germination, exposure of PAMPs enables the activation of host LC3-associated phagocytosis (LAP), which promotes the killing and eradication of fungi. Shielding PAMPs from detection by pattern recognition receptors is a predominant strategy adopted by fungi for evasion from host immunity. Recent studies with *A. fumigatus* have demonstrated that LAP activation also requires the removal of fungal cell wall melanin, based on the genetic, biochemical or biological (germination) data [125]. Fungal melanin is known to inhibit the NADPH oxidase-dependent activation of LAP via excluding the p22phox subunit from the phagosome. Therefore, LAP blockade is a general trait of melanin pigments [126]. Physiologically, melanin-triggered LAP blockade is able to enhance fungal virulence.

5. Medical Relevance of Fungal Pigments

In extreme environments, fungi having pigmented cell walls in their spores and mycelia are able to tolerate dehydration–hydration cycles, thermal fluctuations, toxic metal-contamination, and high ultraviolet light (UV) radiation better than the moniliaceous fungi, which are unpigmented [127]. A remarkable example is the black fungus *Cryomyces antarcticus*, an extremophile initially recovered from Antarctic deserts and later found to be able to survive in the outer part of the International Space Station [38]. The wide geographical distribution of pigmented fungal species and their extraordinary ability to survive in hostile environments suggest that fungal pigments may be of value in a variety of fields, including radioprotection and biomedical applications.

In addition to their broad distribution and their ability to synthesize secondary metabolites such as medically significant pigments, many fungi are readily grown in lab conditions. The prospect of large-scale production attracts interest from both the pharmaceutical industry and fundamental science. The scope of focus includes species from a wide range of environments of marine origin, soil, as well as endophytic fungi from terrestrial and marine flora and endolichenic origin [127,128].

5.1. Medical Roles of Melanins

Recent studies with fungi capable of producing melanin (also called melanotic fungi) have revealed a wide spectrum of functions for this kind of pigment, ranging from drought and radiation resistance to increased virulence in fungal pathogens. For example, melanotic fungi are present in diverse environments including but not limited to radiation-contaminated soils (e.g., the Chernobyl nuclear power station and the surrounding soils), Antarctic deserts, ancient cave paintings, and spacecraft, etc. In fact, melanized fungi, including spores of *Aspergillus niger*, *Cladosporium herbarum*, *Ulocladium chartatum*, *Basipetospora halophile*, etc., have been reported to be able to grow on the Mir Space Station and International Space Station. Some edible mushrooms (Basidiomycetes) are rich in melanins as well. Melanins contain persistent free radical centers, as detected by electron paramagnetic resonance (EPR) spectroscopy. These free radical centers allow melanins acting as antioxidants to scavenge free radicals and to protect the fungi from oxidative stress and damage by ionizing radiation in both atmospheric and terrestrial environments [129]. In a recent study [130], Pacelli et al. compared the effects of densely-ionizing deuterons and sparsely ionizing X-rays on two melanized fungal species, namely the fast-growing pathogenic basiodiomycete *Cryptococcus neoformans* and the slow-growing Antarctic rock-inhabiting *Cryomyces antarcticus* to their non-melanized counterparts. *C. antarcticus* showed more resistance to deuterons than *C. neoformans*, and similar resistance to X-rays was observed for both species. Melanin offered protection against high-dose (1.5 kGy) deuterons for both *C. neoformans* and *C. antarcticus* ($p < 10^{-4}$). In addition, melanin protected *C. antarcticus* ($p < 10^{-4}$) and probably *C. neoformans* against X-rays (0.3 kGy). The use of both XTT (2,3-bis(2-methoxy-4-nitro-5-sulfophenyl)25-[(phenyl-amino)carbonyl]-2H-tetrazolium hydroxide) and MTT (2-(4,5-dimethyl-2-thiazolyl)23,5-diphenyl-2H-tetrazolium bromide) assays in parallel helped define the location of the melanin-mediated electron transfer in the cells. Deuterons increased XTT activity in melanized strains of both species, while the activity in non-melanized cells remained unchanged or decreased. The opposite was observed with levels of ATP (the indicator of the metabolic activity of cells): upon exposure to deuterons, ATP decreased in melanized strains, but not in non-melanized ones. Larger and more distinct differences in both XTT and ATP were found in *C. neoformans* than in *C. antarcticus*. Further research by Pacelli et al. revealed that *C. antarcticus* could produce both 1,8-dihydroxynaphthalene (DHN) and L-3,4- dihydroxyphenylalanine (L-DOPA) melanins, which likely contribute to its ability to thrive under diverse stress conditions [131]. Together, these results indicate for the first time that melanin could protect both fast- and slow-growing fungi from high-dose radiation (deuterons in this case) under physiological conditions [130]. Indeed, the remarkable ability of melanin to confer resistance to radiation led to further research that revealed that melanized fungi could harvest radiation energy for their growth and reproduction. Broadly speaking, such an ability has both fundamental ecological significance and practical implications. Ecologically, the ability of melanotic fungi to make use of electromagnetic radiation for their physiological processes helps us understand energy flows in the biosphere. Practically, the understanding paves the way for potentially developing melanin-based radiation resistant materials and equipment. In astrobiology and space travel, melanin represents a new type of material with which to experiment for space travel [132].

The unique stress-resistance properties of *C. antarcticus*, a heavily melanized black fungus in the class Dothideomycetes of the Ascomycota phyla, were further demonstrated by its ability to survive not only the Antarctic desert but also the harsh outer space environment and cosmic radiation exposure [38,130]. Indeed, the mechanisms for how fungal melanin molecules confer resistance to extreme heat/cold stresses and high-dosage radiations have gained wide attention from scientists [133,134]. One of the protection mechanisms is the chelating properties of melanins towards biologically damaging free radicals generated by the physical and chemical stresses. These results are inspiring biomedical engineering in healthcare and in radiation- and thermal-protective applications [135].

Aside from melanized microscopic fungi such as yeasts and molds, melanized macroscopic mushrooms are also attracting increasing attentions. One such mushroom is the medicinal mushroom *Inonotus obliquus* (also called Chaga), a basidiomycete. This mushroom grows mainly on birch trees and is broadly distributed across the northern hemisphere in Europe, Asia, and North America. Long used in folk medicine, Chaga contains massive amounts of melanin pigments. Several research groups have reported the therapeutic potential of Chaga's bioactive pigments in countering the proliferation of cancer cells, and diabetes mellitus [136,137]. The pigments extracted from *I. obliquus* contain both phenolic compounds such as melanins, and triterpenoids such as inotodiol. Interestingly, the water-soluble component of melanin complexes of *I. obliquus* possesses insulin-sensitizing and hypoglycemic activity [138], consistent with their anti-diabetic functions. In addition, aqueous extracts of *I. obliquus* lowered the viability of human hepatoma (HepG2) cell lines in a dose-dependent fashion in vitro [139]. Subsequent studies revealed that the intraperitoneal administration of aqueous extracts of *I. obliquus* could inhibit tumor growth in vivo in mice implanted with melanoma B16-F10 cells at a dose of 20 mg/kg/d for 10 d [140]. Similarly, another study of melanin from *I. obliquus* showed its suppressive effect on the proliferation of HeLa 229 tumor cells [141]. Together with the above beneficial effects, the discovery of water-soluble melanin from *I. obliquus* in protecting DNA from carcinogenic damage [142] is also accelerating the preclinical studies of *I. obliquus* pigments. Results from such studies could generate significant medical and economic benefits in an environmentally friendly way (green chemistry).

As shown above, melanins have many potential benefits to humans. However, their presence in pathogenic fungi can also cause detrimental effects to humans both directly as human fungal pathogens or indirectly as plant and animal fungal pathogens. For example, in the human fungal pathogens *C. neoformans* [143], *Aspergillus fumigatus* [144], *Paracoccidioides brasiliensis* [145], *Penicillium marneffei* [146], *Fonsecaea pedrosoi* [147], and *Sporothrix schenckii* [148], melanin is an important virulence factor and contributes to drug resistance. Recently identified as a part of host innate immunity, platelets could be activated by exposure to *Aspergillus fumigatus*. In *A. fumigatus*, a mutant lacking melanin exhibited decreased platelet-stimulating activity, and the platelet activating effect can be mimicked by "melanin ghosts" [149], large isolated melanin molecules. Interestingly, melanins in *Fonsecaea monophora*, a common causal agent of chromoblastomycosis, were able to reduce Th1 cytokines and to elevate Th2 cytokines secreted by macrophages [150], likely by regulating the MAPK signalling pathway of macrophages [151], pointing out that melanins interacting with the macrophages might contribute to the immune evasion of fungal pathogen. In these species, melanin protects the invading fungal pathogen against oxidative stress within phagocytes, interferes with host cell signaling and autophagy, and masks recognition of fungal cells by the host immune system. Our understanding should benefit the development of therapeutic targets for treating mycosis in humans.

5.2. Medical Roles of Other Pigments

5.2.1. Anti-Tumor Activities

Carcinogenesis is a prolonged and multi-phase process that involves initiation, promotion, progression, and metastasis as an outcome of an unbalanced cell proliferation and cell apoptosis /death [152]. Anti-proliferative agents are often associated with anti-mutagenic and hence anti-tumor activities.

The majority of the anticancer chemotherapeutics prescribed nowadays, such as alkylating agents, antibiotics, compounds targeting the cell cycle (microtubules, G1/S/G/M checkpoint proteins), and topoisomerase inhibitors, are cytotoxic compounds. These compounds are developed to kill carcinoma cells/tissues more effectively than normal cells/tissues because they generally target the more rapidly dividing tumor cells. However, this may bring about side effects on some actively dividing normal healthy cells, like bone marrow cells, hair follicles and gastrointestinal epithelial cells. Notwithstanding the problems associated with the application of cytotoxic drugs, bioassays testing cytotoxicity toward a panel of cancer cell types represent a reliable approach to screen natural products with anticancer potential, which has led to the successful discovery of anti-cancer pharmaceuticals like paclitaxel and camptothecin [153]. Table 1 summarizes the known fungal pigments with antitumor activities. These pigments are produced by fungi from a variety of species and ecological niches, including species in the genus *Monascus*, endophytic fungi, marine fungi, mushrooms, and fungi residing in special habitats. Together, they have demonstrated activity against a variety of cancer cell lines.

Table 1. Fungal pigments with antitumor activities.

Fungal Sources	Fungal Species/Strain	Isolated Compound	Chemical Nature	Tumor Model/Cell Lines/Target Enzyme	Activity/Active Concentration	References
Monascus-fermented red rice	*Monascus pilosus*	Monascin & ankaflavin rubropunctamine & monascorubramine	Azaphilones	Ames test and Peroxynitrite-and UVB-induced mouse skin carcinogenesis model	Accelerate the mutagen decomposition	Ho et al. [154] Hsu et al. [155] Akihisa et al. [156]
	Monascus sp.	Ankaflavin		HepG2, A549 #/IC_{50}	15 μg/mL	Su et al. [157]
	Monascus purpureus	Monaphilone A Monoaphilone B		HEp-2, WiDr #/IC_{50} HEp-2, WiDr/IC_{50}	72.1, 55.8 μM 77.6, 55.3 μM	Hsu et al. [158]
		Rubropunctatin		BGC-823 #/IC_{50} and in vivo mouse model	12.57 μM	Zheng et al. [159]
	M. purpureus	Monascopyridine C & D		IHKE (kidney epithelial cell) CCK8 assay/EC_{50}	20.7–43.2 μmol/L	Knecht et al. [160]
	Monascus sp.	Glutamic acid derivative of *Monascus* orange pigments† (S)-(+)-1-amino-2-propanol derivative of the above orange pigments†		B16F10 (mouse melanoma cells) tyrosinase expression	30% inhibition 35% inhibition	Jo et al. [161]
	Monascus pilosus	Monascuspiloin	Monascin analog	PC-3 tumors of nude mice	42.5% inhibition (in vivo)	Chiu et al. [162]
Endophytic fungi	A fungus endophytic to *Mimosops elengi*	Ergoflavin	Xanthenes	ACHN (renal cell carcinoma), H460 (non-small-cell lung carcinoma), Panc1(pancreas), HCT116 (colon cancer), and Calu-1 (lung carcinoma)	1.2, 4.0, 2.4, 8.0, 1.5μM/IC_{50}	Deshmukh et al. [163]
	Phomopsis longicolla, endophytic to *Dicerandra frutescens*	Dicerandrol A, Dicerandrol B, Dicerandrol C	Xanthenes	A549, HCT116 #/IC_{50} A549, HCT116 A549, HCT116	7.0, 7.0 μg/mL 1.8, 1.8 μg/mL 1.8, 7.0 μg/mL	Wagenaar & Clardy [164]
	Chaetomium globosum endophytic to *Ginkgo biloba*	Chaetomugilides A–C	Azaphilone alkaloids	HepG-2#	1.7–3.4 μM/IC_{50}	Li et al. [165]
	Penicillium sp. CR1642D endophytic to Costa Rican rainforest	Penexanthone A Penexanthone B Dicerandrols B	Xanthones	A panel of cancer cell lines (Myeloma, lymphoma, leukemia, breast, prostate), also showing enhanced effects regarding tumor-stromal interaction	1–17 μM/IC_{50} IC_{50} of 1.2 μM (+stroma) *vs.* 2.4 μM (-stroma) in RPMI8226; 3.4 μM (+ stroma) *vs* 10.2 μM (-stroma) in H929	Cao et al. [166]
	Chaetomium globosum endophytic to marine fish *Mugil cephalus*	Chaetomugilin A Chaetomugilin C Chaetomugilin F	Azaphilone alkaloids	P388(murine), HL-60 (human) leukemia	8.7, 7.3 μM/IC_{50} 3.6, 2.7 μM/IC_{50} 3.3, 1.3 μM/IC_{50}	Yasuhide et al. [167]

Table 1. *Cont.*

Marine fungi	*Aspergillus tubingensis* GX1-5E	TMC 256 A1	Naphtho-γ-pyrone	MCF-7 & MDA-MB-435 (breast carcinoma), Hep3B & Huh7 (hepatoma), SNB19 & U87 MG (glioblastoma)	19.92–47.98 μM/IC_{50}	Sakurai et al. [168] Huang et al. [169]
	Penicillium pinophilum Hedgcok	Pinophilin A Pinophilin B	Hydrogenated azaphilones	Mammalian DNA polymerases (pols)A, B, Y	48.6–55.6 μM/IC_{50}	Myobatake et al. [170]
	Diaporthe sp. SCSIO 41011	*epi*-isochromophilone II isochromophilone D	Chloroazaphilones	ACHN, 786-O, OS-RC-2 (three renal carcinoma) 786-O renal carcinoma	4.4, 3.0, 3.9 μM/IC_{50} 8.9 μM/IC50	Luo et al. [171]
	Chaetomium globosum HDN151398	*N*-glutarylchaetoviridin C Chaetomugilin A Chaetomugilin C	chloroazaphilones Azaphilone Alkaloids	MGC-803, HO8910 # HL-60, HCT-116 # HL-60, HO8910	6.6, 9.7 μM/IC50 6.4, 6.1 μM/IC50 6.6, 8.8 μM/IC50	Sun et al. [172]
	Nigrospora sp. strain 1403	Bostrycindeoxybostrycin	Anthraquinones	A549, HepG2 # A549, HepG2	2.64, 5.90 μg/mL 2.44, 4.41 μg/mL	Xia et al. [173]
	Penicillium sp.	(+)-formylanserinone B anserinones B	Pentaketides	MDA-MB-435 #	2.90 μg/mL 3.60 μg/mL	Gautschi et al. [174]
Fungi in special habitats	*Pleurostomophora* sp. from a copper mine of North America	Berkchaetoazaphilones A, C Berkchaetorubramine berkchaetoazaphilone B	Azaphilones	Caspase-1 MMP-3 ξ Y79 # LOX IMVI #	150,25,50 μM/IC_{50} 130,15, 45 μM/IC_{50} 1.1 μM /IC_{50} 10 μM/IC_{50}	Stierle et al. [175]
	Coniella fragariae from goose dung	Coniellin A Coniellin A, D, E	Azaphilones	MDA-MB-231#	4.4 μM /IC50 and suppress tumor migration by 98% at 10 μM	Yu et al. [176]
Macrofungi (mushroom)	*Lactarius subvellereus*	Subvellerolactone B, Subvellerolactone D, Subvellerolactone E	Sesquiterpene hydroxylactones	A549, SK-MEL-2 #, HCT-15 A549, HCT-15 A549, HCT-15	26.5, 18.3, 14.2 μM/IC50 25.1,17.8μM/IC50 19.6, 28.7μM/IC50	Kim et al. [177]
	Boletus pseudocalopus	Grifolin derivatives 1–3	Phenolic compounds	A549, B16F1 (mouse melanoma)	5.0–9.0 μg/mL 3.5–7.3 μg/mL	Song et al. [178]
	Albatrellus confluens	Albatrellin	Meroterpenoid	HepG2	1.55 μg/mL	Yang et al. [179]
	Albatrellus flettii	Grifolin, neogrifolin, confluentin	Phenolic compounds	SW480 & HT29 (two colon cancer lines)	35.4, 30.7μM/IC50 34.6, 30.1μM/IC50 33.5, 25.8μM/IC50	Yaqoob et al. [180]

† rubropunctatin & monascorubin. ξ Matrix metalloproteinase-3 (MMP-3). # a variety of human cancer cell lines: A549 (lung adenocarcinoma), HepG2 (heptoblasoma), HEp-2 (laryngeal carcinoma), WiDr (colon adenocarcinoma), MGC-803 (gastric adenocarcinoma), HO8910 (ovarian cancer), HL-60 (promyelocytic leukemia), HCT-116 (colon cancer), MDA-MB-435 (breast cancer), Y79 (retinoblastoma), LOX IMVI (melanoma), SK-MEL-2 (skin melanoma), HCT-15 (colon adenocarcinoma).

5.2.2. Anti-Biofilm Activity

Melanin from the wood-ear edible fungus *Auricularia auricula* has been demonstrated to display distinct anti-biofilm effects towards *Escherichia coli* K-12, *Pseudomonas aeruginosa* PAO1, and *P. fluorescens* P-3, respectively, while no inhibitory activities were observed on bacterial growth [181]. Zhu et al. demonstrated that pigments from *A. auricula* can repress the production of violacein, a quorum-sensing signal for biofilm formation in the reporter strain *Chromobacterium violaceum* CV026 [182]. It is believed that the ability to form biofilm constitutes the predominant virulence factor for bacterial pathogens, as the biofilm formation is known to account for 80% of bacterial infections in humans [183]. Furthermore, biofilms represent a major cause of nosocomial infections, especially related to the emergence of multi-drug resistant strains, thereby contributing to refractory infectious diseases in humans. Thus, the quest of novel compounds that may effectively inhibit these biofilm-forming bacteria has been a hot topic in both pharmaceutical and clinical settings.

5.2.3. Photosensitizers

Perylenequinones (PQs) represent a class of fungal pigments with a signature 3,10-dihydroxy-4,9-perylene-quinone chromophore. PQs are known as reactive oxygen species (ROS)-generating photosensitizers in medical and agricultural settings. Consequently, PQs have attracted significant attention [184]. Indeed, PQs-producing bambusicolous fungal species *Shiraia bambusicola* has long been used in Chinese folk medicine to treat stomach ache, rheumatic pain, as well as some dermatologic disorders like vitiligo and psoriasis [185]. Interestingly, PQs have also been extracted from fungal fruiting bodies, where a diverse microbial community typically resides. At present, how the PQs are produced in fungal fruiting bodies is not known [186].

Due to the complexity and difficulty of the chemical synthesis of PQs [187], mushroom fruiting bodies have been the major source for the supply of PQs. However, recent work by Ma et al. revealed that *Pseudomonas fulva* SB1 isolated from the fruiting body of the fungus *Shiraia bambusicola* was able to boost the production of fungal PQs, including hypocrellins A, C (HA and HC), and elsinochromes A–C (EA, EB and EC). The results revealed that following two days of co-cultures, *Shiraia* mycelial cultures exhibited the highest production of HA, about 3.2-fold of that in axenic culture. The co-culture might have led to the elicitation of fungal conidiation and the formation of compacted fungal pellets, as compared to axenic culture. Furthermore, the bacterial co-culture might have up-regulated the expression of polyketide synthase gene (PKS) and activated genes of the ATP-binding cassette (ABC) transporter as well as major facilitator superfamily transporter (MFS) for PQ exudation [186].

5.2.4. Cholesterol-Lowering and/or Anti-Atherosclerotic Agents

In view of the association of obesity with an elevated risk of developing diabetes as well as cardiovascular disorders, the search for the cholesterol-lowering compounds has attracted much attention from chemists, biologists, pharmacists and medical practitioners. Monascin and ankaflavin, the two yellow pigments from *Monascus* spp., are found to have a remarkable antiobesity activity in a 3T3-L1 preadipocyte model of rat [71]. Their hypolipidemic effects were attributed to: (i) the inhibition of the differentiation and lipogenesis of preadipocytes by downregulating CCAT/enhancer-binding protein β (C/EBPβ) expression and its downstream peroxisome proliferator-activated receptor γ (PPARγ) and CCAT/enhancer-binding protein α (C/EBPα) expressions, and (ii) the inhibition of lipogenesis by increasing lipase activity and decreasing heparin releasable lipoprotein lipase (HR-LPL) activity [72]. Notably, the actions of monascin and ankaflavin do not resemble that of monacolin K (viz. lovastatin), a well-known cholesterol-lowering drug, which elevates creatine phosphokinase (CPK) activity, known to be a rhabdomyolysis marker [72].

Epidemiological studies have revealed a potential role of carotenoids in preventing cardiovascular disease (CVD) [8]. Reductions of low density lipoprotein (LDL) oxidation and of oxidative stress during plaque formation were thought to account for their effects. An association has been found between low serum lycopene level and an increased risk of atherosclerotic vascular events in middle-aged men [188]. Recent studies have demonstrated that fungal species such as *Blakesleea trispora* and *Rhodotorula* can produce lycopenes [189], thereby indicating their potential roles in the production of pharmaceuticals, food additives, or nutritional supplements that could be beneficial for patients with CVD.

5.2.5. Promising Anti-Alzheimer Agents

As one of the most common forms of dementia, Alzheimer's disease (AD) results in huge economic, emotional, and healthcare costs to individuals, families, and societies throughout the world. At present, there is a relative shortage of therapies available to treat AD. The aggregation of the microtubule-associated protein tau is suspected to be a seminal event in AD, thereby reducing/eliminating tau aggregation represents a potential therapeutic target for AD treatment and prevention. Paranjape et al. screened *Aspergillus nidulans* secondary metabolites for their capability to inhibit tau aggregation in vitro using an arachidonic acid polymerization protocol [190]. An aggregation inhibitor, asperbenzaldehyde, was identified. Asperbenzaldehyde is an intermediate in azaphilone biosynthesis. The authors further examined 11 azaphilone derivatives to determine their inhibitory activities against tau aggregation in vitro. All compounds examined were able to inhibit tau filament assembly to some degree, and four of the eleven compounds exhibited a pronounced activity of disassembling pre-formed tau aggregates in a dose-dependent fashion.

Similarly, a detailed investigation into the metabolites of the Mediterranean sponge *Tethya aurantium*-associated fungus *Bartalinia robillardoides* strain LF550 led to the isolation and identification of new chloroazaphilones, helicusin E, isochromophilone X and isochromophilone XI, along with the known pigment helicusin A [191]. Based on the bioassays of these chloroazaphilones, helicusin A (Figure 12) was found to be a potent acetylcholinesterase inhibitor with an IC_{50} value of 2.1 μM. Given that inhibition of acetylcholinesterase is widely considered a therapeutic target for ameliorating AD [192], helicusin A is seen as a promising lead compound to develop a potential novel pharmaceutical regime against AD.

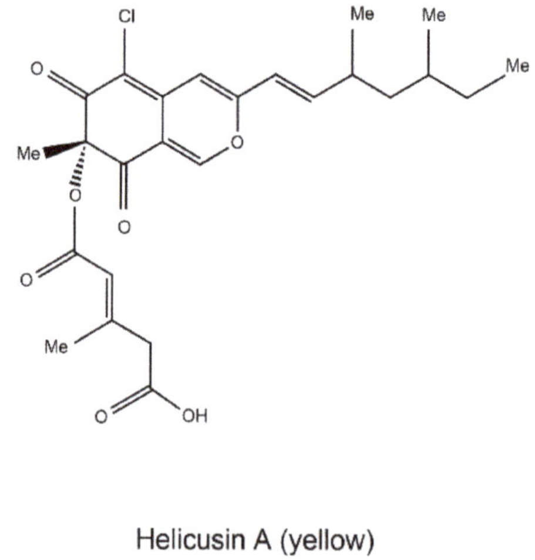

Figure 12. Structure of helicusin A, a pigment from marine fungus *Bartalinia robillardoides*.

We should note that the anti-Alzheimer activities of the above-mentioned pigments have only come from in vitro studies. Such in vitro results are different from the fungal pigments already in clinical use, such as the cholesterol-lowering and/or anti-atherosclerotic agent monascin as described in 5.2.4. However, the preliminary results as demonstrated by fungal pigments are showing great promise as anti-Alzheimer's disease drugs.

5.2.6. Anti-Inflammatory Activity

Hsu et al. (2011) reported the isolation of an azaphilonidal derivative monaphilone A as well as ankaflavin from the fermented products of *Monascus purpureus* NTU 568 [193]. Further studies showed that ankaflavin and monaphilone A decreased lipopolysaccharide (LPS)-induced inflammatory responses including production of nitrite, and expressions of inducible nitric oxide synthase (iNOS) and cyclooxygenase 2 (COX-2) in murine macrophage RAW 264.7 cells, pinpointing the anti-inflammatory effects of these two fungal pigments.

5.2.7. Antimicrobial Activities

A number of fungal pigments have shown significant antimicrobial activities [194,195]. Table 2 summaries the results from a few representative studies over the last decade. These pigments were from a variety of fungi living in diverse ecological niches, including marine environments, soil, and plants. The pigment-producing fungi are also broad, including both unicellular forms such as *Rhodotorula glutinis* and filamentous fungi such as *Monascus ruber* and species in the genera *Aspergillus, Fusarium,* and *Penicillium* (Table 2). Furthermore, most of these fungal pigments demonstrated broad activities against different groups of microorganisms, including plant and human fungal pathogens as well as both Gram-positive and Gram-negative bacterial pathogens. Of special note are pigments chaetoviridide A and B from the deep sea-derived filamentous fungus *Chaetomium* sp. NA-S0-R1 and pigments penicilones B–D from *Penicillium janthinellum* strain HK106 isolated from mangrove soil that showed strong activities against the methicillin-resistant *Staphylococcus aureus* (MRSA) [196,197].

5.2.8. Others

Aside from the above-described roles of fungal pigments in human health, other human health related roles have also been reported. Those included antioxidative, cytotoxic, and immunosuppressive roles. Below are brief descriptions of those studies.

Anti-oxidative properties are commonly found among fungal pigments. The conjugated polyene chain in the carotenoids provides chemical reactivity against oxidizing compounds and free radicals that may otherwise result in damage to cellular functions. Apart from the anti-oxidative actions of carotenoids previously reviewed elsewhere [15], investigation of the Chinese medicinal fungus *Phellinus igniarius* has led to the isolation and identification of three fungal pigments derived from pyrano[4,3-c]isochromen-4-one, phelligridins H, I, and J [198]. These orange or yellow pigments all showed antioxidant activity, as evidenced by inhibiting rat liver microsomal lipid peroxidation with IC_{50} values of 4.8, 3.7, and 6.5μM, respectively, for phelligridins H, I, and J. In addition, phelligridins H and I inhibited protein tyrosine phosphatase 1B (PTP1B), while phelligridin J exhibited cytotoxic activity against four human cancer cell lines, A2708 (ovary cancer), A549 (lung cancer), Bel-7402 (hepatoma) and HCT-8 (colon cancer) [198]. As PTP1B is an important player in cell proliferation, differentiation, and malignancy, being involved, for instance, in the development of breast cancer, lung cancer, and esophageal squamous cell carcinoma [199,200]. The above findings highlight the potential of these novel pigments as antitumor agents.

Table 2. Antimicrobial activities of fungal pigments.

Fungal Sources	Fungal Species/Strain	Bioactive Component	Target Microbes [1]	Antimicrobial Assay [2]	Reference
Marine sponge-associated, Indonesia	*Trichoderma parareesei*	Yellow pigment	*Salmonella typhi, Escherichia coli,* multi-drug resistant strain	MIC: 1000 µg/mL (weak)	Sibero et al. [201]
Deep sea, West Pacific Ocean	*Chaetomium* sp. NA-S0-R1	Chaetoviridide A, B	*Vibrio rotiferianus, Vibrio vulnificus* and MRSA (*Staphylococcus aureus* ATCC 43300 & CGMCC 1.12409)	MIC: 7.3–7.8µg/mL	Wang et al. [197]
Spoiled onion	*Penicillium purpurogenum*	Red exopigment	*S. aureus, Salmonella typhi, E. coli, Corynebacterium diptheriae, Pseudomonas aeruginosa*	Agar diffusion assay showing inhibition zone (diameter 1.5–2.3 cm)	Patil et al. [202]
Tropical Culture Collection André Tosello (Campinas, SP, Brazil).	*Monascus ruber* CCT 3802	Orange pigments (monascorubrin, rubropunctatin) Red pigments (monascorubramine, rubropunctamine)	Foodborne bacterium *S. aureus* ATCC 25923, *S. aureus* ATCC 25923, *E. coli* ATCC 25922	Radial diffusion assay showing inhibition zone (diameter 0.15 cm) Inhibition zone (diameter 0.35, 0.63 cm, respectively)	Vendruscolo et al. [203]
Stressed environment	*Fusarium sp.*	Reddish orange pigment	*Klebsiella pnuemoniae, E. coli, Shigella sp.* (bacteria) *Aspergillus niger, Candida albicans* (fungi)	Well diffusion assay showing inhibition zone (diameter 1.6–2.9 cm)	Mani et al. [204]
Western Ghats forest, India	*Penicillium* sp. MF5	Yellow pigments	*Bacillus subtilis*	MIC: 12.5 µg/mL	Saravanan & Radhakrishnan [205]
Persian type culture collection (PTCC), Tehran, Iran	*Rhodotorula glutinis* PTCC 5256	Carotenoid pigments	*S. aureus, Bacillus cereus, Streptococcus pyogenes, E. coli, Salmonella enteritidis, Enterococcus faecalis, Listeria monocytogenes*	Disk diffusion assay showing inhibition zone (diameter 0.9–1.1 cm)	Yolmeh et al. [206]
Endophyte on marine brown algae, eastern China	*Aspergillus versicolor*	Asperversin, brevianamide M	*E. coli, S. aureus*	Disk diffusion assay showing inhibition zone (diameter 2.0–2.2 cm)	Miao et al. [207]
Endophyte on leaves of *Panax notoginseng*	*Emericella* sp. XL029	14-hydroxyltajixanthone 14-hydroxyltajixanthone, its hydrate, chloride derivative as well as epitajixanthone hydrate	Fungus- *Drechslera maydis, Rhizoctonia cerealis, Fusarium oxysporum* and *Physalospora piricola* Effective against all tested bacteria (except for drug resistant *Staphylococcus aureus*)	MIC: 25 µg/mL MIC: 12.5–50 µg/mL	Wu et al. [195]
Mangrove rhizosphere soil	*Penicillium janthinellum* HK1-6	Penicilones B–D	MRSA (*S. aureus* ATCC 43300, ATCC 33591)	MIC: 3.13–6.25 µg/mL	Chen et al. [196]

[1], MRSA: Methicillin-resistant *Staphylococcus aureus;* [2], MIC: minimum inhibitory concentrations.

The macula of human eye contains two carotenoids, lutein and zeaxanthin. Based on the NIH Eye Disease Case-Control Study [208], the dietary intake of antioxidants such as carotenoids is highly correlated to the incidence of age-related macular degeneration (AMD), demonstrating a statistically significant linear trend associating the reduction of risk with increasing consumption of carotenoids. It is widely recognized that these carotenoids may protect the macula from light-induced damage and scavenge free radicals generated in the photoreceptors [8].

Four new azaphilones, namely longirostrerones A–D, along with three known sterols, have been isolated from extracts of the Thai soil fungus *Chaetomium longirostre* [209]. Longirostrerones A–D exhibited potent cytotoxicity against KB (human epidermoid carcinoma of the mouth) cell lines with IC_{50} of 1.04, 1.52, 0.23, and 6.38 μM, respectively. Among these azaphilones under study, longirostrerone C showed significant cytotoxicity against KB (IC_{50} = 0.23 μM), close to the control drugs doxorubicine and ellipticine. Moreover, longirostrerones A showed strong inhibitory effects on human breast adenocarcinoma MCF7 and small-cell lung cancer NCI-H187 cancer cell lines with IC_{50} values of 0.24 and 3.08 μM, respectively. Apart from the antitumor activities shown above, longirostrerones A–C also displayed antimalarial activity against *Plasmodium falciparum* with the IC_{50} values ranging from 0.62 to 3.73 μM.

An early report revealed the pronounced immunosuppressive activities of monascin and ankaflavin, two yellow oligoketide pigments isolated from the mycelium of *Monascus purpureus* [210]. By inhibiting murine T-splenocyte proliferation, monascin and ankaflavin could interact with regulatory mechanisms of the immune system and thus suppress its function, indicating that the two *Monascus* pigments might serve as promising immunosuppressants.

6. Structure–Activity Relationship (SAR) Studies of Fungal Pigments

The structure–activity relationships have been investigated for several groups of fungal pigments. For example, the presence of a halogen atom at C-5, a proton at C-8 and a diene moiety in the C-3 side chain of the 6-oxoisochromane ring in azaphilones are necessary for gp120-CD_4 binding, as exemplified by isochromophilones and their derivatives [211]. Furthermore, (an) electrophilic ketone(s) and/or enone(s) at both C-6 and C-8 in the 6-oxoisochromane ring are indispensable for the activity of cholesteryl ester transfer protein (CETP) inhibition, which is found in the azaphilone pigments such as chaetoviridin B, sclerotiorin, and rotiorin, etc. [78].

In addition to the above-mentioned bioactivities, studies with murine macrophage cells RAW 264.7 have demonstrated that the inhibition of NO production relies on the structure of azaphilones [212]. For instance, spiro-derivatives daldinins C, E, and F (structures shown in Figure 13) only exert weak inhibition on NO production. In contrast, azaphilones bearing a lactone ring such as multiformin D and sassafrins A–C (Figure 13), are shown to exert stronger inhibitory actions on NO production than daldinins. Entonaemin A, rubiginosins A and B (Figure 13) are all azaphilones where an orsellinic acid moiety is attached to the bicyclic core via an ester linkage. However, since only rubiginosin A displays strong activity in this subclass of azaphilones, the location of the orsellinic acid moiety must not influence the activity of the compounds. Nonetheless, the presence of an acetyl group (such as in the case of rubiginosin A) is indispensable for an elevation of activity. Azaphilones with a fatty acid side chain linked to a bicyclic azaphilone core via an ester bond, like rubiginosin C, cohaerins A and B (Figure 13), demonstrated only weak activities. The most potent inhibitors of NO production seem to be dimeric azaphilones, such as rutilins A and B (Figure 13). Overall, while the acetyl group is necessary for inhibition, the location of orsellinic acid does not seem to change their activities [212]. Taken together, these studies showed that the inhibitory effects of azaphilones are substantially fortified by the number of orsellinic acid moieties in the molecule and the presence of conjugated double bonds in dimeric compounds [212].

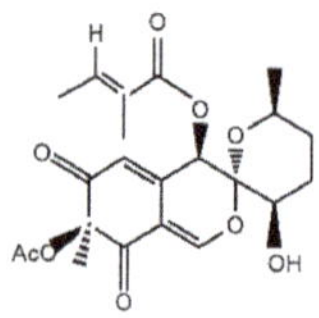
Daldinin C

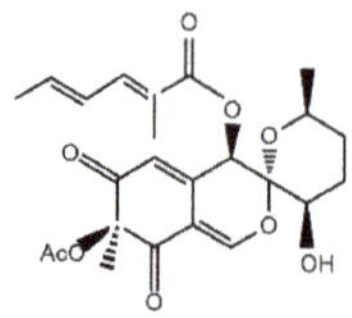
Daldinin E

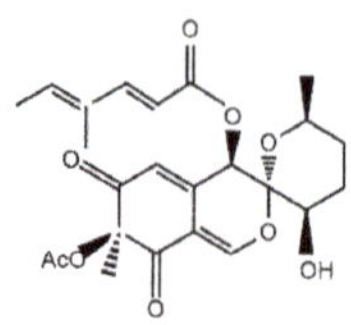
Daldinin F

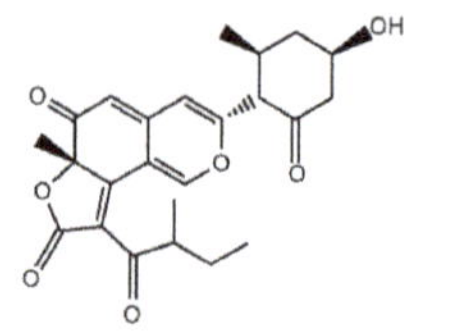
Multiformin D

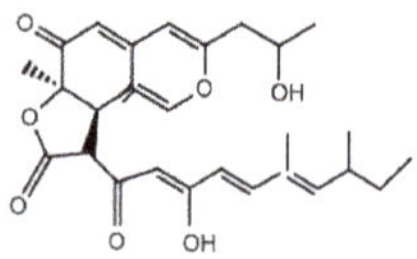
Sassafrin A

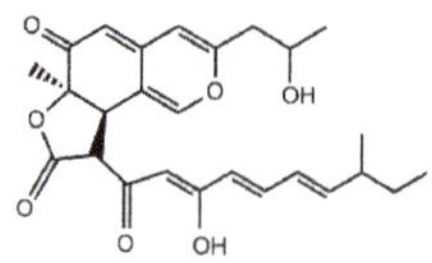
Sassafrin B

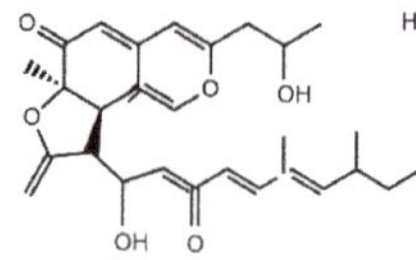
Sassafrin C

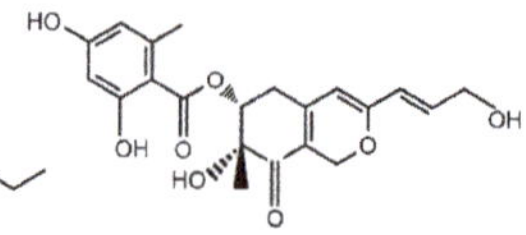
Entonacmin A

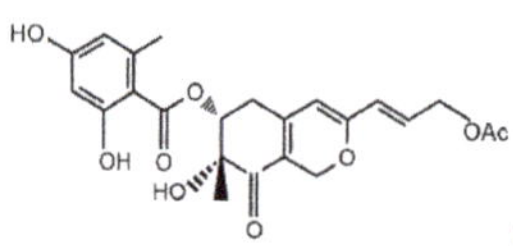
Rubiginosin A

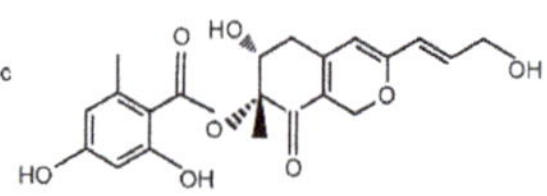
Rubiginosin B

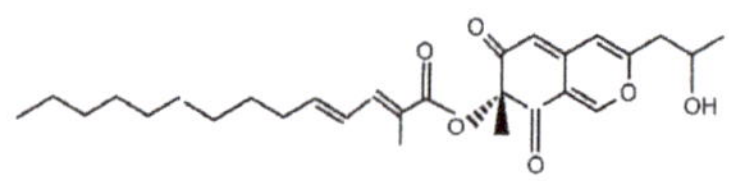
Rubiginosin C

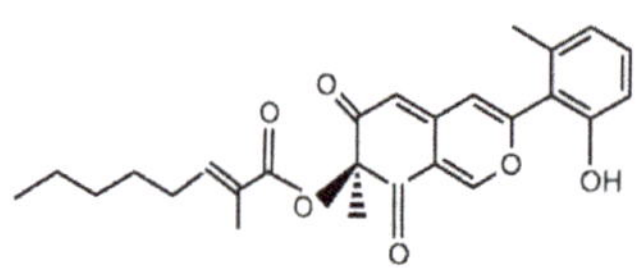
Cohaerin A

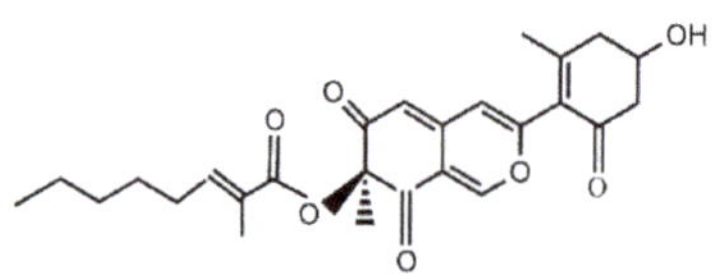
Cohaerin B

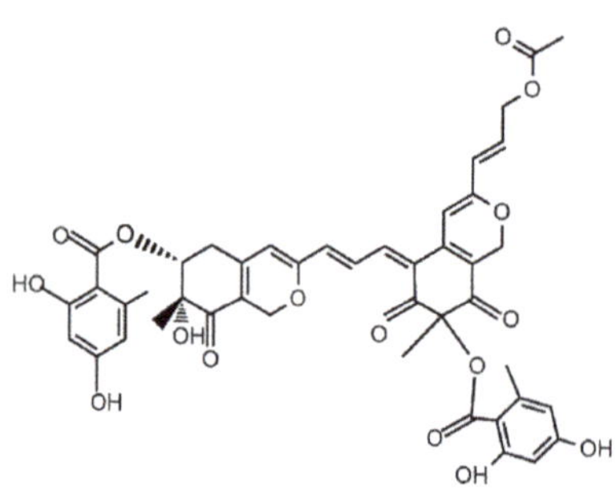
Rutilin A

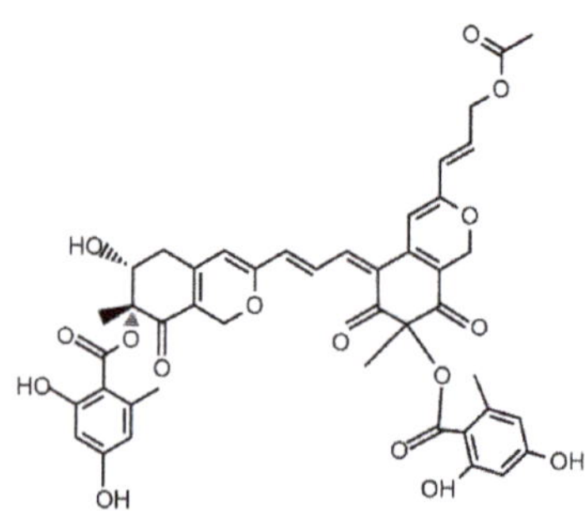
Rutilin B

Figure 13. Structures of azaphilones.

Previous studies have shown that the antioxidant properties of the carotenoids are tightly correlated with their chemical structure. Both the polyene chain and these other structural features influence the chemical properties (e.g., redox properties) of the carotenoids [213]. For example, the structure–activity relationship was investigated by Rodrigues and his colleagues for fifteen carotenoids, showing that the opening of the β-ionone ring and the increase of chromophore extension in the carotenoid structure could be the predominant factors giving rise to the elevated capacity of peroxyl radical scavenging [214]. Specifically, the conjugated polyene chromophore gives rise to not only the light absorption properties, and thus hues, but also the chemical properties of the compound and hence light-harvesting and photoprotective activity. In addition, the polyene chain is the main determinant responsible for antioxidant role of carotenoids due to its chemical reactivity with free radicals, and oxidizing agents. Furthermore, carotenoids are found to be in precise locations and orientations within subcellular structures in vivo. Thus, their chemical and physical traits are strongly affected by other molecules in their vicinity (particularly proteins as well as membrane lipids), and vice versa [215]. Indeed, structural characteristics, such as size, shape, and polarity, are critical to the capability of a carotenoid to fit precisely into its molecular milieu to allow for its functionality. The role of carotenoids in influencing cell membrane-associated molecular processes by modifying the structure, properties, and stability of these membranes may constitute a significant aspect of their beneficial effects on human health.

Apart from their antioxidant actions, studies with marine carotenoids have unveiled the relationship between structure and anti-obesity activity. The key structure responsible for anti-obesity activity is thought to be the carotenoid end of the polyene chromophore bearing an allenic bond and two hydroxyl groups [216].

7. Conclusions and Further Prospects

Fungal pigments have primarily been known for their extensive application as colorants in the food, cosmetic and textiles industries for the past several decades. However, their applications other than as colorants are very broad, including as cholesterol-lowering drugs, exemplified by monacolin K, a clinically used drug also called lovastatin [217], and various anti-microbial and antitumor agents. Unexpected functions of new pigments as well as known ones are continuously being revealed. Fungal pigments have attracted significant interests from the pharmaceutical industry as sources of potential drugs to fight many life-threatening diseases such as cardiovascular disorders, Alzheimer's disease, human carcinomas (hepatoma, breast, lung, colorectal, gastric, pancreatic, leukemia, hematopoietic, renal cell, and other cancers), infectious diseases, and parasitic diseases such as malaria. Since many fungal pigments have shown potential relevance to human health, there is increasing interest in understanding their mechanisms of action to pave the way for the future development of therapeutics against both acute and chronic diseases. Among the fungal pigments, melanin stands out for two reasons: (i) melanin confers excellent protection from such stressors as radiation, thermal fluctuations and drought, thereby giving an insight into the development of radiation-protective agents and devices for the human exploration into the outer space; (ii) melanin is known to contribute to pathogenesis and drug resistance in pathogenic fungi, and consequently represents an important target for the treatment of recalcitrant fungal infection in humans.

Thanks to advancements in microbiological, biochemical, genetic and genomics technologies and the development of high-throughput bioassay systems as well as the bioinformatics analyses, there are several potentially fruitful areas of research on fungal pigments. First, there is a diversity of fungal pigments produced by evolutionarily distant fungi. At present, the origins for the genes and the regulatory pathways governing their synthesis are largely unknown. Phylogenomic and transcriptomic studies should provide novel insights in this area. Second, a standardized test system and/or protocol should be established for screening the bioactivity of fungal pigments for both in vitro and in vivo assays, including appropriate positive and negative controls. Third, the structure–activity relationships are unknown for most fungal pigments. Finally, while the focus of our review is on colored fungal pigments, fungal metabolites that impact pigment productionunpub in both fungi and other organisms, including precursors and regulators for and metabolic products of pigments but they themselves are not colored, represent an additional source of fungal metabolites that have also been explored for human health benefits. One such metabolite in this category is kojic acid, a compound obtained from the filamentous fungus *Aspergillus oryzae* [218]. While colorless itself, kojic acid can inhibit tyrosinase which is essential for the synthesis of melanin in human skin and, consequently, it has been broadly used in dermatological applications [219]. With the increasing demand from the pharmaceutical industry, fungal pigments and their related products are expected to be an excellent resource for the design and development of novel therapeutic molecules in the future.

Author Contributions: L.L. and J.X. conceived and delineated manuscript organization; L.L. wrote the original draft; J.X. revised the manuscript. All authors have read and agreed to the published version of the manuscript.

Funding: This work was supported by the Natural Science and Engineering Research Council of Canada (2020-531998) and McMaster University (to J.X.), and by a China Scholarship Council fellowship (to L.L.).

Acknowledgments: We thank Heather Yoell for helpful suggestion and proofreading.

Conflicts of Interest: The authors declare no conflict of interest.

References

1. Demain, A.L.; Vandamme, E.J.; Collins, J.; Buchholz, K. History of industrial biotechnology. In *Industrial Biotechnology: Microorganisms*; Wittmann, C., Liao, J.C., Eds.; Wiley-VCH: Weinheim, Germany, 2017; pp. 3–84.
2. Boral, H.; Metinb, B.; Döğenc, A.; Seyedmousavi, S.; Ilkit, M. Overview of selected virulence attributes in *Aspergillus fumigatus*, *Candida albicans*, *Cryptococcus neoformans*, *Trichophyton rubrum* and *Exophiala dermatitidis*. *Fungal Genet. Biol.* **2018**, *111*, 92–107. [CrossRef] [PubMed]
3. Gunasekaran, S.; Poorniammal, R. Optimization of fermentation conditions for red pigment production from *Penicillium* sp. under submerged cultivation. *Afr. J. Biotechnol.* **2008**, *7*, 1894–1898. [CrossRef]
4. Méndez, A.; Pérez, C.; Monta˜éz, J.C.; Martínez, G.; Aguilar, C.N. Red pigment production by *Penicillium purpurogenum* GH2 is influenced by pH and temperature. *J. Zhejiang Univ. Sci. B (Biomed. Biotechnol.)* **2011**, *12*, 961–968. [CrossRef] [PubMed]
5. Akilandeswari1, P.; Pradeep, B.V. Exploration of industrially important pigments from soil fungi. *Appl. Microbiol. Biotechnol.* **2016**, *100*, 1631–1643. [CrossRef] [PubMed]
6. Dos Reis Celestino, J.; De Carvalho, L.E.; Da Paz Lima, M.; Lima, A.M.; Ogusku, M.M.; De Souza, J.V.B. Bioprospecting of amazon soil fungi with the potential for pigment production. *Process Biochem.* **2014**, *49*, 569–575. [CrossRef]
7. Dufosse, L.; Galaup, P.; Yaron, A.; Arad, S.M.; Blanc, P.; Murthy, K.N.C.; Ravishanka, G.A. Microorganisms and microalgae as source of pigments for use: A scientific oddity or an industrial reality? *Trends. Food Sci. Technol.* **2005**, *16*, 389–406. [CrossRef]
8. Fraser, P.D.; Bramley, P.M. The biosynthesis and nutritional uses of carotenoids. *Prog. Lipid Res.* **2004**, *43*, 228–265. [CrossRef]
9. Lee, D.; Jang, E.-H.; Lee, M.; Kim, S.-W.; Lee, Y.; Lee, K.-T.; Bahn, Y.-S. Unraveling melanin biosynthesis and signaling networks in *Cryptococcus neoformans*. *mBio* **2019**, *10*, e02267-19. [CrossRef]
10. Chang, P.-K.; Cary, J.W.; Lebar, M.D. Biosynthesis of conidial and sclerotial pigments in *Aspergillus* species. *Appl. Microbiol. Biotech.* **2020**, *104*, 2277–2286. [CrossRef]
11. Brilhantea, R.S.N.; da Rochab, M.G.; de Oliveiraa, J.S.; Pereira-Neto, W.A.; de Melo Guedes, G.M.; de Aguiar Cordeiro, R.; Sidrim, J.J.; Rocha, M.F.; Castelo, D.D. *Cryptococcus neoformans/Cryptococcus gattii* species complex melanized by epinephrine: Increased yeast survival after amphotericin B exposure. *Microb. Pathog.* **2020**, *143*, 104123. [CrossRef]
12. Santos, L.A.; Grisolia, J.C.; Burger, E.; de Araujo Paula, F.B.; Dias, A.L.T.; Malaquias, L.C.C. Virulence factors of *Paracoccidioides brasiliensis* as therapeutic targets: A review. *Antonie Leeuwenhoek Int. J. Gen.* **2020**, *113*, 593–604. [CrossRef] [PubMed]
13. Avalos, J.; Díaz-Sánchez, V.; García-Martínez, J.; Castrillo, M.; Ruger-Herreros, M.; Limón, M.C. Carotenoids. In *Biosynthesis and Molecular Genetics of Fungal Secondary Metabolites*; Martín, J.F., García-Estrada, C., Zeilinger, S., Eds.; Springer: New York, NY, USA, 2014; pp. 149–185.
14. Barredo, J.L.; García-Estrada, C.; Kosalkova, K.; Barreiro, C. Biosynthesis of astaxanthin as a main carotenoid in the heterobasidiomycetous yeast *Xanthophyllomyces dendrorhous*. *J. Fungi* **2017**, *3*, 44. [CrossRef] [PubMed]
15. Avalos, J.; Limón, M.C. Biological roles of fungal carotenoids. *Curr. Genet.* **2015**, *61*, 309–324. [CrossRef] [PubMed]
16. Mata-Gómez, L.; Montañez, J.C.; Méndez-Zavala, A.; Aguilar, C.N. Biotechnological production of carotenoids by yeasts: An overview. *Microb. Cell Fact.* **2014**, *13*, 12. [CrossRef]

17. Aksu, Z.; Eren, A.T. Carotenoids production by the yeast *Rhodotorula mucilaginosa*: Use of agricultural wastes as a carbon source. *Process Biochem.* **2005**, *40*, 2985–2991. [CrossRef]
18. Dufosse, L. Microbial production of food grade pigments. *Food Technol. Biotechnol.* **2006**, *44*, 313–321.
19. Panesar, R.; Kaur, S.; Panesar, P.S. Production of microbial pigments utilizing agro-industrial waste: A review. *Curr. Opin. Food. Sci.* **2015**, *1*, 70–76. [CrossRef]
20. Prado-Cabrero, A.; Scherzinger, D.; Avalos, J.; Al-Babili, S. Retinal biosynthesis in fungi: Characterization of the carotenoid oxygenase CarX from *Fusarium fujikuroi*. *Eukaryot. Cell* **2007**, *6*, 650–657. [CrossRef]
21. Xu, F.; Yuan, Q.-P.; Zhu, Y. Improved production of lycopene and β-carotene by *Blakeslea trispora* with oxygen-vectors. *Process Biochem.* **2007**, *42*, 289–293. [CrossRef]
22. Hernández-Almanza, A.; Montaneza, J.C.; Aguilar-González, M.A.; Martínez-Ávila, C.; Rodríguez-Herrera, R.; Aguilar, C.N. *Rhodotorula glutinis* as source of pigments and metabolites for food industry. *Food Biosci.* **2014**, *5*, 64–72. [CrossRef]
23. Phaff, H.; Miller, M.; Yoneyama, M.; Soneda, M. A comparative study of the yeast florae associated with trees on the Japanese Islands and on the west coast of North America. In Proceedings of the 4th International Fermentation Symposium: Fermentation Technology Today, Kyoto, Japan, 19–25 March 1972; pp. 759–774.
24. Andrewes, A.G.; Starr, M.P. (3R,3′R)-Astaxanthin from the yeast *Phaffia rhodozyma*. *Phytochemistry* **1976**, *15*, 1009–1011. [CrossRef]
25. Rodriguez-Saiz, M.; de la Fuente, J.L.; Barredo, J.L. *Xanthophyllomyces dendrorhous* for the industrial production of astaxanthin. *Appl. Microbiol. Biotechnol.* **2010**, *88*, 645–658. [CrossRef]
26. Schmidt, I.; Schewe, H.; Gassel, S.; Jin, C.; Buckingham, J.; Hümbelin, M.; Sandmann, G.; Schrader, J. Biotechnological production of astaxanthin with *Phaffia rhodozyma/Xanthophyllomyces dendrorhous*. *Appl. Microbiol. Biotechnol.* **2010**, *8*, 555–571. [CrossRef]
27. Mathewaroth, M.M. Carotenoids in Erythropoietic protoporphyria and other photosensitivity diseases. *Ann. N. Y. Acad. Sci.* **1993**, *691*, 127–138. [CrossRef] [PubMed]
28. Sandmann, G. Carotenoids of biotechnological importance. In *Biotechnology of Isoprenoids, Advances in Biochemical Engineering/Biotechnology*; Schrader, J., Bohlmann, J., Eds.; Springer: Berlin/Heidelberg, Germany, 2014; Volume 148, pp. 449–467.
29. Camacho, E.; Vij, R.; Chrissian, C.; Prados-Rosales, R.; Gil, D.; O'Meally, R.N.; Cordero, R.J.B.; Cole, R.N.; McCaffery, J.M.; Stark, R.E.; et al. The structural unit of melanin in the cell wall of the fungal pathogen *Cryptococcus neoformans*. *J. Biol. Chem.* **2019**, *294*, 10471–10489. [CrossRef]
30. Wheeler, M.H.; Bell, A.A. Melanins and their importance in pathogenic fungi. *Curr. Top. Med. Mycol.* **1988**, *2*, 338–387. [PubMed]
31. Perez-Cuesta, U.; Aparicio-Fernandez, L.; Guruceaga, X.; Martin-Souto, L.; Abad-Diaz-de-Cerio, A.; Antoran, A.; Buldain, I.; Hernando, F.L.; Ramirez-Garcia, A.; Rementeria, A. Melanin and pyomelanin in *Aspergillus fumigatus*: From its genetics to host interaction. *Int. Microbiol.* **2020**, *23*, 55–63. [CrossRef] [PubMed]
32. Heinekamp, T.; Thywißen, A.; Macheleidt, J.; Keller, S.; Valiante, V.; Brakhage, A.A. *Aspergillus fumigatus* melanins: Interference with the host endocytosis pathway and impact on virulence. *Front. Microbiol.* **2012**, *3*, 440. [CrossRef]
33. Morris-Jones, R.; Gomez, B.L.; Diez, S.; Uran, M.; Morris-Jones, S.D.; Casadevall, A.; Nosanchuk, J.D.; Hamilton, A.J. Synthesis of melanin pigment by *Candida albicans* in vitro and during infection. *Infect. Immun.* **2005**, *73*, 6147–6150. [CrossRef]
34. Walker, C.A.; Gomez, B.L.; Mora-Montes, H.M.; Mackenzie, K.S.; Munro, C.A.; Brown, A.J.; Gow, N.A.; Kibbler, C.C.; Odds, F.C. Melanin externalization in *Candida albicans* depends on cell wall chitin structures. *Eukaryot. Cell* **2010**, *9*, 1329–1342. [CrossRef]
35. Solano, F. Melanins: Skin pigments and much More—Types, structural models, biological functions, and formation routes. *New J. Sci.* **2014**, *2014*, 498276. [CrossRef]
36. Weijn, A.; van den Berg-Somhorst, D.B.P.M.; Slootweg, J.C.; Vincken, J.-P.; Gruppen, H.; Wichers, H.J.; Mes, J.J. Main phenolic compounds of the melanin biosynthesis pathway in bruising-tolerant and bruising-sensitive button mushroom (*Agaricus bisporus*) strains. *J. Agric. Food Chem.* **2013**, *61*, 8224–8231. [CrossRef] [PubMed]
37. Belozerskaya, T.A.; Gessler, N.N.; Averýanov, A.A. Melanin pigments of fungi. In *Fungal Metabolites. Reference Series in Phytochemistry*; Merillon, J.M., Ramawat, K., Eds.; Springer: Cham, Switzerland, 2015; pp. 1–29.

38. Pacelli, C.; Cassaro, A.; Maturilli, A.; Timperio, A.M.; Gevi, F.; Cavalazzi, B.; Stefan, M.; Ghica, D.; Onofri, S. Multidisciplinary characterization of melanin pigments from the black fungus *Cryomyces antarcticus*. *Appl. Microbiol. Biotechnol.* **2020**, *104*, 6385–6395. [CrossRef] [PubMed]
39. Ambrico, M. Special issue: Melanin, a long lasting history bridging natural pigments and organic bioelectronics. *Polym. Int.* **2016**, *65*, 1249–1250. [CrossRef]
40. Kogej, T.; Wheeler, M.H.; Lanišnik RiŽner, T.; Gunde-Cimerman, N. Evidence for 1,8-dihydroxynaphthalene melanin in three halophilic black yeasts grown under saline and non-saline conditions. *FEMS Microbiol. Lett.* **2004**, *232*, 203–209. [CrossRef]
41. Kalra, R.; Conlan, X.A.; Goel, M. Fungi as a potential source of pigments: Harnessing filamentous fungi. *Front. Chem.* **2020**, *8*, 369. [CrossRef]
42. Mapari, S.A.S.; Meyer, A.S.; Thrane, U.; Risvad, J.C.F. Identification of potentially safe promising fungal cell factories for the production of polyketide natural food colorants using chemotaxonomic rationale microbial cell factories. *Microb. Cell Fact.* **2009**, *8*, 1–15. [CrossRef]
43. Mapari, S.A.S.; Thrane, U.; Meyer, A.S. Fungal polyketide azaphilone pigments as future natural food colorants. *Trends Biotechnol.* **2010**, *28*, 300–307. [CrossRef]
44. Hobson, D.K.; Wales, D.S. Green dyes. *J. Stud. Dyn. Chang.* **1998**, *114*, 42–44. [CrossRef]
45. Durán, N.; Teixeira, M.F.S.; Conti, R.D.; Esposito, E. Ecological-friendly pigments from fungi. *Crit. Rev. Food Sci. Nutr.* **2002**, *42*, 53–66. [CrossRef]
46. Yadav, A.N.; Kour, D.; Rana, K.L.; Yadav, N.; Singh, B.; Chauhan, V.S.; Rastegari, A.A.; Hesham, A.E.; Gupta, V.K. Metabolic engineering to synthetic biology of secondary metabolites production. In *New and Future Developments in Microbial Biotechnology and Bioengineering*; Gupta, V.K., Pandey, A., Eds.; Elsevier: Amsterdam, The Netherlands, 2019; pp. 279–320.
47. Venil, C.K.; Velmurugan, P.; Devi, P.R.; Dufossé, L.; Ravi, A.V. Fungal pigments: Potential coloring compounds for wide ranging applications in textile dyeing. *J. Fungi* **2020**, *6*, 68. [CrossRef] [PubMed]
48. Gesslera, N.N.; Egorovaa, A.S.; Belozerskaya, T.A. Fungal anthraquinones. *Appl. Biochem. Microbiol.* **2013**, *49*, 85–99. [CrossRef]
49. Augustin, N.; Nuthakki, V.K.; Abdullaha, M.; Hassan, Q.P.; Gandhi, S.G.; Bharate, S.B. Discovery of helminthosporin, an anthraquinone isolated from *Rumex abyssinicus* Jacq as a dual cholinesterase inhibitor. *ACS Omega* **2020**, *5*, 1616–1624. [CrossRef]
50. Caro, Y.; Anamale, L.; Fouillaud, M.; Laurent, P.; Petit, T.; Dufossé, L. Natural hydroxyanthraquinoid pigments as potent food grade colorants: An overview. *Nat. Prod. Bioprospect.* **2012**, *2*, 174–193. [CrossRef]
51. Mapari, S.A.S.; Nielsen, K.F.; Larsen, T.O.; Frisvad, J.C.; Meyer, A.S.; Thrane, U. Exploring fungal biodiversity for the production of water-soluble pigments as potential natural food colorants. *Curr. Opin. Biotechnol.* **2005**, *16*, 231–238. [CrossRef] [PubMed]
52. Wang, W.; Zhu, T.; Tao, H.; Lu, Z.; Fang, Y.; Gu, Q.; Zhu, W. Two new cytotoxic quinone type compounds from the halotolerant fungus *Aspergillus variecolor*. *J. Antibiot.* **2007**, *60*, 603–607. [CrossRef] [PubMed]
53. Wang, P.L.; Li, D.Y.; Xie, L.R.; Wu, X.; Hua, H.M.; Li, Z.L. Two new compounds from a marine-derived fungus *Penicillium oxalicum*. *Nat. Prod. Res.* **2014**, *28*, 290–293. [CrossRef]
54. Li, S.-W.; Yang, T.-C.; Lai, C.-C.; Huang, S.-H.; Liao, J.-M.; Wan, L.; Lin, Y.-J.; Lin, C.-W. Antiviral activity of aloe-emodin against influenza A virus via galectin-3 up-regulation. *Eur. J. Pharm.* **2014**, *738*, 125–132. [CrossRef]
55. Huang, Z.; Nong, X.; Ren, Z.; Wang, J.; Zhang, X.; Qi, S. Anti-HSV1, antioxidant and antifouling phenolic compounds from the deep sea derived fungus *Aspergillus versicolor* SCSIO41502. *Bioorg. Med. Chem. Lett.* **2017**, *15*, 787–791. [CrossRef]
56. Miliani, C.; Romani, A.; Favaro, G. Acidichromic effects in 1,2-di- and 1,2,4-tri-hydroxyanthraquinones. A spectrophotometric and fluorimetric study. *J. Phys. Org. Chem.* **2000**, *13*, 141–150. [CrossRef]
57. Babula, P.; Adam, V.; Havel, L.; Kizek, R. Noteworthy secondary metabolites naphthoquinones—Occurrence, pharmacological properties and analysis. *Curr. Pharm. Anal.* **2009**, *5*, 47–68. [CrossRef]
58. Lebeau, J.; Petit, T.; Dufossé, L.; Caro, Y. Putative metabolic pathway for the bioproduction of bikaverin and intermediates thereof in the wild *Fusarium oxysporum* LCP531 strain. *AMB Express* **2019**, *9*, 186. [CrossRef] [PubMed]
59. Zhan, J.; Burns, A.M.; Liu, M.X.; Faeth, S.H.; Gunatilaka, A.A. Search for cell mobility and angiogenesis inhibitors with potential anticancer activity: Beauvericin and other constituents of two endophytic strains of *Fusarium oxysporum*. *J. Nat. Prod.* **2007**, *70*, 227–232. [CrossRef] [PubMed]

60. Son, S.W.; Kim, H.Y.; Choi, G.J.; Lim, H.K.; Jang, K.S.; Lee, S.O.; Lee, S.; Sung, N.D.; Kim, J.C. Bikaverin and fusaric acid from *Fusarium oxysporum* show antioomycete activity against *Phytophtora infestans*. *J. Appl Microbiol.* **2008**, *104*, 692–698. [CrossRef]
61. Limón, M.C.; Rodríguez-Ortiz, R.; Avalos, J. Bikaverin production and applications. *Appl. Microbiol. Biotechnol.* **2010**, *87*, 21–29. [CrossRef] [PubMed]
62. Nirmaladevi, D.; Venkataramana, M.; Chandranayaka, S.; Ramesha, A.; Jameel, N.M.; Srinivas, C. Neuroprotective effects of bikaverin on H_2O_2-induced oxidative stress mediated neuronal damage in SH-SY5Y cell line. *Cell Mol. Neurobiol.* **2014**, *34*, 973–985. [CrossRef]
63. Sturdikova, M.; Slugen, D.; Lesova, K.; Rosenberg, M. Microbial production of coloured azaphiolone metabolites. *Chem. Listy* **2000**, *94*, 105–110.
64. Zhu, J.; Nicholas, P.; Grigoriadis, N.P.; Lee, J.P.; Porco, J.A. Synthesis of the azaphilones using copper-mediated enantioselective oxidative dearomatization. *J. Am. Chem. Soc.* **2005**, *127*, 9342–9343. [CrossRef]
65. Dong, J.; Zhou, Y.; Li, R.; Zhou, W.; Li, L.; Zhu, Y.; Huang, R.; Zhang, K. New nematicidal azaphilonesfrom the aquatic fungus *Pseudohalonectria adversaria* YMF1.01019. *FEMS Microbiol. Lett.* **2006**, *264*, 65–69. [CrossRef]
66. Gao, J.-M.; Yang, S.-X.; Qin, J.-C. Azaphilones: Chemistry and biology. *Chem. Rev.* **2013**, *113*, 4755–4811. [CrossRef]
67. Nakanishi, K. Studies in microbial and insect natural products chemistry. *J. Nat. Med.* **2006**, *60*, 2–20. [CrossRef]
68. Gill, M. Pigments of fungi (Macromycetes). *Nat. Prod. Rep.* **2003**, *20*, 615–639. [CrossRef] [PubMed]
69. Mapari, S.A.S.; Meyer, A.S.; Thrane, U. Colorimetric characterization for comparative analysis of fungal pigments and natural food colorants. *J. Agric. Food Chem.* **2006**, *54*, 7027–7035. [CrossRef] [PubMed]
70. Jou, P.C.; Ho, B.Y.; Hsu, Y.W.; Pan, T.M. The effect of *Monascus* secondary polyketide metabolites, monascin and ankaflavin, on adipogenesis and lipolysis activity in 3T3-L1. *J. Agric. Food Chem.* **2010**, *58*, 12703–12709. [CrossRef] [PubMed]
71. Lee, C.L.; Kung, Y.H.; Wu, C.L.; Hsu, Y.W.; Pan, T.M. Monascin and ankaflavin act as novel hypolipidemic and high-density lipoprotein cholesterol-raising agents in red mold dioscorea. *J. Agric. Food Chem.* **2010**, *59*, 8199–8207. [CrossRef] [PubMed]
72. Lee, C.-L.; Wen, J.-Y.; Hsu, Y.-W.; Pan, T.-M. Monascus-fermented yellow pigments monascin and ankaflavin showed antiobesity effect via the suppression of differentiation and lipogenesis in obese rats fed a high-fat diet. *J. Agric. Food Chem.* **2013**, *61*, 1493–1500. [CrossRef] [PubMed]
73. Lee, C.-L.; Hung, Y.-P.; Hsu, Y.-W.; Pan, T.-M. Monascin and ankaflavin have more anti-atherosclerosis effect and less side effect involving increasing creatinine phosphokinase activity than monacolin K under the same dosages. *J. Agric. Food Chem.* **2013**, *61*, 143–150. [CrossRef]
74. Cheng, C.-F.; Pan, T.-M. Monascus-fermented red mold dioscorea protects mice against alcohol-induced liver injury, whereas its metabolites ankaflavin and monascin regulate ethanol-induced peroxisome proliferator-activated receptor-γ and sterol regulatory element-binding transcription factor-1 expression in HepG2 cells. *J. Sci. Food Agric.* **2018**, *98*, 1889–1898.
75. Akihisa, T.; Tokuda, H.; Ukiya, M.; Iyota, A.; Yasukawa, K.; Sakamoto, N.; Kimura, Y.; Suzuki, T.; Takayasu, J.; Nishino, H. Anti-tumor-initiating effects of monascin, an azaphilonoid pigment from the extract of *Monascus pilosus* fermented rice (red-mold rice). *Chem. Biodivers.* **2005**, *2*, 1305–1309. [CrossRef]
76. Park, J.-H.; Choi, G.J.; Jang, K.S.; Lim, H.K.; Kim, H.T.; Cho, K.W.; Kim, J.-C. Antifungal activity against plant pathogenic fungi of chaetoviridins isolated from *Chaetomium globosum*. *FEMS Microbiol. Lett.* **2005**, *252*, 309–313. [CrossRef]
77. Yasukawa, K.; Takahashi, M.; Natori, S.; Kawai, K.; Yamazaki, M.; Takeuchi, M.; Takido, M. Azaphilones inhibit tumor promotion by 12-O-tetradecanoylphorbol-13-acetate in two-stage carcinogenesis in mice. *Oncology* **1994**, *51*, 108–112. [CrossRef] [PubMed]
78. Tomoda, H.; Matsushima, C.; Tabata, N.; Namatame, I.; Tanaka, H.; Bamberger, M.J.; Arai, H.; Fukazawa, M.; Inoue, K.; Omura, S. Structure-specific inhibition of cholesteryl ester transfer protein by azaphilones. *J. Antibiot.* **1999**, *52*, 160–170. [CrossRef] [PubMed]
79. Chidananda, C.; Rao, L.J.M.; Sattur, A.P. Sclerotiorin, from *Penicillium frequentans*, a potent inhibitor of aldose reductase. *Biotechnol. Lett.* **2006**, *28*, 1633–1636. [CrossRef] [PubMed]
80. Pairet, L.; Wrigley, S.K.; Chetland, I.; Reynolds, I.E.; Hayes, M.A.; Holloway, J.; Ainsworth, A.M.; Katzer, W.; Cheng, X.-M.; Hupe, D.J.; et al. Azaphilones with endothelin receptor binding activity produced by *Penicillium sclerotiorum*: Taxonomy, fermentation, isolation, structure elucidation and biological activity. *J. Antibiot.* **1995**, *48*, 913–923. [CrossRef] [PubMed]

81. Lucas, E.M.F.; de Castro, M.C.M.; Takahashi, J.A. Antimicrobial properties of sclerotionrin, isochromophilone VI and pencolide, metabolites from a Brazilian cerrado isolate of *Penicillum sclerotiorum* van beyma. *Braz. J. Microbiol.* **2007**, *38*, 785–789. [CrossRef]
82. Giridharan, P.; Verekar, S.A.; Khanna, A.; Mishra, P.D.; Deshmukh, S.K. Anticancer activity of sclerotiorin, isolated from an endophytic fungus *Cephalotheca faveolata* Yaguchi, Nishim & Udagawa. *Indian J. Exp. Biol.* **2012**, *50*, 464–468. [PubMed]
83. Arunpanichlert, J.; Rukachaisirikul, V.; Sukpondma, Y.; Phongpaichit, S.; Tewtrakul, S.; Rungjindamai, N.; Sakayaroj, J. Azaphilone and isocoumarin derivatives from the endophytic fungus *Penicillium sclerotiorum* PSU-A13. *Chem. Pharm. Bull.* **2010**, *58*, 1033–1036. [CrossRef]
84. Salo, O.; Guzmán-Chávez, F.; Ries, M.I.; Lankhorst, P.P.; Bovenberg, R.A.L.; Vreeken, R.J.; Driessen, A.J.M. Identification of a polyketide synthase involved in sorbicillin biosynthesis by *Penicillium chrysogenum*. *Appl. Environ. Microbiol.* **2016**, *82*, 3971–3978. [CrossRef]
85. Chen, G.; Chu, J. Characterization of two polyketide synthases involved in sorbicillinoid biosynthesis by *Acremonium chrysogenum* using the CRISPR/Cas9 system. *Appl. Biochem. Biotechnol.* **2019**, *188*, 1134–1144. [CrossRef]
86. Pastre, R.; Marinho, A.M.R.; Rodrigues-Filho, E.; Souza, A.Q.L.; Pereira, J.O. Diversity of polyketides produced by *Penicillium* species isolated from Melia azedarach and Murraya paniculata. *Quim. Nova* **2007**, *30*, 1867–1871. [CrossRef]
87. Teixeira, M.F.S.; Martins, M.S.; Da Silva, J.C.; Kirsch, L.S.; Fernandes, O.C.C.; Carneiro, A.L.B.; Da Conti, R.; Durán, N. Amazonian biodiversity: Pigments from *Aspergillus* and *Penicillium*–characterizations, antibacterial activities and their toxicities. *Curr. Trends Biotechnol. Pharm.* **2012**, *6*, 300–311.
88. Velíšek, J.; Davídek, J.; Cejpek, K. Biosynthesis of food constituents: Natural pigments. Part 1—A review. *Czech J. Food Sci.* **2007**, *25*, 291–315. [CrossRef]
89. Velíšek, J.; Cejpek, K. Pigments of higher fungi: A review. *Czech J. Food Sci.* **2011**, *29*, 87–102. [CrossRef]
90. Fujii, I.; Watanabe, A.; Sankawa, U.; Ebizuka, Y. Identification of claisen cyclase domain in fungal polyketide synthase WA, a naphthopyrone synthase of *Aspergillus nidulans*. *Chem. Biol.* **2001**, *8*, 189–197. [CrossRef]
91. Wohlert, S.E.; Wendt-Pienkowski, E.; Bao, W.L.; Hutchinson, C.R. Production of aromatic minimal polyketides by the daunorubicin polyketide synthase genes reveals the incompatibility of the heterologous DpsY and JadI cyclases. *J. Nat. Prod.* **2001**, *64*, 1077–1080. [CrossRef]
92. Schümann, J.; Hertweck, C. Advances in cloning, functional analysis and heterologous expression of fungal polyketide synthase genes. *J. Biotechnol.* **2006**, *124*, 690–703. [CrossRef]
93. Hajjaj, H.; Klaébé, A.; Goma, G.; Blanc, P.J.; Barbier, E.; Francois, J. Medium-chain fatty acids Affect citrinin production in the filamentous fungus *Monascus ruber*. *Appl. Envriron. Microbiol.* **2000**, *66*, 1120–1125. [CrossRef]
94. Velíšek, J.; Davídek, J.; Cejpek, K. Biosynthesis of food constituents: Natural pigments. Part 2—A review. *Czech J. Food Sci.* **2008**, *26*, 73–98. [CrossRef]
95. Jensen, R.A.; Pierson, D.A. Evolutionary implications of different types of microbial enzymology for L-tyrosine biosynthesis. *Nature* **1975**, *254*, 667–671. [CrossRef]
96. Musso, H. The pigments of fly agaric, *Amanita muscaria*. *Tetrahedron* **1979**, *35*, 2843–2853. [CrossRef]
97. Von Ardenne, R.; Döpp, H.; Musso, H.; Steiglich, W. Über das vorkommen von muscaflavin bei hygrocyben (Agaricales) und seine dihydroazepin-strukture. *Z. Nat. C* **1974**, *29*, 637–639.
98. Hallsworth, J.E.; Magan, N. Culture age, temperature, and pH affect the polyol and trehalose contents of fungal propagules. *Appl. Environ. Microbiol.* **1996**, *62*, 2435–2442. [CrossRef]
99. Bhatia, S.; Sharma, K.; Sharma, A.; Garg, A.; Kumar, S.; Purohit, A.P. Mycosporine and mycosporine-like amino acids: A paramount tool against ultraviolet irradiation. *Pharmacogn. Rev.* **2011**, *5*, 138–146. [CrossRef]
100. Ghazaei, C. Molecular insights into pathogenesis and infection with *Aspergillus fumigatus*. *Malays. J. Med. Sci.* **2017**, *24*, 10–20. [CrossRef]
101. Marcos, C.M.; de Oliveira, H.C.; de Melo, W.D.; da Silva, J.D.; Assato, P.A.; Scorzoni, L.; Rossi, S.A.; de Paula e Silva, A.C.; Mendes-Giannini, M.J.; Almeida, A.M. Anti-immune strategies of pathogenic fungi. *Front. Cell. Infect. Microbiol.* **2016**, *6*, 142. [CrossRef]
102. Hawkins, A.R.; Lamb, H.K.; Moore, J.D.; Charles, I.G.; Roberts, C.F. The pre-chorismate (shikimate) and quinate pathways in filamentous fungi: Theoretical and practical aspects. *J. Gen. Microbiol.* **1993**, *139*, 2891–2899. [CrossRef]

103. Kögl, F.; Becker, H.; de Voss, G.; Wirth, E.L. Untersuchungen über Pilzfarbstoffe. VII. Die Synthese des Atromentins. Zur Kenntnis der Atromentinsäure. *Ann. Chem.* **1928**, *465*, 243. [CrossRef]
104. Liu, J. Natural terphenyls: Developments since 1877. *Chem. Rev.* **2006**, *106*, 2209–2223. [CrossRef]
105. Dewick, P.M. The biosynthesis of shikimate metabolites. *Nat. Prod. Rep.* **1994**, *11*, 173–203. [CrossRef]
106. Tauber, J.P.; Gallegos-Monterrosa, R.; Kovacs, A.T.; Shelest, E.; Hoffmeister, D. Dissimilar pigment regulation in *Serpula lacrymans* and *Paxillus involutus* during inter-kingdom interactions. *Microbiology* **2018**, *164*, 65–77. [CrossRef]
107. Sullivan, G.; Garrett, R.D.; Lenehan, R.F. Occurrence of atromentin and thelephoric acid in cultures of *Clitocybe subilludens*. *J. Pharm. Sci.* **1971**, *60*, 1727. [CrossRef] [PubMed]
108. Britton, G.; Liaaen-Jensen, S.; Pfander, H. *Carotenoids: Handbook*; Birkhauser: Basel, Switzerland, 2004.
109. Britton, G.; Liaaen-Jensen, S.; Pfander, H. *Carotenoids*; Birkhäuser: Basel, Switzerland, 1998; Volume 1–2.
110. Spiteller, P.; Hamprecht, D.; Steglich, W. Biosynthesis of the 2H-azepine alkaloid chalciporone. *J. Am. Chem. Soc.* **2001**, *123*, 4837–4838. [CrossRef] [PubMed]
111. Eggert, C. Laccase-catalyzed formation of cinnabarinic acid is responsible for antibacterial activity of *Pycnoporus cinnabarinus*. *Microbiol. Res.* **1997**, *152*, 315–318. [CrossRef]
112. Kim, W.-G.; Lee, I.-K.; Kim, J.-P.; Ryoo, I.-J.; Koshino, H.; Yoo, I.-D. New indole derivatives with free radical scavenging activity from *Agrocybe cylindracea*. *J. Nat. Prod.* **1997**, *60*, 721–723. [CrossRef] [PubMed]
113. Koshcheenko, K.A.; Baklashova, T.G.; Kozlovskii, A.G.; Arinbasarov, M.U.; Skryabin, G.K. Hydroxylation of indolyl-3-acetic acid by the fungus *Aspergillus niger* IBFM-F-12. *Prikl. Biokhim. Mikrobiol.* **1977**, *13*, 248–254.
114. Ringø, E.; Olsen, R.E.; Mayhew, T.M.; Myklebust, R. Electron microscopy of the intestinal microflora of fish. *Aquaculture* **2003**, *227*, 395–415. [CrossRef]
115. Bogusławska-Wąs, E.; Dłubała, A.; Laskowska, M. The role of *Rhodotorula mucilaginosa* in selected biological process of wild fish. *Fish Physiol. Biochem.* **2019**, *45*, 511–521. [CrossRef]
116. Pérez-Rodríguez, L. Carotenoids in evolutionary ecology: Re-evaluating the antioxidant role. *BioEssays* **2009**, *31*, 1116–1126. [CrossRef]
117. Garbe, A.; Buck, J.; Hämmerling, U. Retinoids are important cofactors in T cell activation. *J. Exp. Med.* **1992**, *176*, 109–117. [CrossRef]
118. Geissmann, F.; Revy, P.; Brousse, N.; Lepelletier, Y.; Folli, C.; Durandy, A.; Chambon, P.; Dy, M. Retinoids regulate survival and antigen presentation by immature dendritic cells. *J. Exp. Med.* **2003**, *198*, 623–634. [CrossRef]
119. Chew, B.P.; Park, J.S. Carotenoid action on the immune response. *J. Nutr.* **2004**, *134*, 257S–261S. [CrossRef]
120. Cornet, S.; Biard, C.; Moret, Y. Is there a role for antioxidant carotenoids in limiting self-harming immune response in invertebrates? *Biol. Lett.* **2007**, *3*, 284–288. [CrossRef]
121. El-Sheekh, M.M.; Mahmoud, Y.A.; Abo-Shady, A.M.; Hamza, W. Efficacy of *Rhodotorula glutinis* and *Spirulina platensis* carotenoids in immunopotentiation of mice infected with *Candida albicans* SC5314 and *Pseudomonas aureginosa* 35. *Folia Microbiol.* **2010**, *55*, 61–67. [CrossRef]
122. Pais, P.; Costa, C.; Cavalheiro, M.; Romão, D.; Teixeira, M.C. Transcriptional control of drug resistance, virulence and immune system evasion in pathogenic fungi: A cross-species comparison. *Front. Cell. Infect. Microbiol.* **2016**, *6*, 131. [CrossRef]
123. Bayry, J.; Beaussart, A.; Dufrêne, Y.F.; Sharma, M.; Bansal, K.; Kniemeyer, O.; Aimanianda, V.; Brakhage, A.A.; Kaveri, S.V.; Kwon-Chung, K.J.; et al. Surface structure characterization of *Aspergillus fumigatus* conidia mutated in the melanin synthesis pathway and their human cellular immune response. *Infect. Immun.* **2014**, *82*, 3141–3153. [CrossRef]
124. Huang, J.; Canadien, V.; Lam, G.Y.; Steinberg, B.E.; Dinauer, M.C.; Magalhaes, M.A.O.; Glogauer, M.; Grinstein, S.; Brumell, J.H. Activation of antibacterial autophagy by NADPH oxidases. *Proc. Natl. Acad. Sci. USA* **2009**, *106*, 6226–6231. [CrossRef]
125. Akoumianaki, T.; Kyrmizi, I.; Valsecchi, I.; Gresnigt, M.S.; Samonis, G.; Drakos, E.; Boumpas, D.; Muszkieta, L.; Prevost, M.C.; Kontoyiannis, D.P.; et al. *Aspergillus* cell wall melanin blocks LC3-associated phagocytosis to promote pathogenicity. *Cell Host Microbe* **2016**, *19*, 79–90. [CrossRef]
126. Chamilos, G.; Akoumianaki, T.; Kyrmizi, I.; Brakhage, A.; Beauvais, A.; Latge, J.-P. Melanin targets LC3-associated phagocytosis (LAP): A novel pathogenetic mechanism in fungal disease. *Autophagy* **2016**, *12*, 888–889. [CrossRef]

127. Dufossé, L.; Fouillaud, M.; Caro, Y.; Mapari, S.A.; Sutthiwong, N. Filamentous fungi are large-scale producers of pigments and colorants for the food industry. *Curr. Opin. Biotechnol.* **2014**, *26*, 56–61. [CrossRef]
128. Lagashetti, A.C.; Dufossé, L.; Singh, S.K.; Singh, P.N. Fungal pigments and their prospects in different industries. *Microorganisms* **2019**, *7*, 604. [CrossRef]
129. Sarna, T.; Plonka, P.M. Biophysical studies of melanin. In *Paramagnetic, Ion-Exchange, and Redox Properties of Melanin Pigments and Their Photoreactivity*; Eaton, S.R., Eaton, G.R., Berliner, L.J., Eds.; Springer: Boston, MA, USA, 2005; pp. 125–146.
130. Pacelli, C.; Bryan, R.A.; Onofri, S.; Selbmann, L.; Shuryak, I.; Dadachova, E. Melanin is effective in protecting fast and slow growing fungi from various types of ionizing radiation. *Environ. Microbiol.* **2017**, *19*, 1612–1624. [CrossRef] [PubMed]
131. Casadevall, A.; Cordero, R.; Bryan, R.; Nosanchuk, J.; Dadachova, E. Melanin, radiation, and energy transduction in fungi. In *The Fungal Kingdom*; Heitman, J., Howlett, B., Crous, P., Stukenbrock, E., James, T., Gow, N., Eds.; ASM Press: Washington, DC, USA, 2017; pp. 509–514.
132. Selbmann, L.; Pacelli, C.; Zucconi, L.; Dadachova, E.; Moeller, R.; de Vera, J.; Onofri, S. Resistance of an Antarctic cryptoendolithic black fungus to radiation gives new insights of astrobiological relevance. *Fungal Biol.* **2018**, *122*, 546–554. [CrossRef] [PubMed]
133. Dadachova, E.; Bryan, R.A.; Howell, R.C.; Schweitzer, A.D.; Aisen, P.; Nosanchuk, J.D.; Casadevall, A. The radioprotective properties of fungal melanin are a function of its chemical composition, stable radical presence and spatial arrangement. *Pigment. Cell Melanoma Res.* **2008**, *21*, 192–199. [CrossRef] [PubMed]
134. Khajo, A.; Bryan, R.A.; Friedman, M.; Burger, R.M.; Levitsky, Y.; Casadevall, A.; Magliozzo, R.S.; Dadachova, E. Protection of melanized *Cryptococcus neoformans* from lethal dose gamma irradiation involves changes in melanin's chemical structure and paramagnetism. *PLoS ONE* **2011**, *6*, e25092. [CrossRef] [PubMed]
135. Cordero, R.J.; Casadevall, A. Functions of fungal melanin beyond virulence. *Fungal Biol. Rev.* **2017**, *31*, 99–112. [CrossRef] [PubMed]
136. Duru, K.C.; Kovaleva, E.G.; Danilova, I.G.; van der Bijl, P. The pharmacological potential and possible molecular mechanisms of action of *Inonotus obliquus* from preclinical studies. *Phytother. Res.* **2019**, *33*, 1966–1980. [CrossRef]
137. Géry, A.; Dubreule, C.; André, V.; Rioult, J.-P.; Bouchart, V.; Heutte, N.; de Pécoulas, P.E.; Krivomaz, T.; Garon, D. Chaga (*Inonotus obliquus*), a future potential medicinal fungus in oncology? A chemical study and a comparison of the cytotoxicity against human lung adenocarcinoma cells (A549) and human bronchial epithelial cells (BEAS-2B). *Integr. Cancer Ther.* **2018**, *17*, 832–843. [CrossRef]
138. Lee, J.H.; Hyun, C.K. Insulin-sensitizing and beneficial lipid metabolic effects of the water-soluble melanin complex extracted from *Inonotus obliquus*. *Phytother. Res.* **2014**, *28*, 1320–1328. [CrossRef]
139. Youn, M.J.; Kim, J.K.; Park, S.Y.; Kim, Y.; Kim, S.J.; Lee, J.S.; Park, R. Chaga mushroom (*Inonotus obliquus*) induces G0/G1 arrest and apoptosis in human hepatoma HepG2 cells. *World J. Gastroenterol.* **2008**, *14*, 511–517. [CrossRef]
140. Youn, M.J.; Kim, J.K.; Park, S.Y.; Kim, Y.; Park, C.; Kim, E.S.; Park, R. Potential anticancer properties of the water extract of *Inonotus obliquus* by induction of apoptosis in melanoma B16-F10 cells. *J. Ethnopharmacol.* **2009**, *121*, 221–228. [CrossRef]
141. Zheng, W.; Zhao, Y.; Zheng, X.; Liu, Y.; Pan, S.; Dai, Y.; Liu, F. Production of antioxidant and antitumor metabolites by submerged cultures of *Inonotus obliquus* cocultured with *Phellinus punctatus*. *Appl. Microbiol. Biotechnol.* **2011**, *89*, 157–167. [CrossRef] [PubMed]
142. Babitskaya, V.G.; Shcherba, V.V.; Ikonnikova, N.V. Melanin Complex of the Fungus *Inonotus obliquus*. *Appl. Biochem. Microbiol.* **2000**, *36*, 377–381. [CrossRef]
143. Salas, S.D.; Bennett, J.E.; Kwon-Chung, K.J.; Perfect, J.R.; Williamson, P.R. Effect of the laccase gene CNLAC1, on virulence of *Cryptococcus neoformans*. *J. Exp. Med.* **1996**, *184*, 377–386. [CrossRef] [PubMed]
144. Tsai, H.F.; Wheeler, M.H.; Chang, Y.C.; Kwon-Chung, K.J. A developmentally regulated gene cluster involved in conidial pigment biosynthesis in *Aspergillus fumigatus*. *J. Bacteriol.* **1999**, *81*, 6469–6477. [CrossRef] [PubMed]
145. Taborda, C.P.; da Silva, M.B.; Nosanchuk, J.D.; Travassos, L.R. Melanin as a virulence factor of *Paracoccidioides brasiliensis* and other dimorphic pathogenic fungi: A minireview. *Mycopathologia* **2008**, *165*, 331–339. [CrossRef]
146. Liu, D.; Wei, L.; Guo, T.; Tan, W. Detection of DOPA-melanin in the dimorphic fungal pathogen *Penicillium marneffei* and its effect on macrophage phagocytosis in vitro. *PLoS ONE* **2014**, *9*, e92610. [CrossRef]

147. Cunha, M.M.; Franzen, A.J.; Seabra, S.H.; Herbst, M.H.; Vugman, N.V.; Borba, L.P.; de Souza, W.; Rozental, S. Melanin in *Fonsecaea pedrosoi*: A trap for oxidative radicals. *BMC Microbiol.* **2010**, *10*, 80. [CrossRef]
148. Romero-Martinez, R.; Wheeler, M.; Guerrero-Plata, A.; Rico, G.; Torres-Guerrero, H. Biosynthesis and functions of melanin in *Sporothrix schenckii*. *Infect. Immun.* **2000**, *68*, 3696–3703. [CrossRef]
149. Rambach, G.; Blum, G.; Latge, J.P.; Fontaine, T.; Heinekamp, T.; Hagleitner, M.; Jeckstrom, H.; Weigel, G.; Wurtinger, P.; Pfaller, K.; et al. Identification of *Aspergillus fumigatus* surface components that mediate interaction of conidia and hyphae with human platelets. *J. Infect. Dis.* **2015**, *212*, 1140–1149. [CrossRef]
150. Zhang, J.; Wang, L.; Xi, L.; Huang, H.; Hu, Y.; Li, X.; Huang, X.; Lu, S.; Sun, J. Melanin in a meristematic mutant of *Fonsecaea monophora* inhibits the production of nitric oxide and Th1 cytokines of murine macrophages. *Mycopathologia* **2013**, *175*, 515–522. [CrossRef]
151. Shi, M.; Sun, J.; Lu, S.; Qin, J.; Xi, L.; Zhang, J. Transcriptional profiling of macrophages infected with *Fonsecaea monophora*. *Mycoses* **2019**, *62*, 374–383. [CrossRef]
152. Hsu, W.H.; Pan, T.M. *Monascus purpureus*-fermented products and oral cancer: A review. *Appl. Microbiol. Biotechnol.* **2012**, *93*, 1831–1842. [CrossRef] [PubMed]
153. Wu, X.-Z. A new classification system of anticancer drugs—Based on cell biological mechanisms. *Med. Hypotheses* **2006**, *66*, 883–887. [CrossRef] [PubMed]
154. Ho, B.Y.; Wu, Y.M.; Hsu, Y.W.; Hsu, L.C.; Kuo, Y.H.; Chang, K.J.; Pan, T.M. Effects of *Monascus*-fermented rice extract on malignant cell-associated neovascularization and intravasation determined using the chicken embryo chorioallantoic membrane model. *Integr. Cancer Ther.* **2010**, *9*, 204–212. [CrossRef] [PubMed]
155. Hsu, W.H.; Lee, B.H.; Pan, T.M. Effects of red mold dioscorea on oral carcinogenesis in DMBA-induced hamster animal model. *Food Chem. Toxicol.* **2011**, *49*, 1292–1297. [CrossRef] [PubMed]
156. Akihisa, T.; Tokuda, H.; Yasukawa, K.; Ukiya, M.; Kiyota, A.; Sakamoto, N.; Suzuki, T.; Tanabe, N.; Nishino, H. Azaphilones, furanoisophthalides, and amino acids from the extracts of *Monascus pilosus*-fermented rice (red-mold rice) and their chemopreventive effects. *J. Agric. Food Chem.* **2005**, *53*, 562–565. [CrossRef] [PubMed]
157. Su, N.W.; Lin, Y.L.; Lee, M.H.; Ho, C.Y. Ankaflavin from *Monascus*-fermented red rice exhibits selective cytotoxic effect and induces cell death on Hep G2 cells. *J. Agric. Food Chem.* **2005**, *53*, 1949–1954. [CrossRef]
158. Hsu, Y.W.; Hsu, L.C.; Liang, Y.H.; Kuo, Y.H.; Pan, T.M. Monaphilones A–C, three new antiproliferative azaphilone derivatives from *Monascus purpureus* NTU 568. *J. Agric. Food Chem.* **2010**, *58*, 8211–8216. [CrossRef]
159. Zheng, Y.; Xin, Y.; Shi, X.; Guo, Y. Anti-cancer effect of rubropunctatin against human gastric carcinoma cells BGC-823. *Appl. Microbiol. Biotechnol.* **2010**, *88*, 1169–1177. [CrossRef]
160. Knecht, A.; Cramer, B.; Humpf, H.U. New *Monascus* metabolites: Structure elucidation and toxicological properties studied with immortalized human kidney epithelial cells. *Mol. Nutr. Food Res.* **2006**, *50*, 314–321. [CrossRef]
161. Jo, D.; Choe, D.; Nam, K.; Shin, C.S. Biological evaluation of novel derivatives of the orange pigments from *Monascus* sp. as inhibitors of melanogenesis. *Biotechnol. Lett.* **2014**, *36*, 1605–1613. [CrossRef] [PubMed]
162. Chiu, H.-W.; Fang, W.-H.; Chen, Y.-L.; Wu, M.-D.; Yuan, G.-F.; Ho, S.-Y.; Wang, Y.-J. Monascuspiloin enhances the radiation sensitivity of human prostate cancer cells by stimulating endoplasmic reticulum stress and inducing autophagy. *PLoS ONE* **2012**, *7*, e40462. [CrossRef] [PubMed]
163. Deshmukh, S.K.; Mishra, P.D.; Kulkarni-Almeida, A.; Verekar, S.; Sahoo, M.R.; Periyasamy, G.; Goswami, H.; Khanna, A.; Balakrishnan, A.; Vishwakarma, R. Anti-inflammatory and anticancer activity of ergoflavin isolated from an endophytic fungus. *Chem. Biodivers.* **2009**, *6*, 784–789. [CrossRef] [PubMed]
164. Wagenaar, M.M.; Clardy, J. Dicerandrols, new antibiotic and cytotoxic dimers produced by the fungus *Phomopsis longicolla* isolated from an endangered mint. *J. Nat. Prod.* **2001**, *64*, 1006–1009. [CrossRef]
165. Li, X.; Sheng, Y.T.; Zhang, Y.-M.; Qin, J.-C. Cytotoxic azaphilone alkaloids from *Chaetomium globosum* TY1. *Bioorg. Med. Chem. Lett.* **2013**, *23*, 2945–2947. [CrossRef]
166. Cao, S.; McMillin, D.W.; Tamayo, G.; Delmore, J.; Mitsiades, C.S.; Clardy, J. Inhibition of tumor cells interacting with stromal cells by xanthones isolated from a Costa Rican *Penicillium* sp. *J. Nat. Prod.* **2012**, *75*, 793–797. [CrossRef]
167. Yasuhide, M.; Yamada, T.; Numata, A.; Tanaka, R. Chaetomugilins, new selectively cytotoxic metabolites, produced by a marine fish-derived *Chaetomium* species. *J. Antibiot.* **2008**, *61*, 615–622. [CrossRef]
168. Sakurai, M.; Kohno, J.; Yamamoto, K.; Okuda, T.; Nishio, M.; Kawano, K.; Ohnuki, T. TMC-256A1 and C1, New inhibitors of IL-4 signal transduction produced by *Aspergillus niger* var niger TC 1629. *J. Antibiot.* **2002**, *55*, 685–692. [CrossRef]

169. Huang, H.B.; Xiao, Z.E.; Feng, X.J.; Huang, C.-H.; Zhu, X.; Ju, J.-H.; Li, M.-F.; Lin, Y.-C.; Liu, L.; She, Z.-G. Cytotoxic naphtho-γ-pyrones from the mangrove endophytic fungus *Aspergillus tubingensis* (GX1-5E). *Helv. Chim. Acta* **2011**, *94*, 1732–1740. [CrossRef]

170. Myobatake, Y.; Takeuchi, T.; Kuramochi, K.; Kuriyama, I.; Ishido, T.; Hirano, K.; Sugawara, F.; Yoshida, H.; Mizushina, Y. Pinophilins A and B, inhibitors of mammalian A-, B-, and Y-family DNA polymerases and human cancer cell proliferation. *J. Nat. Prod.* **2012**, *75*, 135–141. [CrossRef]

171. Luo, X.; Lin, X.; Tao, H.; Wang, J.; Li, J.; Yang, B.; Zhou, X.; Liu, Y. Isochromophilones A-F, cytotoxic chloroazaphilones from the marine mangrove endophytic fungus *Diaporthe* sp. SCSIO 41011. *J. Nat. Prod.* **2018**, *81*, 934–941. [CrossRef] [PubMed]

172. Sun, C.; Ge, X.; Mudassir, S.; Zhou, L.; Yu, G.; Che, Q.; Zhang, G.; Peng, J.; Gu, Q.; Zhu, T.; et al. New Glutamine-containing azaphilone alkaloids from deep-sea-derived fungus *Chaetomium globosum* HDN151398. *Mar. Drugs* **2019**, *17*, 253. [CrossRef] [PubMed]

173. Xia, X.; Li, Q.; Li, J.; Shao, C.; Zhang, J.; Zhang, Y.; Liu, X.; Lin, Y.; Liu, C.; She, Z. Two new derivatives of griseofulvin from the mangrove endophytic Fungus *Nigrospora* sp. (Strain No.1403) from *Kandelia candel* (L.) druce. *Planta Med.* **2011**, *77*, 1735–1738. [CrossRef] [PubMed]

174. Gautschi, J.T.; Amagata, T.; Amagata, A.; Valeriote, F.A.; Mooberry, S.L.; Crews, P. Expanding the strategies in natural product studies of marine-derived fungi: A chemical investigation of *Penicillium* obtained from deep water sediment. *J. Nat. Prod.* **2004**, *67*, 362–367. [CrossRef]

175. Stierle, A.A.; Stierle, D.B.; Girtsman, T.; Mou, T.C.; Antczak, C.; Djaballah, H. Azaphilones from an acid mine extremophile strain of a *Pleurostomophora* sp. *J. Nat. Prod.* **2015**, *78*, 2917–2923. [CrossRef]

176. Yu, H.; Sperlich, J.; Mádi, A.; Kurtá, T.; Dai, H.; Teusch, N.; Guo, Z.-Y.; Zou, K.; Liu, Z.; Proksch, P. Azaphilone derivatives from the fungus *Coniella fragariae* inhibit NF-κB activation and reduce tumor cell migration. *J. Nat. Prod.* **2018**, *81*, 2493–2500. [CrossRef]

177. Kim, K.H.; Noh, H.J.; Choi, S.U.; Park, K.M.; Seok, S.-J.; Lee, K.R. Lactarane sesquiterpenoids from Lactarius subvellereus and their cytotoxicity. *Bioorg. Med. Chem. Lett.* **2010**, *20*, 5385–5388. [CrossRef]

178. Song, J.; Manir, M.M.; Moon, S.-S. Cytotoxic grifolin derivatives isolated from the wild mushroom *Boletus pseudocalopus* (Basidiomycetes). *Chem. Biodivers.* **2009**, *6*, 1435–1440. [CrossRef]

179. Yang, X.-L.; Qin, C.; Wang, F.; Dong, Z.-J.; Liu, J.-K. A New meroterpenoid pigment from the Basidiomycete *Albatrellus confluens*. *Chem. Biodivers.* **2008**, *5*, 484–489. [CrossRef]

180. Yaqoob, A.; Li, W.M.; Liu, V.; Wang, C.; Mackedenski, S.; Tackaberry, L.E.; Massicotte, H.B.; Egger, K.N.; Reimer, K.; Lee, C.H. Grifolin, neogrifolin and confluentin from the terricolous polypore *Albatrellus flettii* suppress KRAS expression in human colon cancer cells. *PLoS ONE* **2020**, *15*, e0231948. [CrossRef]

181. Li, B.; Li, W.; Chen, X.H.; Jiang, M.; Dong, M.S. In vitro antibiofilm activity of the melanin from *Auricularia auricula*, an edible jelly mushroom. *Ann. Microbiol.* **2012**, *62*, 1523–1530.

182. Zhu, H.; He, C.-C.; Chu, Q.-H. Inhibition of quorum sensing in *Chromobacterium violaceum* by pigments extracted from *Auricularia auricular*. *Lett. Appl. Microbiol.* **2011**, *52*, 269–274. [CrossRef] [PubMed]

183. Rasmussen, T.B.; Givskov, M. Quorum sensing inhibitors: A bargain of effects. *Microbiology* **2006**, *152*, 895–904. [CrossRef] [PubMed]

184. Mulrooey, C.A.; O'Brien, E.M.; Morgan, B.J.; Kozlowski, M.C. Perylenequinones: Isolation, synthesis, and biological activity. *Eur. J. Org. Chem.* **2012**, *21*, 3887–3904. [CrossRef] [PubMed]

185. Zhen, J.; Wu, D. Novel therapeutically and diagnostic applications of hypocrellins and hypericins. *Photochem Photobiol.* **1995**, *61*, 529–539.

186. Ma, Y.J.; Zheng, L.P.; Wang, J.W. Inducing perylenequinone production from a bambusicolous fungus *Shiraia* sp. S9 through co-culture with a fruiting body-associated bacterium *Pseudomonas fulva* SB1. *Microb. Cell Fact.* **2019**, *18*, 121. [CrossRef]

187. O'Brien, E.M.; Morgan, B.J.; Mulrooney, C.A.; Carroll, P.J.; Kozlowski, M.C. Perylene-quinone natural products: Total synthesis of hypocrellin A. *J. Org. Chem.* **2010**, *75*, 57–68. [CrossRef]

188. Rissanen, T.; Voutilainen, S.; Nyyssonen, K.; Salonen, J.T. Lycopene, atherosclerosis, and coronary heart disease. *Exp. Biol. Med.* **2002**, *227*, 900–907. [CrossRef]

189. Hernández-Almanza, A.; Montanez, J.; Martínez, G.; Aguilar-Jimenez, A.; Contreras-Esquivel, J.C.; Aguilar, C.N. Lycopene: Progress in microbial production. *Trends Food Sci. Technol.* **2016**, *56*, 142–148. [CrossRef]

190. Paranjape, S.R.; Riley, A.P.; Somoza, A.D.; Oakley, C.E.; Wang, C.C.C.; Prisinzano, T.E.; Oakley, B.R.; Gamblin, T.C. Azaphilones inhibit Tau aggregation and dissolve Tau aggregates in vitro. *ACS Chem. Neurosci.* **2015**, *6*, 751–760. [CrossRef]
191. Jansen, N.; Ohlendorf, B.; Arlette Erhard, A.; Bruhn, T.; Bringmann, G.; Imhoff, J.F. Helicusin E, isochromophilone X and isochromophilone XI: New chloroazaphilones produced by the fungus *Bartalinia robillardoides* strain LF550. *Mar. Drugs* **2013**, *11*, 800–816. [CrossRef] [PubMed]
192. McGleenon, B.M.; Dynan, K.B.; Passmore, A.P. Acetylcholinesterase inhibitors in Alzheimer's disease. *Br. J. Clin. Pharmacol.* **1999**, *48*, 471–480. [CrossRef]
193. Hsu, L.-C.; Hsu, Y.-W.; Liang, Y.-H.; Kuo, Y.-H.; Pan, T.-M. Anti-tumor and anti-inflammatory properties of ankaflavin and monaphilone A from *Monascus purpureus* NTU 568. *J. Agric. Food Chem.* **2011**, *59*, 1124–1130. [CrossRef] [PubMed]
194. Narsing Rao, M.P.; Xiao, M.; Li, W.-J. Fungal and bacterial pigments: Secondary metabolites with wide applications. *Front. Microbiol.* **2017**, *8*, 1113. [CrossRef] [PubMed]
195. Wu, X.; Fang, L.-Z.; Liu, F.-L.; Pang, X.-J.; Qin, H.-L.; Zhao, T.; Xu, L.-L.; Yang, D.-F.; Yang, X.-L. New prenylxanthones, polyketide hemiterpenoid pigments from the endophytic fungus *Emericella* sp. XL029 and their anti-agricultural pathogenic fungal and antibacterial activities. *RSC Adv.* **2017**, *7*, 31115. [CrossRef]
196. Chen, M.; Shen, N.; Chen, Z.-Q.; Zhang, F.-M.; Chen, Y. Penicilones A–D, anti-MRSA azaphilones from the marine-derived fungus *Penicillium janthinellum* HK1-6. *J. Nat. Prod.* **2017**, *80*, 1081–1086. [CrossRef]
197. Wang, W.; Liao, Y.; Chen, R.; Hou, Y.; Ke, W.; Zhang, B.; Gao, M.; Shao, Z.; Chen, J.; Li, F. Chlorinated azaphilone pigments with antimicrobial and cytotoxic activities isolated from the deep sea derived fungus *Chaetomium* sp. NA-S01-R1. *Mar. Drugs.* **2018**, *16*, 61. [CrossRef]
198. Wang, Y.; Shang, X.-Y.; Wang, S.-J.; Mo, S.-Y.; Li, S.; Yang, Y.-C.; Ye, F.; Shi, J.-G.; He, L. Structures, biogenesis, and biological activities of pyrano[4,3-c]isochromen-4-one derivatives from the fungus *Phellinus igniarius*. *J. Nat. Prod.* **2007**, *70*, 296–299. [CrossRef]
199. Kostrzewa, T.; Styszko, J.; Gorska-Ponikowska, M.; Sledzinski, T.; Kuban-Jan-kowska, A. Inhibitors of protein tyrosine phosphatase PTP1B with anticancer potential. *Anticancer Res.* **2019**, *39*, 3379–3384. [CrossRef]
200. Pan, B.-Q.; Xie, Z.-H.; Hao, J.-J.; Zhang, Y.; Xu, X.; Cai, Y.; Wang, M.-R. PTP1B up-regulates EGFR expression by dephosphorylating MYH9 at Y1408 to promote cell migration and invasion in esophageal squamous cell carcinoma. *Biochem. Biophys. Res. Commun.* **2020**, *522*, 53–60. [CrossRef]
201. Sibero, M.T.; Triningsih, D.W.; Radjasa, O.K.; Sabdono, A.; Trianto, A. Evaluation of antimicrobial activity and identification of yellow pigmented marine sponge associated fungi from Teluk Awur, Jepara, Central Java. *Indones. J. Biotechnol.* **2016**, *21*, 1–11. [CrossRef]
202. Patil, S.A.; Sivanandhan, G.; Thakare, D.B. Effect of physical and chemical parameters on the production of red exopigment from *Penicillium purpurogenum* isolated from spoilt onion and study of its antimicrobial activity. *Int. J. Curr. Microbiol. Appl. Sci.* **2015**, *4*, 599–609.
203. Vendruscolo, F.; Tosin, I.; Giachini, D.J.; Schmidell, W.; Ninow, J.L. Antimicrobial activity of *Monascus* pigments produced in submerged fermentation. *J. Food Process. Preserv.* **2014**, *38*, 1860–1865. [CrossRef]
204. Mani, V.M.; Priya, M.S.; Dhaylini, S.; Preethi, K. Antioxidant and antimicrobial evaluation of bioactive pigment from *Fusarium* sp. isolated from stressed environment. *Int. J. Curr. Microbiol. Appl. Sci.* **2015**, *4*, 1147–1158.
205. Saravanan, D.; Radhakrishnan, M. Antimicrobial activity of pigments produced by fungi from Western Ghats. *J. Chem. Pharm. Res.* **2016**, *8*, 634–638.
206. Yolmeh, M.; Hamedi, H.; Khomeiri, M. Antimicrobial activity of pigments extracted from *Rhodotorula glutinis* against some bacteria and fungi. *Zahedan J. Res. Med Sci.* **2016**, *18*, e4954. [CrossRef]
207. Miao, F.-P.; Li, X.-D.; Liu, X.-H.; Cichewicz, R.H.; Ji, N.-Y. Secondary metabolites from an algicolous *Aspergillus versicolor* strain. *Mar. Drugs* **2012**, *10*, 131–139. [CrossRef]
208. Seddon, J.M.; Ajani, A.; Sperduto, R.D.; Hiller, R.; Blair, N.; Burton, T.C.; Farber, M.D.; Gragoudas, E.S.; Haller, J.; Miller, D.T.; et al. Dietary carotenoids, vitamins A, C, and E, and advanced age-related macular degeneration. *JAMA* **1994**, *272*, 1413–1420. [CrossRef]
209. Panthama, N.; Kanokmedhakul, S.; Kanokmedhakul, K.; Soytong, K. Cytotoxic and antimalarial azaphilones from *Chaetomium longirostre*. *J. Nat. Prod.* **2011**, *74*, 2395–2399. [CrossRef]

210. Martínková, L.; Patáková-Juzlová, P.; Krên, V.; Kucêrová, Z.; Havlícêk, V.; Olsŏ ovský, P.; Hovorka, O.; Ríhová, B.; Veselý, D.; Veselá, D.; et al. Biological activities of oligoketide pigments of *Monascus purpureus*. *Food Addit. Contam.* **1999**, *16*, 15–24. [CrossRef]
211. Matsuzaki, K.; Takara, H.; Inokoshi, J.; Tanaka, H.; Masuma, R.; Omura, S. New brominated and halogen-less derivatives and structure-activity relationship of azaphilones inhibiting gp120-CD4 binding. *J. Antibiot.* **1998**, *51*, 1004–1011. [CrossRef] [PubMed]
212. Quang, D.N.; Harinantenaina, L.; Nishizawa, T.; Hashimoto, T.; Kohchi, C.; Soma, G.; Asakawa, Y. Inhibition of nitric oxide production in RAW 264.7 cells by azaphilones from *Xylariaceous* fungi. *Biol. Pharm. Bull.* **2006**, *29*, 34–37. [CrossRef]
213. El-Agamey, A.; Lowe, G.M.; McGarvey, D.J.; Mortensen, A.; Phillip, D.M.; George, T.G.; Young, A.J. Carotenoid radical chemistry and antioxidant/pro-oxidant properties. *Arch. Biochem. Biophys.* **2004**, *430*, 37–48. [CrossRef]
214. Rodrigues, E.; Mariutti, L.R.B.; Chisté, R.C.; Mercadante, A.Z. Development of a novel micro-assay for evaluation of peroxyl radical scavenger capacity: Application to carotenoids and structure–activity relationship. *Food Chem.* **2012**, *135*, 2103–2111. [CrossRef]
215. Britton, G. Structure and properties of carotenoids in relation to function. *FASEB J.* **1995**, *9*, 1551. [CrossRef]
216. Miyashita, K. Function of Marine Carotenoids. In *Food Factors for Health Promotion*; Yoshikawa, T., Ed.; Forum Nutr; Karger AG: Basel, Switzerland, 2009; Volume 61, pp. 136–146.
217. Endo, A. Monacolin K, a new hypocholesterolemic agent produced by a *Monascus* species. *J. Antibiot.* **1979**, *32*, 852–854. [CrossRef]
218. Yamada, R.; Yoshie, T.; Wakai, S.; Asai-Nakashima, N.; Okazaki, F.; Ogino, C.; Hisada, H.; Tsutsumi, H.; Hata, Y.; Kondo, A. *Aspergillus oryzae*-based cell factory for direct kojic acid production from cellulose. *Microb. Cell Fact.* **2014**, *13*, 71. [CrossRef]
219. Saeedi, M.; Eslamifar, M.; Khezri, K. Kojic acid applications in cosmetic and pharmaceutical preparations. *Biomed. Pharmacother.* **2019**, *110*, 582–593. [CrossRef]

Journal of *Fungi*

MDPI

Review

Fungal Pigments: Potential Coloring Compounds for Wide Ranging Applications in Textile Dyeing

Chidambaram Kulandaisamy Venil [1,*], Palanivel Velmurugan [2], Laurent Dufossé [3,*], Ponnuswamy Renuka Devi [1] and Arumugam Veera Ravi [2]

1 Department of Biotechnology, Anna University, Regional Campus – Coimbatore, Coimbatore 641046, Tamil Nadu, India; renukadev@gmail.com
2 Department of Biotechnology, Alagappa University – Science Campus, Karaikudi 630003, Tamil Nadu, India; palanivelmurugan2008@gmail.com (P.V.); aveeraravi@rediffmail.com (A.V.R.)
3 CHEMBIOPRO Chimie et Biotechnologie des Produits Naturels, ESIROI Département agroalimentaire, Université de la Réunion, F-97490 Sainte-Clotilde, Ile de La Réunion, Indian Ocean, France
* Correspondence: ckvenil1@gmail.com (C.K.V.); laurent.dufosse@univ-reunion.fr (L.D.); Tel.: +91-95-6659-0251 (C.K.V.); +262-2-6221-7544 (L.D.)

Received: 30 April 2020; Accepted: 17 May 2020; Published: 20 May 2020

Abstract: Synthetic pigments/non-renewable coloring sources used normally in the textile industry release toxic substances into the environment, causing perilous ecological challenges. To be safer from such challenges of synthetic colorants, academia and industries have explored the use of natural colorants such as microbial pigments. Such explorations have created a fervent interest among textile stakeholders to undertake the dyeing of textile fabrics, especially with fungal pigments. The biodegradable and sustainable production of natural colorants from fungal sources stand as being comparatively advantageous to synthetic dyes. The prospective scope of fungal pigments has emerged in the opening of many new avenues in textile colorants for wide ranging applications. Applying the biotechnological processes, fungal pigments like carotenoids, melanins, flavins, phenazines, quinones, monascins, violacein, indigo, etc. could be extracted on an industrial scale. This review appraises the studies and applications of various fungal pigments in dyeing textile fabrics and is furthermore shedding light on the importance of toxicity testing, genetic manipulations of fungal pigments, and their future perspectives under biotechnological approaches.

Keywords: fungal pigments; textile dyeing; toxicity testing; biotechnological approaches; challenges; limits

1. Introduction

Rapid industrialization in modern times has pressed the swift formulation and use of synthetic colorants in increased volumes in the food, medical, textile, and other industries regardless of their carcinogenic, immunosuppressive, and non-eco-friendly effects. Presently, many studies are focusing on finding alternatives to synthetic colorants, thereby improving the quality of the environment that affects various life forms. Nevertheless, advancements in biotechnology and the widespread awareness in ecological conservation, environmental protection, healthcare, etc. have generated a fervent interest among the public, industry, and researchers for exploring colorants from natural resources as an alternative to synthetic colorants.

Color plays a vibrant role in product acceptability in several industrial segments [1]. Dyes and pigments provide coloring ingredients and have been exploited by man for their artistic value. Dyes, being much smaller than pigments, are easily soluble during application and lose their nature whereas pigments, being about 1–2 μm, are insoluble. Moreover, pigments and dyes differ only by physical characteristics and not by chemical characteristics [2]. The dyes adhere to the surface and form covalent bonds with salts or metals by either physical adsorption or mechanical retention. Dyes used in the textile industries are carcinogenic and may trigger allergic reactions, and they have an environmental limitation because these dyes require strong acids, alkalis, solvents, and heavy metal catalysts, leading to toxic reactions.

Textiles is the key industry, among other industries, that uses diversified dyestuff. Lebeau et al. [3] reports that the anguish dyestuff industry requires higher natural pigments that have led to the emergence of natural colorants from microbes [4]. The increasing demand for pigments promotes multiplying research activities to explore filamentous fungi for pigment production. Research studies have shown that eco-friendly natural pigments from fungi are the best alternative to synthetic pigments because of their fast growth, easy processing, and important roles in transcription and intercellular communication. Fungal pigments have resistance and protect against biotic and abiotic agents (antagonistic microbes and UV radiation) and possess various biological applications [5]. Fungal pigments also have possible opportunities for commercializing their pigmented biomolecules for their application in food, cosmetics, textile fabrics, etc., because of their versatility, structure, and ease for large scale cultivation [6].

The present review appraises the research status of fungal pigments and their applicability in textile fabrics across broad characteristics like the ecology and diversity of fungal pigments, toxicity testing, biotechnological approaches for enhanced production, etc. Moreover, it aims to draw the attention of the academic and textile industries toward the challenges and the application of fungal pigments for textile dyeing, besides the prospective of genetic engineering research avenues in this field.

2. Historical Note on Pigments

The art of coloring has spread right from the development of human civilization. The use of pigments as coloring agents has been in vogue since prehistoric times. Europe and China, more than 5000 years ago, practiced dyeing with plants, leaves, barks, and insects. The Indus Valley-era, as early as 2600–1900 BCE (Before the Common Era or Before the Current Era), used madder dye to color garments at Mohenjodaro and Harappa. Roman centurions extracted red colorants from marine molluscs, *Murex* sp. to dye tunics and Egyptians used natural indigo from the plant *Isatis tinctoria.* Chinese Yanghai used alizarin, purpurin, and indigo in textiles in the Late Bronze Age (1700 BC). During the same Bronze Age, Phoenicians extracted Tyrian purple from the murex shellfish; though it was expensive, it was highly looked-for and so this colorant was the first initiation to global trade [7]. Generally, natural dyes have a strong tradition in India, Turkey, Mexico, Morocco, Europe, China, and countries of West Africa. Thus, natural dyes and dyestuffs are closely associated right from human civilization and are as old as textiles.

Dyes from insects (cochineal and kermes) were common in the 15th century. Cochineal (crimson) dye from cactus insects was in use to dye clothes and as an artist's pigment, and later in food items. France, Holland, and Germany started using plant dyes in industry in the 16th century, and England used wood to dye clothes in the 17th century. Following the gradual application of natural dyes in different parts of the world, quercitron, the pigment extracted from the inner bark of oak, was patented as a dyeing material in America in 1775 [8].

Later, in the 18th century, Swedish chemist, Scheele discovered that chlorine destroyed vegetable colors, and following that, indigo began to develop in England. A natural dye, cudbear, which was extracted from various lichen sources, was later patented. The use of saffron from plants and cochineal from animals to dye clothes have been reported, and natural colorants from plants and animals were used until the middle of the 19th century [9]. About 95% of plants were characterized for plant pigments in America and Europe, however, due to several drawbacks to extract color from plants (stability, shelf life, etc.), researchers in the dye stuff industries looked forward to other alternate sources for colorants with better stability and shelf life.

In 1856, as an alternative to natural colorants, William Henry Perkin, a British chemist, discovered the first synthetic dye 'mauve' from artificial quinine. It was a favorite color of the royal family and so its importance consequently promoted the bloom of the innovative synthetic colorant industries. Thereafter, the industrial revolution rapidly propelled the production of synthetic colorants, which attracted various markets due to the easy manufacturing processes and production of superior coloring properties with less cost. Commercial dyeing industries, appreciating the undercurrent of the industrial revolution, switched over to synthetic dyes due to their production advantages and market potential, and consequently to date, synthetic colorants rule these industries, especially in the textile industry.

3. Ecology of Fungal Pigments

Fungi are diverse and abundant eukaryotic organisms on earth and their presence, even in extreme ecosystems, make it possible for them to produce novel secondary metabolites. Fungi inhabiting a plethora of ecosystems from terrestrial milieus to marine environments are omnipresent. They are spread across various eco zones, from polar to tropical regions and from aerial to deep-sea environments [10]. Current studies have reported the production of new molecules, mainly new pigments from a marine environment. Several polyketide compounds with novel biological activity have been isolated from fungi in deep-sea environments [11]. Marine ecological niches are still mostly unexplored and characteristics of marine ecosystems like salinity, low temperature, and dark induce microbes to produce novel metabolites. Tropical ecosystems are potential niches, and it is mainly mangroves that have the highest diversity of marine fungi because of their rich organic matter, which favors the production of valuable metabolites. Researchers have found that in extreme conditions, pigmented fungi could tolerate hydration/dehydration cycles and high radiation better than non-pigmented fungi. For instance, fungal melanin produced by many filamentous fungi has antioxidant activity, thereby protecting the structures, and provide durability to survive in aggressive environments. Hirot et al. [12] reported a novel metabolite, a green pigment, terphenylquinone, from *A. niger* isolated from Mediterranean sponge, *Axinella damicornis.*

4. Fungal Pigments

Filamentous fungi produce amazing pigments like carotenoids, melanins, flavins, phenazines, quinones, and monascins from different chemical classes [13]. Carotenoids and polyketides come under the classification of fungal pigments [14]. Most fungi produce pigments that are water soluble and ideal for industrial production since it is easy to scale up in industrial fermenters and could be extracted easily without organic solvents (Figure 1).

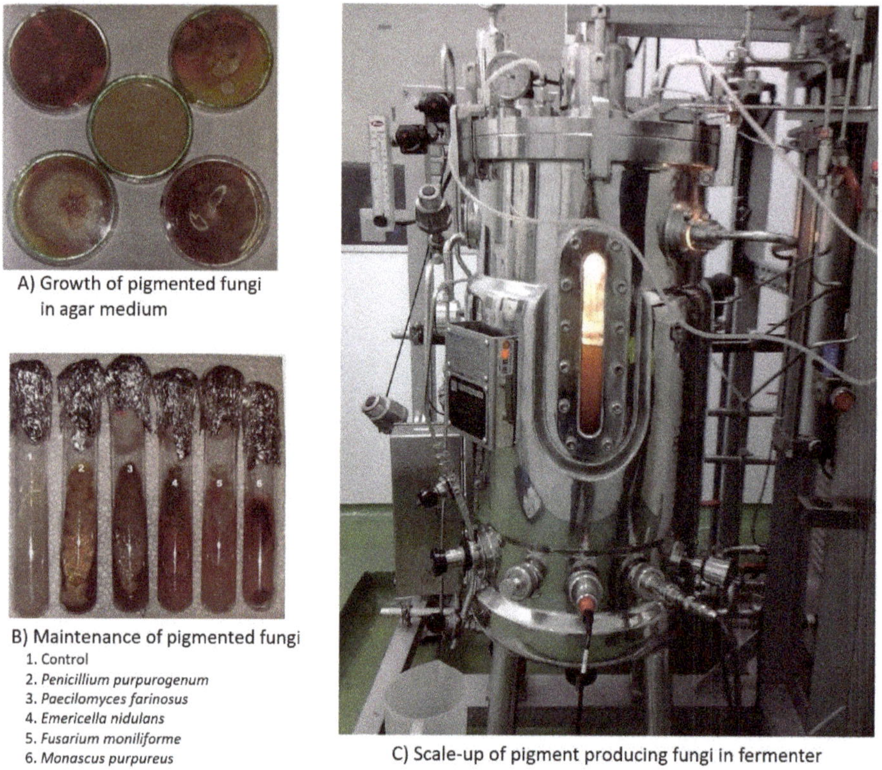

Figure 1. From Petri dish to fermenter: scale-up of pigment producing fungi. (**A**) Growth of pigmented fungi in agar medium; (**B**) Maintenance of pigmented fungi; (**C**) Scale-up of pigment producing fungi in fermenter.

4.1. Carotenoids

Carotenoids are formed through the isoprenoid pathway and produce a striking color with enormous biological activities like antioxidants, antimicrobials, membrane stabilizers, and precursors to vitamin A. Carotenoids are synthesized as carotenes (hydrocarbons) and xanthophylls (oxy-derivatives of carotenes). Several filamentous fungi produce different types of carotenoids (β-carotene, lycopene) and xanthophylls (astaxanthin, lutein, zeaxanthin, and violaxanthin), which are used in the animal feed industry for coloration and serve important roles as precursors of vitamin A.

Carotenoids are comprised mainly of C_{40} isoprenoids containing a polyene chain of conjugated double bonds [15]. They are lipid soluble pigments and have an aliphatic polyene chain that includes a conjugated double bond that acts as a chromophore and gives a yellow to red color. Avalos and Limon [16] reported that this conjugated polyene chain gives a chemical reaction against oxidizing agents and damage cell components. The carotene molecule is formed by head-to-head condensation of two geranylgeranyl precursors with acyclic $C_{40}H_{56}$ structures. Carotenoids possess ecological functions and protect against lethal photo-oxidation [17].

Fungal carotenoids are biosynthesized by the mevalonate pathway by 5-carbon isopentenyl pyrophosphate (IPP) as a precursor, synthesized from hydroxymethylglutaryl coenzyme A (HMG-CoA) or from derivatives of 1-deoxy-D-xylulose 5-phosphate (DXP) or 2-C-methyl-D-erythritol (MEP) generated from the condensation of pyruvate and glyceraldehyde 3-phosphate (non-mevalonate pathway) [18]. The IPP is condensed to form geranylgeranyl pyrophosphate (GGPP), which induces the formation of cis-phytoene. The introduction of α-/β ionone of the polyene chain gives the characteristic carotenoid structures. The MEP pathway starts from the condensation of pyruvate and glyceraldehyde-3-phosphate and is catalyzed by DXP synthase to produce DXP, consequently reduced into MEP by DXP reductase [19]. Avalos and Limon [16] reported that during the growth phase of *Penicillium* and *Phycomyces*, carotenoids are produced to protect from photo-oxidation. Bhosale et al. [20] reported that the cell functions are satisfied by sterols, dolichols, and ubiquinones, whereas as a response to environmental stress, astaxanthin and canthaxanthin are accumulated.

Ogbonna [21] reported that *Blakeslea trispora* produced β-carotene and lycopene; *Ashbya gossypii* gives lactoflavin (riboflavin), and *Penicillium oxalicum* produces a commercial colorant Arpink Red. β-carotene has been described in *Rhodosporidium* sp., *Penicillium* sp., *Aspergillus giganteus*, *Sclerotium rolfsii*, *Sclerotinia sclerotiorum*, and *Sporidiobolus pararoseus*.

4.2. Polyketides

Filamentous fungi produce more remarkable stable pigments than any other natural pigments [22]. Fungal polyketides are synthesized by polyketide synthases (PKS) from acetyl coA and malonyl CoA. Fungal polyketides are tetraketides and octoketides possessing eight C_2 units. Polyketides represent anthraquinones, hydroxyanthraquinones, naphthoquinones, and azaphilones. Polyketide based fungal pigments produce a wide spectrum of colors ranging from red, yellow, orange, brown, and black. These polyketides contain valuable bioactive properties like anticancer, immunosuppressors, antimicrobials, antibiotics, etc. [23]. Many fungal polyketide pigments including anthraquinone, hydroxyanthraquinones, and naphthoquinones produce a wide range of colors for industrial applications. Anthraquinones are produced by *Fusarium* sp., *Trichoderma*, *Aspergillus*, *Eurotium* sp., *Penicillium* sp., etc. Arpink Red^{TM} (*Penicillium oxalicum*) was the first commercial product marketed by Ascolor Biotech, Czech Republic [13].

The derivative of anthraquinone is hydroxyanthraquinone with the replacement of one hydrogen atom by hydroxyl groups. *Aspergillus* sp., *Fusarium* sp., *Penicillium* sp., etc. produce hydroxyanthraquinone as intermediate metabolites [24]. Naphthoquinones are mainly produced by *Fusarium* sp., possessing yellow, orange, and brown colors [25]. Azaphilones are synthesized via the polyketide pathway and are produced by *Aspergillus*, *Chaetomium*, *Monascus*, and *Penicillium* [26]. Azaphilones have similar chemical and molecular structures like *Monascus* pigments [27]. *Monascus* produce six types of polyketide pigments: monascin (yellow), ankaflavin (orange), monascorubin, rubropunctatin, monascorubramine, rubropuntamine (red), and these pigments are sensitive to heat, light, and pH. *Monascus* pigments react with chemicals in the medium (proteins, amino acids, etc.) and form water-soluble pigments [27]. *Monascus* sp. produce edible pigments like monacolins, dimerumic acid, ergosterol, and γ-aminobutyric acid [28]. *Monascus* azaphilone pigments belong to yellow (monascin 1 and ankaflavin), orange (rubropunctatin 3 and monascorubrin 4), and red (rubropunctamine 5 and monascorubramine 6), respectively, and are produced in the genera *Monascus* and *Talaromyces* [29,30]. *Monascus* azaphilone pigments were used as additives in cosmetics because of their pleasant color and excellent capability to absorb harmful UV rays [28] and as dyes for printers, textile yarn, and also to improve the efficiency of solar panels when this pigment is applied as a novel sensitizing dye in solar cells [28,29]. *Monascus* azaphilone pigments have biological activities like anti-diabetic, anti-inflammatory, anti-cancer, anti-microbial, and anti-obesity properties [29,31]. *Monascus* sp. also produces monacolin K (lovastatin) and are used as serum cholesterol lowering drugs because of their inhibition toward 3-hydroxy-3-methylglutaryl-coenzyme A reductase, which controls the biosynthesis of cholesterol [32].

In spite of its tremendous economic potential, the biosynthetic pathway of *Monascus* azaphilone pigments is proposed. The orange pigments (rubropunctatin 3 and monascorubrin 4) were formed by the esterification of a β-ketoacid (fattyacid synthase pathway) to chromophore (polyketide synthase pathway) pathway, reducing the orange pigments to yellow pigments (monascin 1 and ankaflavin). On the contrary, the amination of the orange pigments with NH_3 gives a red pigment (rubropunctamine

5 and monascorubramine 6) [33]. In addition to the typical *Monascus* azaphilone pigments, more than 100 of its congeners have been identified recently, which includes yellow (49), red (47), orange (8), and purple (1) pigments isolated from *Monascus* sp. and *Talaromyces (Penicillium)* sp. [30]. The structural diversity of these pigments is based on the number of carbon skeletons produced by core enzymes (polyketide synthases and non-ribosomal peptide synthetases) [34]. These carbon skeletons are modified by tailoring enzymes for various industrial applications, and metabolic engineering will ensure the safety and productivity of the fermentations.

4.3. Anthraquinones

Anthraquinones are found in *Aspergillus* sp., *Eurotium* sp., *Emericella* sp., *Fusarium* sp., *Penicillium* sp., *Mycosphaerella* sp., *Microsporum* sp., and exhibit a wide range of biological activities including antimicrobial, herbicidal, and insecticidal properties [35]. Anthraquinones produce a yellow color, whereas the substituents produce various hues of the molecules ranging from yellow, orange, red, brown, and violet. Anthraquinones are composed of three benzene rings with 1,10 dioxoanthracene ($C_{14}H_8O_2$) and have two ketone groups in the center. Most widespread fungal anthraquinones are 1,8 dihydroxy and 1, 5,8 trihydroxy anthraquinone derivatives. Anthraquinones are present either in free form or glycoside attached to the O- or C- bond in the side chain, which makes them water-soluble. The characteristics of fungal anthraquinones are dimeric structures formed by C–C bonds. The three ring structure suggests that these compounds can intercalate with DNA and are used in small doses [36]. Another important characteristic is their absorption spectra at 405 nm [37]. They exhibited a wide range of colors with chromatic properties and is of interest for dyeing in the most requested industries like cosmetics, textile dyeing, printing, and food industries. Anthraquinones are quite complex, with a great diversity of chemical structure and parameters (light, pH, temperature, oxygen transfer, carbon and nitrogen sources, inoculum concentration, etc.), which have a large impact on pigment production.

4.4. Hydroxyanthraquinone

Hydroxyanthraquinone isolated from *Haloresellinia* sp. (marine fungus) showed cytotoxic activities [38]. Another hydroxyanthraquinone, aspergillus H and I from *Aspergillus versicolor* have exhibited antiviral activity toward HSV-1 [39]. Shi et al. [40] reported two new hydroxyanthraquinones, namely harzianumnones A (1) and B (2), from the coral fungus *T. harzianum*, possessing cytotoxic activity against the HeLa cell line.

Anthraquinones represent a class of the quinone family with a basic structure of 9,10-dioxoanthracene containing two ketone groups on a central ring. The diversity of these compounds depends on the position of the substituents replacing H atoms on the basic structure. When *n*-hydrogen atoms are replaced by hydroxyl groups, the molecule is called hydroxyanthraquinone. 1,3,6,8-tetra-hydroxyanthraquinones were isolated from *Microsphaeropsis* sp. (associated with marine sponge), *Geosmithia* sp., *Trichoderma* sp., and *Verticicladiella* sp. [37]. Hydroxyanthraquinone derivatives were isolated from the mangrove fungi *Guignardia* sp. and *Halorosellinia* sp., possessing cytotoxicity against the cancer cell line [41]. Fouillaud et al. [37] reported that the 1-hydroxy-3-methylanthraquinone containing only one hydroxyl group on the R1 position had excellent cytotoxic activity against cancer cells, whereas dihydroxyanthraquinones with 1-hydroxy decreased the anticancer activity. 1,4,6,8-tetrahydroxyanthraquinones from *Aspergillus glaucus* showed excellent antibacterial activity against *Bacillus brevis*. The hydroxyanthraquinone derivatives like 1,3,8-trihydroxy-6-methyl-anthraquinone, aloe-emodin 8-O-glucopyranoside, 1,8-dihydroxy-3-methoxy-6-methyl-anthraquinone, and 1,4,5-trihdroxy-7-ethoxy-2-methyl-anthraquinone were isolated from *Drechslera rostrate* and *Eurotium tonpholium*, possessing anti-leishmanial activity [42].

Hydroxyanthraquinone pigments have excellent light fastness properties and no metallization required for dyeing and are considered as reactive dyes. They form complexes with various metals viz aluminum, barium, calcium, copper, iron, etc. and display excellent brightness compared to azo dyes. The ability to form this type of complex is of great concern in textile industries because they easily form covalent bonds with many fibers like cotton, wool, and nylon [37].

4.5. Naphthoquinones

Among the quinones, naphthoquinones are important secondary metabolites and more than 100 naphthoquinones have been identified with different structures in filamentous fungi [43,44], exhibiting various activities like anti-microbial, anti-inflammatory, and anti-cancer properties because of their tendency to prevent DNA damage [45,46]. Newman and Townsend [44] reported that naphthoquinones are considered as the model system to study the synthesis of polyketides in filamentous fungi. On the contrary, Manicilla et al. [45] demonstrated that a significant difference exists in naphthoquinone biosynthesis, representing the pathway as polyphyletic and not polyketide. The 1, 4-naphthoquinones congeners with *Monascus* azaphilone pigments are referred to as *Monascus* naphthoquinones and are detectable in trace quantities only while fermenting *Monascus* azaphilone pigments by the wild-type of *Monascus* sp. [46].

Naphthoquinones are widespread in fungi with phytotoxic, antimicrobials, insecticidal, anti-carcinogneic, cytostatic activities, etc. Naphthoquinones are produced by *Chlorociboria* sp. and *Arthrographis cuboidea* and they differ from naphthoquinones produced by *Trichoderma* sp. and *Fusarium* sp., possessing antagonistic properties toward insects, bacteria, and fungi [47]. Naphthoquinone-like pigments are produced by wood spalting fungi *Sctalidium cuboideum* (draconin-red, red pigment) and *Chlorociboria* sp. (xylindein, blue-green pigment) [48]. Among these naphthoquinone-like pigments, xylindein has been researched for solar energy and textile dyeing applications and when used in the textile and paint industry as dyestuffs, draconin-red crystals formed with excellent color stability.

More than 100 types of naphthoquinone metabolites are produced by 63 fungal species [49]. Naphthoquinones are closely related to chemical defense and have biological roles in electron transfer in various oxidative process (photosynthesis, oxidative phosphorylation) [50]. They possess antifungal activity against all *Candida* sp. (*C. albicans, C. krusei, C. kefyr, C. parapsilosis*), a multi drug resistant pathogen, among bone marrow patients and is resistant to fluconazole [51]. Ferreira et al. [52] reported that the chemistry behind the mechanism of action should be understood.

4.6. Azaphilones

Azaphilones are interesting secondary metabolites with a structurally diverse class, having pyrone-quinone structures possessing high-oxygenated bicyclic core and chiral quaternary center. The biosynthesis includes the fatty acid synthesis pathway and polyketide pathway. The first pathway assembles the polyketide chain from acetic acid and five malonic acid to form the chromphore structures. The second pathway produces medium chain fatty acids by the transesterification reaction.

Azaphilones, a fungal polyketide containing a pyrone-quinone bicyclic core and quaternary carbon center, have various activities like anti-microbial, anti-oxidative, anti-cancer, and anti-inflammatory properties [53–55]. Penicilones A-D, novel azophilones with different configurations at the quaternary carbon center, has remarkable anti-MRSA activity [56]. Citrifurans A-D, unusual dimers of azaphilones, has an inhibitory activity against LPS-induced NO production [57]. Recently, 13 different types of azaphilones are classified based on their chemical structure, which includes the austdiol-type, bulgarialactone-type, citrinin-type, deflectin-type, hydrogenated spiro-azaphilones, and O-containing *Monascus* pigments [54] obtained from *Penicillium* sp. and *Talaromyces* sp. with a wide range of biological activities.

Austdiol-type azaphilones are characterized by an austdiol core with 19 members that include fusaraisochromenone, nemanecins A-C, and perangustols A and B [58,59]. Felinone A from *Beauveria feline*, xylariphilone from *Xylariales* sp., and aspergillusone C from *Aspergillus clavatus* showed cytotoxicity against various cancer cell lines [60–62]. Bulgarialactone-type azaphilones, 5,6-dihydroxyacetosellin and monakaocinol, have a conjugated and linear γ-lactone ring. These types of azaphilones have been isolated from marine fungus, namely *Epicoccum nigrum* and *Monascus kaoliang* [63,64]. Citrinin-type azaphilones are monomeric citrinin derivatives like annulohypoxylomans A-C, annulohypoxylomanols A-B, annulohypoxyloside, and pentaketide [65]. The novel citrinin analogues (3S,4R)-6-hydroxy-8-methoxy-3,5-dimethyl isochromanol and (3S)-6-hydroxy-8-methoxy-3-

methylisochroman were isolated from *Penicillium* sp. [66]. The isochroman glycoside metabolites, monascuspilorin and monascupurpurin were produced by *Monascus pilosus* BCRC38072 and *Monascus purpureus* BCRC 31499 [67,68]. Penicitol A, a citrinin dimer from mangrove fungus *Penicillium chrysogenum* HND-11 has cytotoxic activity against HeLa, HEK-293, HCT-116, and A549 cell lines [69]. Deflectin-type azaphilones were characterized with a γ-lactone ring together with a ketone aliphatic chain [70]. Six deflectins from *Aspergillus deflectus* NCC0415 showed inhibitory activity against protein tyrosine phosphatases and SHP2. Colletotrichones from endophytic fungus *Colletotrichum* sp. exhibit remarkable antibacterial activity [71]. Coniellins from goose dung fungus, *Coniella fragariae,* showed inhibition of NF-kB activation in MDA-MB-231, thereby reducing the migration of tumor cells with more than 60% inhibition [70]. Hydrogenated spiro-azaphilones have a five- or six-membered ring spiroketal system on an azaphilone skeleton. Monascuspirolides A and B from *Monascus purpureus* BCRC 38110 showed stronger NO inhibitory activity with the IC_{50} value of 17.5 and 23.5 μM, respectively [72]. Peniazaphilin A from *Penicillium* sp. CPCC400786 has anti-HIV activity [73]. O-containing *Monascus* pigments has four azaphilones obtained from *Monascus* sp. Monascusazaphilones A and B from *Monascus purpureus* BCRC inhibited NO production by macrophages and it was stronger than the positive control [53].

Azaphilones are produced mainly by *Penicillium, Monascus, Chaetomium*, and *Talaromyces* exhibiting wide color tones as yellow, orange, and red. Xiong et al. [74] reported that the red color has been associated by reacting yellow pigments with amine groups by exchanging pyran oxygen for nitrogen. Azaphilones have a broad spectrum of applications in medicine like decreasing blood pressure, antioxidants, anti-microbials, etc. [75]. Osmanova et al. [55] reported that the activities are related to the reactions of azaphilones with the amino group in the formation of vinylogous γ pyridines.

5. Fungal Dyes for Textile Applications

The textile industry is the largest industry by economic contribution and employment generation. Fungal pigments have excellent colorfastness and staining properties and they are of great interest to textile industries; they warrant production under controlled conditions, with no seasonal fluctuations, and are biodegradable [14]. The burden of reducing the hazards to the environment rests with the dyeing industries. In spite of the positive features of fungal dyes, it has not met the expectations of the textile industry because of irregular fixation [76]. In addition, there are no standardized methods for the industrial dyeing of fungal dyes [48,77]. Therefore, a proper method to standardize the industrial dyeing of fungal dyes and many novel fungal pigments should be taken forward for standardized industrial applications.

Fungal pigments, due to their stability and consistency, have been reported for their use as alternatives to synthetic dyes in the textile industry [47]. These pigments can absorb light in the ultraviolet region and when applied as textile dyes, can protect human skin from harmful UV radiation [78] (Nambela et al. 2020). The potential of fungal pigments in the textile industries has been investigated by various researchers [79–82]. Through a biotechnological approach, anthraquinones are produced by fungi, namely *Trichoderma* sp., *Drechslera* sp., *Aspergillus* sp., and *Curvularia* sp. Cynodontin extracted from *Curvularia lunata* successfully produced two anthraquinone dyes similar to Disperse blue 7 and Acid green 28, the characteristics of which are similar to conventional dyes [83].

Anthraquinones have been reported to be produced from various fungi with antimicrobial activities. Anthraquinones from *Sclerotinia* sp. produce a pink shade and dyed cotton yarns with chemical and natural mordants. The yarns showed excellent stability to heat, light, pH, and temperature [84]. Osmanova et al. [55] reported the existence of water-soluble red and yellow pigments from *T. australis* and *P. murcianum* with an affinity toward wool. Results of this study indicated that the dyes were suitable for industrial conditions as they could withstand temperature and pH. Chemical analysis reported that the red dyes were similar to *Monascus* type pigments [85]. *Penicillium* sp. produces ankaflavin (*Monascus* type pigments) and the ionic nature has a strong affinity of dyes with wool [86]. Nagia and El-Mohamedy [87] reported that anthraquinones from *Fusarium oxysporum* could be used for dyeing wool with excellent color fastness properties and high dye uptake. Anthraquinones produce bright hues with excellent fastness properties and chemical modification may be interesting if it could facilitate the synthesis of dye molecules.

Morales-Oyervides et al. [88] reported the potential application of pigment from *Talaromyces* sp. in the textile industry. The results confirmed the pigment to be a novel source to dye wool textiles. High pigment uptake by the fabric was observed, and the dyeing rate constant and half time dyeing with kinetic behavior well matched those of natural dyes used in the textile industry. The *Talaromyces* pigment has potent antimicrobial properties coupled with the absence of toxicity, which makes this pigment a valuable alternative as a natural dye in textile dyeing [89]. Further characterization of the molecule and study on the interactions between the dye and fabric are worth future research. Celestino et al. [90] reported pigmented fungi *Penicillium sclerotiorum* 2AV2, *Penicillium sclerotiorum* 2AV6, *Aspergillus calidoustus* 4BV13, *Penicillium citrinum* 2AV18, and *Penicillium purpurogenum* 2BV41. *P. sclerotiorum* 2AV2 from soil from the Amazon produces intense color pigments, which could be used for textile applications.

Hinsch et al. [91] reported that the fungal pigments isolated from rotting hardwood logs in Canada produced xylindein (green pigment from *C. aeruginosa*), draconin red (red pigment from *S. cuboideum*), and yellow pigment (*S. ganodermophthorum*) and were able to dye multi fabric test strips. Their results indicated that all these pigments could be used to dye fabrics without the need for additional chemicals. Xylindein exhibited good potential to dye garment fabrics and draconin red for second layer garment fabrics. Spalting (wood-rotting) fungi have been the hot topic of research to extract novel pigments for textile applications. Awkwardly, dichloromethane (DCM) is used to extract colorants from spalting fungi and causes environmental problems and health issues, which is one of the major hurdles holding back spalting fungi from commercialization. Researchers have found that natural oils could be used to transfer pigments from *Chlorociboria* sp., *Scytalidum cuboideum*, and *Scytalidium ganodermophthorum* onto host substrates [92].

Gupta et al. [93] attempted to isolate pigment from *Trichoderma* sp. and explored the possibility for dyeing silk and wool fabrics. The dye from *Trichoderma* was non-toxic to human skin and possesses antimicrobial properties. The color value for the dyed sample was higher for wool when compared to silk fabric, and showed excellent fastness properties to washing and rubbing. Poorniammal et al. [94] evaluated the yellow pigment from *Thermomyces* sp. for the textile dyeing process. Natural mordants and the yellow pigment reduced the influence of pathogens. The overall color fastness properties for the dyed silk fabric was moderate. Due to these antimicrobial properties, this can be used especially in medical applications like bandages, masks, wound dressings, etc. Devi and Karuppan [95] reported on reddish brown pigments from *Alternaria alternata* for their efficiency to dye cotton fabrics. The pigments showed good color fastness properties to perspiration and rubbing. This was the first report to study the reddish brown pigments for *Alternaria alternata* for dyeing cotton fabrics. Velmurugan et al. [96] assessed different water-soluble fungal pigments from *Monascus purpureus*, *Isaria farinosa*, *Emericella nidulans*, *Fusarium verticillioides*, and *Penicillium purpurogenum* for dyeing cotton yarns (Figure 2). This study testified that pre-mordanting with alum and ferrous sulfate achieved variation in shade and color.

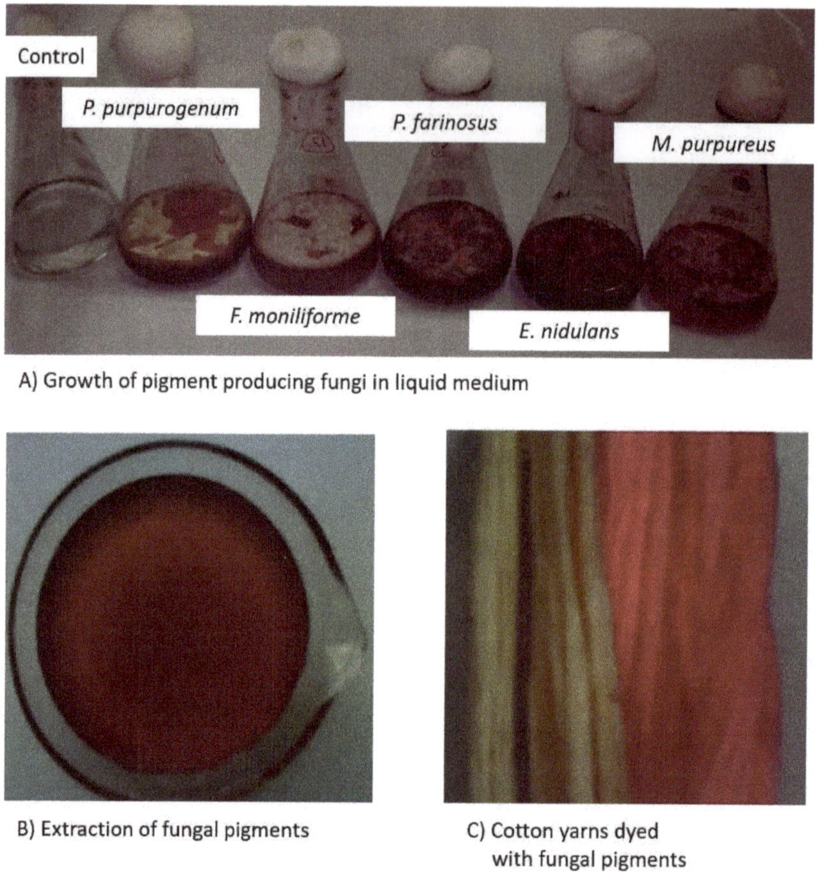

Figure 2. Water-soluble fungal pigments for dyeing cotton yarns. (**A**) Growth of pigment producing fungi in liquid medium; (**B**) Extraction of fungal pigments; (**C**) Cotton yarns dyed with fungal pigments.

The reddish-brown pigment from *Phymatotrichum* sp. (NRC 151) was produced using H-acid (1-naphthol-8 amino, 3,6-disulfonic acid) as a precursor in the medium that showed better fastness to washing, perspiration and light and can be used to dye various fabrics [24]. Sharma et al. [77] isolated *Trichoderma virens, Alternaria alternata,* and *Curvularia lunata* from different habitats for pigment production. The pigments applied to silk and wool showed good fastness properties. The optimum condition for maximum pigment production should be standardized for commercialization in an eco-friendly manner with a cost reduction.

Weber et al. [48] isolated three eco-friendly fungal pigments from *S. cuboideum, C. aeruginosa,* and *S. ganodermophthorum,* which showed their textile dyeing capability without using water and thermal energy. The pigment showed no fading over one week's time and further research is required to study the relationship between the pigments and fabrics, skin sensitivity, toxicity, and stability (UV, time, wash and wearing). Different fungal species with their active pigment for application in the textile industry are shown in Table 1 and Figure 3.

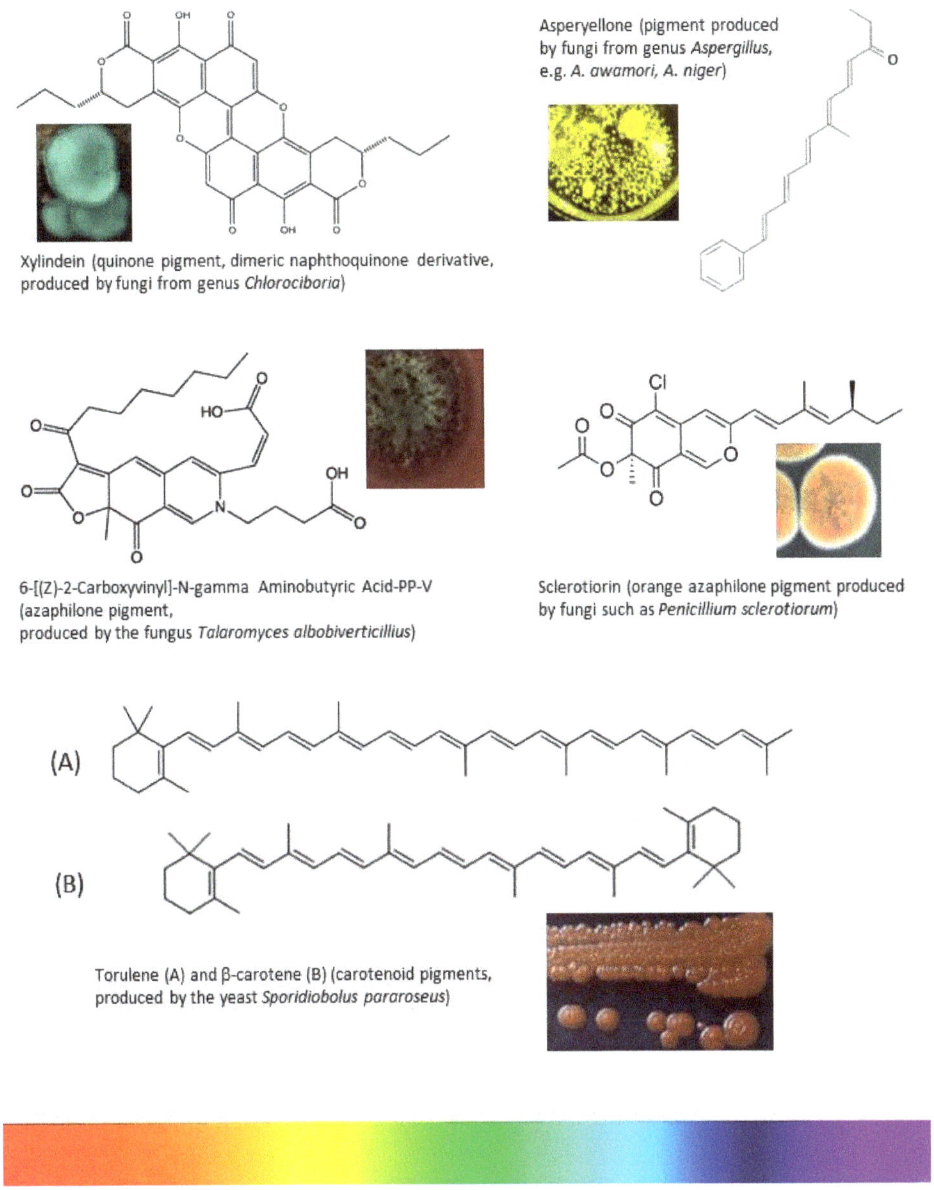

Figure 3. Chemical structures of fungal pigments with potential coloring properties that could be used in textile dyeing.

Table 1. Fungal pigments and their application in the textile industry.

Fungi	Pigment	Color	Fabrics	References
Penicillium oxalicum	Anthraquinones	Arpink Red	Wool	Sardaryan et al. [97]
Trichoderma virens	Anthraquinones	Yellow		
Alternaria alternata	Anthraquinones	Reddish-Brown	Silk, Wool	Sharma et al. [77]
Curvularia lunata	Anthraquinones			
Alternaria alternata	Anthraquinones	Reddish-Brown	Cotton	Devi and Karuppan [95]
Thermomyces sp.	Anthraquinones	Yellow	Cotton, Silk, Wool	Poorniammal et al. [94]
Trichoderma sp.	Anthraquinones	Yellow	Cotton, Silk, Silk cotton	Devi [98]
Trichoderma sp.	Anthraquinones	Yellow	Silk, Wool	Gupta et al. [93]
Penicillium oxalicum (NRC M25)	Anthraquinones	Faint Reddish- Brown	Wool	Mabrouk et al. [24]
Sclerotinia sp.	Anthraquinones	Pinkish-Red	Cotton	Perumal et al. [84]
Aspergillus sp. AN01	Asperyellone	Yellow	Silk, Cotton, Synthetic and Wool fabrics	Iswarya et al. [99]
Monascus purpureus		Red		
Penicillium purpurogenum		Yellow		
Isaria farinosa	Azaphilones	Pink	Cotton	Velmurugan et al. [96]
Fusarium verticillioides		Reddish- Brown		
Emericella nidulans		Red		
Penicillium murcianum	Carotenoids	Yellow	Wool	Hernandez et al. [100]
Talaromyces australis		Red		
Talaromyces australis	2, 4-Di-tert-butylphenol	Red	Cotton fabric	Shibila and Nanthini [101]
Phoma herbarum	Magenta pigment	Magenta	Nylon	Chiba et al. [102]
Monascus purpureus	Monascorubramine	Red	Wool	De santis et al. [103]
Talaromyces verruculosus	Polyketide	Red	Cotton fabric	Chadni et al. [89]
Monascus purpureus	Rubropunctamine	Red	Wool	De santis et al. [103]
Chlorociboria aeruginosa		Green	Bleached cotton, Spun polyamide, Spun polyester, Spun polyacrylic, Worsted wool	Weber et al. [48]; Hinsch et al. [91]
Scytalidium cuboideum	Quinones	Red		
Scytalidium ganodermophthorum		Yellow		
Aspergillus sp.	Quinones	Brown	Cotton, Silk, Silk cotton	Devi [98]
Alternaria alternata	Quinones	Reddish-Brown	Cotton	Gokarneshan [104]
Acrostalagmus (NRC 90)		Brown		
Alternaria alternata (NRC17)		Reddish-Brown		
Alternaria sp. (NRC 97)		Brown		
Aspergillus niger (NRC 95)		Brown		
Bisporomyces sp. (NRC 63)		Deep Brown		
Cunninghamella (NRC 188)	Quinones	Faint Reddish-Brown	Wool	Mabrouk et al. [24]
Penicillium chrysogenum (NRC 74)		Deep Brown		
Penicillium italicum (NRC E11)		Brown		
Penicillium regulosum (NRC 50)		Brown		
Phymatotrichum sp. (NRC 151)		Reddish-Brown		

6. Toxicity Testing for Fungal Pigments

Heo et al. [105] studied the toxicity of fungal pigments extracted from *Penicillium miczynskii, Sanghuangporus baumii, Trichoderma* sp. 1, and *Trichoderma afroharzianum*. The pigments exhibited high radical-scavenging activity. Moderate toxicity was observed in *S. baumii* by the acute toxicity test limiting the applications of this pigment in industry. *P. miczynskii, Trichoderma* sp. 1, and *T. afroharzianum* were reported to be the best fungal pigment for producing strains for safe, water-soluble pigments in the industry.

The cytotoxic activity of fungal pigments from *F. oxysporum, T. verruculosus*, and *Chaetomium* sp. has been tested using various methods such as the yeast toxicity test (YTT), brine shrimp lethality bioassay, or cell counting kit-8 assay. This method of cytotoxicity assay warrants the application of pigments in various industries, especially in the health and pharmaceutical sectors [87]. Poorniammal et al. [106] studied the dermal toxicity of pigments from *Thermomyces* sp. and *P. purpurogenum* in Wistar rats and their results showed that the pigments were non-toxic and have broad scope in the dyeing, printing, and cosmetics industries.

The deep sea fungus *Chaetomium* sp. AN-S01-R1 produces chlorinated azaphilone pigments like chaephilone C (compound **1**) and chaetoviridides A-C (compounds **2–4**). Compound **2** exhibited potent cytotoxic activities toward HepG2 cells with IC_{50} below 5 μM, whereas compounds **1** and **3** showed stronger cytotoxic activities against HeLa cells [107]. To study the cytotoxic mechanism of pigments and their possible industrial applications, further research should be conducted.

The red pigment from *Talaromyces verruculosus* was used to dye cotton fabrics with excellent dye uptake and color fastness properties. The cytotoxicity assay conducted using the brine shrimp lethality test revealed insignificant toxicity and was harmless to use. Hence, researchers should focus on obtaining pigments from this strain for industrial applications and further chemical characterization will open new avenues in the dyeing industry that will reduce the adverse effects of synthetic dyes [104].

Pandiyarajan et al. [108] reported the presence of water-soluble yellow pigments from *Aspergillus* sp. The pigment possessed maximum dyeing ability of the fungus with the hydrothermal method for textile fabrics without mordants and showed better dye uptake in comparison with synthetic dyes. The toxicity of the pigment was tested using a zebra fish model system with an IC_{50} value of 710 μg/mL. This novel pigment can be used an alternative to synthetic dyes for applications in the textile industrial sector.

7. Biotechnology Ways for Enhanced Production

The improvement of fungal pigment production and the knowledge to improve the yield have been gradually studied and extended based on the following major criteria: (1) Through genetic manipulation, the pathways for pigment production and the molecules are better understood as many fungal genomes have been sequenced and many others are still in progress; (2) molecular screening techniques help to improve the gene expression and secretion of unusual metabolites; and (3) the use of optimization strategies such as the use of artificial intelligence to improve the fermentation conditions for pigment production.

Today, the higher production of synthetic dye based industries makes it challenging to use microbes at the industrial level. The major reason associated with this is the higher production cost for instruments, chemicals, and processing. The same can be sorted out using agro-industrial residues for the cheap production with same metabolites at the industrial level. Numerous research on fungal pigments has led to the robustness and tolerance against possible stress at the industrial level, which warrants the efficient production of fungal pigments.

7.1. Genetic Manipulation

The biosynthesis of pigment producing fungi has not been studied in detail and hence genomic screening for pigmented fungi is not possible at the premature stage. The classical methods such as taxonomy, biochemistry, and physiology will be very active in screening pigment producing fungi. The major breakthrough is to genetically modify the strain by genetic transformation, enabling scientists/researchers to modify the targeted gene for unusual metabolites. The genetic transformation of fungi is tedious because of its complex cell wall structures and lack of genetic markers. Hence, species-specific transformation must be required and optimized for every potent strain.

Using genetic engineering approaches like genetic modification, cloning of genes, and exclusion of non-essential genes (toxins), various research studies have been undertaken for enhancing fungal pigment production. Fungal carotenoids from native carotenogenic fungi, *Blakeslea trispora* and *Phycomyces blakesleeanus* are produced up to the industrial scale (17 grams of β-carotene per liter in some cases). Several attempts have been made to introduce the biosynthetic pathway of lycopene or β-carotene in non-carotenogenic strains using metabolic engineering methods. The enhancement of isopentenyl diphosphate (IPP) and dimethylallyl diphosphate (DMAPP) precursors of the MEP or MVA pathways have demonstrated the improved production of lycopene and β-carotene [109].

Fungal strains producing polyketides are not easily cultured in liquid fermentations due to their filamentous growth and dense mycelial growth, which result in increased viscosity and reduced oxygen transfer. The transfer of the polyketide biosynthetic pathway to an industrial strain such as *Saccharomyces cerevisiae* offers an attractive alternative as it allows for easier fermentation and process optimization by metabolic engineering strategies [110]. Type 1 polyketide synthases (PKS) play a vital role in the synthesis of fungal polyketides. The PKS are proteins and are related to eukaryotic fatty acid synthases. Acetyl coenzyme A and malonyl CoA condense to form carbon chains whereas ketoacyl CoA synthase, acyltransferase, acyl carrier domains are also required for the synthesis of polyketides [4]. In *Monascus*, three pigmented or not polyketides are known to be produced, namely citrinin, red pigment, and monacolin K [23]. Many approaches have been used to decrease the production of citrinin (mycotoxin), thereby increasing the production of red pigment. The optimization of various parameters like nitrogen sources, pH, dissolved oxygen, and important genetic alterations were employed to reduce the production of citrinin. In the industrial strain, *M. purpureus* SM001, the polyketide synthase gene *pksCT* disruption, successfully eliminated citrinin production [111].

The investigation about polyketide pigments produced by *F. graminearum, F. decemcellulare,* and *F. bulbigenum* and characterized as naphthoquinones explained that their biosynthesis was the response to environmental stress [112]. Aurofusarin, the main naphthoquinone produced by *Fusarium* sp., exhibited that the gene cluster *PKS12* was responsible for aurofusarin biosynthesis under the control of local transcription factor *AurR1* [113]. The *PKS12* gene cluster is also responsible for the production of rubrofusarin and fuscofusarin [114]. Rubrofusarin is an orange-brown pigment reported to inhibit human DNA topoisomerase II-α and has antibiotic effects on *Mycobacterium tuberculosis.* The polyketide core scaffold differs in their tailoring enzymes, resulting in the production of end products like rubrofusarin B, aurofusarin, nigerone, and nigerasperone A.

The yellow pigment chrysogine by *Penicillium chrysogenum* protects the fungi from ultraviolet radiation and lacks antimicrobial and anticancer activities [115]. The putative biosynthetic gene cluster identified in *P. chrysogenum* includes non-ribosomal peptide synthetase (NRPS) [116], and lately, Wollenberg et al. [117] proved that NRPS is responsible for chrysogine biosynthesis in *Fusarium graminearum.* Anthraquinones have been genetically modified using *Aspergillus* sp., expressing genes related to monodictyphenone and atrochrysone biosynthesis. The metabolic gene cluster for the biosynthesis of red pigment bikaverin from *Fusarium* sp. includes gene encoding PKS (*bik1*) and two genes (*bik2* and *bik3*) encoding tailoring enzymes as well as general transcriptional activator (*bik4*), specific transcriptional activator (*bik5*), and a transporter (*bik6*) [118].

Sen et al. [119] reported that various methods decreased the production of citrinin and mycotoxin, thereby increasing the production of red pigments. The polyketide synthase gene has been studied extensively for the synthesis of citrinin by *Monascus purpureus.* The polyketide synthase gene pksCT was effectively cloned to remove citrinin, thereby enhancing red pigment production by the industrial strain *M. purpureus* SM001 [111]. Lee et al. [120] enhanced the production of monacolin K by *M. pilosus* by *laeA* overexpression using an *A. nidulans alcA* promoter.

The metabolic engineering in fungi is extremely tedious due to the lack of genetic markers and low gene targeting frequencies. CRISPR (Clustered Regularly Interspaced Short Palindromic Repeats) has been successfully used to produce compounds of industrial importance. It consists of enzyme Cas9, the molecular scissors that makes a cut in the target location, enabling the addition or removal of pieces of DNA and guide RNA located inside a longer RNA scaffold. This scaffold binds to target DNA whereas the guide RNA directs the Cas9 enzyme to make a cut at the right point, activating the DNA repair mechanism, and can be used to introduce changes to one or more genomes [121]. Hence, this system is successfully used in genetic engineering to make cell factories for the cost efficient production of natural pigments [122].

Nielson et al. [123] studied the CRISPR cas9 system for *Aspergillus nidulans*, which can also be applied to other fungal types. This has been used for *Talaromyces atroroseus*, a red pigment producer targeted for the food industry. Pohl et al. [116] reported the use of CRISPR cas9 in *Penicillium chrysogenum* with improved production. Limited research is available for pigmented microbes and more research is needed to optimize this system for future industrial applications.

Jia et al. [111] studied the elimination of the mycotoxin citrinin by metabolic engineering. A binary vector system was constructed that disrupts the polyketide synthase gene pks CT in *M. purpureus* SM001 by *Agrobacterium tumefaciens* mediated transformation. The established system was evaluated and showed a high efficiency to improve industrial *Monascus* strains. Westphal et al. [113] reported the enhanced production of aurofusarin by *Fusarium graminearum* by examining the transcription factor AurR1 on the aurofusarin gene cluster in the strain. The overexpression of Aur R1 increased five proteins of the aurofusarin pathway, leading to the 3-fold increase in the production compared to the wild strain.

7.2. Agro-Waste for Clean-Up Production

Agricultural activities produce waste (straw, stem, stalk, husk, peel, legumes, seeds, bagasse, spent grains etc.) throughout the year. There is a great interest in re-using these nutrient rich materials for bioprocessing. The use of this waste can act as a substrate for low cost raw materials to make the process for the production of value added products. Recently, raw materials and byproducts of agro-industrial residues have been considered as low cost substrates for pigment production by microbes (Figure 4). Lopes et al. [124] reported that different strains of filamentous fungi produced pigments on a cheese whey and soya protein medium. Similarly, Kaur et al. [125] studied the enhancement of yellowish pink pigment by *Rhodotorula rubra* MTCC 1446 in a whey and coconut water medium.

Figure 4. Growth of *Monascus* sp. in different agro substrates. (**A**) Submerged fermentation in peanut medium; (**B**) Solid state fermentation in boiled rice; (**C**) Solid state fermentation in peanut husk.

Subhasree et al. [126] stated that red pigment by *Monascus purpureus* was produced by utilizing jackfruit seed as a substrate, and another study reported the production of red pigment by utilizing durian seeds [127]. By utilizing grape waste as a growth substrate, red pigment production by *Monascus purpureusin* was optimized using the response surface methodology [128]. The production of carotenoids by *Rhodotorula glutinis* 1151 was optimized by RSM using a tomato waste medium. Sugarcane and corn bagasse have also been used for pigment production by *Monascus* sp. [129]. Taskin et al. [130] prepared a cheap peptone source by utilizing chicken feather waste through acid hydrolysis and investigated the suitability of this peptone source as a substrate for carotenoid production by *Rhodotorula glutinis* MT-5. Interestingly, 92 mg/L of carotenoids were produced, and this source could be used as a cheap substrate for carotenoid production.

The industries, as a beneficial measure to them, may utilize agro-industrial residues as inexpensive growth substrates for microorganisms. Consuming residues of agro-industries in a bioprocess reduces the costs of production, also solving pollution problem associated with their disposal. Lopes et al. [124] reported the readily available source for pigments produced from *Penicillium chrysogenum* IFL1 and IFL2, *Fusarium graminearum* IFL3, *Monascus purpureus NRRL* 1992, *Penicillium vasconiae* IFL4. All of the fungi produced the water-soluble pigments monascorubin, rubropunctatin, and mycotoxin citrinin on agro-industrial residues (feather meal, fishmeal, cheese whey, grape waste, soybean meal, and rice husk). Sanchez [131] reported the use of agro-industrial residues for low cost alternatives for pigment production, thereby reducing 73% of the total production cost. The main aim is to lower the cost by reducing the cost of raw materials and reducing environmental pollution.

8. Limits, Challenges, and Future Scope for the Dyeing of Fungal Pigments

The reporting of synthetic textile colorants as carcinogenic because of the presence of dioxins (polychlorinated dibenzo-p-dioxins and polychlorinated dibenzofurans) has resulted in the development of eco-friendly, non-toxic colorants. Fungal colorants from a cost effective process that display high colorfastness properties are promising. The authorized usage of fungal pigments varies in diverse parts of the world with respect to its local and traditional usage [13]. For example, the applications of *Monascus* like pigments from *Penicillium* sp. on textile, cotton, wool, leather, paper, paint, etc. have been patented by Mapari et al. [132]. Added important features include the enhanced production of pigments, solubility in water, and stable pigments [133]. The pigment yield can be enhanced by improving the fungal growth or by increasing the accumulation of pigments in the cells [134].

The main challenge is the difficulty in increasing both pigment and biomass yield, since both are negatively correlated. Hence, the relationship between biomass and pigment production needs to be thoroughly studied in order to produce pigments in a controlled way. Pigment production can be controlled by using genetic engineering techniques. Recombinant DNA technologies have been used to alter the activity of enzymes involved in the biosynthesis of carotenoids [135]. The challenges associated with the commercial production of fungal pigments (*Monascus* pigment) is its low solubility in water and poor stability to pH, heat, and light. Dufossé [136] reported that methods have been developed to address these challenges by replacing oxygen with nitrogen from the amino group in the pigment's structure. Furthermore, the major problem is the co-production of toxic compounds with the pigment limiting its applications and preventing regulatory approval.

Most fungi produce a mixture of pigments and the challenge is to produce a pigment with one specific color tone. This can be achieved by optimizing the parameters like substrates, pH, temperature, dissolved oxygen either in solid state or in submerged fermentation. The prospect of developing fungal pigments at the industrial level by selecting the strain with safety measures is imperative. Cost-effective fungal pigments can be produced at lab scale and the technical hurdles that arise when designing the industrial plant need attention. Another important issue is the stability of the pigment over time.

The valuable metabolites from fungi are limited for its commercialization due to the production of mycotoxins together with pigments. Furthermore, toxic worries of fungal pigments have arisen. Toxicological tests are required depending on the final applications of the pigments in either food, cosmetics, textiles, etc. Leading companies have filed numerous patents for *Monascus*-like pigments and the products are in the market. For centuries, *Monascus* pigments have been used in Asia. Most importantly, red pigments from *P. oxalicum* (Natural Red TM), β-carotene, and lycopene from *Blakesla trispora* have been authorized in Europe.

The use of microbes has tremendous advantages as it does not require petroleum based raw materials and this influences the price of the pigment. The qualitative and quantitative research on fungal pigments should be intensified with negative ecological impact. Hence, there is a necessity to explore novel pigments producing fungi from different taxonomic groups to meet the existing demand for natural pigments. The chemical diversity of fungi should be explored for identifying the potent pigments and toxicological testing must be carried out to be accepted by consumers.

9. Conclusions

The growing universal concern is promoting the needs of eco-friendly natural dyes from sustainable resources to counter the hazardous effects of synthetic colorants on the environment and human health. In recent times, the potential of fungal pigment as one of the microbial pigments is explored for textile dyeing in addition to various other applications. The fungi produced an extraordinary range of pigments from different chemical classes. Fungal pigments proved to be nontoxic and achieved adequate color stability to withstand temperature, pH, and additives with color fastness properties when applied to textile fabrics. The advances in biotechnology, genetic engineering strategies for strain improvement, and immense fungal diversity have boosted the use of fungal pigments in the textile industry. However, for commercial applicability, fungal pigments should be tested for toxicity and quality to obtain regulatory approval before entering the market.

Author Contributions: C.K.V.: conceptualization, original draft preparation, writing; P.V., P.R.D., A.V.R.: writing and review; L.D.: writing, review and editing. All authors have read and agreed to the published version of the manuscript.

Funding: This work was supported and funded by the University Grants Commission (UGC), New Delhi under the D.S. Kothari Post-doctoral Fellowship (BL/17-18/0479) dated 25 September 2018.

Acknowledgments: C.K.V. thanks the UGC for awarding the D.S. Kothari Postdoctoral Fellowship (BL/17-18/0479). Additionally, the authors thank Anna University, Regional Campus – Coimbatore for providing the necessary facilities to carry out the project work. Laurent Dufossé deeply thanks the Conseil Régional de La Réunion, Réunion island, Indian Ocean, for continuous financial support of research activities dedicated to microbial pigments.

Conflicts of Interest: The authors declare no conflicts of interest.

References

1. Wrolstad, R.E.; Culver, C.A. Alternatives to those artificial FDC food colorants. *Annu. Rev. Food Sci. Technol.* **2012**, *3*, 59–77. [CrossRef] [PubMed]
2. Patel, S.K.; Saurabh, K.K.; Kamlesh, K.; Mali, M.K. Studies on synthesis of novel low molecular weight anthraquinone disperse dyes and their application on polyester and nylon. *J. Ind. Chem. Soc.* **2012**, *89*, 789–795.
3. Lebeau, J.; Venkatachalam, M.; Fouillaud, M.; Petit, T.; Vinale, F.; Dufossé, L.; Yanis, C. Production and new extraction method of polyketide red pigments produced by Ascomycetes fungi from terrestrial and marine habitats. *J. Fungi* **2017**, *3*, 34. [CrossRef] [PubMed]
4. Mukherjee, G.; Mishra, T.; Deshmukh, S.K. Fungal pigments: Overview. In *Developments in Fungal Biology and Applied Mycology*; Satyanarayana, T., Deshmukh, S.K., Johri, B.N., Eds.; Springer: Singapore, 2017.
5. Eisenman, H.C.; Casadevall, A. Synthesis and assembly of fungal melanin. *Appl. Microbiol. Biotechnol.* **2012**, *93*, 931–940. [CrossRef] [PubMed]
6. Robinson, S.C.; Tudor, D.; Zhang, W.R.; Ng, S.; Copper, P.A. Ability of three yellow pigment producing fungi to colour wood under controlled conditions. *Int. Wood Prod. J.* **2014**, *5*, 103–107. [CrossRef]
7. Bechtold, T.; Mussak, R. *Hanbook of Natural Colorants*; John Wiley and Sons: Hoboken, NJ, USA, 2009.
8. Yusof, M.; Shabbir, M.; Mohammad, F. Natural colorants: Historical, processing and sustainable prospects. *Nat. Prod. Bioprospect.* **2017**, *7*, 123–145. [CrossRef] [PubMed]
9. Hill, D.J. Is there a future for natural dyes? *Color. Technol.* **2008**, *27*, 18–25. [CrossRef]
10. Ramesh, C.; Vinithkumar, N.V.; Kirubagaran, R.; Venil, C.K.; Dufossé, L. Multifaceted applications of microbial pigments: Current knowledge, challenges and future directions for public health implications. *Microorganisms* **2019**, *7*, 186. [CrossRef]
11. Arifeen, M.Z.; Ma, Y.N.; Xue, Y.R.; Liu, C.H. Deep-sea fungi could be the new arsenal for bioactive compounds. *Mar. Drug* **2020**, *18*, 9. [CrossRef]
12. Hirot, J.; Maksimenka, K.; Reichert, M.; Perovic-Ottstadt, S.; Lin, W.H.; Wray, V.; Steube, K.; Schaumann, K.; Weber, H.; Proksch, P.; et al. New natural products from the sponge derived fungus Aspergillus niger. *J. Nat. Prod.* **2004**, *67*, 1532–1543.
13. Dufossé, L.; Fouillaud, M.; Caro, Y.; Mapari, S.A.S.; Sutthiwong, N. Filamentous fungi are large scale producers of pigments and colorants for the food industry. *Curr. Opin. Biotechnol.* **2014**, *26*, 56–61. [CrossRef] [PubMed]
14. Mapari, S.A.S.; Thrane, U.; Meyer, A.S. Fungal polyketide azaphilone pigments as future natural food colorants? *Trends Biotechnol.* **2010**, *28*, 300–307. [CrossRef] [PubMed]
15. Pfander, H. Carotenoids: An overview. *Methods Enzymol.* **1992**, *213*, 3–13.
16. Avalos, J.; Limon, M.C. Biological roles of fungal carotenoids. *Curr. Genet.* **2015**, *61*, 309–324. [CrossRef] [PubMed]
17. Mapari, S.A.S.; Nielsen, K.F.; Larsen, T.O.; Frisvad, J.C.; Meyer, A.S.; Thrane, U. Exploring fungal biodiversity for the production of water soluble pigments as potential natural colorants. *Curr. Opin. Biotechnol.* **2005**, *16*, 231–238. [CrossRef] [PubMed]
18. Rohdich, F.; Kis, K.; Bacher, A.; Eisenreich, W. The non-mevalonate pathway of isoprenoids: Genes, enzymes and intermediates. *Curr. Opin. Chem. Biol.* **2001**, *5*, 535–540. [CrossRef]
19. Zhang, C. Biosynthesis of carotenoids and apocarotenoids by microorganisms and their industrial potential. In *Progress in Carotenoid Research*; BioTrans: Singapore, 2018.

20. Bhosale, P.; Larson, A.J.; Bernstein, P.S. Factorial analysis of tricarboxylic acid cycle intermediates for optimization of zeaxanthin production from Flavobacterium multivorum. *J. Appl. Microbiol.* **2004**, *96*, 623–629. [CrossRef]
21. Ogbonna, C. Production of food colourants by filamentous fungi. *Afr. J. Microbiol. Res.* **2016**, *10*, 960–971.
22. Rao, M.P.N.; Xiao, M.; Li, W.-J. Fungal and bacterial pigments: Secondary metabolites with wide applications. *Front. Microbiol.* **2017**, *8*, 1113.
23. Mapari, S.A.S.; Meyer, A.S.; Thrane, U.; Frisvad, J.C. Identification of potentially safe promising fungal cell factories for the production of polyketide natural food colorants using chemotaxonomic rationale. *Microb. Cell Factories* **2009**, *8*, 24. [CrossRef]
24. Atalla, M.M.; Elkhrisy, E.A.M.; Asem, M.A. Production of textile reddish brown dyes by fungi. *Malays. J. Microbiol.* **2011**, *33*, 40.
25. Babula, P.; Adam, V.; Havel, L.; Kizek, R. Noteworthy secondary metabolites naphthoquinones–their occurrence, pharmacological properties and analysis. *Curr. Pharm. Anal.* **2009**, *5*, 47–67. [CrossRef]
26. Yu, M.L.; Li, Y.X.; Banakar, S.P.; Liu, L.; Shao, C.L.; Li, Z.Y.; Wang, C.Y. New metabolites from the co-culture of marine-derived actinomycete *Streptomyces rochei* MB037 and fungus *Rhinocladiella similis* 35. *Front. Microbiol.* **2019**, *10*, 915. [CrossRef] [PubMed]
27. Dufossé, L. Pigments, microbial. In *Encyclopedia of Microbiology*, 3rd ed.; Elsevier: Amsterdam, The Netherlands, 2009.
28. Chen, W.; He, Y.; Zhou, Y.; Shao, Y.; Feng, Y.; Li, M.; Chen, F. Edible filamentous fungi from the species *monascus*: Early traditional fermentations, modern molecular biology, and future genomics. *Compr. Rev. Food Sci. Food Saf.* **2015**, *14*, 555–567. [CrossRef]
29. Feng, Y.L.; Shao, Y.C.; Chen, F.S. Monascus pigments. *Appl. Microbiol. Biotechnol.* **2012**, *96*, 1421–1440. [CrossRef]
30. Chen, W.; Feng, Y.; Molnar, I.; Chen, F. Nature and nurture: Confluence of pathway determinism with metabolic and chemical serendipity diversifies Monascus azaphilone pigments. *Nat. Prod. Rep.* **2019**, *36*, 561–572. [CrossRef]
31. Lee, C.L.; Wen, J.Y.; Hsu, Y.W.; Pan, T.M. The blood lipid regulation of Monascus-produced monascin and ankaflavin via the suppression of low-density lipoprotein cholesterol assembly and stimulation of apolipoprotein A1 expression in the liver. *J. Microbiol. Immunol.* **2018**, *51*, 27–37. [CrossRef]
32. Lachenmeier, D.W.; Monakhova, Y.B.; Kuballa, T.; LobellBehrends, S.; Maixner, S.; Kohl-Himmelseher, M.; Waldner, A.; Steffen, C. NMR evaluation of total statin content and HMG-CoA reductase inhibition in red yeast rice (Monascus spp.) food supplements. *Chin. Med.* **2012**, *7*, 8. [CrossRef]
33. Chen, W.; Chen, R.; Liu, Q.; He, Y.; He, K.; Ding, X.; Kang, L.; Guo, X.; Xie, N.; Zhou, Y.; et al. Orange, red, yellow: Biosynthesis of azaphilone pigments in monascus fungi. *Chem. Sci.* **2017**, *8*, 4917–4925. [CrossRef]
34. Schor, R.; Cox, R. Classical fungal natural products in the genomic age: The molecular legacy of Harold Raistrick. *Nat. Prod. Rep.* **2018**, *35*, 230–256. [CrossRef]
35. Gessler, N.N.; Egorova, A.S.; Belozerskaya, T.A. Fungal anthraquinones. *Appl. Biochem. Microbiol.* **2013**, *49*, 85–99. [CrossRef]
36. Yang, X.M.; Li, J.S.; Li, Q.Q.; Huang, G.X.; Yan, L.J. Evaluation of the potential toxicity of anthraquinone derivatives in Chinese herbal medicines by the resonance light scattering spectrum. *Asian J. Chem.* **2011**, *23*, 3631–3634.
37. Fouillaud, M.; Venkatachalam, M.; Girard-Valenciennes, E.; Caro, Y.; Dufossé, L. Anthraquinones and derivatives from marine derived fungi: Structural diversity and selected biological activities. *Mar. Drugs* **2016**, *14*, 64. [CrossRef] [PubMed]
38. Zhu, X.; He, Z.; Wu, J.; Yuan, J.; Wen, W.; Hu, Y.; Jiang, Y.; Lin, C.; Zhang, Q.; Lin, M.; et al. A marine anthraquinone SZ-685C overrides Adriamycin resistance in breast cancer cells through suppressing Akt signalling. *Mar. Drugs* **2012**, *10*, 694–711. [CrossRef] [PubMed]
39. Huang, Z.; Nong, X.; Ren, Z.; Wang, J.; Zhang, X.; Qi, S. Anti-HSV1, antioxidant and antifouling phenolic compounds from the deep sea derived fungus Aspergillus versicolor SCSIO41502. *Bioorganic Med. Chem. Lett.* **2017**, *15*, 787–791. [CrossRef] [PubMed]
40. Shi, T.; Hou, X.M.; Li, Z.Y.; Cao, F.; Zhang, Y.H.; Yu, J.Y.; Zhao, D.L.; Shao, C.L.; Wang, C.Y. Harzianumnones A and B: Two hydroxyanthraquinones from the coral derived fungus Trichoderma harzianum. *RSC Adv.* **2018**, *8*, 27596–27601. [CrossRef]

41. Zhang, J.Y.; Tao, L.Y.; Liang, Y.J.; Chen, L.M.; Mi, Y.J.; Zheng, L.S.; Wang, F.; She, Z.G.; Lin, Y.C. Anthracenedione derivatives as anticancer agents isolated from secondary metabolites of the mangrove endophytic fungi. *Mar. Drugs* **2010**, *8*, 1469–1481. [CrossRef]
42. Awaad, A.S.; Al-Zaylaee, H.M.; Alqasoumi, S.I.; Zain, M.E.; Aloyan, E.M.; Alafeefy, A.M.; Awad, E.S.; El-Meligy, R.M. Anti-leishmanial activities of extracts and isolated compounds from *Drechslera rostrata* and *Eurotium tonpholium*. *Phytother. Res.* **2014**, *28*, 774–780. [CrossRef]
43. Wellington, K.W. Understanding cancer and the anticancer activities of naphthoquinones—A review. *RSC Adv.* **2015**, *5*, 20309–20338. [CrossRef]
44. Newman, A.G.; Townsend, C.A. Molecular characterization of the cercosporin biosynthetic pathway in the fungal plant pathogen Cercospora nicotianae. *J. Am. Chem. Soc.* **2016**, *138*, 4219–4228. [CrossRef]
45. Macicella, I.A.; Coatti, G.C.; Biazi, B.I.; Zanetti, T.A.; Baranoski, A.; Marques, L.A.; Corveloni, A.C.; Lepri, S.R.; Mantovani, M.S. Molecular pathways related to the control of proliferation and cell death in 786-O cells treated with plumbagin. *Mol. Biol. Rep.* **2019**, *46*, 6071–6078.
46. Li, M.; Kang, L.; Ding, X.; Liu, J.; Liu, Q.; Shao, Y.; Molnar, I.; Chen, F. Monasone naphthoquinone biosynthesis and resistance in Monascus fungi. *Mol. Biol. Physiol.* **2020**, *11*, e02676-19. [CrossRef] [PubMed]
47. Robinson, S.C.; Tudor, D.; Cooper, P.A. Utilizing pigment-producing fungi to add commercial value to American beech (Fagus grandifolia). *Appl. Microbiol. Biotechnol.* **2012**, *93*, 1041–1048. [CrossRef] [PubMed]
48. Weber, G.; Chen, H.L.; Hinsch, E.; Freitas, S.; Robinson, S. Pigments extracted from the wood staining fungi Chlorociboria aeruginosa, Scytalidium cuboideium and S. ganodermophthorum show potential for use as textile dyes. *Color. Technol.* **2014**, *130*, 445–452. [CrossRef]
49. Medentsev, A.G.; Akimentko, V.K. Naphthoquinone metabolites of the fungi. *Phytochemistry* **1998**, *47*, 935–959. [CrossRef]
50. Coates, C.S.; Ziegler, J.; Manz, K.; Good, J.; Kang, B.; Milikisiyants, S.; Chatterjee, R.; Hao, S.; Golbeck, J.H.; Lakshmi, K.V.J.; et al. The structure and function of quinones in biological solar energy transduction: A cyclic voltammetry, EPR, and hyperfine sub-level correlation (HYSCORE) spectroscopy study of model naphthoquinones. *J. Phys. Chem. B* **2013**, *117*, 7210–7220.
51. Freire, C.P.V.; Ferreira, S.B.; Oliveira, N.S.M.; Matsuura, A.B.J.; Gama, I.L.; Silva, F.C.; Souza, M.C.B.V.; Lima, E.S.; Ferreira, V.F. Synthesis and biological evaluation of substituted α- and β-2,3-dihydrofuran naphthoquinones as potent anticandidal agents. *Med. Chem. Commun.* **2010**, *1*, 229–232. [CrossRef]
52. Ferreira, M.P.S.B.C.; Cardoso, M.F.C.; Silva, F.C.; Ferreira, V.F.; Lima, E.S.; Souza, J.V.B. Antifungal activity of synthetic naphthoquinones against dermatophytes and opportunistic fungi: Preliminary mechanism-of-action tests. *Ann. Clin. Microbiol. Antimicrob.* **2014**, *13*, 26–31. [CrossRef]
53. Chen, C.; Tao, H.; Chen, W.; Yang, B.; Zhou, X.; Luo, X.; Liu, Y. Recent advancements in the chemistry and biology of azaphilones. *RSC Adv.* **2020**, *10*, 10197–10220. [CrossRef]
54. Gao, J.M.; Yang, S.X.; Qin, J.C. Azaphilones: Chemistry and biology. *Chem. Rev.* **2013**, *113*, 4755–4811. [CrossRef]
55. Osmanova, N.; Schultze, W.; Ayoub, N. Azaphilones: A class of fungal metabolites with diverse biological activities. *Phytochem. Rev.* **2010**, *9*, 315–342. [CrossRef]
56. Chen, M.; Shen, N.X.; Chen, Z.Q.; Zhang, F.M.; Chen, Y. Penicilones A-D, anti-MRSA azaphilones from the marine derived fungus Penicillium janthillum HK1-6. *J. Nat. Prod.* **2017**, *80*, 1081–1086. [CrossRef] [PubMed]
57. Yin, G.P.; Wu, Y.R.; Yang, M.H.; Li, T.X.; Wang, X.B.; Zhou, M.M.; Lei, J.L.; Kong, L.Y. Citrifurans A-D four dimeric aromatic polyketides with new carbon skeletons from the fungus Aspergillus sp. *Org. Lett.* **2017**, *19*, 4058–4061. [CrossRef] [PubMed]
58. Kornsakulkarn, J.; Saepua, S.; Suvannakad, R.; Supothina, S.; Boonyuen, N.; Isaka, M.; Prabpai, S.; Kongsaeree, P.; Thongpanchang, C. Cytotoxic tropolones from the fungus Nemania sp. BCC30850. *Tetrahedron* **2017**, *73*, 3505–3512. [CrossRef]
59. Fan, Z.; Sun, Z.H.; Liu, H.X.; Chen, Y.C.; Li, H.H.; Zhang, W.M. Perangustols A and B, a pair of new azaphilone epimers from a marine sediment derived fungus Cladosporium perangustm FS62. *J. Asian Nat. Prod. Res.* **2016**, *18*, 1024–1029. [CrossRef]
60. Arunpanichlert, J.; Rukachaisirikul, V.; Phongpaichit, S.; Supaphon, O.; Sakayaroj, J. Xylariphilone: A new azaphilone derivative from the seagrass derived fungus Xylariales sp. PSU-ES163. *Nat. Prod. Res.* **2016**, *30*, 46–51. [CrossRef]

61. Chen, Y.S.; Cheng, M.J.; Hsiao, Y.; Chan, H.Y.; Hsieh, S.Y.; Chang, C.W.; Liu, T.W.; Chang, H.S.; Chen, I.S. Chemical constituents of the endophytic fungus Hypoxylon sp. 12F0687 isolated from Taiwanese Ilex formosana. *Helv. Chim. Acta* **2015**, *98*, 1167–1176. [CrossRef]
62. Wang, J.; Bai, G.; Liu, Y.; Wang, H.; Li, Y.; Yin, W.; Wang, Y.; Lu, F. Cytotoxic metabolites produced by the endophytic fungus Aspergillus clavatus. *Chem. Lett.* **2015**, *44*, 1148–1149. [CrossRef]
63. Hufendiek, P.; Stoelben, S.S.M.; Kehraus, S.; Merten, N.; Harms, N.; Cruesemann, M.; Arslan, I.; Guetschow, M.; Schneider, T.; Koenig, G.M.; et al. Biosynthetic studies on Acetosellin and structure elucidation of a new acetosellin derivative. *Planta Med.* **2017**, *83*, 1044–1052. [CrossRef]
64. Cheng, M.J.; Wu, M.D.; Chan, H.Y.; Chang, H.S.; Wu, H.C.; Chen, J.J.; Yuan, G.F.; Weng, J.R.; Chang, C.T.; Lin, H.C.; et al. A new azaphilone derivative from the Monasucs kaoliang fermented rice. *Chem. Nat. Compd.* **2019**, *55*, 79–81. [CrossRef]
65. Li, W.; Lee, C.; Bang, S.H.; Ma, J.Y.; Kim, S.; Koh, Y.S.; Shim, S.H. Isochromans and Related constituents from the endophytic fungus *annulohypoxylon truncatum* of *zizania caduciflora* and their anti-inflammatory effects. *J. Nat. Prod.* **2017**, *80*, 205–209. [CrossRef] [PubMed]
66. Orfali, R.S.; Aly, A.H.; Ebrahim, W.; Rudiyansyah, R.; Proksch, P. Isochroman and isocoumarin derivatives from hypersaline lake sediment derived fungus Penicillium sp. *Phytochem. Lett.* **2015**, *13*, 234–238. [CrossRef]
67. Cheng, M.; Wu, M.; Chan, H.; Cheng, Y.C.; Chen, J.J.; Chen, I.S.; Su, Y.S.; Yuan, G.F. New metabolite isolated from the fungus *Monascus pilosus*. *Chem. Nat. Compd.* **2017**, *53*, 44–47. [CrossRef]
68. Li, X.H.; Han, X.H.; Qin, L.L.; He, J.L.; Cao, Z.X.; Guo, D.L.; Deng, Y.; Gu, Y.C. Isochromanes from Aspergillus fumigatus, an endophytic fungus from cordyceps sinensis. *Nat. Prod. Res.* **2019**, *33*, 1870–1875. [CrossRef] [PubMed]
69. Guo, W.; Li, D.; Peng, J.; Zhu, T.; Gu, Q.; Li, D. Penicitols A-C and Penixanacid from the mangrove derived Penicillium chryogenum HDN11-24. *J. Nat. Prod.* **2015**, *78*, 306–310. [CrossRef] [PubMed]
70. Yu, H.; Sperlich, J.; Hofert, S.P.; Janiak, C.; Teusch, N.; Stuhldreier, F.; Wesselborg, S.; Wang, C.; Kassack, M.U.; Dai, H.; et al. Azaphilone pigments and macrodiolides from the coprophilus fungus Coniella fragariae. *Fitoterapia* **2019**, *137*, 104249. [CrossRef]
71. Wang, W.X.; Kusari, S.; Laatsch, H.; Golz, C.; Kusari, P.; Strohmann, C.; Kayser, O.; Spiteller, M. Antibacterial azaphilones from an endophytic fungus, Colletotrichum sp.BS4. *J. Nat. Prod.* **2016**, *79*, 704–710. [CrossRef]
72. Wu, H.C.; Cheng, M.J.; Wu, M.D.; Chen, J.J.; Chen, Y.L.; Chang, H.S. Three new constituents from the fungus of Monascus purpureus and their anti-inflammatory activity. *Phytochem. Lett.* **2019**, *31*, 242–248. [CrossRef]
73. Zhang, D.; Zhao, J.; Wang, X.; Zhao, L.; Liu, H.; Wei, Y.; You, X.; Cen, S.; Yu, L. Peniazaphilin A, a new azaphilone derivative produced by Penicillium sp. CPCC400786. *J. Antibiot.* **2018**, *71*, 905–907. [CrossRef]
74. Xiong, X.; Zhang, X.; Wu, Z.; Wang, Z. Coupled aminophilic reaction and directed metabolic channeling to red *Monascus* pigments by extractive fermentation in nonionic surfactant micelle aqueous solution. *Process. Biochem.* **2015**, *50*, 180–187. [CrossRef]
75. Hu, Z.; Zhang, X.; Wu, Z.; Qi, H.; Wang, Z. Perstraction of intracellular pigments by submerged cultivation of *Monascus* in nonionic surfactant micelle aqueous solution. *Appl. Microbiol. Biotechnol.* **2012**, *94*, 81–89. [CrossRef]
76. Hinsch, E.; Robinson, S.C. Mechanical color reading of wood staining fungal pigment textile dyes: An alternative method for determining color fastness. *Coating* **2016**, *6*, 25. [CrossRef]
77. Sharma, D.; Gupta, C.; Aggarwal, S.; Nagpal, N. Pigment extraction from fungus for textile dyeing. *Indian J. Fibre Text. Res.* **2012**, *37*, 68–73.
78. Nambela, L.; Haule, L.V.; Mgani, Q. A review on source, chemistry, green synthesis and application of textile colorants. *J. Clean. Prod.* **2020**, *246*, 119036. [CrossRef]
79. Lagashetti, A.C.; Dufossé, L.; Singh, S.K.; Singh, P.N. Fungal pigments and their prospects in different industries. *Microorganisms* **2019**, *7*, 604. [CrossRef]
80. Sajid, S.; Akber, N. Applications of fungal pigments in biotechnology. *Pure Appl. Biol.* **2018**, *7*, 922–930. [CrossRef]
81. Kumar, A.; Vishwakarma, H.S.; Singh, J.; Dwivedi, S.; Kumar, M. Microbial pigments: Production and their applications in various industries. *Int. J. Pharm. Chem. Biol. Sci.* **2015**, *5*, 203–212.
82. Caro, Y.; Venkatachalam, M.; Lebeau, J.; Fouillaud, M.; Dufossé, L. Pigments and colorants from filamentous fungi. In *Fungal Metabolites*; Merillon, J.M., Ramawat, K.G., Eds.; Springer International Publishing: Cham, Switzerland, 2017; pp. 499–568.

83. Räisänen, R. *Handbook of Natural Colorants*; John Wiley & Sons: Chichester, UK, 2009; Chapter 11; pp. 183–200.
84. Perumal, K.; Stalin, V.; Chandrasekarenthiran, S.; Sumathi, E.; Saravanakumar, A. Extraction and characterization of pigment from Sclerotinia sp. and its use in dyeing cotton. *Text. Res. J.* **2009**, *79*, 1178–1187. [CrossRef]
85. Visagie, C.M.; Houbraken, J.; Frisvad, J.C. Identification and nomenclature of the genus *Penicillium*. *Stud. Mycol.* **2014**, *78*, 343–371. [CrossRef]
86. Afshari, M.; Shahidi, F.; Mortazavi, S.; Tabatabai, F.; Es'haghi, Z. Investigating the influence of pH, temperature and agitation speed on yellow pigment production by *Penicillium aculeatum* ATCC 10409. *Nat. Prod. Res.* **2015**, *29*, 1300–1306. [CrossRef]
87. Nagia, F.A.; El-Mohamedy, R. Dyeing of wool with natural anthraquinone dyes from Fusarium oxysporum. *Dyes Pigment.* **2007**, *75*, 550–555. [CrossRef]
88. Morales-Oyervides, L.; Oliveira, J.; Sousa-Gallagher, M.; Méndez-Zavala, A.; Montañez, J.C. Assessment of the dyeing properties of the pigments produced by *Talaromyces* spp. *J. Fungi* **2017**, *3*, 38. [CrossRef] [PubMed]
89. Chadni, Z.; Rahaman, M.H.; Jerin, I.; Hoque, K.M.F.; Reza, M.A. Extraction and optimization of red pigment production as secondary metabolites from Talaromyces verruculosus and its potential use in textile industries. *Mycology* **2017**, *8*, 48–57. [CrossRef]
90. Celestino, J.R.; Carvalho, L.; Lima, M.P.; Lima, A.M.; Ogusku, M.M.; de Souze, J.V.B. Bioprospecting of Amazon foil fungi with the potential pigment production. *Process Biochem.* **2014**, *49*, 569–575. [CrossRef]
91. Hinsch, E.M.; Chen, H.L.; Weber, G.; Robinson, S.C. Colorfastness of extracted wood staining fungal pigments on fabrics: A new potential for textile dyes. *J. Text. Appar. Technol. Manag.* **2015**, *3*.
92. Agurto, M.E.P.; Gutierrez, S.M.V.; Chen, H.L.; Robinson, S.C. Wood rotting fungal pigments as colorant coatings on oil based textile paints. *Coating* **2017**, *7*, 152. [CrossRef]
93. Gupta, C.; Sharma, D.; Aggarwal, S.; Nagpal, N. Pigment production from Trichoderma sp. for dyeing of silk and wool. *Int. J. Sci. Nat.* **2013**, *4*, 351–355.
94. Poorniammal, R.; Parthiban, M.; Gunasekaran, S.; Murugesan, R.; Thilagavathi, G. Natural dye production from Thermomyces sp fungi for textile application. *Indian J. Fibre Text. Res.* **2013**, 276–279.
95. Devi, S.; Karuppan, P. Reddish brown pigments from Alternaria alternate for textile dyeing and printing. *Indian J. Fibre Text. Res.* **2015**, *40*, 315–319.
96. Velmurugan, P.; Kim, M.J.; Park, J.S.; Karthikeyan, K.; Lakshmanaperumalsamy, P.; Lee, K.J.; Park, Y.J.; Oh, B.T. Dyeing of cotton yarn with five water soluble fungal pigments obtained from five fungi. *Fibre Polym.* **2010**, *11*, 598–605. [CrossRef]
97. Sardaryan, E.; Zihlova, H.; Strnad, R.; Cermakova, Z. Arpink Red-meet a new natural red food colorant of microbial origin. In *Pigments in Food, More than Colours*; Dufossé, L., Ed.; Elsevier: Quimper, France, 2004; pp. 207–208.
98. Devi, A. Extraction of natural dyes from fungus—An alternate for textile dyeing. *J. Nat. Sci. Res.* **2014**, *4*, 1–6.
99. Iswarya, S.; Shanuja, S.K.; Giri Dev, V.R.; Gnanamani, A. Asperyellone—A suitable coloring agent for protein based textile fabrics: An approach on production, characterization and application. *J. Text. Eng. Fash. Technol.* **2019**, *5*, 73–79.
100. Hernandez, V.; Galleguillos, F.; Thibaut, R.; Muller, A. Fungal dyes for textile applications: Testing of industrial conditions for wool fabrics dyeing. *J. Text. Inst.* **2018**, *110*, 1–6. [CrossRef]
101. Shibila, S.D.; Nanthini, A.U.R. Extraction and characterization of red pigment from Talaromyces australis and its application in dyeing cotton yarn. *Int. Arch. Appl. Sci. Technol.* **2019**, *10*, 81–91.
102. Chiba, S.; Tsuyoshi, N.; Fudou, R.; Ojika, M.; Murakami, Y.; Ogoma, Y.; Oguchi, M.; Yamanaka, S. Magenta pigment produced by fungus. *J. Gen. Appl. Microbiol.* **2006**, *52*, 201–207. [CrossRef] [PubMed]
103. De Santis, D.; Moresi, M.; Gallo, A.M.; Petruccioli, M. Assessment of the dyeing properties of pigments from Monascus purpureus. *J. Chem. Tech. Biotechnol.* **2005**, *80*, 1072–1079. [CrossRef]
104. Gokarneshan, N. Advances in textile printing. *Int. J. Text. Sci. Eng.* **2018**, 01.
105. Heo, Y.M.; Kim, K.; Kwon, S.L.; Na, J.; Lee, H.; Jang, S.; Kim, C.H.; Jung, J.; Kim, J.J. Investigation of filamentous fungi producing safe, functional water soluble pigmnets. *Mycobiology* **2018**, *46*, 269–277. [CrossRef]
106. Poorniammal, R.; Prabhu, S.; Sakthi, A.R.; Gunasekaran, S. Subacute dermal toxicity of Thermomyces sp. and Penicillium purpurogenum pigments in wistar rats. *Int. J. Chem. Stud.* **2019**, *7*, 630–634.

107. Wang, W.; Liao, Y.; Chen, R.; Hou, Y.; Ke, W.; Zhang, B.; Gao, M.; Shao, Z.; Chen, J.; Li, F. Chlorinated azaphilone pigments with antimicrobial and cytotoxic activities isolated from the deep sea derived fungus *Chaetomium* sp. NA-S01-R1. *Mar. Drugs* **2018**, *16*, 61. [CrossRef]
108. Pandiyarajan, S.; Premasudha, P.; Kadirvelu, K. Bio-production of novel water soluble yellow pigment from Aspergillus sp. and exploring its sustainable textile applications. *3 Biotech* **2018**, *8*, 398. [CrossRef] [PubMed]
109. Li, Q.; Fan, F.; Gao, X.; Yang, C.; Bi, C.; Tang, J.; Liu, T.; Zhang, X. Balanced activation of IspG and IspH to eliminate MEP intermediate accumulation and improve isoprenoids production in Escherichia coli. *Metab. Eng.* **2017**, *44*, 13–21. [CrossRef] [PubMed]
110. Rugbjerg, P.; Naesby, M.; Mortensen, U.H.; Frandsen, R.J.N. Reconstruction of the biosynthetic pathway for the core fungal polyketide scaffold rubrofusarin in Saccharomyces cerevisiae. *Microb. Cell Fact.* **2013**, *12*, 31. [CrossRef]
111. Jia, X.Q.; Xu, Z.N.; Zhou, L.P.; Sung, C.K. Elimination of the mycotoxin citrinin production in the industrial important strain Monascus purpureus SM001. *Metab. Eng.* **2010**, *12*, 1–7. [CrossRef]
112. Lebeau, J.; Petit, T.; Clerc, P.; Dufossé, L.; Caro, Y. Isolation of two novel purple naphthoquinone pigmnets concomitant with the bioactive red bikaverin and derivatives thereof produced by F.oxysporum. *Biotechnol. Prog.* **2019**, *35*, 2738. [CrossRef]
113. Westphal, K.R.; Wollenberg, R.D.; Herbst, F.-A.; Sørensen, J.L.; Sondergaard, T.E.; Wimmer, R. Enhancing the production of the fungal pigment aurofusarin in *Fusarium graminearum*. *Toxins* **2018**, *10*, 485. [CrossRef]
114. Klitgaard, A.; Frandsen, R.J.N.; Holm, D.K.; Knudsen, P.B.; Frisvad, J.C.; Nielsen, K.F. Combining UHPLC-high resolution MS and feeding of stable isotope labeled polyketide intermediates for linking precursors to end products. *J. Nat. Prod.* **2015**, *78*, 1518–1525. [CrossRef]
115. Viggiano, A.; Salo, O.; Ali, H.; Szymanski, W.; Lankhorst, P.P.; Nygard, Y.; Bovenberg, R.A.L.; Driessen, A.J.M. Pathway for the biosynthesis of the pigment Chrysogine by Penicillium chrysogenum. *Appl. Environ. Microbiol.* **2018**, *84*. [CrossRef]
116. Pohl, C.; Kiel, J.A.; Driessen, A.J.; Bovenberg, R.A.; Ngard, Y. CRISPR/Cas9 based genome editing of Penicillium chrysogenum. *ACS Synth. Biol.* **2016**, *5*, 754–764. [CrossRef]
117. Wollenberg, R.D.; Saei, W.; Westphal, K.R.; Klitgaard, C.S.; Nielsen, L.K.; Lysøe, E.; Gardiner, D.M.; Wimmer, R.; Sondergaard, T.E.; Sørensen, J.L.; et al. Chrysogine biosynthesis is mediated by a two-module nonribosomal peptide synthetase. *J. Nat. Prod.* **2017**, *80*, 2131–2135. [CrossRef]
118. Rokas, A.; Wisecaver, J.H.; Lind, A.L. The birth, evolution and death of metabolic gene clusters in fungi. *Nat. Rev. Microbiol.* **2018**, *16*, 731–744. [CrossRef]
119. Sen, T.; Barow, C.J.; Deshmukh, S.K. Microbial pigments in the food industry—Challenges and the way forward. *Front. Nutr.* **2019**, *6*, 7. [CrossRef]
120. Lee, S.S.; Lee, J.H.; Lee, I. Strain improvement by overexpression of the laeA gene in monascus pilosus for the production of monascus-fermented rice. *J. Microbiol. Biotechnol.* **2013**, *23*, 959–965. [CrossRef]
121. Doudna, J.A.; Charpentier, E. The new frontier of genome engineering with CRISPR-Cas9. *Science* **2014**, *346*, 1258096. [CrossRef]
122. Donohoue, P.D.; Barrangou, R.; May, A.P. Advances in industrial biotechnology using CRISPR-Cas systems. *Trends Biotechnol.* **2017**, *36*, 134–146. [CrossRef]
123. Nielsen, M.L.; Isbrandt, T.; Rasmussen, K.B.; Thrane, U.; Hoof, J.B.; Larsen, T.O.; Mortensen, U.H. Genes linked to production of secondary metabolites in Talaromyces atroroseus revealed using CRISPR-cas9. *PLoS ONE* **2017**, *12*, 2017. [CrossRef]
124. Lopes, F.C.; Tichota, D.M.; Pereira, J.Q.; Segalin, J.; Rios Ade, O.; Brandelli, A. Pigment production by filamentous fungi on agro-industrial byproducts: An eco-friendly alternative. *Appl. Biochem. Biotechnol.* **2013**, *171*, 616–625. [CrossRef]
125. Kaur, B.; Chakraborty, D.; Kaur, H. Production and stability analysis of yellowish pink pigments from Rhodotorula rubra MTCC 1446. *Int. J. Microbiol.* **2008**, *7*. [CrossRef]
126. Subhasree, R.S.; Babu, D.; Mohan, V.P.C. Effect of carbon and nitrogen sources on stimulation of pigment production by monascus purpureus on jackfruit seeds. *Int. J. Microbiol. Res.* 2011.
127. Srianta, I.; Novita, Y.; Kusumawati, N. Production of monascus pigments on durian seed. Effect of supplementation of carbon source. *J. Pure Appl. Microbiol.* **2012**, *6*, 59–63.
128. Silveira, S.T.; Daroit, D.J.; Brandelli, A. Pigment production by *Monascus purpureus* in grape waste using factorial design. *LWT Food Sci. Technol.* **2008**, *41*, 170–174. [CrossRef]

129. Rajeshwari, T.R.; Ponnusami, V.; Sugumaran, K.R. Production of monascus pigment in low cost fermentation. *Int. J. Chem. Tech. Res.* **2014**, *6*, 2929–2932.
130. Taskin, M.; Sisman, T.; Erdal, S.; Kurbanoglu, E.B. Use of waste chicken feathers as peptone for production of carotenoids in submerged culture of *Rhodotorula glutinis* MT-5. *Eur. Food Res. Technol.* **2011**, *233*, 657–665. [CrossRef]
131. Sanchez, C. Lignocellulosic residues: Biodegradation and bioconversion by fungi. *Biotechnol. Adv.* **2009**, *27*, 185–194. [CrossRef]
132. Mapari, S.A.S.; Meyer, A.S.; Frisvad, J.C.; Thrane, U. Production of Monascus Like Azaphilone Pigment. US Patent No. US20110250656A1, 2008.
133. Pagano, M.C.; Dhar, P.P. Fungal pigments: An overview. In *Fungal Bio-Molecules: Sources, Applications and Recent Developments*, 1st ed.; Gupta, V.K., Mach, R.L., Sreenivasaprasad, S., Eds.; Wiley: London, UK, 2015; ISBN 978-1-118-95829-2.
134. Loto, I.; Gutierrez, M.S.; Barahona, S.; Sepulveda, D.; Martinez-Moya, P.; Baeza, M.; Cifuentes, V.; Alcaino, J. Enhancement of carotenoid production by disrupting the C22-sterol desaturase gene (CYP61) in Xanthophyllomyces dendrorhous. *BMC Microbiol.* **2012**, *12*, 235. [CrossRef]
135. Gmoser, R.; Ferreira, J.A.; Lennartsson, P.R.; Taherzadeh, M.J. Filamentous ascomycetes fungi as a source of natural pigments. *Fungal Biol. Biotechnol.* **2017**, *4*, 4. [CrossRef]
136. Dufossé, L. Microbial production of food grade pigments. *Food Technol. Biotechnol.* **2006**, *44*, 313–323.

MDPI
St. Alban-Anlage 66
4052 Basel
Switzerland
Tel. +41 61 683 77 34
Fax +41 61 302 89 18
www.mdpi.com

Journal of Fungi Editorial Office
E-mail: jof@mdpi.com
www.mdpi.com/journal/jof

www.ingramcontent.com/pod-product-compliance
Lightning Source LLC
Chambersburg PA
CBHW061508220926
51850CB00004B/202
9783036514666